Essentials of Geology

NINTH EDITION

Essentials of Geology

Frederick K. Lutgens · Edward J. Tarbuck

Illustrated by

Dennis Tasa

PEARSON
Prentice Hall

Upper Saddle River, New Jersey 07458

Library of Congress Cataloging-in-Publication Data

Lutgens, Frederick K.
Essentials of geology / Frederick K. Lutgens, Edward J. Tarbuck;
illustrated by Dennis Tasa.—9th ed.
p. cm.
Includes index
ISBN 0-13-149749-9
1. Geology—Textbooks. I. Tarbuck, Edward J. II. Title.
QE26.3.L87 2006
550—dc22 2005047648

Executive Editor: Patrick Lynch
Executive Managing Editor: Kathleen Schiaparelli
Assistant Managing Editor: Beth Sweeten
Production Editor: Edward Thomas
Managing Editor, Science Media: Nicole M. Jackson
Assistant Managing Editor, Science Supplements: Becca Richter
Senior Media Editor: Chris Rapp
Project Manager: Dorothy Marrero
Director of Marketing: Linda Taft-Mackinnon
Manufacturing Manager: Alexis Heydt-Long
Manufacturing Buyer: Alan Fischer
Director of Creative Services: Paul Belfanti
Art Director: Heather Scott
Interior and Cover Design: Tamara Newnam
Copy Editor: Barbara Booth
Editorial Assistant: Sean Hale
Photo Research Administrator: Melinda Reo
Photo Researcher: Yvonne Gerin
Proofreader: Alison Lorber
Image Permission Coordinator: Debbie Hewitson
Color Scanning Supervisor: Joseph Conti
Production Assistant: Nancy Bauer
Composition: Pine Tree Composition
Senior Managing Editor, Art Production and Management: Patty Burns
Assistant Manager, Art Production: Ronda Whitson
Manager, Production Technologies: Matt Haas
Managing Editor, Art Management: Abigail Bass
Art Production Editor: Jessica Einsig
Illustrations: ESM Art Production, Lead Illustrator: Mark Landis
Cover Photo: Bill Hatcher/National Geographic Image Collection
Title Page Photo: Horseshoe Overlook, above the Colorado River. Marc Muench/David Muench Photography Inc

PEARSON Prentice Hall
Pearson Prentice Hall
Pearson Education, Inc.
Upper Saddle River, New Jersey 07458

Printed in the United States of America
10 9 8 7 6 5 4 3 2 1

ISBN 0-13-149749-9

Pearson Education Ltd., *London*
Pearson Education Australia Pty., Limited, *Sydney*
Pearson Education *Singapore*, Pte. Ltd
Pearson Education North Asia Ltd., *Hong Kong*
Pearson Education Canada, Ltd., *Toronto*
Pearson Educación de Mexico, S.A. de C.V.
Pearson Education—Japan, *Tokyo*
Pearson Education Malaysia, Pte. Ltd

To Nancy and Joanne

Brief Contents

1 An Introduction to Geology 1

2 Minerals: Building Blocks of Rocks 32

3 Igneous Rocks 56

4 Volcanoes and Other Igneous Activity 80

5 Weathering and Soils 114

6 Sedimentary Rocks 136

7 Metamorphic Rocks 162

8 Mass Wasting: The Work of Gravity 184

9 Running Water 200

10 Groundwater 222

11 Glaciers and Glaciation 242

12 Deserts and Wind 266

13 Shorelines 284

14 Earthquakes and Earth's Interior 308

15 Plate Tectonics: A Scientific Theory Unfolds 334

16 Origin and Evolution of the Ocean Floor 366

17 Crustal Deformation and Mountain Building 390

18 Geologic Time 414

19 Earth History: A Brief Summary 440

Appendix A Metric and English Units Compared 462

Appendix B Topographic Maps 463

Appendix C Landforms of the Conterminous United States 467

Glossary 469

Index 479

GEODe: Essentials of Geology

A copy of the *GEODe: Essentials of Geology CD-ROM* is packaged with each copy of *ESSENTIALS OF GEOLOGY,* Ninth Edition. This dynamic learning aid reinforces key geologic concepts by using tutorials, animations, and interactive exercises.

Chapter 1 **An Introduction to Geology**

Chapter 2 **Minerals: Building Blocks of Rocks**

Chapter 3 **Igneous Rocks**

Chapter 4 **Volcanoes and Other Igneous Activity**

Chapter 5 **Weathering and Soil**

Chapter 6 **Sedimentary Rocks**

Chapter 7 **Metamorphic Rocks**

Chapter 8 **Mass Wasting: The Work of Gravity**

Chapter 9 **Running Water**

Chapter 10 **Groundwater**

Chapter 11 **Glaciers and Glaciation**

Chapter 12 **Deserts and Winds**

Chapter 13 **Shorelines**

Chapter 14 **Earthquakes and Earth's Interior (Part A)**

Chapter 14 **Earthquakes and Earth's Interior (Part B)**

Chapter 15 **Plate Tectonics: A Scientific Revolution Unfolds**

Chapter 16 **Origin and Evolution of the Ocean Floor**

Chapter 17 **Crustal Deformation and Mountain Building (Part A)**

Chapter 17 **Crustal Deformation and Mountain Building (Part B)**

Chapter 18 **Geologic Time**

This *GEODe: Essentials of Geology* icon appears throughout the book wherever a text discussion has a corresponding activity on the CD-ROM.

Contents

Preface xiii

1 An Introduction to Geology 1

The Science of Geology 2
Geology, People, and the Environment 3
Historical Notes About Geology 3
Catastrophism 3; The Birth of Modern Geology 4
Geologic Time 4
The Nature of Scientific Inquiry 6
Hypothesis 6; Theory 7; Scientific Methods 7
Earth's Spheres 8
Hydrosphere 9; Atmosphere 10; Biosphere 10; Geosphere 10
Earth as a System 11
Earth System Science 11; The Earth System 12
The Rock Cycle: One of Earth's Subsystems 13
The Basic Cycle 16; Alternative Paths 16
Early Evolution of Earth 16
Origin of Planet Earth 16; Formation of Earth's Layered Structure 18
Earth's Internal Structure 19
Layers Defined by Composition 19; Layers Defined by Physical Properties 19
The Face of Earth 21
Features of the Continents 21; Features of the Ocean Basins 22
Dynamic Earth 24
The Theory of Plate Tectonics 24; Plate Boundaries 25
Box 1.1 Do Glaciers Move? An Application of the Scientific Method 8
The Chapter in Review 29 Key Terms 30 Questions for Review 30 Online Study Guide 31

2 Minerals: Building Blocks of Rocks 32

Minerals: Building Blocks of Rocks 34
Elements: Building Blocks of Minerals 35
Why Atoms Bond 37
Ionic Bonds: Electrons Transferred 38; Covalent Bonds: Electrons Shared 39; Other Bonds 39
Isotopes and Radioactive Decay 39
Properties of Minerals 40
Primary Diagnostic Properties 40; Other Properties of Minerals 43
Mineral Groups 43
The Silicates 44
The Silicon–Oxygen Tetrahedron 44; Other Silicate Structures 44; Joining Silicate Structures 45
Common Silicate Minerals 46
The Light Silicates 46; The Dark Silicates 49
Important Nonsilicate Minerals 49
Mineral Resources 50
Box 2.1 Asbestos: What Are the Risks? 51
The Chapter in Review 53 Key Terms 54 Questions for Review 54 Online Study Guide 55

3 Igneous Rocks 56

Magma: The Parent Material of Igneous Rock 58
The Nature of Magma 58; From Magma to Crystalline Rock 59
Igneous Textures 60
Factors Affecting Crystal Size 60; Types of Igneous Textures 61
Igneous Compositions 62
Granitic versus Basaltic Compositions 63; Other Compositional Groups 64; Silica Content as an Indicator of Composition 64
Naming Igneous Rocks 64
Felsic (Granitic) Igneous Rocks 64; Intermediate (Andesitic) Igneous Rocks 67; Mafic (Basaltic) Igneous Rocks 68; Pyroclastic Rocks 68
Origin of Magma 69
Generating Magma from Solid Rock 69
How Magmas Evolve 71
Bowen's Reaction Series and the Composition of Igneous Rocks 71; Magmatic Differentiation 73; Assimilation and Magma Mixing 73; Partial Melting and Magma Formation 74
Mineral Resources and Igneous Processes 75
Box 3.1 A Closer Look at Bowen's Reaction Series 72
The Chapter in Review 77 Key Terms 78 Questions for Review 78 Online Study Guide 79

4 Volcanoes and Other Igneous Activity 80

The Nature of Volcanic Eruptions 82
Factors Affecting Viscosity 82; Importance of Dissolved Gases 84
Materials Extruded During an Eruption 86
Lava Flows 86; Gases 87; Pyroclastic Materials 88
Volcanic Structures and Eruptive Styles 89
Anatomy of a Volcano 89; Shield Volcanoes 90; Cinder Cones 91; Composite Cones 92
Living in the Shadow of a Composite Cone 94
Eruption of Vesuvius A.D. 79 94; Nuée Ardente: A Deadly Pyroclastic Flow 95; Lahars: Mudflows on Active and Inactive Cones 96
Other Volcanic Landforms 97
Calderas 97; Fissure Eruptions and Basalt Plateaus 98; Volcanic Pipes and Necks 99
Intrusive Igneous Activity 100
Nature of Plutons 100; Dikes 100; Sills and Laccoliths 101; Batholiths 102

Plate Tectonics and Igneous Activity 104
Igneous Activity at Convergent Plate Boundaries **105**; Igneous Activity at Divergent Plate Boundaries **105**; Intraplate Igneous Activity **109**
Box 4.1 Anatomy of an Eruption 85
Box 4.2 Can Volcanoes Change Earth's Climate? 108
The Chapter in Review **111** Key Terms **111** Questions for Review **112** Online Study Guide **113**

5 Weathering and Soils 114

Earth's External Processes 116
Weathering 116
Mechanical Weathering 116
Frost Wedging **117**; Unloading **118**; Biological Activity **119**
Chemical Weathering 119
Water and Carbonic Acid **119**; How Granite Weathers **120**; Weathering of Silicate Minerals **121**; Spheroidal Weathering **121**
Rates of Weathering 122
Rock Characteristics **122**; Climate **122**; Differential Weathering **123**
Soil 123
An Interface in the Earth System **123**; What Is Soil? **124**
Controls of Soil Formation 124
Parent Material **124**; Time **124**; Climate **125**; Plants and Animals **125**; Topography **127**
The Soil Profile 127
Classifying Soils 128
Soil Erosion 129
How Soil Is Eroded **129**; Rates of Erosion **130**; Sedimentation and Chemical Pollution **132**
Weathering: The Cause of Ore Deposits 132
Bauxite **132**; Other Deposits **133**
Box 5.1 Clearing the Tropical Rain Forest—The Impact on Its Soils 126
The Chapter in Review **133** Key Terms **134** Questions for Review **134** Online Study Guide **135**

6 Sedimentary Rocks 136

What Is a Sedimentary Rock? 138
Turning Sediment into Sedimentary Rock: Diagenesis and Lithification 139
Types of Sedimentary Rocks 139
Detrital Sedimentary Rocks 140
Shale **140**; Sandstone **141**; Conglomerate and Breccia **143**
Chemical Sedimentary Rocks 144
Limestone **144**; Dolostone **145**; Chert **145**; Evaporites **147**; Coal **147**
Classification of Sedimentary Rocks 149
Sedimentary Rocks Represent Past Environments 150
Sedimentary Structures 151
Nonmetallic Mineral Resources from Sedimentary Rocks 154
Energy Resources from Sedimentary Rocks 155
Coal **156**; Oil and Natural Gas **158**
Box 6.1 The Carbon Cycle and Sedimentary Rocks 146
The Chapter in Review **159** Key Terms **160** Questions for Review **160** Online Study Guide **161**

7 Metamorphic Rocks 162

Metamorphism 164
Agents of Metamorphism 165
Heat as a Metamorphic Agent **165**; Pressure and Differential Stress **166**; Chemically Active Fluids **167**
Metamorphic Textures 168
Foliation **168**; Foliated Textures **169**; Other Metamorphic Textures **171**
Common Metamorphic Rocks 172
Foliated Rocks **172**; Nonfoliated Rocks **174**
Metamorphic Environments 175
Contact or Thermal Metamorphism **175**; Hydrothermal Metamorphism **176**; Regional Metamorphism **176**; Other Metamorphic Environments **177**
Metamorphic Zones 178
Textural Variations **178**; Index Minerals and Metamorphic Grade **178**
The Chapter in Review **181** Key Terms **182** Questions for Review **182** Online Study Guide **183**

8 Mass Wasting: The Work of Gravity 184

A Landslide Disaster in Peru 186
Mass Wasting and Landform Development 187
The Role of Mass Wasting **187**; Slopes Change Through Time **187**
Controls and Triggers of Mass Wasting 188
The Role of Water **188**; Oversteepened Slopes **188**; Removal of Vegetation **188**; Earthquakes as Triggers **189**; Landslides Without Triggers? **189**
Classification of Mass-Wasting Processes 190
Type of Material **190**; Type of Motion **190**; Rate of Movement **191**
Slump 192
Rockslide 192
Debris Flow 193
Debris Flows in Semiarid Regions **193**; Lahars **193**
Earthflow 196
Slow Movements 196
Creep **196**; Solifluction **197**

Box 8.1 Debris Flows on Alluvial Fans: A Case Study from Venezuela 195
Box 8.2 The Sensitive Permafrost Landscape 197
The Chapter in Review 198 Key Terms 198 Questions for Review 198 Online Study Guide 199

9 Running Water 200

Earth as a System: The Hydrologic Cycle 202
Running Water 203
Drainage Basins 203; River Systems 204
Streamflow 204
Gradient and Channel Characteristics 205; Discharge 205; Changes from Upstream to Downstream 206
The Work of Running Water 206
Erosion 207; Transportation 207; Deposition 208
Stream Channels 208
Bedrock Channels 208; Alluvial Channels 208
Base Level and Stream Erosion 210
Shaping Stream Valleys 210
Valley Deepening 212; Valley Widening 212; Changing Base Level and Incised Meanders 212
Depositional Landforms 213
Deltas 213; Natural Levees 213; Alluvial Fans 214
Drainage Patterns 214
Floods and Flood Control 216
Causes and Types of Floods 217; Flood Control 218
Box 9.1 Coastal Wetlands Are Vanishing on the Mississippi Delta 216
The Chapter in Review 220 Key Terms 220 Questions for Review 221 Online Study Guide 221

10 Groundwater 222

Importance of Underground Water 224
Distribution of Underground Water 224
The Water Table 226
Variations in the Water Table 226; Interaction Between Groundwater and Streams 226
Factors Influencing the Storage and Movement of Groundwater 227
Porosity 227; Permeability, Aquitards, and Aquifers 228
How Groundwater Moves 228
Springs 228
Wells 229
Artesian Wells 230
Environmental Problems Associated with Groundwater 231
Treating Groundwater as a Nonrenewable Resource 231; Land Subsidence Caused by Groundwater Withdrawal 232; Groundwater Contamination 232
Hot Springs and Geysers 234
Geothermal Energy 234
The Geologic Work of Groundwater 236
Caverns 236; Karst Topography 237
Box 10.1 Measuring Groundwater Movement 229
Box 10.2 The Case of the Disappearing Lake 239
The Chapter in Review 240 Key Terms 241 Questions for Review 241 Online Study Guide 241

11 Glaciers and Glaciation 242

Glaciers: A Part of Two Basic Cycles 244
Valley (Alpine) Glaciers 245; Ice Sheets 245; Other Types of Glaciers 246
How Glaciers Move 247
Rates of Glacial Movement 247; Budget of a Glacier 248
Glacial Erosion 249
Landforms Created by Glacial Erosion 250
Glaciated Valleys 251; Arêtes and Horns 252; Roches Moutonnées 253
Glacial Deposits 254
Types of Glacial Drift 254; Moraines, Outwash Plains, and Kettles 255; Drumlins, Eskers, and Kames 257
Glaciers of the Ice Age 257
Some Indirect Effects of Ice Age Glaciers 259
Causes of Glaciation 260
Plate Tectonics 260; Variations in Earth's Orbit 261; Other Factors 262
Box 11.1 The Collapse of Antarctic Ice Shelves 246
Box 11.2 Glacial Ice—A Storehouse of Climate Data 263
The Chapter in Review 264 Key Terms 264 Questions for Review 265 Online Study Guide 265

12 Deserts and Wind 266

Distribution and Causes of Dry Lands 268
Low-Latitude Deserts 269; Middle-Latitude Deserts 269
Geologic Processes in Arid Climates 270
Weathering 271; The Role of Water 271
Basin and Range: The Evolution of a Mountainous Desert Landscape 273
Transportation of Sediment by Wind 274
Bed Load 275; Suspended Load 275
Wind Erosion 275
Deflation, Blowouts, and Desert Pavement 275; Wind Abrasion 276
Wind Deposits 277
Sand Deposits 277; Types of Sand Dunes 278; Loess (Silt) Deposits 279
Box 12.1 The Disappearing Aral Sea 272
Box 12.2 Dust Bowl—Soil Erosion in the Great Plains 276
The Chapter in Review 281 Key Terms 282 Questions for Review 282 Online Study Guide 283

13 Shorelines 284

The Shoreline: A Dynamic Interface 286
The Coastal Zone 286
Waves 288
Wave Characteristics 288; Circular Orbital Motion 288; Waves in the Surf Zone 290
Wave Erosion 290
Sand Movement on the Beach 291
Movement Perpendicular to the Shoreline 291; Wave Refraction 291; Beach Drift and Longshore Currents 292
Shoreline Features 293
Erosional Features 294; Depositional Features 295; The Evolving Shore 296
Stabilizing the Shore 296
Hard Stabilization 298; Alternatives to Hard Stabilization 299; Erosion Problems Along U.S. Coasts 299
Coastal Classification 302
Emergent Coasts 302; Submergent Coasts 302
Tides 303
Causes of Tides 303; Monthly Tidal Cycle 304; Tidal Currents 304
Box 13.1 The Move of the Century—Relocating the Cape Hatteras Lighthouse 289
Box 13.2 Coastal Vulnerability to Sea-Level Rise 301
The Chapter in Review 305 Key Terms 306 Questions for Review 306 Online Study Guide 307

14 Earthquakes and Earth's Interior 308

What Is an Earthquake? 311
Earthquakes and Faults 311; Discovering the Cause of Earthquakes 312; Foreshocks and Aftershocks 312
San Andreas Fault: An Active Earthquake Zone 312
Seismology: The Study of Earthquake Waves 314
Locating an Earthquake 316
Measuring the Size of Earthquakes 317
Intensity Scales 317; Magnitude Scales 317
Destruction from Earthquakes 320
Destruction from Seismic Vibrations 321; Tsunami 322; Landslides and Ground Subsidence 323; Fire 324
Can Earthquakes Be Predicted? 324
Short-Range Predictions 324; Long-Range Forecasts 325
Earthquake's and Earth's Interior 327
Layers Defined by Composition 327; Layers Defined by Physical Properties 328; Discovering Earth's Major Layers 329; Discovering Earth's Composition 330
Box 14.1 Damaging Earthquakes East of the Rockies 318
The Chapter in Review 331 Key Terms 332 Questions for Review 332 Online Study Guide 333

15 Plate Tectonics: A Scientific Theory Unfolds 334

Continental Drift: An Idea Before Its Time 336
Evidence: The Continental Jigsaw Puzzle 337; Evidence: Fossils Match Across the Seas 337; Evidence: Rock Types and Structures Match 339; Evidence: Ancient Climates 340
The Great Debate 342
Plate Tectonics: The New Paradigm 342
Earth's Major Plates 342; Plate Boundaries 343
Divergent Boundaries 343
Oceanic Ridges and Seafloor Spreading 346; Continental Rifting 346
Convergent Boundaries 347
Oceanic–Continental Convergence 349; Oceanic–Oceanic Convergence 350; Continental–Continental Convergence 350
Transform Fault Boundaries 351
Testing the Plate Tectonics Model 352
Evidence: Ocean Drilling 352; Evidence: Hot Spots 354; Evidence: Paleomagnetism 355
Measuring Plate Motion 358
What Drives Plate Motion? 359
Forces that Drive Plate Motion 359; Models of Plate-Mantle Convection 361
Plate Tectonics into the Future 362
Box 15.1 The Breakup of Pangaea 338
The Chapter in Review 363 Key Terms 364 Questions for Review 364 Online Study Guide 365

16 Origin and Evolution of the Ocean Floor 366

An Emerging Picture of the Ocean Floor 368
Mapping the Seafloor 368; Viewing the Ocean Floor from Space 370; Provinces of the Ocean Floor 370
Continental Margins 370
Passive Continental Margins 370; Active Continental Margins 372
Features of Deep-Ocean Basins 373
Deep-Ocean Trenches 373; Abyssal Plains 373; Seamounts, Guyots, and Oceanic Plateaus 374
Anatomy of the Oceanic Ridge 375
Origin of Oceanic Lithosphere 377
Seafloor Spreading 378; Why Are Oceanic Ridges Elevated? 379; Spreading Rates and Ridge Topography 379
The Structure of Oceanic Crust 380
Formation of Oceanic Crust 380; Interactions Between Seawater and Oceanic Crust 381
Continental Rifting: The Birth of a New Ocean Basin 381
Evolution of an Ocean Basin 381

Destruction of Oceanic Lithosphere 383
Why Oceanic Lithosphere Subducts **383**; Subducting Plates: The Demise of an Ocean Basin **384**
Opening and Closing Ocean Basins: The Supercontinent Cycle 385
Before Pangaea **385**
Box 16.1 Explaining Coral Atolls—Darwin's Hypothesis 376
The Chapter in Review **387** Key Terms **388** Questions for Review **388** Online Study Guide **389**

17 Crustal Deformation and Mountain Building 390

Rock Deformation 392
Temperature and Confining Pressure **392**; Rock Type **393**; Time **393**
Folds 393
Types of Folds **393**; Domes and Basins **395**
Faults 395
Dip-Slip Faults **396**; Strike-Slip Faults **398**
Joints 401
Mountain Building 401
Mountain Building at Subduction Zones 402
Island Arcs **402**; Mountain Building Along Andean-Type Margins **403**
Collisional Mountain Ranges 405
Terranes and Mountain Building **405**; Continental Collisions **407**
Fault-Block Mountains 408
Vertical Movements of the Crust 409
Isostasy **409**; How High Is Too High? **410**
Box 17.1 The San Andreas Fault System 400
The Chapter in Review **411** Key Terms **412** Questions for Review **412** Online Study Guide **413**

18 Geologic Time 414

Geology Needs a Time Scale 416
Relative Dating—Key Principles 416
Law of Superposition **417**; Principle of Original Horizontality **417**; Principle of Cross-Cutting Relationships **418**; Inclusions **418**; Unconformities **418**; Using Relative Dating Principles **419**
Correlation of Rock Layers 419
Fossils: Evidence of Past Life 420
Types of Fossils **420**; Conditions Favoring Preservation **422**; Fossils and Correlation **422**
Dating with Radioactivity 426
Reviewing Basic Atomic Structure **426**
Radioactivity **426**; Half-Life **428**; Radiometric Dating **428**; Dating with Carbon-14 **430**; Importance of Radiometric Dating **431**
The Geologic Time Scale 431
Structure of the Time Scale **431**; Precambrian Time **432**
Difficulties in Dating the Geologic Time Scale 433
Box 18.1 Radon 429
Box 18.2 Using Tree Rings to Date and Study the Recent Past 432
The Chapter in Review **436** Key Terms **437** Questions for Review **438** Online Study Guide **439**

19 Earth History: A Brief Summary 440

Early Evolution of Earth 440
Earth's Primitive Atmosphere **444**; Earth's Atmosphere Evolves **444**
Precambrian Time: Vast and Enigmatic 444
Precambrian History **445**; Precambrian Fossils **446**
Paleozoic Era: Life Explodes 447
Paleozoic History **447**; Early Paleozoic Life **449**; Late Paleozoic Life **450**
Mesozoic Era: Age of the Dinosaurs 452
Mesozoic History **452**; Mesozoic Life **452**
Cenozoic Era: Age of Mammals 456
Cenozoic North America **456**; Cenozoic Life **457**
Box 19.1 Demise of the Dinosaurs 454
The Chapter in Review **459** Key Terms **460** Questions for Review **460** Online Study Guide **461**

Appendix A
Metric and English Units Compared 462

Appendix B
Topographic Maps 463

Appendix C
Landforms of the Conterminous United States 467

Glossary 469

Index 479

Preface

The Ninth Edition of *Essentials of Geology,* like its predecessors, is a college-level text for students taking their first and perhaps only course in geology. The book is intended to be a meaningful, nontechnical survey for people with little background in science. Usually students are taking this class to meet a portion of their college or university's general requirements.

In addition to being informative and up-to-date, a major goal of *Essentials of Geology* is to meet the need of beginning students for a readable and user-friendly text, a book that is a highly usable tool for learning the basic principles and concepts of geology.

New Design

An all-new design has been developed for the Ninth Edition of *Essentials of Geology.* Wider pages allow for greater flexibility and effectiveness in art and photo placement. In addition, new margin photos add to the book's visual appeal as well as to the effectiveness of text discussions. The new design also allowed us to include a new feature—*Did You Know?* This involves the placement of interesting facts and ideas at various places throughout each chapter that are intended to add interest and relevance to text discussions.

Distinguishing Features

Readability

The language of this book is straightforward and *written to be understood.* Clear, readable discussions with a minimum of technical language are the rule. The frequent headings and subheadings help students follow discussions and identify the important ideas presented in each chapter. In the Ninth Edition, improved readability was achieved by examining chapter organization and flow, and writing in a more personal style. Large portions of several chapters were substantially rewritten in an effort to make the material more understandable.

Illustrations and Photographs

Geology is highly visual. Therefore, photographs and artwork are a very important part of an introductory book. *Essentials of Geology,* Ninth Edition contains dozens of new high-quality photographs that were carefully selected to aid understanding, add realism, and heighten the interest of the reader.

Moreover, there has been substantial revision and improvement of the art program. Clearer, easier to understand line drawings show greater color and shading contrasts. More figures combine the use of diagrams or maps *and* photos together. The result is an art program that illustrates ideas and concepts more clearly than ever before. As in earlier editions, we are grateful to Dennis Tasa, a gifted artist and respected geological illustrator for his outstanding work.

Focus on Learning

To assist student learning, every chapter opens with a series of questions. Each question alerts the reader to an important idea or concept in the chapter. When a chapter has been completed, three useful devices help students review. First, a helpful summary—*The Chapter in Review*—recaps all of the major points. Next is a checklist of *Key Terms* with page references. Learning the language of geology helps students learn the material. This is followed by *Questions for Review,* which helps students examine their knowledge of significant facts and ideas. Each chapter closes with a reminder to visit the Website for *Essentials of Geology,* Ninth Edition (http://www.prenhall.com/lutgens). It contains many excellent opportunities for review and exploration.

Maintaining a Focus on Basic Principles and Instructor Flexibility

The main focus of the Ninth Edition remains the same as in the first eight—to foster student understanding of basic geological principles. As much as possible, we have attempted to provide the reader with a sense of the observational techniques and reasoning processes that constitute the discipline of geology.

The organization of the text remains intentionally traditional. Following the overview of geology in Chapter 1, we turn to a discussion of Earth materials and the related processes of volcanism and weathering. Next, we explore the geological work of gravity, water, wind, and ice in modifying and sculpting landscapes. After this look at external processes, we examine Earth's internal structure and the processes that deform rocks and give rise to mountains. Finally, the text concludes with chapters on geologic time and Earth history. This organization accommodates the study of minerals and rocks in the laboratory, which usually comes early in the course.

Realizing that some instructors may prefer to structure their courses somewhat differently, we made each of the chapters self-contained so that they may be taught in a different sequence. Thus, the instructor who wishes to discuss earthquakes, plate tectonics, and mountain building prior to dealing with erosional processes may do so without difficulty. We also chose to introduce plate tectonics in Chapter 1 so that this important and basic theory could be incorporated in appropriate places throughout the text.

More About the Ninth Edition

The Ninth Edition of *Essentials of Geology* represents a thorough revision. *Every* part of the book was examined carefully with the dual goals of keeping topics current and improving the clarity of text discussions.

Here are some examples of what is new in the Ninth Edition of *Essentials of Geology.*

- *GEODe: Essentials of Geology* CD-ROM. Each copy of *Essentials of Geology* 9e comes with this significantly revised and expanded student learning tool. Organizationally, *GEODe* now has a chapter structure to match chapters 1 through 18 in the book. In addition, the treatment of plate tectonics has been completely revised and significantly expanded. Also, all-new chapters on "Weathering and Soil" (Chapter 5) and "Mass Wasting" (Chapter 8) have been added. Each *GEODe* chapter ends with a review quiz consisting of randomly-generated questions to help students review basic concepts.
- Chapter 1, "An Introduction to Geology," offers an expanded section on "Earth as a System," including new material on open and closed systems and feedback mechanisms. The chapter also includes an expanded discussion of the rock cycle and a new section that describes the "Early Evolution of Earth." A new section on "The Face of Earth" introduces students to the major surface features of the continents and ocean basins.
- Much is new in Chapter 2, "Minerals: Building Blocks of Rocks." Changes include a completely rewritten section on the geologic definition of minerals and an all-new discussion of elements that more clearly explains atomic structure and bonding. Many new and redrawn art pieces reinforce the revision.
- Among the changes in Chapter 5 is an all-new section on "Classifying Soils" that includes a new table on world soil orders and a large new world map showing global soil regions.
- Chapter 7 "Metamorphic Rocks," contains revised and rewritten discussions on "Heat as a Metamorphic Agent," "Pressure and Differential Stress," and "Regional Metamorphism."
- Chapter 8, "Mass Wasting," includes an expanded discussion of "Controls and Triggers of Mass Wasting" and a new special interest box on "Debris Flows on Alluvial Fans: A Case Study from Venezuela."
- Chapter 9, "Running Water," has been reorganized and almost entirely rewritten so that the discussion of streams progresses in a manner that is clearer and more logical for the beginning student.
- Among the numerous changes to Chapter 13, "Shorelines," is a new section on the "Coastal Zone" which deals with the "anatomy" of this dynamic interface.
- Chapter 15 on plate tectonics has been made even better! This chapter has been *extensively* reorganized, revised, and rewritten. Now more than ever, the chapter clearly summarizes and explains the most important unifying theory in the Earth sciences. The new title "A Scientific Theory Unfolds" serves to highlight the fact that a significant emphasis involves tracing the historical development of the theory of plate tectonics, as a way of providing students with insights into how science and scientists work.
- Chapter 16, "Origin and Evolution of the Ocean Floor," is all-new to this edition. Students are asked to examine how the seafloor is generated, why it is continually being destroyed, and what clues it can provide about events that occurred in Earth's history.

The Teaching and Learning Package

The challenge is fundamental and too often overlooked in what seems to have become a weapons race of resources supplemental to textbooks: *instructors need more time, students need more preparation.* With this as a credo, Prentice Hall has produced for this edition perhaps the best set of instructor and student resources ever assembled to support an introductory geology textbook. Not only are they of the highest quality, they are the most *useful.*

Instructor Resources

Prentice Hall continues to improve the instructor resources in this edition with the goal of saving you time in preparing for your classes.

Instructor's Resource Center (IRC) on DVD: The IRC puts your lecture resources all in one easy-to-reach place:

- Two **PowerPoint® presentations** for each chapter—see below (Are illustrations central to your lecture? Check out the *Student Lecture Notebook.*)
- **84 animations** of geologic processes (see below)
- **All of the line art, tables and photos** from the text in .jpg files
- ***Images of Earth*** photo gallery (see below)
- ***Instructor's Manual*** in Microsoft Word
- ***Test Item File*** in Microsoft Word
- **TestGenEQ** test generation and management software

PowerPoints®: Found on the IRC are two PowerPoint® files for each chapter. Cut down on your preparation time, no matter what your lecture needs:

1. **Art and Animations**—All of the line art, tables and photos from the text, along with the animation library, pre-loaded into PowerPoint® slides for easy integration into your presentation.
2. **Lecture Outline**—Authored by Stanley Hatfield of Southwestern Illinois College, this set averages 35 slides per chapter and includes customizable lecture outlines with supporting art.

Animations: Found on the IRC, the *Prentice Hall Geoscience Animation Library* includes 84 animations illuminating many difficult-to-visualize geological topics. Created through a unique collaboration among five of Prentice Hall's leading geoscience authors, these animations represent a truly significant leap forward in lecture presentation aids. Available on the IRC on DVD, each animation is mapped to its corresponding chapter. They are provided both as Flash files and, for your convenience, pre-loaded into PowerPoint® slides.

- Convergent Margins
- Stream Processes
- Faults
- Transform Faults
- Angular Unconformity and Nonconformity
- Beach Drift
- Beach Drift and Longshore Currents
- Folding
- Seismograph Operations
- Breakup of Pangaea
- Nebular Hypothesis
- Oxbow Lake Formation
- Seafloor Spreading
- Glacial Processes and Budget
- Relative Dating
- Tectonic Settings and Volcanic Activity
- Glacial Processes—Plucking and Moraines
- Coastal Stabilization—Jetties, Groins, Breakwaters
- Ocean Circulation
- Global Atmospheric Circulation Model
- Wind Pattern Development
- Collapse of Mt. St. Helens
- Convection in a Lava Lamp
- Convection and Tectonics
- Correlating Processes and Plate Boundaries
- Density and Magma Movement
- Dimensions of the Mantle and Core
- Earthquake Waves
- Earth's Water and Hydrologic Cycle
- Hydrologic Cycle
- Groins and Jetties
- Elastic Rebound
- Erosion of Deformed Sedimentary Rock
- Exposing Metamorphic Rock
- Fault Motions
- Natural Levees
- Glacier Ice
- Folding Rock
- Crater Lake
- Cone of Depression
- Divergent Boundary
- Cross-Beds
- Foliation
- Igneous Features and Landforms
- Stream Terraces
- Volcanoes
- Fractional Crystallization
- Advance and Retreat
- Glacial Isostacy
- Global Wind Patterns
- Graphing Stress and Strain
- Hot Spots
- Calderas
- How Streams Move Sediment
- How Tides Work
- Intrusive Igneous Features
- Earth's Age
- Liquefaction
- Mass Movement
- Meandering Streams
- Metamorphic Rock Foliation
- Tidal Cycle
- Plate Boundaries
- Transform Boundaries
- Plate Motions
- Plate Tectonics and Magma Generation
- Weathering
- Waves
- Radioactive Decay
- Relative and Absolute Motion
- Geologic Dating
- Sea Floor Spreading
- Seismic Wave Motions
- Terrane Formation
- Unconformities
- Inclination and Declination
- Tuttle and Bowen
- Uplift and Mass Movement
- Mantle Melting
- Wave Motion and Refraction
- Water Wave Motion and Refraction
- Water Table
- Wind Pattern Development
- Tsunami

"Images of Earth" Photo Gallery Supplement your personal and text-specific slides with this amazing collection of over 300 geologic photos contributed by Marli Miller (University of Oregon) and other professionals in the field. The photos are grouped by geologic concept and available on the IRC on DVD.

Transparencies Every Dennis Tasa illustration in *Essentials of Geology* 9e is available on full-color, projection enhanced transparency—over 300 in all. (Are illustrations central to your lecture? Check out the *Student Lecture Notebook.*)

Instructor's Manual with Tests Authored by Stanley Hatfield (Southwestern Illinois College), the *Instructor's Manual* contains: learning objectives, chapter outlines, answers to end-of-chapter questions and suggested, short demonstrations to enhance your lecture. The *Test Item File* incorporates art and averages 75 multiple-choice, true/false, short-answer and critical-thinking questions per chapter.

TestGenEQ Available on the IRC, use this electronic version of the *Test Item File* to customize and manage your tests. Create multiple versions, add or edit questions, add illustrations—your customization needs are easily addressed by this powerful software.

Test Item File in Blackboard and WebCT formats Already have your own website set up? Prentice Hall can provide you with the *Essentials of Geology Test Item File* in formats suitable for importation into your Blackboard or WebCT course. Additional course resources are available on the IRC and are available for your use, with permission (see the ReadMe file on the IRC).

Student Resources

The student resources to accompany *Essentials of Geology 9e* have been further refined with the goal of focusing the students' efforts and improving their understanding of the concepts of geology.

GEODe: Essentials of Geology Somewhere between a text and a tutor, *GEODe: Essentials of Geology* reinforces key concepts using animations, video, narration, interactive exercises, and practice quizzes. A copy of *GEODe: Essentials of Geology* is automatically included in every copy of the text purchased from Prentice Hall.

Online Study Guide www.prenhall.com/lutgens

Authored by Gary Solar (State University of New York, College at Buffalo), the *Online Study Guide* contains numerous chapter review exercises (from which students get immediate feedback). Links to other resources are also included for further study. Professors can utilize the quizzing modules in conjunction with a course management system to assess student progress.

Student Lecture Notebook All of the line art from the text and transparency set are reproduced in this full color notebook, with space for notes. Students can now fully focus on the lecture and not be distracted by attempting to replicate figures. Each page is three-hole punched for easy integration with other course materials.

Safari X Textbooks Online SafariX WebBooks Online is an exciting new choice for students looking to save money on their required textbooks. As an alternative to purchasing *Essentials of Geology* 9e, students can subscribe to the same content online and save up to 50 percent off the suggested list price of the print text. SafariX WebBooks offer study advantages that no print textbook can match. With an internet-enhanced SafariX WebBook, students can search *Essentials of Geology 9e* for key concepts, navigate easily to a page number, reading assignment or chapter; or bookmark important pages or sections for quick review at a later date.

Acknowledgments

Writing a college textbook requires the talents and cooperation of many individuals. Working with Dennis Tasa, who is responsible for all of the text's outstanding illustrations and much of the developmental work on *GEODe: Essentials of Geology* is always special for us. We not only value his outstanding artistic talents and imagination but his friendship as well.

Our sincere appreciation goes to those colleagues who prepared in-depth reviews. Their critical comments and thoughtful input helped guide our work and clearly strengthened the text. Special thanks to

Stephen P. Altaner, *University of Illinois–Urbana*
Dirk Baron, *California State University—Bakersfield*
David M. Best, *Northern Arizona University*
Constantin Cranganu, *Brooklyn College*
Dale Easley, *University of New Orleans*
Martha Eppes, *University of North Carolina—Charlotte*
Mark Feigenson, *Rutgers University*
Jacqueline Gallagher, *Florida Atlantic University*
Lindley Hanson, *Salem State College*
Albert T. Hsui, *University of Illinois*
William Opperman, *Broward Community College*
Steven Ralser, *University of Wisconsin—Madison*
Jeffrey R. Swope, *Indiana University-Purdue University Indianapolis*

We also want to acknowledge the team of professionals at Prentice Hall. We sincerely appreciate the company's continuing strong support for excellence and innovation. Special thanks to our Executive Editor Patrick Lynch. We value his leadership and appreciate his attention to detail, excellent communication skills, and easy-going style. The production team, led by Ed

Thomas, has once again done an outstanding job. The strong visual impact of *Essentials of Geology, Ninth Edition* benefited greatly from the work of photo researcher Yvonne Gerin and image permission coordinator Debbie Hewitson. Thanks also to Barbara Booth for her excellent copyediting skills. All are true professionals with whom we are very fortunate to be associated.

Frederick K. Lutgens
Edward J. Tarbuck

Climbers set up camp below Basin Mountain in California's Sierra Nevada Range. *(Photo by Galen Rowell/CORBIS)*

CHAPTER

1

An Introduction to Geology

Focus on Learning

To assist you in learning the important concepts in this chapter, you will find it helpful to focus on the following questions:

- How does physical geology differ from historical geology?
- What is the fundamental difference between uniformitarianism and catastrophism?
- What is relative dating? What are some principles of relative dating?
- How does a scientific hypothesis differ from a scientific theory?
- What are Earth's four major "spheres"?
- Why can Earth be regarded as a system?
- What is the rock cycle? Which geologic interrelationships are illustrated by the cycle?
- How did Earth and the other planets in our solar system originate?
- What criteria were used to establish Earth's layered structure?
- What are the major features of the continents and ocean basins?
- What is the theory of plate tectonics? How do the three types of plate boundaries differ?

Analyzing cores of ocean-floor sediment in the lab. (Ocean Drilling Program)

The spectacular eruption of a volcano, the terror brought by an earthquake, the magnificent scenery of a mountain valley, and the destruction created by a landslide all are subjects for the geologist (Figure 1.1). The study of geology deals with many fascinating and practical questions about our physical environment. What forces produce mountains? Will there soon be another great earthquake in California? What was the Ice Age like? Will there be another? How were these ore deposits formed? Should we look for water here? Is strip mining appropriate in this area? Will oil be found if a well is drilled at this location?

The Science of Geology

The subject of this text is **geology,** from the Greek *geo,* "Earth," and *logos,* "discourse." It is the science that pursues an understanding of planet Earth. Geology is traditionally divided into two broad areas—physical and historical. **Physical geology** examines the materials composing Earth and seeks to understand the many processes that operate beneath and upon its surface. The aim of **historical geology,** on the other hand, is to understand the origin of Earth and its development through time. Thus, it strives to establish a chronological arrangement of the multitude of physical and biological changes that have occurred in the geologic past. The study of physical geology logically precedes the study of Earth history because we must first understand how Earth works before we attempt to unravel its past.

To understand Earth is challenging because our planet is a dynamic body with many interacting parts and a complex history. Throughout its long existence, Earth has been changing. In fact, it is changing as you read this page and will continue to do so into the foreseeable future. Sometimes the changes are rapid and violent, as when landslides or volcanic eruptions occur. Just as often, change takes place so slowly that it goes unnoticed during a lifetime. Scales of size and space also vary greatly among the phenomena that geologists study. Sometimes they must focus on phenomena that are submicroscopic, and at other times they must deal with features that are continental or global in scale.

Geology is perceived as a science that is done in the out of doors, and rightly so. A great deal of geology is based on measurements, observations, and experiments conducted in the field. But geology is also done in the laboratory where, for example, the study of various Earth materials provides insights into many basic processes. Frequently, geology requires an understanding and application of knowledge and principles from physics, chemistry, and biology. Geology is a science that seeks to expand our knowledge of the natural world and our place in it.

▼ **Figure 1.1** On January 13, 2001, a magnitude 7.6 earthquake caused considerable damage in El Salvador. The damage pictured here was caused by a landslide that was triggered by the earthquake. As many as 1000 people were buried under 8 meters (26 feet) of landslide debris. Geologists seek to understand the processes that create such events. *(Photo by Reuters/STR/Getty Images Inc.-Hulton Archive Photos)*

Geology, People, and the Environment

The study of geology sometimes involves interactions between people and the natural environment. Stated another way, many of the problems and issues addressed by geology are of practical value to people.

Natural hazards are a part of living on Earth. Every day they adversely affect millions of people worldwide and are responsible for staggering damages. Among the hazardous Earth processes studied by geologists are volcanoes, floods, earthquakes, and landslides. Of course, geologic hazards are simply *natural* processes. They become hazards only when people try to live where these processes occur (Figure 1.2).

Resources represent another important focus of geology that is of great practical value to people. They include water and soil, a great variety of metallic and nonmetallic minerals, and energy. Together they form the very foundation of modern civilization. Geology deals not only with the formation and occurrence of these vital resources but also with maintaining supplies and with the environmental impact of their extraction and use.

Complicating all environmental issues is rapid world population growth and everyone's aspiration to a better standard of living. Earth is now gaining about 100 million people each year. This means a ballooning demand for resources and a growing pressure for people to dwell in environments having significant geologic hazards.

Not only do geologic processes have an impact on people, but we humans can dramatically influence geologic processes as well. For example, river flooding is natural, but the magnitude and frequency of flooding can be changed significantly by human activities such as clearing forests, building cities, and constructing dams. Unfortunately, natural systems do not always adjust to artificial changes in ways that we can anticipate. Thus, an alteration to the environment that was intended to benefit society often has the opposite effect.

▼ **Figure 1.2** This is an image of Italy's Mount Vesuvius in September 2000. This major volcano is surrounded by the city of Naples and the Bay of Naples. In 79 A.D. Vesuvius explosively erupted, burying the towns of Pompeii and Herculanaeum in volcanic ash. Will it happen again? Geologic hazards are *natural* processes. They become hazards only when people try to live where these processes occur. *(Image courtesy of NASA)*

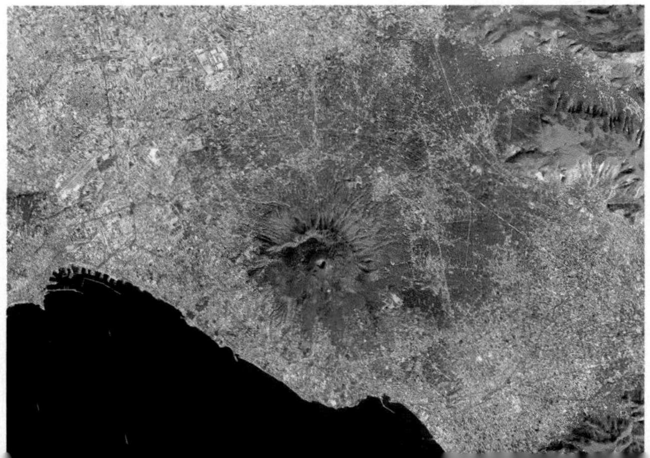

Historical Notes About Geology

The nature of our Earth—its materials and its processes—has been a focus of study for centuries. Writings about fossils, gems, earthquakes, and volcanoes date back to the Greeks, more than 2300 years ago. Certainly, the most influential Greek philosopher was Aristotle. Unfortunately, Aristotle's explanations about the natural world were not derived from keen observations and experiments, as is modern science. Instead, they were arbitrary pronouncements based on the limited knowledge of his day. He believed that rocks were created under the "influence" of the stars and that earthquakes occurred when air in the ground was heated by central fires and escaped explosively! When confronted with a fossil fish, he explained that "a great many fishes live in the earth motionless and are found when excavations are made." Although Aristotle's explanations may have been adequate for his day, they unfortunately continued to be expounded for many centuries, thus thwarting the acceptance of more up-to-date ideas.

Did You Know?

Each year an average American requires huge quantities of Earth materials. Imagine receiving your annual share in a single delivery. A large truck would pull up to your home and unload 12,965 lbs. of stone, 8945 lbs. of sand and gravel, 895 lbs. of cement, 395 lbs. of salt, 361 lbs. of phosphate, and 974 lbs. of other nonmetals. In addition, there would be 709 lbs. of metals, including iron, aluminum, and copper.

Strip mining of coal at Black Mesa, Arizona. (Richard W. Brooks/Photo Researchers, Inc.)

Catastrophism

In the mid-1600s James Ussher, Anglican Archbishop of Armagh, Primate of all Ireland, published a major work that had immediate and profound influences. A respected scholar of the Bible, Ussher constructed a chronology of human and Earth history in which he determined that Earth was only a few thousand years old, having been created in 4004 B.C. Ussher's treatise earned widespread acceptance among Europe's scientific and religious leaders, and his chronology was soon printed in the margins of the Bible itself.

During the seventeenth and eighteenth centuries the doctrine of **catastrophism** strongly influenced people's thinking about Earth. Briefly stated, catastrophists believed that Earth's landscapes had been shaped primarily by great catastrophes. Features such as mountains and canyons, which today

Did You Know?

It took until about the year 1800 for the world population to reach 1 billion. Since then, the planet has added nearly 6 billion more people.

Did You Know?

Shortly after Archbishop Ussher determined an age for Earth, another biblical scholar, Dr. John Lightfoot of Cambridge, felt he could be even more specific. He wrote that Earth was created "...on the 26th of October 4004 B.C. at 9 o'clock in the morning." (As quoted in William L. Stokes, *Essentials of Earth History,* Prentice Hall, Inc. 1973, p. 20.)

we know take great periods of time to form, were explained as having been produced by sudden and often worldwide disasters produced by unknown causes that no longer operate. This philosophy was an attempt to fit the rates of Earth processes to the then current ideas on the age of Earth.

The relationship between catastrophism and the age of Earth has been summarized nicely:

> That the earth had been through tremendous adventures and had seen mighty changes during its obscure past was plainly evident to every inquiring eye; but to concentrate these changes into a few brief millenniums required a tailor-made philosophy, a philosophy whose basis was sudden and violent change.*

The Birth of Modern Geology

James Hutton (1726–1797), the founder of modern geology. (The Natural History Museum, London)

Against this backdrop of Aristotle's views and an Earth created in 4004 B.C., a Scottish physician and gentleman farmer named James Hutton published *Theory of the Earth* in 1795. In this work Hutton put forth a fundamental principle that is a pillar of geology today: **uniformitarianism.** It states that the *physical, chemical, and biological laws that operate today also operated in the geologic past.* In other words, the forces and processes that we observe shaping our planet today have been at work for a very long time. Thus, to understand ancient rocks, we must first understand present-day processes and their results. This idea is commonly stated as *the present is the key to the past.*

Prior to Hutton's *Theory of the Earth,* no one had effectively demonstrated that geological processes can continue over extremely long periods of time. Hutton persuasively argued that forces that appear small could, over long spans of time, produce effects just as great as those resulting from sudden catastrophic events. Hutton carefully cited verifiable observations to support his ideas.

For example, when he argued that mountains are sculpted and ultimately destroyed by weathering and the work of running water, and that their wastes are carried to the oceans by processes that can be observed, Hutton said, "We have a chain of facts which clearly demonstrates . . . that the materials of the wasted mountains have traveled through the rivers"; and further, "There is not one step in all this progress . . . that is not to be actually perceived." He then went on to summarize this thought by asking a question and immediately providing the answer: "What more can we require? Nothing but time."

Today the basic tenets of uniformitarianism are just as viable as in Hutton's day. We realize more strongly than ever that the present gives us insight into the past and that the physical, chemical, and biological laws that govern geological processes remain unchanged through time. However, we also understand that the doctrine should not be taken too literally. To say that geological processes in the past were the same as those occurring today is not to suggest that they always had the same relative importance or operated at precisely the same rate. Moreover, some important geologic processes are not currently observable, but evidence that they occur is well established. For example, we know that Earth has experienced impacts from large meteorites even though we have no human witnesses. Such events altered Earth's crust, modified its climate, and strongly influenced life on the planet.

The acceptance of uniformitarianism meant the acceptance of a very long history for Earth. Although processes vary in their intensity, they still take a very long time to create or destroy major landscape features (Figure 1.3). For example, geologists have established that mountains once existed in portions of present-day Minnesota, Wisconsin, and Michigan. Today the region consists of low hills and plains. Over great expanses of time, erosional processes that wear land away gradually destroyed these peaks. The rock record contains evidence that shows Earth has experienced many such cycles of mountain building and erosion.

In the chapters that follow, we will be examining the materials that compose our planet and the processes that modify it. It is important to remember that although many features of our physical landscape may seem to be unchanging in terms of the decades over which we observe them, they are nevertheless changing, but on time scales of hundreds, thousands, or even many millions of years. Concerning the ever changing nature of Earth through great expanses of geologic time, James Hutton made a statement that was to become his most famous. In concluding his classic 1788 paper, published in the *Transactions of the Royal Society of Edinburgh,* he stated, "The result, therefore, of our present enquiry is, that we find no vestige of a beginning, no prospect of an end."

Geologic Time

Although Hutton and others recognized that geologic time is exceedingly long, they had no methods to accurately determine the age of Earth. However, in 1896 radioactivity was discovered. Using radioactivity for dating was first attempted in 1905 and has been refined ever since. Geologists are now able to assign fairly accurate dates to events in Earth history. For example, we know that the dinosaurs died out about 65 million years ago. Today the age of Earth is put at about 4.5 billion years.

*H. E. Brown, V. E. Monnett, and J. W. Stovall, *Introduction to Geology* (New York: Blaisdell, 1958).

The concept of geologic time is new to many nongeologists. People are accustomed to dealing with increments of time that are measured in hours, days, weeks, and years. Our history books often examine events over spans of centuries, but even a century is difficult to appreciate fully. For most of us, someone or something that is 90 years old is *very old,* and a 1000-year-old artifact is *ancient.*

By contrast, those who study geology must routinely deal with vast time periods—millions or billions (thousands of millions) of years. When viewed in the context of Earth's 4.5-billion-year history, a geologic event that occurred 100 million years ago may be characterized as "recent" by a geologist, and a rock sample that has been dated at 10 million years may be called "young." An appreciation for the magnitude of geologic time is important in the study of geology because many processes are so gradual that vast spans of time are needed before significant changes occur.

During the nineteenth century, long before the discovery of radioactivity and the establishment of reliable *numerical* dates, a geologic time scale was developed using principles of relative dating. **Relative dating** means that events are placed in their proper sequence or order without knowing their age in years. This is done by applying principles such as the **law of superposition,** which states that in layers of sedimentary rocks or lava flows, the youngest layer is on top, and the oldest is on the bottom (assuming that nothing has turned the layers upside down, which sometimes happens). Arizona's Grand Canyon provides a fine example where the oldest rocks are located in the inner gorge while the youngest rocks are found on the rim (Figure 1.3). So the law of superposition establishes the *sequence* of rock layers but not, of course, their numerical ages. Today such a proposal appears to be elementary, but 300 years ago it amounted to a major breakthrough in scientific reasoning by establishing a rational basis for relative time measurements.

Fossils, the remains or traces of prehistoric life, were also essential to the development of a geologic time scale. Fossils are the basis for the **principle of fossil succession,** which states that fossil organisms succeed one another in a definite and determinable order, and therefore any time period can be recognized by its fossil content. This principle was laboriously worked out over decades by collecting fossils from countless rock layers around the world. Once established, it allowed

Did You Know?

Estimates indicate that erosional processes are lowering the North American continent at a rate of about 3 cm per 1000 years. At this rate, it would take 100 million years to level a 3000-meter- (10,000-foot-) high peak.

Did You Know?

If you were to count to 4.5 billion at the rate of one number per second and continued 24 hours a day, 7 days a week, and never stopped, it would take about 150 years to complete the task!

▼ **Figure 1.3** Over millions of years, weathering, gravity, and the erosional work of the Colorado River and the streams that flow into it have created Arizona's Grand Canyon. Geologic processes often act so slowly that changes may not be visible during an entire human lifetime. The relative ages of the rock layers in the canyon can be determined by applying the law of superposition. The youngest rocks are on top, and the oldest are at the bottom. *(Photo by Marc Muench/Muench Photography, Inc.)*

Fossil fish, Eocene epoch, Kemmerer, Wyoming. (Breck P. Kent)

geologists to identify rocks of the same age in widely separated places and to build the *geologic time scale* shown in Figure 1.4.

Notice that units having the same designations do not necessarily extend for the same number of years. For example, the Cambrian period lasted about 50 million years, whereas the Silurian period spanned only about 26 million years. As we will emphasize again in Chapter 18, this situation exists because the basis for establishing the time scale was not the regular rhythm of a clock, but the changing character of life forms through time. Specific dates were added long after the time scale was established. A glance at Figure 1.4 also reveals that the Phanerozoic eon is divided into many more units than earlier eons even though it encompasses only about 12 percent of Earth history. The meager fossil record for these earlier eons is the primary reason for the lack of detail on this portion of the time scale. Without abundant fossils, geologists lose their primary tool for subdividing geologic time.

The Nature of Scientific Inquiry

All science is based on the assumption that the natural world behaves in a consistent and predictable manner that is comprehensible through careful, systematic study. The overall goal of science is to discover the underlying patterns in nature and then to use this knowledge to make predictions about what should or should not be expected, given certain facts or circumstances. For example, by knowing how oil deposits form, geologists are able to predict the most favorable sites for exploration and, perhaps as important, how to avoid regions having little or no potential.

Fossil ferns, Pennsylvanian period, St. Clair, Pennsylvania. (Breck P. Kent)

The development of new scientific knowledge involves some basic logical processes that are universally accepted. To determine what is occurring in the natural world, scientists collect scientific *"facts"* through observation and measurement. Because some error is inevitable, the accuracy of a particular measurement or observation is always open to question. Nevertheless, these data are essential to science and serve as the springboard for the development of scientific theories.

Hypothesis

Once facts have been gathered and principles have been formulated to describe a natural phenomenon, investigators try to explain how or why things happen in the manner observed. They often do this by constructing a tentative (or untested) explanation, which is called a

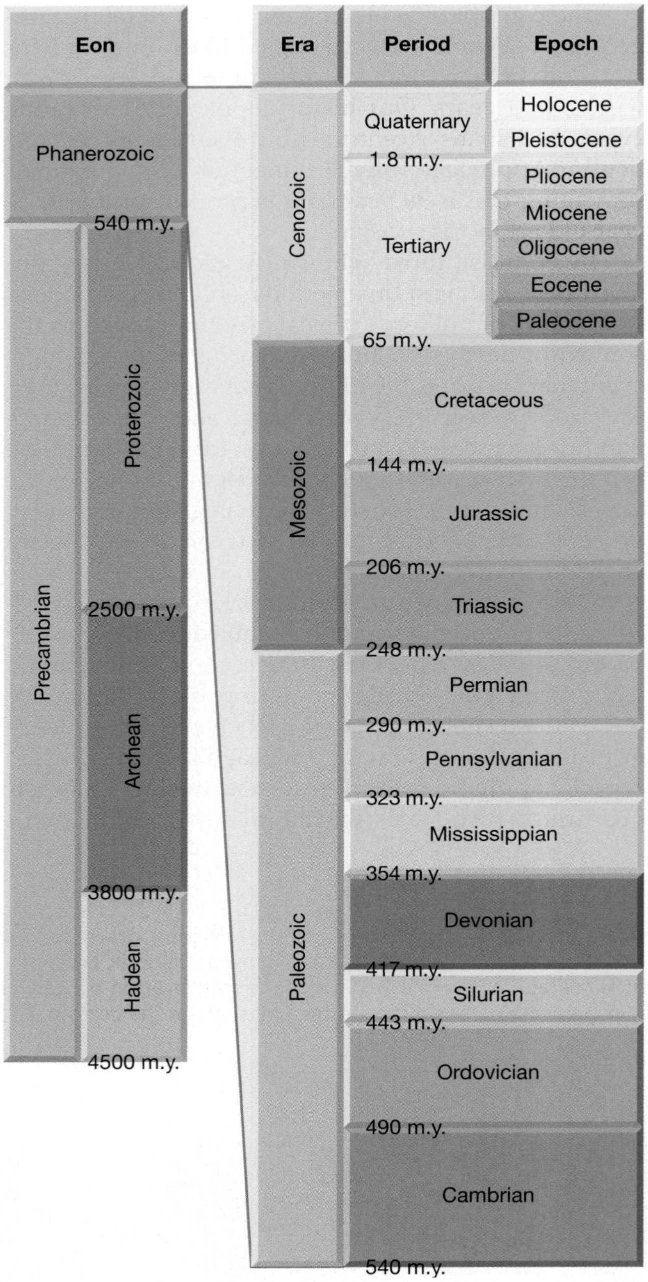

▲ **Figure 1.4** The geologic time scale divides the vast 4.5-billion-year history of Earth into eons, eras, periods, and epochs. We presently live in the Holocene epoch of the Quaternary period. This period is part of the Cenozoic era, which is the latest era of the Phanerozoic eon. Numbers on the time scale represent time in millions of years before the present. These dates were added long after the time scale had been established using relative dating techniques. The Precambrian accounts for more than 88 percent of geologic time. *(Data from Geological Society of America)*

scientific **hypothesis** or **model.** (The term model, although often used synonymously with hypothesis, is a less precise term because it is sometimes used to describe a scientific theory as well.) It is best if an investigator can formulate more than one hypothesis to explain a given set of observations. If an individual scientist is unable to devise multiple models, others in the scientific community will almost always develop alternative explanations. A spirited debate frequently ensues. As a result, extensive research is conducted by proponents of opposing models, and the results are made available to the wider scientific community in scientific journals.

Before a hypothesis can become an accepted part of scientific knowledge, it must pass objective testing and analysis. (If a hypothesis cannot be tested, it is not scientifically useful, no matter how interesting it might seem.) The verification process requires that *predictions* be made based on the model being considered and that the predictions be tested by comparing them against objective observations of nature. Put another way, hypotheses must fit observations other than those used to formulate them in the first place. Those hypotheses that fail rigorous testing are ultimately discarded. The history of science is littered with discarded hypotheses. One of the best known is the Earth-centered model of the universe—a proposal that was supported by the apparent daily motion of the Sun, Moon, and stars around Earth. As the mathematician Jacob Bronowski so ably stated, "Science is a great many things, but in the end they all return to this: Science is the acceptance of what works and the rejection of what does not."

Theory

When a hypothesis has survived extensive scrutiny and when competing models have been eliminated, a hypothesis may be elevated to the status of a scientific **theory.** In everyday language we may say "That's only a theory." But a scientific theory is a well-tested and widely accepted view that the scientific community agrees best explains certain observable facts.

Theories that are extensively documented are held with a very high degree of confidence. Theories of this stature that are comprehensive in scope have a special status. They are called **paradigms** because they explain a large number of interrelated aspects of the natural world. For example, the theory of plate tectonics is a paradigm of the geological sciences that provides the framework for understanding the origin of mountains, earthquakes, and volcanic activity. In addition, plate tectonics explains the evolution of the continents and the ocean basins through time—a topic we will consider later in this chapter.

Scientific Methods

The processes just described, in which scientists gather facts through observations and formulate scientific hypotheses and theories is called the *scientific method.* Contrary to popular belief, the scientific method is not a standard recipe that scientists apply in a routine manner to unravel the secrets of our natural world. Rather, it is an endeavor that involves creativity and insight. Rutherford and Ahlgren put it this way: "Inventing hypotheses or theories to imagine how the world works and then figuring out how they can be put to the test of reality is as creative as writing poetry, composing music, or designing skyscrapers."*

There is no fixed path that scientists always follow that leads unerringly to scientific knowledge. Nevertheless, many scientific investigations involve the following steps: (1) collection of scientific facts (data) through observation and measurement (Figure 1.5); (2) development of one or more working hypotheses or models to explain these facts; (3) development of observations and experiments to test the hypotheses; and (4) the acceptance, modification, or rejection of the hypotheses based on extensive testing (see Box 1.1).

Other scientific discoveries may result from purely theoretical ideas that stand up to extensive examination.

*F. James Rutherford and Andrew Ahlgren, *Science for All Americans* (New York: Oxford University Press, 1990), p. 7.

Surveying a glacier. (Charles D. Winters/Photo Researchers)

▼ **Figure 1.5** This field geologist is checking a seismograph. *(Photo by Andrew Rafkind/Getty Images Inc—Stone Allstock)*

BOX 1.1

Do Glaciers Move? An Application of the Scientific Method

The study of glaciers provides an early application of the scientific method. High in the Alps of Switzerland and France, small glaciers exist in the upper portions of some valleys. In the late eighteenth and early nineteenth centuries, people who farmed and herded animals in these valleys suggested that glaciers in the upper reaches of the valleys had previously been much larger and had occupied downvalley areas. They based their explanation on the fact that the valley floors were littered with angular boulders and other rock debris that seemed identical to the materials that they could see in and near the glaciers at the heads of the valleys.

Although the explanation of these observation seemed logical, others did not accept the notion that masses of ice hundreds of meters thick were capable of movement. The disagreement was settled after a simple experiment was designed and carried out to test the hypothesis that glacial ice can move.

Markers were placed in a straight line completely across an alpine glacier. The position of the line was marked on the valley walls so that if the ice moved, the change in position could be detected. After a year or two the results were clear. The markers on the glacier had advanced down the valley, proving that glacial ice indeed moves. In addition, the experiment demonstrated that ice within a glacier does not move at a uniform rate, because the markers in the center advanced farther than did those along the margins. Although most glaciers move too slowly for direct visual detection, the experiment succeeded in demonstrating that movement nevertheless occurs. In the years that followed, this experiment was repeated many times with greater accuracy using more modern surveying techniques. Each time, the basic relationships established by earlier attempts were verified.

The experiment illustrated in Figure 1.A was carried out at Switzerland's Rhône Glacier later in the nineteenth century. It not only traced the movement of markers within the ice but also mapped the position of the glacier's terminus. Notice that even though the ice within the glacier was advancing, the ice front was retreating. As often occurs in science, experiments and observations designed to test one hypothesis yield new information that requires further analysis and explanation.

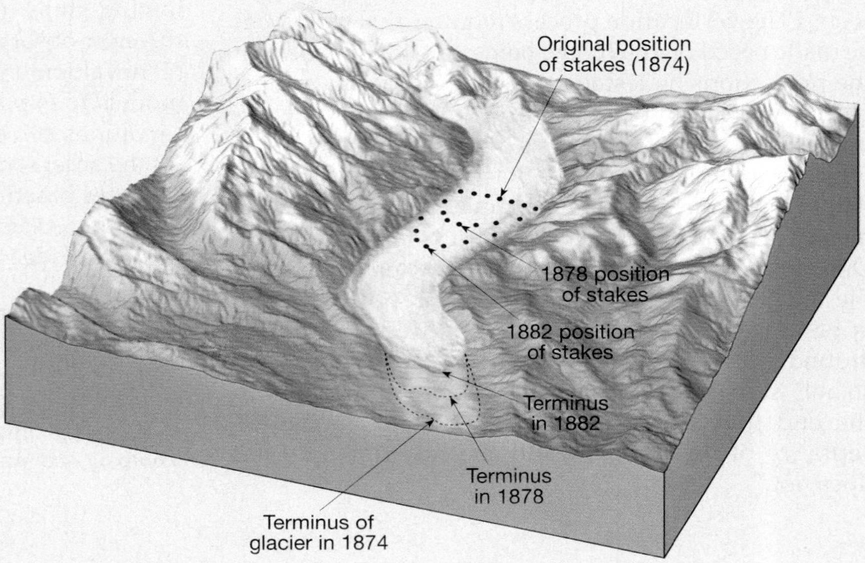

Figure 1.A Ice movement and changes in the terminus at Rhône Glacier, Switzerland. In this classic study of a valley glacier, the movement of stakes clearly show that glacial ice moves and that movement along the sides of the glacier is slower than movement in the center. Also notice that even though the ice front was retreating, the ice within the glacier was advancing.

Carefully examining a rock exposure. (Catherine Ursillo/ Photo Researchers, Inc.)

Some researchers use high-speed computers to simulate what is happening in the "real" world. These models are useful when dealing with natural processes that occur on very long time scales or take place in extreme or inaccessible locations. Still other scientific advancements have been made when a totally unexpected happening occurred during an experiment. These serendipitous discoveries are more than pure luck; for as Louis Pasteur stated, "In the field of observation, chance favors only the prepared mind."

Scientific knowledge is acquired through several avenues, so it might be best to describe the nature of scientific inquiry as the *methods* of science rather than *the* scientific method. In addition, it should always be remembered that even the most compelling scientific theories are still simplified explanations of the natural world.

Earth's Spheres

GEODe — An Introduction to Geology
▼ A View of Earth

A view such as the one in Figure 1.6A provided the *Apollo 8* astronauts as well as the rest of humanity with

A.

B.

▲ **Figure 1.6** **A.** View that greeted the *Apollo 8* astronauts as their spacecraft emerged from behind the Moon. *(NASA Headquarters)* **B.** Africa and Arabia are prominent in this image of Earth taken from *Apollo 17.* The tan cloud-free zones over the land coincide with major desert regions. The band of clouds across central Africa is associated with a much wetter climate that in places sustains tropical rain forests. The dark blue of the oceans and the swirling cloud patterns remind us of the importance of the oceans and the atmosphere. Antarctica, a continent covered by glacial ice, is visible at the South Pole. *(NASA/Science Source/Photo Researchers, Inc.)*

a unique perspective of our home. Seen from space, Earth is breathtaking in its beauty and startling in its solitude. Such an image reminds us that our home is, after all, a planet—small, self-contained, and in some ways even fragile.

As we look more closely at our planet from space, it becomes apparent that Earth is much more than rock and soil. In fact, the most conspicuous features in Figure 1.6A are not continents but swirling clouds suspended above the surface and the vast global ocean. These features emphasize the importance of water to our planet.

The closer view of Earth from space shown in Figure 1.6B helps us appreciate why the physical environment is traditionally divided into three major parts: the water portion of our planet, the hydrosphere; Earth's gaseous envelope, the atmosphere; and, of course, the solid Earth, or geosphere. It needs to be emphasized that our environment is highly integrated and is not dominated by rock, water, or air alone. Rather, it is characterized by continuous interactions as air comes in contact with rock, rock with water, and water with air. Moreover, the biosphere, which is the totality of all plant and animal life on our planet,interacts with each of the three physical realms and is an equally integral part of the planet. Thus, Earth can be thought of as consisting of four major spheres: the hydrosphere, atmosphere, geosphere, and biosphere.

The interactions among Earth's four spheres are incalculable. Figure 1.7 provides us with one easy-to-visualize example. The shoreline is an obvious meeting place for rock, water, and air. In this scene, ocean waves that were created by the drag of air moving across the water are breaking against the rocky shore. The force of the water can be powerful, and the erosional work that is accomplished can be great.

Did You Know?

A scientific *law* is a basic principle that describes a particular behavior of nature that is generally narrow in scope and can be stated briefly—often as a simple mathematical equation.

Hydrosphere

Earth is sometimes called the *blue* planet. Water more than anything else makes Earth unique. The **hydrosphere** is a dynamic mass of liquid that is continually on the move, evaporating from the oceans to the atmosphere, precipitating back to the land, and running back to the ocean again. The global ocean is certainly the most prominent feature of the hydrosphere, blanketing nearly 71 percent of Earth's surface to an average depth of about 3800 meters (12,500 feet). It accounts for about 97 percent of Earth's water. However, the hydrosphere also includes the freshwater found underground and in streams, lakes, and glaciers. Moreover, water is an important component of all living things.

Although these latter sources constitute just a tiny fraction of the total, they are much more important than their meager percentage indicates. Streams, glaciers, and groundwater are responsible for creating many of

Did You Know?

In 1492 when Columbus set sail, many Europeans thought that Earth was flat and that Columbus would sail off the edge. However, more than 2000 years earlier, ancient Greeks realized that Earth was spherical because it always cast a curved shadow on the Moon during a lunar eclipse. In fact, Erastosthenes (276–194 B.C.) calculated Earth's circumference and obtained a value close to the modern measurement of 40,075 km (24,902 mi).

our planet's varied landforms, as well as the freshwater that is so vital to life on land.

Atmosphere

Earth is surrounded by a life-giving gaseous envelope called the **atmosphere.** Compared with the solid Earth, the atmosphere is thin and tenuous. One half lies below an altitude of 5.6 kilometers (3.5 miles), and 90 percent occurs within just 16 kilometers (10 miles) of Earth's surface. By comparison, the radius of the solid Earth (distance from the surface to the center) is about 6400 kilometers (nearly 4000 miles)! Despite its modest dimensions, this thin blanket of air is an integral part of the planet. It not only provides the air that we breathe but also acts to protect us from the Sun's intense heat and dangerous ultraviolet radiation. The energy exchanges that continually occur between the atmosphere and the surface and between the atmosphere and space produce the effects we call weather and climate.

This jet is at about 30,000 feet. More than two-thirds of the atmosphere is below this height. (Warren Faidley/Weatherstock)

If, like the Moon, Earth had no atmosphere, our planet would not only be lifeless but many of the processes and interactions that make the surface such a dynamic place could not operate. Without weathering and erosion, the face of our planet might more closely resemble the lunar surface, which has not changed appreciably in nearly 3 billion years.

Biosphere

The **biosphere** includes all life on Earth. Ocean life is concentrated in the sunlit surface waters of the sea. Most life on land is also concentrated near the surface, with tree roots and burrowing animals reaching a few meters underground and flying insects and birds reaching a kilometer or so above Earth. A surprising variety of life forms are also adapted to extreme environments. For example, on the ocean floor where pressures are extreme and no light penetrates, there are places where vents spew hot, mineral-rich fluids that support communities of exotic life forms. On land, some bacteria thrive in rocks as deep as 4 kilometers (2.5 miles) and in boiling hot springs. Moreover, air currents can carry microorganisms many kilometers into the atmosphere. But even when we consider these extremes, life still must be thought of as being confined to a narrow band very near Earth's surface.

Did You Know?

The volume of ocean water is so large that if Earth's solid mass were perfectly smooth (level) and spherical, the oceans would cover Earth's entire surface to a uniform depth of more than 2000 m (1.2 mi)!

Plants and animals depend on the physical environment for the basics of life. However, organisms do not just respond to their physical environment. Indeed, the biosphere powerfully influences the other three spheres. Without life, the makeup and nature of the geosphere, hydrosphere, and atmosphere would be very different.

Geosphere

Lying beneath the atmosphere and the oceans is the solid Earth, or **geosphere.** The geosphere extends from the surface to the center of the planet, a depth of 6400 kilometers, making it by far the largest of Earth's four spheres. Much of our study of the solid Earth focuses on the more accessible surface features. Fortunately, many of these features represent the outward expressions of the dynamic behavior of Earth's interior. By examining the most prominent surface features and their global extent, we can obtain clues to the dynamic processes that have shaped our planet. A first look at the structure of Earth's interior and at the major surface features of the geosphere will come later in the chapter.

Soil, the thin veneer of material at Earth's surface that supports the growth of plants, may be thought of as part of all four spheres. The solid portion is a mixture of weathered rock debris (geosphere) and organic matter from decayed plant and animal life (biosphere). The decomposed and disintegrated rock debris is the product of weathering processes that require air (atmosphere) and water (hydrosphere). Air and water also occupy the open spaces between the solid particles.

▼ **Figure 1.7** The shoreline is one obvious meeting place for the hydrosphere, atmosphere, and solid Earth. In this scene, along California's Big Sur coastline, ocean waves that were created by the force of moving air break against the rocky shore. The force of the water can be powerful, and the erosional work that is accomplished can be great. *(Photo by David Muench/Muench Photography, Inc.)*

Earth as a System

Anyone who studies Earth soon learns that our planet is a dynamic body with many separate but interacting parts or *spheres.* The hydrosphere, atmosphere, biosphere, and geosphere and all of their components can be studied separately. However, the parts are not isolated. Each is related in some way to the others to produce a complex and continuously interacting whole that we call the *Earth system.*

Earth System Science

A simple example of the interactions among different parts of the Earth system occurs every winter as moisture evaporates from the Pacific Ocean and subsequently falls as rain in the hills of southern California, triggering destructive landslides. The processes that move water from the hydrosphere to the atmosphere and then to the solid Earth have a profound impact on the plants and animals (including humans) that inhabit the affected regions.

Scientists have recognized that in order to more fully understand our planet, they must learn how its individual components (land, water, air, and life forms) are interconnected. This endeavor, called **Earth system science,** aims to study Earth as a *system* composed of numerous interacting parts, or *subsystems.* Rather than looking through the limited lens of only one of the traditional sciences—geology, atmospheric science, chemistry, biology, etc.—Earth system science attempts to integrate the knowledge of several academic fields. Using this interdisciplinary approach, we hope to achieve the level of understanding necessary to comprehend and solve many of our global environmental problems.

What Is a System? Most of us hear and use the term *system* frequently. We may service our car's cooling *system,* make use of the city's transportation *system,* and participate in the political *system.* A news report might inform us of an approaching weather *system.* Further, we know that Earth is just a small part of a larger system known as the *solar system* which in turn is a *subsystem* of the even larger system called the Milky Way Galaxy.

Loosely defined, a **system** can be any size group of interacting parts that form a complex whole. Most natural systems are driven by sources of energy that move matter and/or energy from one place to another. A simple analogy is a car's cooling system, which contains a liquid (usually water and antifreeze) that is driven from the engine to the radiator and back again. The role of this system is to transfer heat generated by combustion in the engine to the radiator, where moving air removes it from the system. Hence, the term cooling system.

Systems like a car's cooling system are self-contained with regard to matter and are called **closed systems.** Although energy moves freely in and out of a closed system, no matter (liquid in the case of our auto's cooling systems) enters or leaves the system. (This assumes you don't get a leak in your radiator.) By contrast, most natural systems are **open systems** and are far more complicated than the foregoing example. In an open system both energy and matter flow into and out of the system. In a weather system such as a hurricane, factors such as the quantity of water vapor available for cloud formation, the amount of heat released by condensing water vapor, and the flow of air into and out of the storm can fluctuate a great deal. At times the storm may strengthen; at other times it may remain stable or weaken.

Did You Know?

Primitive life first appeared in the oceans about 4 billion years ago and has been spreading and diversifying ever since.

These tube worms live in the high pressures and complete darkness of the ocean floor. (Al Giddings Images, Inc.)

Feedback Mechanisms. Most natural systems have mechanisms that tend to enhance change, as well as other mechanisms that tend to resist change and thus stabilize the system. For example, when we get too hot, we perspire to cool down. This cooling phenomenon works to stabilize our body temperature and is referred to as a **negative feedback mechanism.** Negative feedback mechanisms work to maintain the system as it is or, in other words, to maintain the status quo. By contrast, mechanisms that enhance or drive change are called **positive feedback mechanisms.**

Most of Earth's systems, particularly the climate system, contain a wide variety of negative and positive feedback mechanisms. For example, substantial scientific evidence indicates that Earth has entered a period of global warming. One consequence of global warming is that some of the world's glaciers and ice caps have begun to melt. Highly reflective snow- and ice-covered surfaces are gradually being replaced by brown soils, green trees, or blue oceans, all of which are darker, so they absorb more sunlight. Therefore, as Earth warms and some snow and ice melt, our planet absorbs more sunlight. The result is a positive feedback that contributes to the warming.

On the other hand, an increase in global temperature also causes greater evaporation of water from Earth's land-sea surface. One result of having more water vapor in the air is an increase in cloud cover. Because cloud tops are white and highly reflective, more sunlight is reflected back to space, which diminishes the amount of sunshine reaching Earth's surface and thus reduces global temperatures. Further, warmer temperatures tend to promote the growth of vegetation. Plants in turn remove carbon dioxide (CO_2) from the air. Since carbon dioxide is one of the atmosphere's

Did You Know?

During the twentieth century, the global average surface temperature increased by about 0.6°C (1°F). By the end of the twenty-first century, globally averaged surface temperature is projected to increase by an additional 1.4 to 5.8°C (2.5 to 14.5°F).

greenhouse gases, its removal has a negative impact on global warming.*

In addition to natural processes, we must also consider the human element. Extensive cutting and clearing of the tropical rain forests and the burning of fossil fuels (oil, natural gas, and coal) result in an increase in atmospheric CO_2. Such activity appears to have contributed to the increase in global temperature that our planet is experiencing. One of the daunting tasks for Earth system scientists is to predict what the climate will be like in the future by taking into account many variables, including technological changes, population trends, and the overall impact of the numerous competing positive and negative feedback mechanisms.

The Earth System

The Earth system has a nearly endless array of subsystems in which matter is recycled over and over again (Figure 1.8). One example that you will learn about in Chapter 6 traces the movements of carbon among Earth's four spheres. It shows us, for example, that the carbon dioxide in the air and the carbon in living things and in certain sedimentary rocks is all part of a subsystem described by the *carbon cycle.*

Energy from the Sun powers Earth's external processes.

Cycles in the Earth System. A more familiar loop or subsystem is the *hydrologic cycle.* It represents the unending circulation of Earth's water among the hydrosphere, atmosphere, biosphere, and geosphere. Water enters the atmosphere by evaporation from Earth's surface and by transpiration from plants. Water vapor condenses in the atmosphere to form clouds, which in turn produce precipitation that falls back to Earth's surface. Some of the rain that falls onto the land sinks in to be taken up by plants or become groundwater, and some flows across the surface toward the ocean.

Viewed over long time spans, the rocks of the geosphere are constantly forming, changing, and reforming (Figure 1.8). The loop that involves the processes by which one rock changes to another is called the *rock cycle* and will be discussed at some length in the following section. The cycles of the Earth system, such as the hydrologic and rock cycles, are not independent of one another. To the contrary, there are many places where they interface. An **interface** is a common boundary where different parts of a system come in contact and interact. For example, in Figure 1.8, weathering at the surface gradually disintegrates and decomposes solid rock. The work of gravity and running water may eventually move this material to another place and deposit it. Later, groundwater percolating through the debris may leave behind mineral matter that cements the grains together into solid rock (a rock that is often very different from the rock we started with). This changing of one rock into another could not have occurred without the movement of water through the hydrologic cycle. There are many places where one cycle or loop in the Earth system interfaces with and is a basic part of another.

Energy for the Earth System. The Earth system is powered by energy from two sources. The Sun drives external processes that occur in the atmosphere, hydrosphere, and at Earth's surface. Weather and climate, ocean circulation, and erosional processes are driven by energy from the Sun. Earth's interior is the second source of energy. Heat remaining from when our planet formed, and heat that is continuously generated by decay of radioactive elements, powers the internal processes that produce volcanoes, earthquakes, and mountains.

The Parts are Linked. The parts of the Earth system are linked so that a change in one part can produce changes in any or all of the other parts. For example, when a volcano erupts, lava from Earth's interior may flow out at the surface and block a nearby valley. This new obstruction influences the region's drainage system by creating a lake or causing streams to change course. The large quantities of volcanic ash and gases that can be emitted during an eruption might be blown high into the atmosphere and influence the amount of solar energy that can reach Earth's surface. The result could be a drop in air temperatures over the entire hemisphere.

Where the surface is covered by lava flows or a thick layer of volcanic ash, existing soils are buried. This causes the soil-forming processes to begin anew to transform the new surface material into soil (Figure 1.9). The soil that eventually forms will reflect the interactions among many parts of the Earth system—the volcanic parent material, the type and rate of weathering, and the impact of biological activity. Of course, there would also be significant changes in the biosphere. Some organisms and their habitats would be eliminated by the lava and ash, whereas new settings for life, such as the lake, would be created. The potential climate change could also impact sensitive life forms.

The Earth system is characterized by processes that vary on spatial scales from fractions of millimeters to thousands of kilometers. Time scales for Earth's processes range from milliseconds to billions of years. As we learn about Earth, it becomes increasingly clear that despite significant separations in distance or time, many processes are connected, and a change in one component can influence the entire system.

*Greenhouse gases absorb heat energy emitted by Earth and thus help keep the atmosphere warm.

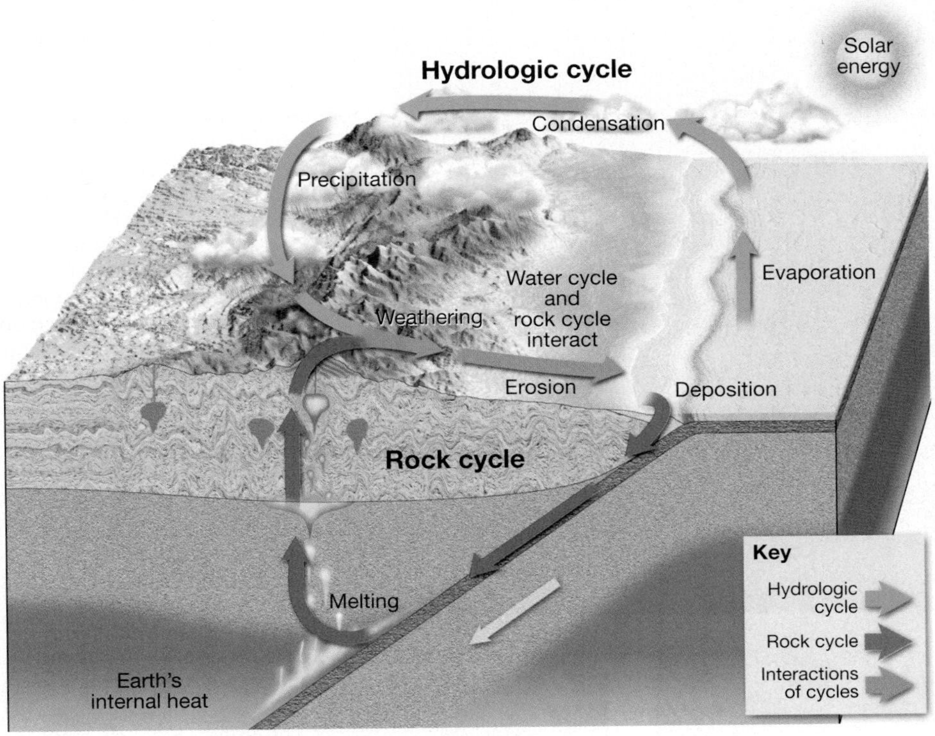

◀ **Figure 1.8** Each part of the Earth system is related to every other part to produce a complex interacting whole. The Earth system involves many cycles, including the hydrologic cycle and the rock cycle. Such cycles are not independent of each other. There are many places where they interface.

Humans are *part of* the Earth system, a system in which the living and nonliving components are entwined and interconnected. Therefore, our actions produce changes in all of the other parts. When we burn gasoline and coal, build breakwaters along the shoreline, dispose of our wastes, and clear the land, we cause other parts of the system to respond, often in unforeseen ways. Throughout this book you will learn about many of Earth's subsystems: the hydrologic system, the tectonic (mountain-building) system, and the rock cycle, to name a few. Remember that these components *and we humans* are all part of the complex interacting whole we call the Earth system.

▼ **Figure 1.9** When Mount St. Helens erupted in May 1980, the area shown here was buried by a volcanic mudflow. Now plants are reestablished and new soil is forming. *(Jack Dykinga Photography)*

The Rock Cycle: One of Earth's Subsystems

GEODe: An Introduction to Geology
▼ The Rock Cycle

Rock is the most common and abundant material on Earth. To a curious traveler, the variety seems nearly endless. When a rock is examined closely, we find that it consists of smaller crystals or grains called minerals. *Minerals* are chemical compounds (or sometimes single elements), each with its own composition and physical properties. The grains or crystals may be microscopically small or easily seen with the unaided eye.

The nature and appearance of a rock is strongly influenced by the minerals that compose it. In addition, a rock's *texture*—the size, shape, and/or arrangement of its constituent minerals—also has a significant effect on its appearance. A rock's mineral composition and texture, in turn, are a reflection of the geologic processes that created it.

The characteristics of the rocks in Figure 1.10 provided geologists with the clues they needed to determine the processes that formed them. This is true of all rocks. Such analyses are critical to an understanding of our planet. This understanding has many practical applications, as in the search for basic mineral and energy resources and the solution of environmental problems.

Geologists divide rocks into three major groups: igneous, sedimentary, and metamorphic. In Figure 1.10,

A.

B.

C.

◄ **Figure 1.10** **A.** This fine-grained rock, called *basalt,* is part of a lava flow from Sunset Crater in northern Arizona. It formed when molten rock erupted at Earth's surface hundreds of years ago and solidified. *(Photo by David Muench)* **B.** This rock is exposed in the walls of southern Utah's Zion National Park. This layer, known as the Navajo Sandstone, consists of durable grains of the glassy mineral quartz that once covered this region with mile after mile of drifting sand dunes. *(Photo by Tom Bean/DRK Photo)* **C.** This rock unit, known as the Vishnu Schist, is exposed in the inner gorge of the Grand Canyon. Its formation is associated with environments far below Earth's surface where temperatures and pressures are high and with ancient mountain-building processes that occurred in Precambrian time. *(Photo by Tom Bean/DRK Photo)*

the lava flow in northern Arizona is classified as igneous, the sandstone in Utah's Zion National Park is sedimentary, and the schist exposed at the bottom of the Grand Canyon is metamorphic.

Texture and mineral composition are important rock properties.

In the preceding section, you learned that Earth is a system. This means that our planet consists of many interacting parts that form a complex whole. Nowhere is this idea better illustrated than when we examine the rock cycle (Figure 1.11). The **rock cycle** allows us to view many of the interrelationships among different parts of the Earth system. Knowledge of the rock cycle will help you more clearly understand the idea that each rock group is linked to the others by the processes that act upon and within the planet. You can consider the rock cycle to be a simplified but useful overview of physical geology. What follows is a brief introduction to the rock cycle. Learn the rock cycle well; you will be examining its interrelationships in greater detail throughout this book.

The Basic Cycle

We begin at the top of Figure 1.11. **Magma** is molten material that forms inside Earth. Eventually magma cools and solidifies. This process, called *crystallization,* may occur either beneath the surface or, following a

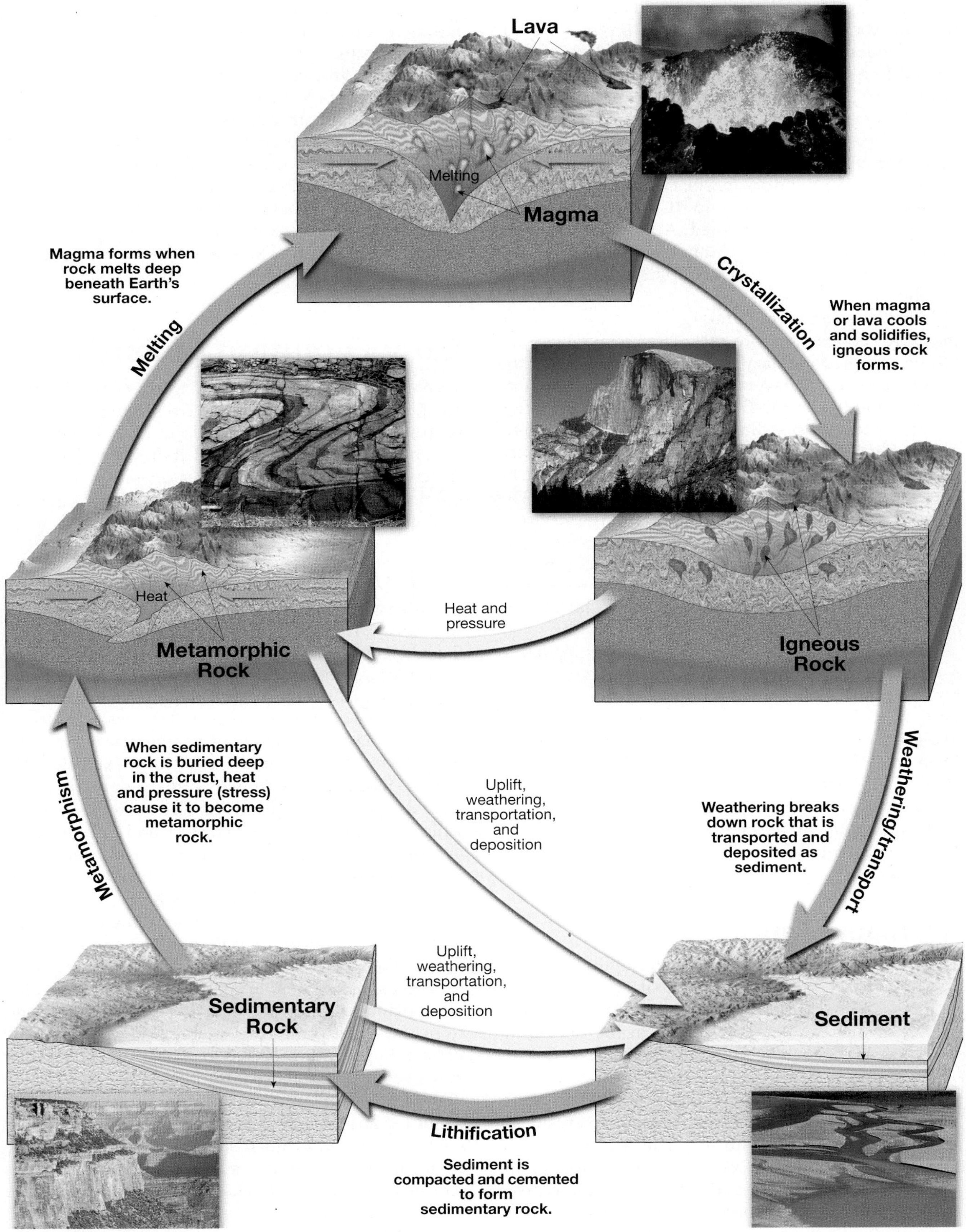
Lava
Melting
Magma
Magma forms when rock melts deep beneath Earth's surface.
Melting
Crystallization
When magma or lava cools and solidifies, igneous rock forms.
Heat
Metamorphic Rock
Heat and pressure
Igneous Rock
Metamorphism
When sedimentary rock is buried deep in the crust, heat and pressure (stress) cause it to become metamorphic rock.
Uplift, weathering, transportation, and deposition
Weathering/transport
Weathering breaks down rock that is transported and deposited as sediment.
Uplift, weathering, transportation, and deposition
Sedimentary Rock
Sediment
Lithification
Sediment is compacted and cemented to form sedimentary rock.

▲ **Figure 1.11** Viewed over long spans, rocks are constantly forming, changing, and reforming. The rock cycle helps us understand the origin of the three basic rock groups. Arrows represent processes that link each group to the others.

An explosive volcanic eruption can influence all spheres of the Earth system. (Guillermo Granja/Reuters/CORBIS)

volcanic eruption, at the surface. In either situation, the resulting rocks are called **igneous rocks** (*ignis* = fire).

If igneous rocks are exposed at the surface, they will undergo *weathering*, in which the day-in and day-out influences of the atmosphere slowly disintegrate and decompose rocks. The materials that result are often moved downslope by gravity before being picked up and transported by any of a number of erosional agents, such as running water, glaciers, wind, or waves. Eventually these particles and dissolved substances, called **sediment,** are deposited. Although most sediment ultimately comes to rest in the ocean, other sites of deposition include river floodplains, desert basins, swamps, and sanddunes.

Next the sediments undergo *lithification*, a term meaning "conversion into rock." Sediment is usually lithified into **sedimentary rock** when compacted by the weight of overlying layers or when cemented as percolating groundwater fills the pores with mineral matter.

If the resulting sedimentary rock is buried deep within Earth and involved in the dynamics of mountain building or intruded by a mass of magma, it will be subjected to great pressures and/or intense heat. The sedimentary rock will react to the changing environment and turn into the third rock type, **metamorphic rock.** If metamorphic rock is subjected to still higher temperatures, it will melt, creating magma, which will eventually crystallize into igneous rock, starting the cycle all over again.

Although rocks may seem to be unchanging masses, the rock cycle shows that they are not. The changes, however, take time—great amounts of time. In addition, the rock cycle is operating all over the world, but in different stages. Today new magma is forming under the island of Hawaii, while the Colorado Rockies are slowly being wom down by weathering and erosion. Some of this weathered debris will eventually be carried to the Gulf of Mexico, where it will add to the already substantial mass of sediment that has accumulated there.

Alternative Paths

The paths shown in the basic cycle are not the only ones that are possible. To the contrary, other paths are just as likely to be followed as those described in the preceding section. These alternatives are indicated by the blue arrows in Figure 1.11.

Igneous rocks, rather than being exposed to weathering and erosion at Earth's surface, may remain deeply buried. Eventually these masses may be subjected to the strong compressional forces and high temperatures associated with mountain building. When this occurs, they are transformed directly into metamorphic rocks.

> **Did You Know?**
>
> The *Light-year* is a unit for measuring distances to stars. Such distances are so large that familiar units such as kilometers or miles are cumbersome to use. One light-year is the distance light travels in one Earth year—about 9.5 trillion km (5.8 trillion mi)!

Metamorphic and sedimentary rocks, as well as sediment, do not always remain buried. Rather, overlying layers may be stripped away, exposing the once buried rock. When this happens, the material is attacked by weathering processes and turned into new raw materials for sedimentary rocks.

Where does the energy that drives Earth's rock cycle come from? Processes driven by heat from Earth's interior are responsible for forming igneous and metamorphic rocks. Weathering and the movement of weathered material are external processes powered by energy from the Sun. External processes produce sedimentary rocks.

Early Evolution of Earth

Recent earthquakes caused by displacements of Earth's crust, along with lavas erupted from active volcanoes, represent only the latest in a long line of events by which our planet has attained its present form and structure. The geologic processes operating in Earth's interior can be best understood when viewed in the context of much earlier events in Earth history.

Origin of Planet Earth

The following scenario describes the most widely accepted views of the origin of our solar system. Although this model is presented as fact, keep in mind that like all scientific hypotheses, this one is subject to revision and even outright rejection. Nevertheless, it remains the most consistent set of ideas to explain what we observe today.

Our scenario begins about 14 billion years ago with the *Big Bang*, an incomprehensibly large explosion that sent all matter of the universe flying outward at incredible speeds. In time, the debris from this explosion, which was almost entirely hydrogen and helium, began to cool and condense into the first stars and galaxies. It was in one of these galaxies, the Milky Way, that our solar system and planet Earth took form.

Earth is one of nine planets that, along with several dozen moons and numerous smaller bodies, revolve around the Sun. The orderly nature of our solar system leads most researchers to conclude that Earth and the other planets formed at essentially the same time and from the same primordial material as the Sun. The **nebular hypothesis** proposes that the bodies of our solar system evolved from an enormous rotating cloud called the *solar nebula* (Figure 1.12). Besides the hydrogen and helium atoms generated during the Big Bang, the solar nebula consisted of microscopic dust grains and the ejected matter of long-dead stars. (Nuclear fusion in stars converts hydrogen and helium into the other elements found in the universe.)

Nearly 5 billion years ago this huge cloud of gases and minute grains of heavier elements began to slowly contract due to the gravitational interactions among its particles. Some external influence, such as a shock wave traveling from a catastrophic explosion (*supernova*), may have triggered the collapse. As this slowly spiraling nebula contracted, it rotated faster and faster for the same reason ice skaters do when they draw their arms toward their bodies. Eventually the inward pull of gravity came into balance with the outward force caused by the rotational motion of the nebula (Figure 1.12). By this time the once vast cloud had assumed a flat disk shape

▲ **Figure 1.12** Formation of the solar system according to the nebular hypothesis. **A.** The birth of our solar system began as dust and gases (nebula) started to gravitationally collapse. **B.** The nebula contracted into a rotating disk that was heated by the conversion of gravitational energy into thermal energy. **C.** Cooling of the nebular cloud caused rocky and metallic material to condense into tiny particles. **D.** Repeated collisions caused the dust-size particles to gradually coalesce into asteroid-size bodies. **E.** Within a few million years these bodies accreted into the planets.

Nebulae such as these are where stars are born. (National Optical Astronomy Observatories)

with a large concentration of material at its center called the *protosun* (pre-Sun). (Astronomers are fairly confident that the nebular cloud formed a disk because similar structures have been detected around other stars.)

During the collapse, gravitational energy was converted to thermal energy (heat), causing the temperature of the inner portion of the nebula to dramatically rise. At these high temperatures, the dust grains broke up into molecules and excited atomic particles. However, at distances beyond the orbit of Mars, the temperatures probably remained quite low. At −200°C, the tiny particles in the outer portion of the nebula were likely covered with a thick layer of ices made of frozen water, carbon dioxide, ammonia, and methane. (Some of this material still resides in the outermost reaches of the solar system in a region called the *Oort cloud.*) The disk-shaped cloud also contained appreciable amounts of the lighter gases hydrogen and helium.

Did You Know?

A growing number of astronomers are of the opinion that Pluto should not be classified as a planet. They think it is best described as a large cometlike body that orbits the outer reaches of the solar system.

The formation of the Sun marked the end of the period of contraction and thus the end of gravitational heating. Temperatures in the region where the inner planets now reside began to decline. The decrease in temperature caused those substances with high melting points to condense into tiny particles that began to coalesce (join together). Materials such as iron and nickel and the elements of which the rock-forming minerals are composed—silicon, calcium, sodium, and so forth—formed metallic and rocky clumps that orbited the Sun (Figure 1.12). Repeated collisions caused these masses to coalesce into larger asteroid-size bodies, called *protoplanets,* which in a few tens of millions of years accreted into the four inner planets we call Mercury, Venus, Earth, and Mars. Not all of these clumps of matter were incorporated into the protoplanets. Those rocky and metallic pieces that remained in orbit are called *meteorites* when they survive an impact with Earth.

As more and more material was swept up by the protoplanets, the high-velocity impact of nebular debris caused the temperature of these bodies to rise. Because of their relatively high temperatures and weak gravitational fields, the inner planets were unable to accumulate much of the lighter components of the nebular cloud. The lightest of these, hydrogen and helium, were eventually whisked from the inner solar system by the solar winds.

At the same time that the inner planets were forming, the larger, outer planets (Jupiter, Saturn, Uranus, and Neptune), along with their extensive satellite systems, were also developing. Because of low temperatures far from the Sun, the material from which these planets formed contained a high percentage of ices—water, carbon dioxide, ammonia, and methane—as well as rocky and metallic debris. The accumulation of ices accounts in part for the large size and low density of the outer planets. The two most massive planets, Jupiter and Saturn, had a surface gravity sufficient to attract and hold large quantities of even the lightest elements—hydrogen and helium.

Formation of Earth's Layered Structure

As material accumulated to form Earth (and for a short period afterward), the high-velocity impact of nebular debris and the decay of radioactive elements caused the temperature of our planet to steadily increase. During this time of intense heating, Earth became hot enough that iron and nickel began to melt. Melting produced liquid blobs of heavy metal that sank toward the center of the planet. This process occurred rapidly on the scale of geologic time and produced Earth's dense iron-rich core.

The early period of heating resulted in another process of chemical differentiation, whereby melting formed buoyant masses of molten rock that rose toward the surface, where they solidified to produce a primitive crust. These rocky materials were enriched in oxygen and "oxygen-seeking" elements, particularly silicon and aluminum, along with lesser amounts of calcium, sodium, potassium, iron, and magnesium. In addition, some heavy metals such as gold, lead, and uranium, which have low melting points or were highly soluble in the ascending molten masses, were scavenged from Earth's interior and concentrated in the developing crust. This early period of chemical segregation established the three basic divisions of Earth's interior—the iron-rich *core;* the thin *primitive crust;* and Earth's largest layer, called the *mantle,* which is located between the core and crust.

An important consequence of this early period of chemical differentiation is that large quantities of gaseous materials were allowed to escape from Earth's interior, as happens today during volcanic eruptions. By this process a primitive atmosphere gradually evolved. It is on this planet, with this atmosphere, that life as we know it came into existence.

Following the events that established Earth's basic structure, the primitive crust was lost to erosion and other geologic processes, so we have no direct record of its makeup. When and exactly how the continental crust—and thus Earth's first landmasses—came into existence is a matter of ongoing research. Nevertheless, there is general agreement that the continental crust formed gradually over the last 4 billion years. (The oldest rocks yet discovered are isolated fragments found in the Northwest Territories of Canada that have radiometric dates of about 4 billion years.) In addition, as you will see in subsequent chapters, Earth is an evolving planet whose continents (and ocean basins) have continually changed shape and even location during much of this period.

Earth's Internal Structure

GEODe ESSENTIALS OF GEOLOGY An Introduction to Geology
▼ Earth's Layered Structure

In the preceding section, you learned that the segregation of material that began early in Earth's history resulted in the formation of three layers defined by their chemical composition—the crust, mantle, and core. In addition to these compositionally distinct layers, Earth can be divided into layers based on physical properties. The physical properties used to define such zones include whether the layer is solid or liquid and how weak or strong it is. Knowledge of both types of layered structures is essential to our understanding of basic geologic processes, such as volcanism, earthquakes, and mountain building (Figure 1.13).

Layers Defined by Composition

Crust. The **crust,** Earth's comparatively thin, rocky outer skin, is generally divided into oceanic and continental crust. The oceanic crust is roughly 7 kilometers (5 miles) thick and composed of dark igneous rocks called *basalt* and *gabbro.* By contrast, the continental crust averages 35–40 kilometers (25 miles) thick but may exceed 70 kilometers (40 miles) in some mountainous regions. Unlike the oceanic crust, which has a relatively homogeneous chemical composition, the continental crust consists of many rock types. The upper crust has an average composition of a *granitic rock* called *granodiorite,* whereas the composition of the lowermost continental crust is more akin to basalt (see Chapter 3). Continental rocks have an average density of about 2.7 g/cm^3, and some have been discovered that are 4 billion years old. The rocks of the oceanic crust are younger (180 million years or less) and more dense (about 3.0 g/cm^3) than continental rocks.*

Mantle. Over 82 percent of Earth's volume is contained in the **mantle,** a solid, rocky shell that extends to a depth of 2900 kilometers (1800 miles). The boundary between the crust and mantle represents a marked change in chemical composition. The dominant rock type in the uppermost mantle is *peridotite,* which has a density of 3.3 g/cm^3. At greater depth peridotite changes by assuming a more compact crystalline structure and hence a greater density.

Core. The composition of the **core** is thought to be an iron-nickel alloy with minor amounts of oxygen, silicon, and sulfur—elements that readily form compounds with iron. At the extreme pressure found in the core, this iron-rich material has an average density of nearly 11 g/cm^3 and approaches 14 times the density of water at Earth's center.

Did You Know?

We have never sampled the mantle or core directly. The structure of Earth's interior is determined by analyzing seismic waves from earthquakes. As these waves of energy penetrate Earth's interior, they change speed and are bent and reflected as they move through zones having different properties. Monitoring stations around the world detect and record this energy.

Layers Defined by Physical Properties

Earth's interior is characterized by a gradual increase in temperature, pressure, and density with depth. Estimates put the temperature at a depth of 100 kilometers at between 1200°C and 1400°C, whereas the temperature at Earth's center may exceed 6700°C. Clearly, Earth's interior has retained much of the energy acquired during its formative years, despite the fact that heat is continuously flowing toward the surface, where it is lost to space. The increase in pressure with depth causes a corresponding increase in rock density.

The gradual increase in temperature and pressure with depth affects the physical properties and hence the mechanical behavior of Earth materials. When a substance is heated, its chemical bonds weaken and its mechanical strength (resistance to deformation) is reduced. If the temperature exceeds the melting point of an Earth material, the material's chemical bonds break and melting ensues. If temperature were the only factor that determined whether a substance melted, our planet would be a molten ball covered with a thin, solid outer shell. However, pressure also increases with depth and tends to increase rock strength. Furthermore, because melting is accompanied by an increase in volume, it occurs at higher temperatures at depth because of greater confining pressure. Thus, depending on the physical environment (temperature and pressure), a particular Earth material may behave like a brittle solid, deform in a puttylike manner, or even melt and become liquid.

Earth can be divided into five main layers based on physical properties and hence mechanical strength—the *lithosphere, asthenosphere, mesosphere (lower mantle), outer core,* and *inner core.*

*Liquid water has a density of 1 g/cm^3; therefore, the density of basalt is three times that of water.

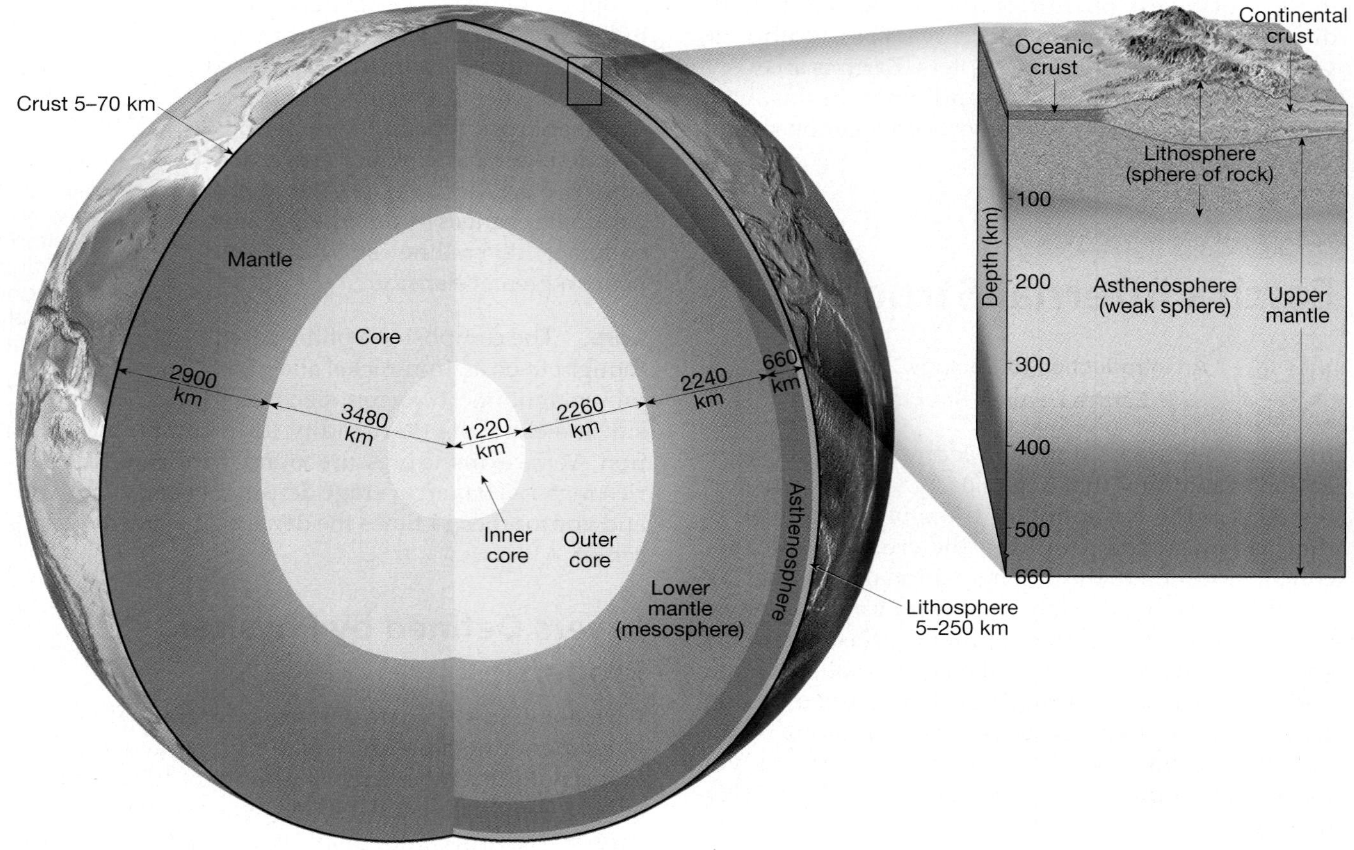

▲ **Figure 1.13** Views of Earth's layered structure. The left side of the large cross section shows that Earth's interior is divided into three different layers based on compositional differences—the crust, mantle, and core. The right side of the large cross section depicts the five main layers of Earth's interior based on physical properties and hence mechanical strength—the lithospere, asthenosphere, lower mantle (mesosphere), outer core, and inner core. The block diagram to the right of the large cross section shows an enlarged view of the upper portion of Earth's interior.

Lithosphere and Asthenosphere. Based on physical properties, Earth's outermost layer consists of the crust and uppermost mantle and forms a relatively cool, rigid shell. Although this layer is composed of materials with markedly different chemical compositions, it tends to act as a unit that exhibits rigid behavior—mainly because it is cool and thus strong. This layer, called the **lithosphere** (*sphere of rock*), averages about 100 kilometers (more than 60 miles) in thickness but may be more than 250 kilometers (150 miles) thick below the older portions of the continents (Figure 1.13). Within the ocean basins the lithosphere is only a few kilometers thick along the oceanic ridges but increases to perhaps 100 kilometers in regions of older and cooler oceanic crust.

Beneath the lithosphere, in the upper mantle lies a soft, comparatively weak layer known as the **asthenosphere** ("weak sphere"). The top portion of the asthenosphere has a temperature/pressure regime that results in a small amount of melting. Within this very weak zone the lithosphere is mechanically detached from the layer below. The result is that the lithosphere is able to move independently of the asthenosphere, a fact we will consider later in the chapter.

It is important to emphasize that the strength of various Earth materials is a function of both their composition and of the temperature and pressure of their environment. You should not get the idea that the entire lithosphere behaves like a brittle solid similar to rocks found on the surface. Rather, the rocks of the lithosphere get progressively hotter and weaker (more easily deformed) with increasing depth. At the depth of the uppermost asthenosphere, the rocks are close enough to their melting temperatures (some melting may actually occur) that they are very easily deformed. Thus, the uppermost asthenosphere is weak because it is near its melting point, just as hot wax is weaker than cold wax.

Mesosphere or Lower Mantle. Below the zone of weakness in the uppermost asthenosphere, increased pressure counteracts the effects of higher temperature, and the rocks gradually strengthen with depth. Between the depths of 660 kilometers and 2900 kilometers a more rigid layer, called the **mesosphere** (*middle sphere*) or **lower mantle,** is found. Despite their strength, the rocks of the mesosphere are still very hot and capable of very gradual flow.

Inner and Outer Core. The core, which is composed mostly of an iron-nickel alloy, is divided into two regions that exhibit very different mechanical strengths. The **outer core** is a *liquid layer* 2270 kilometers (1410 miles) thick. It is the movement of metallic iron within this zone that generates Earth's magnetic field. The **inner core** is a sphere having a radius of 1216 kilometers (754 miles). Despite its higher temperature, the material in the inner core is stronger (because of immense pressure) than the outer core and behaves like a *solid.*

The Face of Earth

The two principal divisions of Earth's surface are the continents and the ocean basins (Figure 1.14). A significant difference between these two areas is their relative levels. The continents are remarkably flat features that have the appearance of plateaus protruding above sea level. With an average elevation of about 0.8 kilometer (0.5 mile), continents lie close to sea level, except for limited areas of mountainous terrain. By contrast, the average depth of the ocean floor is about 3.8 kilometers (2.4 miles) below sea level, or about 4.5 kilometers (2.8 miles) lower than the average elevation of the continents.

The elevation difference between the continents and ocean basins is primarily the result of differences in their respective densities and thicknesses. Recall that the continents average 35–40 kilometers in thickness and are composed of granitic rocks having a density of about 2.7 g/cm^3. The basaltic rocks that comprise the oceanic crust average only 7 kilometers thick and have an average density of about 3.0 g/cm^3. Thus, the thicker and less dense continental crust is more buoyant than the oceanic crust. As a result, continental crust floats on top of the deformable rocks of the mantle at a higher level than oceanic crust for the same reason that a large, empty (less dense) cargo ship rides higher than a small, loaded (denser) one.

Features of the Continents

The largest features of the continents can be grouped into two distinct categories: extensive, flat stable areas that have been eroded nearly to sea level, and uplifted regions of deformed rocks that make up present-day mountain belts.

The most prominent topographic features of the continents are linear mountain belts. Although the distribution of mountains appears to be random, this is not the case. When the youngest mountains are considered (those less than 100 million years old), we find that they are located principally in two major zones. The circum-Pacific belt (the region surrounding the Pacific Ocean) includes the mountains of the western Americas and continues into the western Pacific in the form of volcanic islands such as the Aleutians, Japan, and the Philippines (Figure 1.14).

The other major mountainous belt extends eastward from the Alps through Iran and the Himalayas and then dips southward into Indonesia. Careful examination of mountainous terrains reveals that most are places where thick sequences of rocks have been squeezed and highly deformed, as if placed in a gigantic vise. Older mountains are also found on the continents. Examples include the Appalachians in the eastern United States and the Urals in Russia. Their once lofty peaks are now worn low, the result of millions of years of erosion.

Unlike the young mountain belts, which have formed within the last 100 million years, the interiors of the continents have been relatively stable (undisturbed) for the last 600 million years or even longer. Typically, these regions were involved in mountain-building episodes much earlier in Earth's history.

Within the stable interiors are areas known as **shields** which are expansive, flat regions composed of deformed crystalline rock. Notice in Figure 1.15 that the Canadian Shield is exposed in much of the northeastern part of North America. Age determinations for various shields have shown that they are truly ancient regions. All contain Precambrian-age rocks that are over 1 billion years old, with some samples approaching 4 billion years in age (see Figure 1.4 to review the geologic time scale). These oldest-known rocks exhibit evidence of enormous forces that have folded and faulted them and altered them with great heat and pressure. Thus, we conclude that these rocks were once part of an ancient mountain system that has since been eroded away to produce these expansive, flat regions.

Other flat areas of the stable interior exist in which highly deformed rocks, like those found in the shields, are covered by a relatively thin veneer of sedimentary rocks. These areas are called **stable platforms.** The sedimentary rocks in stable platforms are nearly horizontal except where they have been warped to form large basins or domes. In North America a major portion of the stable platforms is located between the Canadian Shield and the Rocky Mountains (Figure 1.15).

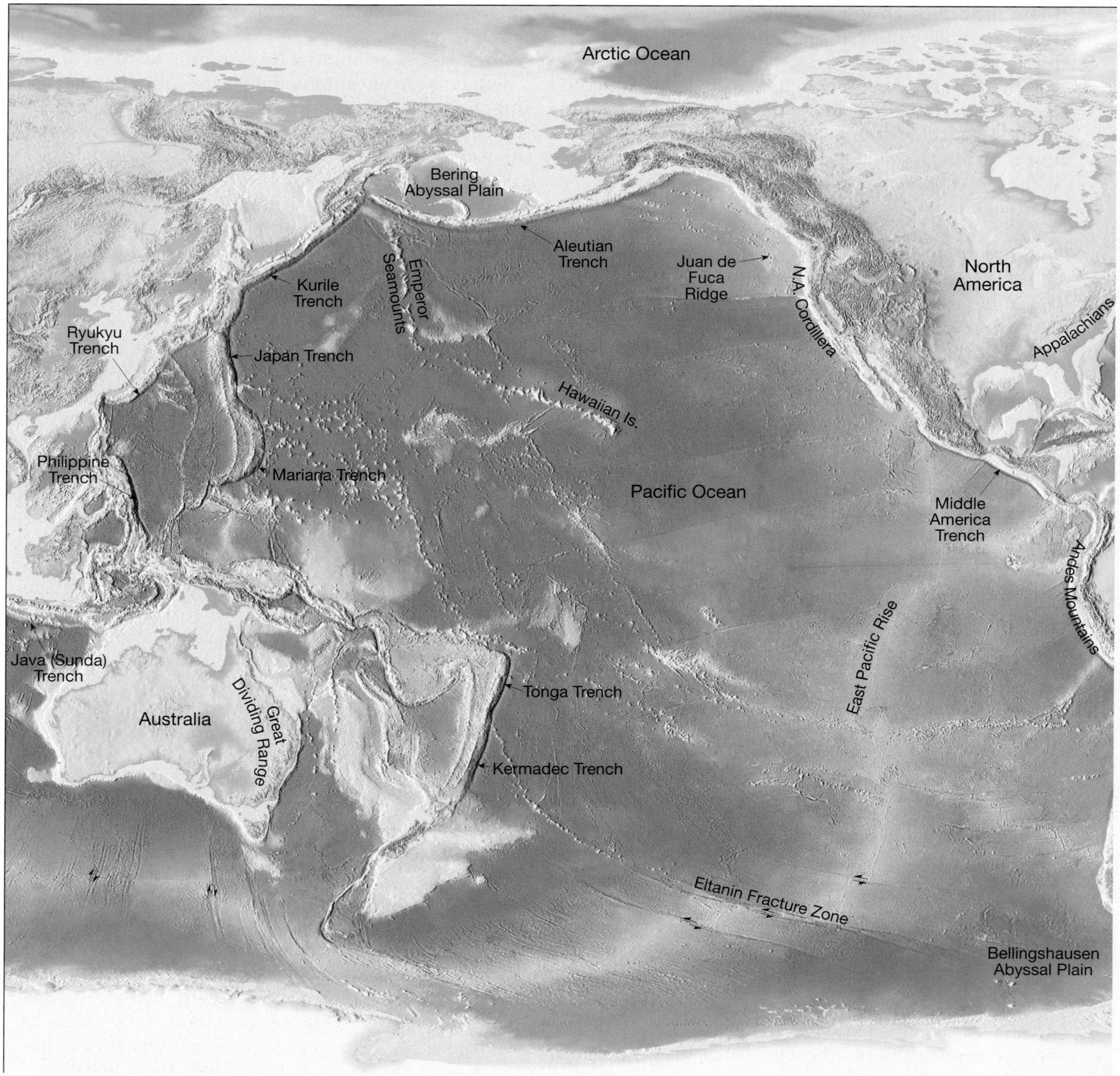

▲ **Figure 1.14** Major surface features of the geosphere.

Features of the Ocean Basins

If all water were drained from the ocean basins, a great variety of features would be seen, including linear chains of volcanoes, deep canyons, plateaus, and large expanses of monotonously flat plains. In fact, the scenery would be nearly as diverse as that on the continents (see Figure 1.14).

During the past 50 years, oceanographers using modern sonar equipment have slowly mapped significant portions of the ocean floor. From these studies they have defined three major regions: *continental margins, deep-ocean basins,* and *oceanic (mid-ocean) ridges.*

The **continental margin** is that portion of the seafloor adjacent to major landmasses. Although land

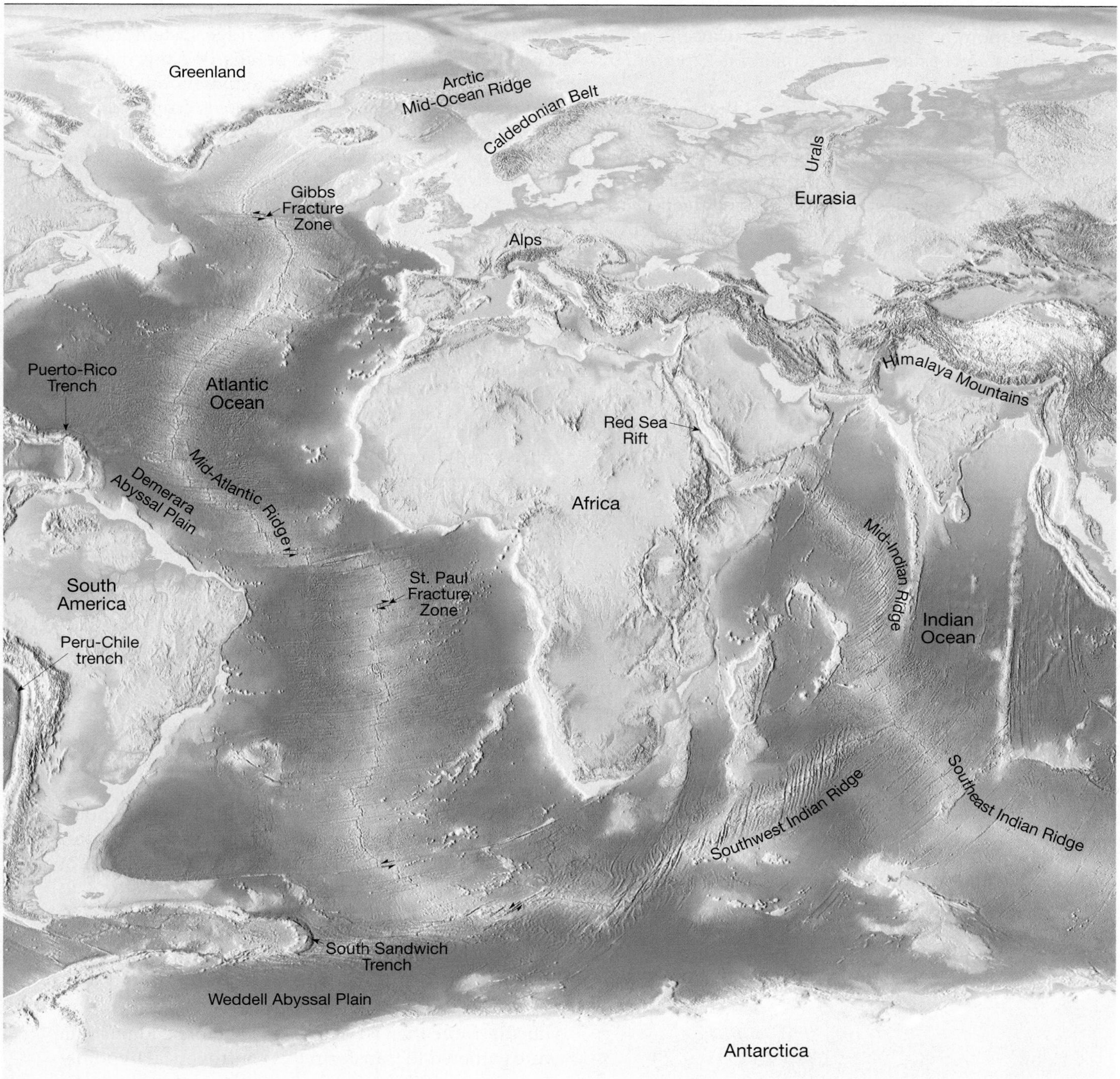

▲ **Figure 1.14** Continued.

and sea meet at the shoreline, this is not the boundary between the continents and the ocean basins. Rather, along most coasts a gently sloping platform of material, called the **continental shelf,** extends seaward from the shore. Because it is underlain by continental crust, it is considered a flooded extension of the continents. A glance at Figure 1.14 shows that the width of the continental shelf is variable. For example, it is broad along the East and Gulf coasts of the United States but relatively narrow along the Pacific margin of the continent.

The boundary between the continents and the deep-ocean basins lies along the **continental slope,** which is a relatively steep dropoff that extends from the outer edge of the continental shelf to the floor of the deep

Mount Everest in the Himalayas. (Galen Rowell)

ocean (Figure 1.14). Using this as the dividing line, we find that about 60 percent of Earth's surface is represented by ocean basins and the remaining 40 percent by continents.

Beyond the continental margin lies the **deep-ocean basin.** Part of this region consists of incredibly flat features called **abyssal plains.** However, the ocean floor also contains extremely deep depressions that are thousands of meters deep. Although these **deep-ocean trenches** are relatively narrow and represent only a small fraction of the ocean floor, they are nevertheless very significant features. Some trenches are located adjacent to young mountains that flank the continents. For example, in Figure 1.14 the Peru-Chile trench off the west coast of South America parallels the Andes Mountains. Other trenches parallel linear island chains called *volcanic island arcs.* The deep-ocean basins are also dotted with submerged volcanic structures called **seamounts,** which sometimes form long narrow chains.

Did You Know?

Ocean depths are often expressed in *fathoms.* One fathom equals 1.8 m or 6 ft, which is about the distance of a person's outstretched arms. The term is derived from how depth-sounding lines were brought back on board a vessel by hand. As the line was hauled in, a worker counted the number of arm lengths collected. By knowing the length of the person's outstretched arms, the amount of line taken in could be calculated. The length of one fathom was later standardized to 6 ft.

The most prominent feature on the ocean floor is the **oceanic** or **mid-ocean ridge.** As shown in Figure 1.14, the Mid-Atlantic Ridge and the East Pacific Rise are parts of this system. This broad elevated feature forms a continuous belt that winds for more than 70,000 kilometers (43,000 miles) around the globe in a manner similar to the seam of a baseball. Rather than consisting of highly deformed rock, such as most of the mountains on the continents, the oceanic ridge system consists of layer upon layer of igneous rock that has been fractured and uplifted.

Understanding the topographic features that comprise the face of Earth is critical to our understanding of the mechanisms that have shaped our planet. What is the significance of the enormous ridge system that extends through all the world's oceans? What is the connection, if any, between young, active mountain belts and deep-ocean trenches? What forces crumple rocks to produce majestic mountain ranges? These are questions that will be addressed in some of the coming chapters as we investigate the dynamic processes that shaped our planet in the geologic past and will continue to shape it in the future.

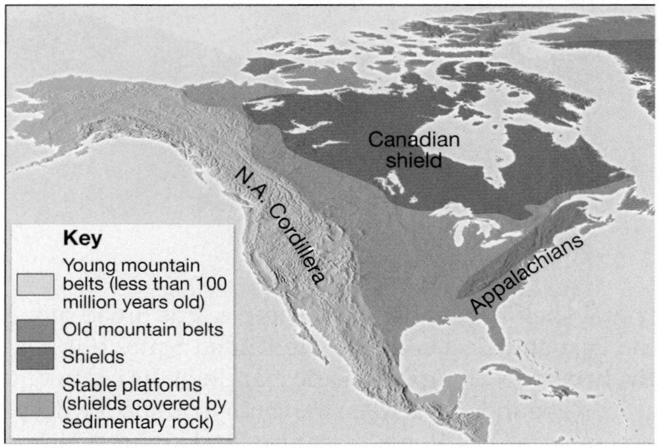

▲ **Figure 1.15** Major features of North America.

Dynamic Earth

GEODe — Essentials of Geology
Plate Tectonics
▼ Introduction

Earth is a dynamic planet! If we could go back in time a few hundred million years, we would find the face of our planet dramatically different from what we see today. There would be no Mount St. Helens, Rocky Mountains, or Gulf of Mexico. Moreover, we would find continents having different sizes and shapes and located in different positions than today's landmasses. In contrast, over the past few billion years the Moon's surface has remained essentially unchanged—only a few craters have been added.

The Theory of Plate Tectonics

Within the past several decades a great deal has been learned about the workings of our dynamic planet. This period has seen an unequaled revolution in our understanding of Earth. The revolution began in the early part of the twentieth century with the radical proposal of *continental drift*—the idea that the continents moved about the face of the planet. This proposal contradicted the established view that the continents and ocean basins are permanent and stationary features on the face of Earth. For that reason, the notion of drifting continents was received with great skepticism and even ridicule. More than 50 years passed before enough data were gathered to transform this controversial hypothesis into a sound theory that wove together the basic processes known to operate on Earth. The theory that finally emerged, called the **theory of plate tectonics,** provided geologists with the first comprehensive model of Earth's internal workings.

According to the plate tectonics model, Earth's rigid outer shell (*lithosphere*) is broken into numerous slabs called **plates,** which are in continual motion. As shown in Figure 1.16, seven major lithospheric plates are recognized. They are the North American, South American, Pacific, African, Eurasian, Australian, and Antarctic plates. Intermediate-size plates include the Caribbean,

Nazca, Philippine, Arabian, Cocos, and Scotia plates. In addition, over a dozen smaller plates have been identified, but are not shown in Figure 1.16. Note that several large plates include an entire continent plus a large area of seafloor (for example, the South American plate). However, none of the plates are defined entirely by the margins of a single continent.

The lithospheric plates move relative to each other at a very slow but continuous rate that averages about 5 centimeters (2 inches) a year. This movement is ultimately driven by the unequal distribution of heat within Earth. Hot material found deep in the mantle moves slowly upward and serves as one part of our planet's internal convective system. Concurrently, cooler, denser slabs of lithosphere descend back into the mantle, setting Earth's rigid outer shell in motion. Ultimately, the titanic, grinding movements of Earth's lithospheric plates generate earthquakes, create volcanoes, and deform large masses of rock into mountains.

Plate Boundaries

Lithospheric plates move as coherent units relative to all other plates. Although the interiors of plates may experience some deformation, all major interactions among individual plates (and therefore most deformation) occurs along their *boundaries.* In fact, the first attempts to outline plate boundaries were made using locations of earthquakes. Later work showed that plates are bounded by three distinct types of boundaries, which are differentiated by the type of relative movement they exhibit. These boundaries are depicted at the bottom of Figure 1.16 and are briefly described here:

1. **Divergent boundaries**—where plates move apart, resulting in upwelling of material from the mantle to create new seafloor (Figure 1.16A).
2. **Convergent boundaries**—where plates move together, resulting in the subduction (consumption) of oceanic lithosphere into the mantle (Figure 1.16B). Convergence can also result in the collision of two continental margins to create a major mountain system.
3. **Transform fault boundaries**—where plates grind past each other without the production or destruction of lithosphere (Figure 1.16C).

If you examine Figure 1.16, you can see that each large plate is bounded by a combination of these boundaries. Movement along one boundary requires that adjustments be made at the others.

Divergent Boundaries. Plate spreading (divergence) occurs mainly at the oceanic ridge. As plates pull apart, the fractures created are immediately filled with molten rock that wells up from the asthenosphere below (Figure 1.17). This hot material slowly cools to hard rock, producing new slivers of seafloor. This happens again and again over millions of years, adding thousands of square kilometers of new seafloor.

This mechanism has created the floor of the Atlantic Ocean during the past 160 million years and is appropriately called **seafloor spreading** (Figure 1.17). The rate of seafloor spreading varies considerably from one spreading center to another. Spreading rates of only 2.5 centimeters (1 inch) per year are typical in the North Atlantic, whereas much faster rates (20 centimeters, 8 inches per year) have been measured along the East Pacific Rise. Even the most rapid rates of spreading are slow on the scale of human history. Nevertheless, the slowest rate of lithosphere production is rapid enough to have created all of Earth's ocean basins over the last 200 million years. In fact, none of the ocean floor that has been dated exceeds 180 million years in age.

Along divergent boundaries where molten rock emerges, the oceanic lithosphere is elevated, because it is hot and occupies more volume than do cooler rocks. Worldwide, this ridge extends for over 70,000 kilometers through all major ocean basins (see Figure 1.15). As new lithosphere is formed along the oceanic ridge, it is slowly yet continually displaced away from the zone of upwelling along the ridge axis. Thus, it begins to cool and contract, thereby increasing in density. This thermal contraction accounts for the greater ocean depths that exist away from the ridge. In addition, cooling causes the mantle rocks below the oceanic crust to strengthen, thereby adding to the plate's thickness. Stated another way, the thickness of oceanic lithosphere is age dependent. The older (cooler) it is, the greater its thickness.

Convergent Boundaries. Although new lithosphere is constantly being added at the oceanic ridges, the planet is not growing in size—its total surface area remains constant. To accommodate the newly created lithosphere, older oceanic plates return to the mantle along *convergent boundaries.* As two plates slowly converge, the leading edge of one slab is bent downward, allowing it to slide beneath the other. The surface expression produced by the descending plate is an ocean *trench,* like the Peru–Chile trench illustrated in Figures 1.14 and 1.17.

> **Did You Know?**
>
> The average rate at which plates move relative to each other is roughly the same rate at which human fingernails grow.

Plate margins where oceanic crust is being consumed are called **subduction zones.** Here, as the subducted plate moves downward, it enters a high-temperature, high-pressure environment. Some subducted materials, as well as more voluminous amounts of the asthenosphere located above the subducting slab, melt and migrate upward into the overriding plate. Occasionally this molten rock may reach the surface, where

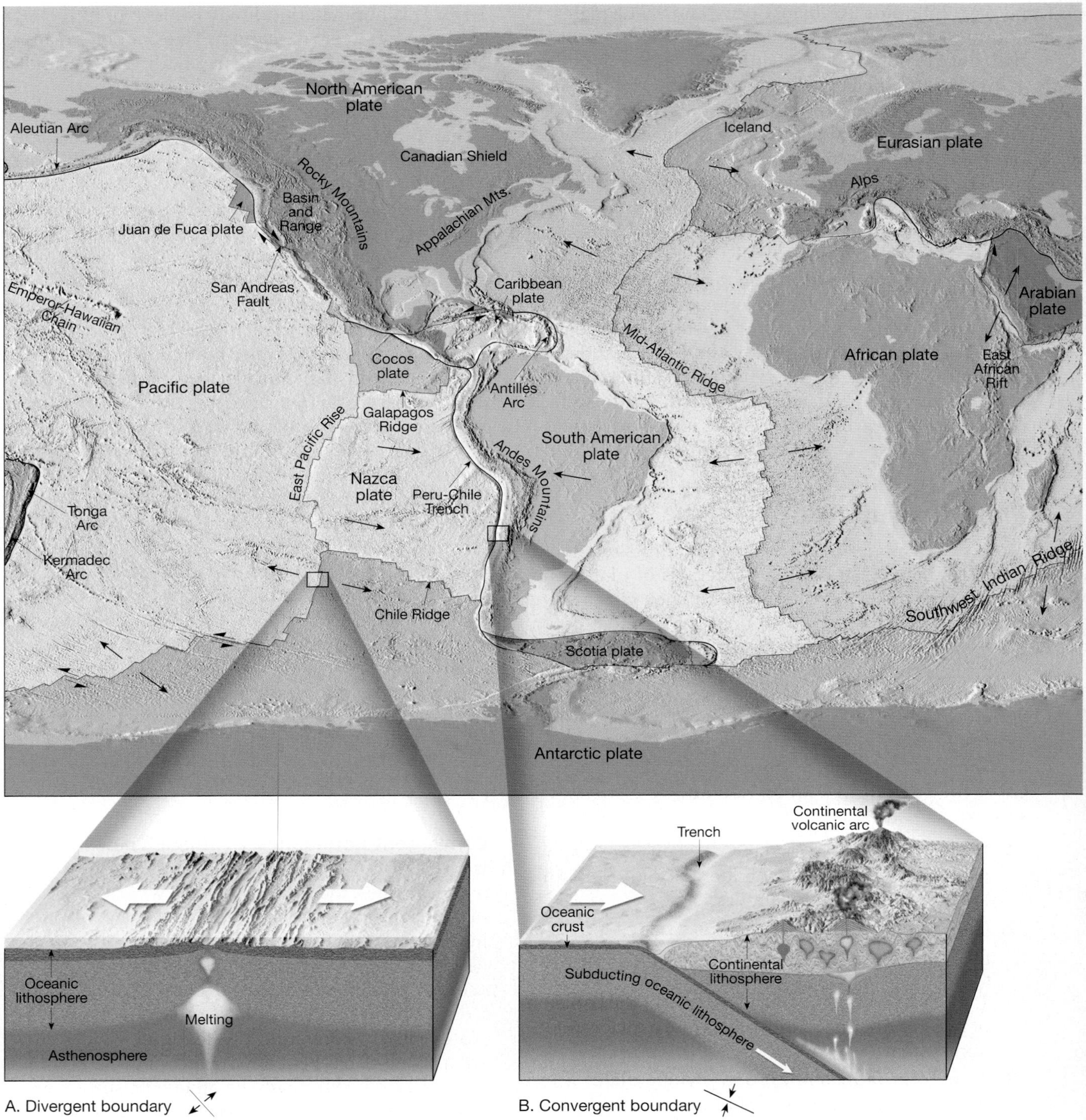

▲ **Figure 1.16** Mosaic of rigid plates that constitute Earth's outer shell. *(After W. B. Hamilton, U.S. Geological Survey)*

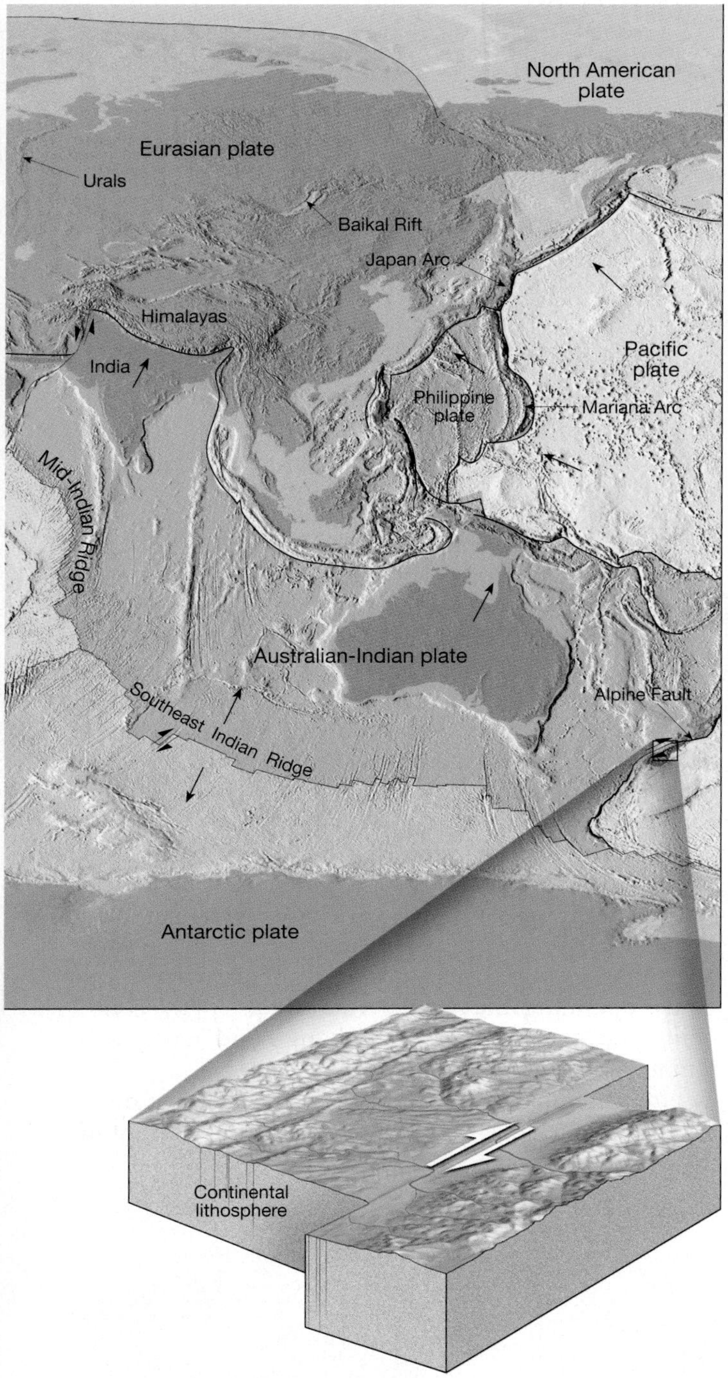

C. Transform fault boundary

▲ **Figure 1.16** Continued.

it gives rise to explosive volcanic eruptions like Mount St. Helens in 1980. However, much of this molten rock never reaches the surface; rather, it solidifies at depth and acts to thicken the crust.

Whenever slabs of continental lithosphere and oceanic lithosphere converge, the continental plate being less dense remains "floating," while the denser oceanic lithosphere sinks into the asthenosphere (Figure 1.17). The classic convergent boundary of this type occurs along the western margin of South America where the Nazca plate descends beneath the adjacent continental block. Here subduction along the Peru–Chile trench gave rise to the Andes Mountains, a linear chain of deformed rocks capped with numerous volcanoes—a number of which are still active.

The simplest type of convergence occurs where one oceanic plate is thrust beneath another. Here subduction results in the production of magma in a manner similar to that in the Andes, except volcanoes grow from the floor of the ocean rather than on a continent. If this activity is sustained, it will eventually build a chain of volcanic structures that emerge from the sea as a *volcanic island arc.* Most volcanic island arcs are found in the Pacific Ocean, as exemplified by the Aleutian, Mariana, and Tonga islands.

As we saw earlier, when an oceanic plate is subducted beneath continental lithosphere, an Andean-type mountain range develops along the margin of the continent. However, if the subducting plate also contains continental lithosphere, continued subduction eventually brings the two continents together (Figure 1.18). Whereas oceanic lithosphere is relatively dense and sinks into the asthenosphere, continental lithosphere is buoyant, which prevents it from being subducted to any great depth. The result is a collision between the two continental blocks (Figure 1.18). Such a collision occurred when the subcontinent of India "rammed" into Asia and produced the Himalayas—the most spectacular mountain range on Earth. During this collision, the continental crust buckled, fractured, and was generally shortened and thickened. In addition to the Himalayas, several other major mountain systems, including the Alps, Appalachians, and Urals, formed during continental collisions along convergent plate boundaries.

Transform Fault Boundaries. Transform fault boundaries are located where plates grind past each other without either generating new lithosphere or consuming old lithosphere. These faults form in the direction of plate movement and were first discovered in association with offsets in oceanic ridges (see Figure 1.16).

Although most transform faults are located along oceanic ridges, some slice through the continents. Two examples are the earthquake-prone San Andreas Fault

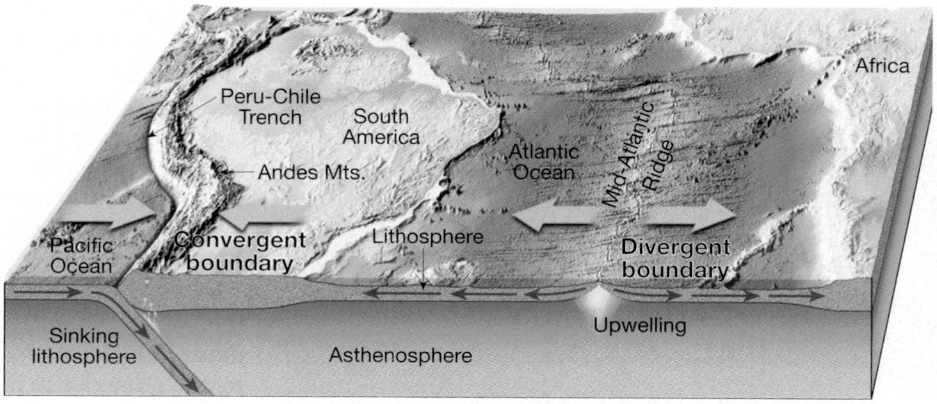

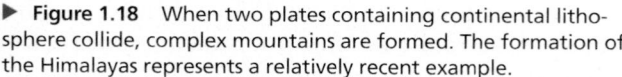

▶ **Figure 1.17** Convergent boundaries occur where two plates move together, as along the western margin of South America. Divergent boundaries are located where adjacent plates move away from one another. The Mid-Atlantic ridge is such a boundary.

of California and the Alpine Fault of New Zealand. Along the San Andreas Fault the Pacific plate is moving toward the northwest, relative to the adjacent North American plate (see Figure 1.16). The movement along this boundary does not go unnoticed. As these plates pass, strain builds in the rocks on opposite sides of the fault. Occasionally the rocks adjust, releasing energy in the form of a great earthquake of the type that devastated San Francisco in 1906.

The San Andreas Fault. (Michael Collier/DRK Photo)

Changing Boundaries. Although the total surface area of Earth does not change, individual plates may diminish or grow in area depending on the distribution of convergent and divergent boundaries. For example, the Antarctic and African plates are almost entirely bounded by spreading centers and hence are growing larger. By contrast, the Pacific plate is being subducted along much of its perimeter and is therefore diminishing in area. At the current rate, the Pacific would close completely in 300 million years—but this is unlikely because changes in plate boundaries will probably occur before that time.

New plate boundaries are created in response to changes in the forces acting on the lithosphere. For example, a relatively new divergent boundary is located in Africa, in a region known as the East African Rift Valleys. If spreading continues in this region, the African plate will split into two plates, separated by a new ocean basin. At other locations plates carrying continental crust are moving toward each other. Eventually these continents may collide and be sutured together. Thus, the boundary that once separated these plates disappears, and two plates become one.

As long as temperatures within the interior of our planet remain significantly higher than those at the surface, material within Earth will continue to circulate. This internal flow, in turn, will keep the rigid outer shell of Earth in motion. Thus, while Earth's internal heat engine is operating, the positions and shapes of the continents and ocean basins will change and Earth will remain a dynamic planet.

In the remaining chapters we will examine in more detail the workings of our dynamic planet in light of the plate tectonics theory.

▶ **Figure 1.18** When two plates containing continental lithosphere collide, complex mountains are formed. The formation of the Himalayas represents a relatively recent example.

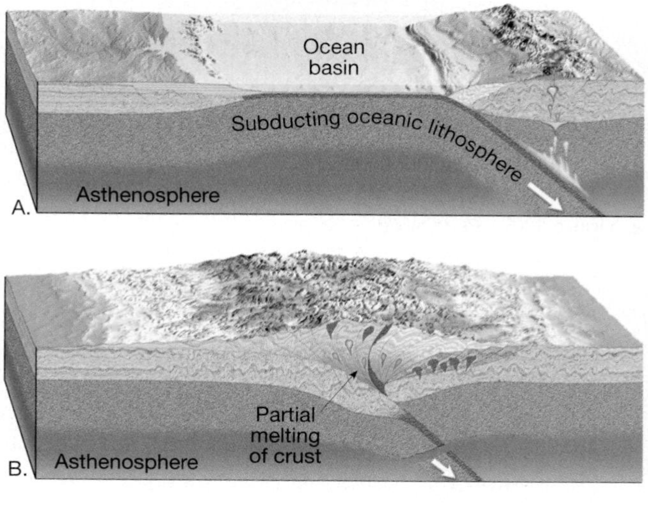

The Chapter in Review

- *Geology* means "the study of Earth." The two broad areas of the science of geology are (1) *physical geology,* which examines the materials composing Earth and the processes that operate beneath and upon its surface; and (2) *historical geology,* which seeks to understand the origin of Earth and its development through time.
- The relationship between people and the natural environment is an important focus of geology. This includes natural hazards, resources, and human influences on geologic processes.
- During the seventeenth and eighteenth centuries, *catastrophism* influenced the formulation of explanations about Earth. Catastrophism states that Earth's landscapes developed over short time spans primarily as a result of great catastrophes. By contrast, *uniformitarianism,* one of the fundamental principles of modern geology advanced by *James Hutton* in the late 1700s, states that the physical, chemical, and biological laws that operate today have also operated in the geologic past. The idea is often summarized as *"The present is the key to the past."* Hutton argued that processes that appear to be slow-acting could, over long spans of time, produce effects that are just as great as those resulting from sudden catastrophic events.
- Using the principles of *relative dating,* the placing of events in their proper sequence or order without knowing their age in years, scientists developed a geologic time scale during the nineteenth century. Relative dates can be established by applying such principles as the *law of superposition* and the *principle of fossil succession.*
- All science is based on the assumption that the natural world behaves in a consistent and predictable manner. The process by which scientists gather facts and formulate scientific *hypotheses* and *theories* is called the *scientific method.* To determine what is occurring in the natural world, scientists often (1) collect facts, (2) develop a scientific hypothesis, (3) construct experiments to test the hypothesis, and (4) accept, modify, or reject the hypothesis on the basis of extensive testing. Other discoveries represent purely theoretical ideas that have stood up to extensive examination. Still other scientific advancements have been made when a totally unexpected happening occurred during an experiment.
- Earth's physical environment is traditionally divided into three major parts: the solid Earth, or *geosphere;* the water portion of our planet, the *hydrosphere;* and Earth's gaseous envelope, the *atmosphere.* In addition, the *biosphere,* the totality of life on Earth, interacts with each of the three physical realms and is an equally integral part of Earth.
- Although each of Earth's four spheres can be studied separately, they are all related in a complex and continuously interacting whole that we call the *Earth system. Earth system science* uses an interdisciplinary approach to integrate the knowledge of several academic fields in the study of our planet and its global environmental problems.
- A *system* is a group of interacting parts that form a complex whole. *Closed systems* are those in which energy moves freely in and out, but matter does not enter or leave the system. In an open system, both energy and matter flow into and out of the system.
- Most natural systems have mechanisms that tend to enhance change, called *positive feedback mechanisms,* and other mechanisms, called *negative feedback mechanisms,* that tend to resist change and thus stabilize the system.
- The two sources of energy that power the Earth system are (1) the Sun, which drives the external processes that occur in the atmosphere, hydrosphere, and at Earth's surface, and (2) heat from Earth's interior that powers the internal processes that produce volcanoes, earthquakes, and mountains.
- The *rock cycle* is one of the many cycles or loops of the Earth system in which matter is recycled. The rock cycle is a means of viewing many of the interrelationships of geology. It illustrates the origin of the three basic rock groups and the role of various geologic processes in transforming one rock type into another.
- The *nebular hypothesis* describes the formation of the solar system. The planets and Sun began forming about 5 billion years ago from a large cloud of dust and gases. As the cloud contracted, it began to rotate and assume a disk shape. Material that was gravitationally pulled toward the center became the *protosun.* Within the rotating disk, small centers, called *protoplanets,* swept up more and more of the cloud's debris. Because of the high temperatures near the Sun, the inner planets were unable to accumulate many of the elements that vaporize at low temperatures. Because of the very cold temperatures existing far from the Sun, the large outer planets consist of huge amounts of ices and lighter materials. These substances account for the comparatively large sizes and low densities of the outer planets.
- Earth's internal structure is divided into layers based on differences in chemical composition and on the basis of changes in physical properties. Compositionally, Earth is divided into a thin outer *crust,* a solid rocky *mantle,* and a dense *core.* Based on physical properties, the layers of Earth are (1) the *lithosphere*—the cool, rigid outermost layer that averages about 100 kilometers thick, (2) the *asthenosphere,* a relatively weak layer located in the mantle beneath the lithosphere, (3) the more rigid *mesosphere,* where rocks are very hot and capable of very gradual flow, (4) the liquid *outer core,* where Earth's magnetic field is generated, and (5) the solid *inner core.*
- Two principal divisions of Earth's surface are the *continents* and *ocean basins.* A significant difference is their relative levels. The elevation differences between continents and ocean

basins is primarily the result of differences in their respective densities and thicknesses.

- The largest features of the continents can be divided into two categories: *mountain belts* and the *stable interior.* The ocean floor is divided into three major topographic units: *continental margins, deep-ocean basins,* and *oceanic ridges.*
- The *theory of plate tectonics* provides a comprehensive model of Earth's internal workings. It holds that Earth's rigid outer lithosphere consists of several segments called *plates* that are slowly and continually in motion relative to one another. Most earthquakes, volcanic activity, and mountain building are associated with the movements of these plates.
- The three distinct types of plate boundaries are (1) *divergent boundaries,* where plates move apart; (2) *convergent boundaries,* where plates move together, causing one to go beneath another, or where plates collide, which occurs when the leading edges are made of continental crust; and (3) *transform fault boundaries,* where plates slide past one another.

Key Terms

abyssal plain (p. 24)
asthenosphere (p. 20)
atmosphere (p. 10)
biosphere (p. 10)
catastrophism (p. 3)
closed system (p. 11)
continental margin (p. 22)
continental shelf (p. 23)
continental slope (p. 23)
convergent boundary (p. 25)
core (p. 19)
crust (p. 19)
deep-ocean basin (p. 24)
deep-ocean trench (p. 24)
divergent boundary (p. 25)
Earth system science (p. 11)
fossil (p. 5)
fossil succession, principle of (p. 5)
geology (p. 2)
geosphere (p. 10)
historical geology (p. 2)
hydrosphere (p. 9)
hypothesis (p. 7)
igneous rock (p. 16)
inner core (p. 21)
interface (p. 12)
lithosphere (p. 20)
lower mantle (p. 21)
magma (p. 14)
mantle (p. 19)
mesosphere (p. 21)
metamorphic rock (p. 16)
model (p. 7)
nebular hypothesis (p. 16)
negative feedback mechanism (p. 11)
oceanic (mid-ocean) ridge (p. 24)
open system (p. 11)
outer core (p. 21)
paradigm (p. 7)
physical geology (p. 2)
plate (p. 24)
plate tectonics, theory of (p. 24)
positive feedback mechanism (p. 11)
relative dating (p. 5)
rock cycle (p. 14)
seafloor spreading (p. 25)
seamount (p. 24)
sediment (p. 16)
sedimentary rock (p. 16)
shield (p. 21)
stable platform (p. 21)
subduction zone (p. 25)
superposition, law of (p. 5)
system (p. 11)
theory (p. 8)
transform fault boundary (p. 25)
uniformitarianism (p. 4)

Questions for Review

1. Geology is traditionally divided into two broad areas. Name and describe these two subdivisions.
2. List at least three phenomena that could be regarded as geologic hazards.
3. Briefly describe Aristotle's influence on the science of geology.
4. How did the proponents of catastrophism perceive the age of Earth?
5. Describe the doctrine of uniformitarianism. How did the advocates of this idea view the age of Earth?
6. About how old is Earth?
7. The geologic time scale was established without the aid of radiometric dating. What principles were used to develop the time scale?
8. How is a scientific hypothesis different from a scientific theory?
9. List and briefly describe the four spheres that constitute our natural environment.
10. How is an open system different from a closed system?
11. Contrast positive feedback mechanisms and negative feedback mechanisms.
12. What are the two sources of energy for the Earth system?
13. Using the rock cycle, explain the statement "One rock is the raw material for another."
14. Briefly describe the events that led to the formation of our solar system.
15. List and briefly describe Earth's compositional layers.
16. Contrast the lithosphere and the asthenosphere.
17. Describe the general distribution of Earth's youngest mountains.
18. Distinguish between shields and stable platforms.
19. List the three basic types of plate boundaries, and describe the relative movement each exhibits.
20. With which type of plate boundary is each of the following associated: subduction zone, San Andreas Fault, seafloor spreading, and Mount St. Helens?

Online Study Guide

The *Essentials of Geology* Web site uses the resources and flexibility of the Internet to aid in your study of the topics in this chapter. Written and developed by geology instructors, this site will help improve your understanding of geology. Visit **www.prenhall.com/lutgens** and click on the cover of *Essentials of Geology 9e* to find:

- Online review quizzes.
- Critical thinking exercises.
- Links to chapter-specific Web resources.
- Internet-wide key-term searches.

http://www.prenhall.com/lutgens

Topaz crystal with Albite. *(Photo by Jeff Scovil)*

CHAPTER 2

Minerals: Building Blocks of Rocks

Focus on Learning

To assist you in learning the important concepts in this chapter, you will find it helpful to focus on the following questions:

- What are minerals, and how are they different from rocks?
- What is the basic structure of an atom? How do atoms bond together?
- How do isotopes of the same element vary from one another, and why are some isotopes radioactive?
- What are some of the physical and chemical properties of minerals? How can these properties be used to distinguish one mineral from another?
- What is the most abundant mineral group? What do all minerals within this group have in common?
- What are some important nonsilicate minerals?
- When is the term ore used in reference to a mineral?

Copper has many important uses. (Getty Images, Inc.)

Earth's crust and oceans are the source of a wide variety of useful and essential minerals. In fact, practically every manufactured product contains materials obtained from minerals. Most people are familiar with the common uses of many basic metals, including aluminum in beverage cans, copper in electrical wiring, and gold and silver in jewelry. But some people are not aware that pencil lead contains the greasy-feeling mineral graphite and that bath powders are made from the mineral talc. Moreover, many do not know that drill bits impregnated with diamonds are employed by dentists to drill through tooth enamel, or that the common mineral quartz is the source of silicon for computer chips (Figure 2.1).

As the mineral requirements of modern society grow, the need to locate additional supplies of useful minerals also grows, becoming more challenging as well. In addition to the economic uses of rocks and minerals, all of the processes studied by geologists are in some way related to the properties of these basic Earth materials. Events such as volcanic eruptions, mountain building, weathering and erosion, and even earthquakes involve rocks and minerals. Consequently, a basic knowledge of Earth materials is essential to an understanding of all geologic phenomena.

Gold on quartz.

Minerals: Building Blocks of Rocks

GEODe — ESSENTIALS OF GEOLOGY

Minerals: Building Blocks of Rocks
▼ Introduction

The term *mineral* is used in several different ways. For example, those concerned with health and fitness extol the benefits of vitamins and minerals. The mining industry typically uses the word when referring to anything taken out of the ground, such as coal, iron ore, or sand and gravel. The guessing game known as *Twenty Questions* usually begins with the question, *Is it Animal, Vegetable or Mineral?* What criteria do geologists use to determine whether something is a mineral?

Did You Know?

Native copper, which is easy to pound into different shapes, was widely used for tools by early humans. Later, craftsmen combined copper with tin to make bronze, a tougher and harder alloy. The Bronze Age lasted from 2200 to 800 B.C., when the ability to extract iron from minerals such as hematite was discovered.

Geologists define a **mineral** as *any naturally occurring inorganic solid that possesses an orderly crystalline structure and a definite chemical composition.* Thus, those Earth materials that are classified as minerals exhibit the following characteristics:

1. **Naturally occurring.** Minerals form by natural, geologic processes. Consequently, synthetic diamonds and rubies, as well as a variety of other useful materials produced in a laboratory, are not considered minerals.
2. **Solid substance.** Minerals are solids within the temperature ranges normally experienced at Earth's surface. Thus, ice (frozen water) is considered a mineral, whereas liquid water and water vapor are not.
3. **Orderly crystalline structure.** Minerals are crystalline substances, which means their atoms are arranged in an orderly, repetitive manner as shown in Figure 2.2. This orderly packing of atoms is reflected in the regularly shaped objects we call crystals (Figure 2.2D). Some naturally occurring solids, such as volcanic glass (obsidian), lack a repetitive atomic structure and are not considered minerals.
4. **Definite chemical composition.** Most minerals are chemical **compounds** made up of two or more elements. The common mineral quartz, for example, consists of two oxygen (O) atoms for every silicon (Si) atom, giving it a chemical composition expressed by the formula SiO_2. However, a few minerals, such as gold, sulfur, and silver, consist of only a single element.
5. **Generally inorganic.** Inorganic crystalline solids, as exemplified by ordinary table salt (halite), that are found naturally in the ground are considered minerals. Organic compounds, on the other hand, are generally not. Sugar, a crystalline solid like salt but which comes from sugarcane or sugar beets, is a common example of such an organic compound. However, many marine animals secrete inorganic compounds, such as calcium carbonate (calcite), in the form of shells and coral reefs. These materials are considered minerals by most geologists.

In contrast to minerals, rocks are more loosely defined. Simply, a **rock** is any solid mass of mineral, or mineral-like, matter that occurs naturally as part of our planet. A few rocks are composed almost entirely of one mineral. A common example is the sedimentary rock *limestone,* which consists of impure masses of the mineral calcite. However, most rocks, like the common rock granite shown in Figure 2.3, occur as aggregates of several kinds of minerals. Here, the term *aggregate* implies that the minerals are joined in such a way that the properties of each mineral are retained. Note that you can easily identify the mineral constituents of the granite in Figure 2.3.

A few rocks are composed of nonmineral matter. These include the volcanic rocks *obsidian* and *pumice,* which are noncrystalline glassy substances, and *coal,* which consists of solid organic debris.

Although this chapter deals primarily with the nature of minerals, keep in mind that most rocks are simply aggregates of minerals. Because the properties of rocks are determined largely by the chemical composition and crystalline structure of the minerals contained within them, we will first consider these Earth materials. Then, in Chapters 3, 6, and 7, we will take a closer look at Earth's major rock groups.

▲ **Figure 2.1** Quartz crystals, variety amethyst, Las Vigas, Mexico. *(Photo by Jeff Scovil)*

Elements: Building Blocks of Minerals

You are probably familiar with the names of many elements, including copper, gold, oxygen, and carbon. An **element** is a substance that cannot be broken down into simpler substances by chemical or physical means. There are about 90 different elements found in nature, and consequently about 90 different kinds of atoms. In addition, scientists have succeeded in creating about 23 synthetic elements.

The elements can be organized into rows so that those with similar properties are in the same column. This arrangement, called the **periodic table**, is shown in Figure 2.4. Notice that symbols are used to provide a

The mineral sulfur consists of only a single element.

▶ **Figure 2.2** This diagram illustrates the orderly arrangement of sodium and chlorine ions in the mineral halite. The arrangement of atoms into basic building blocks having a cubic shape results in regularly shaped cubic crystals. *(Photo by M. Claye/Jacana Scientific Control/Photo Researchers, Inc.)*

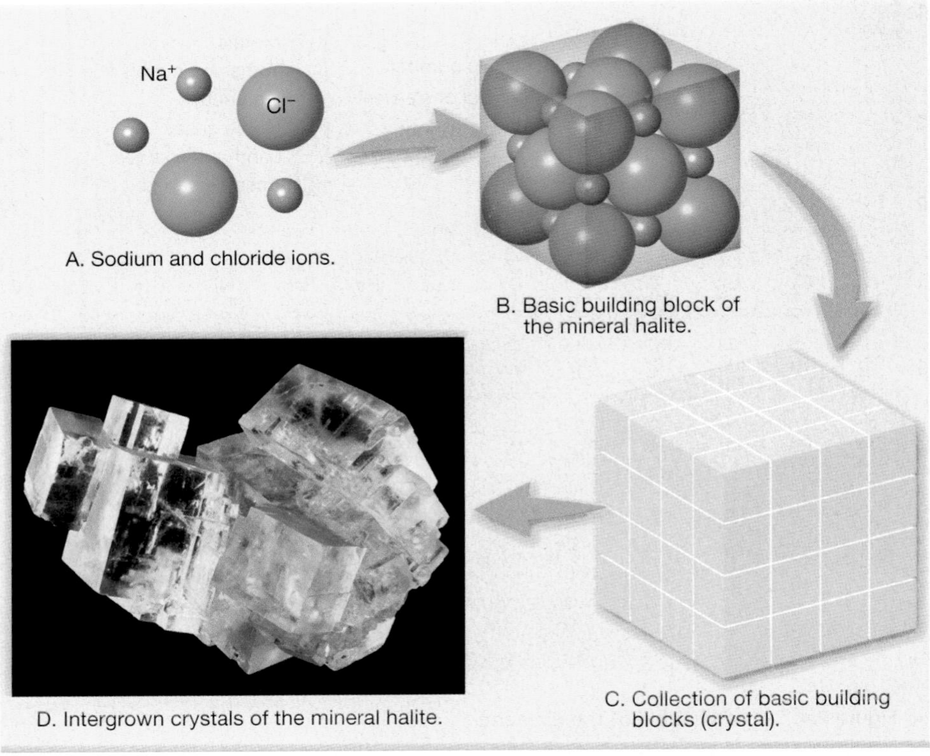

▶ **Figure 2.3** Rocks are aggregates of one or more minerals. *(All photos by E. J. Tarbuck)*

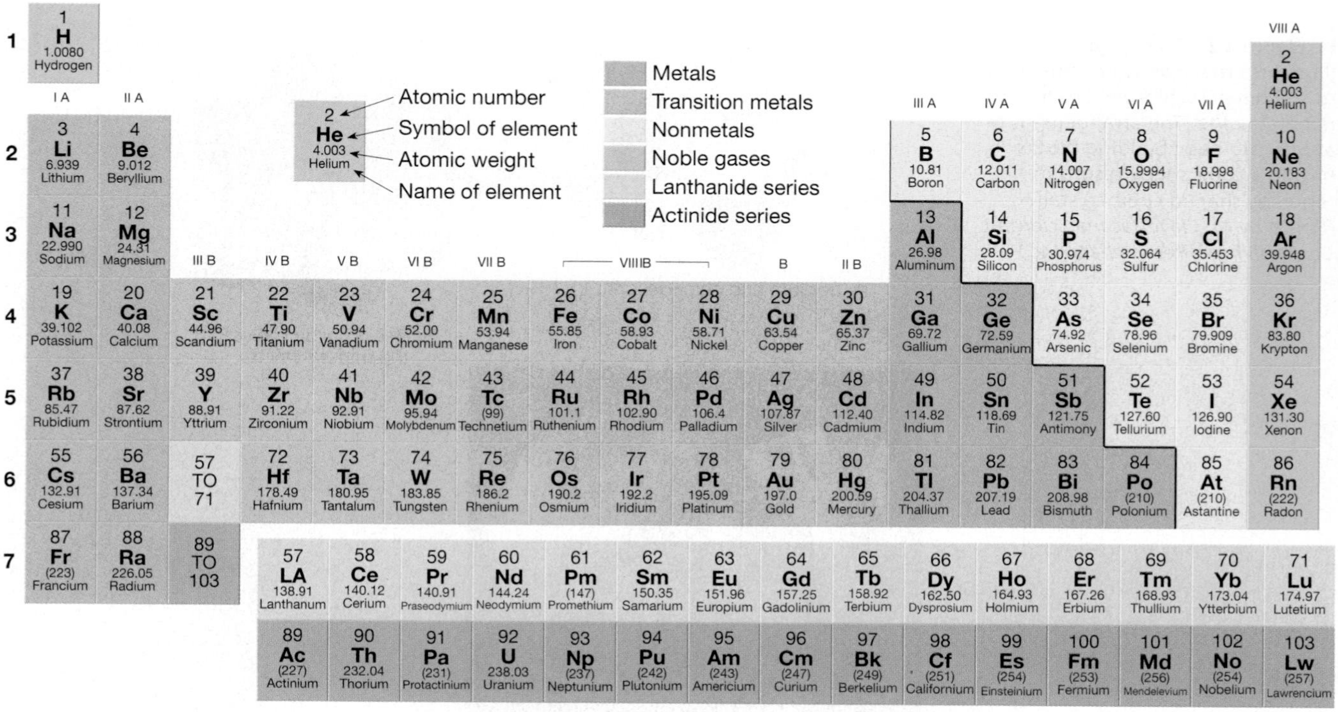

▲ **Figure 2.4** Periodic Table of the Elements.

shorthand way of representing an element. Each element is also known by its atomic number, which is shown above each symbol on the periodic table.

Some elements, such as copper (number 29) and gold (number 79), exist in nature with single atoms as the basic unit. Thus, native copper and gold are minerals made entirely from one element. However, many elements are quite reactive and join together with atoms of one or more other elements to form chemical compounds. As a result, most minerals are chemical compounds consisting of two or more different elements.

To understand how elements combine to form compounds, we must first consider the atom. The **atom** (*a* = not, *tomos* = cut) is the smallest particle of matter that retains the essential characteristics of an element. It is this extremely small particle that does the combining.

The central region of an atom is called the **nucleus.** The nucleus contains protons and neutrons. **Protons** are very dense particles with positive electrical charges. **Neutrons** have the same mass as a proton but lack an electrical charge.

Orbiting the nucleus are **electrons,** which have negative electrical charges. For convenience, we sometimes diagram atoms to show the electrons orbiting the nucleus, like the orderly orbiting of the planets around the Sun (Figure 2.5A). However, electrons move in a less predictable way than planets. As a result, they create a sphere-shaped negative zone around the nucleus. A more realistic picture of the positions of electrons can be obtained by envisioning a cloud of negatively charged electrons surrounding the nucleus (Figure 2.5B).

Studies of electron configurations predict that individual electrons move within regions around the nucleus called **principal shells,** or **energy levels.** Furthermore, each of these shells can hold a specific number of electrons, with the outermost principal shell containing the **valence electrons.** Valence electrons are important because, as you will see later, they are involved in chemical bonding.

Some atoms have only a single proton in their nuclei, while others contain more than 100 protons. The number of protons in the nucleus of an atom is called the **atomic number.** For example, all atoms with six protons are carbon atoms; hence, the atomic number of carbon is 6. Likewise, every atom with eight protons is an oxygen atom.

Atoms in their natural state also have the same number of electrons as protons, so the atomic number also equals the number of electrons surrounding the nucleus. Therefore, carbon has six electrons to match its six protons, and oxygen has eight electrons to match its eight protons. Neutrons have no charge, so the positive charge of the protons is exactly balanced by the negative charge of the electrons. Consequently, atoms in the natural state are neutral electrically and have no overall electrical charge.

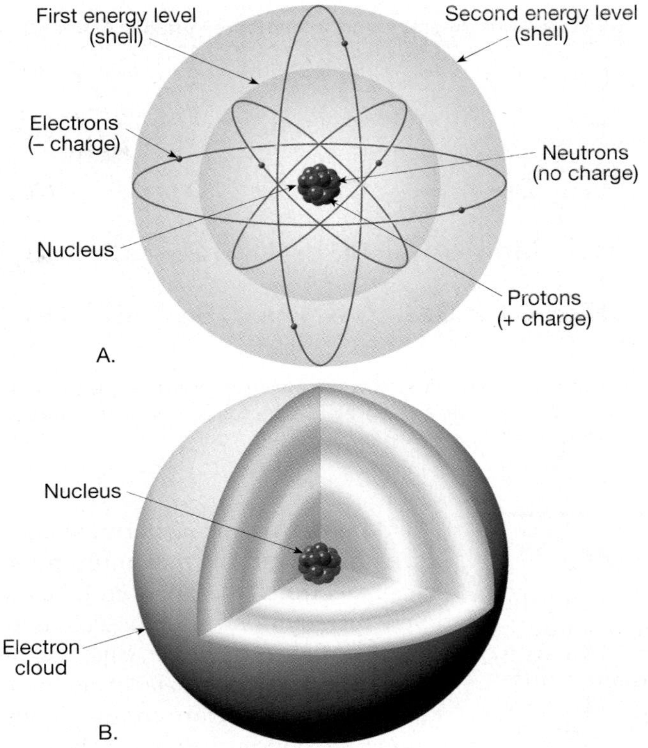

▲ **Figure 2.5** Two models of the atom. **A.** A very simplified view of the atom, which includes a central nucleus, composed of protons and neutrons, encircled by high-speed electrons. **B.** Another model of the atoms showing spherically shaped electron clouds (energy level shells). Note that these models are not drawn to scale.

Why Atoms Bond

Why do elements join together to form compounds? From experimentation it has been learned that the forces holding the atoms together are electrical. Further, it is known that chemical bonding results in a change in the electron configuration of the bonded atoms. As we noted earlier, it is the valence electrons (outer-shell electrons) that are generally involved in chemical bonding. Figure 2.6 shows a shorthand way of representing the number of electrons in the outer principal shell (valence electrons). Notice that the elements in group I have one valence electron, those in group II have two, and so forth up to group VIII, which has eight valence electrons.

Other than the first shell, which can hold a maximum of two electrons, *a stable configuration occurs when the valence shell contains eight electrons.* Only the noble gases, such as neon and argon, have a complete outermost principal shell. Hence, the noble gases are the least chemically reactive and are designated as "inert."

Did You Know?

Although wood pulp is the main ingredient for making newsprint, many higher grades of paper contain large quantities of clay minerals. In fact, each page of this textbook consists of about 25 percent clay (the mineral kaolinite). If rolled into a ball, this much clay would be roughly the size of a tennis ball.

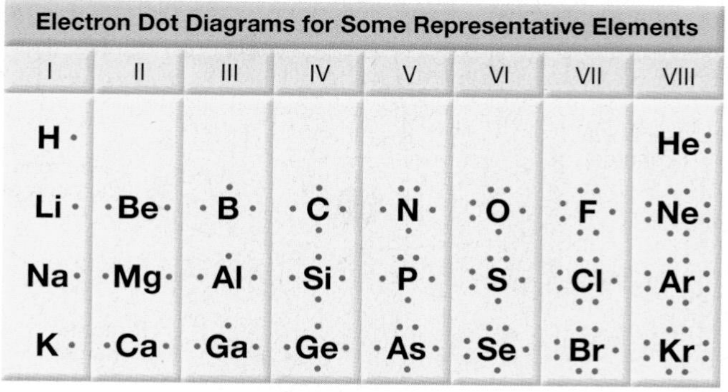

Electron Dot Diagrams for Some Representative Elements
I II III IV V VI VII VIII
H He
Li Be B C N O F Ne
Na Mg Al Si P S Cl Ar
K Ca Ga Ge As Se Br Kr

▲ **Figure 2.6** Dot diagrams for some representative elements. Each dot represents a valence electron found in the outermost principal shell.

> ### Did You Know?
> The purity of gold is expressed by the number of *karats,* where 24 karats is pure gold. Gold less than 24 karats is an alloy (mixture) of gold and another metal, usually copper or silver. For example, 14-karat gold contains 14 parts gold (by weight) mixed with 10 parts of other metals.

When an atom's outermost shell does not contain the maximum number of electrons (8), the atom is likely to chemically bond with one or more other atoms. A *chemical bond* is the sharing or transfer of electrons to attain a stable electron configuration among the bonding atoms. If the electrons are transferred, the bond is an *ionic bond.* If the electrons are shared, the bond is called a *covalent bond.* In either case, the bonding atoms get stable electron configurations, which usually consist of eight electrons in the outer shell.

Ionic Bonds: Electrons Transferred

Perhaps the easiest type of bond to visualize is an **ionic bond.** In ionic bonding, one or more valence electrons are transferred from one atom to another. Simply, one atom gives up its valence electrons, and the other uses them to complete its outer shell. A common example of ionic bonding is sodium (Na) and chlorine (Cl) joining to produce sodium chloride (common table salt). This is shown in Figure 2.7A. Notice that sodium gives up its single valence electron to chlorine. As a result, sodium achieves a stable configuration having eight electrons in its outermost shell. By acquiring the electron that sodium loses, chlorine—which has seven valence electrons—gains the eighth electron needed to complete its outermost shell. Thus, through the transfer of a single electron, both the sodium and chlorine atoms have acquired a stable electron configuration.

> ### Did You Know?
> Pure platinum is an extremely rare and valuable mineral. Platinum's wear and tarnish resistance make it well suited for making fine jewelry. In addition, the automotive industry uses platinum as an oxidation catalyst in catalytic converters that treat automobile exhaust emissions. With a specific gravity of 21.5, platinum is the densest naturally occurring mineral.

Once electron transfer takes place, atoms are no longer electrically neutral. By giving up one electron, a neutral sodium atom becomes *positively charged* (11 protons/10 electrons). Similarly, by acquiring one electron, the neutral chlorine atom becomes *negatively charged* (17 protons/18 electrons). Atoms which have an electrical charge because of the unequal numbers of electrons and protons, are called **ions.** We know that ions with like charges repel, and those with unlike charges attract. Thus, an *ionic bond* is the attraction of oppositely charged ions to one another, producing an electrically neutral compound.

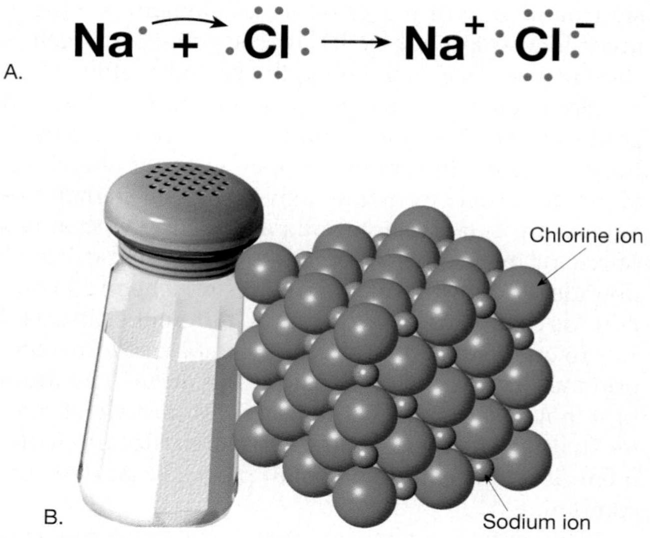

▲ **Figure 2.7** Chemical bonding of sodium and chlorine atoms to produce sodium chloride (table salt). **A.** Through the transfer of one electron in the outer shell of a sodium atom to the outer shell of a chlorine atom, the sodium becomes a positive ion and chlorine a negative ion. **B.** Diagram illustrating the arrangement of sodium and chlorine ions in table salt.

Figure 2.7B illustrates the arrangement of sodium and chlorine ions in ordinary table salt. Notice that salt consists of alternating sodium and chlorine ions, positioned in such a manner that each positive ion is attracted to and surrounded on all sides by negative ions, and vice versa. This arrangement maximizes the attraction between ions with unlike charges while minimizing the repulsion between ions with like charges. Thus, *ionic compounds consist of an orderly arrangement of oppositely charged ions assembled in a definite ratio that provides overall electrical neutrality.*

The properties of a chemical compound are *dramatically different* from the properties of the elements comprising it. For example, sodium, a soft, silvery metal, is extremely reactive and poisonous. If you were to consume even a small amount of elemental sodium, you would need immediate medical attention. Chlorine, a green poisonous gas, is so toxic it was used as a chemical weapon during World War I. Together, however, these elements produce sodium chloride, a harmless flavor enhancer that we call table salt. When elements combine to form compounds, their properties change dramatically.

Covalent Bonds: Electrons Shared

Not all atoms combine by transferring electrons to form ions. Some atoms *share* electrons. For example, the gaseous elements oxygen (O_2), hydrogen (H_2), and chlorine (Cl_2) exist as stable molecules consisting of two atoms bonded together, without a complete transfer of electrons.

Figure 2.8 illustrates the sharing of a pair of electrons between two chlorine atoms to form a molecule of chlorine gas (Cl_2). By overlapping their outer shells, these chlorine atoms share a pair of electrons. Thus, each chlorine atom has acquired, through cooperative action, the needed eight electrons to complete its outer shell. The bond produced by the sharing of electrons is called a **covalent bond.**

A common analogy may help you visualize a covalent bond. Imagine two people at opposite ends of a dimly lit room, each reading under a separate lamp. By moving the lamps to the center of the room, they are able to combine their light sources so each can see better. Just as the overlapping light beams meld, the shared electrons that provide the "electrical glue" in covalent bonds are indistinguishable from each other. The most common mineral group, the silicates, contains the element silicon, which readily forms covalent bonds with oxygen.

Other Bonds

As you might suspect, many chemical bonds are actually hybrids. They consist to some degree of electron sharing, as in covalent bonding, and to some degree of electron transfer, as in ionic bonding. Furthermore, both ionic and covalent bonds may occur within the same compound. This occurs in many silicate minerals, where silicon and oxygen atoms are covalently bonded to form the basic building block common to all silicates. These structures in turn are ionically bonded to metallic ions, producing various electrically neutral chemical compounds.

Another chemical bond exists in which valence electrons are free to migrate from one ion to another. The mobile valence electrons serve as the electrical glue. This type of electron sharing is found in metals such as copper, gold, aluminum, and silver and is called **metallic bonding.** Metallic bonding accounts for the high electrical conductivity of metals, the ease with which metals are shaped, and numerous other special properties of metals.

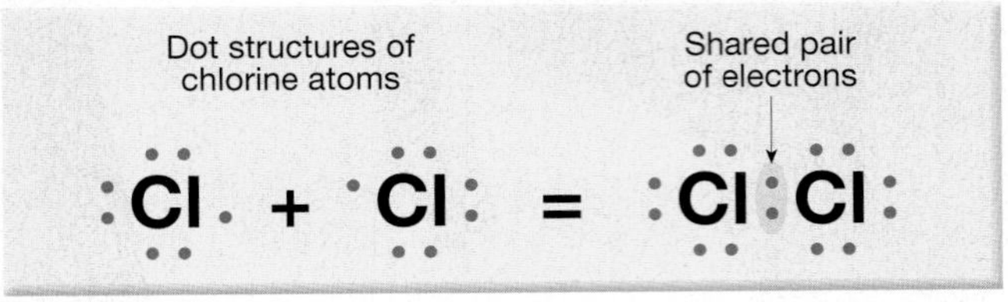

▲ **Figure 2.8** Dot diagrams used to illustrate the sharing of a pair of electrons between two chlorine atoms to form a chlorine molecule. Notice that by sharing a pair of electrons, both chlorine atoms achieve a full outer shell (8 electrons).

Isotopes and Radioactive Decay

Subatomic particles are so incredibly small that a special unit, called an *atomic mass unit*, was devised to express their mass. A proton or a neutron has a mass just slightly more than one atomic mass unit, whereas an electron is only about one two-thousandth of an atomic mass unit. Thus, although electrons play an active role in chemical reactions, they do not contribute significantly to the mass of an atom.

The **mass number** of an atom is simply the total of its neutrons and protons. Atoms of the same element always have the same number of protons. But the number of neutrons for atoms of the same element can vary. Atoms with the same number of protons but different numbers of neutrons are **isotopes** of that element. Isotopes of the same element are labeled by placing the mass number after the element's name or symbol.

Carbon has several different isotopes. Models for two of these, carbon-14 and carbon-12, are shown in Figure 2.9. Since all atoms of the same element have the same number of protons, and carbon has six, carbon-12 also has *six neutrons* to give it a mass number of 12. Likewise, carbon-14 has

Chlorine is a green poisonous gas. (Charles Winters/Photo Researchers)

Did You Know?

The names of precious gems often differ from the names of parent minerals. For example, *sapphire* is one of two gems that are varieties of the same mineral, *corundum.* Tiny amounts of the elements titanium and iron in corundum produce the most prized blue sapphires. When the mineral corundum contains chromium, it exhibits a brilliant red color and the gem is called *ruby.*

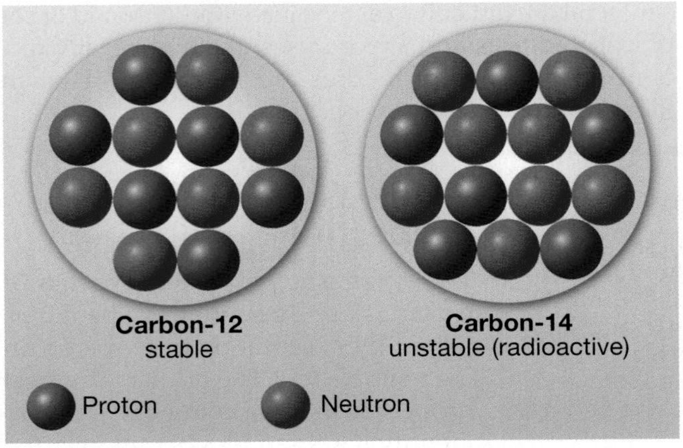

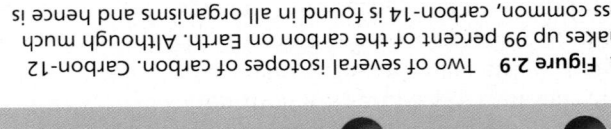

▲ **Figure 2.9** Two of several isotopes of carbon. Carbon-12 makes up 99 percent of the carbon on Earth. Although much less common, carbon-14 is found in all organisms and hence is useful for dating some fossils.

Metallic bonding binds the copper atoms in this sample of native copper.

six protons plus *eight neutrons* to give it a mass number of 14.

In chemical behavior, all isotopes of the same element are nearly identical. To distinguish among them is like trying to differentiate identical twins, with one being slightly heavier. Because isotopes of an element react the same chemically, different isotopes can become parts of the same mineral. For example, when the mineral calcite forms from calcium, carbon, and oxygen, some of its carbon atoms are carbon-12 and some are carbon-14.

The nuclei of most atoms are stable. However, many elements do have isotopes in which the nuclei are unstable. "Unstable" means that the isotopes disintegrate through a process called **radioactive decay.** Radioactive decay occurs when the forces that bind the nucleus are not strong enough.

During radioactive decay, unstable atoms radiate energy and emit particles. Some of this energy powers the movements of Earth's crust and upper mantle. The rates at which unstable atoms decay are measurable. Therefore, certain radioactive atoms can be used to determine the ages of fossils, rocks, and minerals. A discussion of radioactive decay and its applications in dating past geologic events is found in Chapter 18.

Properties of Minerals

GEODe: Minerals: Building Blocks of Rocks
▼ Physical Properties of Minerals

Minerals are solids formed by inorganic processes. Each mineral has an orderly arrangement of atoms (crystalline structure) and a definite chemical composition, which give it a unique set of physical properties. Because the internal structure and chemical composition of a mineral are difficult to determine without the aid of sophisticated tests and apparatuses, the more easily recognized physical properties are frequently used in identification.

Did You Know?

One of the world's heaviest-cut and polished gemstones is the 22,892.5-carat golden-yellow topaz. Currently housed in the Smithsonian Institution, this roughly 10-pound gem is about the size of an automobile headlight and could hardly be used as a piece of jewelry, except perhaps by an elephant.

Primary Diagnostic Properties

The diagnostic physical properties of minerals are those that can be determined by observation or by performing a simple test. The primary physical properties that are commonly used to identify hand samples of minerals are: crystal form, luster, color, streak, hardness, cleavage, fracture, and density or specific gravity. Secondary (or "special") properties that are exhibited by a limited number of minerals include: magnetism, taste, feel, smell, elasticity, malleability, double refraction, and chemical reaction to hydrochloric acid.

Crystal Form. **Crystal form** is the external expression of a mineral's internal orderly arrangement of atoms. Generally, when a mineral forms without space restrictions, it will develop individual crystals with well-formed crystal faces. Figure 2.1 shows the distinctive hexagonal crystals of quartz that form when space and time permit. However, most of the time, crystal growth is severely constrained. It is stunted because of competition for space, resulting in an intergrown mass of small, jammed crystals, none of which exhibits its crystal form. This is what happened to the minerals in the granite in Figure 2.3. Thus, most inorganic solid objects are composed of crystals, but they are not clearly visible to the unaided eye.

Luster. **Luster** is the appearance or quality of light reflected from the surface of a mineral. Minerals that have the appearance of metals, regardless of color, are said to have a *metallic luster.* Minerals with a *nonmetallic luster* are described by various adjectives. These include vitreous (glassy, like the quartz crystals in Figure 2.1), pearly, silky, resinous, and earthy (dull). Some minerals appear somewhat metallic in luster and are said to be *submetallic.*

Color. Although **color** is an obvious feature of a mineral, it is often an unreliable diagnostic property. Slight impurities in the common mineral quartz, for example, give it a variety of colors, including pink, purple (amethyst) milky white, and even black.

Streak. **Streak** is the color of a mineral in its powdered form, which is a much more reliable indication of color. Streak is obtained by rubbing the mineral across a piece of hard, unglazed porcelain, termed a *streak plate* (Figure 2.10). Whereas the color of a mineral may vary from sample to sample, the streak usually does not and is

▲ **Figure 2.10** Although the color of a mineral may not be very helpful in identification, the streak, which is the color of the powdered mineral, can be very useful. *(Photo by Dennis Tasa)*

therefore the more consistent property. Streak can also help to distinguish minerals with metallic lusters from those having nonmetallic lusters. Metallic minerals generally have a dense, dark streak, whereas minerals with nonmetallic lusters do not.

It should be noted that not all minerals produce a streak when using a streak plate. No streak is produced if the mineral is harder than the streak plate.

Hardness. One of the most useful diagnostic properties is **hardness,** a measure of the resistance of a mineral to abrasion or scratching. This property is determined by rubbing the mineral to be identified against another mineral of known hardness. One will scratch the other (unless they have the same hardness).

Geologists use a standard hardness scale, called the **Mohs scale.** It consists of 10 minerals arranged in order from 10 (hardest) to 1 (softest), as shown in Figure 2.11. Any mineral of unknown hardness can be rubbed against these to determine its hardness. In the field, other handy objects work, too. For example, your fingernail has a hardness of 2.5, a copper penny 3.5, and a piece of glass 5.5. The mineral gypsum, which has a hardness of 2, can be easily scratched by your fingernail. Conversely, the similar-looking mineral calcite, which has a hardness of 3, cannot be scratched by your fingernail. Calcite cannot scratch glass, because its hardness is less than 5.5. Quartz, the hardest of the common minerals at 7, will scratch a glass plate. Diamonds, hardest of all, scratch anything.

Pyrite "fools gold" exhibits a metallic luster.

Cleavage. In the crystal structure of a mineral, some bonds are weaker than others. These bonds are where a mineral will break when it is stressed. **Cleavage** is the tendency of a mineral to cleave, or break, along planes of weak bonding. Not all minerals have definite planes of weak bonding, but those that possess cleavage can be identified by the distinctive smooth surfaces that are produced when the mineral is broken.

The simplest type of cleavage is exhibited by minerals called micas (Figure 2.12). Because the micas have weak bonds in one direction, they cleave to form thin, flat sheets. Some minerals have several cleavage planes that produce smooth surfaces when broken, whereas others exhibit poor cleavage and still others have no cleavage at all. When minerals break evenly in more than one direction, cleavage is described by the *number*

Did You Know?

Archaeological finds show that more than 2000 years ago, the Romans used lead for pipes to transport water within their buildings. In fact, Roman smelting of lead and copper ores between 500 B.C. and 300 A.D. caused a small but significant rise in atmospheric pollution, as recorded in Greenland ice cores.

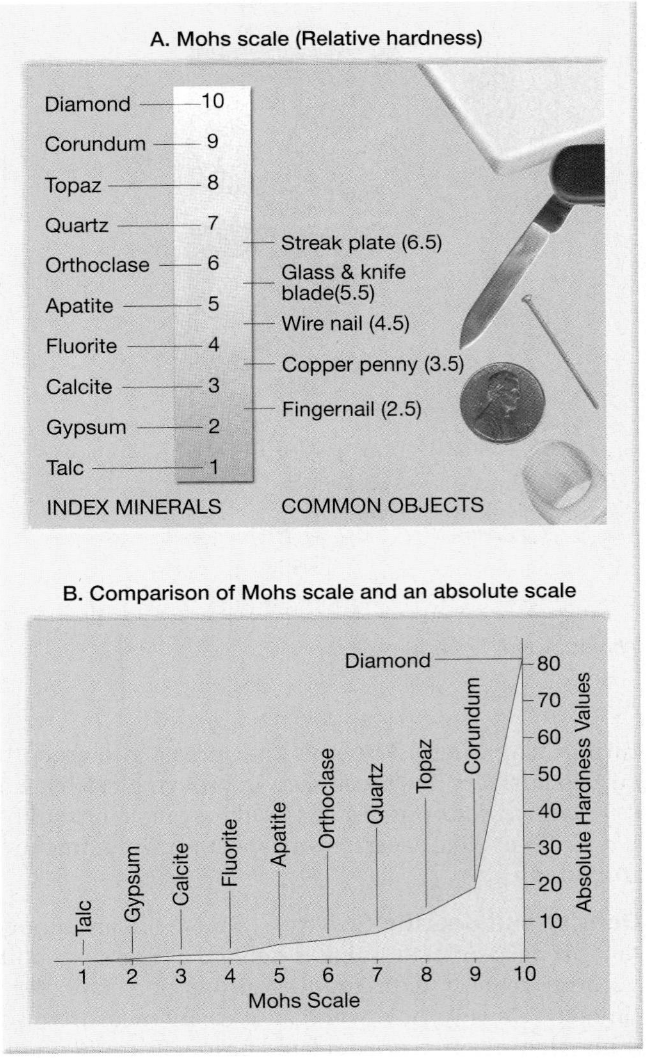

▲ **Figure 2.11** Hardness scales. **A.** Mohs scale of hardness, with the hardness of some common objects. **B.** Relationship between Mohs relative hardness scale and an absolute hardness scale.

▼ **Figure 2.12** The thin sheets shown here were produced by splitting a mica (muscovite) crystal parallel to its perfect cleavage. *(Photo by Breck P. Kent)*

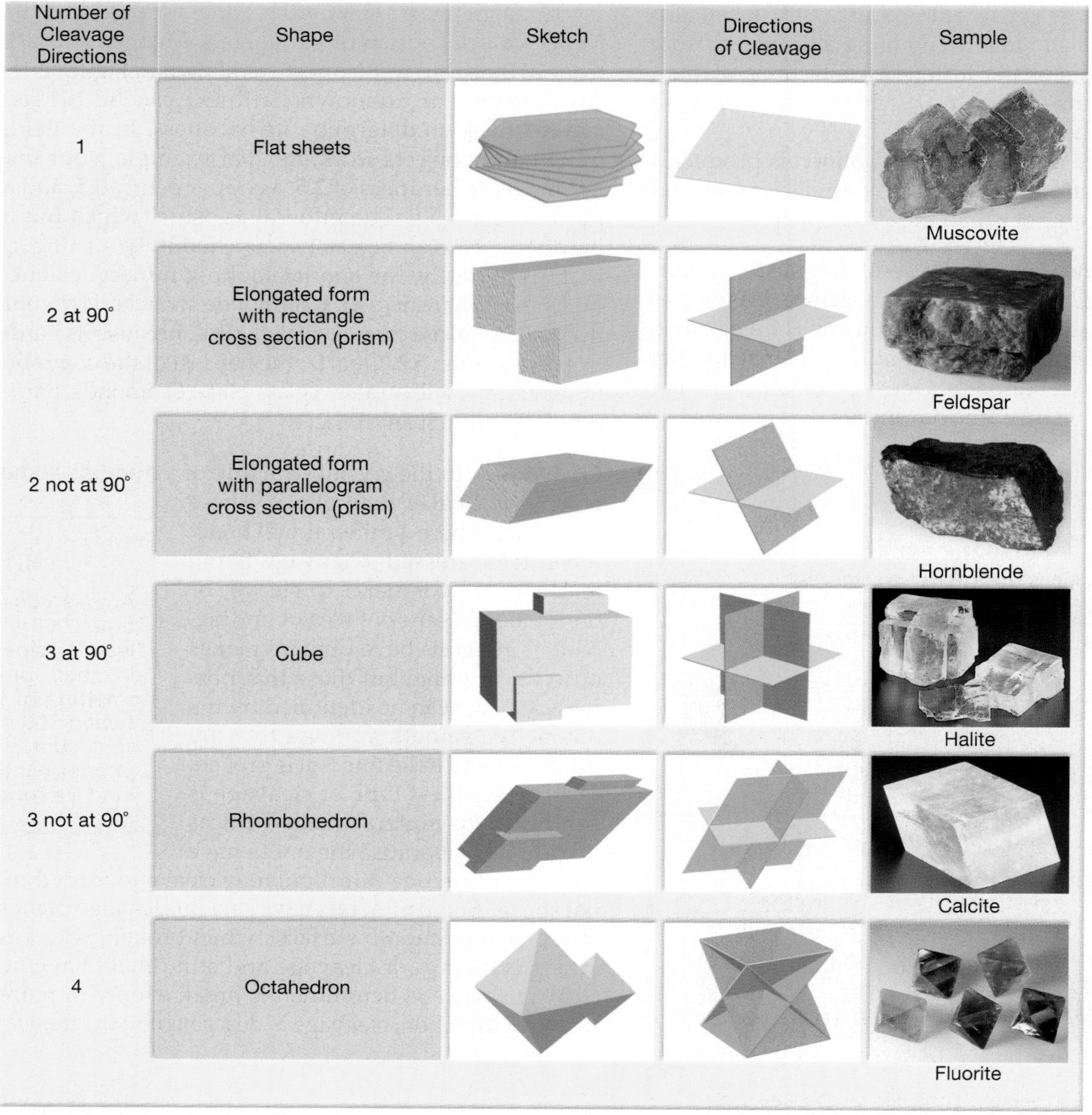

Number of Cleavage Directions
Shape
Sketch
Directions of Cleavage
Sample
1
Flat sheets
Muscovite
2 at 90°
Elongated form with rectangle cross section (prism)
Feldspar
2 not at 90°
Elongated form with parallelogram cross section (prism)
Hornblende
3 at 90°
Cube
Halite
3 not at 90°
Rhombohedron
Calcite
4
Octahedron
Fluorite

▲ **Figure 2.13** Common cleavage directions exhibited by minerals. *(Photos by E. J. Tarbuck)*

of planes exhibited and the *angles at which they meet* (Figure 2.13).

Do not confuse *cleavage* with *crystal form!* When a mineral exhibits cleavage, it will break into pieces *that have the same geometry as each other.* By contrast, the quartz crystals shown in Figure 2.1 do not have cleavage. If broken, they fracture into pieces that do not resemble each other or the original crystals.

Did You Know?

The name crystal is derived from the Greek (*krystallos,* meaning ice) and was applied to quartz crystals. The ancient Greeks thought quartz was water that had crystallized at high pressures deep inside Earth.

Fracture. Minerals that do not exhibit cleavage when broken, such as quartz, are said to **fracture.** Even fracturing has variety: Minerals that break into smooth curved surfaces like those seen in broken glass have a *conchoidal fracture* (Figure 2.14). Others break into splinters or fibers, like asbestos, but most minerals fracture irregularly.

Density and Specific Gravity. **Density** is an important property of matter that is defined as mass per unit volume (usually expressed as grams per cubic centimeter). Density, however, is not easily measured, so mineralogists often use a related property, *specific gravity,* to describe the density of a mineral. **Specific gravity** compares the weight of a mineral to the weight of an equal volume of water. For example, if a cubic centimeter of a mineral weighs three times as much as a

▲ **Figure 2.14** Conchoidal fracture. The smooth curved surfaces result when minerals break in a glasslike manner. *(Photo by E. J. Tarbuck)*

cubic centimeter of water, its specific gravity is 3. (It turns out that the specific gravity of a mineral yields a unitless number, which is numerically equal to density.) With a little practice, you can estimate the specific gravity of minerals by hefting them in your hand. For instance, if a mineral feels as heavy as the common rocks you have handled, its specific gravity will probably be somewhere between 2.5 and 3. Some metallic minerals have a specific gravity noticeably greater than that of common rock-forming minerals. Galena, which is an ore of lead, has a specific gravity of roughly 7.5. The specific gravity of pure 24-karat gold is almost 20.

Other Properties of Minerals

In addition to the properties that were just described, some minerals can be recognized by other distinctive properties. For example, halite is ordinary salt, so it is quickly identified with your tongue. Thin sheets of mica will bend and elastically snap back. Gold is malleable, which means it can be easily hammered or shaped. Talc and graphite both have distinctive feels. Talc feels soapy. Graphite feels greasy (it is a principal ingredient in dry lubricants).

A few minerals, such as magnetite, have a high iron content and can be picked up with a magnet. Some varieties of magnetite (lodestone) are natural magnets and will pick up small iron-based objects such as pins and paper clips. Some minerals exhibit special optical properties. For example, when a transparent piece of calcite is placed over printed material, the letters appear doubled. This optical property is known as *double refraction.* In addition, the streak of many sulfur-bearing minerals smells like rotten eggs.

One very simple chemical test involves placing a drop of dilute hydrochloric acid from a dropper bottle on a freshly broken mineral surface. Certain minerals, called carbonates, will effervesce (fizz) with hydrochloric acid. This test is useful in identifying the mineral calcite, which is a common carbonate mineral.

In summary, a number of special physical and chemical properties are useful in identifying particular minerals. These include taste, smell, elasticity, malleability, feel, magnetism, double refraction, and chemical reaction to hydrochloric acid. Remember that every one of these properties depends on the composition (elements) of a mineral and its structure (how the atoms are arranged).

Diamond is by far the hardest mineral known. (Diane Pendland/Smithsonian Institution)

Mineral Groups

GEODe ESSENTIALS OF GEOLOGY

Minerals: Building Blocks of Rocks
▼ Mineral Groups

Nearly 4000 minerals have been named, and several new ones are identified each year. Fortunately, for students who are beginning to study minerals, no more than a few dozen are abundant! Collectively, these few make up most of the rocks of Earth's crust and, as such, are often referred to as the **rock-forming minerals.** It is also interesting to note that *only eight elements* make up the bulk of these minerals and represent over 98 percent (by weight) of the continental crust (Figure 2.15). These elements, in order of abundance, are oxygen (O), silicon (Si), aluminum (Al), iron (Fe), calcium (Ca), sodium (Na), potassium (K), and magnesium (Mg).

Magnetite is the most common magnetic mineral.

As shown in Figure 2.15, silicon and oxygen are by far the most common elements in Earth's crust. Furthermore, these two elements readily combine to form the framework for the most common mineral group, the **silicates,** which account for more than 90 percent of Earth's crust.

Because other mineral groups are far less abundant in Earth's crust than the silicates, they are often grouped together under the heading **nonsilicates.** Although not as common as the silicates, some nonsilicate minerals are very important economically. They provide us with the iron and aluminum to build our automobiles; gypsum for plaster and dry-wall to construct our homes; and copper for wire to carry electricity and to connect us to the Internet. Some common nonsilicate mineral groups include the carbonates, sulfates, and halides. In addition to their economic importance, these mineral groups include members that are major constituents in sediments and sedimentary rocks.

We will first discuss the most common mineral group, the silicates, and then consider some of the prominent nonsilicate mineral groups.

Double refraction, exhibited by the mineral calcite. (Chip Clark)

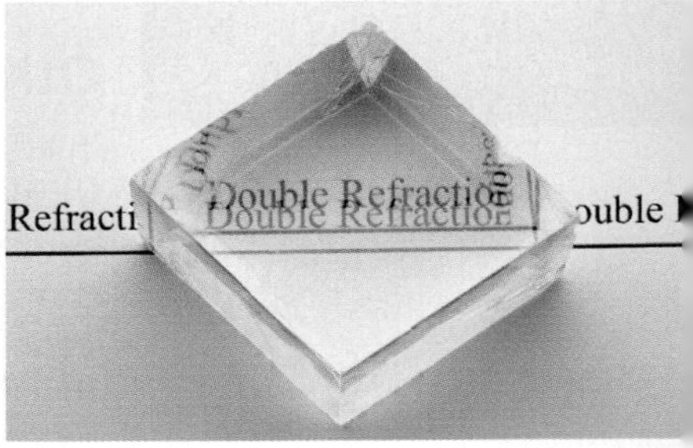

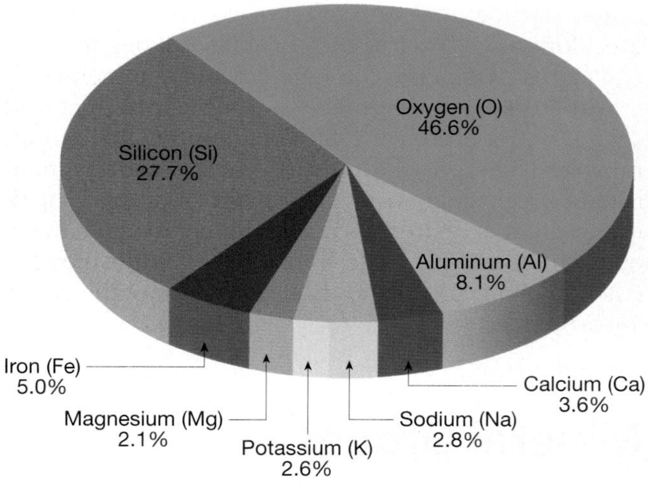

▲ **Figure 2.15** Relative abundance of the eight most abundant elements in the continental crust.

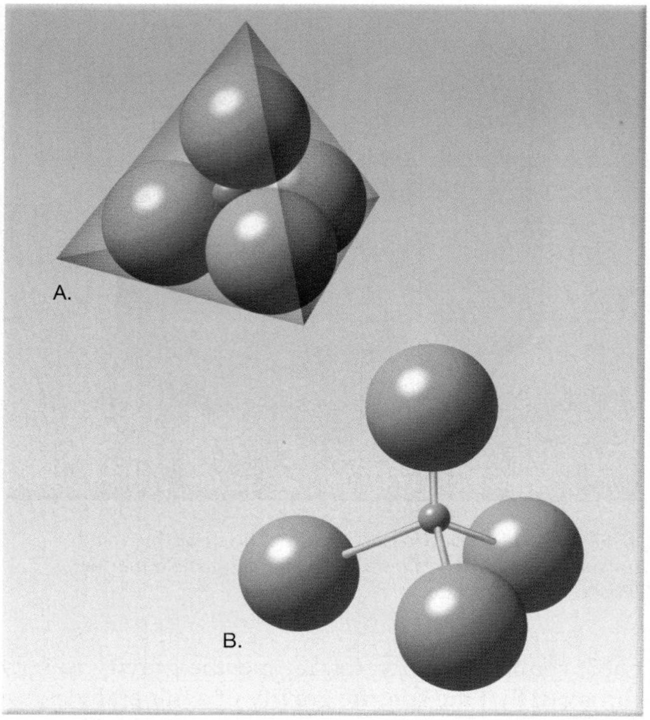

▲ **Figure 2.16** Two representations of the silicon–oxygen tetrahedron. **A.** The four large spheres represent oxygen atoms and the blue sphere represents a silicon atom. The spheres are drawn in proportion to the radii of the atoms. **B.** A model of the tetrahedron using rods to depict the bonds that connect the atoms.

The Silicates

GEODe Minerals: Building Blocks of Rocks
▼ Mineral Groups

Every silicate mineral contains the elements oxygen and silicon. Moreover, except for a few minerals such as quartz, the crystalline structure of most silicate minerals contains one or more of the other common elements of Earth's crust. These elements give rise to the great variety of silicate minerals and their varied properties.

Calcite reacting with a weak acid. (Chip Clark)

The Silicon–Oxygen Tetrahedron

All silicates have the same fundamental building block, the **silicon–oxygen tetrahedron.** This structure consists of four oxygen ions surrounding a much smaller silicon ion (Figure 2.16). The silicon–oxygen tetrahedron is a complex ion (SiO_4^{-4}) with a charge of −4.

In nature, the simplest way in which these tetrahedra join together to become neutral compounds is through the addition of positively charged ions (Figure 2.17). In this way, a chemically stable structure is produced, consisting of individual tetrahedra linked together by various positively charged ions.

Other Silicate Structures

In addition to positive ions providing the opposite electrical charge needed to bind the tetrahedra, the tetrahedra may link with themselves in a variety of configurations. For example, the tetrahedra may be joined to form *single chains, double chains,* or *sheet structures,* as shown in Figure 2.18. The joining of

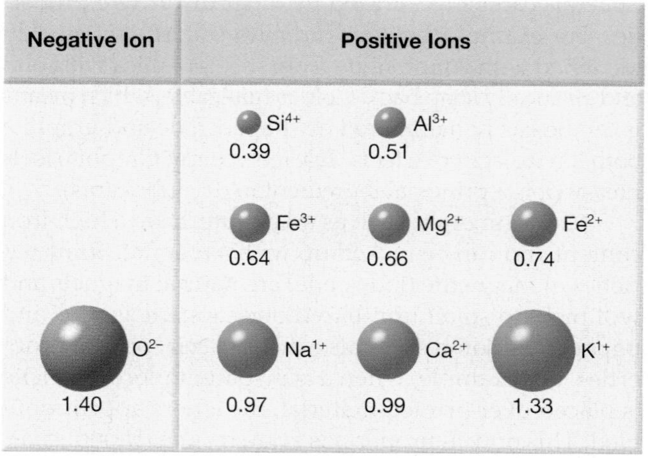

▲ **Figure 2.17** Relative sizes and electrical charges of the eight most common elements in Earth's crust. These are the most common ions in rock-forming minerals. Ionic radii are expressed in angstroms (1 angstrom equals 10^{-8} cm).

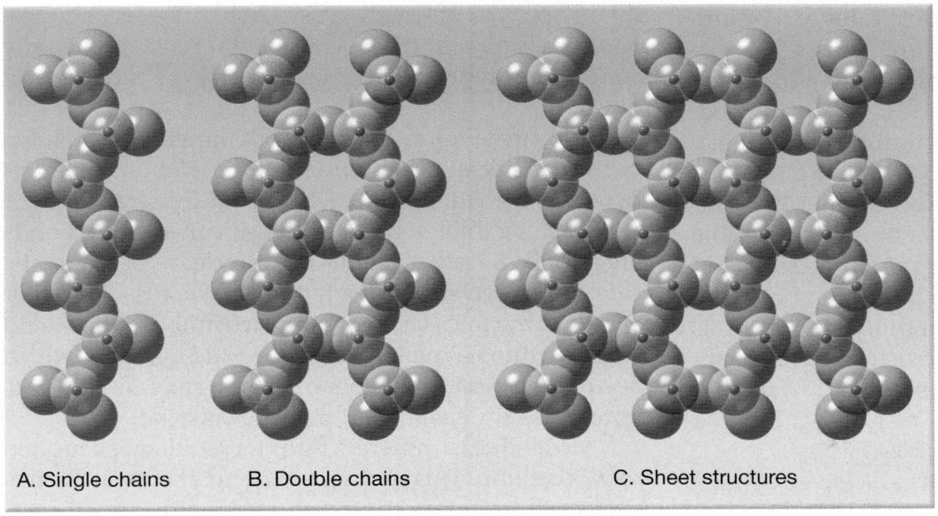

◀ **Figure 2.18** Three types of silicate structures. **A.** Single chains. **B.** Double chains. **C.** Sheet structures.

tetrahedra in each of these configurations results from the sharing of oxygen atoms between silicon atoms in adjacent tetrahedra.

To understand better how this sharing takes place, select one of the silicon ions (small blue spheres) near the middle of the single-chain structure shown in Figure 2.18. Notice that this silicon ion is completely surrounded by four larger oxygen ions (you are looking *through* one of the four to see the blue silicon ion). Also notice that, of the four oxygen ions, two are joined to other silicon ions, whereas the other two are not shared in this manner. *It is the linkage across the shared oxygen ions that joins the tetrahedra into a chain structure.* Now, examine a silicon ion near the middle of the sheet structure and count the number of shared and unshared oxygen ions surrounding it. The increase in the degree of sharing accounts for the sheet structure. Other silicate structures exist, and the most common has all of the oxygen ions shared to produce a complex three-dimensional framework.

By now you can see that the ratio of oxygen ions to silicon ions differs in each of the silicate structures. In the isolated tetrahedron, there are four oxygen ions for every silicon ion. In the single chain, the oxygen-to-silicon ratio is 3:1, and in the three-dimensional framework this ratio is 2:1. Consequently, as more of the oxygen ions are shared, the percentage of silicon in the structure increases. Silicate minerals are therefore described as having a high or low silicon content based on their ratio of oxygen to silicon. This difference in silicon content is important, as we shall see in a later chapter when we consider the formation of igneous rocks.

Joining Silicate Structures

Most silicate structures, including single chains, double chains, and sheets, are not neutral chemical compounds. Thus, like the individual tetrahedra, they are all neutralized by the inclusion of positively charged ions that bond them together into a variety of complex crystalline configurations. The ions that most often link silicate structures are those of the elements iron (Fe), magnesium (Mg), potassium (K), sodium (Na), aluminum (Al), and calcium (Ca).

Notice in Figure 2.17 that each of these ions has a particular atomic size and a particular charge. Generally, ions of approximately the same size are able to substitute freely for one another. For instance, ions of iron (Fe^{2+}) and magnesium (Mg^{2+}) are nearly the same size and substitute for each other without altering the mineral structure. This also holds true for calcium and sodium ions, which can occupy the same site in a crystalline structure. In addition, aluminum (Al) often substitutes for silicon in the silicon–oxygen tetrahedron.

Because of the ability of silicate structures to readily accommodate different ions at a given bonding site, individual specimens of a particular mineral may contain varying amounts of certain elements. A mineral of this type is often expressed by a chemical formula that uses parentheses to show the variable component. A good example is the mineral olivine, $(Mg, Fe)_2SiO_4$, which is magnesium/iron silicate. As you can see from the formula, it is the iron (Fe^{2+}) and magnesium (Mg^{2+}) ions in olivine that freely substitute for each other. At one extreme, olivine may contain iron without any magnesium (Fe_2SiO_4, or iron silicate) and at the other, iron is totally lacking (Mg_2SiO_4, or magnesium silicate). Between these end members, any ratio of iron to magnesium is possible. Thus, olivine, as well as many other silicate minerals, is actually a *family* of minerals with a range of composition between the two end members.

In certain substitutions, the ions that interchange do not have the same electrical charge. For instance, when

Did You Know?

Most "scoopable" kitty litters sold on the market today contain a naturally occurring material called *bentonite*. Bentonite is largely composed of highly absorbent clay minerals that swell and clump up in the presence of moisture. This allows kitty waste to be isolated and scooped out, leaving behind clean litter.

calcium (Ca^{2+}) substitutes for sodium (Na^{1+}), the structure gains a positive charge. In nature, one way in which this substitution is accomplished, while still maintaining overall electrical neutrality, is the simultaneous substitution of aluminum (Al^{3+}) for silicon (Si^{4+}). This particular double substitution occurs in a mineral called *plagioclase feldspar.* It is a member of the most abundant family of minerals found in Earth's crust. The end members of this particular feldspar series are a calcium–aluminum silicate (anorthite, $\mathit{Ca}Al_2Si_2O_8$) and a sodium–aluminum silicate (albite, $\mathit{Na}AlSi_3O_8$).

We are now prepared to review silicate structures in light of what we know about chemical bonding. An examination of Figure 2.17 shows that among the major constituents of the silicate minerals, only oxygen is a negatively charged ion. Because oppositely charged ions attract (and similarly charged ions repel), the chemical bonds that hold silicate structures together form between oxygen and positively charged ions. Thus, the positively charged ions arrange themselves so that they can be as close as possible to oxygen while remaining as far apart from *each other* as possible. Because of its small size and high charge (−4), the silicon (Si) ion forms the strongest bonds with oxygen. Aluminum (Al), although not as strongly bonded to oxygen as silicon, is more strongly bonded than calcium (Ca), magnesium (Mg), iron (Fe), sodium (Na), and potassium (K). In many ways, aluminum plays a role similar to silicon by being the central ion in the basic tetrahedral structure.

Granite, like that found in Yosemite National Park, is composed mostly of light silicate minerals.

Most silicate minerals consist of a basic framework composed of either a single silicon or aluminum ion surrounded by four negatively charged oxygen ions. These tetrahedra often link together to form a variety of other silicate structures (chains, sheets, etc.) through shared oxygen atoms. Finally, positively charged ions bond with the oxygen atoms of these silicate structures to create the more complex crystalline structures that characterize the silicate minerals.

This quartz sample exhibits well-developed hexagonal crystals. (Breck P. Kent)

Common Silicate Minerals

GEODe ESSENTIALS OF GEOLOGY
Minerals: Building Blocks of Rocks
▼ Mineral Groups

To reiterate, the silicates are the most abundant mineral group and have the silicate ion (SiO_4^{4-}) as their basic building block. The major silicate groups and common examples are given in Figure 2.19. The feldspars are by far the most plentiful silicate group, comprising more than 50 percent of Earth's crust. Quartz, the second most abundant mineral in the continental crust, is the only common mineral made completely of silicon and oxygen.

Notice in Figure 2.19 that each mineral group has a particular silicate structure and that a relationship exists between the internal structure of a mineral and the cleavage it exhibits. Because the silicon–oxygen bonds are strong, silicate minerals tend to cleave between the silicon–oxygen structures rather than across them. For example, the micas have a sheet structure and thus tend to cleave into flat plates (see Figure 2.12). Quartz, which has equally strong silicon–oxygen bonds in all directions, has no cleavage but fractures instead.

Most silicate minerals form (crystallize) as molten rock is cooling. This cooling can occur at or near Earth's surface (low temperature and pressure) or at great depths (high temperature and pressure). The environment during crystallization and the chemical composition of the molten rock determine to a large degree which minerals are produced. For example, the silicate mineral olivine crystallizes at high temperatures, whereas quartz crystallizes at much lower temperatures.

In addition, some silicate minerals form at Earth's surface from the weathered products of other silicate minerals. Still other silicate minerals are formed under the extreme pressures associated with mountain building. Each silicate mineral, therefore, has a structure and a chemical composition that *indicate the conditions under which it formed.* Thus, by carefully examining the mineral constituents of rocks, geologists can often determine the circumstances under which the rocks formed.

We will now examine some of the most common silicate minerals, which we divide into two major groups on the basis of their chemical makeup.

The Light Silicates

The **light** (or **nonferromagnesian**) **silicates** are generally light in color and have a specific gravity of about 2.7, which is considerably less than the dark (ferromagnesian) silicates. These differences are mainly attributable to the presence or absence of iron and magnesium. The light silicates contain varying amounts of aluminum, potassium, calcium, and sodium rather than iron and magnesium.

Feldspar Group. *Feldspar,* the most common mineral group, can form under a wide range of temperatures and pressures, a fact that partially accounts for its abundance. All of the feldspars have similar physical properties. They have two planes of cleavage meeting at or near 90-degree angles, are relatively hard (6 on the Mohs scale), and have a luster that ranges from glassy to pearly. As one component in a rock, feldspar crystals

Mineral/Formula	Cleavage	Silicate Structure	Example
Olivine $(Mg, Fe)_2SiO_4$	None	Single tetrahedron 	
Pyroxene group (Augite) $(Mg,Fe)SiO_3$	Two planes at right angles	Single chains 	
Amphibole group (Hornblende) $Ca_2(Fe,Mg)_5Si_8O_{22}(OH)_2$	Two planes at 60° and 120°	Double chains 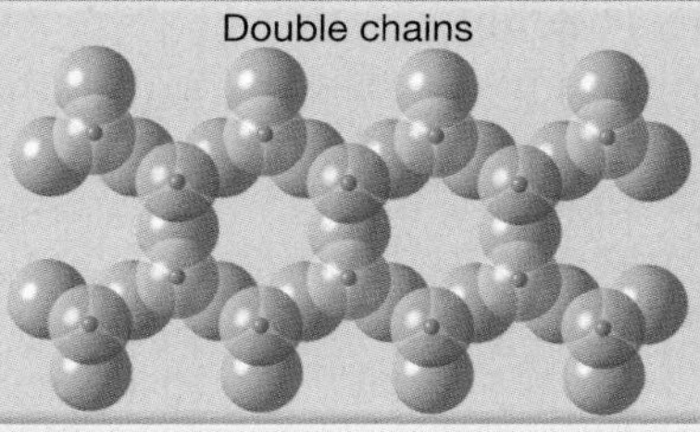	
Micas: Biotite $K(Mg,Fe)_3AlSi_3O_{10}(OH)_2$	One plane	Sheets 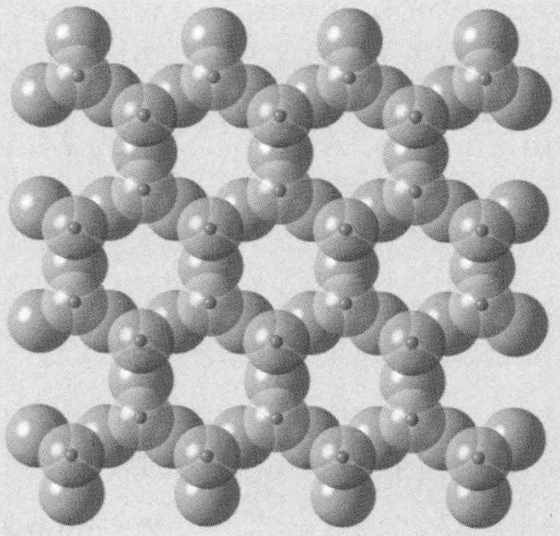	
Micas: Muscovite $KAl_2(AlSi_3O_{10})(OH)_2$	One plane	Sheets	
Feldspars: Potassium feldspar (Orthoclase) $KAlSi_3O_8$	Two planes at 90°	Three-dimensional networks 	
Feldspars: Plagioclase feldspar $(Ca,Na)AlSi_3O_8$	Two planes at 90°	Three-dimensional networks	
Quartz SiO_2	None	Three-dimensional networks	

▲ **Figure 2.19** Common silicate minerals. Note that the complexity of silicate structures increases down the chart. *(Photos by E. J. Tarbuck and Dennis Tasa)*

Did You Know?

Mica crystals have been known to grow to great size. One found in Russia's Ural Mountains was 54 square feet in area and more than 1.5 feet thick.

can be identified by their rectangular shape and rather smooth shiny faces (Figure 2.19).

Two different feldspar structures exist. One group of feldspar minerals contains potassium ions in its structure and is therefore referred to as *potassium feldspar.* (*Orthoclase* and *microcline* are common members of the potassium feldspar group.) The other group, called *plagioclase feldspar,* contains both sodium and calcium ions that freely substitute for one another depending on the environment during crystallization.

Basalt, a common igneous rock, contains a large percentage of dark silicate minerals. (Peterson/USGS)

Potassium feldspar is usually light cream to salmon pink in color. The plagioclase feldspars, on the other hand, range in color from white to medium gray. However, color should not be used to distinguish these groups. The only sure way to distinguish the feldspars physically is to look for a multitude of fine parallel lines, called *striations.* Striations are found on some cleavage planes of plagioclase feldspar but are not present on potassium feldspar (Figure 2.20).

White Sands National Monument, New Mexico, is the world's largest gypsum dune field.

Quartz. *Quartz* is the only common silicate mineral consisting entirely of silicon and oxygen. As such, the term *silica* is applied to quartz, which has the chemical formula SiO_2. As the structure of quartz contains a ratio of two oxygen ions (O^{2-}) for every silicon ion (Si^{4+}), no other positive ions are needed to attain neutrality.

In quartz, a three-dimensional framework is developed through the complete sharing of oxygen by adjacent silicon atoms. Thus, all of the bonds in quartz are of the strong silicon–oxygen type. Consequently, quartz is hard, resistant to weathering, and does not have cleavage. When broken, quartz generally exhibits conchoidal fracture. In a pure form, quartz is clear and if allowed to grow without interference, will form hexagonal crystals that develop pyramid-shaped ends. However, like most other clear minerals, quartz is often colored by inclusions of various ions (impurities) and forms without developing good crystal faces. The most common varieties of quartz are milky (white), smoky (gray), rose (pink), amethyst (purple), and rock crystal (clear) (Figure 2.21).

Muscovite. *Muscovite* is a common member of the mica family. It is light in color and has a pearly luster. Like other micas, muscovite has excellent cleavage in one direction. In thin sheets, muscovite is clear, a property that accounts for its use as window "glass" during the Middle Ages. Because muscovite is very shiny, it can often be identified by the sparkle it gives a rock. If you have ever looked closely at beach sand, you may have seen the glimmering brilliance of the mica flakes scattered among the other sand grains.

Clay Minerals. *Clay* is a term used to describe a variety of complex minerals that, like the micas, have a

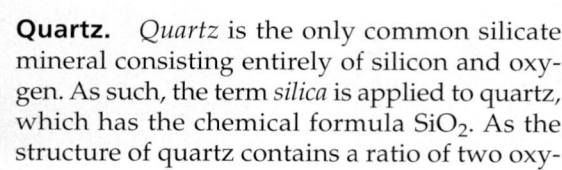

▲ **Figure 2.20** These parallel lines, called *striations,* are a distinguishing characteristic of the plagioclase feldspars. *(Photo by E. J. Tarbuck)*

▼ **Figure 2.21** Quartz. Some minerals, such as quartz, occur in a variety of colors. These samples include crystal quartz (colorless), amethyst (purple), citrine (yellow), and smoky quartz (gray to black). *(Photo by E. J. Tarbuck)*

sheet structure. Unlike other common silicates, such as quartz and feldspar, clays do not form in igneous environments. Rather, most clay minerals originate as products of the chemical weathering of other silicate minerals. Thus, clay minerals make up a large percentage of the surface material we call soil. Because of the importance of soil in agriculture, and because of its role as a supporting material for buildings, clay minerals are extremely important to humans. In addition, clays account for nearly half the volume of sedimentary rocks.

The Dark Silicates

The **dark** (or **ferromagnesian**) **silicates** are those minerals containing ions of iron (iron = *ferro*) and/or magnesium in their structure. Because of their iron content, ferromagnesian silicates are dark in color and have a greater specific gravity, between 3.2 and 3.6, than nonferromagnesian silicates. The most common dark silicate minerals are olivine, the pyroxenes, the amphiboles, dark mica (biotite), and garnet.

Olivine Group. *Olivine* is a family of high-temperature silicate minerals that are black to olive green in color and have a glassy luster and a conchoidal fracture. Rather than developing large crystals, olivine commonly forms small, rounded crystals that give rocks consisting largely of olivine a granular appearance. Olivine is composed of individual tetrahedra, which are bonded together by a mixture of iron and magnesium ions positioned so as to link the oxygen atoms and magnesium atoms together. Because the three-dimensional network generated in this fashion does not have its weak bonds aligned, olivine does not possess cleavage.

Pyroxene Group. The *pyroxenes* are a group of complex minerals thought to be important components of Earth's mantle. The most common member, *augite*, is a black, opaque mineral with two directions of cleavage that meet at nearly a 90-degree angle. Augite is one of the dominant minerals in basalt, a common igneous rock of the oceanic crust and volcanic areas on the continents.

Amphibole Group. *Hornblende* is the most common member of a chemically complex group of minerals called *amphiboles*. Hornblende is usually dark green to black in color, and except for its cleavage angles, which are about 60 degrees and 120 degrees, it is very similar in appearance to augite. This helps distinguish it from pyroxene, which forms rather blocky crystals. Hornblende is found predominantly in continental rocks, where it often makes up the dark portion of an otherwise light-colored rock.

Biotite. *Biotite* is the dark iron-rich member of the mica family. Like other micas, biotite possesses a sheet structure that gives it excellent cleavage in one direction. Biotite also has a shiny black appearance that helps distinguish it from the other ferromagnesian minerals. Like hornblende, biotite is a common constituent of continental rocks, including the igneous rock granite.

Did You Know?

Clay minerals have been used as an additive to thicken milkshakes in fast-food restaurants.

Important Nonsilicate Minerals

GEODe: Minerals: Building Blocks of Rocks
▼ Mineral Groups

Nonsilicate minerals are typically subdivided into *classes*, based on the negatively charged ion or complex ion that the members have in common (Table 2.1). For example, the *oxides* contain the negative oxygen ion (O^{2-}), which is bonded to one or more kinds of positive ions. Thus, within each mineral class, the basic structure and type of bonding is similar. As a result, the minerals in each group have similar physical properties that are useful in mineral identification.

Although the nonsilicates make up only about 8 percent of Earth's crust, some minerals, such as gypsum, calcite, and halite, occur as constituents in sedimentary rocks in significant amounts. Furthermore, many others are important economically. Table 2.1 lists some of the nonsilicate mineral classes and a few examples of each. A brief discussion of a few of the more common nonsilicate minerals follows.

Some of the most common nonsilicate minerals belong to one of three classes of minerals—the carbonates (CO_3^{2-}), the sulfates (SO_4^{2-}), and the halides (Cl^{1-}, F^{1-}, Br^{1-}). The carbonate minerals are much simpler structurally than the silicates. This mineral group is composed of the carbonate ion (CO_3^{2-}) and one or more kinds of positive ions. The two most common carbonate minerals are *calcite*, $CaCO_3$ (calcium carbonate) and *dolomite*, $CaMg(CO_3)_2$ (calcium/magnesium carbonate). Because these minerals are similar both physically and chemically, they are difficult to distinguish from each other. Both have a vitreous luster, a hardness between 3 and 4, and nearly perfect rhombic cleavage. They can, however, be distinguished by using dilute hydrochloric acid. Calcite reacts vigorously with this acid, whereas dolomite reacts much more slowly. Calcite and dolomite are usually found together as the primary constituents in the sedimentary rocks limestone and dolostone. When calcite is the dominant mineral, the rock is called *limestone*, whereas *dolostone* results from a

Galena is an important ore of lead.

Hematite is one of the main ores of iron.

Did You Know?

The material we commonly call mother-of-pearl is a carbonate mineral called aragonite. Mother-of-pearl is produced by a large group of marine organisms commonly called clams (mollusks).

predominance of dolomite. Limestone has many uses, including as road aggregate, as building stone, and as the main ingredient in Portland cement.

Two other nonsilicate minerals frequently found in sedimentary rocks are *halite* and *gypsum*. Both minerals are commonly found in thick layers that are the last vestiges of ancient seas that have long since evaporated (Figure 2.22). Like limestone, both are important nonmetallic resources. Halite is the mineral name for common table salt (NaCl). Gypsum ($CaSO_4 \cdot 2H_2O$), which is calcium sulfate with water bound into the structure, is the mineral of which plaster and other similar building materials are composed.

Did You Know?

The mineral pyrite is commonly known as "fool's gold," because its golden-yellow color can lead to its being mistaken for gold. The name pyrite is derived from the Greek *pyros* ("fire"), because it gives off sparks when struck sharply.

Most nonsilicate mineral classes contain members that are prized for their economic value. This includes the oxides, whose members hematite and magnetite are important ores of iron. Also significant are the sulfides, which are basically compounds of sulfur (S) and one or more metals. Examples of important sulfide minerals include galena (lead), sphalerite (zinc), and chalcopyrite (copper). In addition, native elements, including gold, silver, and carbon (diamonds), plus a host of other nonsilicate minerals—fluorite (flux in making steel), corundum (gemstone, abrasive), and uraninite (a uranium source)—are important economically.

Mineral Resources

Earth's crust and oceans are the source of a wide variety of useful and essential substances (see Box 2.1). In fact, practically every manufactured product contains materials obtained from minerals. Table 2.1 lists important mineral groups in this category.

Mineral resources are the endowment of useful minerals ultimately available commercially. Resources include already identified deposits from which minerals can be extracted profitably, called **reserves,** as well as known deposits that are not yet economically or technologically recoverable. Deposits inferred to exist, but not yet discovered, are also considered mineral resources.

Did You Know?

Gypsum, a white-to-transparent mineral, was first used as a building material in Anatolia (present-day Turkey) around 6000 B.C. It is also found on the interiors of the great pyramids in Egypt, which were erected in about 3700 B.C. Today, an average new American home contains more than 7 metric tons of gypsum in the form of 6,000 square feet of wallboard.

The term **ore** is used to denote those useful metallic minerals that can be mined at a profit. In common usage, the term ore is also applied to some nonmetallic minerals such as fluorite and sulfur. However, materials used for such purposes as building stone, road aggregate, abrasives, ceramics, and fertilizers are not usually called ores; rather, they are classified as industrial rocks and minerals.

Recall that more than 98 percent of Earth's crust is composed of only eight elements, and except for oxygen and silicon, all other elements make up a relatively small fraction of common crustal rocks (see Figure 2.15). Indeed, the natural concentrations of many elements are exceedingly small. A deposit containing the average percentage of a valuable element is worthless because the cost of extracting it greatly exceeds the value of the material recovered.

To be considered of value, an element must be concentrated above the level of its average crustal abundance. Generally, the lower the crustal abundance, the greater the concentration. For example, copper makes up about 0.0135 percent of the crust. However, for a material to be considered as copper ore, it must contain a concentration that is about 100 times this amount. Aluminum, in contrast, represents 8.13 percent of the crust and must be concentrated to only about four times its average crustal percentage before it can be extracted profitably.

It is important to realize that a deposit may become profitable to extract or lose its profitability because of economic changes. If demand for a metal increases and prices rise sufficiently, the status of a previously unprofitable deposit changes, and it becomes an ore. The status of unprofitable deposits may also change if a technological advance allows the ore to be extracted at a lower cost than before.

Conversely, changing economic factors can turn a once profitable ore deposit into an unprofitable deposit

▼ **Figure 2.22** Thick bed of halite (salt) at an underground mine in Grand Saline, Texas. *(Photo by Tom Bochsler)*

BOX 2.1

Asbestos: What Are the Risks?

Once considered safe enough to use in toothpaste, asbestos became one of the most feared contaminants on Earth. The asbestos panic in the United States began in 1986 when the Environmental Protection Agency (EPA) instituted the Asbestos Hazard Emergency Response Act. It required inspection of all pubic and private schools for asbestos. This brought asbestos to public attention and raised parental fears that children could contract asbestos-related cancers because of high levels of airborne fibers in schools.

What is Asbestos? Asbestos is a commercial term applied to a variety of silicate minerals that readily separate into thin, strong fibers that are highly flexible, heat resistant, and relatively inert (Figure 2.A). These properties make asbestos a desirable material for the manufacture of a wide variety of products, including insulation, fireproof fabrics, cement, floor tiles, and car-brake linings. In addition, wall coatings rich in asbestos fibers were used extensively during the U.S. building boom of the 1950s and early 1960s.

The mineral *chrysotile,* marketed as "white asbestos," belongs to the serpentine mineral group and accounts for the vast majority of asbestos sold commercially. All other forms of asbestos are amphiboles and constitute less than 10 percent of asbestos used commercially. The two most common amphibole asbestos minerals are informally called "brown" and "blue" asbestos, respectively.

Exposure and Risk. Health concerns about asbestos stem largely from claims of high death rates among asbestos miners attributed to asbestosis (lung scarring from asbestos fiber inhalation), mesothelioma (cancer of chest and abdominal cavity), and lung cancer. The degree of concern created by these claims is demonstrated by the growth of an entire industry built around asbestos removal from buildings.

The stiff, straight fibers of brown and blue (amphibole) asbestos are known to readily pierce, and remain lodged in, the linings of human lungs. The fibers are physically and chemically stable and are not broken down in the human body. These forms of asbestos are therefore a genuine cause for concern. White asbestos, however, being a different mineral, has different properties. The curly fibers of white asbestos are readily expelled from the lungs and, if they are not expelled, can dissolve within a year.

The U.S. Geological Survey has taken the position that the risks from the most widely used form of asbestos (chrysotile or "white asbestos") are minimal to nonexistent. They cite studies of miners of white asbestos in northern Italy that show mortality rates from mesothelioma and lung cancer differ very little from the general public. Another study was conducted on people living in the area of Thetford Mines, Quebec, once the largest chrysotile mine in the world. For many years there were no dust controls, so these people were exposed to extremely high levels of airborne asbestos. Nevertheless, they exhibited normal levels of the diseases thought to be associated with asbestos exposure.

Despite the fact that over 90 percent of all asbestos used commercially is white asbestos, a number of countries have moved to ban the use of asbestos in many applications, because the different mineral forms of asbestos are not distinguished. Very little of this once exalted mineral is presently used in the United States. Perhaps future studies will determine whether the asbestos panic, in which billions of dollars have been spent on testing and removal, was warranted or not.

Figure 2.A Asbestos. This sample is a fibrous form of the mineral serpentine. (Photo by E. J. Tarbuck)

Table 2.1 Common Nonsilicate Mineral Groups

Mineral Groups [key ion(s) or element(s)]	Mineral Name	Chemical Formula	Economic Use
Carbonates (CO_3^{2-})	Calcite	$CaCO_3$	Portland cement, lime
	Dolomite	$CaMg(CO_3)_2$	Portland cement, lime
Halides (Cl^{1-}, F^{1-}, Br^{1-})	Halite	$NaCl$	Common salt
	Fluorite	CaF_2	Used in steelmaking
	Sylvite	KCl	Fertilizer
Oxides (O^{2-})	Hematite	Fe_2O_3	Ore of iron, pigment
	Magnetite	Fe_3O_4	Ore of iron
	Corundum	Al_2O_3	Gemstone, abrasive
	Ice	H_2O	Solid form of water
Sulfides (S^{2-})	Galena	PbS	Ore of lead
	Sphalerite	ZnS	Ore of zinc
	Pyrite	FeS_2	Sulfuric acid production
	Chalcopyrite	$CuFeS_2$	Ore of copper
	Cinnabar	HgS	Ore of mercury
Sulfates (SO_4^{2-})	Gypsum	$CaSO_4 \cdot 2H_2O$	Plaster
	Anhydrite	$CaSO_4$	Plaster
	Barite	$BaSO_4$	Drilling mud
Native elements (single elements)	Gold	Au	Trade, jewelry
	Copper	Cu	Electrical conductor
	Diamond	C	Gemstone, abrasive
	Sulfur	S	Sulfa drugs, chemicals
	Graphite	C	Pencil lead, dry lubricant
	Silver	Ag	Jewelry, photography
	Platinum	Pt	Catalyst

▼ **Figure 2.23** Aerial view of Bingham Canyon copper mine near Salt Lake City, Utah. Although the amount of copper in the rock is less than 1 percent, the huge volumes of material removed and processed each day (about 200,000 tons) yield enough metal to be profitable. *(Photo by Michael Collier)*

that can no longer be called an ore. This situation was illustrated at the copper mining operation located at Bingham Canyon, Utah, one of the largest open-pit mines on Earth (Figure 2.23). Mining was halted there in 1985 because outmoded equipment had driven the cost of extracting the copper beyond the current selling price. The owners responded by replacing an antiquated 1000-car railroad with conveyor belts and pipelines for transporting the ore and waste. These devices achieved a cost reduction of nearly 30 percent and returned this mining operation to profitability.

Over the years, geologists have been keenly interested in learning how natural processes produce localized concentrations of essential minerals. One well-established fact is that occurrences of valuable mineral resources are closely related to the rock cycle. That is, the mechanisms that generate igneous, sedimentary, and metamorphic rocks, including the processes of weathering and erosion, play a major role in producing concentrated accumulations of useful elements.

Moreover, with the development of the theory of plate tectonics, geologists have added another tool for understanding the processes by which one rock is transformed into another. As these rock-forming processes are examined in the following chapters, we will consider their role in producing some of our important mineral resources.

Bornite, an ore of copper.

The Chapter in Review

- A *mineral* is a naturally occurring inorganic solid possessing a definite chemical structure that gives it a unique set of physical properties. Most *rocks* are aggregates composed of two or more minerals.
- The building blocks of minerals are *elements.* An *atom* is the smallest particle of matter that still retains the characteristics of an element. Each atom has a *nucleus,* which contains *protons* (particles with positive electrical charges) and *neutrons* (particles with neutral electrical charges). Orbiting the nucleus of an atom in regions called *energy levels,* or *principal shells,* are *electrons,* which have negative electrical charges. The number of protons in an atom's nucleus determines its *atomic number* and the name of the element. An element is a large collection of electrically neutral atoms, all having the same atomic number.
- Atoms combine with each other to form more complex substances called *compounds.* Atoms bond together by either gaining, losing, or sharing electrons with other atoms. In an *ionic bond,* one or more electrons are transferred from one atom to another, giving the atom a net positive or negative charge. The resulting electrically charged atom is called an *ion.* Ionic compounds consist of an orderly arrangement of oppositely charged ions assembled in a definite ratio that provides overall electrical neutrality. Another type of bond, the *covalent bond,* is produced when atoms share electrons.
- *Isotopes* are variants of the same element, but with a different *mass number* (the total number of neutrons plus protons found in an atom's nucleus). Some isotopes are unstable and disintegrate naturally through a process called *radioactivity.*
- The properties of minerals include *crystal form, luster, color, streak, hardness, cleavage, fracture,* and *density or specific gravity.* In addition, a number of special physical and chemical properties (*taste, smell, elasticity, malleability, feel, magnetism, double refraction,* and *chemical reaction to hydrochloric acid*) are useful in identifying certain minerals. Each mineral has a unique set of properties that can be used for identification.
- Of the nearly 4000 minerals, no more than a few dozen make up most of the rocks of Earth's crust and, as such, are classified as *rock-forming minerals.* Eight elements (oxygen, silicon, aluminum, iron, calcium, sodium, potassium, and magnesium) make up the bulk of these minerals and represent over 98 percent (by weight) of Earth's continental crust.
- The most common mineral group is the *silicates.* All silicate minerals have the negatively charged *silicon–oxygen tetrahedron* as their fundamental building block. In some silicate minerals the tetrahedra are joined in chains (the pyroxene and amphibole groups); in others, the tetrahedra are arranged into sheets (the micas, biotite, and muscovite), or three-dimensional networks (the feldspars and quartz). Binding the tetrahedra and various silicate structures are often the positive ions of iron, magnesium, potassium, sodium, aluminum, and calcium. Each silicate mineral has a structure and a chemical composition that indicates the conditions under which it was formed.
- The *nonsilicate* mineral groups, which contain several economically important minerals, include the *oxides* (e.g., the mineral hematite, mined for iron), *sulfides* (e.g., the mineral sphalerite, mined for zinc, and the mineral galena, mined for lead), *sulfates, halides,* and *native elements* (e.g., gold and silver). The more common nonsilicate rock-forming minerals include the *carbonate minerals,* calcite and dolomite. Two other nonsilicate minerals frequently found in sedimentary rocks are halite and gypsum.
- *Mineral resources* are the endowment of useful minerals ultimately available commercially. Resources include already identified deposits from which minerals can be extracted profitably, called *reserves,* as well as known deposits that are not yet economically or technologically recoverable. Deposits inferred to exist, but not yet discovered, are also considered mineral resources. The term *ore* is used to denote those useful metallic minerals that can be mined for a profit, as well as some nonmetallic minerals, such as fluorite and sulfur, that contain useful substances.

Key Terms

atom (p. 37)
atomic number (p. 37)
cleavage (p. 41)
color (p. 40)
compound (p. 34)
covalent bond (p. 39)
crystal form (p. 40)
dark silicates (p. 49)
density (p. 42)
electron (p. 47)
element (p. 35)
energy levels (p. 37)
ferromagnesian silicates (p. 49)
fracture (p. 42)
hardness (p. 41)
ion (p. 38)
ionic bond (p. 38)
isotope (p. 39)
light silicates (p. 46)
luster (p. 40)
mass number (p. 39)
metallic bond (p. 39)
mineral (p. 34)
mineral resource (p. 50)
Mohs scale (p. 41)
neutron (p. 37)
nonsilicates (p. 46)
nonferromagnesian silicates (p. 43)
nucleus (p. 37)
ore (p. 50)
periodic table (p. 35)
principal shells (p. 37)
proton (p. 37)
radioactive decay (p. 40)
reserve (p. 50)
rock (p. 34)
rock-forming minerals (p. 43)
silicates (p. 43)
silicon–oxygen tetrahedron (p. 44)
specific gravity (p. 42)
streak (p. 40)
valence electron (p. 37)

Questions for Review

1. List five characteristics an Earth material should have in order to be considered a mineral.
2. Define the term *rock.*
3. List the three main particles of an atom, and explain how they differ from one another.
4. If the number of electrons in an atom is 35 and its mass number is 80, calculate the following:
   a. the number of protons
   b. the atomic number
   c. the number of neutrons
5. What is the significance of valence electrons?
6. Briefly distinguish between ionic and covalent bonding.
7. What occurs in an atom to produce an ion?
8. What is an isotope?
9. Although all minerals have an orderly internal arrangement of atoms (crystalline structure), most mineral samples do not demonstrate their crystal form. Why?
10. Why might it be difficult to identify a mineral by its color?
11. If you found a glassy-appearing mineral while rock hunting and had hopes that it was a diamond, what simple test might help you make a determination?
12. Explain the use of corundum as given in Table 2.1 in terms of the Mohs hardness scale.
13. Gold has a specific gravity of almost 20. If a 25-liter pail of water weighs about 25 kilograms, how much would a 25-liter pail of gold weigh?
14. Explain the difference between the terms *silicon* and *silicate.*
15. What do ferromagnesian minerals have in common? List examples of ferromagnesian minerals.
16. What do muscovite and biotite have in common? How do they differ?
17. Should color be used to distinguish between orthoclase and plagioclase feldspar? What is the best means of distinguishing between the two types of feldspar?
18. Each of the following statements describes a silicate mineral or mineral group. In each case, provide the appropriate name.
    a. The most common member of the amphibole group.
    b. The most common nonferromagnesian member of the mica family.
    c. The only silicate mineral made entirely of silicon and oxygen.
    d. A high-temperature silicate with a name that is based on its color.
    e. Characterized by striations.
    f. Originates as a product of chemical weathering.
19. What simple test can be used to distinguish calcite from dolomite?
20. Contrast *resource* and *reserve.*
21. What might cause a mineral deposit that had not been considered an ore to be reclassified as an ore?

Online Study Guide

The *Essentials of Geology* Web site uses the resources and flexibility of the Internet to aid in your study of the topics in this chapter. Written and developed by geology instructors, this site will help improve your understanding of geology. Visit **www.prenhall.com/lutgens** and click on the cover of *Essentials of Geology 9e* to find:

- Online review quizzes.
- Critical thinking exercises.
- Links to chapter-specific Web resources.
- Internet-wide key-term searches.

http://www.prenhall.com/lutgens

Man rapelling from Glacier Point in California's Yosemite National Park. *(Photo by Don Mason/CORBIS/Bettmann)*

CHAPTER 3

Igneous Rocks

Focus on Learning

To assist you in learning the important concepts in this chapter, you will find it helpful to focus on the following questions:

- How are igneous rocks formed?
- How does magma differ from lava?
- What two criteria are used to classify igneous rocks?
- How does the rate of cooling of magma influence the crystal size of minerals in igneous rocks?
- How is the mineral makeup of an igneous rock related to Bowen's reaction series?
- In what ways are granitic rocks different from basaltic rocks?
- How are economic deposits of gold, silver, and many other metals formed?

▲ **Figure 3.1** Fluid basaltic lava erupting and flowing down the slopes of Hawaii's Kilauea Volcano. *(Photo by G. Brad Lewis/Getty Images, Inc./Liaison)*

Igneous rocks and metamorphic rocks, derived from igneous "parents," make up about 95 percent of Earth's crust. Furthermore, the mantle, which accounts for over 82 percent of Earth's volume, is also composed of igneous rock. Thus, Earth can be described as a huge mass of igneous rocks covered with a thin veneer of sedimentary rocks and having a relatively small iron-rich core. Consequently, a basic knowledge of igneous rocks is essential to our understanding of the structure, composition, and internal workings of our planet.

Mount Rushmore National Monument is carved from intrusive igneous rocks. (March Muench)

Occasionally, magma is explosively ejected from a volcano. (Steve Kaufman/DRK)

Magma: The Parent Material of Igneous Rock

GEODe ESSENTIALS OF GEOLOGY

Igneous Rocks
▼ Introduction

In our discussion of the rock cycle, it was pointed out that **igneous rocks** (from the Latin *ignis,* or fire) form as molten rock cools and solidifies. Abundant evidence supports the fact that the parent material for igneous rocks, called *magma,* is formed by a process called *partial melting.* Partial melting occurs at various levels within Earth's crust and upper mantle to depths of perhaps 250 kilometers (about 150 miles). We will explore the origin of magma later in this chapter.

Once formed, a magma body buoyantly rises toward the surface because it is less dense than the surrounding rocks. Occasionally molten rock breaks through, producing a spectacular volcanic eruption. Magma that reaches Earth's surface is called **lava.** Sometimes magma is explosively ejected from a vent, producing a catastrophic eruption. However, not all eruptions are violent; many volcanoes emit quiet outpourings of very fluid lava (Figure 3.1).

Igneous rocks that form when molten rock solidifies *at the surface* are classified as **extrusive,** or **volcanic** (after the fire god Vulcan). Extrusive igneous rocks are abundant in western portions of the Americas, including the volcanic cones of the Cascade Range and the extensive lava flows of the Columbia Plateau. In addition, many oceanic islands, typified by the Hawaiian chain, are composed almost entirely of volcanic igneous rocks.

Magma that loses its mobility before reaching the surface eventually crystallizes at depth. Igneous rocks that *form at depth* are termed **intrusive,** or **plutonic** (after Pluto, the god of the lower world in classical mythology). Intrusive igneous rocks would never outcrop at the surface if portions of the crust were not uplifted and the overlying rocks stripped away by erosion. (When a mass of crustal rock is exposed—not covered with soil—it is called an *outcrop.*) Exposures of intrusive igneous rocks occur in many places, including Mount Washington, New Hampshire, Stone Mountain, Georgia, the Black Hills of South Dakota, and Yosemite National Park, California.

The Nature of Magma

Magma is completely or partly molten material, which on cooling solidifies to form an igneous rock. Most magmas consist of three distinct parts—a liquid component, a solid component, and a gaseous phase.

The liquid portion, called **melt,** is composed of mobile ions of those elements commonly found in Earth's crust. Melt is made up mostly of ions of silicon and oxygen that readily combine to form silica (SiO_2), as well as lesser amounts of aluminum, potassium, calcium, sodium, iron, and magnesium.

The solid components (if any) in magma are silicate minerals that have already crystallized from the melt. As a magma body cools, the size and number of crystals increase. During the last stage of cooling, a magma body is mostly a crystalline solid with only minor amounts of melt.

Water vapor (H_2O), carbon dioxide (CO_2), and sulfur dioxide (SO_2) are the most common gases found in magma and are confined by the immense pressure exerted by the overlying rocks. These gaseous components, called **volatiles,** are dissolved within the melt. (Volatiles are those materials that will readily vaporize—form a gas—at surface pressures.) Volatiles remain part of the magma until it either moves near the surface (low-pressure environment) or until the magma body is essentially crystallized, at which time any remaining volatiles freely migrate away.

From Magma to Crystalline Rock

In hot magma, ions and groups of ions join together and break apart constantly. As a magma cools, the ions begin to move more slowly and eventually join together into orderly crystalline structures. This process, called **crystallization,** generates various silicate minerals that reside within the remaining melt.

Before we examine how magma crystallizes, let us first examine how a simple crystalline solid melts. In any crystalline solid, the ions are arranged in a closely packed regular pattern. However, they are not without some motion. They exhibit a sort of restricted vibration about fixed points. As the temperature rises, the ions vibrate more rapidly and consequently collide with ever increasing vigor with their neighbors. Thus, heating causes the ions to occupy more space, which in turn causes the solid to expand. When the ions are vibrating rapidly enough to overcome the force of the chemical bonds, the solid begins to melt. At this stage the ions are able to slide past one another, and their orderly crystalline structure disintegrates. Thus, melting converts a solid consisting of tight, uniformly packed ions into a liquid composed of unordered ions moving randomly about.

In the process of crystallization, cooling reverses the events of melting. As the temperature of the liquid drops, the ions pack closer and closer together as they slow their rate of movement. When cooled sufficiently, the forces of the chemical bonds will again confine the ions to an orderly crystalline arrangement.

When magma cools, it is generally the silicon and oxygen atoms that link together first to form silicon–oxygen tetrahedra, the basic building blocks of the silicate minerals. As a magma continues to lose heat to its surroundings, the tetrahedra join with each other and with other ions to form embryonic crystal nuclei. Slowly each nucleus grows as ions lose their mobility and join the crystalline network.

The earliest formed minerals have space to grow and tend to have better developed crystal faces than do the later ones that fill the remaining space. Eventually all of the melt is transformed into a solid mass of interlocking silicate minerals that we call an *igneous rock* (Figure 3.2).

Did you Know?

During the catastrophic eruption of Vesuvius in A.D. 79, the entire city of Pompeii (near Naples, Italy) was completely buried by several meters of pumice and volcanic ash. Centuries passed, and new towns sprang up around Vesuvius. It was not until 1595, during a construction project, that the remains of Pompeii came to light. Today, thousands of tourists stroll amongst the excavated remains of Pompeii's shops, taverns, and villas.

As this lava cools, it is transformed into a solid mass of interlocking minerals we call an igneous rock. (Griggs/USGS)

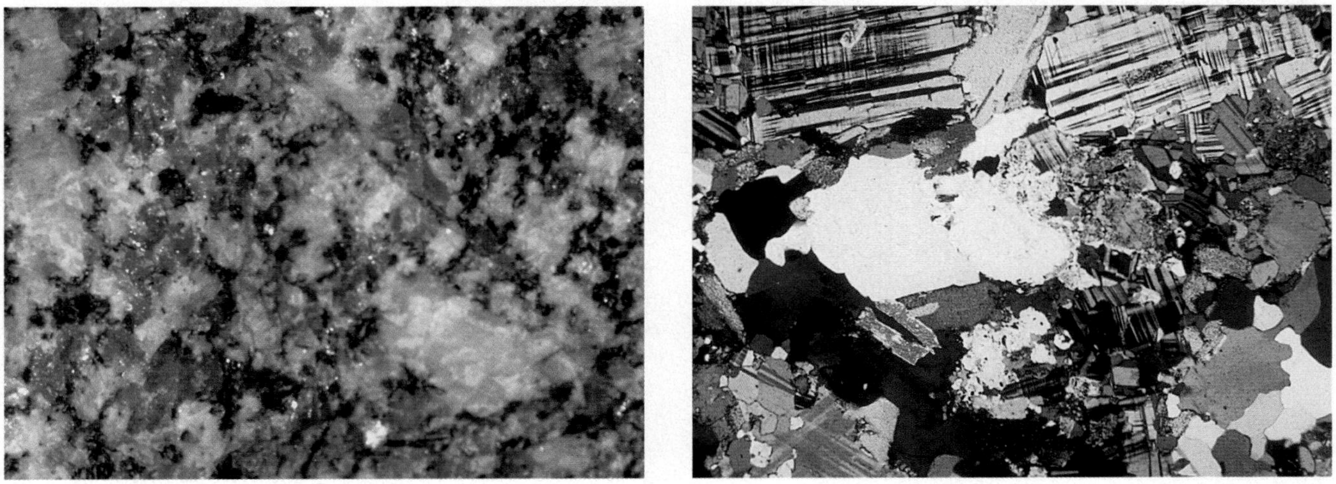

A. B.

▲ **Figure 3.2** **A.** Close-up of interlocking crystals in a coarse-grained igneous rock. The largest crystals are about 1 centimeter in length. **B.** Photomicrograph of interlocking crystals in a coarse-grained igneous rock. *(Photos courtesy of E. J. Tarbuck)*

As you will see later, the crystallization of magma is much more complex than just described. Whereas a single compound, such as water, crystallizes at a specific temperature, solidification of magma with its diverse chemistry spans a temperature range of 200°C. During crystallization, the composition of the melt continually changes as ions are selectively removed and incorporated into the earliest formed minerals. If the melt should separate from the earliest formed minerals, its composition will be different from that of the original magma. Thus, a single magma may generate rocks with widely differing compositions. As a consequence, a great variety of igneous rocks exist. We will return to this important idea later in the chapter.

Although the crystallization of magma is complex, it is nevertheless possible to classify igneous rocks based on their mineral composition and the conditions under which they formed. Their environment during crystallization can be roughly inferred from the size and arrangement of the mineral grains, a property called *texture.* Consequently, *igneous rocks are most often classified by their texture and mineral composition.* We will consider these two rock characteristics in the following sections.

Igneous Textures

Igneous Rocks
▼ Igneous Textures

The term **texture,** when applied to an igneous rock, is used to describe the overall appearance of the rock based on the size, shape, and arrangement of its interlocking crystals (Figure 3.3). Texture is an important characteristic because it reveals a great deal about the environment in which the rock formed. This fact allows geologists to make inferences about a rock's origin while working in the field where sophisticated equipment is not available.

Factors Affecting Crystal Size

Three factors contribute to the textures of igneous rocks: (1) *the rate at which magma cools;* (2) *the amount of silica present;* and (3) *the amount of dissolved gases in the magma.* Of these, the rate of cooling is the dominant factor, but like all generalizations, this one has numerous exceptions.

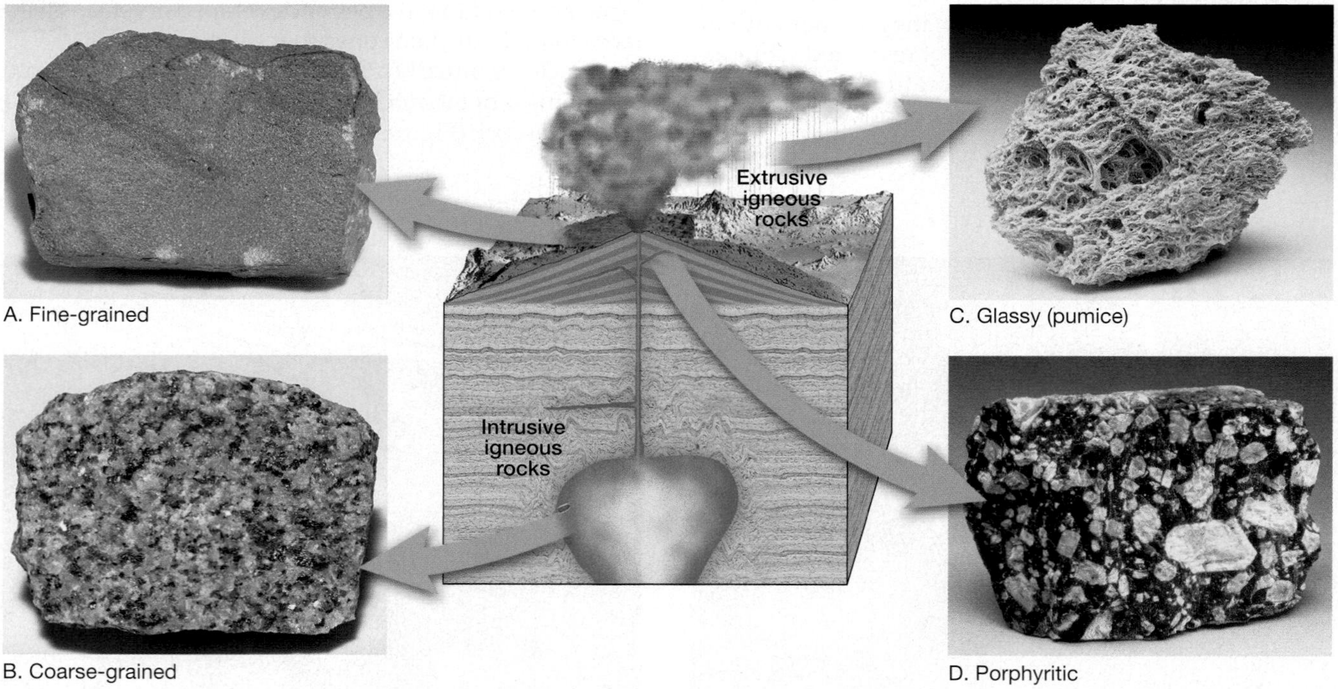

▲ **Figure 3.3** Igneous rock textures. **A.** Igneous rocks that form at or near Earth's surface cool quickly and often exhibit a fine-grained texture. **B.** Coarse-grained igneous rocks form when magma slowly crystallizes at depth. **C.** During a volcanic eruption in which silica-rich lava is ejected into the atmosphere, a frothy glass called pumice may form. **D.** A porphyritic texture results when magma that already contains some large crystals migrates to a new location where the rate of cooling increases. The resulting rock consists of larger crystals embedded within a matrix of smaller crystals. *(Photos courtesy of E. J. Tarbuck)*

As a magma body loses heat to its surroundings, the mobility of its ions decreases. A very large magma body located at great depth will cool over a period of perhaps tens or hundreds of thousands of years. Initially, relatively few crystal nuclei form. Slow cooling permits ions to migrate freely until they eventually join one of the existing crystalline structures. Consequently, slow cooling promotes the growth of fewer but larger crystals.

On the other hand, when cooling occurs more rapidly—for example, in a thin lava flow—the ions quickly lose their mobility and readily combine to form crystals. This results in the development of numerous embryonic nuclei, all of which compete for the available ions. The result is a solid mass of small intergrown crystals.

When molten material is quenched quickly, there may not be sufficient time for the ions to arrange into a crystalline network. Rocks that consist of unordered ions are referred to as **glass.**

▲ **Figure 3.4** Scoria is a volcanic rock that exhibits a vesicular texture. Vesicles are small holes left by escaping gas bubbles. *(Photo courtesy of E. J. Tarbuck)*

Types of Igneous Textures

As you saw, the effect of cooling on rock textures is fairly straightforward. Slow cooling promotes the growth of large crystals, whereas rapid cooling tends to generate smaller crystals. We will consider the other factors affecting crystal growth as we examine the major textural types.

Aphanitic (fine-grained) Texture. Igneous rocks that form at the surface or as small masses within the upper crust where cooling is relatively rapid possess a very fine-grained texture termed **aphanitic.** By definition, the crystals that make up aphanitic rocks are so small that individual minerals can only be distinguished with the aid of a microscope (Figure 3.3A). Because mineral identification is not possible, we commonly characterize fine-grained rocks as being light, intermediate, or dark in color. Using this system of grouping, light-colored aphanitic rocks are those containing primarily light-colored nonferromagnesian silicate minerals and so forth (see the section titled "Common Silicate Minerals" in Chapter 2).

Commonly seen in many aphanitic rocks are the voids left by gas bubbles that escape as lava solidifies. These spherical or elongated openings are called *vesicles,* and the rocks that contain them are said to have a **vesicular texture.** Rocks that exhibit a vesicular texture usually form in the upper zone of a lava flow, where cooling occurs rapidly enough to "freeze" the lava, thereby preserving the openings produced by the expanding gas bubbles (Figure 3.4).

Phaneritic (coarse-grained) Texture. When large masses of magma slowly solidify far below the surface, they form igneous rocks that exhibit a coarse-grained texture described as **phaneritic.** These coarse-grained rocks consist of a mass of intergrown crystals that are roughly equal in size and large enough so that the individual minerals can be identified without the aid of a microscope (Figure 3.3B). (Geologists often use a small magnifying lens to aid in identifying coarse-grained minerals.) Because phaneritic rocks form deep within Earth's crust, their exposure at Earth's surface results only after erosion removes the overlying rocks that once surrounded the magma chamber.

Porphyritic Texture. A large mass of magma located at depth may require tens to hundreds of thousands of years to solidify. Because different minerals crystallize at different temperatures (as well as at differing rates), it is possible for some crystals to become quite large before others even begin to form. If magma containing some large crystals should change environments—for example, by erupting at the surface—the remaining liquid portion of the lava would cool relatively quickly. The resulting rock, which has large crystals embedded in a matrix of smaller crystals, is said to have a **porphyritic texture** (Figure 3.3D). The large crystals in such a rock are referred to as **phenocrysts,** whereas the matrix of smaller crystals is called **groundmass.** A rock with such a texture is termed a **porphyry.**

Glassy Texture. During some volcanic eruptions, molten rock is ejected into the atmosphere, where it is quenched quickly. Rapid cooling of this type may generate rocks having a **glassy texture.** As we indicated earlier, glass results when unordered ions are "frozen"

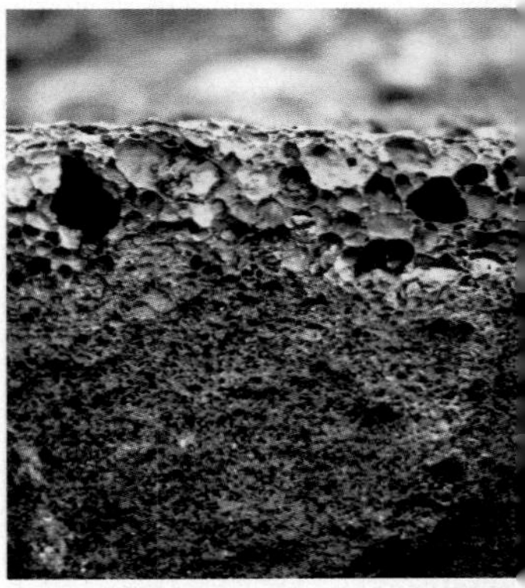

Vesicles are small holes left by escaping gas bubbles.

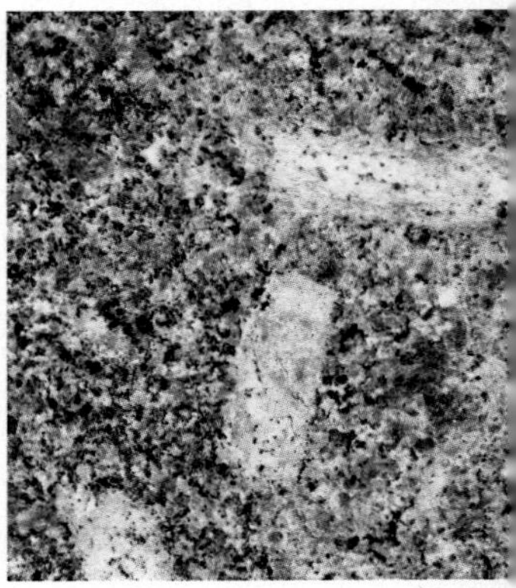

Close-up of rectangular feldspar phenocrysts in a porphyritic igneous rock.

Did you Know?

People have been making glass in roughly the same way for at least 2000 years. The process involves melting certain Earth materials and cooling the liquid quickly before the atoms have time to form an orderly crystalline structure. This is the same way that natural glass, called *obsidian,* is generated from lava. It is possible to produce glass from a variety of materials, but most commercial glass is produced from quartz sand and lesser amounts of carbonate minerals.

before they are able to unite into an orderly crystalline structure. *Obsidian,* a common type of natural glass, is similar in appearance to a dark chunk of manufactured glass (Figure 3.3C).

Lava flows composed of obsidian several tens of feet thick occur in some places (Figure 3.5). Thus, rapid cooling is not the only mechanism by which a glassy texture can form. As a general rule, magmas with a high silica content tend to form long, chainlike structures before crystallization is complete. These structures in turn impede ionic transport and increase the magma's viscosity. (*Viscosity* is a measure of a fluid's resistance to flow.)

▲ **Figure 3.5** This obsidian flow was extruded from a vent along the south wall of Newberry Caldera, Oregon. Note the road for scale. *(Photo by Marli Miller)*

Granitic magma, which is rich in silica, may be extruded as an extremely viscous mass that eventually solidifies to form obsidian. By contrast, basaltic magma, which is low in silica, forms very fluid lavas that upon cooling usually generate fine-grained crystalline rocks. However, the surface of basaltic lava may be quenched rapidly enough to form a thin, glassy skin. Moreover, Hawaiian volcanoes sometimes generate lava fountains, which spray basaltic lava tens of meters into the air. Such activity can produce strands of volcanic glass called *Pele's hair,* after the Hawaiian goddess of volcanoes.

Volcanic glass (obsidian) was used by Native Americans for making sharp tools. (Jeffrey Scovil)

Pyroclastic (fragmental) Texture. Some igneous rocks are formed from the consolidation of individual rock fragments that are ejected during a violent volcanic eruption. The ejected particles might be very fine ash, molten blobs, or large angular blocks torn from the walls of the vent during the eruption. Igneous rocks composed of these rock fragments are said to have a **pyroclastic** or **fragmental texture.**

Because pyroclastic rocks are made of individual particles or fragments rather than interlocking crystals, their textures often appear to be more similar to sedimentary rocks than to other igneous rocks.

Pegmatitic Texture. Under special conditions, exceptionally coarse-grained igneous rocks, called **pegmatites,** may form. These rocks, which are composed of interlocking crystals all larger than a centimeter in diameter, are said to have a **pegmatitic texture.** Most pegmatites are found around the margins of large plutons as small masses or thin veins that commonly extend into the adjacent host rock.

Pegmatites form in the late stages of crystallization, when water and other volatiles, such as chlorine, fluorine, and sulfur, make up an unusually high percentage of the melt. Because ion migration is enhanced in these fluid-rich environments, the crystals that form are abnormally large. Thus, the large crystals in pegmatites are not the result of inordinately long cooling histories; rather, they are the consequence of the fluid-rich environment in which crystallization takes place.

The composition of most pegmatites is similar to that of granite. Thus, pegmatites contain large crystals of quartz, feldspar, and muscovite. However, some contain significant quantities of comparatively rare and hence valuable minerals (see section titled "Mineral Resources and Igneous Processes" at the end of this chapter).

Igneous Compositions

GEODe ESSENTIALS OF GEOLOGY

Igneous Rocks

▼ Igneous Compositions

Igneous rocks are mainly composed of silicate minerals. Furthermore, the mineral makeup of a particular igneous rock is ultimately determined by the chemical composition of the magma from which it crystallizes. Recall that magma is composed largely of the eight elements that are the major constituents of the silicate minerals. Chemical analysis shows that silicon and oxygen (usually expressed as the silica [SiO_2] content of a magma) are by far the most abundant constituents of igneous rocks. These two elements, plus ions of aluminum (Al), calcium (Ca), sodium (Na), potassium (K), magnesium (Mg), and iron (Fe), make up roughly 98 percent by weight of most magmas. In addition, magma contains small amounts of many other elements, including titanium and manganese, and trace amounts of much rarer elements such as gold, silver, and uranium.

As magma cools and solidifies, these elements combine to form two major groups of silicate minerals. The *dark* (or *ferromagnesian*) *silicates* are rich in iron and/or magnesium and comparatively low in silica. *Olivine,*

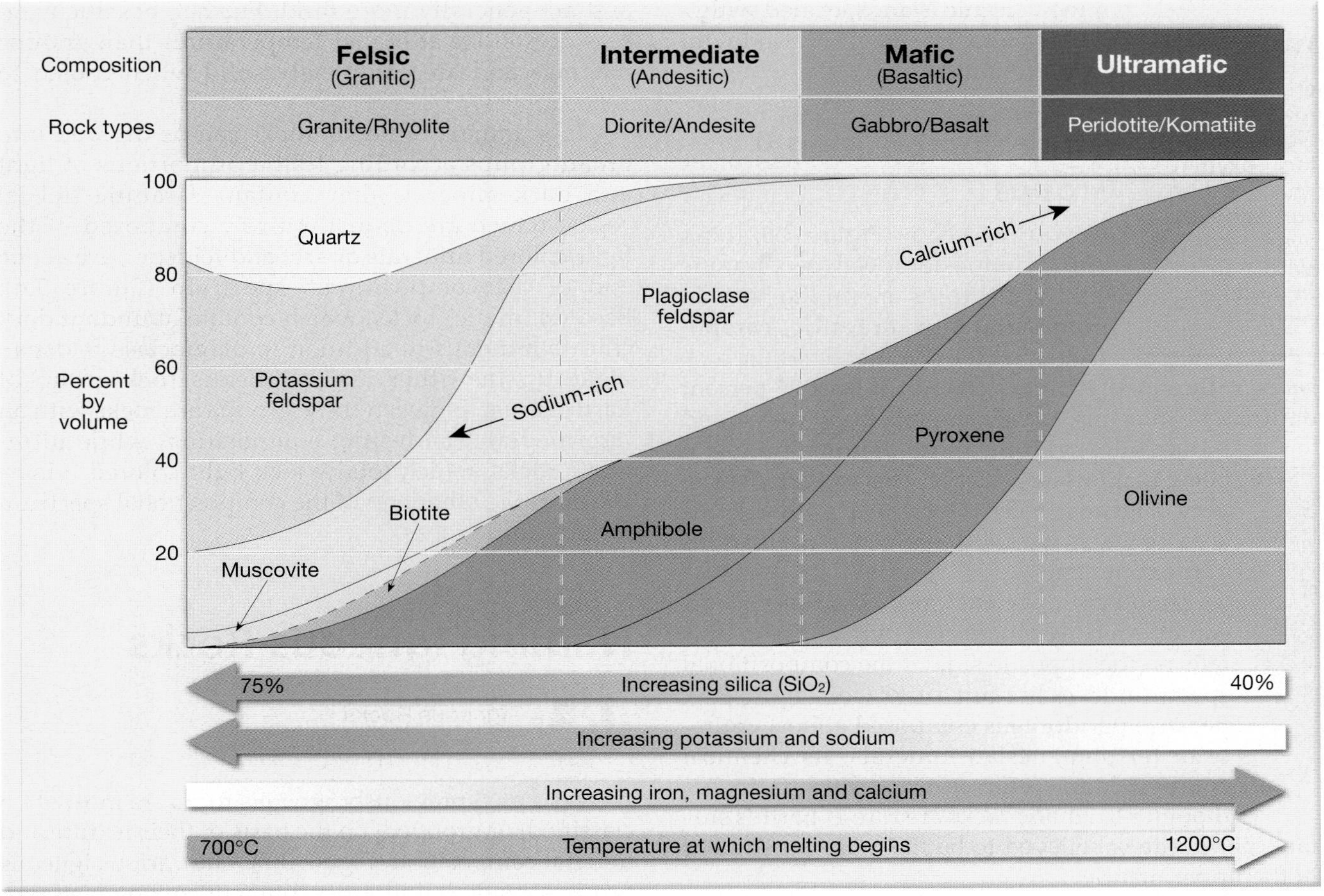

▲ **Figure 3.6** Mineralogy of common igneous rocks and the magmas from which they form. *(After Dietrich, Daily, and Larsen)*

pyroxene, amphibole, and *biotite mica* are the common dark silicate minerals of Earth's crust. By contrast, the *light* (or *nonferromagnesian*) *silicates* contain greater amounts of potassium, sodium, and calcium rather than iron and magnesium. As a group, these minerals are richer in silica than the dark silicates. The light silicates include *quartz, muscovite mica,* and the most abundant mineral group, the *feldspars.* The feldspars make up over 40 percent of most igneous rocks. Thus, in addition to feldspar, igneous rocks contain some combination of the other light and/or dark silicates listed here.

Granitic versus Basaltic Compositions

Despite their great compositional diversity, igneous rocks (and the magmas from which they form) can be divided into broad groups according to their proportions of light and dark minerals (Figure 3.6). Near one end of the continuum are rocks composed almost entirely of light-colored silicates—quartz and feldspar. Igneous rocks in which these are the dominant minerals have a **granitic composition.** Geologists also refer to granitic rocks as being **felsic,** a term derived from *fe*ldspar and *si*lica (quartz). In addition to quartz and feldspar, most granitic rocks contain about 10 percent dark silicate minerals, usually biotite mica and amphibole. Granitic rocks are rich in silica (about 70 percent) and are major constituents of the continental crust.

Rocks that contain substantial dark silicate minerals and calcium-rich plagioclase feldspar (but no quartz) are said to have a **basaltic composition** (Figure 3.6). Because basaltic rocks contain a high percentage of ferromagnesian minerals, geologists also refer to them as **mafic** (from *ma*gnesium and *f*errum, the Latin name for iron). Because of their iron content, mafic rocks are typically darker and have a greater density than granitic rocks. Basaltic rocks make up the ocean floor as well as many

Did you Know?

Enormous crystals have been found in a type of igneous rock called *pegmatite.* For example, single crystals of the mineral feldspar the size of a house have been unearthed in North Carolina and Norway, while huge muscovite (mica) plates weighing 85 tons have been mined in India. In addition, crystals as large as telephone poles (over 40 feet in length) of the lithium-bearing mineral spodumene have been found in the Black Hills of South Dakota.

Did you Know?

The most abundant element in Earth's crust is *oxygen.* It makes up 47 percent of the Earth's crust by weight and 94 percent by volume. In fact, oxygen atoms are so much bigger than most of the other atoms in common minerals that the crust is mainly oxygen atoms packed closely together with smaller atoms, such as silicon, aluminum, and potassium tucked in-between.

of the volcanic islands located within the ocean basins. Basalt is also found on the continents.

Other Compositional Groups

As you can see in Figure 3.6, rocks with a composition between granitic and basaltic rocks are said to have an **intermediate,** or **andesitic, composition** after the common volcanic rock *andesite.* Intermediate rocks contain at least 25 percent dark silicate minerals, mainly amphibole, pyroxene, and biotite mica with the other dominant mineral being plagioclase feldspar. This important category of igneous rocks is associated with volcanic activity that is typically confined to the margins of the continents.

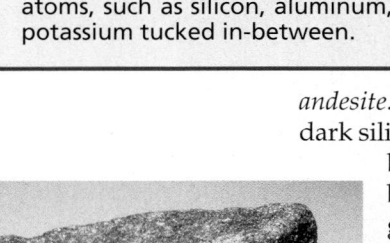

Olivine-rich peridotite.

Another important igneous rock, *peridotite,* contains mostly olivine and pyroxene and thus falls on the opposite side of the compositional spectrum from granitic rocks (see Figure 3.6). Because peridotite is composed almost entirely of ferromagnesian minerals, its chemical composition is referred to as **ultramafic.** Although ultramafic rocks are rare at Earth's surface, peridotite is believed to be the main constituent of the upper mantle.

Silica Content as an Indicator of Composition

An important aspect of the chemical composition of igneous rocks is their silica (SiO_2) content. Recall that silicon and oxygen are the two most abundant elements in igneous rocks. Typically, the silica content of crustal rocks ranges from a low of about 45 percent in ultramafic rocks to a high of over 70 percent in felsic rocks (Figure 3.6). The percentage of silica in igneous rocks actually varies in a systematic manner that parallels the abundance of other elements. For example, rocks comparatively low in silica contain large amounts of iron, magnesium, and calcium. By contrast, rocks high in silica contain very small amounts of those elements but are enriched instead in sodium and potassium. Consequently, the chemical makeup of an igneous rock can be inferred directly from its silica content.

The main ingredient of commercial glass is silica that is obtained from the mineral quartz. (Guy Ryecart/DK)

Further, the amount of silica present in magma strongly influences its behavior. Granitic magma, which has a high silica content, is quite viscous (thick) and exists as a liquid at temperatures as low as 700°C. On the other hand, basaltic magmas are low in silica and are generally more fluid. Further, basaltic magmas crystallize at higher temperatures than granitic magmas and are completely solid when cooled to 1000°C.

In summary, igneous rocks can be divided into broad groups according to the proportions of light and dark minerals they contain. Granitic (felsic) rocks, which are almost entirely composed of the light-colored minerals quartz and feldspar, are at one end of the compositional spectrum (Figure 3.6). Basaltic (mafic) rocks, which contain abundant dark silicate minerals in addition to plagioclase feldspar, make up the other major igneous rock group of Earth's crust. Between these groups are rocks with an intermediate (andesitic) composition, while ultramafic rocks, which totally lack light-colored minerals, lie at the other end of the compositional spectrum from granitic rocks.

Naming Igneous Rocks

GEODe ESSENTIALS OF GEOLOGY

Igneous Rocks
▼ Naming Igneous Rocks

As was stated previously, igneous rocks are most often classified, or grouped, on the basis of their texture and mineral composition (Figure 3.7). The various igneous textures result mainly from different cooling histories, whereas the mineral composition of an igneous rock is the consequence of the chemical makeup of its parent magma. Because igneous rocks are classified on the basis of their mineral composition and texture, two rocks may have similar mineral constituents but have different textures and hence different names. For example, *granite,* a coarse-grained plutonic rock, has a fine-grained volcanic equivalent called *rhyolite.* Although these rocks are mineralogically the same, they have different cooling histories and do not look at all alike (Figure 3.8).

Felsic (Granitic) Igneous Rocks

Granite. Granite is perhaps the best known of all igneous rocks (Figure 3.8A). This is partly because of its natural beauty, which is enhanced when it is polished, and partly because of its abundance in the continental crust. Slabs of polished granite are commonly used for tombstones and monuments and as building stones. Well-known areas in the United States where granite is quarried include Barre, Vermont; Mount Airy, North Carolina; and Saint Cloud, Minnesota.

Granite is a phaneritic rock composed of about 25 percent quartz and roughly 65 percent feldspar, mostly potassium- and sodium-rich varieties. Quartz crystals,

Chemical Composition	Granitic (Felsic)	Andesitic (Intermediate)	Basaltic (Mafic)	Ultramafic
Dominant Minerals	Quartz Potassium feldspar Sodium-rich plagioclase feldspar	Amphibole Sodium- and calcium-rich plagioclase feldspar	Pyroxene Calcium-rich plagioclase feldspar	Olivine Pyroxene
Accessory Minerals	Amphibole Muscovite Biotite	Pyroxene Biotite	Amphibole Olivine	Calcium-rich plagioclase feldspar
TEXTURE: Phaneritic (coarse-grained)	**Granite**	**Diorite**	**Gabbro**	**Peridotite**
TEXTURE: Aphanitic (fine-grained)	**Rhyolite**	**Andesite**	**Basalt**	**Komatiite** (rare)
TEXTURE: Porphyritic	"Porphyritic" precedes any of the above names whenever there are appreciable phenocrysts			Uncommon
TEXTURE: Glassy	**Obsidian** (compact glass) **Pumice** (frothy glass)			
TEXTURE: Pyroclastic (fragmental)	**Tuff** (fragments less than 2 mm) **Volcanic Breccia** (fragments greater than 2 mm)			
Rock Color (based on % of dark minerals)	0% to 25%	25% to 45%	45% to 85%	85% to 100%

▲ **Figure 3.7** Classification of the major groups of igneous rocks based on their mineral composition and texture. Phaneritic (coarse-grained) rocks are plutonic, solidifying deep underground. Aphanitic (fine-grained) rocks are volcanic, or solidify as shallow, thin plutons.

▼ **Figure 3.8** **A.** Granite, one of the most common phaneritic igneous rocks. **B.** Rhyolite, the aphanitic equivalent of granite, is less abundant. *(Photos by E. J. Tarbuck)*

A. Granite

Close up

B. Rhyolite

Close up

▲ **Figure 3.9** Rocks contain information about the processes that produce them. This massive granitic monolith (El Capitán) located in Yosemite National Park, California, was once a molten mass found deep within Earth. *(Photo by Tim Fitzharris/Minden Pictures)*

▼ **Figure 3.10** Igneous rocks that exhibit a glassy texture. **A.** Obsidian, a glassy volcanic rock. **B.** Pumice, a glassy rock containing numerous vesicles. *(Photos courtesy of E. J. Tarbuck)*

A. Obsidian **B. Pumice**

which are roughly spherical in shape, are often glassy and clear to light gray in color. By contrast, feldspar crystals are not as glassy, are generally white to gray or salmon pink in color, and exhibit a rectangular rather than spherical shape. When potassium feldspar is dominant and dark pink in color, granite appears almost reddish. This variety is popular as a building stone. However, the feldspar grains are often white to gray, so that when mixed with lesser amounts of dark silicates, granite appears light gray in color.

Other minor constituents of granite are muscovite and some dark silicates, particularly biotite and amphibole. Although the dark components generally make up less than 10 percent of most granites, dark minerals appear to be more prominent than their percentage would indicate.

Granite may also have a porphyritic texture. These specimens contain feldspar crystals of a centimeter or more in length that are scattered among the coarse-grained groundmass of quartz and amphibole.

Granite and other related crystalline rocks are often the by-products of mountain building. Because granite is very resistant to weathering, it frequently forms the core of eroded mountains. For example, Pikes Peak in the Rockies, Mount Rushmore in the Black Hills, the White Mountains of New Hampshire, Stone Mountain in Georgia, and Yosemite National Park in the Sierra Nevada are all areas where large quantities of granite are exposed at the surface (Figure 3.9).

Granite is a very abundant rock. However, it has become common practice among geologists to apply the term granite to any coarse-grained intrusive rock composed predominantly of light silicate materials. We will follow this practice for the sake of simplicity. You should keep in mind that this use of the term granite covers rocks having a wide range of mineral compositions.

Rhyolite. Rhyolite is the extrusive equivalent of granite and, like granite, is composed essentially of the light-colored silicates (Figure 3.8B). This fact accounts for its color, which is usually buff to pink or occasionally very light gray. Rhyolite is aphanitic and frequently contains glass fragments and voids, indicating rapid cooling in a surface environment. When rhyolite contains phenocrysts, they are small and composed of either quartz or potassium feldspar. In contrast to granite, which is widely distributed as large plutonic masses, occurrences of rhyolite are less common and generally less voluminous. Yellowstone Park is one well-known exception. Here rhyolite lava flows and thick ash deposits of similar composition are extensive.

Obsidian. *Obsidian* is a dark-colored glassy rock that usually forms when silica-rich lava is quenched quickly (Figure 3.10A). In contrast to the orderly arrangement of ions characteristic of minerals, *the ions in glass are unordered.* Consequently, glassy rocks such as obsidian are not composed of minerals in the same sense as most other rocks.

Although usually black or reddish-brown in color, obsidian has a high silica content (Figure 3.10A). Thus, its composition is more akin to the light-colored igneous rocks such as granite than to the dark rocks of basaltic composition. By itself, silica is clear like window glass; the dark color results from the presence of metallic ions. If you examine a thin edge of a piece of obsidian, it will be nearly transparent. Because of its excellent conchoidal fracture and ability to hold a sharp, hard edge, obsidian was a prized material from which Native Americans chipped arrowheads and cutting tools.

Did you Know?

Although Earth's upper mantle is often depicted in diagrams as a reddish-colored layer to show its high temperature, it's color is actually a dark green. But how do we know it is green? On occasion, unmelted fragments of mantle rocks are brought to Earth's surface by rapidly rising magma and are ejected from volcanoes. These mantle rocks, called *peridotite*, are composed mainly of olivine, which, as the name suggests, is olive-green and pyroxene, which is a blackish green.

Pumice. *Pumice* is a volcanic rock that, like obsidian, has a glassy texture. Usually found with obsidian, pumice forms when large amounts of gas escape through lava to generate a gray, frothy mass (Figure 3.10B). In some samples, the voids are quite noticeable, whereas in others the pumice resembles fine shards of intertwined glass. Because of the large percentage of voids, many samples of pumice will float when placed in water. Oftentimes flow lines are visible in pumice, indicating that some movement occurred before solidification was complete. Moreover, pumice and obsidian can often be found in the same rock mass, where they exist in alternating layers.

Polished slabs of granite have many commercial uses. (Andreas von Einsiedel)

Intermediate (Andesitic) Igneous Rocks

Andesite. Andesite is a medium-gray fine-grained rock of volcanic origin. Its name comes from South America's Andes Mountains, where numerous volcanoes are composed of this rock type. In addition to the volcanoes of the Andes, many of the volcanic structures encircling the Pacific Ocean are of andesitic composition. Andesite commonly exhibits a porphyritic texture (Figure 3.11). When this is the case, the phenocrysts are often light, rectangular crystals of plagioclase feldspar or black, elongated amphibole crystals. Andesite often resembles rhyolite, so their identification usually requires microscopic examination to verify the abundance, or lack, of quartz crystals. Andesite contains minor amounts of quartz, whereas rhyolite is composed of about 25 percent quartz.

Diorite. *Diorite* is the plutonic equivalent of andesite It is a coarse-grained intrusive rock that looks somewhat similar to gray granite. However, it can be

▲ **Figure 3.11** Andesite porphyry. **A.** Hand sample of andesite porphyry, a common volcanic rock. **B.** Photomicrograph of a thin section of andesite porphyry to illustrate texture. Notice that the few large crystals (phenocrysts) are surrounded by much smaller crystals (groundmass). *(Photo by E. J. Tarbuck)*

▲ **Figure 3.12** Diorite is a phaneritic igneous rock of intermediate composition. The white crystals are plagioclase feldspar and the black crystals are amphibole and biotite. *(Photo by E. J. Tarbuck)*

Did you Know?

The formation of the most common chemical elements on Earth, such as oxygen, silicon, and iron, occurred billions of years ago inside distant stars. Through various processes of fusion, these stars converted the lightest elements, mostly hydrogen, into these heavier elements. In fact, most heavy elements found in the solar system as well as the atoms in your body are believed to have formed from debris scattered by preexisting stars.

distinguished from granite by the absence of visible quartz crystals and because it contains a higher percentage of dark silicate minerals. The mineral makeup of diorite is primarily sodium-rich plagioclase feldspar and amphibole, with lesser amounts of biotite. Because the light-colored feldspar grains and dark amphibole crystals appear to be roughly equal in abundance, diorite has a salt-and-pepper appearance (Figure 3.12).

Mafic (Basaltic) Igneous Rocks

Basalt. *Basalt* is a very dark green to black fine-grained volcanic rock composed primarily of pyroxene and calcium-rich plagioclase feldspar, with lesser amounts of olivine and amphibole present (Figure 3.13A). When porphyritic, basalt commonly contains small light-colored feldspar phenocrysts or glassy-appearing olivine phenocrysts embedded in a dark groundmass.

Large block of obsidian.

Basalt is the most common extrusive igneous rock. Many volcanic islands, such as the Hawaiian Islands and Iceland, are composed mainly of basalt (see Figure 3.1). Further, the upper layers of the oceanic crust consist of basalt. In the United States, large portions of central Oregon and Washington were the sites of extensive basaltic outpourings (see Figure 4.20, p. 99). At some locations these once fluid basaltic flows have accumulated to thicknesses approaching 3 kilometers.

Gabbro. Gabbro is the intrusive equivalent of basalt (Figure 3.13B). Like basalt, it is very dark green to black in color and composed primarily of pyroxene and calcium-rich plagioclase feldspar. Although gabbro is not a common constituent of the continental crust, it undoubtedly makes up a significant percentage of the oceanic crust. Here large portions of the magma found in underground reservoirs that once fed basalt flows eventually solidified at depth to form gabbro.

Pyroclastic Rocks

Pyroclastic rocks are composed of fragments ejected during a volcanic eruption. One of the most common pyroclastic rocks, called *tuff*, is composed mainly of tiny ash-size fragments that were later cemented together. In situations where the ash particles remained hot enough to fuse, the rock is called *welded tuff*. Although welded tuff consists mostly of tiny glass shards, it may contain walnut-size pieces of pumice and other rock fragments.

Welded tuffs blanket vast portions of once volcanically active areas of the western United States. Some of these tuff deposits are hundreds of feet thick and extend

▲ **Figure 3.13** These dark-colored mafic rocks are composed primarily of pyroxene and calcium-rich plagioclase. **A.** Basalt is aphanitic and a very common extrusive rock. **B.** Gabbro, the phaneritic equivalent of basalt, is less abundant. Photomicrographs are magnified about 27 times. *(Photos by E. J. Tarbuck)*

for tens of miles from their source. Most formed millions of years ago as volcanic ash spewed from large volcanic structures (calderas) in an avalanche style, spreading laterally at speeds approaching 100 kilometers per hour. Early investigators of these deposits incorrectly classified them as rhyolite lava flows. Today we know that this silica-rich lava is too viscous (thick) to flow more than a few miles from a vent.

Unlike most igneous rock names, such as granite and basalt, the term tuff does not imply mineral composition. Thus, it is frequently used with a modifier, as, for example, rhyolite tuff.

Origin of Magma

Although some magmas show evidence of at least some components that were derived from melting of crustal rocks, geologists are now confident that most magma originates by melting in Earth's mantle. It is also clear that plate tectonics plays a major role in the generation of most magma. The greatest amount of igneous activity occurs at divergent plate boundaries in association with seafloor spreading. Substantial amounts of magma are also produced at subduction zones where oceanic lithosphere descends into the mantle. The magma generated here contains components of the mantle as well as subducted crust and subducted sediments. In addition, some magma appears to be generated deep in the mantle where it is not influenced directly by plate motions.

Generating Magma from Solid Rock

Based on available scientific evidence, *Earth's crust and mantle are composed primarily of solid, not molten, rock.* Although the outer core is a fluid, its iron-rich material is very dense and remains deep within Earth. So, what is the source of magma that produces igneous activity?

Geologists conclude that most magma originates when essentially solid rock, located in the crust and upper mantle, melts. The most obvious way to generate magma from solid rock is to raise the temperature above the rock's melting point.

Role of Heat. What source of heat is sufficient to melt rock? Workers in underground mines know that temperatures get higher as they go deeper. Although the rate of temperature change varies from place to place, it *averages* between 20°C and 30°C per kilometer in the *upper* crust. This change in temperature with depth is known as the **geothermal gradient.** Estimates indicate that the temperature at a depth of 100 kilometers ranges between 1200°C and 1400°C. At these high temperatures, rocks in the lower crust and upper mantle are near

Did you Know?

During the Stone Age, volcanic glass (obsidian) was used for making cutting tools. Today, scalpels made from obsidian are being employed for delicate plastic surgery because they leave less scarring. "The steel scalpel has a rough edge, where the obsidian scalpel is smoother and sharper," explains Lee Green, MD, an associate professor at the University of Michigan Medical School.

Pumice is very light weight because it is mostly air. (Chip Clark)

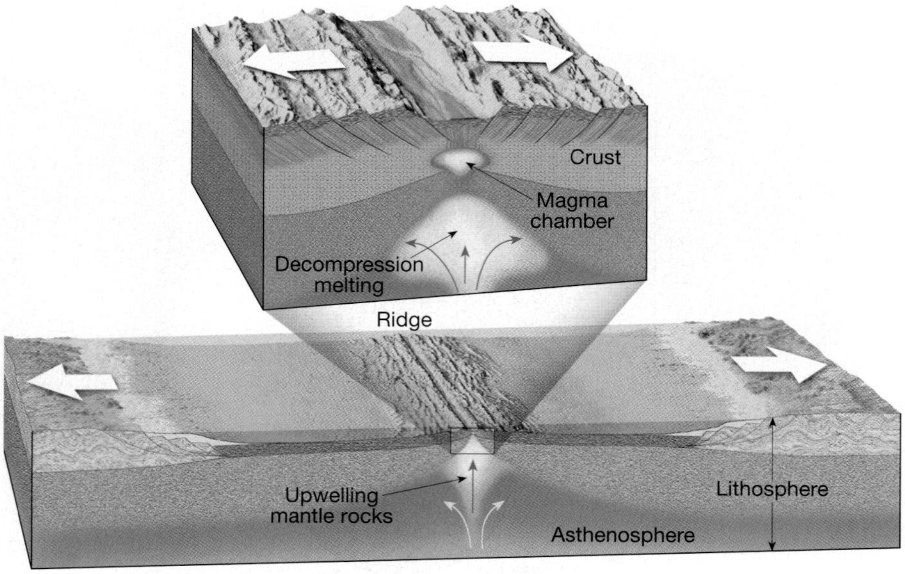

▶ **Figure 3.14** As hot mantle rock ascends, it continually moves into zones of lower pressure. This drop in confining pressure can trigger melting, even without additional heat.

> **Did you Know?**
>
> Since the 1930s *Lava Soap,* which contains powdered pumice, has been used by carpenters, mechanics, and other people working with their hands. Because of the abrasiveness of pumice, lava soap is particularly good at removing grease, dirt, paints, and adhesives.

The valley walls of the Columbia River gorge are composed of ancient basalt flows.

but somewhat below their melting points. Thus, they are very hot but still essentially solid.

One source of heat to melt crustal rocks is basaltic magma that originates in the mantle. As basaltic magma buoyantly rises, it often "ponds" beneath crustal rocks, which have a lower density and are already near their melting temperatures. This results in the partial melting of crustal rocks and the formation of a secondary, silica-rich magma. But what is the source of heat to melt mantle rock? As you will learn, the vast bulk of mantle-derived magma forms without the aid of an additional heat source.

Role of Pressure. If temperature were the only factor that determined whether or not rock melts, our planet would be a molten ball covered with a thin, solid outer shell. This, of course, is not the case. The reason is that pressure also increases with depth.

Melting, which is accompanied by an increase in volume, *occurs at higher temperatures at depth* because of greater confining pressure. Consequently, an increase in confining pressure causes an increase in the rock's melting temperature. Conversely, reducing confining pressure lowers a rock's melting temperature. When confining pressure drops enough, **decompression melting** is triggered. This may occur when mantle rock *ascends* as a result of convective upwelling, thereby moving into zones of lower pressure. Decompression melting is responsible for generating magma along divergent plate boundaries (oceanic ridges) where plates are rifting apart (Figure 3.14). Here, below the ridge crest, hot mantle rocks rise upward to replace the material that shifted horizontally.

Role of Volatiles. Another important factor affecting the melting temperature of rock is its water content. Water and other volatiles act as salt does to melt ice.

▶ **Figure 3.15** As an oceanic plate descends into the mantle, water and other volatiles are driven from the subducting crustal rocks. These volatiles lower the melting temperature of mantle rock sufficiently to generate melt.

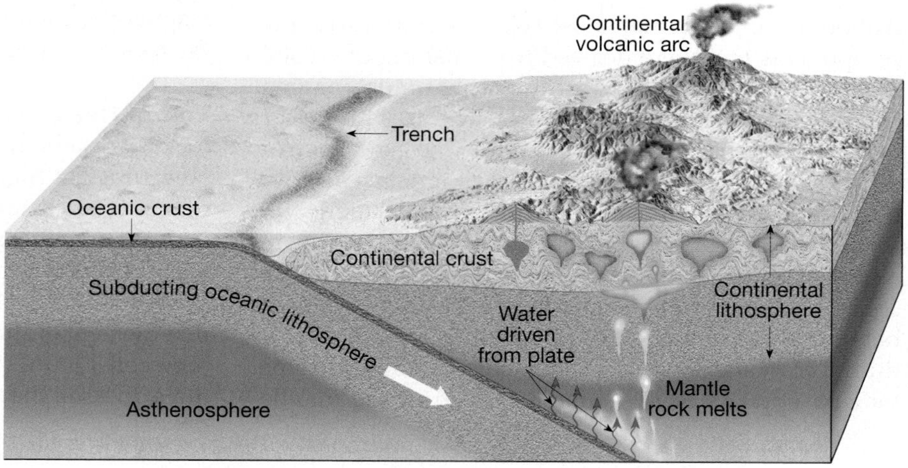

That is, volatiles cause rock to melt at lower temperatures. Further, the effect of volatiles is magnified by increased pressure. Consequently, "wet" rock buried at depth has a much lower melting temperature than does "dry" rock of the same composition and under the same confining pressure. Therefore, in addition to a rock's composition, its temperature, depth (confining pressure), and water content determine whether it exists as a solid or liquid.

Volatiles play an important role in generating magma at convergent plate boundaries where cool slabs of oceanic lithosphere descend into the mantle (Figure 3.15). As an oceanic plate sinks, both heat and pressure drive water from the subducting crustal rocks. These volatiles, which are very mobile, migrate into the wedge of hot mantle that lies above. This process is believed to lower the melting temperature of mantle rock sufficiently to generate some melt. Laboratory studies have shown that the melting point of basalt can be lowered by as much as 100°C by the addition of only 0.1 percent water. When enough mantle-derived basaltic magma forms, it will buoyantly rise toward the surface.

In summary, magma can be generated under three sets of conditions: (1) *heat* may be added; for example, a magma body from a deeper source intrudes and melts crustal rock; (2) *a decrease in pressure* (without the addition of heat) can result in *decompression melting;* and (3) the *introduction of volatiles* (principally water) can lower the melting temperature of mantle rock sufficiently to generate magma.

How Magmas Evolve

Because a large variety of igneous rocks exists, it is logical to assume that an equally large variety of magmas must also exist. However, geologists have observed that a single volcano may extrude lavas exhibiting quite different compositions. Data of this type led them to examine the possibility that magma might change (evolve) and thus become the parent to a variety of igneous rocks. To explore this idea, a pioneering investigation into the crystallization of magma was carried out by N. L. Bowen in the first quarter of the twentieth century.

Welded tuff from Valles Caldera near Los Alamos, New Mexico. (M. Miller)

Bowen's Reaction Series and the Composition of Igneous Rocks

Recall that ice freezes at a single temperature whereas magma crystallizes through at least 200°C of cooling. In a laboratory setting Bowen and his coworkers demonstrated that as a basaltic magma cools, minerals tend to crystallize in a systematic fashion based on their melting points. As shown in Figure 3.16, the first mineral to crystallize from a basaltic magma is the ferromagnesian mineral olivine. Further cooling generates calcium-rich plagioclase feldspar as well as pyroxene, and so forth down the diagram.

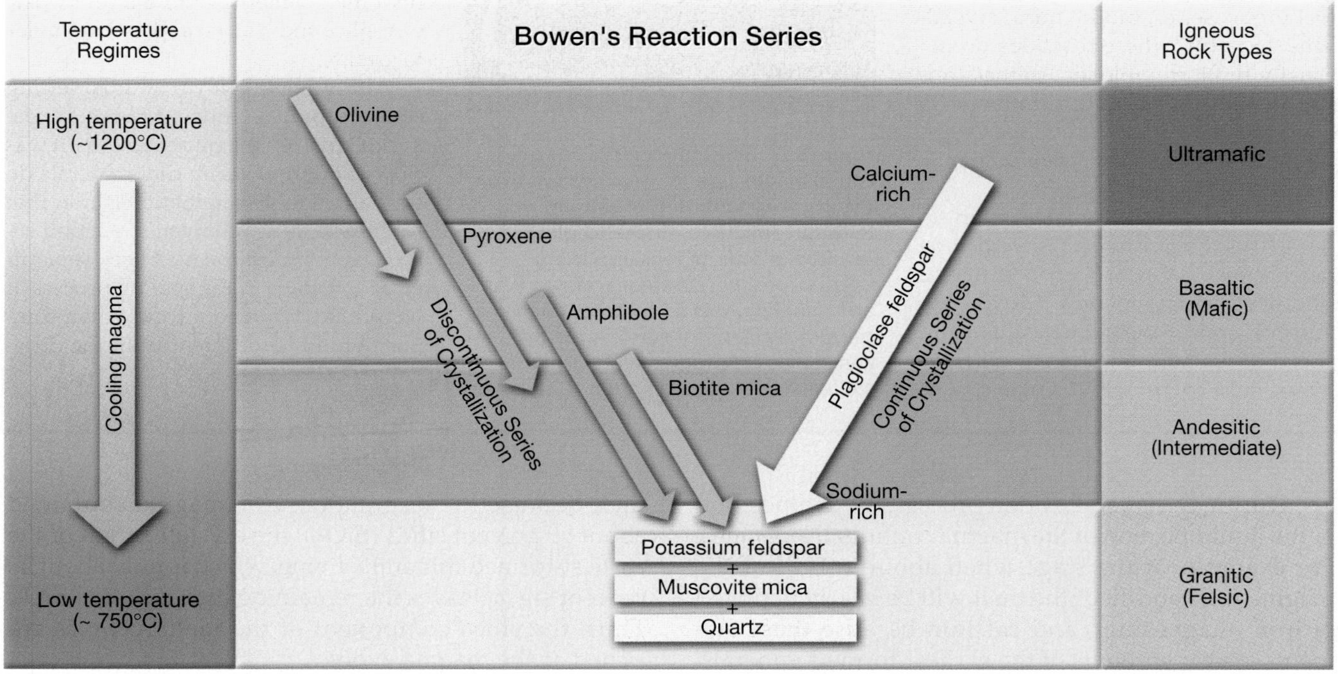

▲ **Figure 3.16** Bowen's reaction series shows the sequence in which minerals crystallize from a magma. Compare this figure to the mineral composition of the rock groups in Figure 3.7. Note that each rock group consists of minerals that crystallize in the same temperature range.

BOX 3.1
A Closer Look at Bowen's Reaction Series

Although it is highly idealized, Bowen's reaction series provides us with a visual representation of the order in which minerals crystallize from a magma of average composition (see Figure 3.16). This model assumes that the magma cools slowly at depth in an otherwise unchanging environment. Notice that Bowen's reaction series is divided into two branches—a discontinuous series and a continuous series.

Discontinuous Reaction Series. The upper left branch of Bowen's reaction series indicates that as a magma cools, olivine is the first mineral to crystallize. Once formed, olivine will chemically react with the remaining melt to form the mineral pyroxene (see Figure 3.16). In this reaction, olivine, which is composed of individual silicon–oxygen tetrahedra, incorporates more silica into its structure, thereby linking its tetrahedra into single-chain structures of the mineral pyroxene. (*Note:* Pyroxene has a lower crystallization temperature than olivine and is more stable at lower temperatures.) As the magma body cools further, the pyroxene crystals will in turn react with the melt to generate the double-chain structure of amphibole. This reaction will continue until the last mineral in this series, biotite mica, crystallizes. In nature, these reactions do not usually run to completion, so that various amounts of each of the minerals in the series may exist at any given time, and some minerals such as biotite may never form.

This branch of Bowen's reaction series is called a *discontinuous reaction series* because at each step a different silicate structure emerges. Olivine, the first mineral in the sequence, is composed of isolated tetrahedra, whereas pyroxene is composed of single chains, amphibole of double chains, and biotite of sheet structures.

Continuous Reaction Series. The right branch of the reaction series, called the *continuous reaction series,* illustrates that calcium-rich plagioclase feldspar crystals react with the sodium ions in the melt to become progressively more sodium-rich (see Figure 3.16). Here the sodium ions diffuse into the feldspar crystals and displace the calcium ions in the crystal lattice. Often, the rate of cooling occurs rapidly enough to prohibit a complete replacement of the calcium ions by sodium ions. In these instances, the feldspar crystals will have calcium-rich interiors surrounded by zones that are progressively richer in sodium (Figure 3.A).

During the last stage of crystallization, after much of the magma has solidified, potassium feldspar forms. (Muscovite will form in pegmatites and other plutonic igneous rocks that crystallize at considerable depth.) Finally, if the remaining melt has excess silica, the mineral quartz will form.

Figure 3.A Photomicrograph of a zoned plagioclase feldspar crystal. After this crystal (composed of calcium-rich feldspar) solidified, further cooling resulted in sodium ions displacing calcium ions. Because replacement was not complete, this feldspar crystal has a calcium-rich interior surrounded by zones that are progressively richer in sodium. (Photo courtesy of E. J. Tarbuck)

Testing Bowen's Reaction Series. During an eruption of Hawaii's Kilauea volcano in 1965, basaltic lava poured into a pit crater, forming a lava lake that became a natural laboratory for testing Bowen's reaction series. When the surface of the lava lake cooled enough to form a crust, geologists drilled into the magma and periodically removed samples that were quenched to preserve the melt, and minerals that were growing within it. By sampling the lava at successive stages of cooling, a history of crystallization was recorded.

As Bowen's reaction series predicts, olivine crystallized early but later ceased to form, and was partly reabsorbed into the cooling melt. (In a larger magma body that cooled more slowly, we would expect most, if not all, of the olivine to react with the melt and change to pyroxene.) Most important, the melt changed composition throughout the course of crystallization. In contrast to the original basaltic lava, which contained about 50 percent silica (SiO_2), the final melt contained more than 75 percent silica and had a composition similar to granite.

Although the lava in this setting cooled rapidly compared to rates experienced in deep magma chambers, it was slow enough to verify that minerals do crystallize in a systematic fashion that roughly parallels Bowen's reaction series. Further, had the melt been separated at any stage in the cooling process, it would have formed a rock with a composition much different from the original lava.

During the crystallization process, the composition of the liquid portion of the magma continually changes. For example, at the stage when about a third of the magma has solidified, the melt will be nearly depleted of iron, magnesium, and calcium because these elements are constituents of the earliest-formed minerals. The removal of these elements from the melt will cause it to become enriched in sodium and potassium. Further, because the original basaltic magma contained about 50 percent silica (SiO_2), the crystallization of the earliest-formed mineral, olivine, which is only about 40 percent silica, leaves the remaining melt richer in SiO_2. Thus, the silica component of the melt becomes enriched as the magma evolves.

Bowen also demonstrated that if the solid components of a magma remain in contact with the

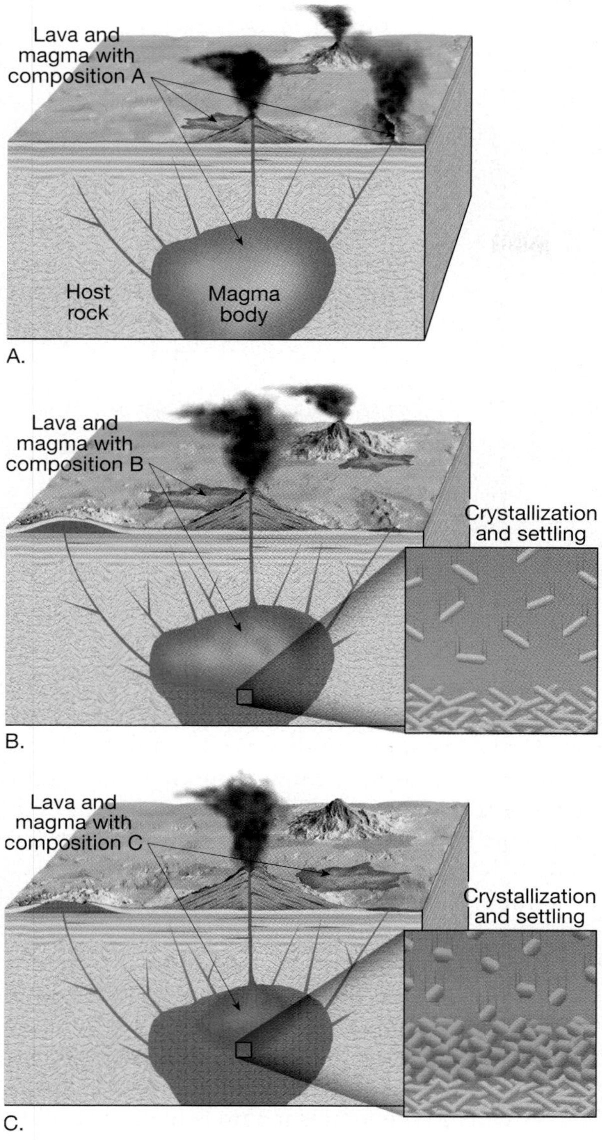

▲ **Figure 3.17** Illustration of how a magma evolves as the earliest-formed minerals (those richer in iron, magnesium, and calcium) crystallize and settle to the bottom of the magma chamber, leaving the remaining melt richer in sodium, potassium, and silica (SiO_2). **A.** Emplacement of a magma body and associated igneous activity generates rocks having a composition similar to that of the initial magma. **B.** After a period of time, crystallization and settling changes the composition of the melt, while generating rocks having a composition quite different than the original magma. **C.** Further magmatic differentiation results in another more highly evolved melt with its associated rock types.

remaining melt, they will chemically react and evolve into the next mineral in the sequence shown in Figure 3.16. For this reason, this arrangement of minerals became known as **Bowen's reaction series** (see Box 3.1). As you will see, in some natural settings the earliest-formed minerals can be separated from the melt, thus halting any further chemical reaction.

The diagram of Bowen's reaction series in Figure 3.16 depicts the sequence that minerals crystallize from a magma of average composition under laboratory conditions. Evidence that this highly idealized crystallization model approximates what can happen in nature comes from the analysis of igneous rocks. In particular, we find that minerals that form in the same general temperature regime on Bowen's reaction series are found together in the same igneous rocks. For example, notice in Figure 3.16 that the minerals quartz, potassium feldspar, and muscovite, which are located in the same region of Bowen's diagram, are typically found together as major constituents of the plutonic igneous rock granite.

Did you Know?

Quartz watches actually contain a quartz crystal to keep time. Before quartz watches, timepieces used some sort of oscillating mass or tuning fork. Cogs and wheels converted this mechanical movement to the movement of the hand. It turns out that if voltage is applied to a quartz crystal, it will oscillate with a consistency that is hundreds of times better for timing than a tuning fork. Because of this property, and modern integrated-circuit technology, quartz watches are now built so cheaply they are sometimes given away in cereal boxes. Modern watches that employ mechanical movements are very expensive indeed.

Magmatic Differentiation

Bowen demonstrated that minerals crystallize from magma in a systematic fashion. But how do Bowen's findings account for the great diversity of igneous rocks? It has been shown that at one or more stages during crystallization, a separation of the solid and liquid components of a magma can occur. One example is called **crystal settling.** This process occurs when the earlier-formed minerals are denser (heavier) than the liquid portion and sink toward the bottom of the magma chamber, as shown in Figure 3.17. When the remaining melt solidifies—either in place or in another location if it migrates into fractures in the surrounding rocks—it will form a rock with a chemical composition much different from the parent magma (Figure 3.17). The formation of one or more secondary magmas from a single parent magma is called **magmatic differentiation.**

At any stage in the evolution of a magma, the solid and liquid components can separate into two chemically distinct units. Further, magmatic differentiation within the secondary melt can generate additional chemically distinct fractions. Consequently, magmatic differentiation and separation of the solid and liquid components at various stages of crystallization can produce several chemically diverse magmas and ultimately a variety of igneous rocks (see Figure 3.17).

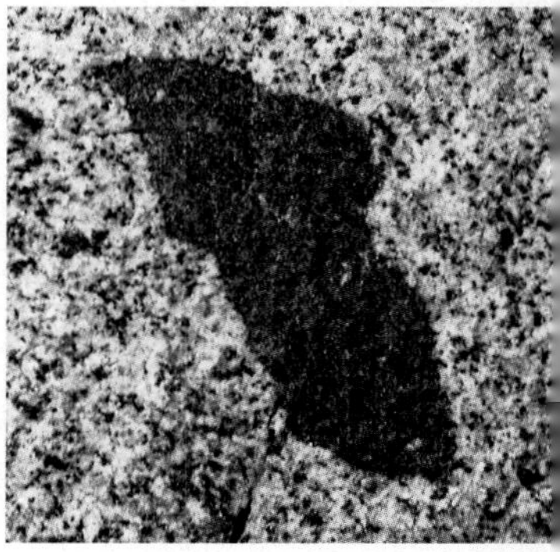

Before it solidifies, a magma body may incorporate some of the surrounding host rock (dark color).

Assimilation and Magma Mixing

Once a magma body forms, its composition can change through the incorporation of foreign material. For example, as magma migrates upward, it may incorporate some of the surrounding host rock, a process called

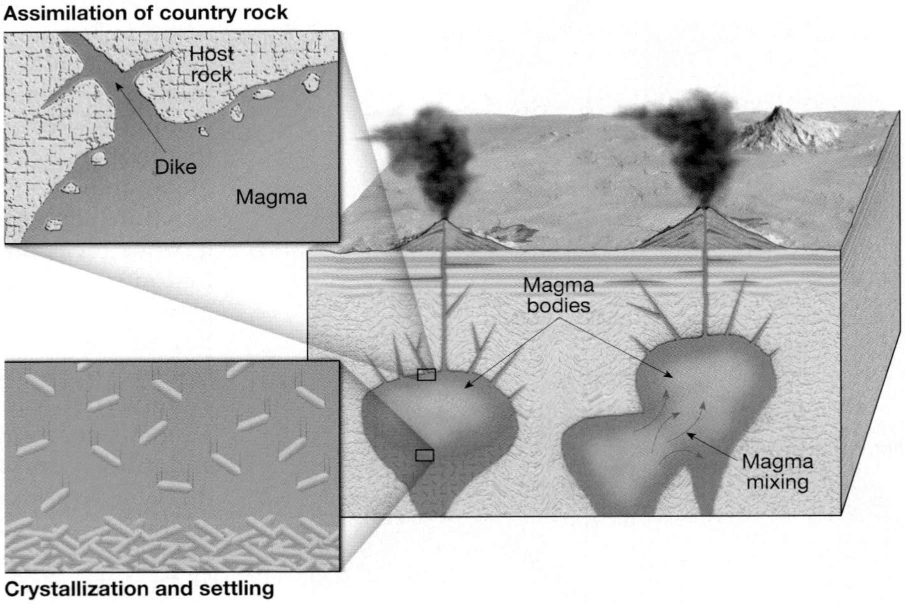

▶ **Figure 3.18** This illustration shows three ways that the composition of a magma body may be altered: magma mixing; assimilation of host rock; and crystallization and settling (magmatic differentiation).

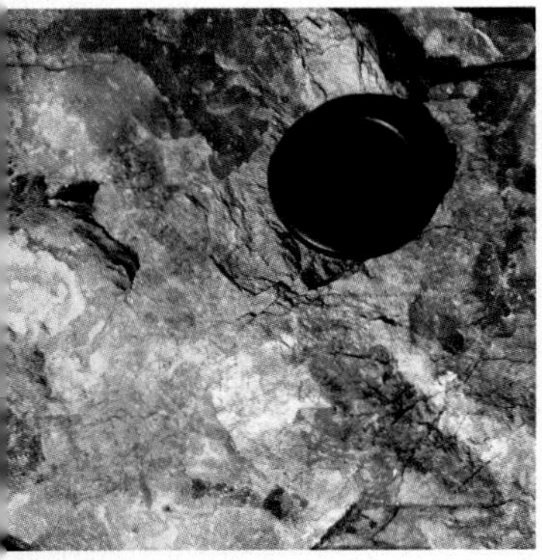

Granite pegmatite composed mainly of feldspar (reddish) and quartz (gray).

assimilation (Figure 3.18). This process may operate in a near-surface environment where rocks are brittle. As the magma pushes upward, stress causes numerous cracks in the overlying rock. The force of the injected magma is often sufficient to dislodge blocks of "foreign" rock and incorporate them into the magma body. In deeper environments, the magma may be hot enough to simply melt and assimilate some of the surrounding host rock, which is near its melting temperature.

Another means by which the composition of a magma body can be altered is called **magma mixing.** This process occurs whenever one magma body intrudes another (Figure 3.18). Once combined, convective flow may stir the two magmas and generate a fluid with an intermediate composition. Magma mixing may occur during the ascent of two chemically distinct magma bodies as the more buoyant mass overtakes the slower moving mass.

▲ **Figure 3.19** This pegmatite in the Black Hills of South Dakota was mined for its large crystals of spodumene, an important source of lithium. Arrows are pointing to impressions left by crystals. Note person in upper center of photo for scale. *(Photo by James G. Kirchner)*

In summary, Bowen successfully demonstrated that through magmatic differentiation, a single parent magma can generate several mineralogically different igneous rocks. Thus, this process, in concert with magma mixing and contamination by crustal rocks, accounts in part for the great diversity of magmas and igneous rocks. We will next look at another important process, partial melting, which also generates magmas having varying compositions.

Partial Melting and Magma Formation

Recall that the crystallization of a magma occurs over a temperature range of at least 200°C. As you might expect, melting, the reverse process, spans a similar temperature range. As rock begins to melt, those minerals with the lowest melting temperatures are the first to melt. Should melting continue, minerals with higher melting points begin to melt and the composition of the magma steadily approaches the overall composition of the rock from which it was derived. Most often, however, melting is not complete. The incomplete melting of rocks is known as **partial melting,** a process that produces most, if not all, magma.

Notice in Figure 3.16 that rocks with a felsic composition are composed of minerals with the lowest melting (crystallization) temperatures—namely, quartz and

Table 3.1 Occurrences of metallic minerals

Metal	Principal Ores	Geological Occurrences
Aluminum	Bauxite	Residual product of weathering
Chromium	Chromite	Magmatic segregation
Copper	Chalcopyrite Bornite Chalcocite	Hydrothermal deposits; contact metamorphism; enrichment by weathering processes
Gold	Native gold	Hydrothermal deposits; placers
Iron	Hematite Magnetite Limonite	Banded sedimentary formations; magmatic segregation
Lead	Galena	Hydrothermal deposits
Magnesium	Magnesite Dolomite	Hydrothermal deposits
Manganese	Pyrolusite	Residual product of weathering
Mercury	Cinnabar	Hydrothermal deposits
Molybdenum	Molybdenite	Hydrothermal deposits
Nickel	Pentlandite	Magmatic segregation
Platinum	Native platinum	Magmatic segregation; placers
Silver	Native silver Argentite	Hydrothermal deposits; enrichment by weathering processes
Tin	Cassiterite	Hydrothermal deposits; placers
Titanium	Ilmenite	Magmatic segregation; placers
Tungsten	Wolframite	Pegmatites; contact metamorphic deposits; placers
Uranium	Uraninite (pitchblende)	Pegmatites; sedimentary deposits
Zinc	Sphalerite	Hydrothermal deposits

potassium feldspar. Also note that as we move up Bowen's reaction series, the minerals have progressively higher melting temperatures and that olivine, which is found at the top, has the highest melting point. When a rock undergoes partial melting, it will form a melt that is enriched in ions from minerals with the lowest melting temperatures. The unmelted crystals are those of minerals with higher melting temperatures. Separation of these two fractions would yield a melt with a chemical composition that is richer in silica and nearer to the felsic (granitic) end of the spectrum than the rock from which it was derived.

In summary, a large number of processes dominated by magmatic differentiation and partial melting, as well as magma mixing and contamination by crustal rocks, account for the great diversity of igneous rocks and the magmas from which they formed.

Mineral Resources and Igneous Processes

Some of the most important accumulations of metals, such as gold, silver, copper, mercury, lead, platinum, and nickel, are produced by igneous processes (Table

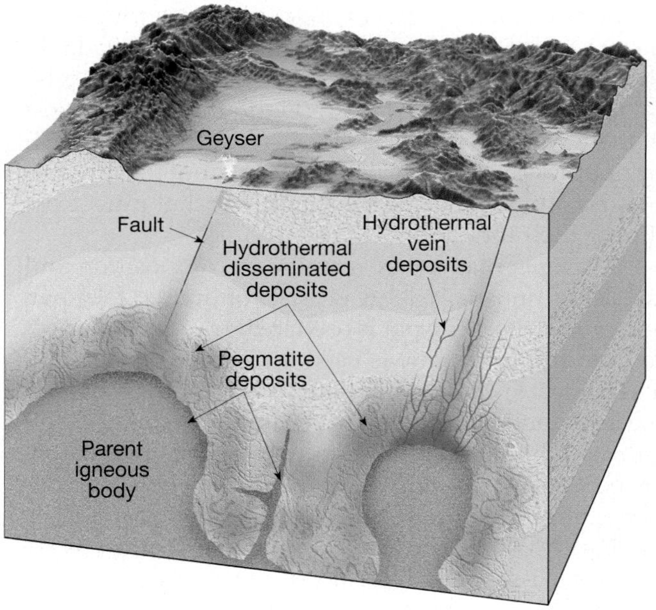

▲ **Figure 3.20** Illustration of the relationship between a parent igneous body and the associated pegmatite and hydrothermal deposits.

3.1). These mineral resources result from processes that concentrate desirable materials to the extent that extraction is economically feasible.

Some pegmatites contain semiprecious gem minerals such as beryl (variety aquamarine).

The igneous processes that generate some of these metal deposits are quite straightforward. For example, as a large basaltic magma body cools, the heavy minerals that crystallize early tend to settle to the lower portion of the magma chamber. This type of magmatic segregation serves to concentrate selected metals producing major deposits of chromite (ore of chromium), magnetite, and platinum. Layers of chromite, interbedded with other heavy minerals, are mined at Montana's Stillwater Complex, whereas the Bushveld Complex in South Africa contains over 70 percent of the world's known reserves of platinum.

Magmatic segregation is also important in the late stages of the magmatic process. This is particularly true of granitic magmas in which the residual melt can become enriched in rare elements and some heavy metals. Further, because water and other volatile substances do not crystallize along with the bulk of the magma body, these fluids make up a high percentage of the melt during the final phase of solidification. Crystallization in a fluid-rich environment, where ion migration is enhanced, results in the formation of crystals several centimeters, or even a few meters, in length. The resulting rocks, called *pegmatites,* are composed of these unusually large crystals (Figure 3.19).

Did you Know?

Although the United States is the largest consumer of industrial diamonds, it has no commercial source of this mineral. It does, however, manufacture large quantities of synthetic diamonds for industrial use.

Feldspar masses the size of houses have been quarried from a pegmatite located in North Carolina. Gigantic hexagonal crystals of muscovite measuring a few meters across have been found in Ontario, Canada. In the Black Hills, spodumene crystals as thick as telephone poles have been mined (Figure 3.19). The largest of these was more than 12 meters (40 feet) long. Not all pegmatites contain such large crystals, but these examples emphasize the special conditions that must exist during their formation.

Most pegmatites are granitic in composition and consist of unusually large crystals of quartz, feldspar, and muscovite. Feldspar is used in the production of ceramics, and muscovite is used for electrical insulation and glitter. Further, pegmatites often contain some of the least abundant elements. Minerals containing the elements lithium, cesium, uranium, and the rare earths are occasionally found. Moreover, some pegmatites contain semiprecious gems such as beryl, topaz, and tourmaline. Most pegmatites are located within large igneous masses or as dikes or veins that cut into the host rock that surrounds the magma chamber (Figure 3.20).

Not all late-stage magmas produce pegmatites, nor do all have a granitic composition. Rather, some magmas become enriched in iron or occasionally copper. For example, at Kirava, Sweden, magma composed of over 60 percent magnetite solidified to produce one of the largest iron deposits in the world.

▲ **Figure 3.21** Light-colored vein deposits emplaced along a series of fractures in dark-colored metamorphic rock. Diablo Lake Overlook, North Cascades National Park, Washington. *(Photo by James E. Patterson)*

Among the best known and most important ore deposits are those generated from hot, ion-rich fluids called **hydrothermal** (hot-water) **solutions.*** Included in this group are the gold deposits of the Homestake mine in South Dakota, the lead, zinc, and silver ores near Coeur d'Alene, Idaho, the silver deposits of the Comstock Lode in Nevada, and the copper ores of Michigan's Keweenaw Peninsula.

The majority of hydrothermal deposits originate from hot, metal-rich fluids that are remnants of late-stage magmatic processes. During solidification, liquids plus various metallic ions accumulate near the top of the magma chamber. Because of their mobility, these ion-rich solutions can migrate great distances through the surrounding rock before they are eventually deposited, usually as sulfides of various metals (Figure 3.20). Some of these fluids move along openings such as fractures or bedding planes, where they cool and precipitate metallic ions to produce *vein deposits* (Figure 3.21). Many of the most productive deposits of gold, silver, and mercury occur as hydrothermal vein deposits.

Another important type of accumulation generated by hydrothermal activity is called a *disseminated deposit.* Rather than being concentrated in narrow veins and dikes, these ores are distributed as minute masses throughout the entire rock mass. Much of the world's copper is extracted from disseminated deposits, in-

* Because these hot, ion-rich fluids tend to chemically alter the host rock, this process is called hydrothermal metamorphism and is discussed in Chapter 7.

cluding those at Chuquicamata, Chile, and the huge Bingham Canyon copper mine in Utah (see Figure 2.22). Because these accumulations contain only 0.4 to 0.8 percent copper, between 125 and 250 kilograms of ore must be mined for every kilogram of metal recovered. The environmental impact of these large excavations, including the problems of waste disposal, is significant.

Another economically important mineral with an igneous origin is diamond. Although best known as gems, diamonds are used extensively as abrasives. Diamonds are thought to originate at depths of nearly 200 kilometers, where the confining pressure is great enough to generate this high-pressure form of carbon. Once crystallized, the diamonds are carried upward through pipe-shaped conduits that increase in diameter toward the surface. In diamond-bearing pipes, nearly the entire pipe contains diamond crystals that are disseminated throughout an ultramafic rock called *kimberlite.* The most productive kimberlite pipes are those in South Africa. The only equivalent source of diamonds in the United States is located near Murfreesboro, Arkansas, but this deposit is exhausted and serves today merely as a tourist attraction.

The Chapter in Review

- *Igneous rocks* form when *magma cools* and solidifies. *Extrusive,* or *volcanic,* igneous rocks result when *lava* cools at the surface. Magma that solidifies at depth produces *intrusive,* or *plutonic,* igneous rocks.

- As magma cools, the ions that compose it arrange themselves into orderly patterns during a process called *crystallization.* Slow cooling results in the formation of rather large crystals. Conversely, when cooling occurs rapidly, the outcome is a solid mass consisting of tiny intergrown crystals. When molten material is quenched instantly, a mass of unordered atoms, referred to as *glass,* forms.

- Igneous rocks are most often classified by their *texture* and *mineral composition.*

- The texture of an igneous rock refers to the overall appearance of the rock based on the size and arrangement of its interlocking crystals. The most important factor affecting texture is the rate at which magma cools. Common igneous rock textures include *aphanitic,* with grains too small to be distinguished without the aid of a microscope; *phaneritic,* with intergrown crystals that are roughly equal in size and large enough to be identified with the unaided eye; *porphyritic,* which has large crystals (*phenocrysts*) interbedded in a matrix of smaller crystals (*groundmass*); and *glassy.*

- The mineral composition of an igneous rock is the consequence of the chemical makeup of the parent magma and the environment of crystallization. Igneous rocks are divided into broad compositional groups based on the percentage of dark and light silicate minerals they contain. *Felsic rocks* (e.g., granite and rhyolite) are composed mostly of the light-colored silicate minerals potassium feldspar and quartz. Rocks of *intermediate* composition (e.g., andesite and diorite) contain plagioclase feldspar and amphibole. *Mafic rocks* (e.g., basalt and gabbro) contain abundant olivine, pyroxene, and calcium feldspar. They are high in iron, magnesium, and calcium, low in silicon, and are dark gray to black in color.

- The mineral makeup of an igneous rock is ultimately determined by the chemical composition of the magma from which it crystallizes. N. L. Bowen discovered that as magma cools in the laboratory, those minerals with higher melting points crystallize before minerals with lower melting points. *Bowen's reaction series* illustrates the sequence of mineral formation within magma.

- During the crystallization of magma, if the earlier-formed minerals are denser than the liquid portion, they will settle to the bottom of the magma chamber during a process called *crystal settling.* Owing to the fact that crystal settling removes the earlier-formed minerals, the remaining melt will form a rock with a chemical composition much different from the parent magma. The process of developing more than one magma type from a common magma is called *magmatic differentiation.*

- Once a magma body forms, its comparison can change through the incorporation of foreign material, a process termed *assimilation,* or by *magma mixing.*

- Magma originates from essentially solid rock of the crust and mantle. In addition to a rock's composition, its temperature, depth (confining pressure), and water content determine whether it exists as a solid or liquid. Thus, magma can be generated by *raising a rock's temperature,* as occurs when a hot mantle plume "ponds" beneath crustal rocks. A *decrease in pressure* can cause *decompression melting.* Further, the *introduction of volatiles* (water) can lower a rock's melting point sufficiently to generate magma. Because melting is generally not complete, a process called *partial melting* produces a melt made of the lowest-melting-temperature minerals, which are higher in silica than the original rock. Thus, magmas generated by partial melting are nearer to the felsic end of the compositional spectrum than are the rocks from which they formed.

- Some of the most important accumulations of metals, such as gold, silver, lead, and copper, are produced by igneous processes. The best-known and most important ore deposits are generated from *hydrothermal* (hot-water) *solutions*. Hydrothermal deposits are thought to originate from hot, metal-rich fluids that are remnants of late-stage magmatic processes. These ion-rich solutions move along fractures or bedding planes, cool, and precipitate the metallic ions to produce *vein deposits*. In a *disseminated deposit* (e.g., much of the world's copper deposits), the ores from hydrothermal solutions are distributed as minute masses throughout the entire rock mass.

Key Terms

andesitic composition (p. 64)
aphanitic texture (p. 61)
assimilation (p. 73)
basaltic composition (p. 63)
Bowen's reaction series (p. 73)
crystallization (p. 59)
crystal settling (p. 73)
decompression melting (p. 70)
extrusive (p. 58)
felsic (p. 63)
fragmental texture (p. 62)
geothermal gradient (p. 69)
glass (p. 61)
glassy texture (p. 61)
groundmass (p. 61)
granitic composition (p. 63)
hydrothermal solutions (p. 76)
igneous rocks (p. 58)
intermediate composition (p. 64)
intrusive (p. 58)
lava (p. 58)
mafic (p. 63)
magma (p. 58)
magma mixing (p. 74)
magmatic differentiation (p. 73)
melt (p. 58)
partial melting (p. 74)
pegmatite (p. 62)
pegmatitic texture (p. 62)
phaneritic texture (p. 61)
phenocryst (p. 61)
plutonic (p. 58)
porphyritic texture (p. 61)
porphyry (p. 61)
pyroclastic texture (p. 62)
texture (p. 60)
ultramafic (p. 64)
vesicular texture (p. 61)
volatiles (p. 59)
volcanic (p. 58)

Questions for Review

1. What is magma?
2. How does lava differ from magma?
3. How does the rate of cooling influence the crystallization process?
4. In addition to the rate of cooling, what two other factors influence the crystallization process?
5. The classification of igneous rocks is based largely on two criteria. Name these criteria.
6. The statements that follow relate to terms describing igneous rock textures. For each statement, identify the appropriate term.
   a. Openings produced by escaping gases.
   b. Obsidian exhibits this texture.
   c. A matrix of fine crystals surrounding phenocrysts.
   d. Crystals are too small to be seen without a microscope.
   e. A texture characterized by two distinctly different crystal sizes.
   f. Coarse-grained, with crystals of roughly equal size.
   g. Exceptionally large crystals exceeding 1 centimeter in diameter.
7. Why are the crystals in pegmatites so large?
8. What does a porphyritic texture indicate about an igneous rock?
9. How are granite and rhyolite different? In what way are they similar?
10. Compare and contrast each of the following pairs of rocks:
    a. granite and diorite
    b. basalt and gabbro
    c. andesite and rhyolite
11. How does tuff differ from other igneous rocks, such as granite and basalt?
12. What is the geothermal gradient?
13. Describe the three conditions that cause rock to melt.
14. What is magmatic differentiation? How might this process lead to the formation of several different igneous rocks from a single magma?
15. Relate the classification of igneous rocks to Bowen's reaction series.
16. What is partial melting?
17. How does the composition of a melt produced by partial melting compare with the composition of the parent rock?
18. List two types of hydrothermal deposits.

Online Study Guide

The *Essentials of Geology* Web site uses the resources and flexibility of the Internet to aid in your study of the topics in this chapter. Written and developed by geology instructors, this site will help improve your understanding of geology. Visit **www.prenhall.com/lutgens** and click on the cover of *Essentials of Geology 9e* to find:

- Online review quizzes.
- Critical thinking exercises.
- Links to chapter-specific Web resources.
- Internet-wide key-term searches.

http://www.prenhall.com/lutgens

A recent eruption of Italy's Mount Etna, one of Europe's most active volcanoes. *(Photo by Art Wolf)*

CHAPTER

4

Volcanoes and Other Igneous Activity

Focus on Learning

To assist you in learning the important concepts in this chapter, you will find it helpful to focus on the following questions:

- What primary factors determine the nature of volcanic eruptions? How do these factors affect a magma's viscosity?
- What materials are associated with a volcanic eruption?
- What are the eruptive patterns and basic characteristics of the three types of volcanoes generally recognized by volcanologists?
- What criteria are used to classify intrusive igneous bodies? What are some of these features?
- What is the relation between volcanic activity and plate tectonics?

On Sunday, May 18, 1980, the largest volcanic eruption to occur in North America in historic times transformed a picturesque volcano into a decapitated remnant (Figure 4.1). On this date in southwestern Washington State, Mount St. Helens erupted with tremendous force. The blast blew out the entire north flank of the volcano, leaving a gaping hole. In one brief moment, a prominent volcano whose summit had been more than 2900 meters (9500 feet) above sea level was lowered by more than 400 meters (1350 feet).

The event devastated a wide swath of timber-rich land on the north side of the mountain (Figure 4.2). Trees within a 400-square-kilometer area lay intertwined and flattened, stripped of their branches and appearing from the air like toothpicks strewn about. The accompanying mudflows carried ash, trees, and water-saturated rock debris 29 kilometers (18 miles) down the Toutle River. The eruption claimed 59 lives, some dying from the intense heat and the suffocating cloud of ash and gases, others from being hurled by the blast, and still others from entrapment in the mudflows.

The eruption ejected nearly a cubic kilometer of ash and rock debris. Following the devastating explosion, Mount St. Helens continued to emit great quantities of hot gases and ash. The force of the blast was so strong that some ash was propelled more than 18,000 meters (over 11 miles) into the stratosphere. During the next few days, this very fine-grained material was carried around Earth by strong upper-air winds. Measurable deposits were reported in Oklahoma and Minnesota, with crop damage into central Montana. Meanwhile, ash fallout in the immediate vicinity exceeded 2 meters in depth. The air over Yakima, Washington (130 kilometers to the east), was so filled with ash that residents experienced midnightlike darkness at noon.

Lava fountain, Kilauea Volcano, Hawaii. (Douglas Peebles)

Not all volcanic eruptions are as violent as the 1980 Mount St. Helens event. Some volcanoes, such as Hawaii's Kilauea volcano, generate relatively quiet outpourings of fluid lavas. These "gentle" eruptions are not without some fiery displays; occasionally fountains of incandescent lava spray hundreds of meters into the air. Such events, however, typically pose minimal threat to human life and property, and the lava generally falls back into a lava pool.

Testimony to the quiet nature of Kilauea's eruptions is the fact that the Hawaiian Volcanoes Observatory has operated on its summit since 1912. This, despite the fact that Kilauea has had more than 50 eruptive phases since record-keeping began in 1823. Further, the longest and largest of Kilauea's eruptions began in 1983 and remains active, although it has received only modest media attention.

Why do volcanoes like Mount St. Helens erupt explosively, whereas others like Kilauea are relatively quiet? Why do volcanoes occur in chains like the Aleutian Islands or the Cascade Range? Why do some volcanoes form on the ocean floor, while others occur on the continents? This chapter will deal with these and other questions as we explore the nature and movement of magma and lava (Figure 4.3).

The Nature of Volcanic Eruptions

GEODe ESSENTIALS OF GEOLOGY Volcanoes and Other Igneous Activity
▼ The Nature of Volcanic Eruptions

Volcanic activity is commonly perceived as a process that produces a picturesque, cone-shaped structure that periodically erupts in a violent manner, like Mount St. Helens (Box 4.1). Although some eruptions may be very explosive, many are not. What determines whether a volcano extrudes magma violently or "gently"? The primary factors include the magma's *composition,* its *temperature,* and the amount of *dissolved gases* it contains. To varying degrees, these factors affect the magma's mobility, or **viscosity.** The more viscous the material, the greater its resistance to flow. (For example, syrup is more viscous than water.) Magma associated with an explosive eruption may be five times more viscous than magma that is extruded in a quiescent manner.

Factors Affecting Viscosity

The effect of temperature on viscosity is easily seen. Just as heating syrup makes it more fluid (less viscous), the mobility of lava is strongly influenced by temperature. As lava cools and begins to congeal, its mobility decreases and eventually the flow halts.

A more significant factor influencing volcanic behavior is the chemical composition of the magma. This was discussed in Chapter 3 with the classification of igneous rocks. Recall that a major difference among various igneous rocks is their silica (SiO_2) content (Table 4.1). Magmas that produce mafic rocks such as basalt contain about 50 percent silica, whereas magmas that produce felsic rocks (granite and its extrusive equivalent, rhyolite) contain over 70 percent silica. The intermediate rock types—andesite and diorite—contain about 60 percent silica.

A magma's viscosity is directly related to its silica content. In general, the more silica in magma, the greater its viscosity. The flow of magma is impeded because silica structures link together into long chains, even before crystallization begins. Consequently, because of their high silica content, rhyolitic (felsic) lavas are very viscous and tend to form comparatively short, thick flows. By contrast, basaltic (mafic) lavas, which contain

▲ **Figure 4.1** Before-and-after photographs show the transformation of Mount St. Helens caused by the May 18, 1980, eruption. The dark area in the "after" photo is debris-filled Spirit Lake, partially visible in the "before" photo. *(Photos courtesy of U.S. Geological Survey)*

Did You Know?

Chile, Peru, and Ecuador boast the highest volcanoes in the world. Here dozens of cones exceed 20,000 feet. In fact, two volcanoes located in Ecuador, named Chimborazo and Cotopaxi, were once considered the world's highest mountains. That distinction remained until the Himalayas were surveyed in the nineteenth century.

less silica, tend to be quite fluid and have been known to travel distances of 150 kilometers (90 miles) or more before congealing.

Importance of Dissolved Gases

The gas content of a magma also affects its mobility. Dissolved gases tend to increase the fluidity of magma. Of far greater consequence is the fact that escaping gases provide enough force to propel molten rock from a volcanic vent.

The summits of volcanoes often begin to inflate months or even years before an eruption occurs. This indicates that magma is migrating into a shallow reservoir located within the cone. During this phase **volatiles** (the gaseous component of magma consisting mostly of water) tend to migrate upward and accumulate near the top of the magma chamber. Thus, the upper portion of a magma body is enriched in dissolved gases.

Lava flowing from a small lava tube. (Griggs/USGS)

When an eruption starts, gas-charged magma moves from the magma chamber and rises through the volcanic conduit or vent. As the magma nears the surface, the confining pressure is greatly reduced. This reduction in pressure allows the dissolved gases to be released suddenly, just as opening a warm soda bottle allows carbon dioxide gas bubbles to escape. At temperatures of 1000°C and low near-surface pressures, these gases will expand to occupy hundreds of times their original volume.

Very fluid basaltic magmas allow the expanding gases to migrate upward and escape from the vent with relative ease. As they escape, the gases may propel incandescent lava hundreds of meters into the air, producing lava fountains. Although spectacular, such fountains are mostly harmless and not generally associated with major explosive events that cause great loss of life and property. Rather, eruptions of fluid basaltic lavas, such as those that occur in Hawaii, are generally quiescent.

At the other extreme, highly viscous magmas explosively expel jets of hot ash-laden gases that evolve into buoyant plumes called **eruption columns** that extend thousands of meters into the atmosphere (Figure 4.4). Prior to an explosive eruption, an extended period of *magmatic differentiation* occurs in which iron-rich minerals crystallize and settle out, leaving the upper part of the magma enriched in silica and dissolved gases. As this volatile-rich magma moves up the volcanic vent toward the surface, these gases begin to collect as tiny bubbles. For reasons that are still poorly understood, at some height in the conduit, this mixture is transformed into a gas jet containing tiny magma fragments that are explosively ejected from the volcano. This type of explosive eruption is exemplified by Mount Pinatubo in the Philippines (1991) and Mount St. Helens (1980).

▲ **Figure 4.2** Douglas fir trees were snapped off or uprooted by the lateral blast of Mount St. Helens on May 18, 1980. Note two scientists, lower left, for scale. *(Photo by Lyn Topinka, U.S. Geological Survey)*

As magma in the upper portion of the vent is ejected, the pressure on the molten rock directly below drops. Thus, rather than a single "bang," volcanic eruptions are really a series of explosions. This process might logically continue until the magma chamber is emptied, much like a geyser empties itself of water (see Chapter 10). However, this generally does not happen. The soluble gases in a viscous magma migrate upward quite slowly. Only within the uppermost portion of the magma body does the gas content build sufficiently to trigger explosive eruptions. Thus, an explosive event is commonly followed by the quiet emission of "degassed" lavas. However, once this eruptive phase ceases, the process of gas buildup begins anew. This time lag may partially explain the

▼ **Figure 4.3** A river of basaltic lava from the January 17, 2002, eruption of Mount Nyiragongo destroyed many homes in Goma, Congo. *(Photo by AFP/Marco Longari/Getty Images/Agence France Presse)*

BOX 4.1

Anatomy of an Eruption

The events leading to the May 18, 1980, eruption of Mount St. Helens began about two months earlier as a series of minor Earth tremors centered beneath the awakening mountain (Figure 4.A, part A). The tremors were caused by the upward movement of magma within the mountain. The first volcanic activity took place a week later, when a small amount of ash and steam rose from the summit. Over the next several weeks, sporadic eruptions of varied intensity occurred. Prior to the main eruption, the primary concern had been the potential hazard of mudflows. These moving lobes of saturated soil and rock are created as ice and snow melt from the heat emitted from magma within the volcano.

The only warning of a potential eruption was a bulge on the volcano's north flank (Figure 4.A, part B). Careful monitoring of this dome-shaped structure indicated a very slow but steady growth rate of a few meters per day. If the growth rate of the bulge changed appreciably, an eruption might quickly follow. Unfortunately, no such variation was detected prior to the explosion. In fact, the seismic activity decreased during the two days preceding the huge blast.

Dozens of scientists were monitoring the mountain when it exploded. "Vancouver, Vancouver, this is it!" was the only warning—and last words from one scientist—that preceded the unleashing of tremendous quantities of pent-up gases. The trigger was a medium-sized earthquake. Its vibrations sent the north slope of the cone plummeting into the Toutle River, removing the overburden that had trapped the magma below (Figure 4.A, part C). With the pressure reduced, the water in the magma vaporized and expanded, causing the mountainside to rupture like an overheated steam boiler. Because the eruption originated around the bulge, several hundred meters below the summit, the initial blast was directed laterally rather than vertically. Had the full force of the eruption been upward, far less destruction would have occurred.

Mount St. Helens is one of 15 large volcanoes and innumerable smaller ones that comprise the Cascade Range, which extends from British Columbia to northern California. Eight of the largest volcanoes have been active in the past few hundred years. Of the remaining seven active volcanoes, the most likely to erupt again are Mount Baker and Mount Rainier in Washington, Mount Shasta and Lassen Peak in California, and Mount Hood in Oregon.

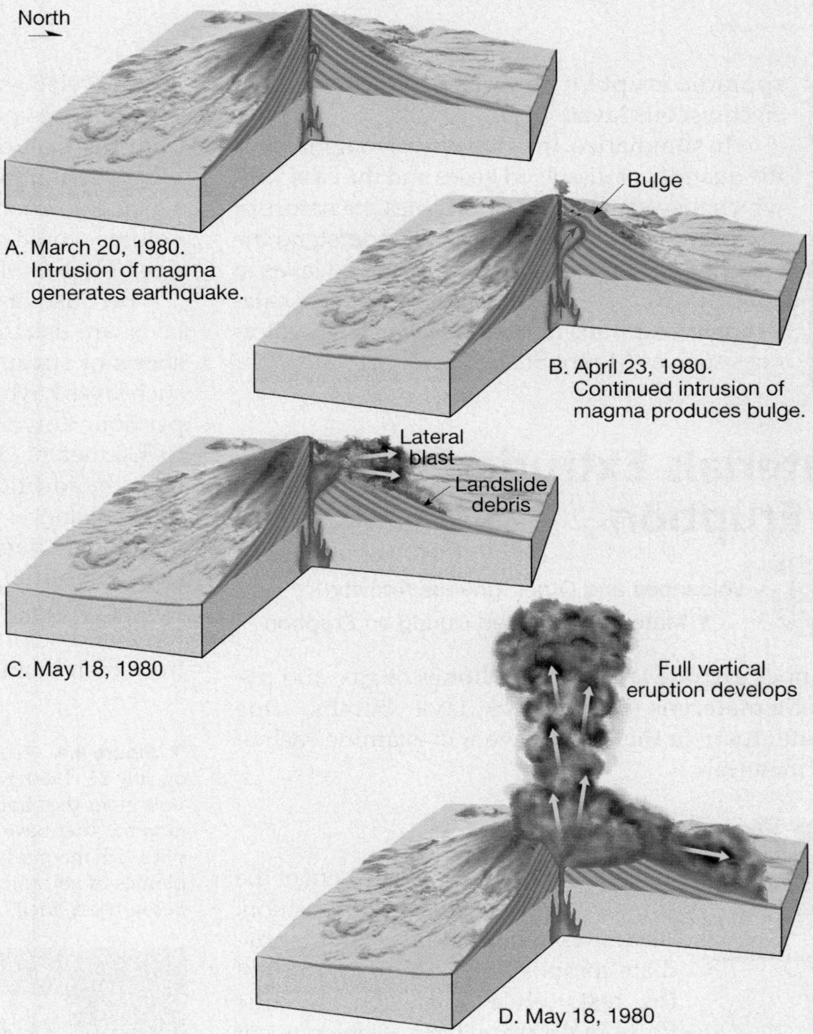

Figure 4.A Idealized diagrams showing the events in the May 18, 1980, eruption of Mount St. Helens. A. First, a sizable earthquake recorded on Mount St. Helens indicates that renewed volcanic activity is possible. B. Alarming growth of a bulge on the north flank suggests increasing magma pressure below. C. Triggered by an earthquake, a giant landslide reduced the confining pressure on the magma body and initiated an explosive lateral blast. D. Within seconds a large vertical eruption sent a column of volcanic ash to an altitude of about 18 kilometers (11 miles). This phase of the eruption continued for over nine hours.

Table 4.1 Magmas have different compositions, which cause their properties to vary

Composition	Silica Content	Viscosity	Gas Content	Tendency to Form Pyroclastics	Volcanic Landform
Mafic (Basaltic) magma	Least (~50%)	Least	Least (1–2%)	Least	Shield Volcanoes Basalt Plateaus Cinder Cones
Intermediate (Andesitic) magma	Intermediate (~60%)	Intermediate (3–4%)	Intermediate	Intermediate	Composite Cones
Felsic (Rhyolitic) magma	Most (~70%)	Greatest	Most (4–6%)	Greatest	Volcanic Domes Pyroclastic Flows

Lava flowing across road, Kilauea, Hawaii. (Griggs/USGS)

sporadic eruptive patterns of volcanoes that eject viscous lavas.

To summarize, the viscosity of magma, plus the quantity of dissolved gases and the ease with which they can escape, determines the nature of a volcanic eruption. We can now understand the "gentle" volcanic eruptions of hot, fluid lavas in Hawaii and the explosive and sometimes catastrophic eruptions of viscous lavas from volcanoes such as Mount St. Helens.

Materials Extruded During an Eruption

GEODe: Essentials of Geology — Volcanoes and Other Igneous Activity ▼ Materials Extruded During an Eruption

Volcanoes extrude lava, large volumes of gas, and pyroclastic materials (broken rock, lava "bombs," fine ash, and dust). In this section we will examine each of these materials.

Lava Flows

The vast majority of lava on Earth, over 90 percent of the total volume, is estimated to be basaltic in composition. Andesites and other lavas of intermediate composition account for most of the rest, while silica-rich rhyolitic flows make up as little as one percent of the total. Recent basaltic flows from two Hawaiian volcanoes, Mauna Loa and Kilauea, had volumes that ranged up to about 0.5 cubic kilometer. One of the largest flows of basaltic lava in historic times came from Iceland's Laki fissure in 1783. The volume of this flow measured 12 cubic kilometers (nearly 3 cubic miles) and some of the lava traveled as far as 88 kilometers (55 miles) from its source. Some prehistoric eruptions, such as those that built the Columbia Plateau in the Pacific Northwest, were even larger. One flow of basaltic lava exceeded 1200 cubic kilometers (almost 300 cubic miles). Such a volume would be sufficient to build three volcanoes the size of Italy's Mount Etna, one of Earth's largest cones.

Because of their lower silica content, hot basaltic lavas are usually very fluid. They flow in thin, broad sheets or streamlike ribbons. On the island of Hawaii, such lavas have been clocked at 30 kilometers (19 miles) per hour down steep slopes. However, flow rates of 10 to 300 meters (30 to 1000 feet) per hour are more common. In addition, basaltic lavas have been known to travel distances of 150 kilometers (90 miles) or more before congealing. By contrast, the movement of silica-rich (rhyolitic) lava may be too slow to perceive. Furthermore, most rhyolitic lavas are comparatively thick and seldom travel more than a few kilometers from their vents. As you might expect, andesitic lavas,

Did You Know?

The eruption of Tambora, Indonesia, in 1815 is the largest-known volcanic event in modern history. About 20 times more ash and rock was explosively ejected during this eruption than was emitted during the 1980 Mount St. Helens event. The sound of the explosion was heard an incredible 3000 miles away, about the distance across the United States.

▼ **Figure 4.4** Eruption column formed by Mount St. Helens on July 22, 1980, two months after the huge May eruption. Volcanoes that border the Pacific Ocean are fed largely by magmas that have intermediate or felsic compositions. These silica-rich magmas often erupt explosively, generating large plumes of volcanic dust and ash. *(David Weintraub/Photo Researchers, Inc.)*

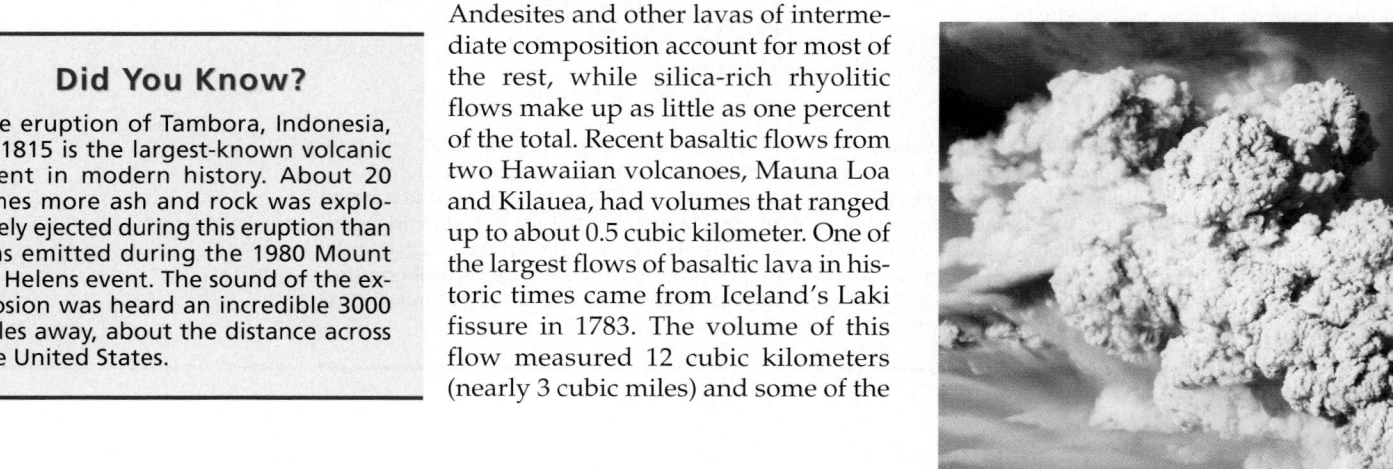

A.

B.

▲ **Figure 4.5** **A.** Typical pahoehoe (ropy) lava flow, Kilauea, Hawaii. *(Photo by Doug Perrine/DRK Photo)* **B.** Typical slow-moving aa flow. *(Photo by J. D. Griggs/U.S. Geological Survey)*

which are intermediate in composition, exhibit characteristics that are between the extremes.

Pahoehoe Flows. When fluid basaltic lavas of the Hawaiian type congeal, they commonly form a relatively smooth skin that wrinkles as the still-molten subsurface lava continues to advance (Figure 4.5A). They are known as **pahoehoe flows** (pronounced pah-hoy-hoy) and resemble the twisting braids in ropes.

Hardened pahoehoe flows commonly contain tunnels that once were horizontal conduits carrying lava from the volcanic vent to the flow's leading edge. These openings develop in the interior of a flow where temperatures remain high long after the surface hardens. Under these conditions, the still-molten lava within the conduits continues its forward motion, leaving behind the cavelike voids called **lava tubes.** Lava tubes are important because they allow fluid lavas to advance great distances from their source. Lava tubes are rare in andesitic and rhyolitic lavas.

Aa Flows. Another common type of basaltic lava, called **aa** (pronounced ah-ah), has a surface of rough, jagged blocks with dangerously sharp edges and spiny projections (Figure 4.5B). Active aa flows are relatively cool and thick and advance at rates from 5 to 50 meters per hour. Moreover, gases escaping from the surface produce numerous voids and sharp spines in the congealing lava. As the molten interior advances, the outer crust is broken further, giving the flow the appearance of an advancing mass of rough, clinkery rubble.

The lava that flowed from the Mexican volcano Parícutin and buried the city of San Juan Parangaricutiro was of the aa type (see Figure 4.12). At times one of the flows from Parícutin moved only 1 meter per day, but it continued to advance day in and day out for more than three months.

Several factors appear to be responsible for the differences between pahoehoe and aa lava flows. In Hawaii, pahoehoe flows are hotter, richer in gases, and faster than aa flows on comparable slopes. Further, most Hawaiian flows begin as pahoehoe but may change to aa flows as they move downslope.

This lava tube once carried molten rock to the toe of a large flow. (Douglas Peebles)

Block Lavas. In contrast to fluid basaltic magmas which typically produce pahoehoe and aa flows, andesitic and rhyolitic magmas tend to generate **block lavas.** Block lava consists largely of detached, blocks with slightly curved surfaces that cover unbroken lava in the interior. Although similar to aa flows, these lavas consist of blocks with comparatively smooth surfaces, rather than having rough, clinkery surfaces.

Pillow Lavas. Recall that much of Earth's volcanic output occurs along oceanic ridges (divergent plate boundaries). When outpourings of lava occur on the ocean floor, or when lava enters the ocean, the flow's outer skin quickly congeals. However, the lava is usually able to move forward by breaking through the hardened surface. This process occurs over and over, as molten basalt is extruded—like toothpaste from a tightly squeezed tube. The result is a lava flow composed of elongated structures resembling large bed pillows stacked one atop the other. These structures, called **pillow lavas,** are useful in the reconstruction of Earth history because whenever they are observed, they indicate that the lava flow formed in an underwater environment.

Ancient pillow lavas at Trinity Bay, Newfoundland. (Geologic Survey of Canada)

Gasses

Magmas contain varying amounts of dissolved gases (*volatiles*) held in the molten rock by confining pressure, just as carbon dioxide is held in soft drinks. As with soft drinks, as soon as the pressure is reduced, the gases begin to escape. Obtaining gas samples from an erupting volcano is difficult and dangerous, so geologists usually estimate the amount of gas originally contained within the magma.

Sulfur-rich gases created these yellow sulfur deposits. (Christian Grzimek/Photo Researchers)

The gaseous portion of most magmas makes up from 1 to 6 percent of the total weight, with most of this in the form of water vapor. Although the percentage may be small, the actual quantity of emitted gas can exceed thousands of tons per day.

The composition of volcanic gases is important because they contribute significantly to the gases that make up our planet's atmosphere. Analyses of samples taken during Hawaiian eruptions indicate that the gases are about 70 percent water vapor, 15 percent carbon dioxide, 5 percent nitrogen, 5 percent sulfur dioxide, and lesser amounts of chlorine, hydrogen, and argon. Sulfur compounds are easily recognized by their pungent odor. Volcanoes are a natural source of air pollution, including sulfur dioxide, which readily combines with water to form sulfuric acid.

In addition to propelling magma from a volcano, gases play an important role in creating the narrow conduit that connects the magma chamber to the surface. First, high temperatures and the buoyant force from the magma body cracks the rock above. Then, hot blasts of high-pressure gases expand the cracks and develop a passageway to the surface. Once the passageway is completed, the hot gases armed with rock fragments erode its walls, producing a larger conduit. Because these erosive forces are concentrated on any protrusion along the pathway, the volcanic pipes that are produced have a circular shape. As the conduit enlarges, magma moves upward to produce surface activity. Following an eruptive phase, the volcanic pipe often becomes choked with a mixture of congealed magma and debris that was not thrown clear of the vent. Before the next eruption, a new surge of explosive gases may again clear the conduit.

Occasionally, eruptions emit colossal amounts of volcanic gases that rise high into the atmosphere, where they may reside for several years. Some of these eruptions may have an impact on Earth's climate, a topic we will consider later in the chapter.

Pyroclastic Materials

When basaltic lava is extruded, dissolved gases escape quite freely and continually. These gases propel incandescent blobs of lava to great heights (Figure 4.6). Some of this ejected material may land near the vent and build a cone-shaped structure, whereas smaller particles will be carried great distances by the wind. By contrast, viscous (rhyolitic) magmas are highly charged with gases, and upon release they expand a thousandfold as they blow pulverized rock, lava, and glass fragments from the vent. The particles produced in both of these situations are referred to as **pyroclastic materials.** These ejected fragments range in size from very fine dust and sand-sized volcanic ash (less than 2 millimeters) to pieces that weigh several tons.

Ash and *dust* particles are produced from gas-laden viscous magma during an explosive eruption (see Figure 4.4). As magma moves up in the vent, the gases rapidly expand, generating a froth of melt that might resemble froth that flows from a just opened bottle of

▼ **Figure 4.6** Volcanic bombs forming during an eruption of Hawaii's Kilauea volcano. Ejected lava fragments take on a streamlined shape as they sail through the air. The bomb in the insert is about 10 centimeters long. *(Photo by Arthur Roy/National Audubon Society/Photo Researchers, Inc.; inset photo courtesy of E. J. Tarbuck)*

champagne. As the hot gases expand explosively, the froth is blown into very fine glassy fragments. When the hot ash falls, the glassy shards often fuse to form a rock called *welded tuff.* Sheets of this material, as well as ash deposits that later consolidate, cover vast portions of the western United States.

Also common are pyroclasts that range in size from small beads to walnuts termed *lapilli* ("little stones"). These ejecta are commonly called *cinders* (2–64 millimeters). Particles larger than 64 millimeters (2.5 inches) in diameter are called *blocks* when they are made of hardened lava, and *bombs* when they are ejected as incandescent lava. Because bombs are semimolten upon ejection, they often take on a streamlined shape as they hurtle through the air (Figure 4.6). Because of their size, bombs and blocks usually fall on the slopes of a cone; however, they are occasionally propelled far from the volcano by the force of escaping gases.

Volcanic Structures and Eruptive Styles

GEODe ESSENTIALS OF GEOLOGY
Volcanoes and Other Igneous Activity
▼ Volcanic Structures and Eruptive Styles

The popular image of a volcano is that of a solitary, graceful, snowcapped cone, such as Mount Hood in Oregon or Japan's Fujiyama. These picturesque, conical mountains are produced by volcanic activity that occurred intermittently over thousands, or even hundreds of thousands, of years. However, many volcanoes do not fit this image. Some volcanoes are only 30 meters (100 feet) high and formed during a single eruptive phase that may have lasted only a few days. Further, numerous volcanic landforms are not "volcanoes" at all. For example, Alaska's Valley of Ten Thousand Smokes is a flat-topped deposit consisting of 15 cubic kilometers of ash that erupted in less than 60 hours and blanketed a section of river valley to a depth of 200 meters (600 feet).

Volcanic landforms come in a wide variety of shapes and sizes, and each structure has a unique eruptive history. Nevertheless, volcanologists have been able to classify volcanic landforms and determine their eruptive patterns. In this section we will consider the general anatomy of a volcano and look at three major volcanic types: shield volcanoes, cinder cones, and composite cones. This discussion will be followed by an overview of other significant volcanic landforms.

Did You Know?

Lava bombs 20 feet long and weighing over 200 tons were thrown 2000 feet (about the length of 7 football fields) from the vent during an eruption of the Japanese volcano Asama.

Anatomy of a Volcano

Volcanic activity frequently begins when a fissure (crack) develops in the crust as magma moves forcefully toward the surface. As the gas-rich magma moves up this linear fissure, its path is usually localized into a circular **conduit,** or **pipe,** that terminates at a surface opening called a **vent** (Figure 4.7). Successive eruptions of lava, pyroclastic material, or frequently a combination of both often separated by long periods of inactivity eventually build the structure we call a **volcano.**

Located at the summit of many volcanoes is a steep-walled depression called a **crater.** Craters are constructional features that were built upward as ejected fragments collected around the vent to form a doughnut-like structure. Some volcanoes have multiple summit craters, whereas others have very large, more or less circular depressions called *calderas.* Calderas are large collapse structures that may or may not form in association with a volcano. (We will consider the formation of various types of calderas later.)

The crater of Mt Vesuvius near Naples, Italy.
(Krafft/Photo Researchers)

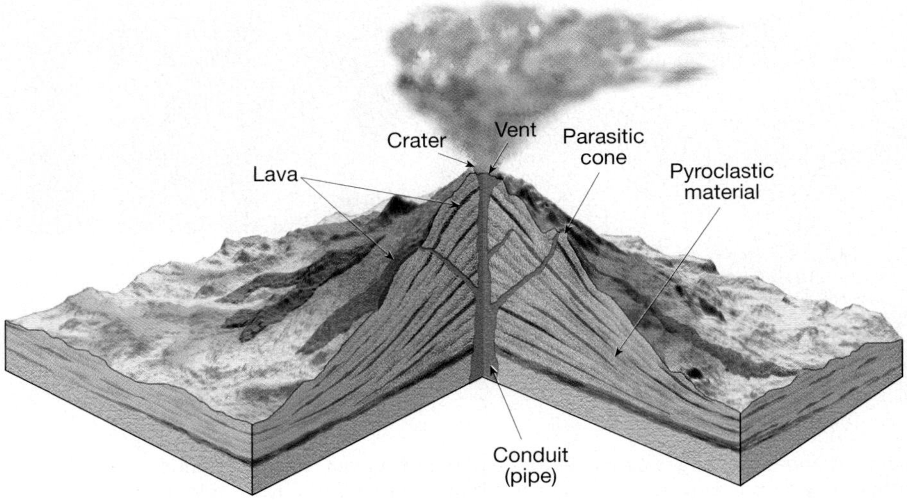

◀ **Figure 4.7** Anatomy of a "typical" composite cone (see also Figures 4.8 and 4.11 for a comparison with a shield and cinder cone, respectively).

During early stages of growth, most volcanic discharges come from a central summit vent. As a volcano matures, material also tends to be emitted from fissures that develop along the flanks, or base, of the volcano. Continued activity from a flank eruption may produce a small **parasitic cone.** Mount Etna in Italy, for example, has more than 200 secondary vents, some of which have built cones. Many of these vents, however, emit only gases and are appropriately called **fumaroles.**

Did You Know?

Kilauea volcano located on the big island of Hawaii is the most continuously active volcano on Earth. Eruptions have regularly led to lava flowing into the ocean at a rate of roughly 5 meters per second for years on end.

The form of a particular volcano is largely determined by the composition of the contributing magma. As you will see, fluid Hawaiian-type lavas tend to produce broad structures with gentle slopes, whereas more viscous silica-rich lavas (and some gas-rich basaltic lavas) tend to generate cones with moderate to steep slopes.

Shastina (left) is a parasitic cone on Mount Shasta, California. (M. Miller)

Shield Volcanoes

Shield volcanoes are produced by the accumulation of fluid basaltic lavas and exhibit the shape of a broad, slightly domed structure that resembles a warrior's shield (Figure 4.8). Most shield volcanoes have grown up from the deep-ocean floor to form islands or seamounts. For example, the islands of the Hawaiian chain, Iceland, and the Galapagos are either a single shield volcano or the coalescence of several shields. Some shield volcanoes, however, occur on continents.

Extensive study of the Hawaiian Islands confirms that each shield was built from a myriad of basaltic lava flows averaging a few meters thick. Also, these islands consist of only about 1 percent pyroclastic ejecta.

Mauna Loa is one of five overlapping shield volcanoes that together comprise the Big Island of Hawaii. From its base on the floor of the Pacific Ocean to its summit, Mauna Loa is over 9 kilometers (6 miles) high, exceeding that of Mount Everest. This massive pile of basaltic rock has a volume of 40,000 cubic kilometers that was extruded over a period of nearly a million years. For comparison, the volume of material composing Mauna Loa is roughly 200 times greater than the amount composing a large composite cone such as Mount Rainier (Figure 4.9). Most shields, however, are more modest in size. For example, the classic Icelandic shield, Skjalbreidur, rises to a height of only about 600 meters (2000 feet) and is 10 kilometers (6 miles) across its base.

Young shields, particularly those located in Iceland, emit very fluid lava from a central summit vent and have sides with gentle slopes that vary from 1 to 5 degrees. Mature shields, as exemplified by Mauna Loa, have steeper slopes in the middle sections (about 10 degrees), while their summits and flanks are comparatively flat. During the mature stage, lavas are discharged from the summit vents, as well as from rift zones that

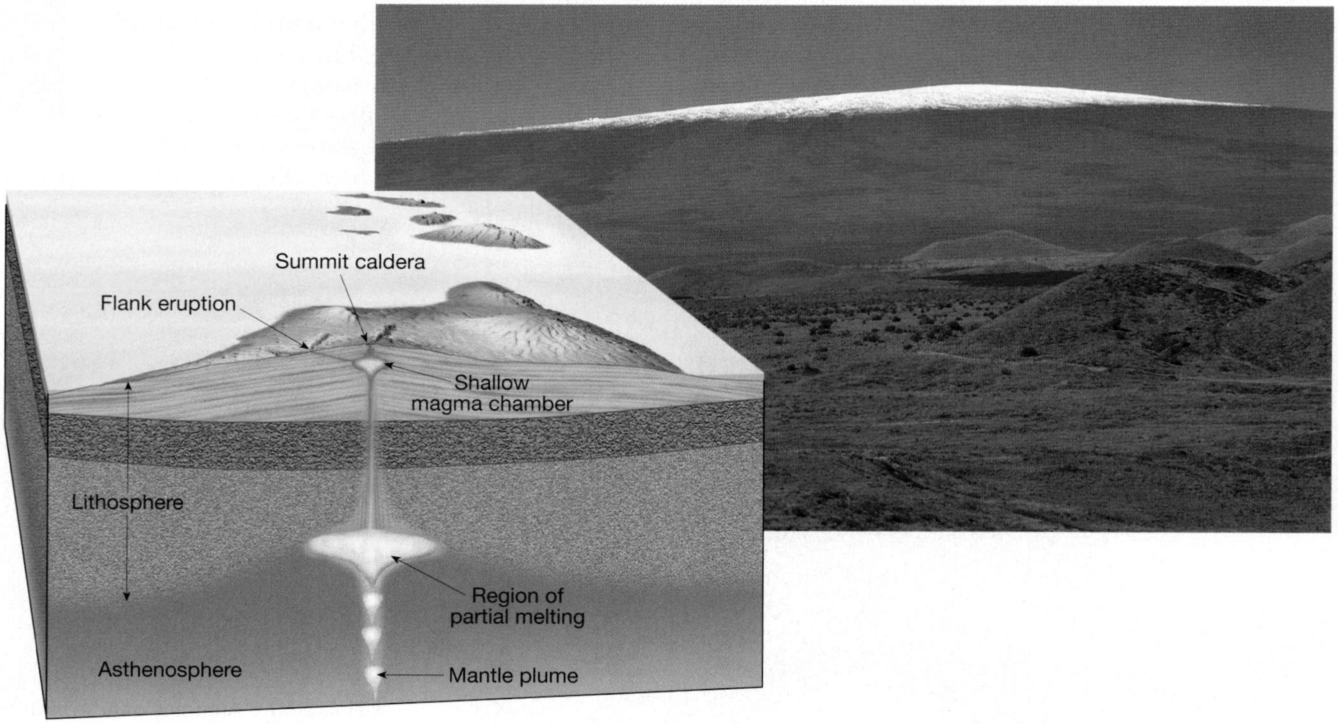

▲ **Figure 4.8** Mauna Loa is one of five shield volcanoes that together make up the island of Hawaii. Shield volcanoes are built primarily of fluid basaltic lava flows and contain only a small percentage of pyroclastic materials. *(Photo by Greg Vaughn)*

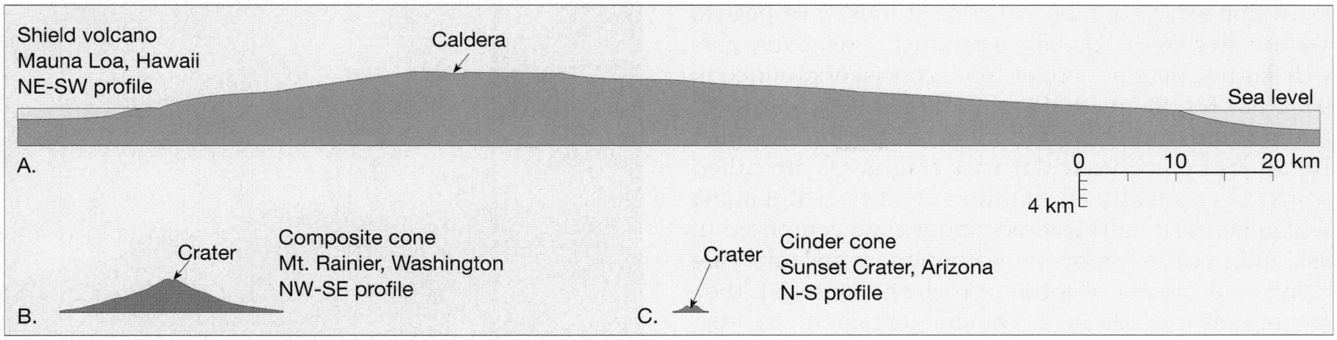

▲ **Figure 4.9** Profiles of volcanic landforms. **A.** Profile of Mauna Loa, Hawaii, the largest shield volcano in the Hawaiian chain. Note size comparison with Mt. Rainier, Washington, a large composite cone. **B.** Profile of Mt. Rainier, Washington. Note how it dwarfs a typical cinder cone. **C.** Profile of Sunset, Arizona, a typical steep-sided cinder cone.

develop along the slopes. Most lava is discharged as the fluid pahoehoe type, but as these flows cool downslope, many change into a clinkery aa flow. Once an eruption is well established, a large fraction of the lava (perhaps 80 percent) flows through a well-developed system of lava tubes. This greatly increases the distance lava can travel before it solidifies. Thus, lava emitted near the summit often reaches the sea, thereby adding to the width of the cone at the expense of its height.

Another feature common to a mature, active shield volcano is a large steep-walled caldera that occupies its summit. Calderas form when the roof of the volcano collapses as magma from the central magma reservoir migrates to the flanks, often feeding fissure eruptions. Mauna Loa's summit caldera measures 2.6 by 4.5 kilometers (1.6 by 2.8 miles) and has a depth that averages about 150 meters (500 feet).

In their final stage of growth, the activity on mature shields is more sporadic and pyroclastic ejections are more prevalent. Further, the lavas increase in viscosity, resulting in thicker, shorter flows. These eruptions tend to steepen the slope of the summit area, which often becomes capped with clusters of cinder cones. This explains why Mauna Kea, a very mature volcano that has not erupted in historic times, has a steeper summit than Mauna Loa, which erupted as recently as 1984. Astronomers are so certain that Mauna Kea is "over the hill" that they have built an elaborate observatory on its summit, housing some of the world's finest (and most expensive) telescopes.

Kilauea, Hawaii: Eruption of a Shield Volcano. Kilauea, the most active and intensely studied shield volcano in the world, is located on the island of Hawaii in the shadow of Mauna Loa. More than 50 eruptions have been witnessed here since record keeping began in 1823. Several months before each eruptive phase, Kilauea inflates as magma gradually migrates upward and accumulates in a central reservoir located a few kilometers below the summit. For up to 24 hours in advance of an eruption, swarms of small earthquakes warn of the impending activity.

Summit caldera, Mauna Loa, Hawaii. The smaller depressions are pit craters. (Peterson/USGS)

Most of the activity on Kilauea during the past 50 years occurred along the flanks of the volcano in a region called the East Rift Zone. A rift eruption here in 1960 engulfed the coastal village of Kapoho, located nearly 30 kilometers (20 miles) from the source. The longest and largest rift eruption ever recorded on Kilauea began in 1983 and continues to this day, with no signs of abating. The first discharge began along a 6-kilometer (4-mile) fissure where a 100-meter- (300-foot-) high "curtain of fire" formed as red-hot lava was ejected skyward (Figure 4.10). When the activity became localized, a cinder and spatter cone given the Hawaiian name *Puu Oo* was built. Over the next three years the general eruptive pattern consisted of short periods (hours to days) when fountains of gas-rich lava sprayed skyward. Each event was followed by nearly a month of inactivity.

By the summer of 1986 a new vent opened up 3 kilometers downrift. Here smooth-surfaced pahoehoe lava formed a lava lake. Occasionally the lake overflowed, but more often lava escaped through tunnels to feed pahoehoe flows that moved down the southeastern flank of the volcano toward the sea. These flows destroyed nearly a hundred rural homes, covered a major roadway, and eventually reached the sea. Lava has been intermittently pouring into the ocean ever since, adding new land to the island of Hawaii.

Did You Know?

According to legend, Pele, the Hawaiian goddess of volcanoes, makes her home at the summit of Kilauea volcano. Evidence for her existence is "Pele's hair," which consists of thin, delicate strands of glass, which are soft and flexible and have a golden-brown color. This threadlike volcanic glass forms when blobs of hot lava are spattered and shredded by escaping gases.

Cinder Cones

As the name suggests, **cinder cones** (also called **scoria cones**) are built from ejected lava fragments that take on the appearance of cinders or clinkers as they begin to harden while in flight. These fragments range in size from fine ash to bombs but consist mostly of pea- to walnut-size lapilli. Usually a product of relatively gas-rich basaltic magma, cinder cones consist of rounded to irregular fragments that are markedly vesicular (containing voids) and have a black to reddish-brown color. Recall that these vesicular rock fragments are called *scoria*. Occasionally an eruption of silica-rich magma will generate a light-colored cinder cone composed of ash and pumice fragments. Although cinder cones are composed mostly of loose pyroclastic material, they sometimes extrude lava. On such occasions the discharges come from vents located at or near the base rather than from the summit crater.

Cinder cones have a very simple, distinctive shape determined by the slope that loose pyroclastic material maintains as it comes to rest (Figure 4.11). Because cinders have a high angle of repose (the steepest angle at which material remains stable), young cinder cones are steep-sided, having *slopes* between 30 and 40 degrees. In addition, cinder cones have large, deep craters in relation to the overall size of the structure. Although relatively symmetrical, many cinder cones are elongated, and higher on the side that was downwind during the eruptions.

A cinder cone in northern Arizona being mined for landscape rock.

Cinder cones are usually the product of a single eruptive episode that sometimes lasts only a few weeks and rarely exceeds a few years. Once this event ceases, the magma in the pipe connecting the vent to the magma chamber solidifies, and the volcano never erupts again. As a consequence of this short life span, cinder cones are small, usually between 30 meters (100 feet) and 300 meters (1000 feet), and rarely exceed 700 meters (2100 feet) in height (see Figure 4.9).

Did You Know?

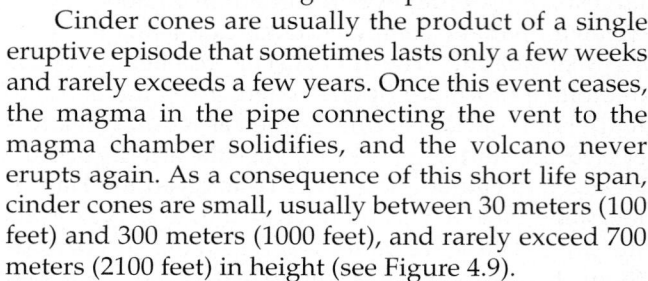

A short distance off the south coast of Hawaii is a submarine volcano called Loihi. Although very active, it has another 3000 feet to go before it breaks the surface and becomes another island in the Hawaiian chain.

Cinder cones are found by the thousands all around the globe. Some are located in volcanic fields like the one located near Flagstaff, Arizona, which consists of about 600 cones. Others are parasitic cones of larger volcanoes. Mount Etna, for example, has dozens of cinder cones dotting its flanks.

Parícutin: Life of a Garden-Variety Cinder Cone. One of the very few volcanoes studied by geologists from beginning to end is the cinder cone called Parícutin, located about 320 kilometers (200 miles) west of Mexico City. In 1943 its eruptive phase began in a cornfield owned by Dionisio Pulido, who witnessed the event as he prepared the field for planting.

For two weeks prior to the first eruption, numerous Earth tremors caused apprehension in the nearby village of Parícutin. Then on February 20 sulfurous gases began billowing from a small depression that had been in the cornfield for as long as people could remember. During the night, hot, glowing rock fragments

▲ **Figure 4.10** Lava extruded along the East Rift Zone, Kilauea, Hawaii. *(Photo by Greg Vaughn)*

were ejected from the vent, producing a spectacular fireworks display. Explosive discharges continued, throwing hot fragments and ash occasionally as high as 6000 meters (20,000 feet) above the crater rim. Larger fragments fell near the crater, some remaining incandescent as they rolled down the slope. These built an aesthetically pleasing cone, while finer ash fell over a much larger area, burning and eventually covering the village of Parícutin. In the first day the cone grew to 40 meters (130 feet), and by the fifth day it was over 100 meters (330 feet) high. Within the first year more than 90 percent of the total ejecta had been discharged.

The first lava flow came from a fissure that opened just north of the cone, but after a few months flows began to emerge from the base of the cone itself. In June 1944 a clinkery aa flow 10 meters (30 feet) thick moved over much of the village of San Juan Parangaricutiro, leaving only the church steeple exposed (Figure 4.12). After nine years of intermittent pyroclastic explosions and nearly continuous discharge of lava from vents at its base, the activity ceased almost as quickly as it had begun. Today, Parícutin is just another one of the scores of cinder cones dotting the landscape in this region of Mexico. Like the others, it will not erupt again.

Composite Cones

Earth's most picturesque yet potentially dangerous volcanoes are **composite cones,** or **stratovolcanoes** (Figure 4.13). Most are located in a relatively narrow zone that rims the Pacific Ocean, appropriately called the *Ring of Fire* (see Figure 4.27). This active zone includes a chain of continental volcanoes that are distributed along the west coast of South and North America, including the large cones of the Andes and the Cascade Range of the western United States and Canada. The latter group includes Mount St. Helens, Mount Rainier, and Mount Garibaldi. The most active regions in the Ring of Fire are located along curved belts of volcanic islands

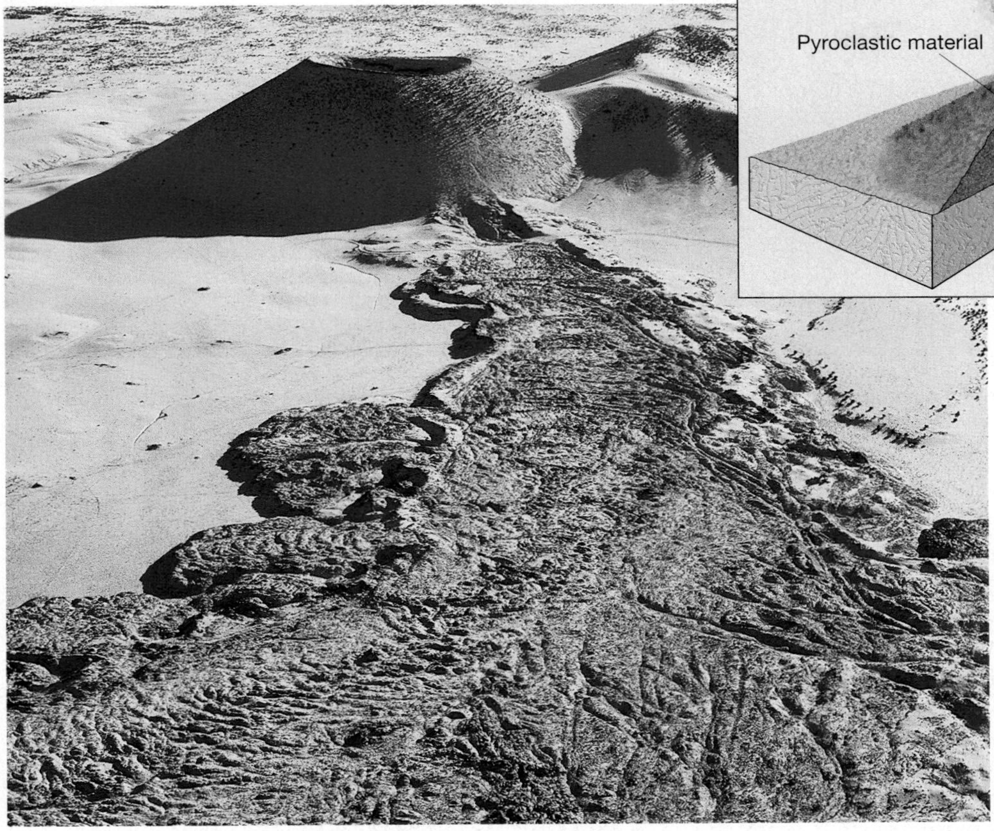

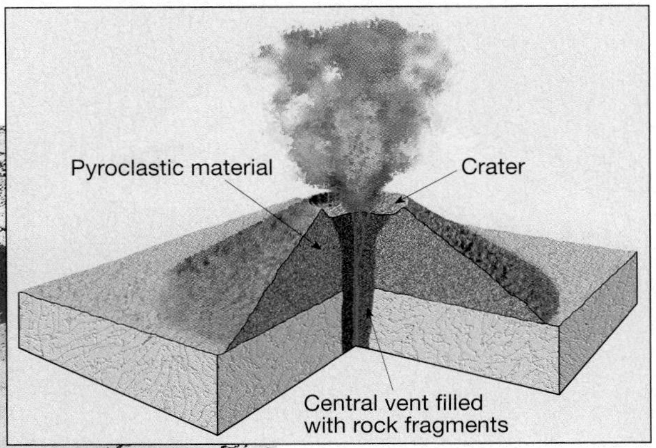

▲ **Figure 4.11** SP Crater, a cinder cone north of Flagstaff, Arizona. Cinder cones are built from ejected lava fragments and are usually less than 300 meters (1000 feet) high. *(Photo by Michael Collier)*

▼ **Figure 4.12** The village of San Juan Parangaricutiro engulfed by lava from Parícutin, shown in the background. Only the church towers remain. *(Photo by Tad Nichols)*

situated adjacent to the deep-ocean trenches of the northern and western Pacific. This nearly continuous chain of volcanoes stretches from the Aleutian Islands to Japan and the Philippines and ends on the North Island of New Zealand.

The classic composite cone is a large, nearly symmetrical structure composed of both lava and pyroclastic deposits. Just as shield volcanoes owe their shape to fluid basaltic lavas, composite cones reflect the nature of the erupted material. For the most part, composite cones are the product of gas-rich magma having an andesitic composition. (Composite cones may also emit various amounts of material having a basaltic and/or rhyolitic composition.) Relative to shields, the silica-rich magmas typical of composite cones generate thick viscous lavas that travel short distances. In addition, composite cones may generate explosive eruptions that eject huge quantities of pyroclastic material.

The growth of a "typical" composite cone begins with both pyroclastic material and lava being emitted from a central vent. As the structure matures, lavas tend to flow from fissures that develop on the lower flanks

▲ **Figure 4.13** These two volcanoes, Pomerape and Parinacota, exhibit the classic shape of a composite cone. Lauca National Park, Chile. *(Photo by Michael Giannechini/Photo Researchers, Inc.)*

Amphitheater-shaped scar left after the 1980 eruption of Mount St. Helens. (C. C. Lockwood/DRK)

of the cone. This activity may alternate with explosive eruptions that eject pyroclastic material from the summit crater. Sometimes both activities occur simultaneously.

A conical shape, with a steep summit area and more gradually sloping flanks, is typical of many large composite cones. This classic profile, which adorns calendars and postcards, is partially a consequence of the way viscous lavas and pyroclastic ejecta contribute to the growth of the cone. Coarse fragments ejected from the summit crater tend to accumulate near their source. Because of their high angle of repose, coarse materials contribute to the steep slopes of the summit area. Finer ejecta, on the other hand, are deposited as a thin layer over a large area. This acts to flatten the flank of the cone. In addition, during the early stages of growth, lavas tend to be more abundant and flow greater distances from the vent than do later lavas. This contributes to the cone's broad base. As the volcano matures, the short flows that come from the central vent serve to armor and strengthen the summit area. Consequently, steep slopes exceeding 40 degrees are sometimes possible. Two of the most perfect cones—Mount Mayon in the Philippines and Fujiyama in Japan—exhibit the classic form we expect of a composite cone, with its steep summit and gently sloping flanks.

Despite their symmetrical form, most composite cones have a complex history. Huge mounds of volcanic debris surrounding many cones provide evidence that, in the distant past, a large section of the volcano slid downslope as a massive landslide. Others develop horseshoe-shaped depressions at their summits as a result of explosive eruptions, or as occurred during the 1980 eruption of Mount St. Helens, a combination of a landslide and the eruption of 0.6 cubic kilometer of magma left a gaping void on the north side of the cone. Often, so much rebuilding has occurred since these eruptions that no trace of the amphitheater-shaped scar remains.

Vesuvius in Italy provides us with another example of the complex history of a volcanic region. This young volcano was built on the same site where an older cone was destroyed by an eruption in A.D. 79. In the following section we will look at another aspect of composite cones: their destructive nature.

Living in the Shadow of a Composite Cone

More than 50 volcanoes have erupted in the United States in the past 200 years (Figure 4.14). Fortunately, the most explosive of these eruptions, except for Mount St. Helens in 1980, occurred in sparsely inhabited regions of Alaska. On a global scale, numerous destructive eruptions have occurred during the past few thousand years, a few of which may have influenced the course of human civilization.

Eruption of Vesuvius A.D. 79

In addition to producing some of the most violent volcanic activity, composite cones can erupt unexpectedly. One of the best documented of these events was the A.D. 79 eruption of the Italian volcano we now call Vesuvius. Prior to this eruption, Vesuvius had been dormant for centuries and had vineyards adorning its sunny slopes. On August 24, however, the tranquility ended, and in less than 24 hours the city of Pompeii (near Naples) and more than 2000 of its 20,000 residents perished. Some were entombed beneath a layer of pumice nearly 3 meters (10 feet) thick, while others were encased within a layer of hardened ash. They remained this way for nearly 17 centuries, until the city was partially excavated,

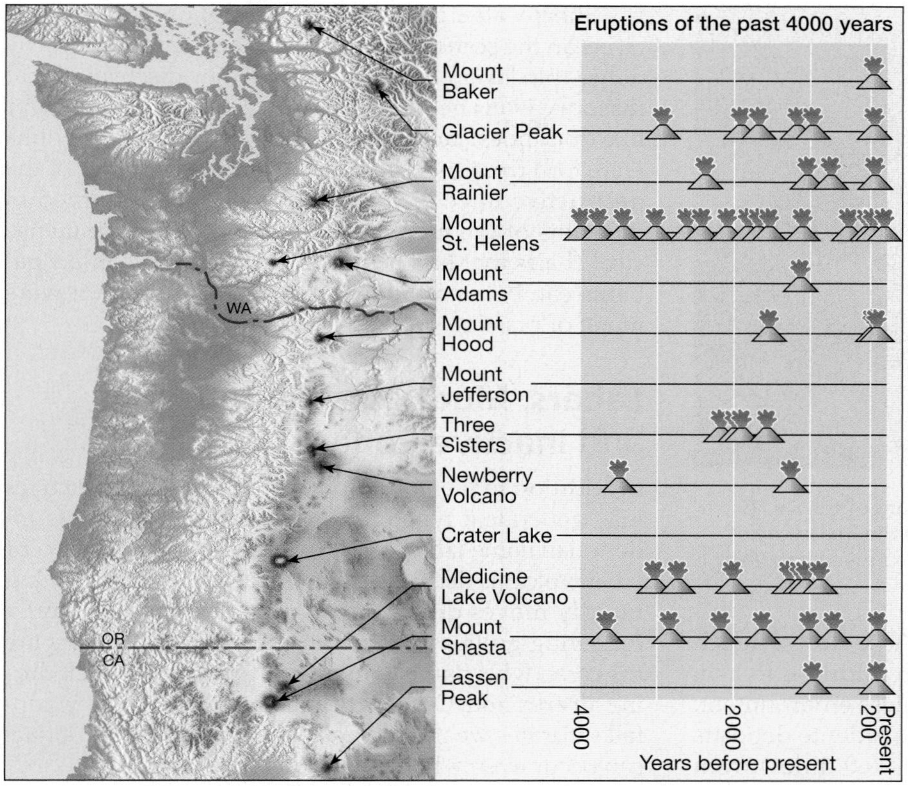

◀ **Figure 4.14** Of the 13 potentially active volcanoes in the Cascade Range, 11 have erupted in the past 4000 years and 7 in just the past 200 years. More than 100 eruptions, most of which were explosive, have occurred in the past 4000 years. Mount St. Helens is the most active volcano in the Cascades. Its eruptions have ranged from relatively quiet outflows of lava to explosive events much larger than that of May 18, 1980. Each eruption symbol in the diagram represents from one to several dozen eruptions closely spaced in time. *(After U.S. Geological Survey)*

giving archaeologists a superbly detailed picture of ancient Roman life.

By reconciling historical records with detailed scientific studies of the region, volcanologists have pieced together the chronology of the destruction of Pompeii. The eruption most likely began as steam discharges on the morning of August 24. By early afternoon fine ash and pumice fragments formed a tall eruptive cloud emanating from Vesuvius. Shortly thereafter, debris from this cloud began to shower Pompeii, located 9 kilometers (6 miles) downwind of the volcano. Undoubtedly, many people fled during this early phase of the eruption. For the next several hours pumice fragments as large as 5 centimeters (2 inches) fell on Pompeii. One historical record of this eruption states that people located more distant than Pompeii tied pillows to their heads in order to fend off the flying fragments.

The pumice fall continued for several hours, accumulating at the rate of 12 to 15 centimeters (5 to 6 inches) per hour. Most of the roofs in Pompeii eventually gave way. Despite the accumulation of more than 2 meters of pumice, many of the people that had not evacuated Pompeii were probably still alive the morning of August 25. Then suddenly and unexpectedly a surge of searing hot dust and gas swept rapidly down the flanks of Vesuvius. This blast killed an estimated 2000 people who had somehow managed to survive the pumice fall. Some may have been killed by flying debris, but most died of suffocation as a result of inhaling ash-laden gases. Their remains were quickly buried by the falling ash, which rain cemented into a hard mass before their bodies had time to decay. The subsequent decomposition of the bodies produced cavities in the hardened ash that replicated the form of the entombed bodies, preserving in some cases even facial expressions. Nineteenth-century excavators found these cavities and created casts of the corpses by pouring plaster of Paris into the voids (Figure 4.15). Some of the plaster casts show victims trying to cover their mouths.

Building uncovered during the excavation of Pompeii, Italy. (C. Havens/CORBIS)

Nuée Ardente: A Deadly Pyroclastic Flow

Among the most devastating phenomena associated with composite cones are **pyroclastic flows** that consist of hot gases infused with incandescent ash and larger rock fragments. The most destructive of these fiery flows, called **nuée ardentes** (also referred to as *glowing avalanches*), are capable of racing down steep volcanic slopes at speeds that can approach 200 kilometers (125 miles) per hour (Figure 4.16).

The ground-hugging portion of a glowing avalanche is rich in particulate matter, which is suspended by jets of buoyant gases passing upward through the flow. Some of these gases have escaped from newly erupted volcanic fragments. In addition, air that is overtaken and trapped by an advancing flow may be heated sufficiently to provide buoyancy to the particulate

▲ **Figure 4.15** Plaster casts of several victims of the A.D. 79 eruption of Mount Vesuvius that destroyed the Italian city of Pompeii. *(Photo by Leonard von Matt/Photo Researchers, Inc.)*

matter of the nuée ardente. Thus, these flows, which can include large rock fragments in addition to ash, travel downslope in a nearly frictionless environment. This helps to explain why some nuée ardente deposits are found more than 100 kilometers (60 miles) from their source.

The pull of gravity is the force that causes these heavier-than-air flows to sweep downslope much like a snow avalanche. Some pyroclastic flows result when a powerful eruption blasts pyroclastic material laterally out the side of a volcano. Probably more often, nuée ardentes form from the collapse of tall eruption columns that form over a volcano during an explosive event. Once gravity overcomes the initial upward thrust provided by the escaping gases, the ejecta begin to fall. Massive amounts of incandescent blocks, ash, and pumice fragments that fall onto the summit area begin to cascade downslope under the influence of gravity. The largest fragments have been observed bouncing down the flanks of a cone, while the finer materials travel rapidly as an expanding tongue-shaped cloud.

The Destruction of St. Pierre. In 1902 an infamous nuée ardente from Mount Pelée, a small volcano on the Caribbean island of Martinique, destroyed the port town of St. Pierre. The destruction happened in moments and was so devastating that almost all of St. Pierre's 28,000 inhabitants were killed. Only one person on the outskirts of town—a prisoner protected in a dungeon—and a few people on ships in the harbor were spared (Figure 4.17).

Did You Know?

In 1902, when Mount Pelée erupted, the U.S. Senate was preparing to vote on the location of a canal to connect the Pacific and Atlantic oceans. The choice was between Panama and Nicaragua, a country that featured a smoking volcano on its postage stamp. Supporters of the Panamanian route used the potential threat of volcanic eruptions in Nicaragua to bolster their argument in favor of Panama. The U.S. Senate subsequently approved the construction of the Panama Canal by eight votes.

Shortly after this calamitous eruption, scientists arrived on the scene. Although St. Pierre was mantled by only a thin layer of volcanic debris, they discovered that masonry walls nearly a meter thick were knocked over like dominoes; large trees were uprooted and cannons were torn from their mounts. A further reminder of the destructive force of this nuée ardente is preserved in the ruins of the mental hospital. One of the immense steel chairs that had been used to confine alcoholic patients can be seen today, contorted, as though it were made of plastic.

Lahars: Mudflows on Active and Inactive Cones

In addition to violent eruptions, large composite cones may generate a type of mudflow referred to by its Indonesian name **lahar.** These destructive mudflows occur when volcanic debris becomes saturated with water and rapidly moves down steep volcanic slopes, generally following gullies and stream valleys. Some lahars are triggered when large volumes of ice and snow melt during an eruption. Others are generated when heavy rainfall saturates weathered volcanic deposits. Thus, lahars can occur even when a volcano is *not* erupting.

When Mount St. Helens erupted in 1980, several lahars formed. These mudflows and accompanying flood waters raced down the valleys of the north and south forks of the Toutle River at speeds exceeding 30 kilometers per hour. Water levels in the river rose to 4 meters (13 feet) above flood stage, destroying or severely damaging nearly all the homes and bridges along the

▼ **Figure 4.16** Nuée ardente races down the slope of Mount St. Helens on August 7, 1980, at speeds in excess of 100 kilometers (60 miles) per hour. *(Photo by Peter W. Lipman, U.S. Geological Survey)*

▲ **Figure 4.17** St. Pierre as it appeared shortly after the eruption of Mount Pelée, 1902. *(Photo by Heilprin/Underwood and Underwood/Library of Congress)*

impacted area. Fortunately, the area was not densely populated.

In 1985 deadly lahars were produced during a small eruption of Nevado del Ruiz, a 5300-meter (17,400-foot) volcano in the Andes Mountains of Colombia. Hot, pyroclastic material melted ice and snow that capped the mountain (Nevado means *snow* in Spanish) and sent torrents of ash and debris down three major river valleys that flank the volcano. Reaching speeds of 100 kilometers (60 miles) per hour, these mudflows tragically took 25,000 lives.

Mount Rainier, Washington, is considered by many to be America's most dangerous volcano because, like Nevado del Ruiz, it has a thick year-round mantle of snow and ice. Adding to the risk is the fact that 100,000 people live in the valleys around Rainier, and many homes are built on lahars that flowed down the volcano hundreds or thousands of years ago. A future eruption, or perhaps just a period of heavy rainfall, may produce lahars that will likely take similar paths.

Other Volcanic Landforms

The most obvious volcanic structure is a cone. But other distinctive and important landforms are also associated with volcanic activity.

Calderas

Calderas are large collapse depressions having a more or less circular form. Their diameters exceed one kilometer, and many are tens of kilometers across. (Those less than a kilometer across are called *collapse pits*.) Most calderas are formed by one of the following processes: 1) the collapse of the summit of a large composite volcano following an explosive eruption of silica-rich pumice and ash fragments; 2) the collapse of the top of a shield volcano caused by subterranean drainage from a central magma chamber; and 3) the collapse of a large area, independent of any preexisting volcanic structures, caused by the discharge of colossal volumes of silica-rich pumice and ash along ring fractures.

A lahar from Japan's Unzen Volcano buried this house. (T. Pierson/USGS)

Crater Lake–Type Calderas. Crater Lake, Oregon, is located in a caldera that has a maximum diameter of 10 kilometers (6 miles) and is 1175 meters (over 3800 feet) deep. This caldera formed about 7000 years ago when a composite cone, later named Mount Mazama, violently extruded 50 to 70 cubic kilometers of pyroclastic material (Figure 4.18). With the loss of support, 1500 meters (nearly a mile) of the summit of this once prominent cone collapsed. After the collapse, rainwater filled the caldera (Figure 4.19). Later volcanic activity built a small cinder cone in the lake. Today this cone, called Wizard Island, provides a mute reminder of past activity.

Hawaiian-Type Calderas. Although most calderas are produced by *collapse following an explosive eruption*, some are

Did You Know?

Some calderas are so large that most people who visit geothermal features, such as those found in Yellowstone National Park, are unaware that they are standing within one of the largest volcanic depressions on Earth.

not. For example, Hawaii's active shield volcanoes, Mauna Loa and Kilauea, both have large calderas at their summits. Kilauea's measures 3.3 by 4.4 kilometers (about 2 by 3 miles) and is 150 meters (500 feet) deep. Each caldera formed by gradual subsidence of the summit as magma slowly drained laterally from the central magma chamber to a rift zone, often producing flank eruptions.

Yellowstone-Type Calderas. Although the 1980 eruption of Mount St. Helens was spectacular, it pales by comparison to what happened 630,000 years ago in the region now occupied by Yellowstone National Park. Here, approximately 1000 cubic kilometers of pyroclastic material erupted, eventually producing a caldera 70 kilometers (43 miles) across. This event produced showers of ash as far away as the Gulf of Mexico. Vestiges of this activity are the many hot springs and geysers in the region.

Did You Know?

Prehistoric caldera eruptions in Yellowstone National Park were about 1000 times larger than the Mount St. Helens episode, strong enough to decimate much of western North America. Volcanologists have compared their destructive forces with that of the impact of a small asteroid. Other caldera eruptions of similar scale occurred in the Jemez Mountains near Santa Fe, New Mexico, Long Valley near Bishop, California, and in Colorado's San Juan Mountains.

Unlike calderas associated with composite cones, these depressions are so large and poorly defined that many remained undetected until high-quality aerial, or satellite, images became available. One of these, the LaGarita caldera, located in the San Juan Mountains of southern Colorado, is about 32 kilometers (20 miles) wide and 80 kilometers (50 miles) long. Despite modern mapping techniques, the entire outline of this structure is still uncertain.

The formation of a large Yellowstone-type caldera begins when a silica-rich (rhyolitic) magma body is emplaced near the surface, upwarping the overlying rocks. Next, ring fractures develop in the roof providing a pathway to the surface for the highly viscous gas-rich magma. This initiates an explosive eruption of colossal proportions ejecting huge volumes (usually exceeding 100 cubic kilometers) of pyroclastic materials, mainly in the form of ash and pumice fragments. Typically, these materials form a pyroclastic flow that spreads across the landscape at speeds that may exceed 100 kilometers (60 miles) per hour, destroying most living things in its path. After coming to rest, the hot fragments of ash and pumice fuse together, forming a welded tuff that closely resembles a solidified lava flow. Finally, with the loss of support, the roof of the magma chamber collapses, generating a large caldera.

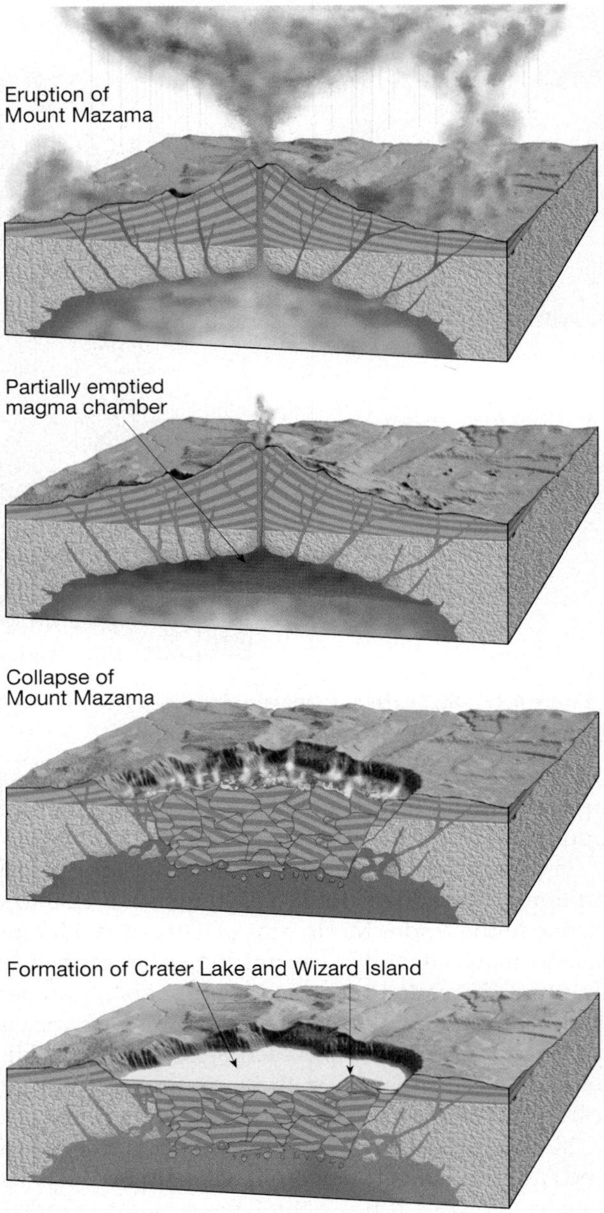

▲ **Figure 4.18** Sequence of events that formed Crater Lake, Oregon. About 7000 years ago a violent eruption partly emptied the magma chamber, causing the summit of former Mount Mazama to collapse. Rainfall and groundwater contributed to form Crater Lake, the deepest lake in the United States. Subsequent eruptions produced the cinder cone called Wizard Island. *(After H. Williams,* The Ancient Volcanoes of Oregon, *p. 47. Courtesy of the University of Oregon)*

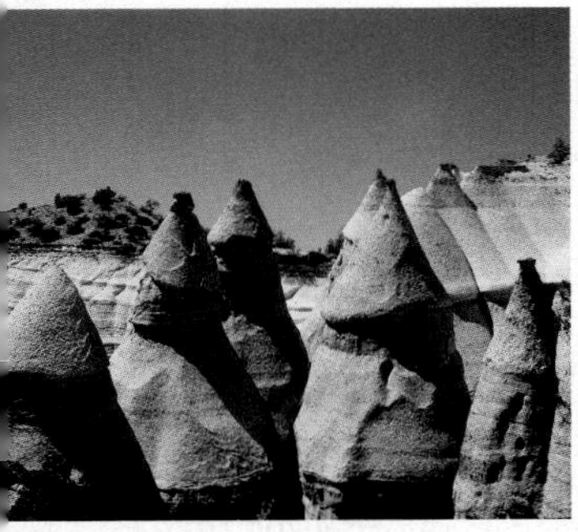

Welded tuffs like these near Santa Fe, New Mexico, often erode to produce wigwam or tent rocks.

Fissure Eruptions and Basalt Plateaus

We think of volcanic eruptions as building a cone or shield from a central vent. But by far the greatest volume of volcanic material is extruded from fractures in the crust called **fissures.** Rather than building a cone, these long, narrow cracks may emit a low-viscosity basaltic lava, blanketing a wide area.

The extensive Columbia Plateau in the northwestern United States was formed this way (Figure 4.20). Here, numerous **fissure eruptions** extruded very fluid basaltic lava (Figure 4.21). Successive flows, some 50 meters (160 feet) thick, buried the existing landscape

▲ **Figure 4.19** Crater Lake in Oregon occupies a caldera 10 kilometers (about 6 miles) in diameter. Wizard Island is a cinder cone located within the caldera. *(Photo by Greg Vaughn/Tom Stack and Associates)*

as they built a lava plateau nearly a mile thick. The fluid nature of the lava is evident, because some remained molten long enough to flow 150 kilometers (90 miles) from its source. The term **flood basalts** appropriately describes these flows. Massive accumulations of basaltic lava, similar to those of the Columbia Plateau, occur worldwide. One of the largest is the Deccan Traps, a thick sequence of flat-lying basalt flows covering nearly 500,000 square kilometers (195,000 square miles) of west central India. When the Deccan Traps formed about 66 million years ago, nearly 2 million cubic kilometers of lava were extruded in less than 1 million years. Another huge deposit of flood basalts, called the Ontong Java Plateau, is found on the floor of the Pacific Ocean. A discussion of the origin of large basalt plateaus is provided later in this chapter in the section on *"Intraplate Igneous Activity."*

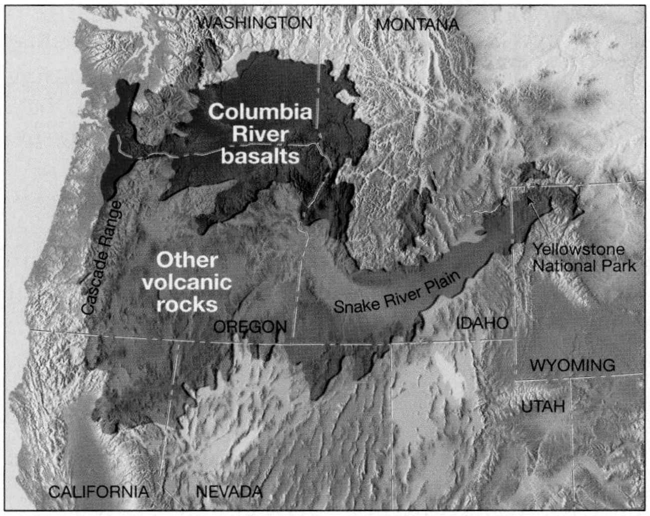

▲ **Figure 4.20** Volcanic areas that compose the Columbia Plateau in the Pacific Northwest. The Columbia River basalts cover an area of nearly 200,000 square kilometers (80,000 square miles). Activity here began about 17 million years ago as lava began to pour out of large fissures, eventually producing a basalt plateau with an average thickness of more than 1 kilometer. *(After U.S. Geological Survey)*

Did You Know?

In 1783 the deadly Laki fissure eruption took place in Iceland along a 15-mile rift. The eruption extruded 12 cubic kilometers of fluid basaltic lava. In addition, sulfurous gases and ash were emitted that destroyed grasslands, killing most of Iceland's livestock. The eruption and ensuing famine killed 10,000 Icelanders—roughly one in five. Recently, a volcanologist who is an expert on the Laki eruption concluded that a similar eruption today would disrupt commercial aviation in the Northern Hemisphere for several months.

Volcanic Pipes and Necks

Most volcanoes are fed magma through short conduits, called *pipes,* that connect a magma chamber to the surface. In rare circumstances, pipes may extend tubelike to depths exceeding 200 kilometers (125 miles). When this occurs, the ultramafic magmas that migrate up these structures produce rocks that are thought to be samples of the mantle that have undergone very little alteration during their ascent. Geologists consider these unusually deep conduits to be "windows" into Earth, for they allow us to view rock normally found only at great depth.

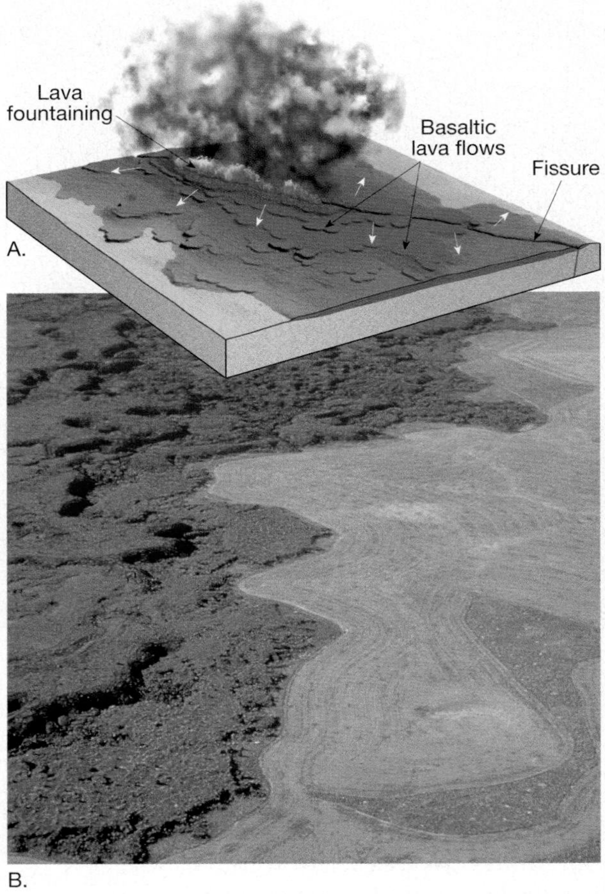

▲ **Figure 4.21** Basaltic fissure eruption. **A.** Lava fountaining from a fissure and formation of fluid lava flows called flood basalts. **B.** Photo of basalt flows near Idaho Falls. *(Photo by John S. Shelton)*

This volcanic pipe at Kimberly, South Africa, was excavated because it contains gem-quality diamonds. (Roger De La Harpe)

The best-known volcanic pipes are the diamond-bearing structures of South Africa. Here, the rocks filling the pipes originated at depths of at least 150 kilometers (90 miles), where pressure is high enough to generate diamonds and other high-pressure minerals. The task of transporting essentially unaltered magma (along with diamond inclusions) through 150 kilometers of solid rock is exceptional. This fact accounts for the scarcity of natural diamonds.

Volcanoes on land are continually being lowered by weathering and erosion. Cinder cones are easily eroded, because they are composed of unconsolidated materials. However, all volcanoes will eventually succumb to relentless erosion over geologic time. As erosion progresses, the rock occupying the volcanic pipe is often more resistant and may remain standing above the surrounding terrain long after most of the cone has vanished. Shiprock, New Mexico, is such a feature and is called a **volcanic neck** (Figure 4.22). This structure, higher than many skyscrapers, is but one of many such landforms that protrude conspicuously from the red desert landscapes of the American Southwest.

Intrusive Igneous Activity

GEODe: Volcanoes and Other Igneous Activity
▼ Intrusive Igneous Activity

Although volcanic eruptions can be among the most violent and spectacular events in nature and therefore worthy of detailed study, most magma is emplaced at depth. Thus, an understanding of intrusive igneous activity is as important to geologists as the study of volcanic events.

The structures that result from the emplacement of igneous material at depth are called **plutons,** named for Pluto, the god of the lower world in classical mythology. Because all plutons form out of view beneath Earth's surface, they can be studied only after uplifting and erosion have exposed them. The challenge lies in reconstructing the events that generated these structures millions or even hundreds of millions of years ago.

For the sake of clarity, we have separated our discussions of volcanism and plutonic activity. Keep in mind, however, that these diverse processes occur simultaneously and involve basically the same materials.

Nature of Plutons

Plutons are known to occur in a great variety of sizes and shapes. Some of the most common types are illustrated in Figure 4.23. Notice that some of these structures have a tabular (tabletop) shape, whereas others are quite massive. Also, observe that some of these bodies cut across existing structures, such as layers of sedimentary rock; others form when magma is injected between sedimentary layers. Because of these differences, intrusive igneous bodies are generally classified according to their shape as either **tabular** or **massive** and by their orientation with respect to the host rock. Plutons are said to be **discordant** if they cut across existing structures, and **concordant** if they form parallel to features such as sedimentary strata. As you can see in Figure 4.23A, plutons are closely associated with volcanic activity. Many of the largest intrusive bodies are the remnants of magma chambers that once fed ancient volcanoes.

Dikes

Dikes are tabular discordant bodies that are produced when magma is injected into fractures. The force exerted by the emplaced magma can be great enough to separate the walls of the fracture further. Once crystallized, these sheetlike structures have thicknesses ranging from less than a centimeter to more than a kilometer. The largest have lengths of hundreds of kilometers. Most dikes, however, are a few meters thick and extend laterally for no more than a few kilometers.

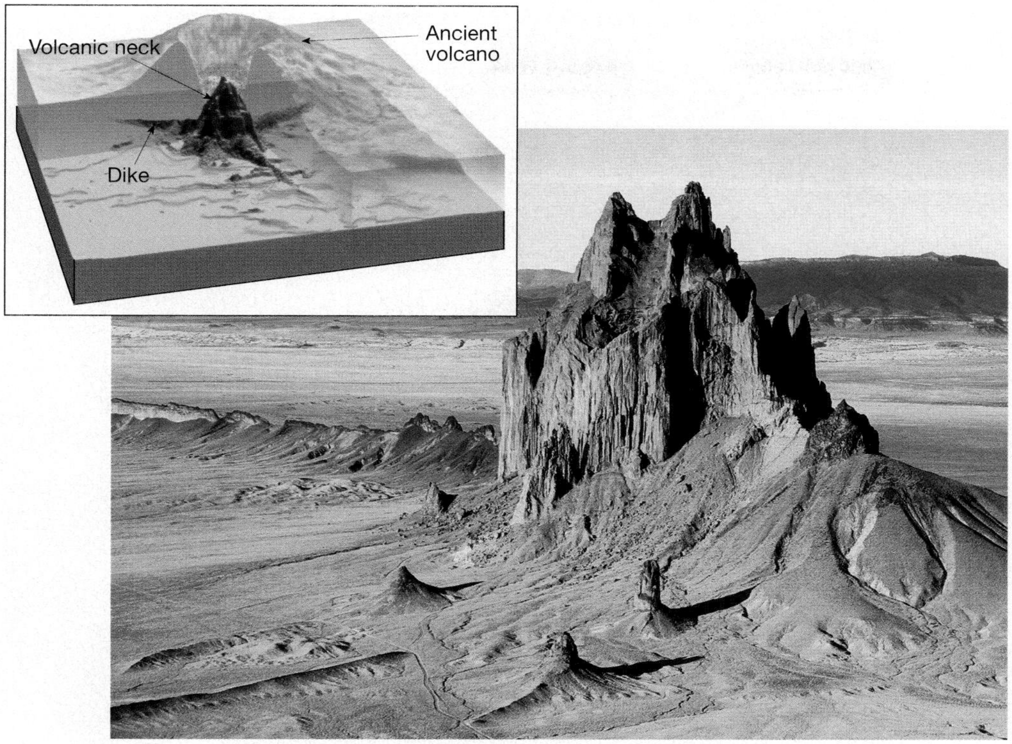

◀ **Figure 4.22** Shiprock, New Mexico, is a volcanic neck. This structure, which stands over 420 meters (1380 feet) high, consists of igneous rock that crystallized in the vent of a volcano that has long since been eroded away. The tabular structure in the background is a dike that served to feed lava flows along the flanks of the once active volcano. *(Photo by Tom Bean)*

Dikes are often found in groups that once served as vertically oriented pathways followed by molten rock that fed ancient lava flows. The parent pluton is generally not observable. Some dikes are found radiating, like spokes on a wheel, from an eroded volcanic neck (see Figure 4.22). In these situations the active ascent of magma is thought to have generated fissures in the volcanic cone out of which lava flowed.

Sills and Laccoliths

Sills and laccoliths are concordant plutons that form when magma is intruded in a near-surface environment. They differ in shape and usually differ in composition.

Sills. **Sills** are tabular plutons formed when magma is injected along sedimentary bedding surfaces (Figure 4.24). Horizontal sills are the most common, although all orientations, even vertical, are known to exist. Because of their relatively uniform thickness and large areal extent, sills are likely the product of very fluid magmas. Magmas having a low silica content are more fluid, so most sills are composed of the rock basalt.

The emplacement of a sill requires that the overlying sedimentary rock be lifted to a height equal to the thickness of the sill. Although this is a formidable task, in shallow environments it often requires less energy than forcing the magma up the remaining distance to the surface. Consequently, sills form only at shallow depths, where the pressure exerted by the weight of overlying rock layers is low. Although sills are intruded between layers, they can be locally discordant. Large sills frequently cut across sedimentary layers and resume their concordant nature at a higher level.

One of the largest and most studied of all sills in the United States is the Palisades Sill. Exposed for 80 kilometers along the west bank of the Hudson River in southeastern New York and northeastern New Jersey, this sill is about 300 meters thick. Because of its resistant nature, the Palisades Sill forms an imposing cliff that can be seen easily from the opposite side of the Hudson.

In many respects, sills closely resemble buried lava flows. Both are tabular and often exhibit columnar jointing (Figure 4.25). **Columnar joints** form as igneous rocks cool and develop shrinkage fractures that produce elongated, pillarlike columns. Further, because sills generally form in near-surface environments and may be only a few meters thick, the emplaced magma often cools quickly enough to generate an aphanitic texture.

Laccoliths. **Laccoliths** are similar to sills because they form when magma is intruded between sedimentary layers in a near-surface environment. However, the magma that generates laccoliths is more viscous. This less fluid magma collects as a lens-shaped mass that arches the overlying strata upward (see Figure 4.23). Consequently, a laccolith can occasionally be detected because of the dome-shaped bulge it creates at the surface.

This dike is more resistant to weathering than the surrounding rocks. (P. Jay Fleisher)

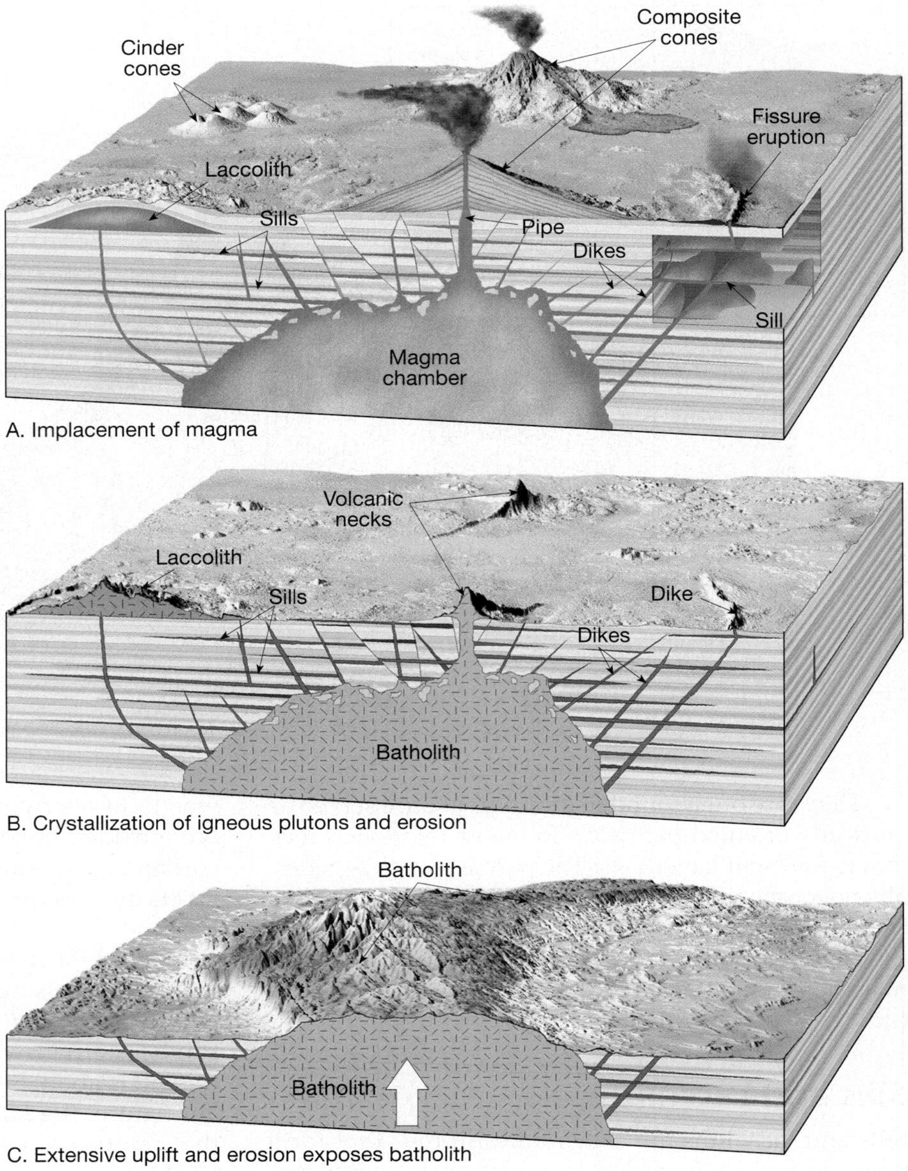

▶ **Figure 4.23** Illustrations showing basic igneous structures. **A.** This block diagram shows the relationship between volcanism and intrusive igneous activity. **B.** This view illustrates the basic intrusive igneous structures, some of which have been exposed by erosion long after their formation. **C.** After millions of years of uplifting and erosion, a batholith is exposed at the surface.

Laccolith exposed in the Henry Mountains of southern Utah. (Michael Collier)

Most large laccoliths are probably not much wider than a few kilometers. The Henry Mountains in southeastern Utah are largely composed of several laccoliths believed to have been fed by a much larger magma body emplaced nearby.

Batholiths

By far the largest intrusive igneous bodies are **batholiths.** Most often, batholiths occur as linear structures several hundreds of kilometers long and up to 100 kilometers wide, as shown in Figure 4.26. The Idaho batholith, for example, encompasses an area of more than 40,000 square kilometers and consists of many plutons. Indirect evidence gathered from gravitational

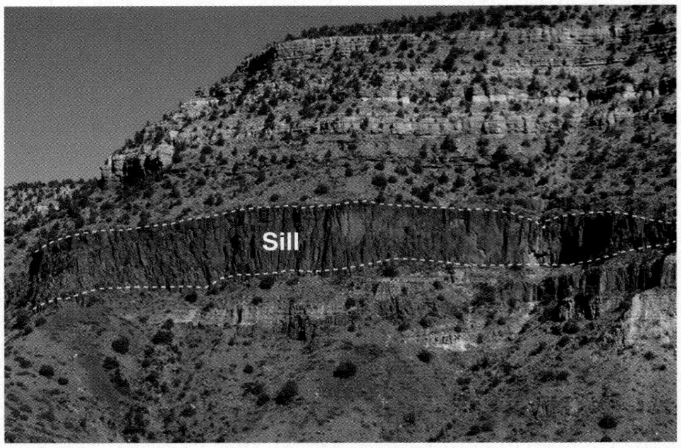

▲ **Figure 4.24** Salt River Canyon, Arizona. The dark, essentially horizontal band is a sill of basaltic composition that intruded horizontal layers of sedimentary rock. *(Photo by E. J. Tarbuck)*

◀ **Figure 4.25** Columnar jointing in basalt, Giants Causeway National Park, Northern Ireland. These five- to seven-sided columns are produced by contraction and fracturing that occurs as a lava flow or sill gradually cools. *(Photo by Tom Till)*

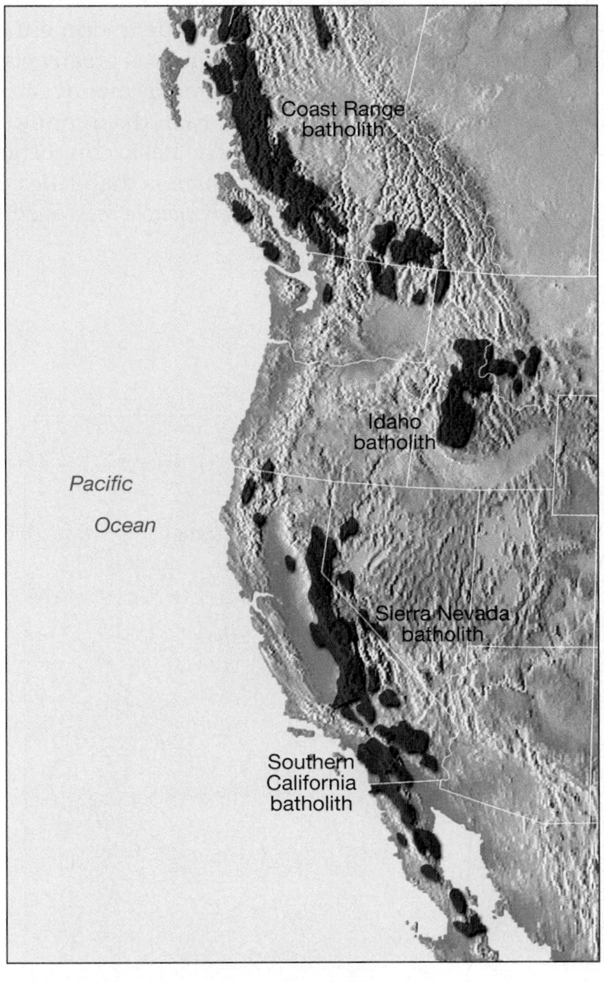

▲ **Figure 4.26** Granitic batholiths that occur along the western margin of North America. These gigantic, elongated bodies consist of numerous plutons that were emplaced during the last 150 million years of Earth history.

studies indicates that batholiths are also very thick, possibly extending dozens of kilometers into the crust.

Batholiths usually consist of rock types having chemical compositions toward the granitic end of the spectrum, although diorite is commonly found. Smaller batholiths can be rather simple structures composed almost entirely of one rock type. However, studies of large batholiths have shown that they consist of a large number of distinct plutons that were intruded over a period of millions of years. The plutonic activity that created the Sierra Nevada batholith, for example, occurred nearly continuously over a 130-million-year span that ended about 80 million years ago during the Cretaceous period.

A portion of the Sierra Nevada batholith. (G. Rowell/CORBIS)

Batholiths may compose the core of mountain systems. Here, uplifting and erosion have removed the surrounding rock, thereby exposing the resistant igneous body. Some of the highest peaks in the Sierra Nevada, such as Mount Whitney, are carved from such a granitic mass.

Large expanses of granitic rock also occur in the stable interiors of the continents, such as the Canadian Shield of North America. These relatively flat exposures are the remains of ancient mountains that have long since been leveled by erosion. Thus, the rocks that make up the batholiths of youthful mountain ranges, such as the Sierra Nevada, were generated near the top of a magma chamber, whereas in shield areas, the roots of former mountains and, thus, the lower portions of batholiths, are exposed.

Did You Know?

In the eastern United States some exposed granitic intrusions have dome-shaped, nearly treeless summits, hence the name "summit balds." Examples include Cadillac Mountain in Maine; Chocurua in New Hampshire; Black Mountain in Vermont; and Stone Mountain, Georgia.

Did You Know?

Although tsunami waves are most often associated with displacement along a fault located on the seafloor, some result from the collapse of a volcanic cone. This was dramatically demonstrated during the 1883 eruption on Krakatau, when the northern half of a volcano plunged into the Sunda Strait, creating a tsunami that exceeded 100 feet in height. Although the island of Krakatau was uninhabited, an estimated 36,000 people were killed along the coastline of the Indonesian islands of Java and Sumatra.

Plate Tectonics and Igneous Activity

Geologists have known for decades that the global distribution of volcanism is not random. Of the more than 800 active volcanoes* that have been identified, most are located along the margins of the ocean basins—most notably within the circum-Pacific belt known as the *Ring of Fire* (Figure 4.27). This group of volcanoes consists mainly of composite cones that emit volatile-rich magma having an intermediate (andesitic) composition that occasionally produces awe-inspiring eruptions.

The volcanoes comprising a second group emit very fluid basaltic lavas and are confined to the deep ocean basins, including well-known examples on Hawaii and Iceland. In addition, this group contains many active submarine volcanoes that dot the ocean floor; particularly notable are the innumerable small seamounts that lay along the axis of the oceanic ridge. At these depths the pressures are so great that seawater does not boil explosively, even in contact with hot lavas. Thus, first-hand knowledge of these eruptions is limited, coming mainly from deep-diving submersibles.

A third group includes those volcanic structures that are irregularly distributed in the interiors of the continents. None are found in Australia nor in the eastern two-thirds of North and South America. Africa is notable because it has many potentially active volcanoes including Mount Kilimanjaro, the highest point on the continent (5895 meters, 19,454 feet). Volcanism on continents is the most diverse, ranging from eruptions of very fluid basaltic lavas, like those that generated the Columbia River Plateau to explosive eruptions of silica-rich rhyolitic magma as occurred in Yellowstone.

Until the late 1960s, geologists had no explanation for the apparently haphazard distribution of continental volcanoes, nor were they able to account for the almost continuous chain of volcanoes that circles the margin of the Pacific basin. With the development of the theory of plate tectonics, the picture was greatly clarified. Recall that most primary (unaltered) magma originates in the upper mantle and that the mantle is essentially solid, *not molten* rock. The basic connection between plate tectonics and volcanism is that *plate motions provide the mechanisms by which mantle rocks melt to generate magma.*

*For our purposes, active volcanoes include those with eruptions that have been dated. At least 700 other cones exhibit geologic evidence that they have erupted within the past 10,000 years and are regarded as potentially active. Further, innumerable active submarine volcanoes are hidden from view in the depths of the ocean and are not counted in these numbers.

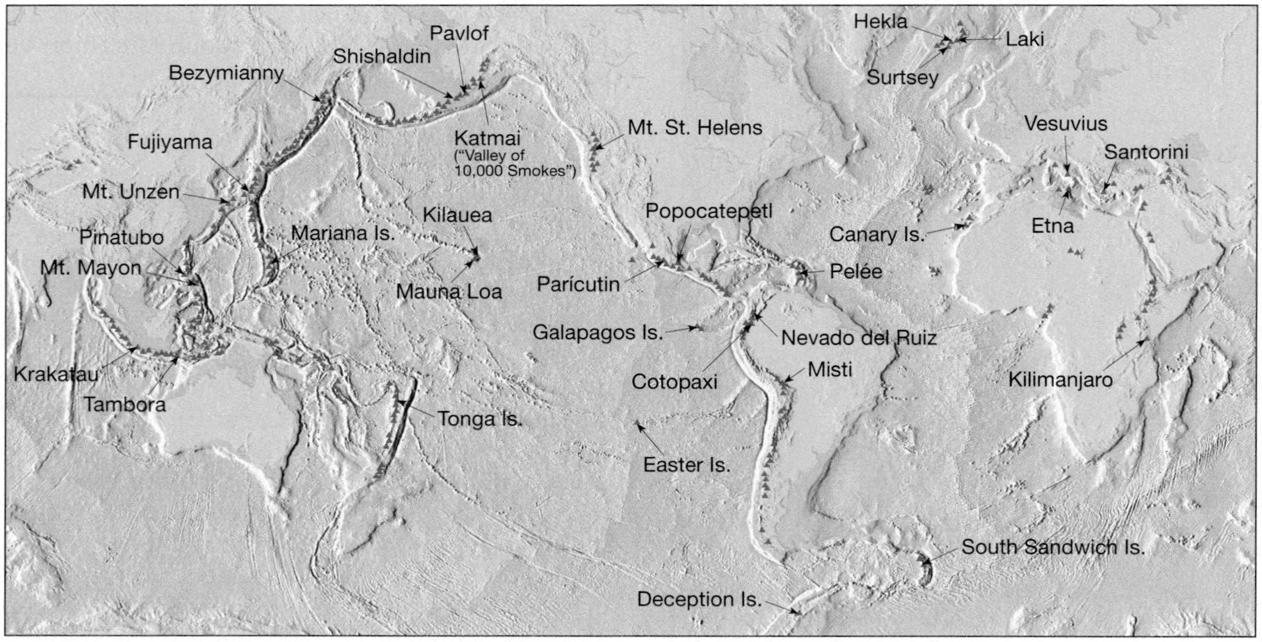

▲ **Figure 4.27** Locations of some of Earth's major volcanoes.

We will examine three zones of igneous activity and their relationship to plate boundaries. These active areas are located 1) along convergent plate boundaries where plates move toward each other and one sinks beneath the other; 2) along divergent plate boundaries, where plates move away from each other and new seafloor is created; and 3) areas within the plates proper that are not associated with any plate boundary. (It should be noted that volcanic activity rarely occurs along transform plate boundaries.) These three volcanic settings are depicted in Figure 4.28. (If you are unclear as to how magma is generated, we suggest that you study the section entitled "Origin of Magma" located at the end of Chapter 3 before proceeding.)

Igneous Activity at Convergent Plate Boundaries

Recall that at convergent plate boundaries, slabs of oceanic crust are bent as they descend into the mantle, generating an oceanic trench. As a slab sinks deeper into the mantle, the increase in temperature and pressure drives volatiles (mostly H_2O) from the oceanic crust. These mobile fluids migrate upward into the wedge-shaped piece of mantle located between the subducting slab and overriding plate (Figure 4.28). Once the sinking slab reaches a depth of about 100 to 150 kilometers, these water-rich fluids reduce the melting point of hot mantle rock sufficiently to trigger some melting. The partial melting of mantle rock (principally peridotite) generates magma with a basaltic composition. After a sufficient quantity of magma has accumulated, it slowly migrates upward.

Volcanism at a convergent plate margin results in the development of a linear or slightly curved chain of volcanoes called a *volcanic arc.* These volcanic chains develop roughly parallel to the associated trench—at distances of 200 to 300 kilometers (100 to 200 miles). Volcanic arcs can be constructed on oceanic, or continental, lithosphere. Those that develop within the ocean and grow large enough for their tops to rise above the surface are labeled *island archipelagos* in most atlases. Geologists prefer the more descriptive term **volcanic island arcs,** or simply **island arcs** (Figure 4.28). Several young volcanic island arcs of this type border the western Pacific basin, including the Aleutians, the Tongas, and the Marianas.

The early stage of island arc volcanism is typically dominated by the eruption of fluid basalts that build numerous shieldlike structures on the ocean floor. Because this activity begins at great depth, volcanic cones must extrude a great deal of lava before their tops rise above the sea to form islands. This cone-building activity, coupled with massive basaltic intrusions as well as magma that is added to the underside of the crust, tends to thicken the arc crust through time. As a result, mature volcanic arcs are underlain by a comparatively thick crust that impedes the upward flow of the mantle-derived basalts. This in turn provides time for magmatic differentiation to occur, in which heavy iron-rich minerals crystallize and settle out, leaving the melt enriched in silica (see Chapter 3). Consequently, as the arc matures, the magmas that reach the surface tend to erupt silica-rich andesites and even some rhyolites. In addition, magmatic differentiation tends to concentrate the available volatiles (water) into the more silica-rich components of these magmas. Because they emit viscous volatile-rich magma, island arc volcanoes typically have explosive eruptions.

Did You Know?

At 14,411 feet in altitude, Washington's Mount Rainier is the tallest of the 15 great volcanoes that make up the backbone of the Cascade Range. Although considered active, its summit is covered by more than 25 alpine glaciers.

Volcanism associated with convergent plate boundaries may also develop where slabs at oceanic lithosphere are subducted under continental lithosphere to produce a **continental volcanic arc** (Figure 4.28). The mechanisms that generate these mantle-derived magmas are essentially the same as those operating at island arcs. The major difference is that continental crust is much thicker and is composed of rocks having a higher silica content than oceanic crust. Hence through the assimilation of silica-rich crustal rocks, plus extensive magmatic differentiation, a mantle-derived magma may become highly evolved as it rises through continental crust. Stated another way, the primary magmas generated in the mantle may change from a comparatively dry, fluid basaltic magma to a viscous andesitic or rhyolitic magma having a high concentration of volatiles as it moves up through the continental crust. The volcanic chain of the Andes Mountains along the western edge of South America is perhaps the best example of a mature continental volcanic arc.

Since the Pacific basin is essentially bordered by convergent plate boundaries (and associated subduction zones), it is easy to see why the irregular belt of explosive volcanoes we call the Ring of Fire formed in this region. The volcanoes of the Cascade Range in the northwestern United States, including Mount Hood, Mount Rainier, and Mount Shasta, are included in this group (Box 4.2).

Igneous Activity at Divergent Plate Boundaries

The greatest volume of magma (perhaps 60 percent of Earth's total yearly output) is produced along the oceanic ridge system in association with seafloor spreading (Figure 4.28). Here, below the ridge axis where the lithospheric plates are being continually pulled apart, the solid yet mobile mantle responds to the decrease in overburden and rises upward to fill in the rift. Recall from Chapter 3 that as rock rises, it experiences a

Mount Augustine, Alaska (Stephen Kaufman)

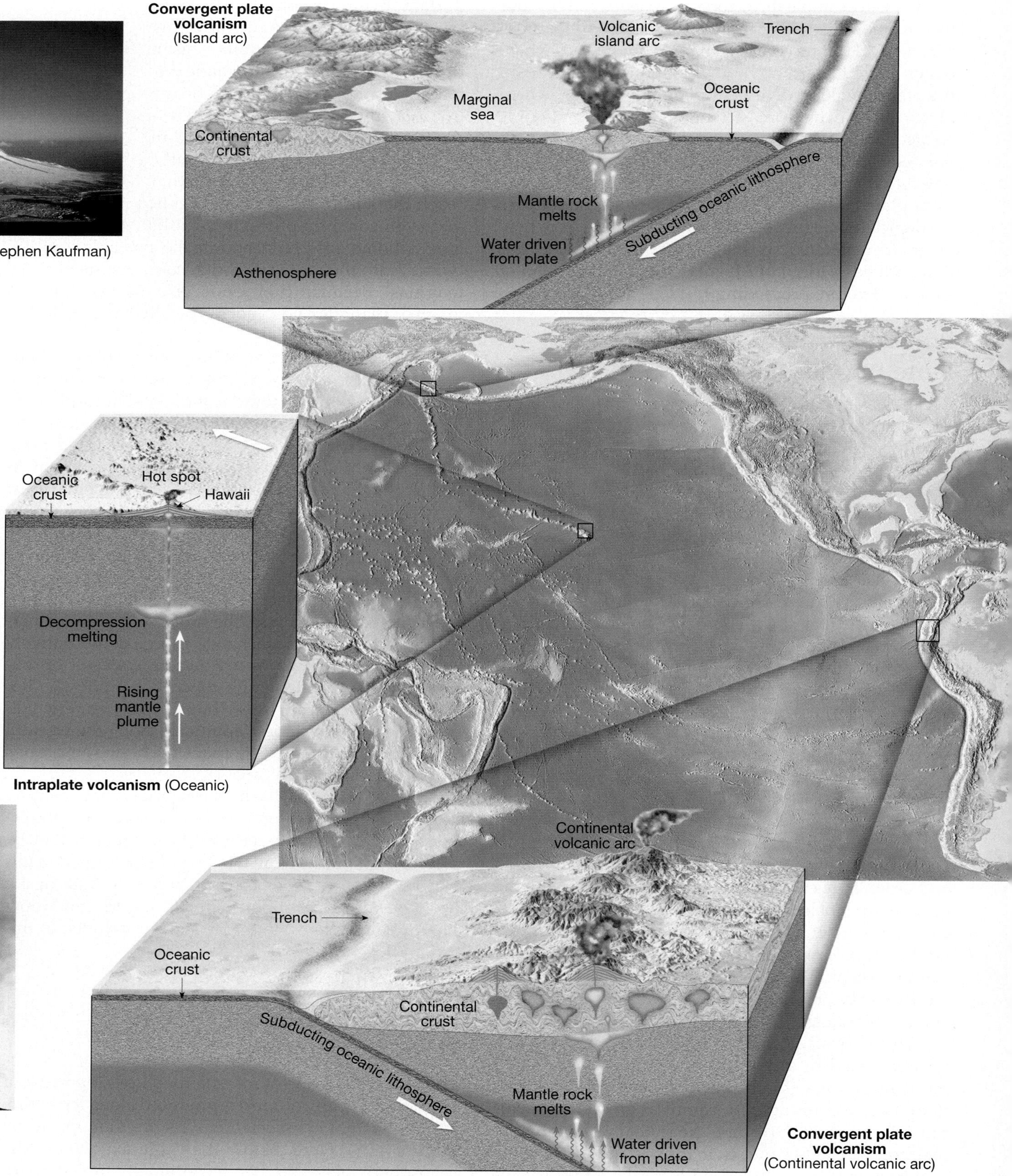

Kilauea, Hawaii (J. D. Griggs/USGS)

▲ **Figure 4.28** Three zones of volcanism. Two of these zones are plate boundaries, and the third is the interior area of the plates.

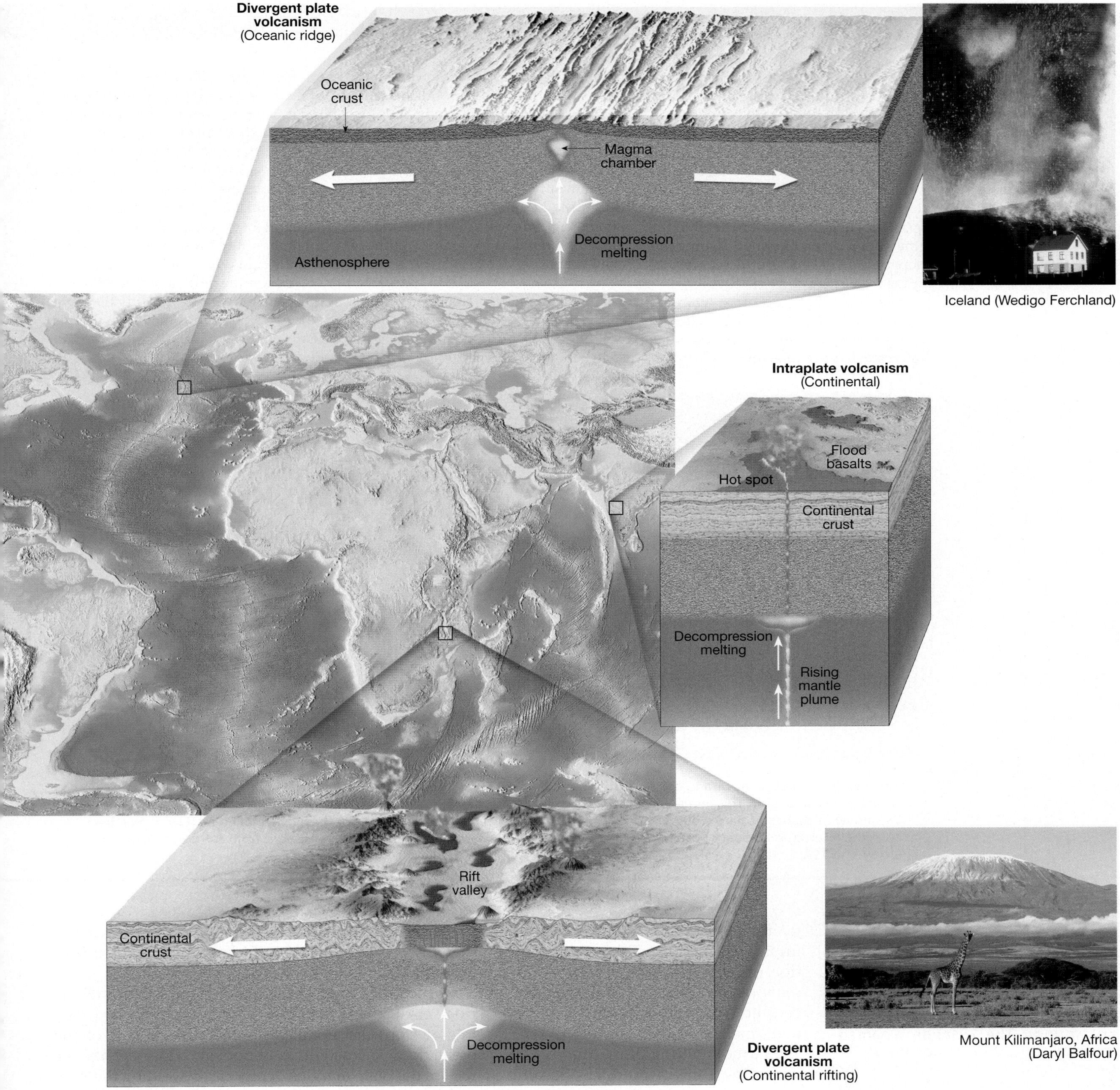
Divergent plate volcanism
(Oceanic ridge)
Oceanic crust
Magma chamber
Decompression melting
Asthenosphere
Iceland (Wedigo Ferchland)
Intraplate volcanism
(Continental)
Flood basalts
Hot spot
Continental crust
Decompression melting
Rising mantle plume
Rift valley
Continental crust
Decompression melting
Divergent plate volcanism
(Continental rifting)
Mount Kilimanjaro, Africa
(Daryl Balfour)

(Figure 4.28, continued)

BOX 4.2

Can Volcanoes Change Earth's Climate?

One example of the interplay between different parts of the Earth system is the relationship between volcanic activity and changes in climate. We know that changes in the composition of the atmosphere can have a significant impact on climate. Moreover, we know that volcanic eruptions can emit large quantities of gases and particles into the atmosphere, thus altering its composition. So do volcanic eruptions actually influence Earth's climate?

The idea that explosive volcanic eruptions might alter Earth's climate was first proposed many years ago. It is still regarded as a plausible explanation for some aspects of climatic variability. Explosive eruptions emit huge quantities of gases and fine-grained debris into the atmosphere (Figure 4.B). The greatest eruptions are sufficiently powerful to inject material high into the stratosphere (an atmospheric layer that extends between the heights of about 10 and 50 kilometers), where it spreads around the globe and remains for many months or even years.

The Basic Premise. The basic premise is that this suspended volcanic material will filter out a portion of the incoming solar radiation, which in turn will drop temperatures in the lowest layer of the atmosphere (this layer, called the *troposphere,* extends from Earth's surface to a height of about 10 kilometers).

More than 200 years ago Benjamin Franklin used this idea to argue that material from the eruption of a large Icelandic volcano could have reflected sunlight back to space and therefore might have been responsible for the unusually cold winter of 1783–1784.

Perhaps the most notable cool period linked to a volcanic event is the "year without a summer" that followed the 1815 eruption of Mount Tambora in Indonesia. The eruption of Tambora is the

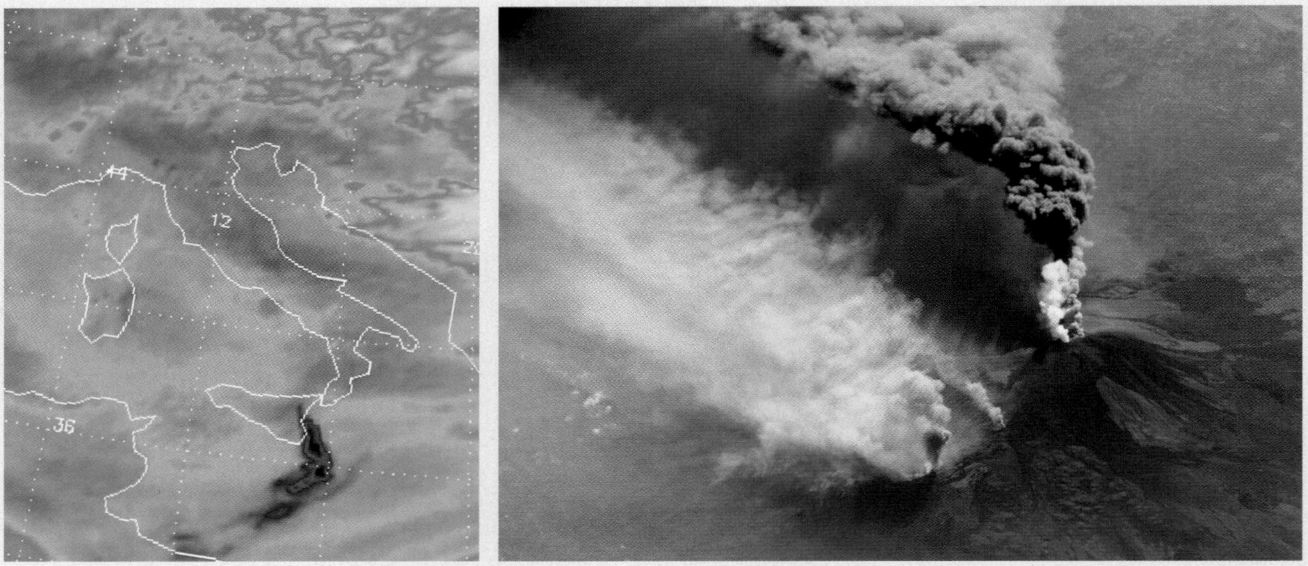

Figure 4.B Mount Etna, a volcano on the island of Sicily, erupting in late October, 2002. Mount Etna is Eurorpe's largest and most active volcano. Left: This image from the Atmospheric Infrared Sounder on NASA's Aqua satellite shows the sulfur dioxide (SO_2) plume in shades of purple and black. Right: This photo of Mount Etna looking southeast was taken by a crew member aboard the International Space Station. It shows a plume of volcanic ash streaming southeastward from the volcano. (Images courtesy of NASA)

Did You Know?

Iceland, which is the largest volcanic island in the world, has over 20 active volcanoes and numerous geysers and hot springs. The Icelanders named their finest gusher of boiling water *Geysir,* a name used to describe similar features around the world, including "Old Faithful Geyser" in Yellowstone National Park.

decrease in confining pressure and undergoes melting without the addition of heat. This process, called *decompression melting,* is the most common process by which mantle rocks melt.

Partial melting of mantle rock at spreading centers produces basaltic magma having a composition that is surprisingly similar to that generated along convergent plate boundaries. Because this newly formed basaltic magma is less dense than the mantle rock from which it was derived, it rises faster than the mantle.

Collecting in reservoirs located just beneath the ridge crest, about 10 percent of this magma eventually migrates upward along fissures to erupt as flows on the ocean floor. This activity continuously adds new basaltic rock to the plate margins, temporarily welding them together, only to break again as spreading continues. Along some ridges, outpourings of bulbous pillow lavas build numerous small seamounts. At other locations largest of modern times. During April 7–12, 1815, this nearly 4000-meter-high (13,000-foot) volcano violently expelled more than 100 cubic kilometers (24 cubic miles) of volcanic debris. The impact of the volcanic aerosols on climate is believed to have been widespread in the Northern Hemisphere. From May through September 1816 an unprecedented series of cold spells affected the northeastern United States and adjacent portions of Canada. There was heavy snow in June and frost in July and August. Abnormal cold was also experienced in much of Western Europe. Similar, although apparently less dramatic, effects were associated with other great explosive volcanoes, including Indonesia's Krakatau in 1883.

Three Modern Examples. Three major volcanic events have provided considerable data and insight regarding the impact of volcanoes on global temperatures. The eruptions of Washington State's Mount St. Helens in 1980, the Mexican volcano El Chichón in 1982, and the Philippines' Mount Pinatubo in 1991 have given scientists an opportunity to study the atmospheric effects of volcanic eruptions with the aid of more sophisticated technology than had been available in the past. Satellite images and remote-sensing instruments allowed scientists to monitor closely the effects of the clouds of gases and ash that these volcanoes emitted.

Mount St. Helens. When Mount St. Helens erupted, there was immediate speculation about the possible effects on our climate. Could such an eruption cause our climate to change? There is no doubt that the large quantity of volcanic ash emitted by the explosive eruption had significant local and regional effects for a short period. Still, studies indicated that any longer-term lowering of hemispheric temperatures was negligible. The cooling was so slight, probably less than 0.1°C (0.2°F), that it could not be distinguished from other natural temperature fluctuations.

El Chichón. Two years of monitoring and studies following the 1982 El Chichón eruption indicated that its cooling effect on global mean temperature was greater than that of Mount St. Helens, on the order of 0.3 to 0.5°C (0.5 to 0.9°F). The eruption of El Chichón was *less explosive* than the Mount St. Helens blast, so why did it have a greater impact on global temperatures? The reason is that the material emitted by Mount St. Helens was largely fine ash that settled out in a relatively short time. El Chichón, on the other hand, emitted far greater quantities of sulfur dioxide gas (an estimated 40 times more) than Mount St. Helens. This gas combines with water vapor in the stratosphere to produce a dense cloud of tiny sulfuric-acid particles. The particles, called *aerosols,* take several years to settle out completely. They lower the troposphere's mean temperature because they reflect solar radiation back to space.

We now understand that volcanic clouds that remain in the stratosphere for a year or more are composed largely of sulfuric-acid droplets and not of dust, as was once thought. Thus, the volume of fine debris emitted during an explosive event is not an accurate criterion for predicting the global atmosphere effects of an eruption.

Mount Pinatubo. The Philippines volcano, Mount Pinatubo, erupted explosively in June 1991, injecting 25 million to 30 million tons of sulfur dioxide into the stratosphere. The event provided scientists with an opportunity to study the climatic impact of a major explosive volcanic eruption using NASA's spaceborne Earth Radiation Budget Experiment. During the next year the haze of tiny aerosols increased the percentage of light reflected by the atmosphere and thus lowered global temperatures by 0.5°C (0.9°F).

It may be true that the impact on global temperature of eruptions like El Chichón and Mount Pinatubo is relatively minor, but many scientists agree that the cooling produced could alter the general pattern of atmospheric circulation for a limited period. Such a change, in turn, could influence the weather in some regions. Predicting or even identifying specific regional effects still presents a considerable challenge to atmospheric scientists.

The preceding examples illustrate that the impact on climate of a single volcanic eruption, no matter how great, is relatively small and short-lived. Therefore, if volcanism is to have a pronounced impact over an extended period, many great eruptions, closely spaced in time, need to occur. If this happens, the stratosphere could become loaded with enough sulfur dioxide and volcanic dust to seriously diminish the amount of solar radiation reaching the surface.

erupted lavas produce fluid flows that create more subdued topography.

Although most spreading centers are located along the axis of an oceanic ridge, some are not. In particular, the East African Rift is a site where continental lithosphere is being pulled apart, forming a continental rift (See Figure 4.28, p. 107, bottom right.). Here, magma is generated by decompression melting in the same manner it is produced along the oceanic ridge system. Vast outpourings of fluid basaltic lavas are common in this region. The East African Rift zone also contains some large composite cones, as exemplified by Mount Kilimanjaro. Like composite cones that form along convergent plate boundaries, these volcanoes form when mantle-derived basalts evolve into volatile-rich andesitic magma as they migrate up through thick silica-rich continental rocks.

Intraplate Igneous Activity

We know why igneous activity is initiated along plate boundaries, but why do eruptions occur in the interiors of

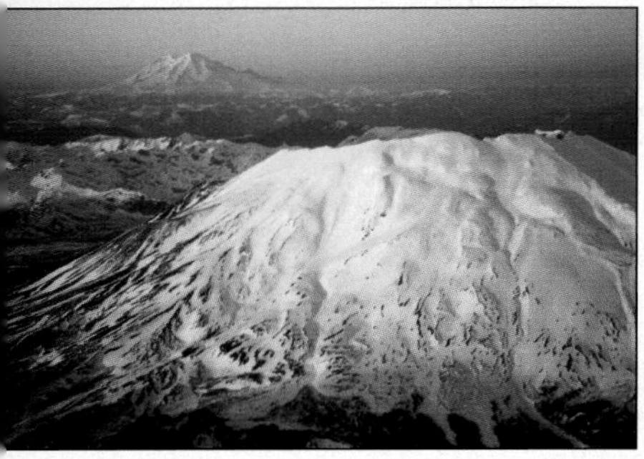

Mount St. Helens (foreground) and Mount Rainier are part of the Cascade Range. (M. Miller)

plates? Hawaii's Kilauea is considered the world's most active volcano, yet it is situated thousands of kilometers from the nearest plate boundary in the middle of the vast Pacific plate (See Figure 4.28, p. 106, left center). Other sites of **intraplate volcanism** (meaning "within the plate") include the Canary Islands, Yellowstone, and several volcanic centers that you may be surprised to learn are located in the Sahara Desert of northern Africa.

We now recognize that most intraplate volcanism occurs where a mass of hotter than normal mantle material called a **mantle plume** ascends toward the surface (Figure 4.29). Although the depth at which (at least some) mantle plumes originate is still hotly debated, many appear to form deep within Earth at the core-mantle boundary. These plumes of solid yet mobile mantle rock rise toward the surface in a manner similar to the blobs that form within a lava lamp. (These are the trendy lamps that contain two immiscible liquids in a glass container. As the base of the lamp is heated, the denser liquid at the bottom becomes buoyant and forms blobs that rise to the top.) Like the blobs in a lava lamp, a mantle plume has a bulbous head that draws out a narrow stalk beneath it as it rises. Once the plume head nears the top of the mantle, decompression melting generates basaltic magma that may eventually trigger volcanism at the surface. The result is a localized volcanic region a few hundred kilometers across called a **hot spot** (Figure 4.29). More than 100 hot spots have been identified, and most have persisted for millions of years. The land surface around hot spots is often elevated, showing that it is buoyed up by a plume of warm low-density material. Furthermore, by measuring the heat flow in these regions, geologists have determined that the mantle beneath hot spots must be 100–150°C hotter than normal.

The volcanic activity on the island of Hawaii, with its outpourings of basaltic lava, is certainly the result of hot-spot volcanism (Figure 4.28). Where a mantle plume has persisted for long periods of time, a chain of volcanic structures may form as the overlying plate moves over it. In the Hawaiian Islands, hot-spot activity is currently centered on Kilauea. However, over the past 80 million years the same mantle plume generated a chain of volcanic islands (and seamounts) that extend thousands of kilometers from the Big Island in a northwestward direction across the Pacific.

Mantle plumes are also thought to be responsible for the vast outpourings of basaltic lava that create large basalt plateaus such as the Columbia Plateau in the northwestern United States, India's Deccan Plateau, and the Ontong Java Plateau in the western Pacific. The most widely accepted explanation for these eruptions, which emit extremely large volumes of basaltic magma over relatively short time intervals, involves a plume with a substantial-sized head. These large structures may have heads that are hundreds of kilometers in diameter connected to a long, narrow tail rising from the core-mantle boundary (Figure 4.29). Upon reaching the base of the lithosphere, the temperature of the material in the plume is estimated to be 200–300°C warmer than the surrounding rock. Thus, as much as 10 to 20 percent of the mantle material making up the plume head rapidly melts. It is this melting that triggers the burst of volcanism that emits voluminous outpourings of lava to form a huge basalt plateau in a matter of a million or so years (Figure 4.29). The comparatively short initial eruptive phase is followed by tens of millions of years of less voluminous activity, as the plume tail slowly rises to the surface. Thus, extending away from most large flood basalt provinces is a chain of volcanic structures, similar to the Hawaiian chain, that terminates over an active hot spot marking the current position of the tail of the plume.

Based on the current state of knowledge, it appears that hot-spot volcanism, with its associated mantle plumes, is responsible for most intraplate volcanism. However, there are some widely scattered volcanic regions located far from any plate boundary that are not

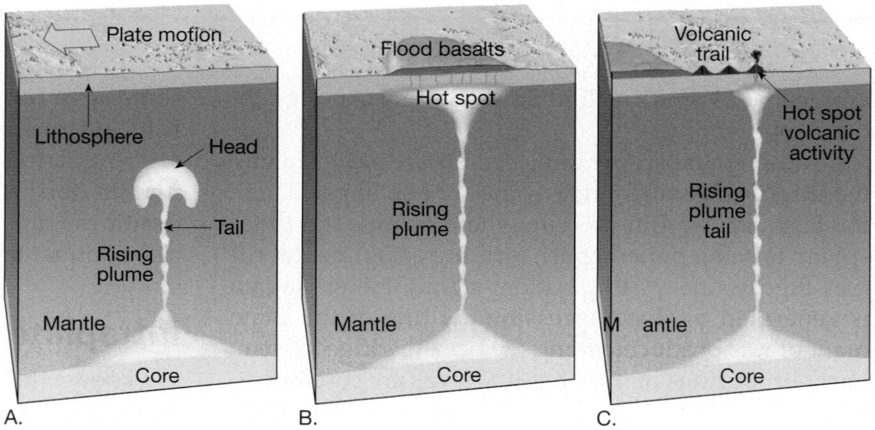

▶ **Figure 4.29** Model of a mantle plume and associated hot-spot volcanism. **A.** A rising mantle plume with large bulbous head and narrow tail. **B.** Rapid decompression melting of the head of a mantle plume produces vast outpourings of basalt. **C.** Less voluminous activity caused by the plume tail produces a linear volcanic chain on the seafloor.

linked to hot-spot volcanism. Well-known examples are found in the Basin and Range province of the western United States and northwestern Mexico. We will consider the cause of volcanism in this region in Chapter 17. A few other volcanic regions defy explanation. Thus, the natural world still has some secrets that remain to be explained by future generations of Earth scientists.

The Chapter in Review

- The primary factors that determine the nature of volcanic eruptions include the magma's *composition*, its *temperature*, and the *amount of dissolved gases* it contains. As lava cools, it begins to congeal, and as *viscosity* increases, its mobility decreases. The *viscosity of magma is directly related to its silica content*. Rhyolitic (felsic) lava, with its high silica content (about 70 percent), is very viscous and forms short, thick flows. Basaltic (mafic) lava, with a lower silica content (about 50 percent), is more fluid and may travel a long distance before congealing. Dissolved gases tend to increase the fluidity of magma and, as they expand, provide the force that propels molten rock from the vent of a volcano.
- The materials associated with a volcanic eruption include (1) *lava flows* (*pahoehoe* flows, which resemble twisted braids; and *aa* flows, consisting of rough, jagged blocks; both form from basaltic lavas); (2) *gases* (primarily *water vapor*); and (3) *pyroclastic material* (pulverized rock and lava fragments blown from the volcano's vent, which include *ash, pumice, lapilli, cinders, blocks,* and *bombs*.)
- Successive eruptions of lava from a central vent result in a mountainous accumulation of material known as a *volcano*. Located at the summit of many volcanoes is a steep-walled depression called a *crater*. *Shield cones* are broad, slightly domed volcanoes built primarily of fluid, basaltic lava. *Cinder cones* have steep slopes composed of pyroclastic material. *Composite cones*, or *stratovolcanoes*, are large, nearly symmetrical structures built of interbedded lavas and pyroclastic deposits. Composite cones produce some of the most violent volcanic activity. Often associated with a violent eruption is a *nuée ardente*, a fiery cloud of hot gases infused with incandescent ash that races down steep volcanic slopes. Large composite cones may also generate a type of mudflow known as a *lahar*.
- Most volcanoes are fed by *conduits* or *pipes*. As erosion progresses, the rock occupying the pipe is often more resistant and may remain standing above the surrounding terrain as a *volcanic neck*. The summits of some volcanoes have large, nearly circular depressions called *calderas* that result from collapse following an explosive eruption. Calderas also form on shield volcanos by subterranean drainage from a central magma chamber, and the largest calderas form by the discharge of colossal volumes of silica-rich pumice along ring fractures. Although volcanic eruptions from a central vent are the most familiar, by far the largest amounts of volcanic material are extruded from cracks in the crust called *fissures*. The term *flood basalts* describes the fluid, waterlike, basaltic lava flows that cover an extensive region in the northwestern United States known as the Columbia Plateau. When silica-rich magma is extruded, *pyroclastic flows* consisting largely of ash and pumice fragments usually result.
- Intrusive igneous bodies are classified according to their *shape* and by their *orientation with respect to the host rock*, generally sedimentary rock. The two general shapes are *tabular* (sheetlike) and *massive*. Intrusive igneous bodies that cut across existing sedimentary beds are said to be *discordant*; those that form parallel to existing sedimentary beds are *concordant*.
- *Dikes* are tabular, discordant igneous bodies produced when magma is injected into fractures that cut across rock layers. Tabular, concordant bodies, called *sills*, form when magma is injected along the bedding surfaces of sedimentary rocks. In many respects, sills closely resemble buried lava flows. *Laccoliths* are similar to sills but form from less fluid magmas that collect as lens-shaped masses that arch the overlying strata upward. *Batholiths*, the largest intrusive igneous bodies, frequently make up the cores of mountains, as exemplified by the Sierra Nevada.
- *Most active volcanoes are associated with plate boundaries.* Active areas of volcanism are found along oceanic ridges where seafloor spreading is occurring (*divergent plate boundaries*), in the vicinity of ocean trenches where one plate is being subducted beneath another (*convergent plate boundaries*), and in the interiors of plates themselves (intraplate volcanism). Rising plumes of hot mantle rock are the source of most intraplate volcanism.

Key Terms

aa flow (p. 87)
batholith (p. 102)
block lavas (p. 87)
caldera (p. 97)
cinder cone (p. 91)
columnar joints (p. 101)
composite cone (p. 92)
concordant (p. 100)
conduit (p. 89)
continental volcanic arc (p. 105)
crater (p. 89)
dike (p. 100)

discordant (p. 100)
eruption column (p. 84)
fissure (p. 98)
fissure eruption (p. 98)
flood basalt (p. 99)
fumarole (p. 90)
hot spot (p. 110)
intraplate volcanism (p. 110)
island arc (p. 105)
laccolith (p. 101)
lahar (p. 96)
lava tube (p. 87)
mantle plume (p. 110)
massive (p. 100)
nuée ardente (p. 95)
pahoehoe flow (p. 77)
parasitic cone (p. 90)
pillow lava (p. 87)
pipe (p. 89)
pluton (p. 100)
pyroclastic flow (p. 95)
pyroclastic material (p. 88)
scoria cone (p. 91)
shield volcano (p. 90)
sill (p. 101)
stratovolcano (p. 92)
tabular (p. 100)
vent (p. 89)
viscosity (p. 82)
volcanic island arc (p. 105)
volcanic neck (p. 100)
volcano (p. 89)
volatiles (p. 84)

Questions for Review

1. What event triggered the May 18, 1980, eruption of Mount St. Helens? (See Box 4.1.)
2. List three factors that determine the nature of a volcanic eruption. What role does each play?
3. Why is a volcano that is fed by highly viscous magma likely to be a greater threat than a volcano supplied with very fluid magma?
4. Describe pahoehoe and aa lava.
5. List the main gases released during a volcanic eruption. Why are gases important in eruptions?
6. How do volcanic bombs differ from blocks of pyroclastic debris?
7. Compare a volcanic crater to a caldera.
8. Compare and contrast the three main types of volcanoes (size, composition, shape, and eruptive style).
9. Name a prominent volcano for each of the three types.
10. Briefly compare the eruptions of Kilauea and Parícutin.
11. Contrast the destruction of Pompeii with the destruction of St. Pierre (time frame, volcanic material, and nature of destruction).
12. Describe the formation of Crater Lake. Compare it to the caldera found on shield volcanoes such as Kilauea.
13. What are the largest volcanic structures on Earth?
14. What is Shiprock, New Mexico, and how did it form?
15. How do the eruptions that created the Columbia Plateau differ from eruptions that create volcanic peaks?
16. Extensive pyroclastic flow deposits are most often associated with which volcanic structures?
17. Describe each of the four intrusive features discussed in the text (dike, sill, laccolith, and batholith).
18. Why might a laccolith be detected at Earth's surface before being exposed by erosion?
19. What is the largest of all intrusive igneous bodies? Is it tabular or massive? Concordant or discordant?
20. Volcanism at divergent plate boundaries is associated with which rock type? What causes rocks to melt in these regions?
21. What is the Ring of Fire?
22. What type of plate boundary is associated with the Ring of Fire?
23. Are volcanoes in the Ring of Fire generally described as quiescent or violent? Name a volcano that would support your answer.
24. Describe the process that generates magma along convergent plate boundaries.
25. What is the source of magma for intraplate volcanism?
26. What is meant by hot-spot volcanism?
27. How do geologists identify hot spots other than volcanism?
28. The Hawaiian Islands and Yellowstone are associated with which of the three zones of volcanism? Cascade Range? Flood basalt provinces?

Online Study Guide

The *Essentials of Geology* Website uses the resources and flexibility of the Internet to aid in your study of the topics in this chapter. Written and developed by geology instructors, this site will help improve your understanding of geology. Visit **www.prenhall.com/lutgens** and click on the cover of *Essentials of Geology 9e* to find:

- Online review quizzes.
- Critical thinking exercises.
- Links to chapter-specific Web resources.
- Internet-wide key-term searches.

http://www.prenhall.com/lutgens

Differential weathering is responsible for much of the spectacular scenery in Utah's Canyonland's National Park.
(Photo by Jack Dykinga Photography)

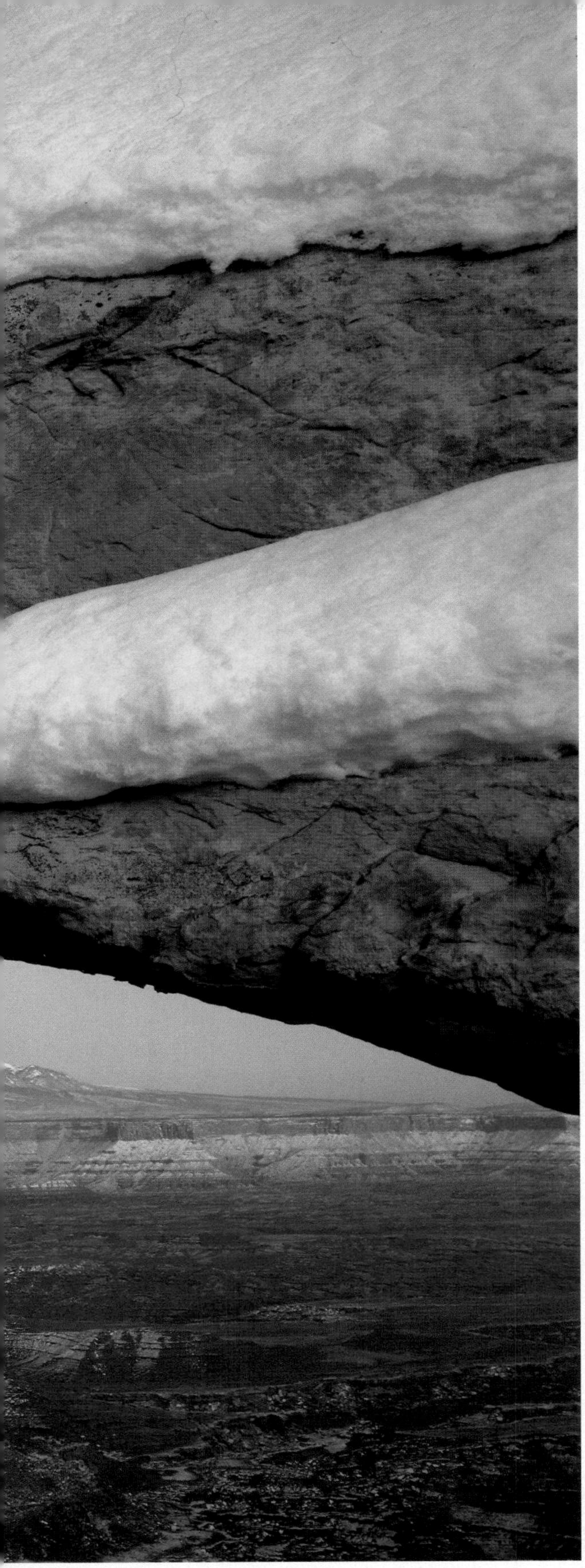

CHAPTER 5

Weathering and Soils

Focus on Learning

To assist you in learning the important concepts in this chapter, you will find it helpful to focus on the following questions:

- What are Earth's external processes, and what roles do they play in the rock cycle?
- What are the two main categories of weathering? In what ways are they different?
- What factors determine the rate at which rock weathers?
- What is soil? What are the factors that control soil formation?
- What factors influence natural rates of soil erosion? What impact have humans had?
- How is weathering related to the formation of ore deposits?

Earth's surface is constantly changing. Rock is disintegrated and decomposed, moved to lower elevations by gravity, and carried away by water, wind, or ice. In this manner Earth's physical landscape is sculptured. This chapter focuses on the first step of this never-ending process—weathering. What causes solid rock to crumble, and why does the type and rate of weathering vary from place to place? Soil, an important product of the weathering process and a vital resource, is also examined.

Earth's External Processes

GEODe: Weathering and Soil
▼ External versus Internal Processes

New Hampshire's famous Old Man of the Mountain, as it appeared before May 3, 2003. (AP Wide World Photo)

The famous granite outcrop after it collapsed. The natural processes that created it ultimately destroyed it. (AP Wide World Photo)

Weathering, mass wasting, and erosion are called **external processes** because they occur at or near Earth's surface and are powered by energy from the Sun. External processes are a basic part of the rock cycle because they are responsible for transforming solid rock into sediment.

To the casual observer, the face of Earth may appear to be without change, unaffected by time. In fact, 200 years ago, most people believed that mountains, lakes, and deserts were permanent features of an Earth that was thought to be no more than a few thousand years old. Today, we know that Earth is 4.5 billion years old and that mountains eventually succumb to weathering and erosion, lakes fill with sediment or are drained by streams, and deserts come and go with changes in climate.

Earth is a dynamic body. Some parts of Earth's surface are gradually elevated by mountain building and volcanic activity. These **internal processes** derive their energy from Earth's interior. Meanwhile, opposing external processes are continually breaking rock apart and moving the debris to lower elevations. The latter processes include:

1. **Weathering**—the physical breakdown (disintegration) and chemical alteration (decomposition) of rocks at or near Earth's surface.
2. **Mass wasting**—the transfer of rock and soil downslope under the influence of gravity.
3. **Erosion**—the physical removal of material by mobile agents such as water, wind, or ice.

In this chapter we will focus on rock weathering and the products generated by this activity (Figure 5.1). However, weathering cannot be easily separated from mass wasting and erosion, because as weathering breaks rocks apart, mass wasting and erosion remove the rock debris. This transport of material by mass wasting and erosion further disintegrates and decomposes the rock.

Weathering

GEODe: Weathering and Soil
▼ Types of Weathering

Weathering goes on all around us, but it seems like such a slow and subtle process that it is easy to underestimate its importance. However, it is worth remembering that weathering is a basic part of the rock cycle and thus a key process in the Earth system.

All materials are susceptible to weathering. Consider, for example, the fabricated product concrete, which closely resembles the sedimentary rock called conglomerate. A newly poured concrete sidewalk has a smooth, fresh, unweathered look. However, not many years later, the same sidewalk will appear chipped, cracked, and rough, with pebbles exposed at the surface. If a tree is nearby, its roots may heave and buckle the concrete as well. The same natural processes that eventually break apart a concrete sidewalk also act to disintegrate rock.

Weathering occurs when rock is mechanically fragmented (disintegrated) and/or chemically altered (decomposed). **Mechanical weathering** is accomplished by physical forces that break rock into smaller and smaller pieces without changing the rock's mineral composition. **Chemical weathering** involves a chemical transformation of rock into one or more new compounds. These two concepts can be illustrated with a piece of paper. The paper can be disintegrated by tearing it into smaller and smaller pieces, whereas decomposition occurs when the paper is set afire and burned.

Why does rock weather? Simply, weathering is the response of Earth materials to a changing environment. For instance, after millions of years of uplift and erosion, the rocks overlying a large intrusive igneous body may be removed, exposing it at the surface. The mass of crystalline rock, which formed deep below ground where temperatures and pressures are much greater than at the surface, is now subjected to a very different and comparatively hostile surface environment. In response, this rock mass will gradually change. This transformation of rock is what we call *weathering*.

In the following sections, we will discuss the various modes of mechanical and chemical weathering. Although we will consider these two processes separately, keep in mind that they usually work simultaneously in nature.

Mechanical Weathering

GEODe: Weathering and Soil
▼ Mechanical Weathering

When a rock undergoes *mechanical weathering,* it is broken into smaller and smaller pieces, each retaining the characteristics of the original material. The end result

▲ **Figure 5.1** Arizona's Monument Valley. When weathering accentuates differences in rocks, spectacular landforms are sometimes created. As the rock gradually disintegrates and decomposes, mass wasting and erosion remove the products of weathering. *(Photo by Michael Collier)*

is many small pieces from a single large one. Figure 5.2 shows that breaking a rock into smaller pieces increases the surface area available for chemical attack. Hence, by breaking rocks into smaller pieces, mechanical weathering increases the amount of surface area available for chemical weathering.

In nature, three physical processes are especially important in breaking rocks into smaller fragments: frost wedging, expansion resulting from unloading, and biological activity. In addition, although the work of erosional agents such as wind, glacial ice, rivers, and waves is usually considered separately from mechanical weathering, it is nevertheless important to point out that as these mobile agents move rock debris, they relentlessly disintegrate these materials.

Frost Wedging

Repeated cycles of freezing and thawing represent an important process of mechanical weathering. Liquid water has the unique property of expanding about 9 percent when it freezes. This increase in volume occurs because, as ice forms, the water molecules arrange themselves into a very open crystalline structure. As a result, when water freezes, it expands and exerts a tremendous outward force. This can be verified by filling a container with water and freezing it. The formation of ice will rupture the container.

In nature, water works its way into cracks in rock and, upon freezing, expands and enlarges these openings. After many freeze-thaw cycles, the rock is broken into angular fragments. This process is appropriately called **frost wedging** (Figure 5.3). Frost wedging is most pronounced in mountainous regions where a daily freeze-thaw cycle often exists. Here, sections of rock are wedged loose and may tumble into large piles called **talus slopes** that often form at the base of steep rock outcrops (Figure 5.3).

Frost wedging also causes great destruction to highways in the northern United States and Canada, particularly in the early spring when the freeze-thaw cycle is well established. Roadways acquire numerous potholes and are occasionally heaved and buckled by this destructive force.

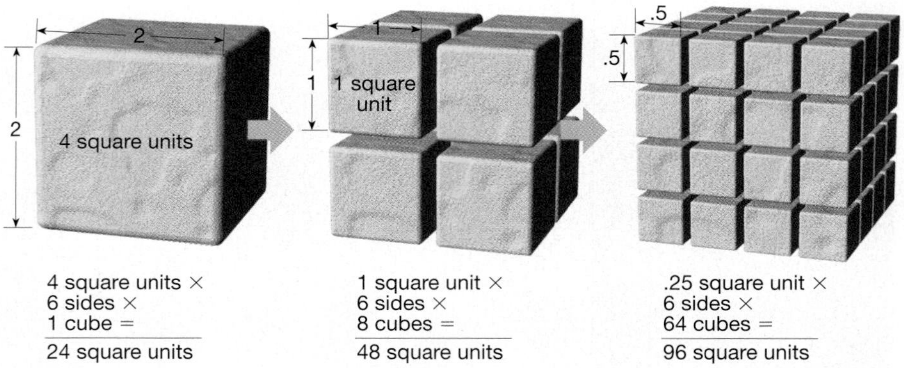

▶ **Figure 5.2** Chemical weathering can occur only to those portions of a rock that are exposed to the elements. Mechanical weathering breaks rock into smaller and smaller pieces, thereby increasing the surface area available for chemical attack.

Unloading

When large masses of igneous rock, particularly granite, are exposed by erosion, concentric slabs begin to break loose. The process generating these onionlike layers is called **sheeting.** It is thought that this occurs, at least in part, because of the great reduction in pressure when the overlying rock is eroded away, a process called *unloading*. Accompanying this unloading, the outer layers expand more than the rock below and thus separate from the rock body (Figure 5.4). Continued weathering eventually causes the slabs to separate and spall off, creating **exfoliation domes.** Excellent examples of exfoliation domes include Stone Mountain, Georgia, and Half Dome (Figure 5.4C) and Liberty Cap in Yosemite National Park.

Deep underground mining provides us with another example of how rocks behave once the confining pressure is removed. Large rock slabs sometimes explode off the walls of newly cut mine tunnels because of the abruptly reduced pressure. Evidence of this type, plus the fact that fracturing occurs parallel to the floor of a quarry when large blocks of rock are removed,

▲ **Figure 5.3** Frost wedging. As water freezes, it expands, exerting a force great enough to break rock. When frost wedging occurs in a setting such as this, the broken rock fragments fall to the base of the cliff and create a cone-shaped accumulation known as talus. *(Photo by Tom & Susan Bean, Inc.)*

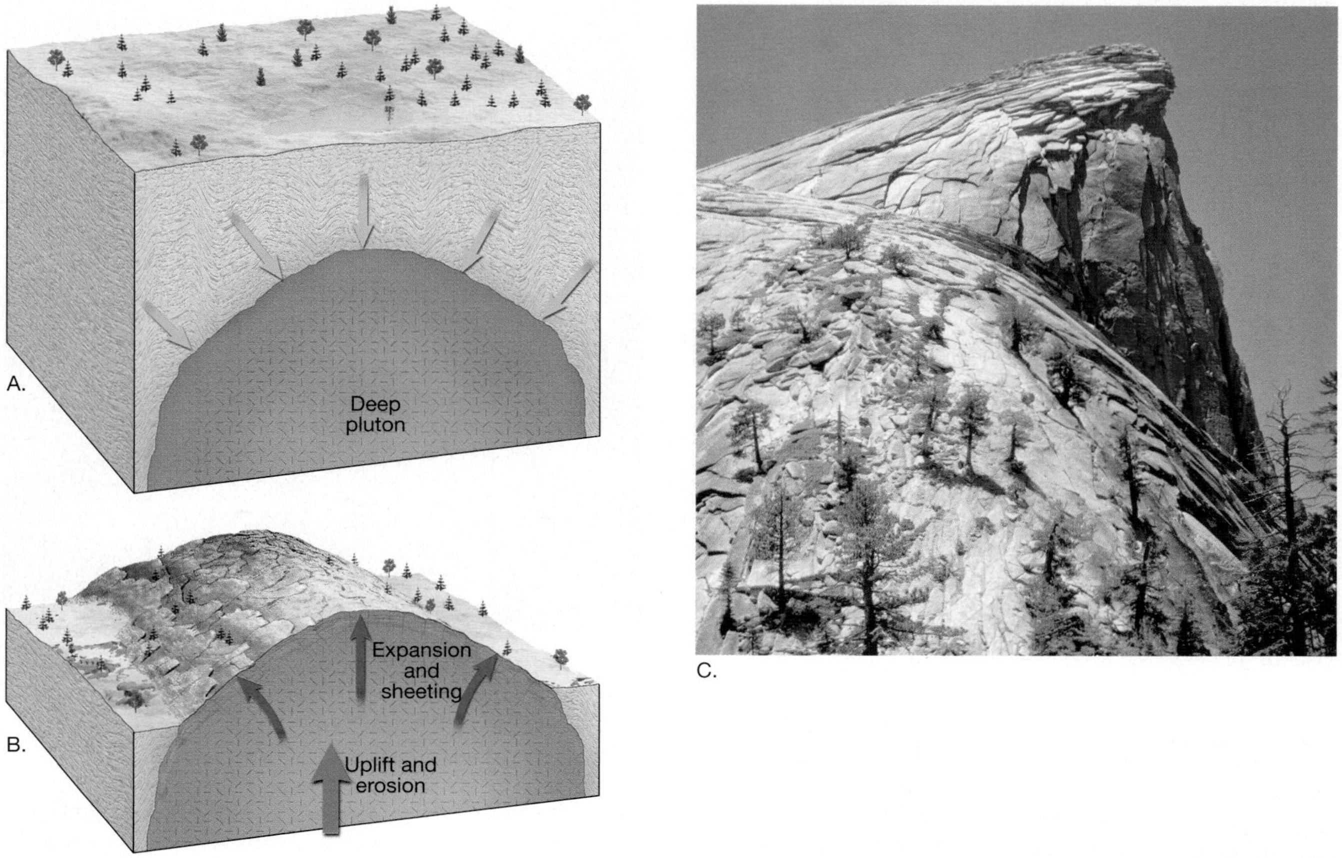

▲ **Figure 5.4** Sheeting is caused by the expansion of crystalline rock as erosion removes the overlying material. When the deeply buried pluton in **A** is exposed at the surface following uplift and erosion in **B,** the igneous mass fractures into thin slabs. The photo in **C** is of the summit of Half Dome in Yosemite National Park, California. It is an exfoliation dome and illustrates the onionlike layers created by sheeting. *(Photo by Breck P. Kent)*

strongly supports the process of unloading as the cause of sheeting.

Biological Activity

Both mechanical and chemical weathering are accomplished by the activities of organisms. Plant roots in search of minerals and water grow into fractures, and as the roots grow, they wedge the rock apart (Figure 5.5). Burrowing animals further break down the rock by moving fresh material to the surface, where physical and chemical processes can more effectively attack it. Of course, where rock has been blasted in search of minerals or for construction, the impact of humans is particularly noticeable.

There are numerous ways that organisms play a role in chemical weathering. For example, plant roots, fungi, and lichens that occupy fractures or may encrust a rock produce acids that promote decomposition. Moreover, some bacteria are capable of extracting compounds from minerals and using the energy from the compound's chemical bonds to supply their life needs. These primitive "mineral-eating" life forms can live at depths as great as a few kilometers.

Chemical Weathering

GEODe — ESSENTIALS OF GEOLOGY

Weathering and Soil
▼ Chemical Weathering

Chemical weathering involves the complex processes that break down rock components and internal structures of minerals. Such processes convert the constituents to new minerals or release them to the surrounding environment. During this transformation, the original rock decomposes into substances that are stable in the surface environment. Consequently, the products of chemical weathering will remain essentially unchanged as long as they remain in an environment similar to the one in which they formed.

Did You Know?

Ions dissolved in underground water can precipitate out of solution to form salts. As the salt crystals grow larger in the open spaces of rock, they can exert a force strong enough to enlarge cracks or push apart surrounding grains.

Water and Carbonic Acid

Water is by far the most important agent of chemical weathering. Although pure water is nonreactive, a small

▲ **Figure 5.5** Root wedging widens fractures in rock and aids the process of mechanical weathering. *(Photo by Tom and Susan Bean, Inc./DRK Photo)*

▲ **Figure 5.6** Iron reacts with oxygen to form iron oxide as seen on these rusted barrels. *(Photo by Stephen J. Krasemann/DRK Photo)*

Did You Know?

The intense heat from a brush or forest fire can cause flakes of rock to spall from boulders or bedrock. As the rock surface becomes overheated, a thin layer expands and shatters.

amount of dissolved material is generally all that is needed to activate it. Oxygen dissolved in water will oxidize some materials. For example, when an iron nail is found in moist soil, it will have a coating of rust (iron oxide), and if the time of exposure has been long, the nail will be so weak that it can be broken as easily as a toothpick. When rocks containing iron-rich minerals oxidize, a yellow to reddish-brown rust will appear on the surface (Figure 5.6).

Carbon dioxide (CO_2) dissolved in water (H_2O) forms carbonic acid (H_2CO_3), the same weak acid produced when soft drinks are carbonated. Rain dissolves some carbon dioxide as it falls through the atmosphere, and additional amounts released by decaying organic matter are acquired as the water percolates through the soil. Carbonic acid ionizes to form the very reactive hydrogen ion (H^+) and the bicarbonate ion (HCO_3^-).

The dissolving power of carbonic acid plays an important part in forming limestone caverns. (David Muench)

How Granite Weathers

To illustrate how rock chemically weathers when attacked by carbonic acid, we will consider the weathering of granite, the most abundant continental rock. Recall that granite consists mainly of quartz and potassium feldspar. The weathering of the potassium feldspar component of granite takes place as follows:

$$\underset{\text{potassium feldspar}}{2\,KAlSi_3O_8} + \underset{\text{carbonic acid}}{2(H^+ + HCO_3^-)} + \underset{\text{water}}{H_2O} \rightarrow$$

$$\underset{\text{clay mineral}}{Al_2Si_2O_5(OH)_4} + \underbrace{\underset{\text{potassium ion}}{2K^+} + \underset{\text{bicarbonate ion}}{2HCO_3^-} + \underset{\text{silica ion}}{4SiO_2}}_{\text{in solution}}$$

In this reaction, the hydrogen ions (H^+) attack and replace potassium ions (K^+) in the feldspar structure, thereby disrupting the crystalline network. Once removed, the potassium is available as a nutrient for plants or becomes the soluble salt potassium bicarbonate ($KHCO_3$), which may be incorporated into other minerals or carried to the ocean in dissolved form by streams.

The most abundant products of the chemical breakdown of feldspar are residual clay minerals. Clay minerals are the end products of weathering and are very stable under surface conditions. Consequently, clay minerals make up a high percentage of the inorganic material in soils. Moreover, the most abundant sedimentary rock, shale, contains a high proportion of clay minerals.

In addition to the formation of clay minerals during the weathering of feldspar, some silica is removed from the feldspar structure and is carried away by groundwater (water beneath Earth's surface). This dissolved silica will eventually precipitate to produce nodules of chert or flint, or it will fill in the pore spaces between sediment grains, or it will be carried to the ocean, where microscopic animals will remove it from the water to build hard silica shells.

A. B.

▲ **Figure 5.7** **A.** Spheroidal weathering is evident in this exposure of granite in California's Joshua Tree National Monument. Because the rocks are attacked more vigorously on the corners and edges, they take on a spherical shape. The lines visible in the rock are called *joints.* Joints are important rock structures that allow water to penetrate and start the weathering process long before the rock is exposed. *(Photo by E. J. Tarbuck)* **B.** Sometimes successive shells are loosened as the weathering process continues to penetrate ever deeper into the rock. *(Photo by Martin Schmidt, Jr.)*

To summarize, the weathering of potassium feldspar generates a residual clay mineral, a soluble salt (potassium bicarbonate), and some silica, which enters into solution.

Quartz, the other main component of granite, is *very resistant* to chemical weathering; it remains substantially unaltered when attacked by weak acidic solutions. As a result, when granite weathers, the feldspar crystals dull and slowly turn to clay, releasing the once interlocked quartz grains, which still retain their fresh, glassy appearance. Although some quartz remains in the soil, much is eventually transported to the sea or to other sites of deposition, where it becomes the main constituent of such features as sandy beaches and sand dunes. In time these quartz grains may become lithified to form the sedimentary rock *sandstone.*

Table 5.1 Products of weathering.

Mineral	Residual Products	Material in Solution
Quartz	Quartz grains	Silica
Feldspars	Clay minerals	Silica, K^+, Na^+, Ca^{2+}
Amphibole (hornblende)	Clay minerals Limonite Hematite	Silica, Ca^{2+}, Mg^{2+}
Olivine	Limonite Hematite	Silica, Mg^{2+}

Weathering of Silicate Minerals

Table 5.1 lists the weathered products of some of the most common silicate minerals. Remember that silicate minerals make up most of Earth's crust and that these minerals are composed essentially of only eight elements. When chemically weathered, silicate minerals yield sodium, calcium, potassium, and magnesium ions that form soluble products, which may be removed by groundwater. The element iron combines with oxygen, producing relatively insoluble iron oxides, which give soil a reddish-brown or yellowish color. Under most conditions the three remaining elements—aluminum, silicon, and oxygen—join with water to produce residual clay minerals. However, even the highly insoluble clay minerals are very slowly removed by subsurface water.

Spheroidal Weathering

In addition to altering the internal structure of minerals, chemical weathering also causes physical changes. For instance, when angular rock masses are attacked by water that enters along joints, the rocks tend to take on a spherical shape. Gradually the corners and edges of the angular blocks become more rounded. The corners are attacked most readily because of their greater surface area, as compared to the edges and faces. This process, called **spheroidal weathering,** gives the weathered rock a more rounded or spherical shape (Figure 5.7A).

Did You Know?

The only common mineral that is very resistant to both mechanical and chemical weathering is quartz.

Sometimes during the formation of spheroidal boulders, successive shells separate from the rock's main body (Figure 5.7B). Eventually the outer shells break off, allowing the chemical-weathering activity to penetrate deeper into the boulder. This spherical scaling results because, as the minerals in the rock weather to clay, they increase in size through the addition of water to their structure. This increased bulk exerts an outward force that causes concentric layers of rock to break loose and fall off. Hence, chemical weathering does produce forces great enough to cause mechanical weathering.

This type of spheroidal weathering, in which shells spall off, should not be confused with the phenomenon of sheeting discussed earlier. In sheeting, the fracturing occurs as a result of unloading, and the rock layers that separate from the main body are largely unaltered at the time of separation.

Rates of Weathering

Several factors influence the type and rate of rock weathering. We have already seen how mechanical weathering affects the rate of weathering. By breaking rock into smaller pieces, the amount of surface area exposed to chemical weathering is increased. Other important factors examined here include rock characteristics and climate.

Rock Characteristics

Rock characteristics encompass all of the chemical traits of rocks, including mineral composition and solubility. In addition, any physical features such as joints (cracks) can be important because they allow water to penetrate rock and start the process of weathering long before the rock is exposed.

The variations in weathering rates due to the mineral constituents can be demonstrated by comparing old headstones made from different rock types. Headstones of granite, which are composed of silicate minerals, are relatively resistant to chemical weathering. We can see this by examining the inscriptions on the headstones shown in Figure 5.8. In contrast, the marble headstone shows signs of extensive chemical alteration over a relatively short period. Marble is composed of calcite (calcium carbonate), which readily dissolves even in a weakly acidic solution.

The silicates, the most abundant mineral group, weather in essentially the same order as their order of crystallization. By examining Bowen's reaction series (see Figure 3.16, p. 71), you can see that olivine crystallizes first and is therefore the least resistant to chemical weathering, whereas quartz, which crystallizes last, is the most resistant.

Climate

Climatic factors, particularly temperature and moisture, are crucial to the rate of rock weathering. One important example from mechanical weathering is that the frequency of freeze-thaw cycles greatly affects the amount of frost wedging. Temperature and moisture also exert a strong influence on rates of chemical weathering and on the kind and amount of vegetation present. Regions with lush vegetation generally have a thick mantle of soil rich in decayed organic matter from which chemically active fluids such as carbonic and humic acids are derived.

The optimum environment for chemical weathering is a combination of warm temperatures and abundant moisture. In polar regions chemical weathering is ineffective because frigid temperatures keep the available moisture locked up as ice, whereas in arid regions

▶ **Figure 5.8** An examination of headstones reveals the rate of chemical weathering on diverse rock types. The granite headstone (left) was erected four years before the marble headstone (right). The inscription date of 1872 on the marble monument is nearly illegible. *(Photos by E. J. Tarbuck)*

there is insufficient moisture to foster rapid chemical weathering.

Human activities can influence the composition of the atmosphere, which in turn can impact the rate of chemical weathering. One well-known example is acid rain (Figure 5.9).

Differential Weathering

Masses of rock do not weather uniformly. Take a moment to look back at the photo of Shiprock, New Mexico, in Figure 4.22 (p. 101). The durable volcanic neck protrudes high above the surrounding terrain. A glance at the chapter-opening photo shows an additional example of this phenomenon, called **differential weathering.** The results vary in scale from the rough, uneven surface of the marble headstone in Figure 5.8 to the boldly sculpted exposures in Bryce Canyon (Figure 5.10).

Many factors influence the rate of rock weathering. Among the most important are variations in the composition of the rock. More resistant rock protrudes as ridges or pinnacles, or as steeper cliffs on an irregular hillside (see Figure 6.3, p. 142). The number and spacing of joints can also be a significant factor (see Figure 5.7A). Differential weathering and subsequent erosion are responsible for creating many unusual and sometimes spectacular rock formations and landforms.

Soil

Soil covers most land surfaces. Along with air and water, it is one of our most indispensable resources. Also like air and water, soil is taken for granted by many of us. The following quote helps put this vital layer in perspective.

> Science, in recent years, has focused more and more on the Earth as a planet, one that for all we know is unique—where a thin blanket of air, a thinner film of water, and the thinnest veneer of soil combine to support a web of life of wondrous diversity in continuous change.*

Soil has accurately been called "the bridge between life and the inanimate world." All life—the entire biosphere—owes its existence to a dozen or so elements that must ultimately come from Earth's crust. Once weathering and other processes create soil, plants carry out the intermediary role of assimilating the necessary elements and making them available to animals, including humans.

Soil is an essential resource that we often take for granted. (Tom Bean/CORBIS)

An Interface in the Earth System

When Earth is viewed as a system, soil is referred to as an interface—a common boundary where different parts of a system interact. This is an appropriate designation because soil forms where the solid Earth, the atmosphere, the hydrosphere, and the biosphere meet. Soil is a material that develops in response to complex environmental interactions among different parts of the Earth system. Over time, soil gradually evolves to a state of equilibrium or balance with the environment. Soil is dynamic and sensitive to almost every aspect of its surroundings. Thus, when environmental changes occur, such as climate, vegetative cover, and animal

*Jack Eddy. "A Fragile Seam of Dark Blue Light," in *Proceedings of the Global Change Research Forum,* U.S. Geological Survey Circular 1086, 1993, p. 15.

▼ **Figure 5.9** Acid rain accelerates the chemical weathering of stone monuments and structures, including this building facade in Leipzig, Germany. As a consequence of burning large quantities of coal and petroleum, more than 40 million tons of sulfur and nitrogen oxides are released into the atmosphere each year in the United States. Through a series of complex chemical reactions, some of these pollutants are converted into acids that then fall to Earth's surface as rain or snow. *(Photo by Doug Plummer/Photo Researchers, Inc.)*

▼ **Figure 5.10** Differential weathering is illustrated by these sculpted rock pinnacles in Bryce National Park, Utah. *(Photo by Ray Mathis/CORBIS/The Stock Market)*

Did You Know?

Sometimes soils become buried and preserved. Later, if these ancient soils, called *paleosols*, are uncovered, they can provide useful clues to climates and the nature of landscapes thousands or millions of years ago.

(including human) activity, the soil responds. Any such change produces a gradual alteration of soil characteristics until a new balance is reached. Although thinly distributed over the land surface, soil functions as a fundamental interface, providing an excellent example of the integration among many parts of the Earth system.

What Is Soil?

With few exceptions, Earth's land surface is covered by **regolith,** the layer of rock and mineral fragments produced by weathering. Some would call this material soil, but soil is more than an accumulation of weathered debris. **Soil** is a combination of mineral and organic matter, water, and air—that portion of the regolith that supports the growth of plants. Although the proportions of the major components in soil vary, the same four components always are present to some extent (Figure 5.11). About one half of the total volume of a good-quality surface soil is a mixture of disintegrated and decomposed rock (mineral matter) and *humus,* the decayed remains of animal and plant life (organic matter). The remaining half consists of pore spaces among the solid particles where air and water circulate.

Although the mineral portion of the soil is usually much greater than the organic portion, humus is an essential component. In addition to being an important source of plant nutrients, humus enhances the soil's ability to retain water. Because plants require air and water to live and grow, the portion of the soil consisting of pore spaces that allow for the circulation of these fluids is as vital as the solid soil constituents.

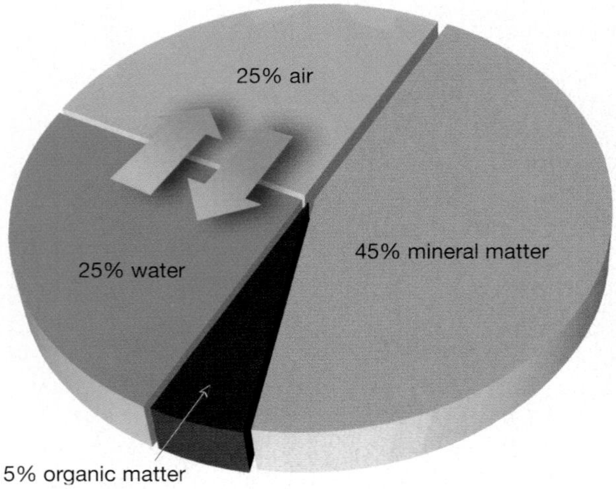

▲ **Figure 5.11** Composition (by volume) of a soil in good condition for plant growth. Although the percentages vary, each soil is composed of mineral and organic matter, water, and air.

Soil water is far from "pure" water; instead, it is a complex solution containing many soluble nutrients. Soil water not only provides the necessary moisture for the chemical reactions that sustain life, it also supplies plants with nutrients in a form they can use. The pore spaces that are not filled with water contain air. This air is the source of necessary oxygen and carbon dioxide for most microorganisms and plants that live in the soil.

Controls of Soil Formation

Soil is the product of the complex interplay of several factors. The most important of these are parent material, time, climate, plants and animals, and topography. Although all of these factors are interdependent, their roles will be examined separately.

Parent Material

The source of the weathered mineral matter from which soils develop is called the **parent material** and is a major factor influencing a newly forming soil. Gradually it undergoes physical and chemical changes as the processes of soil formation progress. Parent material might be the underlying bedrock or it can be a layer of unconsolidated deposits, as in a stream valley. When the parent material is bedrock, the soils are termed *residual soils.* By contrast, those developed on unconsolidated sediment are called *transported soils* (Figure 5.12). Note that transported soils form *in place* on parent materials that have been carried from elsewhere and deposited by gravity, water, wind, or ice.

The nature of the parent material influences soils in two ways. First, the type of parent material affects the rate of weathering and thus the rate of soil formation. (Consider the weathering rates of granite versus marble.) Also, because unconsolidated deposits are already partly weathered and provide more surface area for chemical weathering, soil development on such material usually progresses more rapidly. Second, the chemical makeup of the parent material affects the soil's fertility. This influences the character of the natural vegetation the soil can support.

At one time the parent material was believed to be the primary factor causing differences among soils. Today, soil scientists realize that other factors, especially climate, are more important. In fact, it has been found that similar soils are often produced from different parent materials and that dissimilar soils have developed from the same parent material. Such discoveries reinforce the importance of the other soil-forming factors.

Time

Time is an important component of every geological process, and soil formation is no exception. The nature of soil is strongly influenced by the length of time that

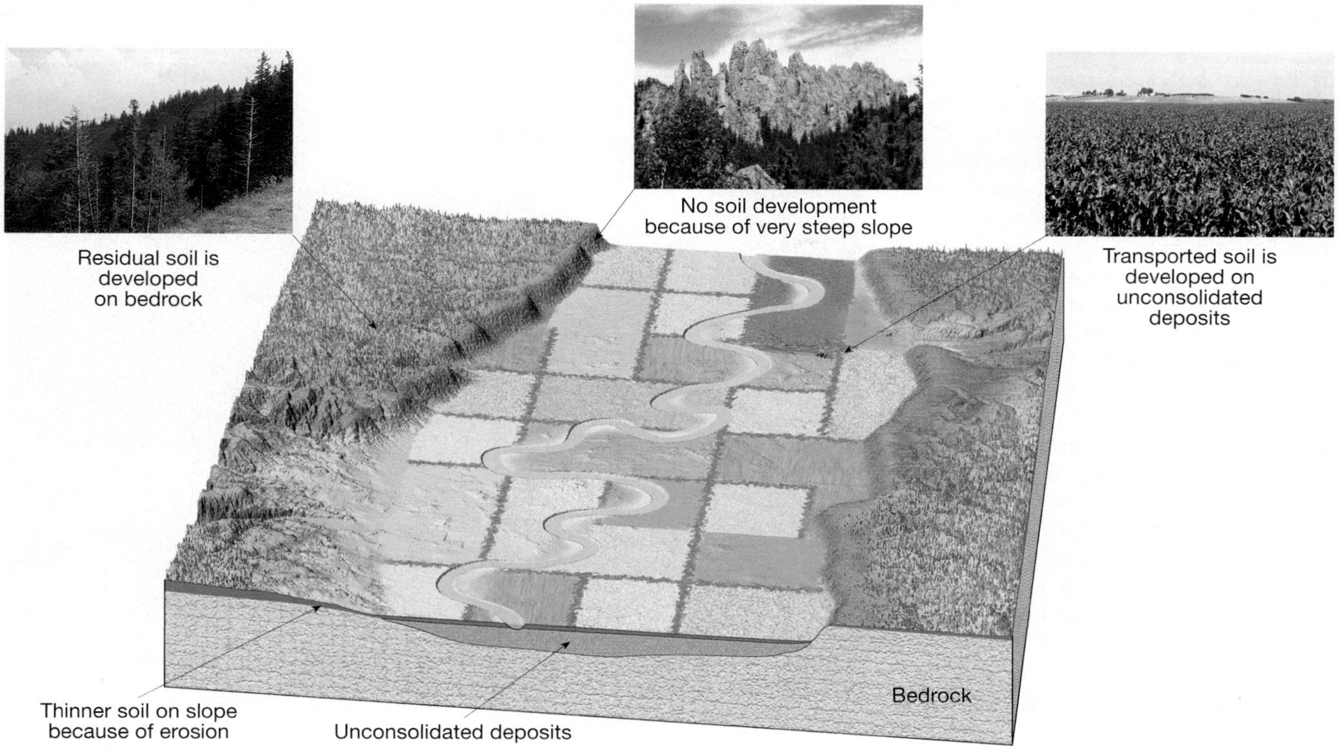
No soil development
because of very steep slope
Residual soil is
developed
on bedrock
Transported soil is
developed on
unconsolidated
deposits
Bedrock
Thinner soil on slope
because of erosion
Unconsolidated deposits

▲ **Figure 5.12** The parent material for residual soils is the underlying bedrock, whereas transported soils form on unconsolidated deposits. Also note that as slopes become steeper, soil becomes thinner.*(Photos by E. J. Tarbuck)*

processes have been operating. If weathering has been going on for a comparatively short time, the parent material determines to a large extent the characteristics of the soil. As weathering processes continue, the influence of parent material on soil is overshadowed by the other soil-forming factors, especially climate. The amount of time required for various soils to evolve cannot be specified, because the soil-forming processes act at varying rates under different circumstances. However, as a rule, the longer a soil has been forming, the thicker it becomes and the less it resembles the parent material.

Climate

Climate is the most influential control of soil formation. Just as temperature and precipitation are the climatic elements that influence people the most, so too are they the elements that exert the strongest impact on soil formation. Variations in temperature and precipitation determine whether chemical or mechanical weathering predominates. They also greatly influence the rate and depth of weathering. For instance, a hot, wet climate may produce a thick layer of chemically weathered soil in the same amount of time that a cold, dry climate produces a thin mantle of mechanically weathered debris. Also, the amount of precipitation influences the degree to which various materials are removed (leached) from the soil, thereby affecting soil fertility. Finally, climatic conditions are important factors controlling the type of plant and animal life present (see Box 5.1)

Plants and Animals

The biosphere plays a vital role in soil formation. The types and abundance of organisms present have a strong influence on the physical and chemical properties of a soil. In fact, for well-developed soils in many regions, the significance of natural vegetation in influencing soil type is frequently implied in the description used by soil scientists. Such phrases as *prairie soil, forest soil,* and *tundra soil* are common.

Plants and animals furnish organic matter to the soil. Certain bog soils are composed almost entirely of organic matter, whereas desert soils may contain only a tiny percentage. Although the quantity of organic matter varies substantially among soils, it is the rare soil that completely lacks it.

The primary source of organic matter is plants, although animals and the uncountable microorganisms also contribute. When organic matter decomposes, important nutrients are supplied to plants, as well as to animals and microorganisms living in the soil. Consequently, soil fertility depends in part on the amount of organic matter present. Furthermore, the decay of plant and animal remains causes the formation of various organic acids. These complex acids hasten the weathering process. Organic matter also has a high water-holding ability and thus aids water retention in a soil.

Did You Know?

It usually takes between 80 and 400 years for soil-forming processes to create 1 cm (less than 1/2 in.) of topsoil.

BOX 5.1

Clearing the Tropical Rain Forest—The Impact on Its Soils

Thick red soils are common in the wet tropics and subtropics. They are the end product of extreme chemical weathering. Because lush tropical rain forests are associated with these soils, we might assume they are fertile and have great potential for agriculture. However, just the opposite is true—they are among the poorest soils for farming. How can this be?

Because rain forest soils develop under conditions of high temperature and heavy rainfall, they are severely leached. Not only does leaching remove the soluble materials such as calcium carbonate but the great quantities of percolating water also remove much of the silica, with the result that insoluble oxides of iron and aluminum become concentrated in the soil. Iron oxides give the soil its distinctive red color. Because bacterial activity is very high in the tropics, rain forest soils contain practically no humus. Moreover, leaching destroys fertility because most plant nutrients are removed by the large volume of downward-percolating water. Therefore, even though the vegetation may be dense and luxuriant, the soil itself contains few available nutrients.

Most nutrients that support the rain forest are locked up in the trees themselves. As vegetation dies and decomposes, the roots of the rain forest trees quickly absorb the nutrients before they are leached from the soil. The nutrients are continuously recycled as trees die and decompose.

Therefore, when forests are cleared to provide land for farming or to harvest the timber, most of the nutrients are removed as well (Figure 5.A). What remains is a soil that contains little to nourish planted crops.

The clearing of rain forests not only removes plant nutrients but also accelerates erosion. When vegetation is present, its roots anchor the soil, and its leaves and branches provide a canopy that protects the ground by deflecting the full force of the frequent heavy rains.

The removal of vegetation also exposes the ground to strong direct sunlight. When baked by the Sun, these tropical soils can harden to a bricklike consistency and become practically impenetrable to water and crop roots. In only a few years, soils in a freshly cleared area may no longer be cultivable.

The term *laterite,* which is often applied to these soils, is derived from the Latin word *latere,* meaning "brick," and was first applied to the use of this material for brickmaking in India and Cambodia. Laborers simply excavated the soil, shaped it, and allowed it to harden in the Sun. Ancient but still well-preserved structures built of laterite remain standing today in the wet tropics (Figure 5.B). Such structures have withstood centuries of weathering because all of the original soluble materials were already removed from the soil by chemical weathering. Laterites are therefore virtually insoluble and very stable.

In summary, we have seen that some rain forest soils are highly leached products of extreme chemical weathering in the warm, wet tropics. Although they may be associated with lush tropical rain forests, these soils are unproductive when vegetation is removed. Moreover, when cleared of plants, these soils are subject to accelerated erosion and can be baked to brick-like hardness by the Sun.

Figure 5.A Clearing the Amazon rain forest in Surinam. The thick lateric soil is highly leached. (Photo by Wesley Bocxe/Photo Researchers)

Figure 5.B This ancient temple at Angor Wat, Cambodia, was built of bricks made of laterite. (Photo by R. Ian Lloyd/CORBIS/The Stock Market)

Microorganisms, including fungi, bacteria, and single-celled protozoa, play an active role in the decay of plant and animal remains. The end product is humus, a material that no longer resembles the plants and animals from which it was formed. In addition, certain microorganisms aid soil fertility because they have the ability to convert atmospheric nitrogen into soil nitrogen.

Earthworms and other burrowing animals act to mix the mineral and organic portions of a soil. Earthworms, for example, feed on organic matter and thoroughly mix soils in which they live, often moving and enriching many tons per acre each year. Burrows and holes also aid the passage of water and air through the soil.

Topography

The lay of the land can vary greatly over short distances. Such variations in topography can lead to the development of a variety of localized soil types. Many of the differences exist because the length and steepness of slopes have a significant impact on the amount of erosion and the water content of soil.

On steep slopes, soils are often poorly developed. In such situations little water can soak in; as a result, soil moisture may be insufficient for vigorous plant growth. Further, because of accelerated erosion on steep slopes, the soils are thin or nonexistent (Figure 5.12).

In contrast, waterlogged soils in poorly drained bottomlands have a much different character. Such soils are usually thick and dark. The dark color results from the large quantity of organic matter that accumulates because saturated conditions retard the decay of vegetation. The optimum terrain for soil development is a flat-to-undulating upland surface. Here we find good drainage, minimum erosion, and sufficient infiltration of water into the soil.

Slope orientation, or the direction a slope is facing, also is significant. In the midlatitudes of the Northern Hemisphere, a south-facing slope receives a great deal more sunlight than a north-facing slope. In fact, a steep north-facing slope may receive no direct sunlight at all. The difference in the amount of solar radiation received causes substantial differences in soil temperature and moisture, which in turn may influence the nature of the vegetation and the character of the soil.

Although we have dealt separately with each of the soil-forming factors, remember that all of them work together to form soil. No single factor is responsible for a soil's character. Rather, it is the combined influence of parent material, time, climate, plants and animals, and topography that determines this character.

The Soil Profile

Because soil-forming processes operate from the surface downward, variations in composition, texture, structure, and color gradually evolve at varying depths. These vertical differences, which usually become more pronounced as time passes, divide the soil into zones or layers known as **horizons.** If you were to dig a trench in soil, you would see that its walls are layered. Such a vertical section through all of the soil horizons constitutes the **soil profile.**

Northern coniferous forest. The type of vegetation strongly influences soil formation. (Carr Clifton)

Figure 5.13 presents an idealized view of a welldeveloped soil profile in which five horizons are identified. From the surface downward, they are designated as *O, A, E, B,* and *C,* respectively. These five horizons are common to soils in temperate regions. The characteristics and extent of development of horizons vary in different environments. Thus, different localities exhibit soil profiles that can contrast greatly with one another.

The *O* horizon consists largely of organic material. This is in contrast to the layers beneath it that consist mainly of mineral matter. The upper portion of the *O* horizon is primarily plant litter such as loose leaves and other organic debris that are still recognizable. By contrast, the lower portion of the *O* horizon is made up of partly decomposed organic matter (humus) in which plant structures can no longer be identified. In addition to plants, the *O* horizon is teeming with microscopic life, including bacteria, fungi, algae, and insects. All of these organisms contribute oxygen, carbon dioxide, and organic acids to the developing soil.

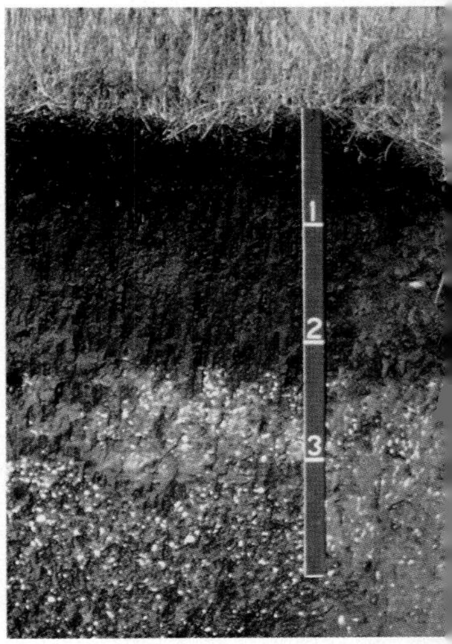

This soil in South Dakota has distinct horizons.

Underlying the organic-rich *O* horizon is the *A* horizon. This zone is largely mineral matter, yet biological activity is high and humus is generally present at up to 30 percent in some instances. Together the *O* and *A* horizons make up what is commonly called *topsoil.* Below the *A* horizon, the *E* horizon is a light-colored layer that contains little organic material. As water percolates downward through this zone, finer particles are carried away. This washing out of the fine soil components is termed **eluviation.** Water percolating downward also dissolves soluble inorganic soil components and carries them to deeper zones. This depletion of soluble materials from the upper soil is termed **leaching.**

Immediately below the *E* horizon is the *B* horizon, or *subsoil.* Much of the material removed from the *E* horizon by eluviation is deposited in the *B* horizon, which is often referred to as the *zone of accumulation.* The accumulation of the fine clay particles enhances water retention in the subsoil. However, in extreme cases clay accumulation can form a very compact and impermeable layer called *hardpan.* The *O, A, E,* and *B* horizons together constitute the **solum,** or true soil. It is in the solum that the soil-forming processes are active and that living roots and other plant and animal life are largely confined.

Below the solum and above the unaltered parent material is the *C* horizon, a layer characterized by partially altered parent material. Whereas the *O, A, E,* and *B* horizons bear little resemblance to the parent material, it is easily identifiable in the *C* horizon. Although this material is undergoing changes that will eventually

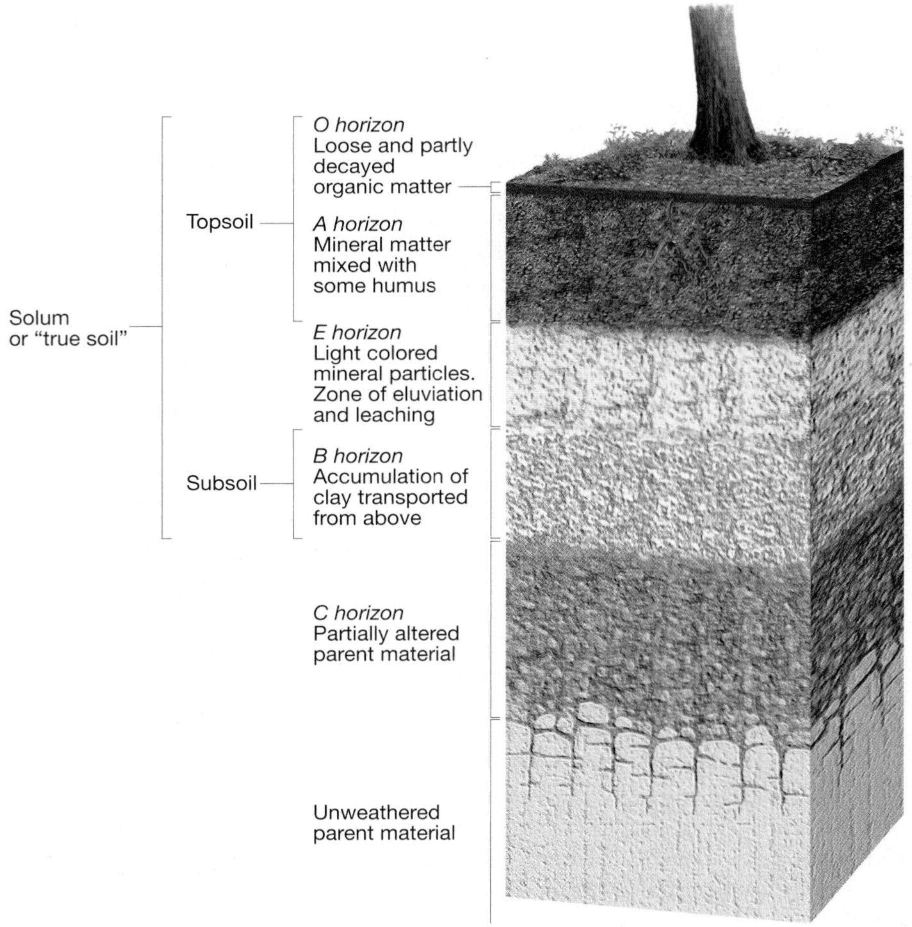

▶ **Figure 5.13** A soil profile is a vertical cross-section from the surface down to the parent material. Well-developed soils show distinct layers called *horizons.* This is an idealized soil profile from a humid climate in the middle latitudes.

transform it into soil, it has not yet crossed the threshold that separates regolith from soil.

The characteristics and extent of development can vary greatly among soils in different environments. The boundaries between soil horizons may be very distinct, or the horizons may blend gradually from one to another. A well-developed soil profile indicates that environmental conditions have been relatively stable over an extended time span and that the soil is mature. By contrast, some soils lack horizons altogether. Such soils are called *immature,* because soil building has been going on for only a short time. Immature soils are also characteristic of steep slopes where erosion continually strips away the soil, preventing full development.

Classifying Soils

There are many variations from place to place and from time to time among the factors that control soil formation. These differences lead to a bewildering variety of soil types. To cope with such variety, it is essential to devise some means of classifying the vast array of data to be studied. By establishing groups consisting of items that have certain important characteristics in common, order and simplicity are introduced. Bringing order to large quantities of information not only aids comprehension and understanding but also facilitates analysis and explanation.

In the United States, soil scientists have devised a system for classifying soils known as the **Soil Taxonomy.** It emphasizes the physical and chemical properties of the soil profile and is organized on the basis of observable soil characteristics. There are six hierarchical categories of classification, ranging from *order,* the broadest category, to *series,* the most specific category. The system recognizes 12 soil orders and more than 19,000 soil series.

The names of the classification units are combinations of syllables, most of which are derived from Latin or Greek. The names are descriptive. For example, soils of the order Aridosol (from the Latin *aridus,* dry, and *solum,* soil) are characteristically dry soils in arid regions. Soils in the order Inceptisols (from Latin *inceptum,* beginning, and *solum,* soil) are soils with only the beginning or inception of profile development.

Brief descriptions of the 12 basic soil orders are provided in Table 5.2. Figure 5.14 shows the complex worldwide distribution pattern of the Soil Taxonomy's 12 soil orders. Like many classification systems, the Soil Taxonomy is not suitable for every purpose. It is especially useful for agricultural and related land-use purposes, but it is not a useful system for engineers who are preparing evaluations of potential construction sites.

Soil Erosion

Soils are just a tiny fraction of all Earth materials, yet they are a vital resource. Because soils are necessary for the growth of rooted plants, they are the very foundation of the human life-support system. Just as human ingenuity can increase the agricultural productivity of soils through fertilization and irrigation, soils can be damaged or destroyed by careless activities. Despite their basic role in providing food, fiber, and other basic materials, soils are among our most abused resources.

Perhaps this neglect and indifference has occurred because a substantial amount of soil seems to remain even where soil erosion is serious. Nevertheless, although the loss of fertile topsoil may not be obvious to the untrained eye, it is a growing problem as human activities expand and disturb more and more of Earth's surface.

How Soil Is Eroded

Soil erosion is a natural process. It is part of the constant recycling of Earth materials that we call the *rock cycle.* Once soil forms, erosional forces, especially water and wind, move soil components from one place to another. Every time it rains, raindrops strike the land with surprising force (Figure 5.15). Each drop acts like a tiny bomb, blasting movable soil particles out of their positions in the soil mass. Then, water flowing across the surface carries away the dislodged soil particles. Because the soil is moved by thin sheets of water, this process is termed *sheet erosion.*

After flowing as a thin, unconfined sheet for a relatively short distance, threads of current typically

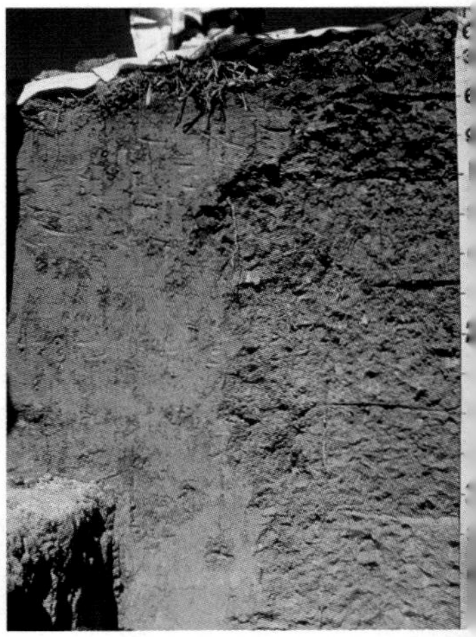

The boundaries between horizons in this soil in Puerto Rico are indistinct. (Soil Science Society of America)

Table 5.2 World Soil Orders

Alfisols	Moderately weathered soils that form under boreal forests or broadleaf deciduous forests, rich in iron and aluminum. Clay particles accumulate in a subsurface layer in response to leaching in moist environments. Fertile, productive soils, because they are neither too wet nor too dry.
Andisols	Young soils in which the parent material is volcanic ash and cinders, deposited by recent volcanic activity.
Aridosols	Soils that develop in dry places; insufficient water to remove soluble minerals, may have an accumulation of calcium carbonate, gypsum, or salt in subsoil; low organic content.
Entisols	Young soils having limited development and exhibiting properties of the parent material. Productivity ranges from very high for some formed on recent river deposits to very low for those forming on shifting sand or rocky slopes.
Gelisols	Young soils with little profile development that occur in regions with permafrost. Low temperatures and frozen conditions for much of the year slow soil-forming processes.
Histosols	Organic soils with little or no climatic implications. Can be found in any climate where organic debris can accumulate to form a bog soil. Dark, partially decomposed organic material commonly referred to as *peat.*
Inceptisols	Weakly developed young soils in which the beginning (inception) of profile development is evident. Most common in humid climates, they exist from the Arctic to the tropics. Native vegetation is most often forest.
Mollisols	Dark, soft soils that have developed under grass vegetation, generally found in prairie areas. Humus-rich surface horizon that is rich in calcium and magnesium. Soil fertility is excellent. Also found in hardwood forests with significant earthworm activity. Climatic range is boreal or alpine to tropical. Dry seasons are normal.
Oxisols	Soils that occur on old land surfaces unless parent materials were strongly weathered before they were deposited. Generally found in the tropics and subtropical regions. Rich in iron and aluminum oxides, oxisols are heavily leached; hence are poor soils for agricultural activity.
Spodosols	Soils found only in humid regions on sandy material. Common in northern coniferous forests and cool humid forests. Beneath the dark upper horizon of weathered organic material lies a light-colored horizon of leached material, the distinctive property of this soil.
Ultisols	Soils that represent the products of long periods of weathering. Water percolating through the soil concentrates clay particles in the lower horizons (argillic horizons). Restricted to humid climates in the temperate regions and the tropics, where the growing season is long. Abundant water and a long frost-free period contribute to extensive leaching, hence poorer soil quality.
Vertisols	Soils containing large amounts of clay, which shrink upon drying and swell with the addition of water. Found in subhumid to arid climates, provided that adequate supplies of water are available to saturate the soil after periods of drought. Soil expansion and contraction exert stresses on human structures.

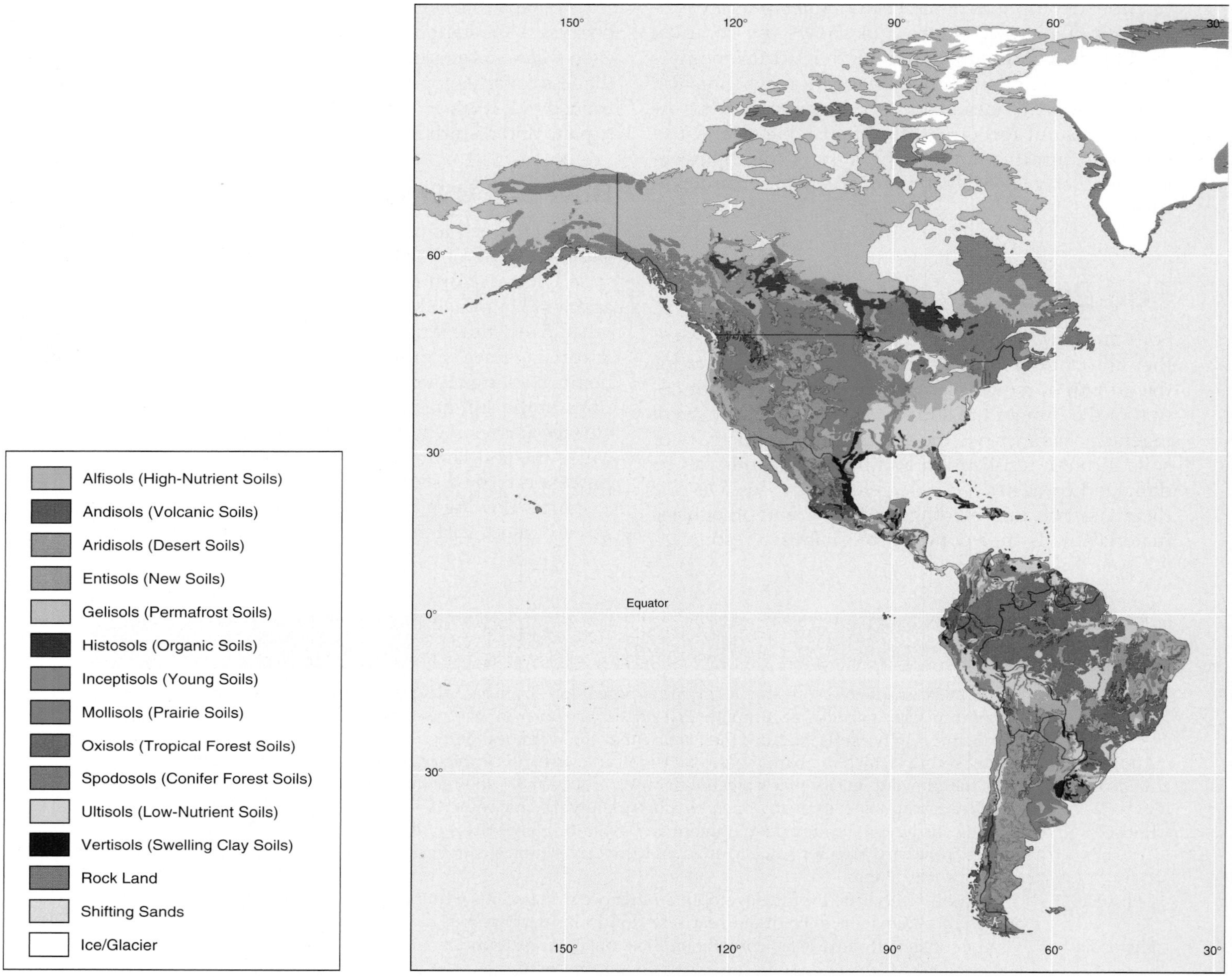

▲ **Figure 5.14** Global soil regions. Worldwide distribution of the Soil Taxonomy's 12 soil orders. (After U.S. Department of Agriculture, Natural Resources Conservation Service, World Soil Resources Staff)

develop, and tiny channels called *rills* begin to form. Still deeper cuts in the soil, known as *gullies,* are created as rills enlarge (Figure 5.16). When normal farm cultivation cannot eliminate the channels, we know the rills have grown large enough to be called gullies. Although most dislodged soil particles move only a short distance during each rainfall, substantial quantities eventually leave the fields and make their way downslope to a stream. Once in the stream channel, these soil particles, which can now be called *sediment,* are transported downstream and eventually deposited.

Rates of Erosion

We know that soil erosion is the ultimate fate of practically all soils. In the past, erosion occurred at slower rates than it does today because more of the land surface was covered and protected by trees, shrubs, grasses, and other plants. However, human activities such as farming, logging, and construction, which remove or disrupt the natural vegetation, have greatly accelerated the rate of soil erosion. Without the stabilizing effect of plants, the soil is more easily swept away by the wind or carried downslope by sheet wash.

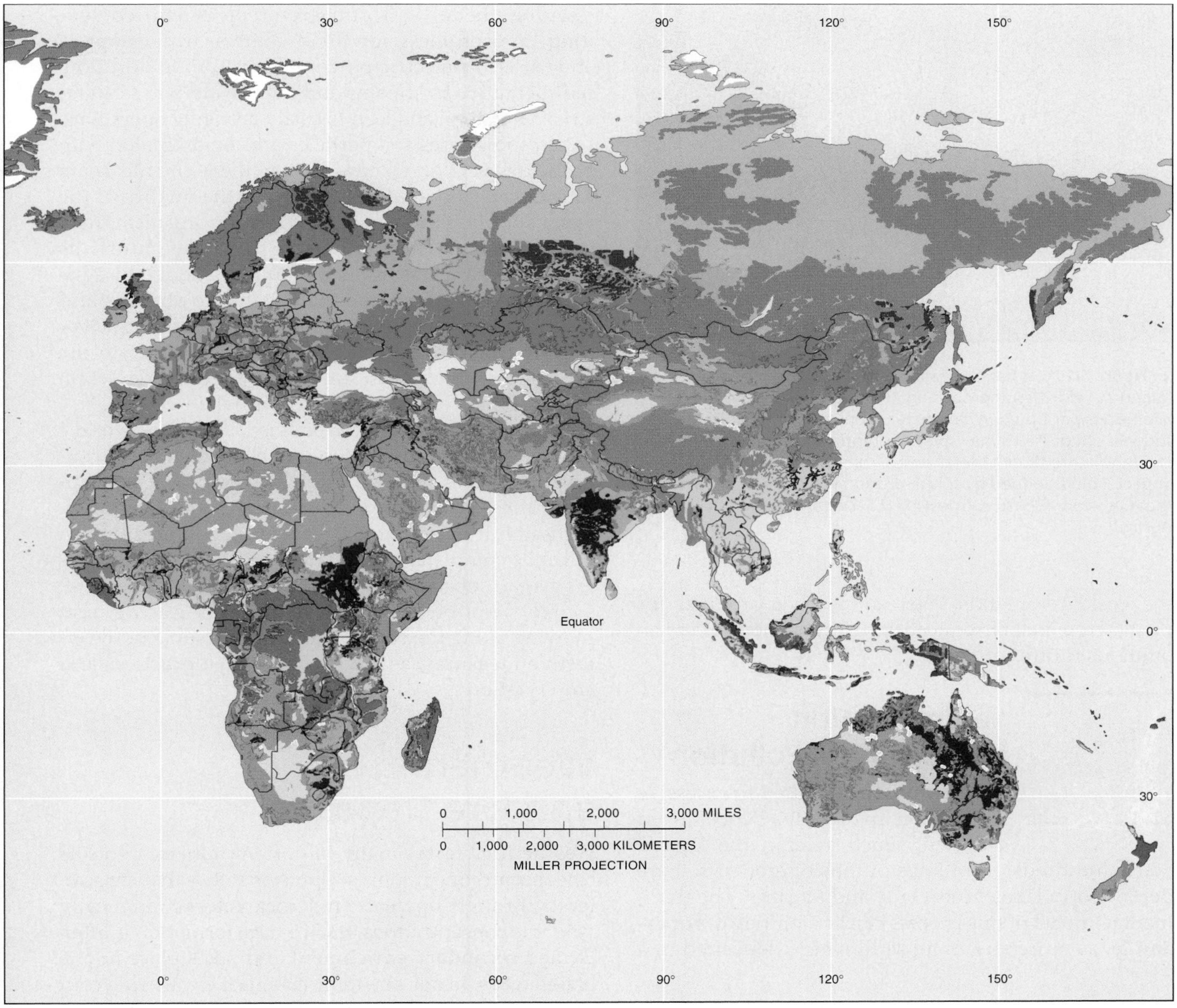

▲ **Figure 5.14** *Continued*

Natural rates of soil erosion vary greatly from one place to another and depend on soil characteristics as well as such factors as climate, topography, and type of vegetation. Over a broad area, erosion caused by surface runoff may be estimated by determining the sediment loads of the streams that drain the region. When studies of this kind were made on a global scale, they indicated that prior to the appearance of humans, sediment transport by rivers to the ocean amounted to just over 9 billion metric tons per year (1 metric ton = 1000 kilograms). By contrast, the amount of material currently transported to the sea by rivers is about 24 billion metric tons per year, or more than two and a half times the earlier rate.

It is more difficult to measure the loss of soil due to wind erosion. However, the removal of soil by wind is generally much less significant than erosion by flowing water except during periods of prolonged drought. When dry conditions prevail, strong winds can remove large quantities of soil from unprotected fields. Such was the case in the 1930s in the portions of the Great Plains that came to be called the *Dust Bowl* (see Box 12.2 and Figure 12.9, both on p. 276).

In many regions the rate of soil erosion is significantly greater than the rate of soil formation. This means that a renewable resource has become nonrenewable in these places. At present, it is estimated that topsoil is eroding faster than it forms on more than one third of

Gully erosion on a Wisconsin farm. (D. P. Burnside/Photo Researchers)

▲ **Figure 5.15** When it is raining, millions of water drops are falling at velocities approaching 10 meters per second (35 kilometers per hour). When water drops strike an exposed surface, soil particles may splash as high as 1 meter into the air and land more than a meter away from the point of raindrop impact. Soil dislodged by splash erosion is more easily moved by sheet erosion. *(Photo courtesy of U.S. Department of Agriculture)*

the world's croplands. The result is lower productivity, poorer crop quality, reduced agricultural income, and an ominous future.

Did You Know?

Just 1 mm of soil lost from a single acre of land amounts to about 5 tons.

Sedimentation and Chemical Pollution

Another problem related to excessive soil erosion involves the deposition of sediment. Each year in the United States hundreds of millions of tons of eroded soil are deposited in lakes, reservoirs, and streams. The detrimental impact of this process can be significant. For example, as more and more sediment is deposited in a reservoir, the capacity of the reservoir is reduced, limiting its usefulness for flood control, water supply, and/or hydroelectric power generation. In addition, sedimentation in streams and other waterways can restrict navigation and lead to costly dredging operations.

In some cases soil particles are contaminated with pesticides used in farming. When these chemicals are introduced into a lake or reservoir, the quality of the water supply is threatened, and aquatic organisms may be endangered. In addition to pesticides, nutrients found naturally in soils as well as those added by agricultural fertilizers make their way into streams and lakes, where they stimulate the growth of plants. Over a period of time, excessive nutrients accelerate the process by which plant growth leads to the depletion of oxygen and an early death of the lake.

The availability of good soils is critical if the world's rapidly growing population is to be fed. On every continent, unnecessary soil loss is occurring because appropriate conservation measures are not being used. Although it is a recognized fact that soil erosion can never be completely eliminated, soil conservation programs can substantially reduce the loss of this basic resource. Windbreaks (rows of trees), terracing, and plowing along the contours of hills are some of the effective measures, as are special tillage practices and crop rotation.

▼ **Figure 5.16** Gully erosion in poorly protected soil, southern Colombia. *(Photo by Carl Purcell/Photo Researchers, Inc.)*

Weathering and Ore Deposits

Weathering creates many important mineral deposits by concentrating minor amounts of metals that are scattered through unweathered rock into economically valuable concentrations. Such a transformation is often termed **secondary enrichment** and takes place in one of two ways. In one situation, chemical weathering coupled with downward-percolating water removes undesired materials from decomposing rock, leaving the desired elements enriched in the upper zones of the soil. The second way is basically the reverse of the first. That is, the desirable elements that are found in low concentrations near the surface are removed and carried to lower zones, where they are redeposited and become more concentrated.

Bauxite

The formation of *bauxite,* the principal ore of aluminum, is one important example of an ore created as a result of enrichment by weathering processes (Figure 5.17). Although aluminum is the third most abundant element in Earth's crust, economically valuable concentrations of this important metal are not common, because most aluminum is tied up in silicate minerals, from which it is extremely difficult to extract.

▲ **Figure 5.17** Bauxite is the principal ore of aluminum and forms as a result of weathering processes under tropical conditions. Its color varies from red or brown to nearly white. *(Photo by E. J. Tarbuck)*

Bauxite forms in rainy tropical climates. When aluminum-rich source rocks are subjected to the intense and prolonged chemical weathering of the tropics, most of the common elements, including calcium, sodium, and potassium, are removed by leaching. Because aluminum is extremely insoluble, it becomes concentrated in the soil (as bauxite, a hydrated aluminum oxide). Thus, the formation of bauxite depends on climatic conditions in which chemical weathering and leaching are pronounced, plus, of course, the presence of aluminum-rich source rock. In a similar manner, important deposits of nickel and cobalt develop from igneous rocks rich in silicate minerals such as olivine.

There is significant concern regarding the mining of bauxite and other residual deposits because they tend to occur in environmentally sensitive areas of the tropics. Mining is preceded by the removal of tropical vegetation, thus destroying rainforest ecosystems. Moreover, the thin moisture-retaining layer of organic matter is also disturbed. When the soil dries out in the hot sun, it becomes bricklike and loses its moisture-retaining qualities. Such soil cannot be productively farmed nor can it support significant forest growth. The long-term consequences of bauxite mining are clearly of concern for developing countries in the tropics, where this important ore is mined.

Did You Know?

In the United States it is estimated that the amount of soil annually eroded from farms exceeds the amount of soil that forms by more than 2 billion tons.

Other Deposits

Many copper and silver deposits result when weathering processes concentrate metals that are dispersed through a low-grade primary ore. Usually such enrichment occurs in deposits containing pyrite (FeS_2) the most common and widespread sulfide mineral. Pyrite is important because when it chemically weathers, sulfuric acid forms, which enables percolating waters to dissolve the ore metals. Once dissolved, the metals gradually migrate downward through the primary ore body until they are precipitated. Deposition takes place because of changes that occur in the chemistry of the solution when it reaches the groundwater zone (the zone beneath the surface where all pore spaces are filled with water). In this manner, the small percentage of dispersed metal can be removed from a large volume of rock and redeposited as a higher-grade ore in a smaller volume of rock.

Did You Know?

Bauxite is a useful indicator of past climates because it records periods of wet tropical climate in the geologic past.

The Chapter in Review

- External processes include (1) *weathering*—the disintegration and decomposition of rock at or near Earth's surface; (2) *mass wasting*—the transfer of rock material downslope under the influence of gravity; and (3) *erosion*—the removal of material by a mobile agent, usually water, wind, or ice. They are called *external processes* because they occur at or near Earth's surface and are powered by energy from the Sun. By contrast, *internal processes,* such as volcanism and mountain building, derive their energy from Earth's interior.
- *Mechanical weathering* is the physical breaking up of rock into smaller pieces. Rocks can be broken into smaller fragments by *frost wedging* (where water works its way into cracks or voids in rock and upon freezing, expands and enlarges the openings), *unloading* (expansion and breaking due to a great reduction in pressure when the overlying rock is eroded away), and *biological activity* (by humans, burrowing animals, plant roots, etc.).
- *Chemical weathering* alters a rock's chemistry, changing it into different substances. Water is by far the most important agent of chemical weathering. Oxygen dissolved in water will *oxidize* iron-rich minerals, while carbon dioxide (CO_2) dissolved in water forms carbonic acid, which attacks and alters rock. The chemical weathering of silicate minerals frequently produces (1) soluble products containing sodium, calcium, potassium, and magnesium ions, and silica in solution; (2) insoluble iron oxides, including limonite and hematite; and (3) clay minerals.
- The rate at which rock weathers depends on such factors as (1) *particle size*—small pieces generally weather faster than large pieces; (2) *mineral makeup*—calcite readily

dissolves in mildly acidic solutions, and silicate minerals that form first from magma are least resistant to chemical weathering; and (3) *climatic factors,* particularly temperature and moisture. Frequently, rocks exposed at Earth's surface do not weather at the same rate. This *differential weathering* of rocks is influenced by such factors as mineral makeup and degree of jointing.

- *Soil* is a combination of mineral and organic matter, water, and air—that portion of the *regolith* (the layer of rock and mineral fragments produced by weathering) that supports the growth of plants. About half of the total volume of a good-quality soil is a mixture of disintegrated and decomposed rock (mineral matter) and *humus* (the decayed remains of animal and plant life); the remaining half consists of pore spaces, where air and water circulate. The most important factors that control soil formation are *parent material, time, climate, plants and animals,* and *topography.*

- Soil-forming processes operate from the surface downward and produce zones or layers in the soil called *horizons.* From the surface downward, the soil horizons are respectively designated as *O* (largely organic matter), *A* (largely mineral matter), *E* (where the fine soil components and soluble materials have been removed by eluviation and leaching), *B* (or *subsoil,* often referred to as the *zone of accumulation*), and *C* (partially altered parent material). Together the *O* and *A* horizons make up what is commonly called the *topsoil.*

- In the United States, soils are classified using a system known as the *Soil Taxonomy.* It is based on physical and chemical properties of the soil profile and includes six hierarchical categories. The system is especially useful for agricultural and related land-use purposes.

- Soil erosion is a natural process. It is part of the constant recycling of Earth materials that we call the *rock cycle.* Once in a stream channel, soil particles, which can now be called *sediment,* are transported downstream and eventually deposited. *Rates of soil erosion* vary from one place to another and depend on the soil's characteristics as well as such factors as climate, slope, and type of vegetation. Human activities have greatly accelerated the rate of soil erosion in many areas.

- Weathering creates ore deposits by concentrating minor amounts of metals into economically valuable deposits. The process, often called *secondary enrichment,* is accomplished by either (1) removing undesirable materials and leaving the desired elements enriched in the upper zones of the soil, or (2) removing and carrying the desirable elements to lower soil zones where they are redeposited and become more concentrated. *Bauxite,* the principal ore of aluminum, is one important ore created as a result of enrichment by weathering processes. In addition, many copper and silver deposits result when weathering processes concentrate metals that were formerly dispersed through low-grade primary ore.

Key Terms

chemical weathering (p. 116)
differential weathering (p. 123)
eluviation (p. 127)
erosion (p. 116)
exfoliation dome (p. 118)
external process (p. 116)
frost wedging (p. 117)
horizon (p. 127)
internal process (p. 116)
leaching (p. 127)
mass wasting (p. 116)
mechanical weathering (p. 116)
parent material (p. 124)
regolith (p. 124)
secondary enrichment (p. 132)
sheeting (p. 118)
soil (p. 124)
soil profile (p. 127)
Soil Taxonomy (p. 123)
solum (p. 127)
spheroidal weathering (p. 121)
talus slope (p. 117)
weathering (p. 116)

Questions for Review

1. Describe the role of external processes in the rock cycle.
2. If two identical rocks were weathered, one mechanically and the other chemically, how would the products of weathering for the two rocks differ?
3. How does mechanical weathering add to the effectiveness of chemical weathering?
4. Describe the formation of an exfoliation dome. Give an example of such a feature.
5. Granite and basalt are exposed at the surface in a hot, wet region.
   a. Which type of weathering will predominate?
   b. Which of the rocks will weather most rapidly? Why?
6. Heat speeds up a chemical reaction. Why then does chemical weathering proceed slowly in a hot desert?
7. How is carbonic acid (H_2CO_3) formed in nature? What results when this acid reacts with potassium feldspar?
8. Relate soil to the Earth system.
9. What factors might cause different soils to develop from the same parent material, or similar soils to form from different parent materials?
10. Which control of soil formation is most important? Explain.
11. How can slope affect the development of soil? What is meant by the term *slope orientation*?

12. List the characteristics associated with each of the horizons in a well-developed soil profile. Which of the horizons constitute the solum? Under what circumstances do soils lack horizons?

13. Is soil erosion a natural process or primarily the result of inappropriate land use by people?

14. List three detrimental effects of soil erosion other than the loss of topsoil from croplands.

15. Name the primary ore of aluminum and describe its formation.

Online Study Guide

The *Essentials of Geology* Web site uses the resources and flexibility of the Internet to aid in your study of the topics in this chapter. Written and developed by geology instructors, this site will help improve your understanding of geology. Visit **www.prenhall.com/lutgens** and click on the cover of *Essentials of Geology 9e* to find:

- Online review quizzes.
- Critical thinking exercises.
- Links to chapter-specific Web resources.
- Internet-wide key-term searches.

http://www.prenhall.com/lutgens

The White Chalk Cliffs of Dover. This prominent chalk deposit underlies large portions of southern England as well as parts of northern France. *(Photo by Jon Arnold/Getty Images, Inc.-Taxi)*

CHAPTER 6

Sedimentary Rocks

Focus on Learning

To assist you in learning the important concepts in this chapter, you will find it helpful to focus on the following questions:

- What is a sedimentary rock? About what percentage of Earth's crust is sedimentary?
- How is sediment turned into sedimentary rock?
- What are the two general types of sedimentary rocks, and how does each type form?
- What is the primary basis for distinguishing among various types of detrital sedimentary rocks? What criterion is used to subdivide the chemical rocks?
- What are sedimentary structures? Why are these structures useful to geologists?
- What are the two broad groups of nonmetallic mineral resources?
- Which energy resources are associated with sedimentary rocks?

Fossils are associated primarily with sedimentary rocks and sediments.

Chapter 5 provided the background you needed to understand the origin of sedimentary rocks. Recall that weathering of existing rocks begins the process. Next, gravity and erosional agents (running water, wind, waves, and ice) remove the products of weathering and carry them to a new location where they are deposited. Usually the particles are broken down further during this transport phase. Following deposition, this material, which is now called **sediment,** becomes lithified (turned to rock). In most cases, the sediment is lithified into solid sedimentary rock by the processes of *compaction* and *cementation.*

What Is a Sedimentary Rock?

GEODe — Essentials of Geology

Sedimentary Rocks

▼ Introduction

The products of mechanical and chemical weathering constitute the raw materials for sedimentary rocks. The word *sedimentary* indicates the nature of these rocks, for it is derived from the Latin *sedimentum,* which means "to settle," a reference to solid material settling out of a fluid (water or air). Most, but not all, sediment is deposited in this fashion.

Did You Know?

About 75 percent of all rock outcrops on the continents are sedimentary.

Weathered debris is constantly being swept from bedrock, carried away, and eventually deposited in lakes, river valleys, seas, and countless other places. The particles in a desert sand dune, the mud on the floor of a swamp, the gravel in a stream bed, and even household dust are examples of this never-ending process. Because the weathering of bedrock and the transport and deposition of the weathering products are continuous, sediment is found almost everywhere. As piles of sediment accumulate, the materials near the bottom are compacted. Over long periods, these sediments become cemented together by mineral matter deposited into the spaces between particles, forming solid rock.

Geologists estimate that sedimentary rocks account for only about 5 percent (by volume) of Earth's outer 16 kilometers (10 miles). However, the importance of this group of rocks is far greater than this percentage would imply. If we were to sample the rocks exposed at the surface, we would find that the great majority are sedimentary (Figure 6.1). Therefore, we may think of sedimentary rocks as comprising a relatively thin and somewhat discontinuous layer in the uppermost portion of the crust. This fact is readily understood when we consider that sediment accumulates at Earth's surface.

Because sediments are deposited at Earth's surface, the rock layers that eventually form contain evidence of past events that occurred at the surface. By their very nature, sedimentary rocks contain within them indications of past environments in which their particles were deposited, and in many cases, clues to the mechanisms involved in their transport. Furthermore, it is sedimentary rocks that contain fossils, which are vital tools in the study of the geologic past. Thus, this group of rocks provides geologists with much of the basic

▼ **Figure 6.1** Sedimentary rocks are exposed at Earth's surface more than igneous and metamorphic rocks. Because they contain fossils and other clues about the geologic past, sedimentary rocks are important in the study of Earth history. The sedimentary rocks shown here are in Canyonlands National Park, Utah. *(Photo by Carr Clifton)*

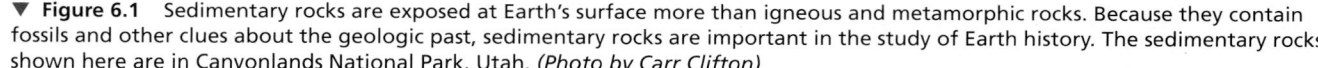

information they need to reconstruct the details of Earth history.

Finally, it should be mentioned that many sedimentary rocks are important economically. Coal, which is burned to provide a significant portion of U.S. electrical energy, is classified as a sedimentary rock. Our other major energy sources—petroleum and natural gas—are associated with sedimentary rocks. So are major sources of iron, aluminum, manganese, and fertilizer, plus numerous materials essential to the construction industry.

Turning Sediment into Sedimentary Rock: Diagenesis and Lithification

A great deal of change can occur to sediment from the time it is deposited until it becomes a sedimentary rock and is subsequently subjected to the temperatures and pressures that convert it to metamorphic rock. The term **diagenesis** is a collective term for all of the chemical, physical, and biological changes that take place after sediments are deposited and during and after lithification.

Burial promotes diagenesis because as sediments are buried, they are subjected to increasingly higher temperatures and pressures. Diagenesis occurs within the upper few kilometers of Earth's crust at temperatures that are generally less than 150° to 200°C. Beyond this somewhat arbitrary threshold, metamorphism is said to occur.

One example of diagenetic change is *recrystallization,* the development of more stable minerals from less stable ones. It is illustrated by the mineral aragonite, the less stable form of calcium carbonate ($CaCO_3$). Aragonite is secreted by many marine organisms to form shells and other hard parts, such as the skeletal structures produced by corals. In some environments, large quantities of these solid materials accumulate as sediment. As burial takes place, aragonite recrystallizes to the more stable form of calcium carbonate, calcite, the main constituent in the sedimentary rock limestone.

Diagenesis includes **lithification,** the processes by which unconsolidated sediments are transformed into solid sedimentary rocks. Basic lithification processes include compaction and cementation.

The most common physical diagenetic change is **compaction.** As sediment accumulates, the weight of overlying material compresses the deeper sediments. The deeper a sediment is buried, the more it is compacted and the firmer it becomes. As the grains are pressed closer and closer, there is considerable reduction in pore space (the open space between particles). For example, when clays are buried beneath several thousand meters of material, the volume of clay may be reduced by as much as 40 percent. As pore space decreases, much of the water that was trapped in the sediments is driven out. Because sands and other coarse sediments are less compressible, compaction is most significant as a lithification process in fine-grained sedimentary rocks.

Modern underground coal mine. (Melvin Grubb)

Cementation is the most important process by which sediments are converted to sedimentary rock. It is a chemical diagenetic change that involves the precipitation of minerals among the individual sediment grains. The cementing materials are carried in solution by water percolating through the open spaces between particles. Through time, the cement precipitates onto the sediment grains, fills the open spaces, and joins the particles. Just as the amount of pore space is reduced during compaction, the addition of cement into a sedimentary deposit reduces its porosity as well.

Did You Know?

Coal is responsible for slightly more than half the electricity generated in the United States.

Calcite, silica, and iron oxide are the most common cements. It is often a relatively simple matter to identify the cementing material. Calcite cement will effervesce with dilute hydrochloric acid. Silica is the hardest cement and thus produces the hardest sedimentary rocks. An orange or dark red color in a sedimentary rock means that iron oxide is present.

Iron oxides are the most important "pigments" that color sedimentary rocks.

Most sedimentary rocks are lithified by means of compaction and cementation. However, some initially form as solid masses of intergrown crystals rather than beginning as accumulations of separate particles that later become solid. Other crystalline sedimentary rocks do not begin that way but are transformed into masses of interlocking crystals sometime after the sediment is deposited.

For example, with time and burial, loose sediment consisting of delicate calcareous skeletal debris may be recrystallized into a relatively dense crystalline limestone. Because crystals grow until they fill all the available space, pore spaces are frequently lacking in crystalline sedimentary rocks. Unless the rocks later develop joints and fractures, they will be relatively impermeable to fluids like water and oil.

Types of Sedimentary Rocks

GEODe ESSENTIALS OF GEOLOGY
Sedimentary Rocks
▼ Types of Sedimentary Rocks

Sediment has two principal sources. First, sediment may be an accumulation of material that originates and is transported as solid particles derived from both mechanical and chemical weathering. Deposits of this type are termed *detrital,* and the sedimentary rocks that they form are called **detrital sedimentary rocks.** The second major source of sediment is soluble material produced

Close-up of a sedimentary rock made of intergrown crystals.

largely by chemical weathering. When these dissolved substances are precipitated by either inorganic or organic processes, the material is known as chemical sediment, and the rocks formed from it are called **chemical sedimentary rocks.**

We will now look at each type of sedimentary rock and some examples of each.

Detrital Sedimentary Rocks

GEODe ESSENTIALS OF GEOLOGY

Sedimentary Rocks
▼ Types of Sedimentary Rocks

Though a wide variety of minerals and rock fragments may be found in detrital rocks, clay minerals and quartz are the chief constituents of most sedimentary rocks in this category. Recall from Chapter 5 that clay minerals are the most abundant product of the chemical weathering of silicate minerals, especially the feldspars. Clays are fine-grained minerals with sheetlike crystalline structures similar to the micas. The other common mineral, quartz, is abundant because it is extremely durable and very resistant to chemical weathering. Thus, when igneous rocks such as granite are attacked by weathering processes, individual quartz grains are freed.

Other common minerals in detrital rocks are feldspars and micas. Because chemical weathering rapidly transforms these minerals into new substances, their presence in sedimentary rocks indicates that erosion and deposition were fast enough to preserve some of the primary minerals from the source rock before they could be decomposed.

Particle size is the primary basis for distinguishing among various detrital sedimentary rocks. Table 6.1 presents the size categories for particles making up detrital rocks. Note that in this context the term *clay* refers only to a particular size and not to the minerals of the same name. Although most clay minerals are of clay size, not all clay-sized sediment consists of clay minerals.

Particle size is not only a convenient method of dividing detrital rocks, but the sizes of the component grains also provide useful information about environments of deposition. Currents of water or air sort the particles by size—the stronger the current, the larger the particle size carried. Gravels, for example, are moved by swiftly flowing rivers as well as by landslides and glaciers. Less energy is required to transport sand; thus, it is common to such features as windblown dunes and some river deposits and beaches. Very little energy is needed to transport clay, so it settles very slowly. Accumulations of these tiny particles are generally associated with the quiet waters of a lake, lagoon, swamp, or certain marine environments.

Common detrital sedimentary rocks, in order of increasing particle size, are shale, sandstone, and conglomerate or breccia. We will now look at each type and how it forms.

Shale

Shale is a sedimentary rock consisting of silt- and clay-size particles (Figure 6.2). These fine-grained detrital rocks account for well over half of all sedimentary rocks. The particles in these rocks are so small that they cannot be readily identified without great magnification and for this reason make shale more difficult to study and analyze than most other sedimentary rocks.

Much of what can be learned is based on particle size. The tiny grains in shale indicate that deposition occurs as the result of gradual settling from relatively quiet, nonturbulent currents. Such environments include lakes, river floodplains, lagoons, and portions of the deep-ocean basins. Even in these quiet environments, there is usually enough turbulence to keep clay-size particles suspended almost indefinitely. Consequently, much of the clay is deposited only after the individual particles coalesce to form larger aggregates.

Sometimes the chemical composition of the rock provides additional information. One example is black

Table 6.1 Particle Size Classification for Detrital Rocks

Size Range (millimeters)	Particle Name	Common Sediment Name	Detrital Rock
>256	Boulder		
64–256	Cobble	Gravel	Conglomerate or breccia
4–64	Pebble		
2–4	Granule		
1/16–2	Sand	Sand	Sandstone
1/256–1/16	Silt	Mud	Shale, mudstone, or siltstone
<1/256	Clay		

0 10 20 30 40 50 60

(Scale in mm)

▲ **Figure 6.2** Shale is a fine-grained detrital rock that is by far the most abundant of all sedimentary rocks. Dark shales containing plant remains are relatively common. *(Photo courtesy of E. J. Tarbuck)*

shale, which is black because it contains abundant organic matter (carbon). When such a rock is found, it strongly implies that deposition occurred in an oxygen-poor environment such as a swamp, where organic materials do not readily oxidize and decay.

As silt and clay accumulate, they tend to form thin layers commonly referred to as *laminae*. Initially the particles in the laminae are oriented randomly. This disordered arrangement leaves a high percentage of open space (called *pore space*) that is filled with water. However, this situation usually changes with time as additional layers of sediment pile up and compact the sediment below.

During this phase the clay and silt particles take on a more parallel alignment and become tightly packed. This rearrangement of grains reduces the size of the pore spaces and forces out much of the water. Once the grains are pressed closely together, the tiny spaces between particles do not readily permit solutions containing cementing material to circulate. Therefore, shales are often described as being weak because they are poorly cemented and therefore not well lithified. The inability of water to penetrate its microscopic pore spaces explains why shale often forms barriers to the subsurface movement of water and petroleum. Indeed, rock layers that contain groundwater are commonly underlain by shale beds that block further downward movement.* The opposite is true for underground reservoirs of petroleum. They are often capped by shale beds that effectively prevent oil and gas from escaping to the surface.

*The relationship between impermeable beds and the occurrence and movement of groundwater is examined in Chapter 10.

It is common to apply the term *shale* to all fine-grained sedimentary rocks, especially in a nontechnical context. However, be aware that there is a more restricted use of the term. In this narrower usage, shale must exhibit the ability to split into thin layers along well-developed, closely spaced planes. This property is termed *fissility*. If the rock breaks into chunks or blocks, the name *mudstone* is applied. Another fine-grained sedimentary rock that, like mudstone, is often grouped with shale but lacks fissility is *siltstone*. As its name implies, siltstone is composed largely of silt-size particles and contains less clay-size material than shale and mudstone.

Although shale is far more common than other sedimentary rocks, it does not usually attract as much notice as other less abundant members of this group. The reason is that shale does not form prominent outcrops as sandstone and limestone often do. Rather, shale crumbles easily and usually forms a cover of soil that hides the unweathered rock below. This is illustrated nicely in the Grand Canyon, where the gentler slopes of weathered shale are quite inconspicuous and overgrown with vegetation, in sharp contrast with the bold cliffs produced by more durable rocks (Figure 6.3).

Although shale beds may not form striking cliffs and prominent outcrops, some deposits have economic value. Certain shales are quarried to obtain raw material for pottery, brick, tile, and china. Moreover, when mixed with limestone, shale is used to make Portland cement. In the future, one type of shale, called oil shale, may become a valuable energy resource.

Sandstone

Sandstone is the name given rocks in which sand-sized grains predominate (Figure 6.4). After shale, sandstone is the most abundant sedimentary rock, accounting for approximately 20 percent of the entire group. Sandstones form in a variety of environments and often contain significant clues about their origin, including sorting, particle shape, and composition.

Sorting is the degree of similarity in particle size in a sedimentary rock. For example, if all the grains in a sample of sandstone are about the same size, the sand is considered *well sorted*. Conversely, if the rock contains mixed large and small particles, the sand is said to be *poorly sorted*. By studying the degree of sorting, we can learn much about the depositing current. Deposits of windblown sand are usually better sorted than deposits sorted by wave activity (Figure 6.5). Particles washed by waves are commonly better sorted than materials deposited by streams. Sediment accumulations that exhibit poor sorting usually result when particles are transported for only a relatively short time and then rapidly deposited. For example, when a turbulent stream reaches the gentler slopes at the base of a steep

Beaches are one common environment in which sand accumulates. (Marc Muench)

▲ **Figure 6.3** Sedimentary rock layers exposed in the walls of the Grand Canyon, Arizona. Beds of resistant sandstone and limestone produce bold cliffs. By contrast, weaker, poorly cemented shale crumbles and produces a gentler slope of weathered debris in which some vegetation is growing. *(Photo by Tom & Susan Bean, Inc.)*

mountain, its velocity is quickly reduced, and poorly sorted sands and gravels are deposited.

The shapes of sand grains can also help decipher the history of a sandstone. When streams, winds, or waves move sand and other sedimentary particles, the grains lose their sharp edges and corners and become more rounded as they collide with other particles

▲ **Figure 6.4** Quartz sandstone. After shale, sandstone is the most abundant sedimentary rock. *(Photos by E. J. Tarbuck)*

during transport. Thus, rounded grains likely have been airborne or waterborne. Further, the degree of rounding indicates the distance or time involved in the transportation of sediment by currents of air or water. Highly rounded grains indicate that a great deal of abrasion and hence a great deal of transport has occurred.

Very angular grains, on the other hand, imply two things: that the materials were transported only a short distance before they were deposited, and that some other medium may have transported them. For example, when glaciers move sediment, the particles are usually made more irregular by the crushing and grinding action of the ice.

In addition to affecting the degree of rounding and the amount of sorting that particles undergo, the length of transport by turbulent air and water currents also influences the mineral composition of a sedimentary deposit. Substantial weathering and long transport lead to the gradual destruction of weaker and less stable minerals, including the feldspars and ferromagnesians. Because quartz is very durable, it is usually the mineral that survives the long trip in a turbulent environment.

The preceding discussion has shown that the origin and history of a sandstone can often be deduced by examining the sorting, roundness, and mineral composition of its constituent grains. Knowing this information allows us to infer that a well-sorted, quartz-rich sandstone consisting of highly rounded grains must be the result of a great deal of transport. Such a rock, in fact, may represent several cycles of weathering,

◀ **Figure 6.5** Sorting is the degree of similarity in particle size. The wind-transported sand grains in these dunes are well sorted because they are all practically the same size. Great Sand Dunes National Park, Colorado. *(Photo by David Muench/David Muench Photography, Inc.)*

transport, and deposition. We may also conclude that a sandstone containing a significant amount of feldspar and angular grains of ferromagnesian minerals underwent little chemical weathering and transport and was probably deposited close to the source area of the particles.

Owing to its durability, quartz is the predominant mineral in most sandstones. When this is the case, the rock may simply be called *quartz sandstone.* When a sandstone contains appreciable quantities of feldspar, the rock is called *arkose.* In addition to feldspar, arkose usually contains quartz and sparkling bits of mica. The mineral composition of arkose indicates that the grains were derived from granitic source rocks. The particles are generally poorly sorted and angular, which suggests short-distance transport, minimal chemical weathering in a relatively dry climate, and rapid deposition and burial.

A third variety of sandstone is known as *graywacke.* Along with quartz and feldspar, this dark-colored rock contains abundant rock fragments and matrix. *Matrix* refers to the silt- and clay-size particles found in spaces between larger sand grains. More than 15 percent of graywacke's volume is matrix. The poor sorting and angular grains characteristic of graywacke suggest that the particles were transported only a relatively short distance from their source area and then rapidly deposited. Before the sediment could be reworked and sorted further, it was buried by additional layers of material. Graywacke is frequently associated with submarine deposits made by dense sediment-choked torrents called turbidity currents.

Conglomerate and Breccia

Conglomerate consists largely of gravels (Figure 6.6). As Table 6.1 indicates, these particles may range in size from large boulders to particles as small as garden peas. The particles are commonly large enough to be identified as distinctive rock types; thus, they can be valuable in identifying the source areas of sediments. More often than not, conglomerates are poorly sorted because the openings between the large gravel particles contain sand or mud.

Gravels accumulate in a variety of environments and usually indicate the existence of steep slopes or very turbulent currents. The coarse particles in a conglomerate may reflect the action of energetic mountain streams or result from strong wave activity along a rapidly eroding coast. Some glacial and landslide deposits also contain plentiful gravel.

Did You Know?

The most important and common material used for making glass is silica, which is usually obtained from the quartz in "clean," well-sorted sandstones.

▲ **Figure 6.6** Conglomerate is composed primarily of rounded gravel-size particles. *(Photo by E. J. Tarbuck)*

If the large particles are angular rather than rounded, the rock is called *breccia* (Figure 6.7). Because large particles abrade and become rounded very rapidly during transport, the pebbles and cobbles in a breccia indicate that they did not travel far from their source area before they were deposited. Thus, as with many other sedimentary rocks, conglomerates and breccias contain clues to their history. Their particle sizes reveal the strength of the currents that transported them, whereas the degree of rounding indicates how far the particles traveled. The fragments within a sample identify the source rocks that supplied them.

Chemical Sedimentary Rocks

GEODe: Sedimentary Rocks
▼ Types of Sedimentary Rocks

Travertine is the rock that decorates many caverns.

In contrast to detrital rocks, which form from the solid products of weathering, chemical sediments derive from material that is carried *in solution* to lakes and seas. This material does not remain dissolved in the water indefinitely, however. Some of it precipitates to form chemical sediments. These become rocks such as limestone, chert, and rock salt.

This precipitation of material occurs in two ways. *Inorganic* processes such as evaporation and chemical activity can produce chemical sediments. *Organic* (life) processes of water-dwelling organisms also form chemical sediments, said to be of **biochemical origin.**

One example of a deposit resulting from inorganic chemical processes is the dripstone that decorates many caves (Figure 6.8). Another is the salt left behind as a body of seawater evaporates. In contrast, many water-dwelling animals and plants extract dissolved mineral matter to form shells and other hard parts. After the organisms die, their skeletons collect by the millions on the floor of a lake or ocean as biochemical sediment (Figure 6.9).

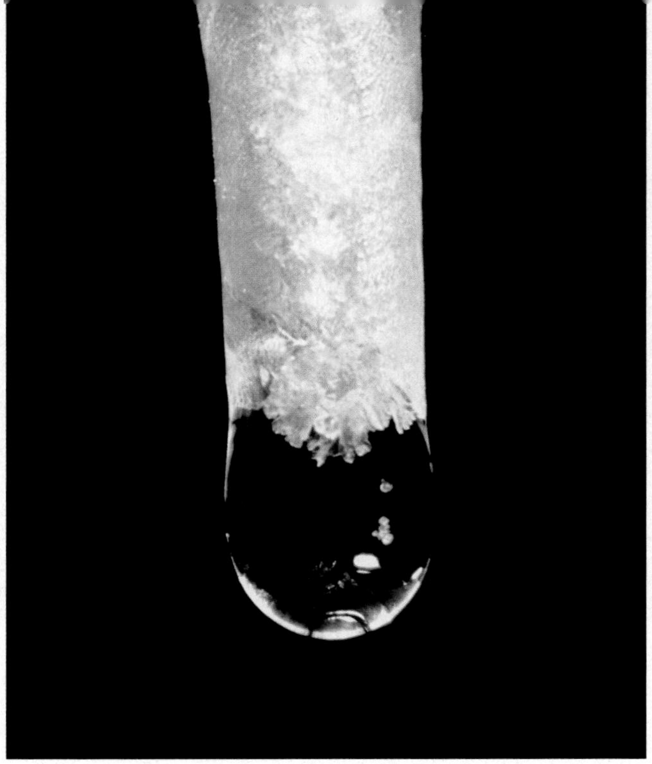

▲ **Figure 6.8** Because many cave deposits are created by the seemingly endless dripping of water over long time spans, they are commonly called *dripstone.* The material being deposited is calcium carbonate ($CaCO_3$) and the rock is a form of limestone called *travertine.* The calcium carbonate is precipitated when some dissolved carbon dioxide escapes from a water drop. *(Photo by Clifford Stroud, National Park Service)*

▲ **Figure 6.7** When the gravel-size particles in a detrital rock are angular, the rock is called breccia. *(Photo by E. J. Tarbuck)*

Limestone

Representing about 10 percent of the total volume of all sedimentary rocks, *limestone* is the most abundant chemical sedimentary rock. It is composed chiefly of the mineral calcite ($CaCO_3$) and forms either by inorganic means or as the result of biochemical processes. Regardless of its origin, the mineral composition of all limestone is similar, yet many different types exist. This is true because limestones are produced under a variety of conditions. Those forms having a marine biochemical origin are by far the most common (see Box 6.1).

Coral Reefs. Corals are one important example of organisms that are capable of creating large quantities of marine limestone. These relatively simple invertebrate animals secrete a calcareous (calcite-rich) external skeleton. Although they are small, corals are capable of creating massive structures called *reefs* (Figure 6.10A). Reefs consist of coral colonies made up of great numbers of individuals that live side by side on a calcite

▲ **Figure 6.9** This rock, called coquina, consists of shell fragments; therefore, it has a biochemical origin. *(Photos by E. J. Tarbuck)*

structure secreted by the animals. In addition, calcium carbonate-secreting algae live with the corals and help cement the entire structure into a solid mass. A wide variety of other organisms also live in and near reefs.

Certainly the best-known modern reef is Australia's Great Barrier Reef, 2000 kilometers (1240 miles) long, but many lesser reefs also exist. They develop in the shallow, warm waters of the tropics and subtropics equatorward of about 30° latitude. Striking examples exist in the Bahamas and Florida Keys.

Of course, not only modern corals build reefs. Corals have been responsible for producing vast quantities of limestone in the geologic past as well. In the United States, reefs of Silurian age are prominent features in Wisconsin, Illinois, and Indiana. In west Texas and adjacent southeastern New Mexico, a massive reef complex formed during the Permian period is strikingly exposed in Guadalupe Mountains National Park (Figure 6.10B).

Coquina and Chalk. Although most limestone is the product of biological processes, this origin is not always evident, because shells and skeletons may undergo considerable change before becoming lithified into rock. However, one easily identified biochemical limestone is *coquina,* a coarse rock composed of poorly cemented shells and shell fragments (Figure 6.9). Another less obvious but nevertheless familiar example is *chalk,* a soft, porous rock made up almost entirely of the hard parts of microscopic marine organisms smaller than the head of a pin. Among the most famous chalk deposits are those exposed along the southeast coast of England (see chapter-opening photo).

Inorganic Limestones. Limestones having an inorganic origin form when chemical changes or high water temperatures increase the concentration of calcium carbonate to the point that it precipitates. *Travertine,* the type of limestone commonly seen in caves, is an example (Figure 6.8). When travertine is deposited in caves, groundwater is the source of the calcium carbonate. As water droplets become exposed to the air in a cavern, some of the carbon dioxide dissolved in the water escapes, causing calcium carbonate to precipitate.

Another variety of inorganic limestone is *oolitic limestone.* It is a rock composed of small spherical grains called *ooids.* Ooids form in shallow marine waters as tiny seed particles (commonly small shell fragments) are moved back and forth by currents. As the grains are rolled about in the warm water, which is supersaturated with calcium carbonate, they become coated with layer upon layer of the precipitate.

Jasper is the red variety of chert.

Dolostone

Closely related to limestone is *dolostone,* a rock composed of the calcium-magnesium carbonate mineral dolomite. Although dolostone can form by direct precipitation from seawater, it is thought that most originates when magnesium in seawater replaces some of the calcium in limestone. The latter hypothesis is reinforced by the fact that there is practically no young dolostone. Rather, most dolostones are ancient rocks in which there was ample time for magnesium to replace calcium.

Chert

Chert is a name used for a number of very compact and hard rocks made of microcrystalline silica (SiO_2). One well-known form is *flint,* whose dark color results from the organic matter it contains. *Jasper,* a red variety, gets its bright color from iron oxide. The banded form is usually referred to as *agate* (Figure 6.11).

Chert deposits are commonly found in one of two situations: as irregularly shaped nodules in limestone and as layers of rock. The silica composing many chert nodules may have been deposited directly from water. Such nodules have an inorganic origin. However, it is unlikely that a very large percentage of chert layers was precipitated directly from seawater, because seawater is seldom saturated with silica. Hence, beds of chert are thought to have originated largely as biochemical sediment.

Flint is a dark variety of chert.

Most water-dwelling organisms that produce hard parts make them of calcium carbonate. But some, such as diatoms and radiolarians, produce glasslike silica skeletons. These tiny organisms are able to extract silica even though seawater contains only tiny quantities of the dissolved material. It is from their remains that most chert beds are believed to originate.

BOX 6.1

The Carbon Cycle and Sedimentary Rocks

To illustrate the movement of material and energy in the Earth system, let us take a brief look at the *carbon cycle* (Figure 6.A). Pure carbon is relatively rare in nature. It is found predominantly in two minerals: diamond and graphite. Most carbon is bonded chemically to other elements to form compounds such as carbon dioxide, calcium carbonate, and the hydrocarbons found in coal and petroleum. Carbon is also the basic building block of life because it readily combines with hydrogen and oxygen to form the fundamental organic compounds that compose living things.

In the atmosphere, carbon is found mainly as carbon dioxide (CO_2). Atmospheric carbon dioxide is significant because it is a greenhouse gas, which means it is an efficient absorber of energy emitted by Earth and thus influences the heating of the atmosphere. Because many of the processes that operate on Earth involve carbon dioxide, this gas is constantly moving into and out of the atmosphere. For example, through the process of photosynthesis, plants absorb carbon dioxide from the atmosphere to produce the essential organic compounds needed for growth. Animals that consume these plants (or consume other animals that eat plants) use these organic compounds as a source of energy and, through the process of respiration, return carbon dioxide to the atmosphere. (Plants also return some CO_2 to the atmosphere via respiration.) Further, when plants die and decay or are burned, this biomass is oxidized, and carbon dioxide is returned to the atmosphere.

Not all dead plant material decays immediately back to carbon dioxide. A small percentage is deposited as sediment. Over long spans of geologic time, considerable biomass is buried with sediment. Under the right conditions, some of these carbon-rich deposits are converted to fossil fuels—coal, petroleum, or natural gas. Eventually some of the fuels are recovered (mined or pumped from a well) and burned to run factories and to fuel our transportation system. One result of fossil-fuel combustion is the release of huge quantities of CO_2 into the atmosphere. Certainly, one of the most active parts of the carbon cycle is the movement of CO_2 from the atmosphere to the biosphere and back again.

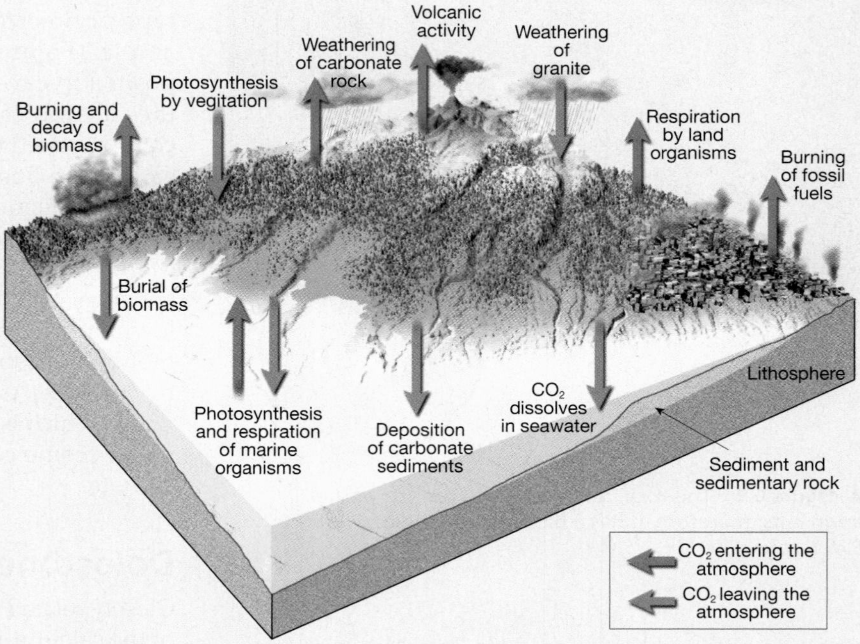

Figure 6.A Simplified diagram of the carbon cycle, with emphasis on the flow of carbon between the atmosphere and the hydrosphere, geosphere and biosphere. The colored arrows show whether the flow of carbon is into or out of the atmosphere.

Carbon also moves from the lithosphere and hydrosphere to the atmosphere and back again. For example, volcanic activity early in Earth's history is thought to be the source of much of the carbon dioxide found in the atmosphere. One way that carbon dioxide makes its way back to the hydrosphere and then to the solid Earth is by first combining with water to form carbonic acid (H_2CO_3), which then attacks the rocks that compose the lithosphere. One product of this chemical weathering of solid rock is the soluble bicarbonate ion ($2\ HCO_3^-$), which is carried by groundwater and streams to the ocean. Here water-dwelling organisms extract this dissolved material to produce hard parts of calcium carbonate ($CaCO_3$). When the organisms die, these skeletal remains settle to the ocean floor as biochemical sediment and become sedimentary rock. In fact, the lithosphere is by far Earth's largest depository of carbon, where it is a constituent of a variety of rocks, the most abundant being limestone. Eventually the limestone may be exposed at Earth's surface, where chemical weathering will cause the carbon stored in the rock to be released to the atmosphere as CO_2.

In summary, carbon moves among all four of Earth's major spheres. It is essential to every living thing in the biosphere. In the atmosphere, carbon dioxide is an important greenhouse gas. In the hydrosphere, carbon dioxide is dissolved in lakes, rivers, and the ocean. In the lithosphere, carbon is contained in carbonate sediments and sedimentary rocks, and is stored as organic matter dispersed through sedimentary rocks and as deposits of coal and petroleum.

A.

B.

▲ **Figure 6.10** **A.** This modern coral reef is at Bora-Bora in French Polynesia. *(Photo by Nancy Sefton/Photo Researchers)* **B.** El Capitán Peak, a massive limestone cliff in Guadalupe Mountains National Park, Texas. The rocks here are an exposed portion of a large reef that formed during the Permian period. *(Photo by Steve Elmore Photography, Inc.)*

Some bedded cherts occur in association with lava flows and layers of volcanic ash. For these occurrences it is probable that the silica was derived from the decomposition of the volcanic ash and not from biochemical sources. Note that when a hand specimen of chert is being examined, there are few reliable criteria by which the mode of origin (inorganic versus biochemical) can be determined.

Like glass, most chert has a conchoidal fracture. Its hardness, ease of chipping, and ability to hold a sharp edge made chert a favorite of Native Americans for fashioning points for spears and arrows. Because of chert's durability and extensive use, "arrowheads" are found in many parts of North America.

Evaporites

Very often, evaporation is the mechanism triggering deposition of chemical precipitates. Minerals commonly precipitated in this fashion include halite (sodium chloride, NaCl), the chief component of *rock salt,* and gypsum (hydrous calcium sulfate, $CaSO_4 \cdot 2\,H_2O$), the main ingredient of *rock gypsum.* Both have significant commercial importance. Halite is familiar to everyone as the common salt used in cooking and for seasoning foods. Of course, it has many other uses—from melting ice on roads to making hydrochloric acid—and has been considered important enough that people have sought, traded, and fought over it for much of human history. Gypsum is the basic ingredient of plaster of Paris. This material is used most extensively in the construction industry for wallboard and interior plaster.

▼ **Figure 6.11** Chert is a name used for a number of dense, hard rocks made of microcrystalline quartz. Agate is the banded variety. *(Photo by Jeffrey A. Scovil)*

Did You Know?

Diatoms have many uses. Examples include filters for straining yeast from beer and filtering swimming-pool water. They are also used as mild abrasives in toothpaste and in household cleaning and polishing products.

In the geologic past, many areas that are now dry land were basins, submerged under shallow arms of a sea that had only narrow connections to the open ocean. Under these conditions, seawater continually moved into the bay to replace water lost by evaporation. Eventually the waters of the bay became saturated and salt deposition began. Such deposits are called **evaporites.**

When a body of seawater evaporates, the minerals that precipitate do so in a sequence determined by their solubility. Less soluble minerals precipitate first, and more soluble minerals precipitate later as salinity increases. For example, gypsum precipitates when about two-thirds to three-quarters of the seawater has evaporated, and halite settles out when nine-tenths of the water has been removed. During the last stages of this process, potassium and magnesium salts precipitate. One of these last-formed salts, the mineral *sylvite,* is mined as a significant source of potassium ("potash") for fertilizer.

Harvesting salt from the evaporation of seawater. (William E. Townsend Jr.)

▲ **Figure 6.12** The Bonneville salt flats in Utah are a well-known example of evaporite deposits. *(Photo by Tom & Susan Bean, Inc.)*

On a smaller scale, evaporite deposits can be seen in many desert basins in the western United States. Here, following rains or periods of snowmelt in the mountains, streams flow from the surrounding mountains into an enclosed basin. As the water evaporates, **salt flats** form when dissolved materials are precipitated as a white crust on the ground (Figure 6.12).

Coal

Coal is quite different from other chemical sedimentary rocks. Unlike limestone and chert, which are calcite- or silica-rich, coal is made mostly of organic matter. Close examination of a piece of coal under a microscope or magnifying glass often reveals plant structures such as leaves, bark, and wood that have been chemically altered but are still identifiable. This supports the conclusion that coal is the end product of large amounts of plant material buried for millions of years.

The initial stage in coal formation is the accumulation of large quantities of plant remains. However, special conditions are required for such accumulations, because dead plants normally decompose when exposed to the atmosphere. An ideal environment that allows for the buildup of plant material is a swamp. Because stagnant swamp water is oxygen-deficient, complete decay (oxidation) of the plant material is not possible. At various times during Earth's history, such environments have been common. Coal undergoes successive stages of formation. With each successive stage, higher temperatures and pressures drive off impurities and volatiles, as shown in Figure 6.13.

Large quantities of plant material can accumulate in stagnant swamp water. This is the first step in the formation of coal. (Carr Clifton)

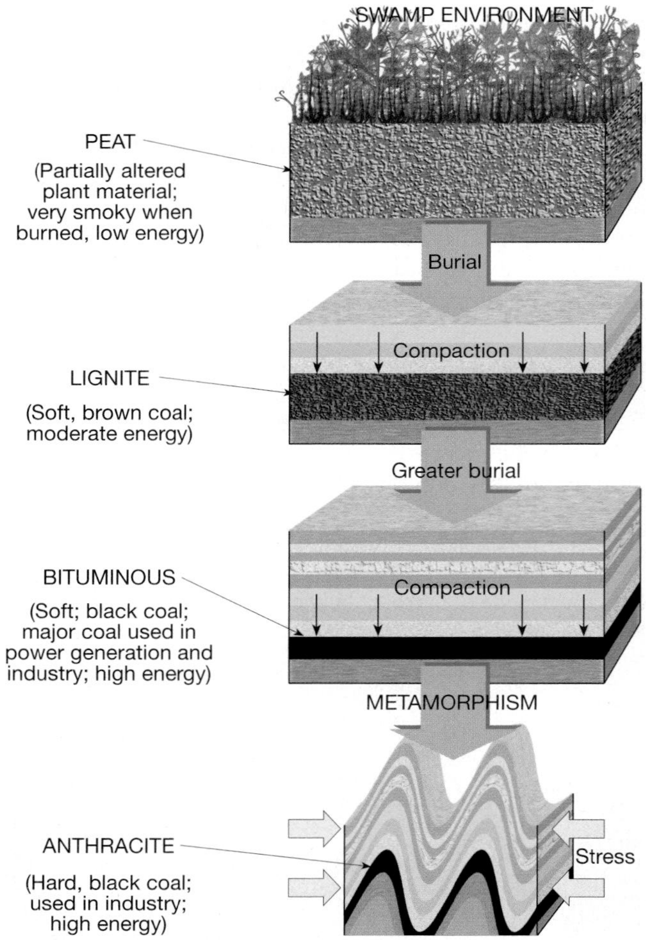

▲ **Figure 6.13** Successive stages in the formation of coal.

Lignite and bituminous coals are sedimentary rocks, but anthracite is a metamorphic rock. Anthracite forms when sedimentary layers are subjected to the folding and deformation associated with mountain building.

Classification of Sedimentary Rocks

The classification scheme in Figure 6.14 divides sedimentary rocks into two major groups: detrital and chemical. Further, we can see that the main criterion for subdividing the detrital rocks is particle size, whereas the primary basis for distinguishing among different rocks in the chemical group is their mineral composition.

As is the case with many (perhaps most) classifications of natural phenomena, the categories presented in Figure 6.14 are more rigid than the actual state of nature. In reality, many of the sedimentary rocks classified into the chemical group also contain at least small quantities of detrital sediment. Many limestones, for example, contain varying amounts of mud or sand, giving them a "sandy" or "shaly" quality. Conversely, because practically all detrital rocks are cemented with material that was originally dissolved in water, they too are far from being "pure."

As was the case with the igneous rocks examined in Chapter 3, *texture* is a part of sedimentary rock classification. Two major textures are used in the classification of sedimentary rocks: clastic and nonclastic. The term **clastic** is taken from a Greek word meaning "broken." Rocks that display a clastic texture consist of discrete fragments and particles that are cemented and compacted together. Although cement is present in the spaces between particles, these openings are rarely filled completely. All detrital rocks have a clastic texture. In

Did You Know?

Each year about 30 percent of the world's supply of salt is extracted from seawater. The seawater is pumped into ponds and allowed to evaporate, leaving behind "artificial evaporites," which are harvested.

Detrital Sedimentary Rocks

Clastic Texture (particle size)	Sediment Name	Rock Name
Coarse (over 2 mm)	Gravel (Rounded particles)	**Conglomerate**
Coarse (over 2 mm)	Gravel (Angular particles)	**Breccia**
Medium (1/16 to 2 mm)	Sand (If abundant feldspar is present the rock is called **Arkose**)	**Sandstone**
Fine (1/16 to 1/256 mm)	Mud	**Siltstone**
Very fine (less than 1/256 mm)	Mud	**Shale**

Chemical Sedimentary Rocks

Composition	Texture	Rock Name	
Calcite, $CaCO_3$	Nonclastic: Fine to coarse crystalline	**Crystalline Limestone**	
Calcite, $CaCO_3$	Nonclastic: Fine to coarse crystalline	**Travertine**	
Calcite, $CaCO_3$	Clastic: Visible shells and shell fragments loosely cemented	**Coquina**	**Biochemical Limestone**
Calcite, $CaCO_3$	Clastic: Various size shells and shell fragments cemented with calcite cement	**Fossiliferous Limestone**	**Biochemical Limestone**
Calcite, $CaCO_3$	Clastic: Microscopic shells and clay	**Chalk**	**Biochemical Limestone**
Quartz, SiO_2	Nonclastic: Very fine crystalline	**Chert (light colored) Flint (dark colored)**	
Gypsum $CaSO_4 \cdot 2H_2O$	Nonclastic: Fine to coarse crystalline	**Rock Gypsum**	
Halite, NaCl	Nonclastic: Fine to coarse crystalline	**Rock Salt**	
Altered plant fragments	Nonclastic: Fine-grained organic matter	**Bituminous Coal**	

▲ **Figure 6.14** Identification of sedimentary rocks. Sedimentary rocks are divided into two major groups—detrital and chemical—based on their source of sediment. The main criterion for naming detrital sedimentary rocks is particle size, whereas the primary basis for distinguishing among chemical sedimentary rocks is their mineral composition.

▲ **Figure 6.15** Like other evaporites, this sample of rock salt is said to have a nonclastic (crystalline) texture because it is composed of intergrown crystals. *(Photo by E. J. Tarbuck)*

addition, some chemical sedimentary rocks exhibit this texture. For example, coquina, the limestone composed of shells and shell fragments, is obviously as clastic as conglomerate or sandstone. The same applies for some varieties of oolitic limestone.

Some chemical sedimentary rocks have a **nonclastic** or **crystalline texture** in which the minerals form a pattern of interlocking crystals. The crystals may be microscopically small or large enough to be visible without magnification. Common examples of rocks with nonclastic textures are evaporites (Figure 6.15). The materials that make up many other nonclastic rocks may actually have originated as detrital deposits. In these instances, the particles probably consisted of shell fragments and other hard parts rich in calcium carbonate or silica. The clastic nature of the grains was subsequently obliterated or obscured because the particles recrystallized when they were consolidated into limestone or chert.

Nonclastic rocks consist of intergrown crystals, and some may resemble igneous rocks, which are also crystalline. The two rock types are usually easy to distinguish because the minerals that make up nonclastic sedimentary rocks are different from those found in most igneous rocks. For example, rock salt, rock gypsum, and some forms of limestone consist of intergrown crystals, but the minerals contained within these rocks (halite, gypsum, and calcite) are seldom associated with igneous rocks.

Did You Know?

These terms are often used when describing the mineral makeup of sedimentary rocks: *siliceous rocks* contain abundant quartz, *argillaceous* rocks are clay-rich, and *carbonate* rocks contain calcite or dolomite.

Sedimentary Rocks Represent Past Environments

GEODe ESSENTIALS OF GEOLOGY
Sedimentary Rocks
▼ Sedimentary Environments

Sedimentary rocks are important in the interpretation of Earth history. By understanding the conditions under which sedimentary rocks form, geologists can often deduce the history of a rock, including information about the origin of its component particles, the method and length of sediment transport, and the nature of the place where the grains eventually came to rest; that is, the environment of deposition. An **environment of deposition** or **sedimentary environment** is simply a geographic setting where sediment is accumulating. Each site is characterized by a particular combination of geologic processes and environmental conditions. Thus, when a series of sedimentary layers are studied, we can see the successive changes in environmental conditions that occurred at a particular place with the passage of time.

At any given time, the geographic setting and environmental conditions of a sedimentary environment determine the nature of the sediments that accumulate. Therefore, geologists carefully study the sediments in present-day depositional environments because the features they find can also be observed in ancient sedimentary rocks. By applying a thorough knowledge of present-day conditions, geologists attempt to reconstruct the ancient environments and geographical relationships of an area at the time a particular set of sedimentary layers were deposited. Such analyses often lead to the creation of maps depicting the geographic distribution of land and sea, mountains and river valleys, deserts and glaciers, and other environments of deposition. The foregoing description is an excellent example of the application of a fundamental principle of modern geology, namely that "the present is the key to the past."*

Sedimentary environments are commonly placed into one of three categories: continental, marine, or transitional (shoreline). Each category includes many specific subenvironments. Figure 6.16 is an idealized diagram illustrating a number of important sedimentary environments associated with each category. Later, Chapters 9 through 13 will describe these environments in detail. Each is an area where sediment accumulates and where organisms live and die. Each produces a characteristic

*For more on this idea, see "The Birth of Modern Geology" in Chapter 1.

sedimentary rock or assemblage that reflects prevailing conditions.

Sedimentary Structures

In addition to variations in grain size, mineral composition, and texture, sediments exhibit a variety of structures. Some, such as graded beds, are created when sediments are accumulating and are a reflection of the transporting medium. Others, such as mud cracks, form after the materials have been deposited and result from processes occurring in the environment. When present, sedimentary structures provide additional information that can be useful in the interpretation of Earth history.

Sedimentary rocks form as layer upon layer of sediment accumulates in various depositional environments. These layers, called **strata,** or **beds,** are probably *the single most common and characteristic feature of sedimentary rocks.* Each stratum is unique. It may be a coarse sandstone, a fossil-rich limestone, a black shale, and so on. When you look at Figure 6.17, or look back through this chapter at Figures 6.1, 6.3, and 6.10B, you will see many such layers, each different from the others. The variations in texture, composition, and thickness reflect the different conditions under which each layer was deposited.

The thickness of beds ranges from microscopically thin to tens of meters thick. Separating the strata are **bedding planes,** flat surfaces along which rocks tend to separate or break. Changes in the grain size or in the composition of the sediment being deposited can create bedding planes. Pauses in deposition can also lead to layering because chances are slight that newly deposited material will be exactly the same as previously deposited sediment. Generally, each bedding plane marks the end of one episode of sedimentation and the beginning of another.

Because sediments usually accumulate as particles that settle from a fluid, most strata are originally deposited as horizontal layers. There are circumstances, however, when sediments do not accumulate in horizontal beds. Sometimes when a bed of sedimentary rock is examined, we see layers within it that are inclined to the horizontal. When this occurs, it is called **cross-bedding** and is most characteristic of sand dunes, river deltas, and certain stream channel deposits (Figure 6.18).

Graded beds represent another special type of bedding. In this case the particles within a single sedimentary layer gradually change from coarse at the bottom to fine at the top. Graded beds are most characteristic of rapid deposition from water containing sediment of varying sizes. When a current experiences a rapid energy loss, the largest particles settle first, followed by successively smaller grains. The deposition of a graded bed is most often associated with a turbidity current, a mass of sediment-choked water that is denser than clear water and that moves downslope along the bottom of a lake or ocean (Figure 6.19).

As geologists examine sedimentary rocks, much can be deduced. A conglomerate, for example, may indicate a high-energy environment, such as a surf zone or rushing stream, where only coarse materials settle out and finer particles are kept suspended (Figure 6.20). If the rock is arkose, it may signify a dry climate where little chemical alteration of feldspar is possible. Carbonaceous shale is a sign of a low-energy, organic-rich environment, such as a swamp or lagoon.

Arkose contains abundant feldspar.

Other features found in some sedimentary rocks also give clues to past environments. Ripple marks are such a feature. **Ripple marks** are small waves of sand that develop on the surface of a sediment layer by the action of moving water or air (Figure 6.21A). The ridges form at right angles to the direction of motion. If the ripple marks were formed by air or water moving in essentially one direction, their form will be asymmetrical. These *current ripple marks* will have steeper sides in the downcurrent direction and more gradual slopes on the upcurrent side. Ripple marks produced by a stream flowing across a sandy channel or by wind blowing over a sand dune are two common examples of current ripples. When present in solid rock, they may be used to determine the direction of movement of ancient wind or water currents. Other ripple marks have a symmetrical form. These features, called *oscillation ripple marks,* result from the back-and-forth movement of surface waves in a shallow nearshore environment.

Ripple marks in a river channel. (Galen Rowell)

Mud cracks (Figure 6.21B) indicate that the sediment in which they were formed was alternately wet and dry. When exposed to air, wet mud dries out and shrinks, producing cracks. Mud cracks are associated with such environments as tidal flats, shallow lakes, and desert basins.

Fossils, the remains or traces of prehistoric life, are important inclusions in sediment and sedimentary rock. They are important tools for interpreting the geologic past. Knowing the nature of the life forms that existed at a particular time helps researchers decipher past environmental conditions. Further, fossils are important time indicators and play a key role in correlating rocks that are of similar ages but are from different places. Fossils will be examined in more detail in Chapter 18.

Fossil of an amphibian from Permian-age rocks in Texas. (T. A. Wiewandt/DRK)

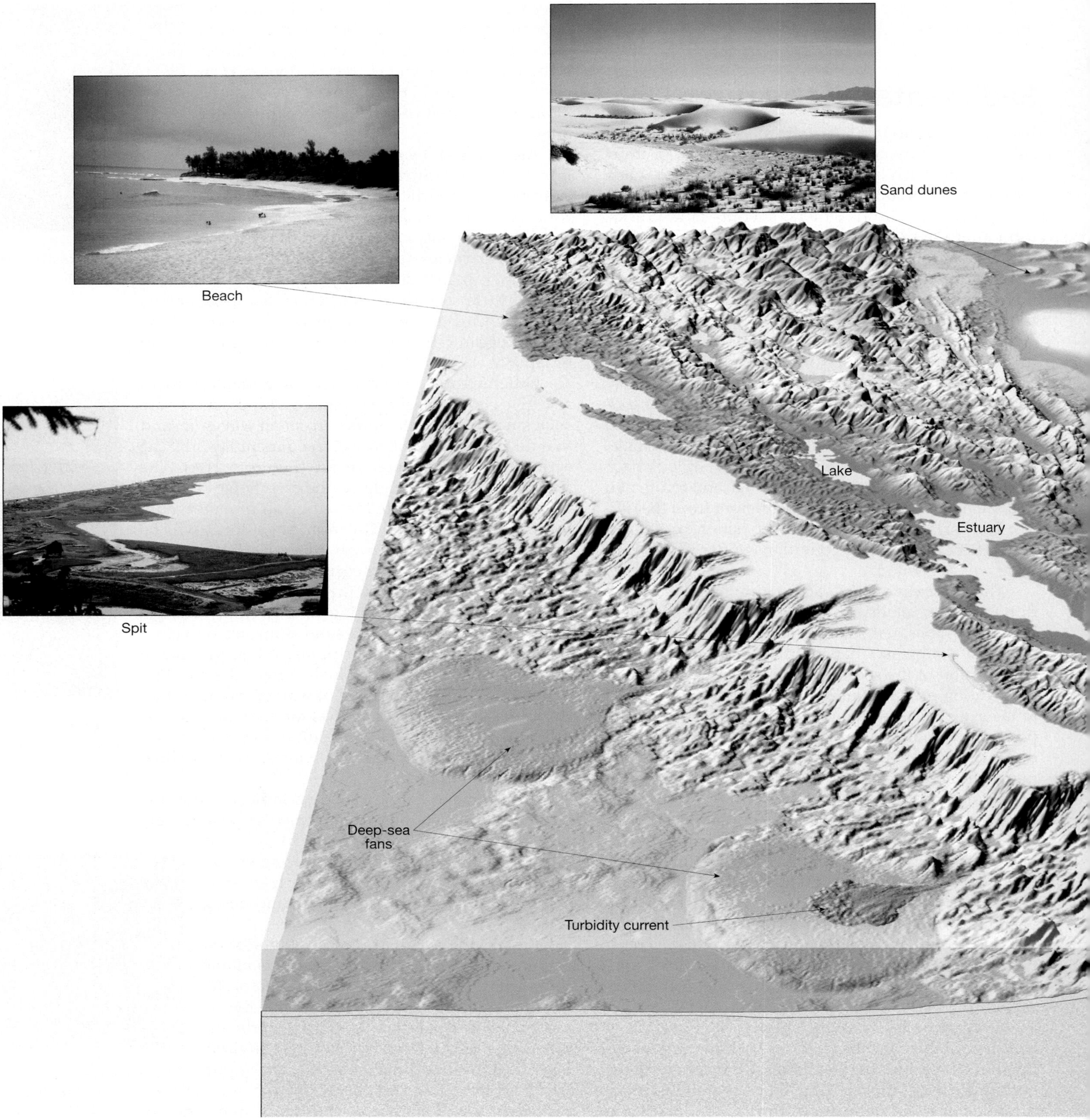

▲ **Figure 6.16** Sedimentary environments are those places where sediment accumulates. Each is characterized by certain physical, chemical, and biological conditions. Because each sediment contains clues about the environment in which it was deposited, sedimentary rocks are important in the interpretation of Earth history. A number of important continental, transitional, and marine sedimentary environments are represented in this idealized diagram. *(Photos by E. J. Tarbuck, except alluvial fan, by Marli Miller)*

Alluvial fans
Glacial deposits
Salt flat
Playa lake
Stream
Swamp
Floodplain
Delta
Lagoon
Reef
Barrier island

▲ **Figure 6.17** This outcrop of sedimentary strata illustrates the characteristic layering of this group of rocks. Minnewaska State Park, New York. *(Photo by Carr Clifton)*

Nonmetallic Mineral Resources from Sedimentary Rocks

Earth materials that are not used as fuels or processed for the metals they contain are referred to as **nonmetallic mineral resources.** Realize that use of the word "mineral" is very broad in this economic context and is quite different from the geologist's strict definition of mineral found in Chapter 2. Nonmetallic mineral resources are extracted and processed either for the nonmetallic elements they contain or for the physical and chemical properties they possess (Table 6.2). Although these resources have diverse origins, many are sediments or sedimentary rocks.

People often do not realize the importance of nonmetallic minerals, because they see only the products that resulted from their use and not the minerals themselves. That is, many nonmetallics are used up in the process of creating other products. Examples include the fluorite and limestone that are part of the steelmaking process, the abrasives required to make a piece of machinery, and the fertilizers needed to grow a food crop.

The quantities of nonmetallic minerals used each year are enormous. Per capita consumption of nonfuel resources in the United States is nearly 11 metric tons, of which over 95 percent are nonmetallics (Figure 6.22). Nonmetallic mineral resources are commonly divided into two broad groups: *building materials* and *industrial minerals.* Because some substances have many different uses, they are found in both categories. Limestone, perhaps the most versatile and widely used rock of all, is the best example. As a building material, it is used not only as crushed rock and building stone but also in making cement. Moreover, as an industrial mineral, limestone is an ingredient in the manufacture of steel and is used in agriculture to neutralize acidic soils.

Did You Know?

Producing one ton of steel requires about one-third ton of limestone and between 2 and 20 lbs. of fluorspar (fluorite).

Other important building materials include cut stone, aggregate (sand, gravel, and crushed rock), gypsum for plaster and wallboard, clay for tile and bricks, and cement, which is made from limestone and shale. Cement and aggregate go into the making of concrete, a material that is essential to practically all construction.

A wide variety of resources are classified as industrial minerals. In some instances these materials are important because they are sources of specific chemical elements or compounds. Such minerals are used in the manufacture of chemicals and the production of fertilizers. In other cases their importance is related to the physical properties they exhibit. Examples include minerals such as corundum and garnet, which are used as abrasives. Although supplies are generally plentiful, most industrial minerals are not nearly as abundant as building materials.

Moreover, deposits are far more restricted in distribution and extent. As a result, many of these nonmetallic resources must be transported considerable distances, which of course adds to their cost. Unlike most building materials, which need a minimum of processing before they are ready to use, many industrial minerals require considerable processing to extract the desired substance at the proper degree of purity for its ultimate use.

A.

B.

▲ **Figure 6.18** **A.** The cutaway section of this sand dune shows cross-bedding. *(Photo by John S. Shelton)* **B.** The cross-bedding of this sandstone indicates it was once a sand dune. *(Photo by David Muench Photography, Inc.)*

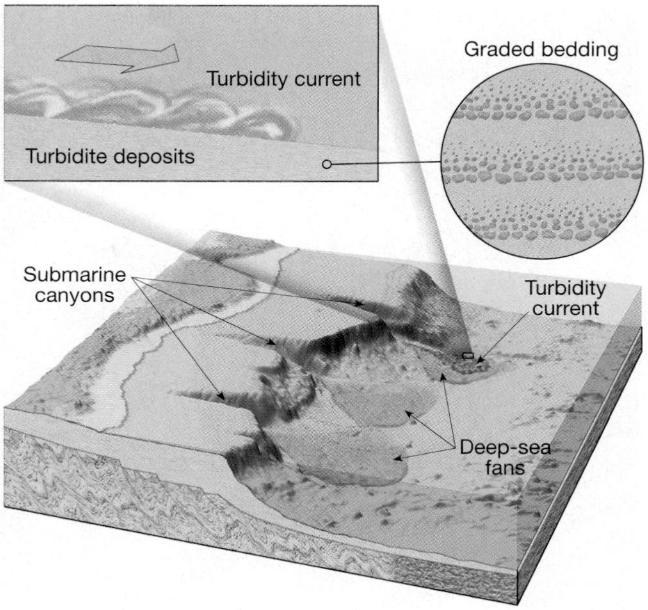

▲ **Figure 6.19** Turbidity currents are downslope movements of dense, sediment-laden water. They are created when sand and mud on the continental shelf and slope are dislodged and thrown into suspension. Because such mud-choked water is denser than normal seawater, it flows downslope, eroding and accumulating more sediment. Beds deposited by these currents are called *turbidites.* Each event produces a single bed characterized by a decrease in sediment size from bottom to top, a feature known as a *graded bed.*

▲ **Figure 6.20** In a turbulent stream channel, only large particles settle out. Finer sediments remain suspended and continue their downstream journey. *(Photo by E. J. Tarbuck)*

Energy Resources from Sedimentary Rocks

Coal, petroleum, and natural gas are the primary fuels of our modern industrial economy. About 86 percent of the energy consumed in the United States today comes from these basic fossil fuels (Figure 6.23). Although major shortages of oil and gas will not occur for many years, proven reserves are declining. Despite new exploration, even in very remote regions and severe environments, new sources of oil are not keeping pace with consumption.

A.

B.

▲ **Figure 6.21** **A.** Ripple marks are produced by currents of water or wind. *(Photo by Stephen Trimble)* **B.** Mud cracks form when wet mud or clay dries out and shrinks. *(Photo by Gary Yeowell/Getty Images, Inc-Stone Allstock)*

Did You Know?

Just 2 kilometers of four-lane highway require more than 85 tons of aggregate. The United States produces about 2 billion tons of aggregate per year. This represents about one half of the entire nonenergy mining volume in the country.

Unless large, new petroleum reserves are discovered (which is possible but not likely), a greater share of our future needs will eventually have to come from coal and/or from alternative energy sources such as nuclear, solar, wind, tidal, and hydroelectric power. Two fossil-fuel alternatives—oil sands and oil shale—are sometimes mentioned as promising new sources of liquid fuels. In the following sections, we will briefly examine the fuels that have traditionally supplied our energy needs.

Coal

Along with oil and natural gas, coal is commonly called a **fossil fuel.** Such a designation is appropriate, because each time we burn coal we are using energy from the Sun that was stored by plants many millions of years ago. We are indeed burning a "fossil."

Coal has been an important fuel for centuries. In the nineteenth and early twentieth centuries, cheap and plentiful coal powered the Industrial Revolution. By 1900, coal was providing 90 percent of the energy used in the United States. Although still important, coal currently provides only about 23 percent of the energy needs of the United States.

Table 6.2 Uses of nonmetallic minerals.

Mineral	Uses
Apatite	Phosphorus fertilizers
Asbestos (chrysotile)	Incombustible fibers
Calcite	Aggregate; steelmaking; soil conditioning; chemicals; cement; building stone
Clay minerals (kaolinite)	Ceramics; china
Corundum	Gemstones; abrasives
Diamond	Gemstones; abrasives
Fluorite	Steelmaking; aluminum refining; glass; chemicals
Garnet	Abrasives; gemstones
Graphite	Pencil lead; lubricant; refractories
Gypsum	Plaster of Paris
Halite	Table salt; chemicals; ice control
Muscovite	Insulator in electrical applications
Quartz	Primary ingredient in glass
Sulfur	Chemicals; fertilizer manufacture
Sylvite	Potassium fertilizers
Talc	Powder used in paints, cosmetics, etc.

▲ **Figure 6.22** The annual per capita consumption of nonmetallic and metallic mineral resources for the United States. About 94 percent of the materials used are nonmetallic.

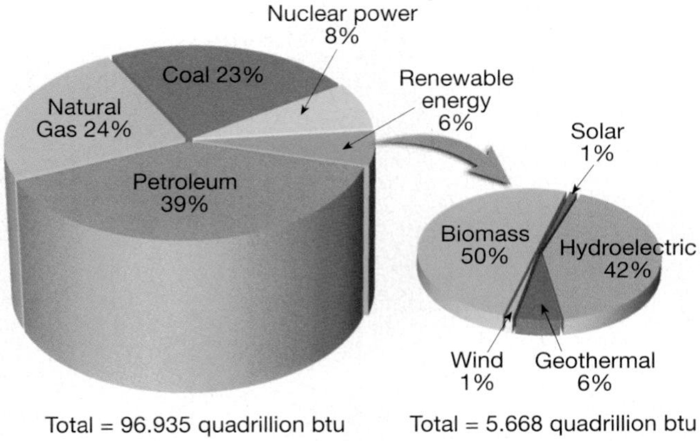

▲ **Figure 6.23** U.S. energy consumption, 2001. The total was nearly 97 quadrillion Btu. A quadrillion, by the way, is 10 raised to the 12th power, or a million million—A quadrillion Btu is a convenient unit for referring to U.S. energy use as a whole. *(Source: U.S. Department of Energy, Energy Information Administration)*

More than 90 percent of present-day coal usage is for the generation of electricity. As oil reserves gradually diminish in the years to come, the use of coal may actually increase. Expanded coal production is possible because the world has enormous reserves and the technology to mine coal efficiently. In the United States, coal fields are widespread and contain supplies that should last for hundreds of years (Figure 6.24).

Although coal is plentiful, its recovery and its use present a number of problems. Surface mining can turn the countryside into a scarred wasteland if careful (and costly) reclamation is not carried out to restore the land. (Today all U.S. surface mines must reclaim the land.) Although underground mining does not scar the landscape to the same degree, it has been costly in terms of human life and health. Underground mining long ago ceased to be a pick-and-shovel operation and is today a highly mechanized and computerized process. Strong federal safety regulations have made U.S. mining quite safe. However, the hazards of roof falls, gas explosions, and working with heavy equipment remain.

Did You Know?

The biomass category in Figure 6.23 is a relatively new name for the oldest fuels used by people. Examples include firewood, charcoal, crop residues, and animal wastes.

Air pollution is a major problem associated with the burning of coal. Much coal contains significant quantities of sulfur. Despite efforts to remove sulfur before the coal is burned, some remains. When the coal is burned, the sulfur is converted into noxious sulfur-oxide gases. Through a series of complex chemical reactions in the atmosphere, the sulfur oxides are converted to sulfuric acid, which then falls to Earth's surface as rain or snow. This *acid precipitation* can have detrimental ecological effects over widespread areas.

Because none of the problems just mentioned are likely to prevent the increased use of this important and abundant fuel, stronger efforts must be made to correct the problems associated with the mining and use of coal.

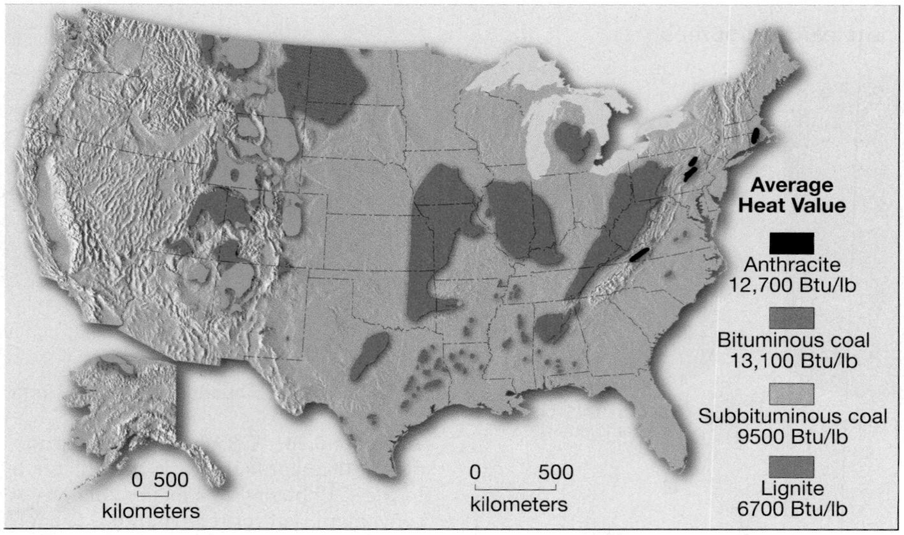

▲ **Figure 6.24** Coal fields of the United States. *(Courtesy of the Bureau of Mines, U.S. Department of the Interior)*

Oil and Natural Gas

Petroleum and natural gas obviously are not rocks, but we class them as mineral resources because they *come from* sedimentary rocks. They are found in similar environments and typically occur together. Both consist of various hydrocarbon compounds (compounds consisting of hydrogen and carbon) mixed together. They may also contain small quantities of other elements, such as sulfur, nitrogen, and oxygen. Like coal, petroleum and natural gas are biological products derived from the remains of organisms. However, the environments in which they form are very different, as are the organisms. Coal is formed mostly from plant material that accumulated in a swampy environment above sea level. Oil and gas are derived from the remains of both plants and animals having a marine origin.

Did You Know?

The first successful oil well was completed by Edwin Drake on August 27, 1859, along Oil Creek near Titusville, Pennsylvania. Oil-bearing strata were encountered at a depth of just 69 ft.

Edwin Drake (right) in front of the first successful oil well. (CORBIS/Bettmann)

Petroleum formation is complex and not completely understood. Nevertheless, we know that it begins with the accumulation of sediment in ocean areas that are rich in plant and animal remains. These accumulations must occur where biological activity is high, such as nearshore areas. However, most marine environments are oxygen-rich, which leads to the decay of organic remains before they can be buried by other sediments. Therefore, accumulations of oil and gas are not as widespread as the marine environments that support abundant biological activity. This limiting factor notwithstanding, large quantities of organic matter are buried and protected from oxidation in many offshore sedimentary basins. With increasing burial over millions of years, chemical reactions gradually transform some of the original organic matter into the liquid and gaseous hydrocarbons we call petroleum and natural gas.

Unlike the solid organic matter from which they formed, the newly created petroleum and natural gas are *mobile.* These fluids are gradually squeezed from the compacting, mud-rich layers where they originate into adjacent permeable beds such as sandstone, where openings between sediment grains are larger. Because all of this occurs underwater, the rock layers containing the oil and gas are already saturated with water. But oil and gas are less dense than water, so they migrate upward through the water-filled pore spaces of the enclosing rocks. Unless something acts to halt this upward migration, the fluids will eventually reach the surface. There the volatile components will evaporate.

Sometimes the upward migration is halted. A geologic environment that allows for economically significant amounts of oil and gas to accumulate underground is termed an **oil trap.** Several geologic structures can act as oil traps. All have two basic conditions in common: a porous, permeable **reservoir rock** that will yield petroleum and natural gas in sufficient quantities to make drilling worthwhile; and a **cap rock,** such as shale, that is virtually impermeable to oil and gas. The cap rock keeps the upwardly mobile oil and gas from escaping at the surface.

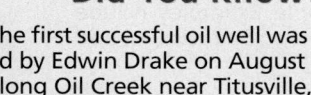

The Chapter in Review

- *Sedimentary rock* consists of *sediment* that has been *lithified* into solid rock. Sediment has two principal sources: (1) as *detrital material*, which originates and is transported as solid particles from both mechanical and chemical weathering, which, when lithified, forms detrital sedimentary rocks; and (2) from soluble material produced largely by chemical weathering, which, when precipitated, forms *chemical sedimentary rocks.*

- *Diagenesis* refers to all of the physical, chemical, and biological changes that occur after sediments are deposited and during and after the time they are turned into sedimentary rock. Burial promotes diagenesis. Diagenesis includes lithification.

- *Lithification* refers to the processes by which unconsolidated sediments are transformed into solid sedimentary rock. Most sedimentary rocks are lithified by means of *compaction* and/or *cementation.* Compaction occurs when the weight of overlying materials compresses the deeper sediments. Cementation, the most important process by which sediments are converted to sedimentary rock, occurs when soluble cementing materials, such as *calcite, silica,* and *iron oxide,* are precipitated onto sediment grains, fill open spaces, and join the particles. Although most sedimentary rocks are lithified by compaction or cementation, certain chemical rocks, such as the evaporites, initially form as solid masses of intergrown crystals.

- *Particle size* is the primary basis for distinguishing among various detrital sedimentary rocks. The size of the particles in a detrital rock indicates the energy of the medium that transported them. For example, gravels are moved by swiftly flowing rivers, whereas less energy is required to transport sand. Common detrital sedimentary rocks include *shale* (silt- and clay-size particles), *sandstone,* and *conglomerate* (rounded gravel-size particles) or *breccia* (angular gravel-size particles).

- Precipitation of chemical sediments occurs in two ways: (1) by *inorganic processes,* such as evaporation and chemical activity; or (2) by *organic processes* of water-dwelling organisms that produce sediments of *biochemical origin. Limestone,* the most abundant chemical sedimentary rock, consists of the mineral calcite ($CaCO_3$) and forms either by inorganic means or as the result of biochemical processes. Inorganic limestones include *travertine,* which is commonly seen in caves, and *oolitic limestone,* consisting of small spherical grains of calcium carbonate. Other common chemical sedimentary rocks include *dolostone* (composed of the calcium-magnesium carbonate mineral dolomite), *chert* (made of microcrystalline quartz), *evaporites* (such as rock salt and rock gypsum), and *coal* (lignite and bituminous).

- Sedimentary rocks can be divided into two main groups: *detrital* and *chemical.* All detrital rocks have a *clastic texture,* which consists of discrete fragments and particles that are cemented and compacted together. The main criterion for subdividing the detrital rocks is particle size. Common detrital rocks include *conglomerate, sandstone,* and *shale.* The primary basis for distinguishing among different rocks in the chemical group is their mineral composition. Some chemical rocks, such as those deposited when seawater evaporates, have a *nonclastic texture* in which the minerals form a pattern of interlocking crystals. However, in reality, many of the sedimentary rocks classified into the chemical group also contain at least small quantities of detrital sediment. Common chemical rocks include *limestone, rock gypsum,* and *coal* (e.g., lignite and bituminous).

- Sedimentary environments are those places where sediment accumulates. They are grouped into continental, marine, and transitional (shoreline) environments. Each is characterized by certain physical, chemical, and biological conditions. Because sediment contains clues about the environment in which it was deposited, sedimentary rocks are important in the interpretation of Earth's history.

- Layers, called *strata,* or *beds,* are probably the single most characteristic feature of sedimentary rocks. Other features found in some sedimentary rocks, such as *ripple marks, mud cracks, cross-bedding, graded bedding,* and *fossils,* also give clues to past environments.

- Earth materials that are not used as fuels or processed for the metals they contain are referred to as *nonmetallic resources.* Many are sediments or sedimentary rocks. The two broad groups of nonmetallic resources are *building materials* and *industrial minerals.* Limestone, perhaps the most versatile and widely used rock of all, is found in both groups.

- *Coal, petroleum,* and *natural gas,* the *fossil fuels* of our modern economy, are all associated with sedimentary rocks. Coal originates from large quantities of plant remains that accumulate in an oxygen-deficient environment, such as a swamp. More than 70 percent of present-day coal usage is for the generation of electricity. Air pollution from the sulfur-oxide gases that form from burning most types of coal is a significant environmental problem.

- Oil and natural gas, which commonly occur together in the pore spaces of some sedimentary rocks, consist of various *hydrocarbon compounds* (compounds made of hydrogen and carbon) mixed together. Petroleum formation is associated with the accumulation of sediment in ocean areas that are rich in plant and animal remains that become buried and isolated in an oxygen-deficient environment. As the mobile petroleum and natural gas form, they migrate and accumulate in adjacent permeable beds such as sandstone. If the upward migration is halted by an impermeable rock layer, referred to as a *cap rock,* a geologic environment develops that allows for economically significant amounts of oil and gas to accumulate underground in what is termed an *oil trap.*

Key Terms

bedding plane (p. 151)
beds (strata) (p. 151)
biochemical origin (p. 144)
cap rock (p. 158)
cementation (p. 139)
chemical sedimentary rock (p. 140)
clastic (p. 149)
compaction (p. 139)
cross-bedding (p. 151)
crystalline texture (p. 150)
detrital sedimentary rock (p. 139)
diagenesis (p. 139)
environment of deposition (p. 150)
evaporite (p. 147)
fossil (p. 151)
fossil fuel (p. 156)
graded bed (p. 151)
lithification (p. 139)
mud crack (p. 151)
nonclastic (p. 150)
nonmetallic mineral resource (p. 154)
oil trap (p. 158)
reservoir rock (p. 158)
ripple mark (p. 151)
salt flat (p. 148)
sediment (p. 138)
sedimentary environment (p. 150)
sorting (p. 141)
strata (beds) (p. 151)

Questions for Review

1. How does the volume of sedimentary rocks in Earth's crust compare with the volume of igneous rocks in the crust? Are sedimentary rocks evenly distributed throughout the crust?

2. What is *diagenesis*? Give an example.

3. Compaction is most important as a lithification process with which sediment size?

4. List three common cements for sedimentary rocks. How might each be identified?

5. What minerals are most common in detrital sedimentary rocks? Why are these minerals so abundant?

6. What is the primary basis for distinguishing among various detrital sedimentary rocks?

7. Why does shale usually crumble quite easily?

8. Distinguish between *conglomerate* and *breccia*.

9. Distinguish between the two categories of chemical sedimentary rocks.

10. What are evaporite deposits? Name a rock that is an evaporite.

11. How is bituminous coal different from lignite? How is anthracite different from bituminous?

12. Each of the following statements describes one or more characteristics of a particular sedimentary rock. For each statement, name the sedimentary rock that is being described.

a. An evaporite used to make plaster.
b. A fine-grained detrital rock that breaks into chunks or blocks.
c. Dark-colored sandstone containing angular rock particles as well as clay, quartz, and feldspar.
d. The most abundant chemical sedimentary rock.
e. A dark-colored, dense, hard rock made of microcrystalline quartz.
f. A variety of limestone composed of small spherical grains.

13. What is the primary basis for distinguishing among different chemical sedimentary rocks?

14. Distinguish between *clastic* and *nonclastic* textures. What type of texture is common to all detrital sedimentary rocks?

15. What is probably the single most characteristic feature of sedimentary rocks?

16. Distinguish between *cross-bedding* and *graded bedding*.

17. Nonmetallic resources are commonly divided into two broad groups. List the two groups and some examples of materials that belong to each.

18. Coal enjoys the advantage of being plentiful. What are some disadvantages associated with the production and use of coal?

19. What is an *oil trap?* List two conditions common to all oil traps.

Online Study Guide

The *Essentials of Geology* Web site uses the resources and flexibility of the Internet to aid in your study of the topics in this chapter. Written and developed by geology instructors, this site will help improve your understanding of geology. Visit **www.prenhall.com/lutgens** and click on the cover of *Essentials of Geology 9e* to find:

- Online review quizzes.
- Critical thinking exercises.
- Links to chapter-specific Web resources.
- Internet-wide key-term searches.

http://www.prenhall.com/lutgens

The French Alps are highly deformed and metamorphosed. *(Photo by J. P. Delobelle/Peter Arnold, Inc.)*

CHAPTER 7

Metamorphic Rocks

Focus on Learning

To assist you in learning the important concepts in this chapter, you will find it helpful to focus on the following questions:

- What are metamorphic rocks, and how do they form?
- In which three geologic settings is metamorphism most likely to occur?
- What are the agents of metamorphism?
- What are the two textural divisions of metamorphic rocks and the conditions associated with the occurrence of each?
- What are the names, textures, and compositions of the common metamorphic rocks?
- How is the intensity, or degree, of metamorphism reflected in the texture and mineralogy of metamorphic rocks?

The folded and metamorphosed rocks shown in Figure 7.1 were once flat-lying sedimentary strata. Compressional forces of unimaginable magnitude and temperatures hundreds of degrees above surface conditions prevailed for perhaps thousands or millions of years to produce the deformation displayed by these rocks. Under such extreme conditions, solid rock responds by folding, fracturing, and often by flowing. This chapter looks at the tectonic forces that forge metamorphic rocks and how these rocks change in appearance, mineralogy, and sometimes even in overall chemical composition.

Ancient metamorphic rocks of the Canadian Shield. (Robert Hildebrand)

Extensive areas of metamorphic rocks are exposed on every continent in the relatively flat regions known as **shields.** These metamorphic regions are found in eastern Canada, Brazil, Africa, India, Australia, and Greenland. Moreover, metamorphic rocks are an important component of many mountain belts, including the Alps and the Appalachians, where they make up a large portion of a mountain's crystalline core. Even those portions of the stable continental interiors that are covered by sedimentary rocks are underlain by metamorphic basement rocks. In these settings the metamorphic rocks are highly deformed and intruded by large igneous masses. Indeed, significant parts of Earth's continental crust are composed of metamorphic and associated igneous rocks.

Unlike some igneous and sedimentary processes that take place in surface or near-surface environments, metamorphism invariably occurs deep within Earth, beyond our direct observation. Notwithstanding this significant obstacle, geologists have developed techniques that have allowed them to learn a great deal about the conditions under which metamorphic rocks form. In turn, the study of metamorphic rocks provides important insights into the geologic processes that operate within Earth's crust and upper mantle.

Metamorphism

GEODe: Metamorphic Rocks ▼ Introduction

Recall from the section on the rock cycle in Chapter 1 that metamorphism is the transformation of one rock type into another. Metamorphic rocks are produced from preexisting igneous, sedimentary, or even from other metamorphic rocks. Thus, every metamorphic rock has a **parent rock**—the rock from which it was formed.

Metamorphism, which means to "change form," is a process that leads to changes in the mineralogy, texture, and often the chemical composition of rocks. Metamorphism takes place where preexisting rock is subjected to a physical or chemical environment that is significantly different from that in which it initially formed. These include changes in temperature, pressure (stress), and the introduction of chemically active fluids. In response to these new conditions, the rock gradually changes until a state of equilibrium with the new environment is reached. Most metamorphic changes

▼ **Figure 7.1** Deformed and metamorphosed rocks, Anza-Borrego Desert State Park, California. *(Photo by A. P. Trujillo/APT Photos)*

occur at the elevated temperatures and pressures that exist in the zone beginning a few kilometers below Earth's surface and extending into the upper mantle.

Metamorphism often progresses incrementally, from slight changes (*low-grade metamorphism*) to substantial changes (*high-grade metamorphism*). For example, under low-grade metamorphism, the common sedimentary rock *shale* becomes the more compact metamorphic rock called *slate.* Hand samples of these rocks are sometimes difficult to distinguish, illustrating that the transition from sedimentary to metamorphic is often gradual and the changes can be subtle.

In more extreme environments, metamorphism causes a transformation so complete that the identity of the parent rock cannot be determined. In high-grade metamorphism, such features as bedding planes, fossils, and vesicles that may have existed in the parent rock are obliterated. Further, when rocks at depth (where temperatures are high) are subjected to directed pressure, they slowly deform to produce a variety of textures as well as large-scale structures such as folds.

In the most extreme metamorphic environments, the temperatures approach those at which rocks melt. However, *during metamorphism the rock must remain essentially solid,* for if complete melting occurs, we have entered the realm of igneous activity.

Most metamorphism occurs in one of three settings:

1. When rock is intruded by a magma body, *contact* or *thermal metamorphism* may take place. Here, change is driven by a rise in temperature within the host rock surrounding an igneous intrusion.
2. *Hydrothermal metamorphism* involves chemical alterations that occur as hot ion-rich water circulates through fractures in rock. This type of metamorphism is usually associated with igneous activity that provides the heat required to drive chemical reactions and circulate these fluids through rock.
3. During mountain building, great quantities of rock are subjected to directed pressures and high temperatures associated with large-scale deformation called *regional metamorphism.*

After considering the agents of metamorphism and some common metamorphic rocks, we will examine these metamorphic environments as well as others.

Agents of Metamorphism

GEODe ESSENTIALS OF GEOLOGY
Metamorphic Rocks
▼ Agents of Metamorphism

The agents of metamorphism include *heat, pressure (stress),* and *chemically active fluids.* During metamorphism, rocks are usually subjected to all three metamorphic agents simultaneously. However, the degree of metamorphism and the contribution of each agent vary greatly from one environment to another.

Heat as a Metamorphic Agent

The most important agent of metamorphism is *heat* because it provides the energy to drive chemical reactions that result in the recrystallization of existing minerals and/or the formation of new minerals. Recall from the discussion of igneous rocks that an increase in temperature causes the ions within a mineral to vibrate more rapidly. Even in a crystalline *solid,* where ions are strongly bonded, this elevated level of activity allows individual atoms to migrate more freely between sites in the crystalline structure.

Did You Know?

Some low-grade metamorphic rocks actually contain fossils. When fossils are present in metamorphic rocks, they provide useful clues for determining the original rock type and its depositional environment. In addition, fossils whose shapes have been distorted during metamorphism provide insight into the extent to which the rock has been deformed.

Changes Caused by Heat. Heat affects Earth materials, especially those that form in low temperature environments, in two ways. First, it promotes recrystallization of individual mineral grains. This is particularly true of clays, fine-grained sediments, and some chemical precipitates. Higher temperatures promote recrystallization where fine particles tend to coalesce into larger grains of the same mineralogy.

Second, heat may raise the temperature of a rock to the point where one, or more, of the minerals are no longer chemically stable. In such cases, the constituent ions tend to arrange themselves into crystalline structures that are more stable in the new high-energy environment. Such chemical reactions result in the creation of new minerals with stable configurations that have an overall composition roughly equivalent to that of the original minerals. (In some environments, ions may migrate into or out of a rock unit, thereby changing its overall chemical composition.)

In summary, if we were to traverse a region of metamorphic rocks (now uplifted and exposed) while traveling in the direction of increasing metamorphism, we would expect to observe two changes largely attributable to increased temperature. The grain size of the rocks would increase and the mineralogy would gradually change.

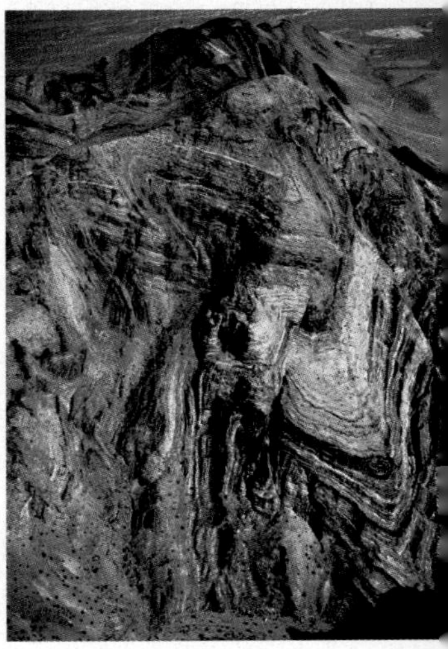

Deformed strata, Death Valley, California. (Michael Collier)

Sources of Heat. The heat to metamorphose rocks comes mainly from energy released by radioactive decay and thermal energy stored in Earth's interior. Recall that temperatures increase with depth at a rate known as the *geothermal gradient.* In the upper crust, this increase in temperature averages between 20°C and 30°C per kilometer (Figure 7.2). Thus, rocks that formed at Earth's surface will experience a gradual increase in temperature as they are transported (subducted) to greater depths (Figure 7.2). When buried to a depth of about 8 kilometers (5 miles), where temperatures are

Did You Know?

Glacial ice is a metamorphic rock that exhibits ductile flow much like hot rocks buried deep within Earth's crust. Although we think of glacial ice as being cold, it is in fact "hot," relative to its melting temperature. Therefore, we should not be surprised that the ice within a glacier gradually flows downslope in response to the force of gravity.

about 200°C, clay minerals tend to become unstable and begin to recrystallize into minerals, such as chlorite and muscovite, that are stable in this environment. (Chlorite is a micalike mineral formed by the metamorphism of dark silicate minerals.) However, many silicate minerals, particularly those found in crystalline igneous rocks—quartz and feldspar for example—remain stable at these temperatures. Thus, metamorphic changes in these minerals generally occur at much greater depths.

Environments where rocks may be carried to great depths and heated include convergent plate boundaries where slabs of sediment-laden oceanic crust are being subducted. In addition, rocks may become deeply buried in large basins where gradual subsidence results in very thick accumulations of sediment (Figure 7.2). Such locations, exemplified by the Gulf of Mexico, are known to develop metamorphic conditions near the base of the pile.

Metamorphism can occur at shallow crustal depths in regions where hot springs and geysers are active.

Furthermore, continental collisions, which result in crustal thickening, cause rocks to become deeply buried where elevated temperatures may trigger partial melting.

Heat may also be transported from the mantle into even the shallowest layers of the crust. Rising mantle plumes, upwelling at mid-ocean ridges, and magma generated by partial melting of mantle rock at subduction zones are three examples (Figure 7.2). Anytime magma forms and buoyantly rises toward the surface, metamorphism generally occurs. When magma intrudes relatively cool rocks at shallow depths, the host rock is "baked." This process, called *contact metamorphism,* will be considered later in this chapter.

Pressure and Differential Stress

Pressure, like temperature, also increases with depth as the thickness of the overlying rock increases. Buried rocks are subjected to **confining pressure,** which is analogous to water pressure, where the forces are applied equally in all directions (Figure 7.3A). The deeper you go in the ocean, the greater the confining pressure. The same is true for rock that is buried. Confining pressure causes the spaces between mineral grains to close, producing a more compact rock having a greater density (Figure 7.3A). Further, at great depths, confining pressure may cause minerals to recrystallize into new minerals that display a more compact crystalline form. Confining pressure does *not,* however, fold and deform rocks like those shown in Figure 7.1.

In addition to confining pressure, rocks may be subjected to directed pressure. This occurs, for example, at convergent plate boundaries where slabs of lithosphere collide. Here the forces that deform rock are unequal in different directions and are referred to as **differential stress.**

Unlike confining pressure, which "squeezes" rock equally in all directions, differential stresses are greater in one direction than in others. As shown in Figure 7.3B, rocks subjected to differential stress are shortened in the direction of greatest stress and elongated, or lengthened, in the direction perpendicular to that stress. As a result, the rocks involved are often *folded* or *flattened* (similar to stepping on a rubber ball). Along convergent plate boundaries the greatest differential stress is directed roughly horizontal in the direction of plate motion, and the least pressure is in the vertical direction. Consequently, in these settings the crust is greatly shortened (horizontally) and thickened (vertically).

In surface environments where temperatures are comparatively low, rocks are *brittle* and tend to fracture when subjected to differential stress. Continued deformation grinds and pulverizes the mineral grains into

Figure 7.2 Illustration of the geothermal gradient and its role in metamorphism. Notice how the geothermal gradient is lowered by the subduction of comparatively cool oceanic lithosphere. By contrast, thermal heating is evident where magma intrudes the upper crust.

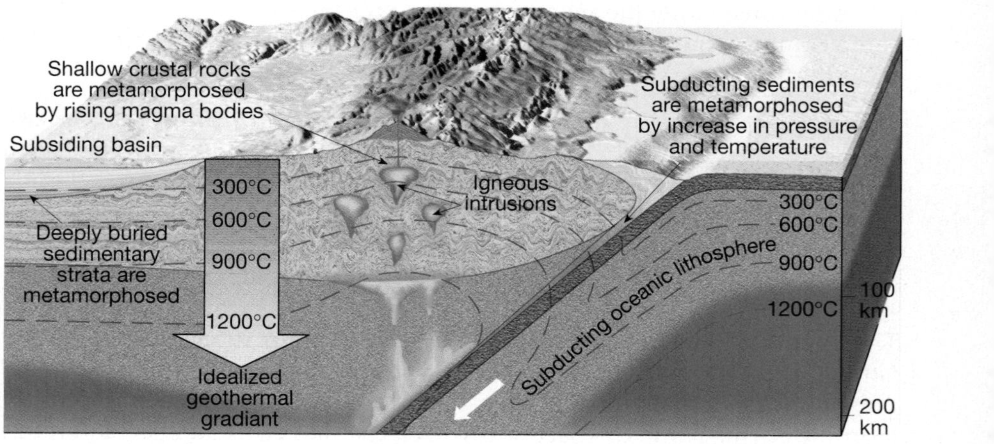

small fragments. By contrast, in high-temperature environments rocks are *ductile.* When rocks exhibit ductile behavior, their mineral grains tend to flatten and elongate when subjected to differential stress. This accounts for their ability to deform by flowing (rather than fracturing) to generate intricate folds. As you will see, differential stress also plays a major role in developing metamorphic textures.

Chemically Active Fluids

Fluids composed mainly of water and other volatile components, including carbon dioxide, are believed to play an important role in some types of metamorphism. Fluids that surround mineral grains act as catalysts to promote recrystallization by enhancing ion migration. In progressively hotter environments these ion-rich fluids become correspondingly more reactive. When two mineral grains are squeezed together, the part of their crystalline structures that touch are the most highly stressed. Ions located at these sites are readily dissolved by the hot fluids and migrate along the surface of the grain to the pore spaces located between individual grains. Thus, hydrothermal fluids aid in the recrystallization of mineral grains by dissolving material from regions of high stress and then precipitating (depositing) this material in areas of low stress. As a result, *minerals tend to recrystallize and grow longer in a direction perpendicular to compressional stresses.*

Where hot fluids circulate freely through rocks, ionic exchange may occur between two adjacent rock layers,

Did You Know?

The temperature of Earth's crust increases with depth, an idea that can be expressed as *the deeper one goes, the hotter it gets.* This causes considerable problems for underground mining efforts. In the Western Deep Levels mine in South Africa, which is 2.5 miles deep, the temperature of the rock is hot enough to scorch human skin. Here, the miners work in groups of two: one to mine the rock, and the other to operate the fan that keeps them cool.

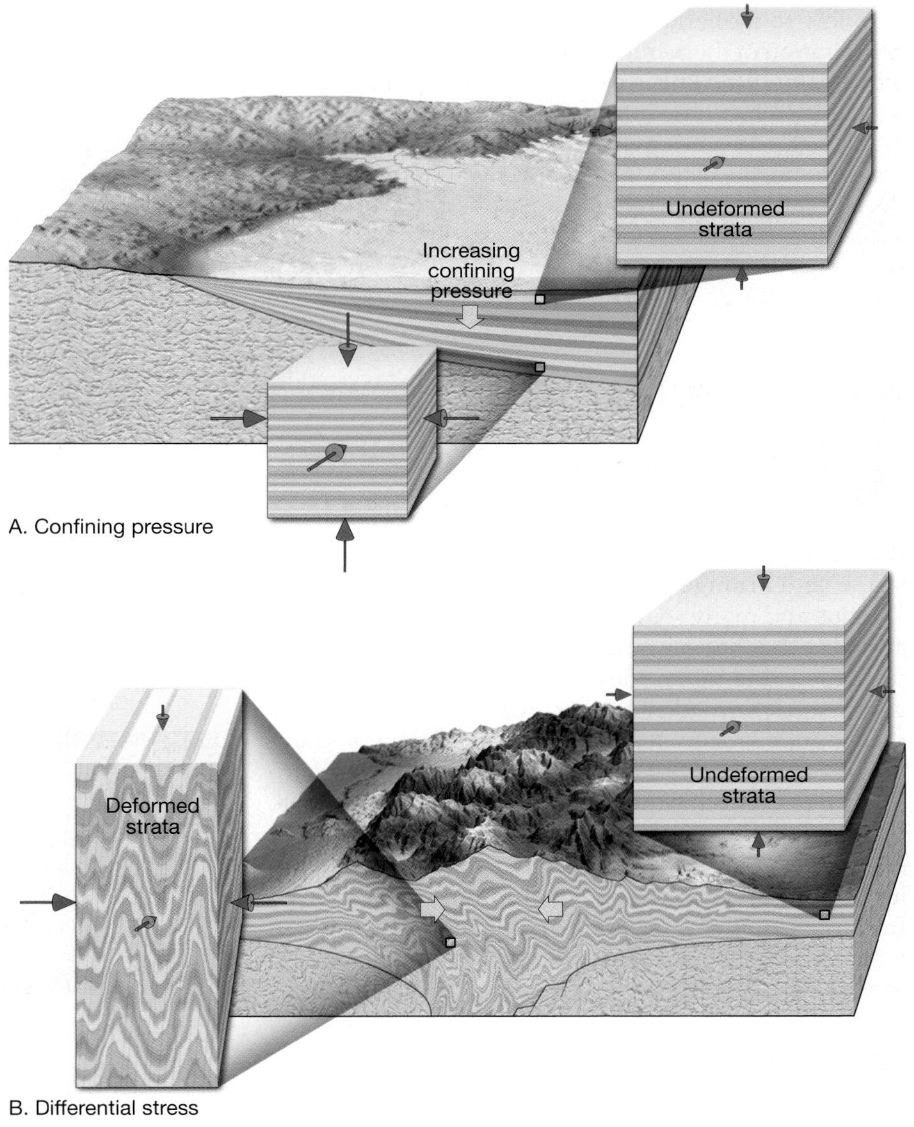

▶ **Figure 7.3** Pressure (stress) as a metamorphic agent. **A.** In a depositional environment, as confining pressure increases, rocks deform by decreasing in volume. **B.** During mountain building, rocks subjected to differential stress are shortened in the direction that pressure is applied, and lengthened in the direction perpendicular to that force.

Did You Know?

Although rare, metamorphic rocks that contain microscopic diamonds have been discovered on Earth's surface. The existence of diamonds in metamorphic rocks indicates that these rocks formed at very high pressures found at crustal depths of at least 100 kilometers (60 miles). How these metamorphic rocks were buried to these great depths and later returned to the surface is still unsolved.

or ions may migrate great distances before they are finally deposited. The latter situation is particularly common when we consider hot fluids that escape during the crystallization of an igneous pluton. If the rocks that surround the pluton differ markedly in composition from that of the invading fluids, there may be a substantial exchange of ions between the fluids and host rocks. When this occurs, a change in the overall composition of the surrounding rock results. In these instances the metamorphic process is called **metasomatism.**

What is the source of these chemically active fluids? Water is plentiful in the pore spaces of most sedimentary rocks, as well as in fractures in igneous rocks. In addition, many minerals, such as clays, micas, and amphiboles, are *hydrated* and thus contain water in their crystalline structures. Elevated temperatures associated with low to moderate-grade metamorphism cause the dehydration of these minerals. Once expelled, the water moves along the surfaces of individual grains and is available to facilitate ion transport. However, in high-grade metamorphic environments, where temperatures are extreme, these fluids may be driven from the rocks. Recall that when oceanic crust is subducted to depths exceeding 100 kilometers, water expelled from these slabs migrates into the mantle wedge above, where it triggers melting.

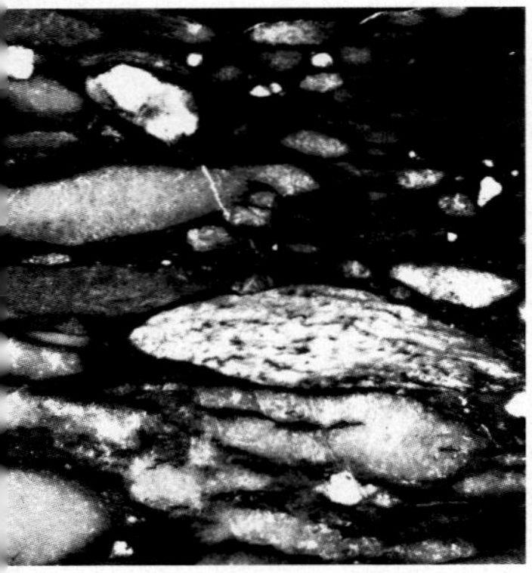

Deformed metaconglomerate exhibits pebbles that have been stretched and flattened.

Metamorphic Textures

GEODe — ESSENTIALS OF GEOLOGY

Metamorphic Rocks
▼ Textural and Mineralogical Changes

Recall that the term **texture** is used to describe the size, shape, and arrangement of grains within a rock. Most igneous and many sedimentary rocks consist of mineral grains that have a random orientation and thus appear the same when viewed from any direction. By contrast, deformed metamorphic rocks that contain platy minerals (micas) and/or elongated minerals (amphiboles), typically display some kind of *preferred orientation* in which the mineral grains exhibit a parallel to subparallel alignment. Like a fistful of pencils, rocks containing elongated minerals that are oriented parallel to each other will appear different when viewed from the side than when viewed head-on. A rock that exhibits a preferred orientation of its minerals is said to possess *foliation.*

Foliation

The term **foliation** refers to any planar (nearly flat) arrangement of mineral grains or structural features within a rock. Although foliation occurs in some sedimentary and even a few types of igneous rocks, it is a fundamental characteristic of regionally metamorphosed rocks—that is, rock units that have been strongly folded and distorted. In metamorphic environments, foliation is ultimately driven by compressional stresses that shorten rock units, causing mineral grains in preexisting rocks to develop parallel, or nearly parallel, alignments. Examples of foliation include the parallel alignment of platy and/or elongated minerals; the parallel alignment of flattened mineral grains and pebbles; compositional banding where the separation of dark and light minerals generate a layered appearance; and slaty cleavage where rocks can be easily split into thin, tabular slabs along parallel surfaces. These diverse types of foliation can form in many different ways, including:

1. Rotation of platy and/or elongated mineral grains into a new orientation.
2. Recrystallization of minerals to form new grains growing in the direction of preferred orientation.
3. Changing the shape of equidimensional grains into elongated shapes that are aligned in a preferred orientation.

The rotation of existing mineral grains is the easiest of these mechanisms to envision. Figure 7.4 illustrates the mechanics by which platy or elongated minerals are rotated. Note that the new alignment is perpendicular to the direction of maximum shortening.

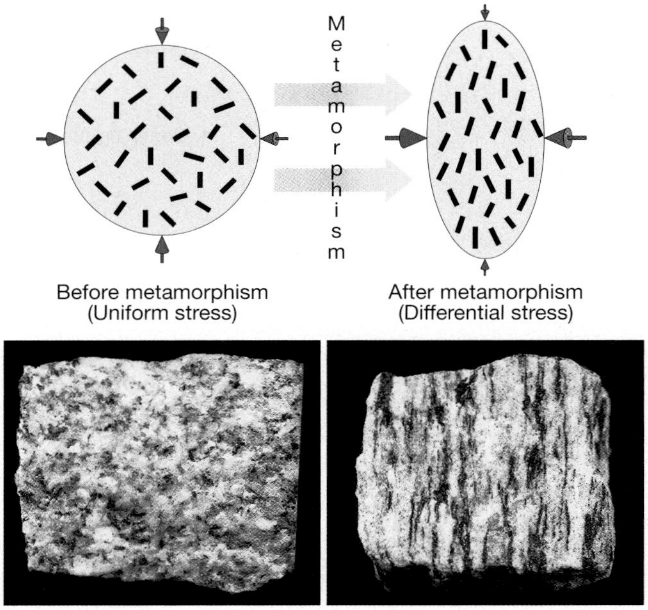

▲ **Figure 7.4** Mechanical rotation of platy or elongated mineral grains. Existing mineral grains keep their random orientation if force is uniformly applied. As differential stress causes rocks to flatten, mineral grains rotate toward the plane of flattening. *(Photos by E. J. Tarbuck)*

Although physical rotation of platy minerals contributes to the development of foliation in low-grade metamorphism, other mechanisms dominate in more extreme environments.

Recall that recrystallization is the creation of new mineral grains out of old ones. During the transformation of shale to slate, minute clay minerals (stable at the surface) recrystallize into microscopic flakes of chlorite and mica (stable at higher temperatures and pressures). In some other settings, old grains are dissolved and migrate to a different site, where they precipitate to form new mineral grains. The growth of new mineral grains tends to develop on old crystals of similar structure and grow in the same orientation as the older crystals. In this way, the new growth "mimics" the growth of old grains and enhances any preexisting preferred orientation. However, recrystallization that accompanies deformation often results in a *new* preferred orientation. As rock units are folded and generally shortened during metamorphism, elongated and platy minerals tend to recrystallize perpendicular to the direction of maximum stress.

Mechanisms that change the shapes of individual grains are especially important for the development of preferred orientations in rocks that contain minerals such as quartz, calcite, and olivine. When these minerals are stressed, they develop elongated grains that align in a direction parallel to maximum flattening (Figure 7.5). This type of deformation occurs in high-temperature environments where ductile deformation is dominant (as opposed to brittle fracturing).

A change in grain shape can occur as units of a mineral's crystalline structure slide relative to one another along discrete planes, thereby distorting the grain as shown in Figure 7.5B. This type of gradual solid-state, plastic flow involves slippage that disrupts the crystal lattice as the positions of atoms change. This typically involves breaking the existing chemical bonds and the formation of new ones. In addition, the shape of a mineral may change as ions move from a point along the margin of the grain that is highly stressed to a less-stressed position on the same grain (Figure 7.5C). This type of deformation occurs by mass transfer of material from one location to another. As you might expect, this mechanism is aided by chemically active fluids and is a type of recrystallization.

Foliated Textures

Various types of foliation exist, depending largely upon the grade of metamorphism and the mineralogy of the parent rock. We will look at three; *rock* or *slaty cleavage; schistosity;* and *gneissic texture.*

Rock or Slaty Cleavage. **Rock cleavage** refers to closely spaced planar surfaces along which rocks split into thin, tabular slabs when hit with a hammer. Rock cleavage is developed in various metamorphic rocks but is best displayed in slates that exhibit an excellent splitting property called **slaty cleavage.**

Depending on the metamorphic environment and the composition of the parent rock, rock cleavage develops in a number of different ways. In a low-grade metamorphic environment, rock cleavage is known to develop where beds of shale (and related sedimentary rocks) are strongly folded and metamorphosed to form slate. The

Did You Know?

The reason high-quality billiard tables are so heavy is that they have surfaces made of a thick slab of the metamorphic rock, slate. Because slate splits easily into slabs, it is highly prized for use as billiard-table surfaces as well as building materials such as floor and roof tiles.

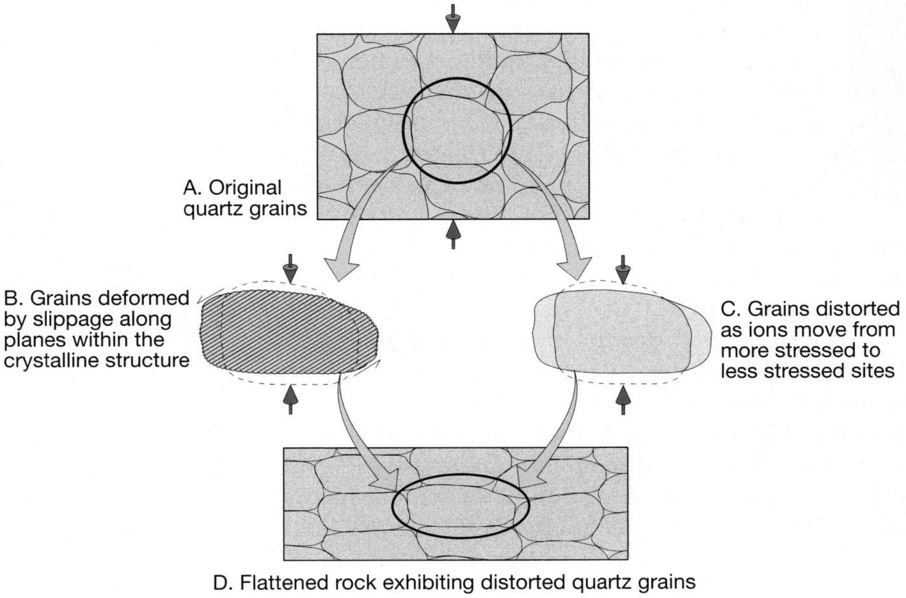
A. Original quartz grains
B. Grains deformed by slippage along planes within the crystalline structure
C. Grains distorted as ions move from more stressed to less stressed sites
D. Flattened rock exhibiting distorted quartz grains

▶ **Figure 7.5** Development of preferred orientations of minerals such as quartz, calcite, and olivine. **A.** Ductile deformation (flattening) of these roughly equidimensional mineral grains can occur in one of two ways. **B.** One mechanism is a solid-state plastic flow that involves intracrystalline gliding of individual units within each grain. **C.** Another mechanism involves dissolving material from areas of high stress and depositing that material in locations of low stress. **D.** Both mechanisms change the shape of the grains, but the volume and composition of each grain remains essentially unchanged.

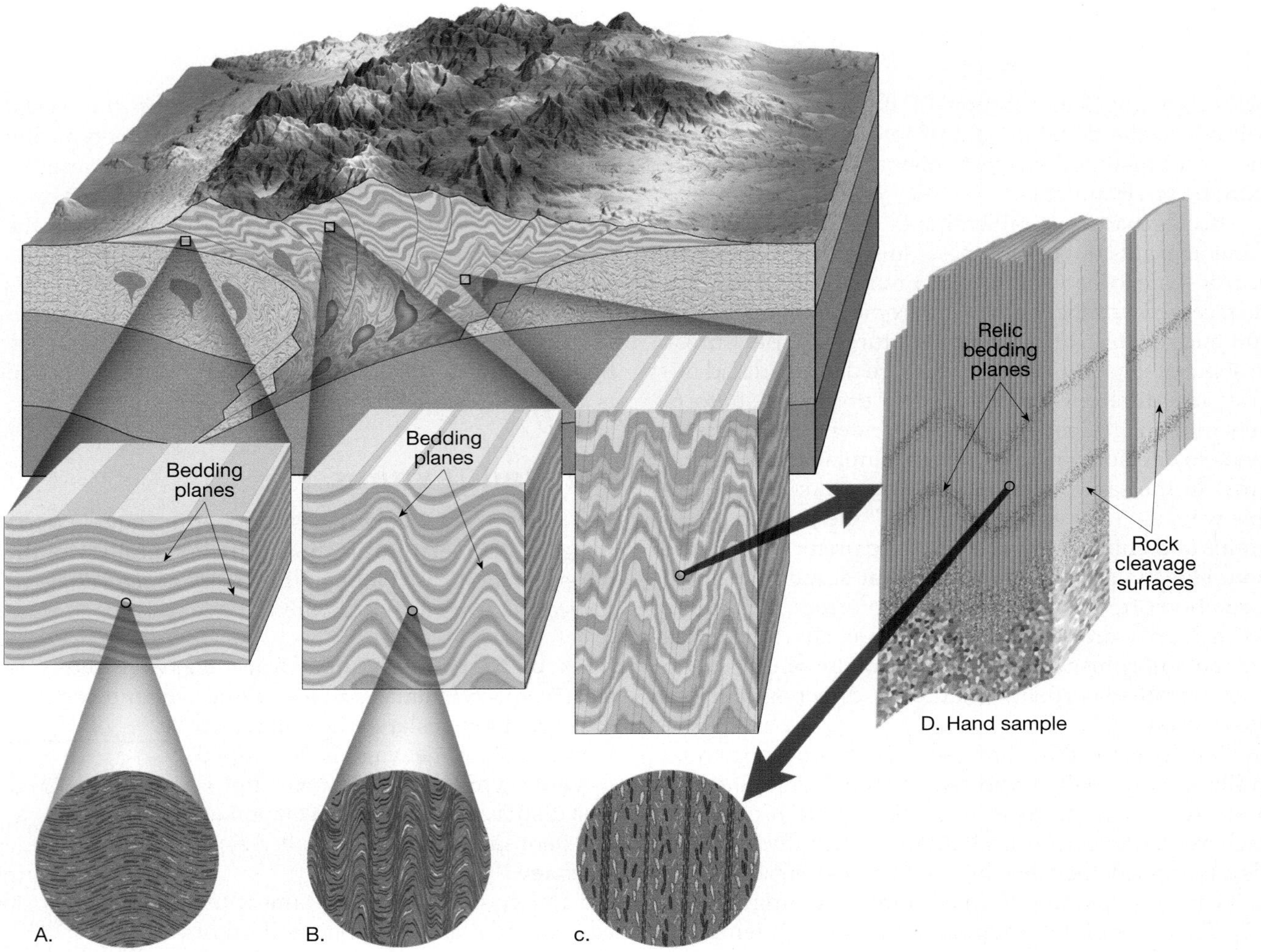

▲ **Figure 7.6** Development of one type of rock cleavage. As shale is strongly folded **(A, B)** and metamorphosed to form slate, the developing mica flakes are bent into microfolds. **C.** Further metamorphism results in the recrystallization of mica grains along the limbs of these folds to enhance the foliation. **D.** Hand sample of slate illustrates rock cleavage and its orientation to relic bedding surfaces.

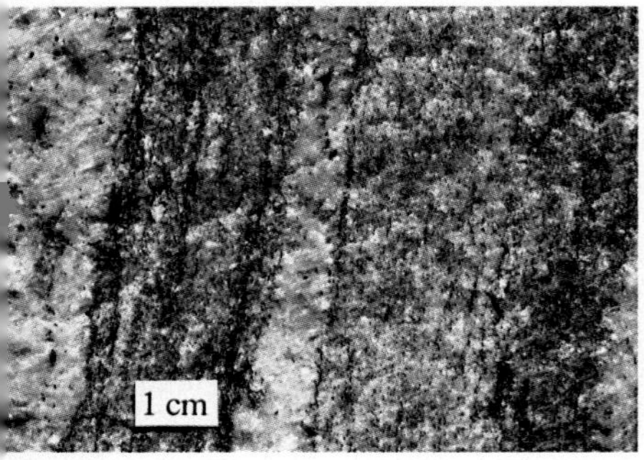

Close-up of gneissic foliation. (M. Miller)

process begins as platy grains are kinked and bent—generating microscopic folds having limbs (sides) that are roughly aligned (Figure 7.6). With further deformation, this new alignment is enhanced as old grains break down and recrystallize preferentially in the direction of the newly developed orientation. In this manner the rock develops narrow parallel zones where mica flakes are concentrated. These planar features alternate with zones containing quartz and other mineral grains that do not exhibit a pronounced linear orientation. It is along these very thin zones, where platy minerals display a parallel alignment, that slate splits (Figure 7.7).

Because slate typically forms during low-grade metamorphism of shale, evidence of the original sedimentary bedding planes is often preserved. However, as shown in Figure 7.6D, the orientation of slate's cleavage usually develops at an oblique angle to the original sedimentary layering. Thus, unlike shale, which splits along bedding planes, slate often splits across them. Other metamorphic rocks, such as schists and gneisses, may also split along planar surfaces and thus exhibit rock cleavage.

Schistosity. Under more extreme temperature-pressure regimes, the minute mica and chlorite grains in slate begin to grow many times larger. When these platy minerals grow large enough to be discernible with the unaided eye and exhibit a planar or layered structure, the rock is said to exhibit a type of foliation called **schistosity.** Rocks having this texture are referred to as *schist.* In addition to platy minerals, schist often contains deformed quartz and feldspar grains that appear as flat, or lens-shaped, grains hidden among the mica grains.

Gneissic Texture. During high-grade metamorphism, ion migrations can result in the segregation of minerals as shown in Figure 7.8. Notice that the dark biotite crystals and light silicate minerals (quartz and

▲ **Figure 7.7** Excellent slaty cleavage is exhibited by the rock in this slate quarry near Alta, Norway. The parallel mineral alignment in this rock allows it to split easily into the flat slabs visible in the photo. *(Photo by Fred Bruemmer/DRK Photo)*

▲ **Figure 7.9** Garnet-mica schist. The dark red garnet crystals (porphyroblasts) are embedded in a matrix of fine-grained micas. *(Photo by E. J. Tarbuck)*

feldspar) have separated, giving the rock a banded appearance called **gneissic texture.** A metamorphic rock with this texture is called *gneiss* (pronounced "nice"). Although foliated, gneisses will not usually split as easily as slates and some schists. Gneisses that do cleave tend to break parallel to their foliation and expose mica-rich surfaces that resemble schist.

Other Metamorphic Textures

Not all metamorphic rocks exhibit a foliated texture. Those that *do not* are referred to as **nonfoliated.** Nonfoliated metamorphic rocks typically develop in environments where deformation is minimal and the parent rocks are composed of minerals that exhibit equidimensional crystals, such as quartz or calcite. For example, when a fine-grained limestone (made of calcite) is metamorphosed by the intrusion of a hot magma body, the small calcite grains recrystallize to form larger interlocking crystals. The resulting rock, *marble,* displays large, equidimensional grains that are randomly oriented similar to those of a coarse-grained igneous rock.

▲ **Figure 7.8** This rock displays a gneissic texture. Notice that the dark biotite flakes and light silicate minerals are segregated, giving the rock a banded or layered appearance. *(Photo by E. J. Tarbuck)*

Another texture common to metamorphic rocks consists of particularly large grains, called *porphyroblasts,* that are surrounded by a fine-grained matrix of other minerals (Figure 7.9). **Porphyroblastic textures** develop in a wide range of rock types and metamorphic environments when minerals in the parent rock recrystallize to form new minerals. During recrystallization certain metamorphic minerals, including garnet, staurolite, and andalusite, invariably develop *a small number of very large crystals.* By contrast, minerals such as muscovite, biotite, and quartz typically form *a large number of very small grains.* As a result, when metamorphism generates the minerals garnet, biotite, and muscovite in the same setting, the rock will contain large crystals (porphyroblasts) of garnet embedded in a fine-grained matrix consisting of biotite and muscovite (Figure 7.9).

The exterior of the Washington Monument is composed of white marble. (Alan Schein/CORBIS)

Common Metamorphic Rocks

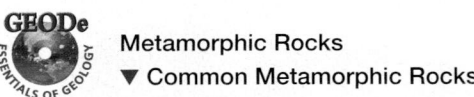

GEODe: Essentials of Geology
Metamorphic Rocks
▼ Common Metamorphic Rocks

Recall that metamorphism causes many changes in rocks, including increased density, change in grain size, reorientation of mineral grains into a planar arrangement known as foliation, and the transformation of low-temperature minerals into high-temperature minerals. Moreover, the introduction of ions may generate new minerals, some of which are economically important.

The major characteristics of some common metamorphic rocks are summarized in Figure 7.10. Notice that metamorphic rocks can be broadly classified by the type of foliation exhibited and to a lesser extent on the chemical composition of the parent rock.

Foliated Rocks

Slate. *Slate* is a very fine-grained (less than 0.5 millimeter) foliated rock composed of minute mica flakes that are too small to be visible (Figure 7.11). Thus, slate generally appears dull and closely resembles shale. A noteworthy characteristic of slate is its excellent rock

Rock Name		Texture	Grain Size	Comments	Parent Rock
Slate	Increasing Metamorphism	Foliated	Very fine	Excellent rock cleavage, smooth dull surfaces	Shale, mudstone, or siltstone
Phyllite			Fine	Breaks along wavy surfaces, glossy sheen	Slate
Schist			Medium to Coarse	Micaceous minerals dominate, scaly foliation	Phyllite
Gneiss			Medium to Coarse	Compositional banding due to segregation of minerals	Schist, granite, or volcanic rocks
Migmatite			Medium to Coarse	Banded rock with zones of light-colored crystalline minerals	Gneiss, schist
Mylonite		Weakly Foliated	Fine	When very fine-grained, resembles chert, often breaks into slabs	Any rock type
Metaconglomerate			Coarse-grained	Stretched pebbles with preferred orientation	Quartz-rich conglomerate
Marble		Nonfoliated	Medium to coarse	Interlocking calcite or dolomite grains	Limestone, dolostone
Quartzite			Medium to coarse	Fused quartz grains, massive, very hard	Quartz sandstone
Hornfels			Fine	Usually, dark massive rock with dull luster	Any rock type
Anthracite			Fine	Shiny black rock that may exhibit conchoidal fracture	Bituminous coal
Fault breccia			Medium to very coarse	Broken fragments in a haphazard arrangement	Any rock type

▲ **Figure 7.10** Classification of common metamorphic rocks.

▲ **Figure 7.11** Slate, a very fine-grained metamorphic rock. *(Photo by E. J. Tarbuck)*

cleavage, or tendency to break into flat slabs. This property traditionally made slate a most useful rock for roof and floor tile, blackboards, and billiard tables.

Slate is most often generated by the low-grade metamorphism of shale, mudstone, or siltstone. Less frequently it is produced when volcanic ash is metamorphosed. Slate's color depends on its mineral constituents. Black (carbonaceous) slate contains organic material, red slate gets its color from iron oxide, and green slate usually contains chlorite.

Phyllite. Phyllite represents a gradation in the degree of metamorphism between slate and schist. Its constituent platy minerals are larger than those in slate but not yet large enough to be readily identifiable with the unaided eye. Although phyllite appears similar to slate, it can be easily distinguished from slate by its glossy sheen and wavy surface (Figure 7.12). Phyllite usually exhibits rock cleavage and is composed mainly of very fine crystals of either muscovite, chlorite, or both.

Schist. Schists are medium- to coarse-grained metamorphic rocks in which platy minerals predominate (Figure 7.13). These flat components commonly include

▲ **Figure 7.12** Phyllite can be distinguished from slate by its glossy sheen and wavy surface. *(Photo by E. J. Tarbuck)*

▲ **Figure 7.13** Mica schist. This sample of schist is composed mostly of muscovite and biotite. *(Photo by E. J. Tarbuck)*

the micas (muscovite and biotite), which display a planar alignment that gives the rock its foliated texture. In addition, schists contain smaller amounts of other minerals, often quartz and feldspar. Schists composed mostly of dark minerals (amphiboles) are known. Like slate, the parent rock of many schists is shale, which has undergone medium- to high-grade metamorphism during major mountain-building episodes.

Slate roof, Switzerland.

The term *schist* describes the texture of a rock, and as such, it is used to describe rocks having a wide variety of chemical compositions. To indicate the composition, mineral names are used. For example, schists composed primarily of muscovite and biotite are called *mica schists.* Depending upon the degree of metamorphism and composition of the parent rock, mica schists often contain *accessory minerals,* some of which are unique to metamorphic rocks. Some common accessory minerals that occur as porphyroblasts include *garnet, staurolite,* and *sillimanite,* in which case the rock is called *garnet-mica schist, staurolite-mica schist,* and so forth (see Figure 7.9).

In addition, schists may be composed largely of the minerals chlorite or talc, in which case they are called *chlorite schist* and *talc schist,* respectively. Both chlorite and talc schists can form when rocks with a basaltic composition undergo metamorphism. Others contain the mineral *graphite,* which is used as pencil "lead," graphite fibers (used in fishing rods), and lubricant (commonly for locks).

Did You Know?

One of America's worst civil-engineering disasters occurred in 1928 when the St. Francis dam in southern California failed. Huge torrents of water washed down San Francesquito Canyon, destroying 900 buildings and taking nearly 500 lives. The eastern part of the dam was built on highly foliated mica schist that was prone to slippage, as evidenced by an earlier landslide in that area. The immense water pressure at the base of the dam and the weak schist to which it was anchored, are thought to have contributed to the failure.

This banded gneiss exhibits evidence of deformation.

Gneiss. *Gneiss* is the term applied to medium- to coarse-grained banded metamorphic rocks in which granular and elongated (as opposed to platy) minerals predominate. The most common minerals in gneiss are quartz, potassium feldspar, and sodium-rich plagioclase feldspar. Most gneisses also contain lesser amounts of biotite, muscovite, and amphibole that develop a preferred orientation. Some gneisses will split along the layers of platy minerals, but most break in an irregular fashion.

Recall that during high-grade metamorphism, the light and dark components separate, giving gneisses their characteristic banded or layered appearance. Thus, most gneisses consist of alternating bands of white or reddish feldspar-rich zones and layers of dark ferromagnesian minerals (see Figure 7.8). These banded gneisses often exhibit evidence of deformation, including folds and faults.

Most gneisses have a felsic composition and are derived from granite or its aphanitic equivalent, rhyolite. However, many form from the high-grade metamorphism of shale. In this instance, gneiss represents the last rock in the sequence of shale, slate, phyllite, schist, and gneiss. Like schists, gneisses may also include large crystals of accessory minerals such as garnet and staurolite. Gneisses made up primarily of dark minerals such as those that compose basalt also occur. For example, an amphibole-rich rock that exhibits a gneissic texture is called *amphibolite.*

▲ **Figure 7.15** Quartzite is a nonfoliated metamorphic rock formed from quartz sandstone. The photomicrograph shows the interlocking quartz grains typical of quartzite. *(Photos by E. J. Tarbuck)*

Nonfoliated Rocks

Marble. *Marble* is a coarse, crystalline metamorphic rock whose parent was limestone or dolostone (Figure 7.14). Pure marble is white and composed essentially of the mineral calcite. Because of its relative softness (hardness of 3), marble is easy to cut and shape. White marble is particularly prized as a stone from which to create monuments and statues, such as the famous statue of *David* by Michelangelo. Unfortunately, since marble is composed of calcium carbonate, it is readily attacked by acid rain.

Replica of David *by Michelangelo, created from white marble.* (Andrew Ward/Getty Images)

▲ **Figure 7.14** Marble, a crystalline rock formed by the metamorphism of limestone. Photomicrograph shows interlocking calcite crystals using polarized light. *(Photos by E. J. Tarbuck)*

The parent rock from which marble forms often contains impurities that tend to color the stone. Thus, marble can be pink, gray, green, or even black and may contain a variety of accessory minerals (chlorite, mica, garnet, and wollastonite). When marble forms from limestone interbedded with shales, it will appear banded and exhibit visible foliation. When deformed, these banded marbles may develop highly contorted mica-rich folds that give the rock a rather artistic design. Hence, these decorative marbles have been used as a building stone since prehistoric times.

Quartzite. *Quartzite* is a very hard metamorphic rock formed from quartz sandstone (Figure 7.15). Under moderate- to high-grade metamorphism, the quartz grains in sandstone fuse together (inset in Figure 7.15). The recrystallization is so complete that when broken, quartzite will split through the quartz grains rather than along their boundaries. In some instances, sedimentary features such as crossbedding are preserved and give the rock a banded appearance. Pure quartzite is white,

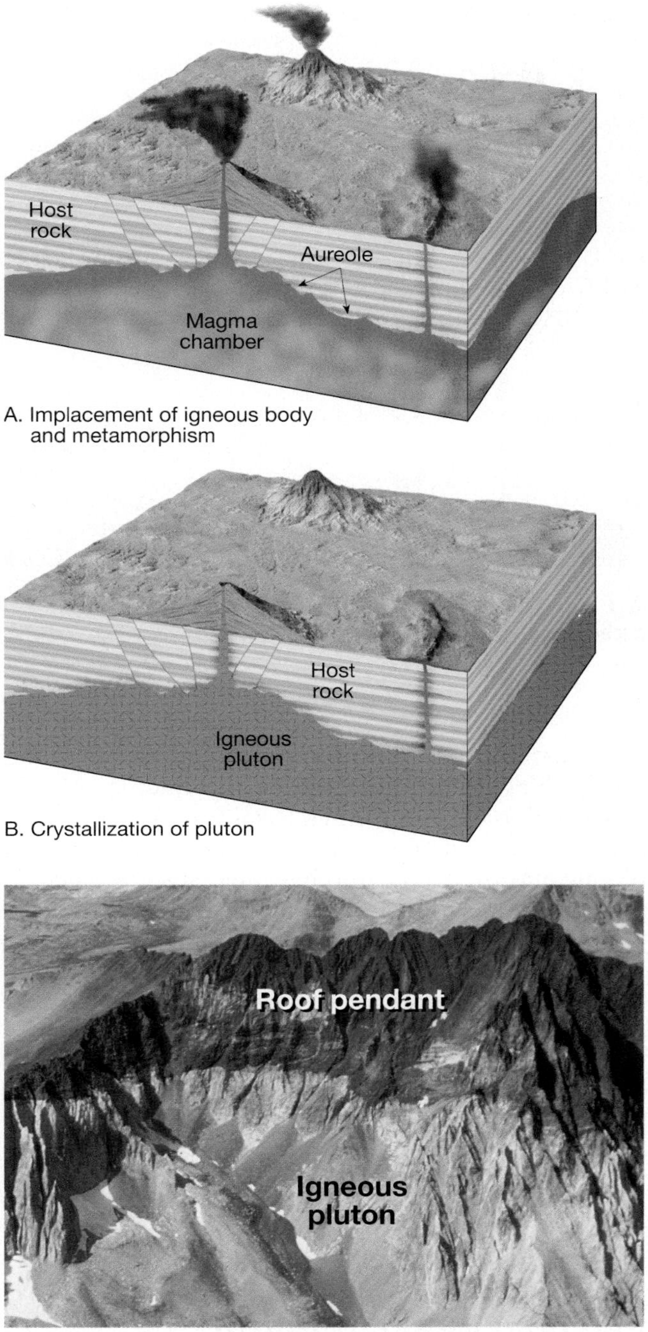

▲ **Figure 7.16** Contact metamorphism produces a zone of alteration called an *aureole* around an intrusive igneous body. In the photo, the dark layer, called a *roof pendant,* consists of metamorphosed host rock adjacent to the upper part of the light-colored igneous pluton. The term *roof pendant* implies that the rock was once the roof of a magma chamber. Sierra Nevada, near Bishop, California. *(Photo by John S. Shelton)*

but iron oxide may produce reddish or pinkish stains, while dark mineral grains may impart a gray color.

Metamorphic Environments

There are a number of environments in which metamorphism occurs. Most are in the vicinity of plate margins, and many are associated with igneous activity. We will consider the following types of metamorphism: 1) *contact* or *thermal metamorphism;* 2) *hydrothermal metamorphism;* 3) *regional metamorphism;* 4) *burial metamorphism;* 5) *impact metamorphism;* and 6) *metamorphism along faults.*

With the exception of shock metamorphism, there is considerable overlap among the other types. Recall that regional metamorphism occurs where lithospheric plates collide to generate mountains. Here, large segments of Earth's crust are folded and faulted while often being intruded by magma rising from the mantle. Hence, the rocks that are deformed and metamorphosed in a regional setting exhibit metamorphic features common to other types of metamorphism.

Did You Know?

Because marble can be carved readily, it has been used for centuries for buildings and memorials. Examples of important structures whose exteriors are clad in marble include the Parthenon in Greece, the Taj Mahal in India, and the Washington Monument in the United States.

Contact or Thermal Metamorphism

Contact or **thermal metamorphism** occurs when rocks immediately surrounding a molten igneous body are "baked" and therefore altered from their original state. The altered rocks occur in a zone called an **aureole** (Figure 7.16). While small intrusions such as dikes and sills typically form aureoles only a few centimeters thick, large intrusions such as batholiths can produce aureoles that extend outward for several kilometers.

Although contact metamorphism is not entirely restricted to shallow crustal depths, it is most easily recognized when it occurs in this setting. Here, the temperature contrast between an intrusion and the surrounding rock is greatest and confining pressure is low. Because contact metamorphism does not involve directed pressure, crystals within the metamorphic aureole are more or less randomly oriented.

During contact metamorphism of mudstones and shales, the clay minerals are baked as if placed in a kiln. The result is a very hard, fine-grained metamorphic rock called *hornfels* (Figure 7.17). Because directed pressure is not a factor in forming these rocks, they are characteristically nonfoliated. Hornfels can form from a variety of parent materials during contact metamorphism, including volcanic ash and basalts. In some cases, large grains of metamorphic minerals, such as garnet and staurolite, may form, giving the hornfels a porphyroblastic texture (see Figure 7.9).

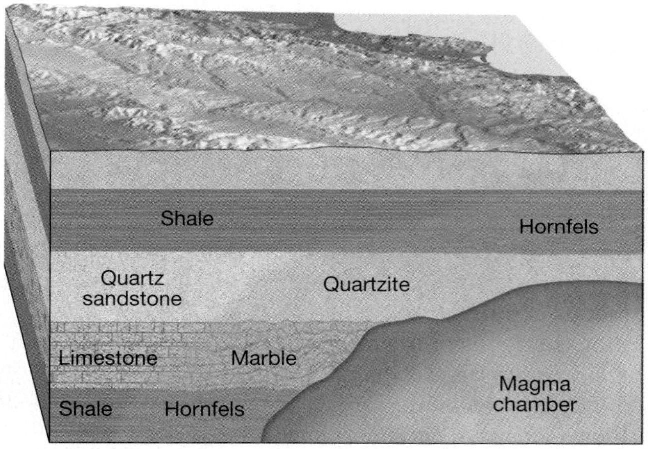

▲ **Figure 7.17** Contact metamorphism of shale yields hornfels, while contact metamorphism of quartz sandstone and limestone produces quartzite and marble, respectively.

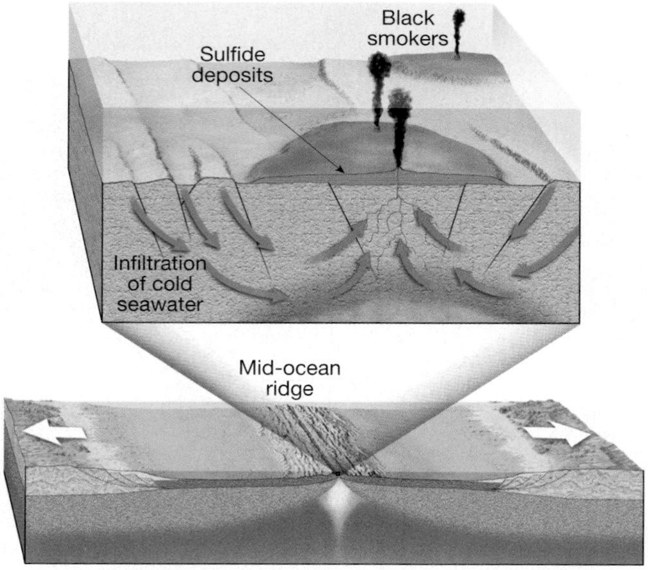

▲ **Figure 7.18** Hydrothermal metamorphism along a mid-ocean ridge.

Did You Know?

Most pigment, which gives paint its color and its ability to hide previous coats, comes from various natural Earth materials. For example, *umber*, which produces a rich chestnut-brown color, is mined on the island of Cyprus. This material was formed on an ancient seafloor by the hydrothermal alteration of iron-and manganese-rich basaltic lavas.

Two other metamorphic rocks produced in association with hornfels are quartzite and marble (Figure 7.17). Their parent rocks are quartz sandstone, and limestone, respectively. Although quartzite and marble are commonly associated with contact metamorphism, unlike hornfels, they also form in other metamorphic environments.

Hydrothermal Metamorphism

When hot, ion-rich fluids circulate through fissures and cracks that develop in rock, a chemical alteration called **hydrothermal metamorphism** occurs. This type of metamorphism is closely associated with igneous activity, since it provides the heat required to circulate these ion-rich solutions. Thus, hydrothermal metamorphism often occurs simultaneously with contact metamorphism in regions where large plutons are emplaced.

As these large magma bodies cool and solidify, the ions that are not incorporated into the crystalline structures of the newly formed silicate minerals, as well as any remaining volatiles (water), are expelled. These ion-rich fluids are called **hydrothermal solutions.** In addition to chemically altering the host rocks, the ions contained in hydrothermal solutions sometimes precipitate to form a variety of economically important mineral deposits.

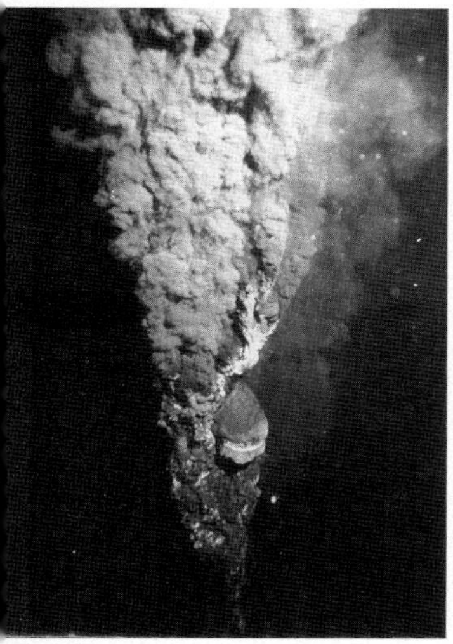

Black smoker spewing hot, mineral-rich seawater.
(R. Ballard/Woods Hole)

As our understanding of plate tectonics expanded, it became clear that the most widespread occurrence of hydrothermal metamorphism is along the axis of the oceanic ridge system (Figure 7.18). Here, as plates move apart, upwelling magma from the mantle generates new seafloor. As seawater percolates through the young, hot oceanic crust, it is heated and chemically reacts with the newly formed basaltic rocks. The result is the conversion of ferromagnesian minerals, such as olivine and pyroxene, into hydrated silicates, including serpentine, chlorite, and talc. In addition, large amounts of metals, such as iron, cobalt, nickel, silver, gold and copper, are also dissolved from the newly formed crust. These hot, metal-rich fluids eventually rise along fractures and gush from the seafloor at temperatures of about 350°C, generating particle-filled clouds called *black smokers.* Upon mixing with the cold seawater, sulfides and carbonate minerals containing these heavy metals precipitate to form metallic deposits, some of which are economically valuable. This is believed to be the origin of the copper ores mined today on the island of Cyprus.

Regional Metamorphism

Most metamorphic rocks are produced during the process of **regional metamorphism** in association with mountain building. During these dynamic events, large segments of Earth's crust are intensely deformed along convergent plate boundaries (Figure 7.19). This activity occurs most often where oceanic lithosphere is subducted to produce island arcs or continental volcanic arcs and during continental collisions.

Metamorphism associated with continental collisions involves the convergence of an active plate margin with a passive continental margin as shown in Figure 7.19. Such collisions generally result in large

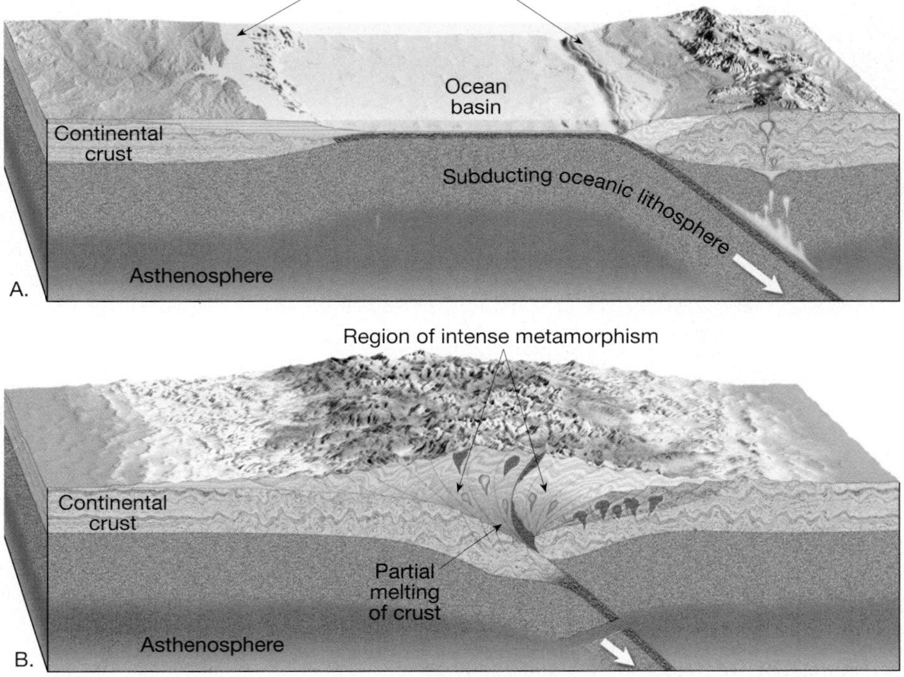

▶ **Figure 7.19** Regional metamorphism occurs where rocks are squeezed between two converging lithospheric plates during mountain building.

segments of Earth's crust being intensely deformed by compressional forces associated with convergent plate motion. Sediments and crustal rocks that form the margins of the colliding continental blocks are folded and faulted, causing them to shorten and thicken like a rumpled carpet (Figure 7.19). This event often involves crystalline continental basement rocks, as well as slices of ocean crust that once floored the intervening ocean basin.

The general thickening of the crust results in buoyant lifting where deformed rocks are elevated high above sea level to form mountainous terrain. Likewise, crustal thickening results in the deep burial of large quantities of rock as crustal blocks are thrust one beneath another. Here, in the roots of mountains, elevated temperatures caused by deep burial are responsible for the most productive and intense metamorphic activity within a mountain belt. Often, these deeply buried rocks become heated to the point of melting. As a result, magma collects until it forms bodies large enough to buoyantly rise and intrude the overlying metamorphic and sedimentary rocks (Figure 7.19). Consequently, the cores of many mountain ranges consist of folded and faulted metamorphic rocks, often intertwined with intruded igneous bodies. Over time, these deformed rock masses are uplifted, and erosion removes the overlying material to expose the igneous and metamorphic rocks that comprise the central core of the mountain range.

Other Metamorphic Environments

Other types of metamorphism occur that generate comparatively smaller amounts of metamorphic rock in localized concentrations.

Deformed metamorphic rocks in the Appalachians. (Phil Dombrowski)

Burial Metamorphism. **Burial metamorphism** occurs in association with very thick accumulations of sedimentary strata in a subsiding basin. Here, low-grade metamorphic conditions may be attained within the lowest layers. Confining pressure and geothermal heat drive the recrystallization of the constituent minerals to change the texture and/or mineralogy of the rock without appreciable deformation.

The depth required for burial metamorphism varies from one location to another, depending on the prevailing geothermal gradient. Low-grade metamorphism often begins at depths of about 8 kilometers (5 miles), where temperatures are about 200°C. However, in areas that exhibit steep geothermal gradients, such as near the Salton Sea in California and in northern New Zealand, drilling operations have collected metamorphic minerals from a depth of only a few kilometers.

Metamorphism Along Fault Zones. Near the surface, rock behaves like a brittle solid. Consequently, movement along a fault zone fractures and pulverizes rock. The result is a loosely coherent rock called *fault breccia,*

Fault breccia, California.
(A. P. Trujillo)

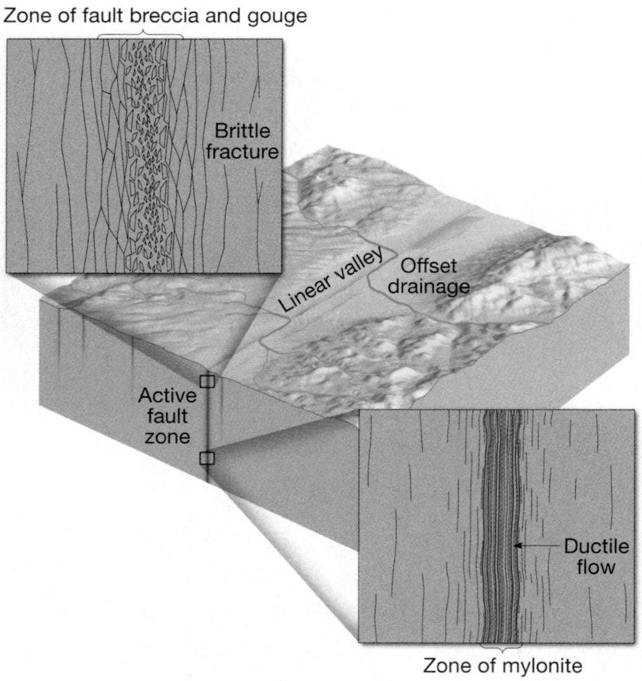

▲ **Figure 7.20** Metamorphism along a fault zone.

Did You Know?

Rocks of the Acasta Gneiss Complex, exposed just east of Great Slave Lake in the Northwest Territories of Canada, have yielded radiometric dates of up to 4.03 billion years. Currently, these are the oldest-known rocks in the world.

which is composed of broken and crushed rock fragments (Figure 7.20). Displacements along California's San Andreas Fault have created a zone of fault breccia and related rock types over 1000 kilometers long and up to 3 kilometers wide.

In some shallow fault zones, a soft, uncemented claylike material called *fault gouge* is also produced. Fault gouge is formed by the crushing and grinding of rock material during fault movement. The resulting crushed material is further altered by groundwater that infiltrates the porous fault zone.

Did You Know?

Although most of central North America has a rather subdued topography, the underlying bedrock is similar to rocks found in the cores of metamorphosed mountain belts. This evidence strongly supports the view that these rocks once formed the roots of ancient mountain chains that may have risen as high as the present-day Himalayas.

Much of the deformation associated with fault zones occurs at great depth and thus at high temperatures. In this environment preexisting minerals deform by ductile flow. As large slabs of rock move in opposite directions, the minerals in the fault zone between them tend to form elongated grains that give the rock a foliated or lineated appearance. Rocks formed in these zones of intense ductile deformation are termed *mylonites.*

Impact Metamorphism. **Impact** (or *shock*) **metamorphism** occurs when high-speed projectiles called *meteorites* (fragments of comets or asteroids) strike Earth's surface. Upon impact the energy of the rapidly moving meteorite is transformed into heat energy and shock waves that pass through the surrounding rocks. The result is pulverized, shattered, and sometimes melted rock. The products of these impacts, called *impactiles,* include mixtures of fused fragmented rock plus glass-rich ejecta that resemble volcanic bombs. In some cases, a very dense form of quartz (*coesite*) and minute *diamonds* are found. These high-pressure minerals provide convincing evidence that pressures and temperatures at least as great as those existing in the upper mantle must have been attained at least briefly at Earth's surface.

Metamorphic Zones

In areas affected by metamorphism, there usually exist systematic variations in the mineralogy and texture of the rocks that can be observed as we traverse the region. These differences are clearly related to variations in the degree of metamorphism experienced in each metamorphic zone.

Textural Variations

If we begin with a clay-rich sedimentary rock such as shale, a gradual increase in metamorphic intensity is accompanied by a general coarsening of the grain size. Thus, we observe shale changing to a fine-grained slate, which then forms phyllite and through continued recrystallization generates a coarse-grained schist (Figure 7.21). Under more intense conditions a gneissic texture that exhibits layers of dark and light minerals may develop. This systematic transition in metamorphic textures can be observed as we approach the Appalachian Mountains from the west. Beds of shale, which once extended over large areas of the eastern United States, still occur as nearly flat-lying strata in Ohio. However, in the broadly folded Appalachians of central Pennsylvania, the rocks that once formed flat-lying beds are folded and display a preferred orientation of platy mineral grains as exhibited by well-developed slaty cleavage. As we move farther eastward into the intensely deformed crystalline Appalachians, we find large outcrops of schists. The most intense zones of metamorphism are found in Vermont and New Hampshire, where gneissic rocks outcrop.

Index Minerals and Metamorphic Grade

In addition to textural changes, we encounter corresponding changes in mineralogy as we shift from regions of low-grade metamorphism to regions of high-grade metamorphism. An idealized transition in mineralogy that results from the regional metamorphism of shale is shown in Figure 7.22. The first new mineral to

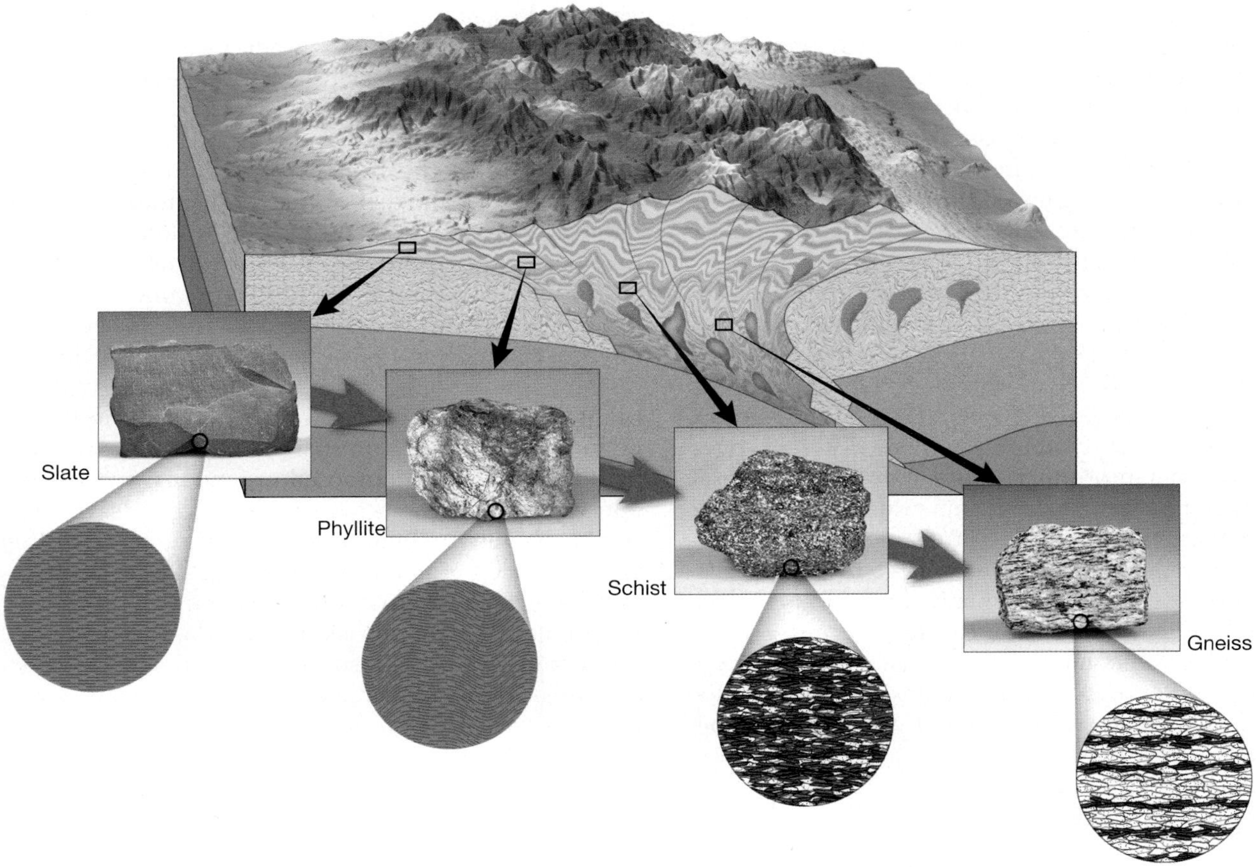

▲ **Figure 7.21** Idealized illustration of progressive regional metamorphism. From left to right, we progress from low-grade metamorphism (slate) to high-grade metamorphism (gneiss). *(Photos by E. J. Tarbuck)*

▼ **Figure 7.22** The typical transition in mineralogy that results from progressive metamorphism of shale.

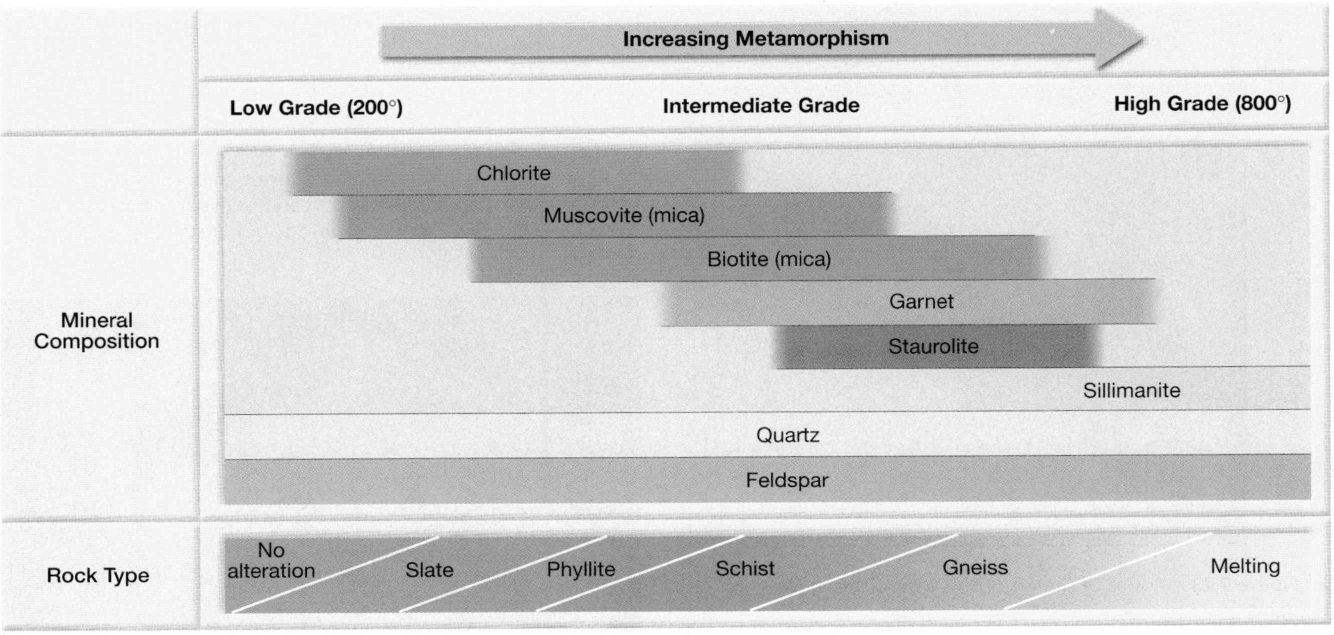

Shock-melted ejecta (tektites) produced when a comet or asteroid impacted Earth. (Brian Mason, Smithsonian)

form as shale changes to slate is chlorite. At higher temperatures flakes of muscovite and biotite begin to dominate. Under more extreme conditions, metamorphic rocks may contain garnet and staurolite crystals. At temperatures approaching the melting point of rock, sillimanite forms. Sillimanite is a high-temperature metamorphic mineral used to make refractory porcelains such as those used in spark plugs.

Through the study of metamorphic rocks in their natural settings (called *field studies*) and through experimental studies, researchers have learned that certain minerals are good indicators of the metamorphic environment in which they formed. Using these **index minerals,** geologists distinguish among different zones of regional metamorphism. For example, the mineral chlorite begins to form when temperatures are relatively low, less than 200°C (Figure 7.23). Thus, rocks that contain chlorite (usually slates) are referred to as *low-grade.* By contrast, the mineral sillimanite only forms in very extreme environments where temperatures exceed 600°C, and rocks containing it are considered *high-grade.* By mapping the occurrences of index minerals, geologists are in effect mapping zones of varying metamorphic grade. *Grade* is a term used in a relative sense to refer to the conditions of temperature (or sometimes pressure) to which a rock has been subjected.

Migmatites. In the most extreme environments, even the highest-grade metamorphic rocks undergo change. For example, gneissic rocks may be heated sufficiently to cause melting to begin. However, recall from our discussion of igneous rocks that different minerals melt at different temperatures. The light-colored silicates, usually quartz and potassium feldspar, have the lowest melting temperatures and begin to melt first, whereas the mafic silicates, such as amphibole and biotite, remain solid. When this partially melted rock cools, the light bands will be composed of igneous, or igneous-appearing components, while the dark bands will consist of unmelted metamorphic material. Rocks of this type are called **migmatites** (Figure 7.24). The light-colored bands in migmatites often form tortuous folds and may contain tabular inclusions of the dark components. Migmatites serve to illustrate the fact that some rocks are transitional and do not clearly belong to any one of the three basic rock groups.

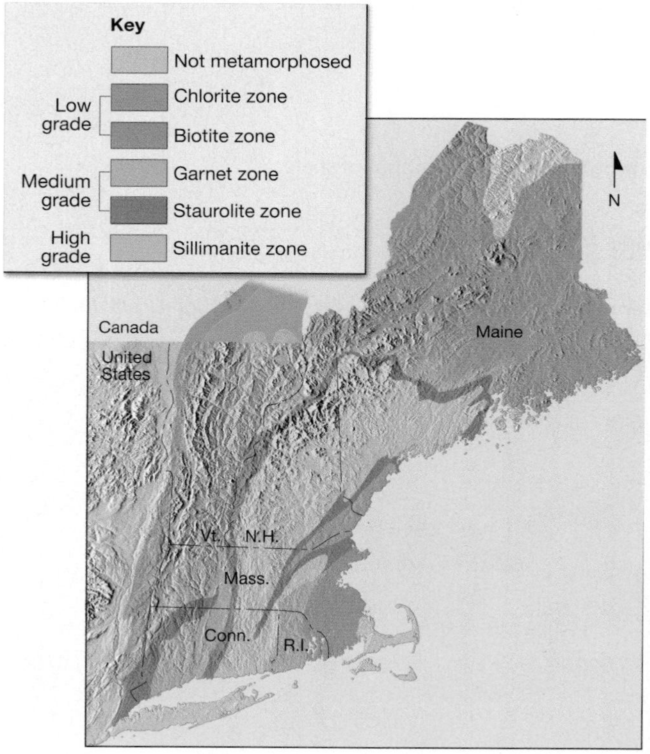

▲ **Figure 7.23** Zones of metamorphic intensities in New England.

▼ **Figure 7.24** Migmatite. The lightest-colored layers are igneous rock composed of quartz and feldspar, whereas the darker layers have a metamorphic origin. *(Photo by Harlan H. Roepke)*

The Chapter in Review

- *Metamorphism* is the transformation of one rock type into another. *Metamorphic rocks* form from preexisting rocks (either igneous, sedimentary, or other metamorphic rocks) that have been altered by the agents of metamorphism, which include *heat, pressure (stress),* and *chemically active fluids.* During metamorphism the material essentially remains solid. The changes that occur in metamorphosed rocks are textural as well as mineralogical.

- Metamorphism most often occurs in one of three settings: (1) when rock is in contact with magma, *contact* or *thermal metamorphism* occurs; (2) where hot, ion-rich water circulates through rock, chemical alteration occurs by a process called *hydrothermal metamorphism;* or (3) during mountain building, where extensive areas of rock undergo *regional metamorphism.* The greatest volume of metamorphic rock is produced during regional metamorphism.

- The mineral makeup of the parent rock determines, to a large extent, the degree to which each metamorphic agent will cause change. Heat is the most important agent because it provides the energy to drive chemical reactions that result in the recrystallization of minerals. Pressure, like temperature, also increases with depth. When subjected to *confining pressure,* minerals may recrystallize into more compact forms. During mountain building, rocks are subjected to *differential stress,* which tends to shorten them in the direction pressure is applied and lengthen them in the direction perpendicular to that force. At depth, rocks are warm and *ductile,* which accounts for their ability to deform by flowing when subjected to differential stresses. Chemically active fluids, most commonly water containing ions in solution, also enhance the metamorphic process by dissolving minerals and aiding the migration and precipitation of this material at other sites.

- *The grade of metamorphism is reflected in the texture and mineralogy of metamorphic rocks.* During regional metamorphism, rocks typically display a *preferred orientation* called *foliation* in which their platy and elongated minerals are aligned. Foliation develops as platy or elongated minerals are rotated into parallel alignment, recrystallize to form new grains that exhibit a preferred orientation, or are plastically deformed into flattened grains that exhibit a planar alignment. *Rock cleavage* is a type of foliation in which rocks split cleanly into thin slabs along surfaces where platy minerals are aligned. *Schistosity* is a type of foliation defined by the parallel alignment of medium- to coarse-grained platy minerals. During high-grade metamorphism, ion migrations can cause minerals to segregate into distinct layers or bands. Metamorphic rocks with a banded texture are called *gneiss.* Metamorphic rocks composed of only one mineral forming equidimensional crystals often appear *nonfoliated. Marble* (metamorphosed limestone) is often nonfoliated. Further, metamorphism can cause the transformation of low-temperature minerals into high-temperature minerals and, through the introduction of ions from *hydrothermal solutions,* generate new minerals, some of which form economically important metallic ore deposits.

- Common foliated metamorphic rocks include *slate, phyllite,* various types of *schists* (e.g., garnet-mica schist), and *gneiss.* Nonfoliated rocks include *marble* (parent rock: limestone) and *quartzite* (most often formed from quartz sandstone).

- The three geologic environments in which metamorphism commonly occurs are (1) *contact* or *thermal metamorphism,* (2) *hydrothermal metamorphism* and (3) *regional metamorphism.* Contact metamorphism occurs when rocks are in contact with an igneous body, resulting in the formation of zones of alteration around the magma called *aureoles.* Most contact metamorphic rocks are fine-grained, dense, tough rocks of various chemical compositions. Because directional pressure is not a major factor, these rocks are not generally foliated. Hydrothermal metamorphism occurs where hot, ion-rich fluids circulate through rock and cause chemical alteration of the constituent minerals. Most hydrothermal alteration occurs along the oceanic ridge system where seawater migrates through hot oceanic crust and chemically alters newly formed basaltic rocks. Metallic ions that are removed from the crust are eventually carried to the floor of the ocean where they precipitate from black smokers to form metallic deposits, some of which may be economically important. Regional metamorphism takes place at considerable depths over an extensive area and is associated with the process of mountain building. A gradation in the degree of change usually exists in association with regional metamorphism, in which the intensity of metamorphism (low- to high-grade) is reflected in the texture and mineralogy of the rocks. In the most extreme metamorphic environments, rocks called *migmatites* fall into a transition zone *somewhere between* "true" igneous rocks and "true" metamorphic rocks.

Key Terms

aureole (p. 175)
burial metamorphism (p. 177)
confining pressure (p. 166)
contact metamorphism (p. 175)
differential stress (p. 166)
foliation (p. 168)
gneissic texture (p. 171)
hydrothermal metamorphism (p. 176)
hydrothermal solution (p. 176)
impact metamorphism (p. 178)
index mineral (p. 180)
metamorphism (p. 164)
metasomatism (p. 168)
migmatite (p. 180)
nonfoliated texture (p. 171)
parent rock (p. 164)
porphyroblastic texture (p. 171)
regional metamorphism (p. 176)
rock cleavage (p. 169)
schistosity (p. 170)
shield (p. 164)
slaty cleavage (p. 169)
texture (p. 168)
thermal metamorphism (p. 175)

Questions for Review

1. What is *metamorphism?* What are the agents that change rocks?
2. Why is heat considered the most important agent of metamorphism?
3. How is confining pressure different than differential stress?
4. What role do chemically active fluids play in metamorphism?
5. In what two ways can the parent rock affect the metamorphic process?
6. What is foliation? Distinguish between *slaty cleavage, schistosity,* and *gneissic* textures.
7. Briefly describe the three mechanisms by which minerals develop a preferred orientation.
8. List some changes that might occur to a rock in response to metamorphic processes.
9. Slate and phyllite resemble each other. How might you distinguish one from the other?
10. Each of the following statements describes one or more characteristics of a particular metamorphic rock. For each statement, name the metamorphic rock that is being described.
    a. calcite-rich and nonfoliated
    b. loosely coherent rock composed of broken fragments that formed along a fault zone
    c. represents a grade of metamorphism between slate and schist
    d. very fine-grained and foliated; excellent rock cleavage
    e. foliated and composed predominately of platy minerals
    f. composed of alternating bands of light and dark silicate minerals
    g. hard, nonfoliated rock resulting from contact metamorphism
11. Distinguish between *contact metamorphism* and *regional metamorphism.* Which creates the greatest quantity of metamorphic rock?
12. Where does most hydrothermal metamorphism occur?
13. Describe *burial metamorphism.*
14. How do geologists use index minerals?
15. Briefly describe the textural changes that occur in the transformation of slate to phyllite to schist and then to gneiss.
16. How are gneisses and migmatites related?

Online Study Guide

The *Essentials of Geology* Web site uses the resources and flexibility of the Internet to aid in your study of the topics in this chapter. Written and developed by geology instructors, this site will help improve your understanding of geology. Visit **www.prenhall.com/lutgens** and click on the cover of *Essentials of Geology 9e* to find:

- Online review quizzes.
- Critical thinking exercises.
- Links to chapter-specific Web resources.
- Internet-wide key-term searches.

http://www.prenhall.com/lutgens

Landslides triggered by heavy rains destroyed these homes in Laguna Niguel, California. *(Photo by Grantpix/Index Stock Imagery)*

CHAPTER

8

Mass Wasting: The Work of Gravity

Focus on Learning

To assist you in learning the important concepts in this chapter, you will find it helpful to focus on the following questions:

- What is the process of mass wasting?
- What role does mass wasting play in the development of valleys?
- What are the controls and triggers of mass wasting?
- What criteria are used to divide and describe the various types of mass wasting?
- What are the general characteristics of slump, rockslide, debris flow, earthflow, and creep?

Peruvian valley before the landslide. (Iris Lozier)

Earth's surface is never perfectly flat but instead consists of slopes. Some are steep and precipitous; others are moderate or gentle. Some are long and gradual; others are short and abrupt. Slopes can be mantled with soil and covered by vegetation or consist of barren rock and rubble. Taken together, slopes are the most common elements in our physical landscape. Although most slopes may appear to be stable and unchanging, the force of gravity causes material to move downslope. At one extreme, the movement may be gradual and practically imperceptible. At the other extreme, it may consist of a roaring debris flow or a thundering rock avalanche. Landslides are a worldwide natural hazard (Figure 8.1). When these hazardous processes lead to loss of life and property, they become natural disasters.

A Landslide Disaster in Peru

Occasionally, news media report the terrifying and often grim details of landslides. For example, on May 31, 1970, a gigantic rock avalanche buried more than 20,000 people in Yungay and Ranrahirca, Peru. There was little warning of the impending disaster; it began and ended in just a matter of a few minutes. The avalanche started about 14 kilometers (9 miles) from Yungay, near the summit of the 6700-meter (22,000-foot) Nevados Huascaran, the loftiest peak in the Peruvian Andes. Triggered by the ground motion from a strong offshore earthquake, a huge mass of rock and ice broke free from the precipitous north face of the mountain. After plunging nearly a kilometer, the material pulverized on impact and immediately began rushing down the mountainside, made fluid by trapped air and melted ice.

The falling debris ripped loose millions of tons of additional debris as it roared downhill. Although the material followed a previously eroded gorge, a portion of the debris jumped a 200 to 300-meter (650- to 1000-foot) bedrock ridge that had protected Yungay from similar events in the past and buried the entire city. After destroying another town in its path, Ranrahirca, the mass of debris finally reached the bottom of the valley. There, its momentum carried it across the Rio Santa and tens of meters up the valley wall on the opposite side.

This was not the first such disaster in the region and will probably not be the last. Just eight years earlier a less spectacular, but nevertheless devastating, rock avalanche took the lives of an estimated 3500 people on the heavily populated valley floor at the base of the mountain. Fortunately, mass movements such as the one just described are infrequent and only occasionally affect large numbers of people.

As with many geologic hazards, the tragic rock avalanche in Peru was triggered by a natural event—in this case, an earthquake. In fact, most mass-wasting events, whether spectacular or subtle, are the result of circumstances that are completely independent of

▼ **Figure 8.1** This rockslide occurred in January 1997 on Highway 140 near the Arch Rock entrance to Yosemite National Park, California. *(Photo by Roger J. Wyan/AP/Wide World Photos)*

human activities. In places where mass wasting is a recognized threat, steps can often be taken to control downslope movements or limit the damages that such movements can cause. If the potential for mass wasting goes unrecognized or is ignored, the results can be costly and dangerous. It should also be pointed out that, although most downslope movements occur whether people are present or not, many occurrences each year are aggravated or even triggered by human actions.

Peruvian valley after the landslide. (Iris Lozier)

Mass Wasting and Landform Development

Landslides are spectacular examples of a common geologic process called mass wasting. **Mass wasting** refers to the downslope movement of rock, regolith, and soil under the direct influence of gravity. It is distinct from the erosional processes that are examined in subsequent chapters because mass wasting does not require a transporting medium such as water, wind, or glacial ice.

The Role of Mass Wasting

In the evolution of most landforms, mass wasting is the step that follows weathering. By itself weathering does not produce significant landforms. Rather, landforms develop as the products of weathering are removed from the places where they originate. Once weathering weakens and breaks rock apart, mass wasting transfers the debris downslope, where a stream, acting as a conveyor belt, usually carries it away. Although there may be many intermediate stops along the way, the sediment is eventually transported to its ultimate destination: the sea.

The combined effects of mass wasting and running water produce stream valleys, which are the most common and conspicuous of Earth's landforms. If streams alone were responsible for creating the valleys in which they flow, valleys would be very narrow features. However, the fact that most river valleys are much wider than they are deep is a strong indication of the significance of mass-wasting processes in supplying material to streams. This is illustrated by the Grand Canyon (Figure 8.2). The walls of the canyon extend far from the Colorado River, owing to the transfer of weathered debris downslope to the river and its tributaries by mass-wasting processes. In this manner, streams and mass wasting combine to modify and sculpt the surface. Of course, glaciers, groundwater, waves, and wind are also important agents in shaping landforms and developing landscapes.

Slopes Change Through Time

It is clear that if mass wasting is to occur, there must be slopes down which rock, soil, and regolith can move. It is Earth's mountain building and volcanic processes that produce these slopes through sporadic changes in the elevations of landmasses and the ocean floor. If dynamic internal processes did not continually produce regions having higher elevations, the system that moves debris to lower elevations would gradually slow and eventually cease.

Did You Know?

Although many people, including geologists, frequently use the word *landslide*, the term has no specific definition in geology. Rather, it is a popular non-technical term used to describe all relatively rapid forms of mass wasting.

Most rapid and spectacular mass-wasting events occur in areas of rugged, geologically young mountains. Newly formed mountains are rapidly eroded by rivers and glaciers into regions characterized by steep and unstable slopes. It is in such settings that massive destructive landslides, such as the Yungay disaster, occur. As mountain building subsides, mass wasting

◀ **Figure 8.2** The walls of the Grand Canyon extend far from the channel of the Colorado River. This results primarily from the transfer of weathered debris downslope to the river and its tributaries by mass-wasting processes. *(Photo by Tom and Susan Bean, Inc.)*

and erosional processes lower the land. Through time, steep and rugged mountain slopes give way to gentler, more subdued terrain. Thus, as a landscape ages, massive and rapid mass-wasting processes give way to smaller, less dramatic downslope movements.

Controls and Triggers of Mass Wasting

GEODe Mass Wasting
▼ Controls and Triggers of Mass Wasting

Gravity is the controlling force of mass wasting, but several factors play an important role in overcoming inertia and creating downslope movements. Long before a landslide occurs, various processes work to weaken slope material, gradually making it more and more susceptible to the pull of gravity. During this span, the slope remains stable but gets closer and closer to being unstable. Eventually, the strength of the slope is weakened to the point that something causes it to cross the threshold from stability to instability. Such an event that initiates downslope movement is called a *trigger.* Remember that the trigger is not the sole cause of the mass-wasting event, but just the last of many causes. Among the common factors that trigger mass-wasting processes are saturation of material with water, oversteepening of slopes, removal of anchoring vegetation, and ground vibrations from earthquakes.

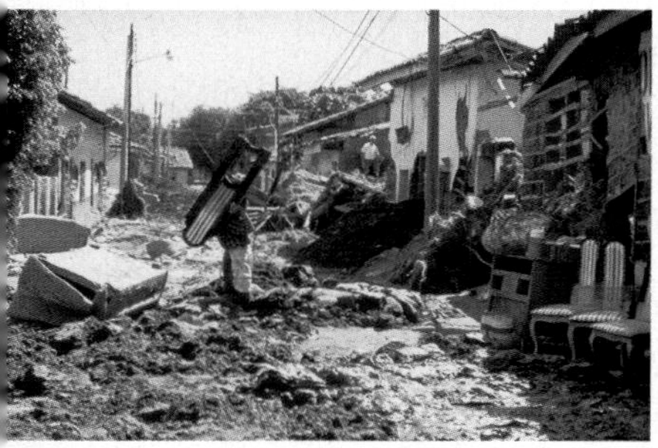

Heavy rains from Hurricane Mitch in 1998 triggered mudflows in Central America. (Noel Quidu/Getty-Liaison)

The Role of Water

Mass wasting is sometimes triggered when heavy rains or periods of snowmelt saturate surface materials. This was the case in October 1998 when torrential downpours associated with Hurricane Mitch triggered devastating mudflows in Central America.

When the pores in sediment become filled with water, the cohesion among particles is destroyed, allowing them to slide past one another with relative ease. For example, when sand is slightly moist, it sticks together quite well. However, if enough water is added to fill the openings between the grains, the sand will ooze out in all directions (Figure 8.3). Thus, saturation reduces the internal resistance of materials, which are then easily set in motion by the force of gravity. When clay is wetted, it becomes very slick—another example of the lubricating effect of water. Water also adds considerable weight to a mass of material. The added weight in itself may be enough to cause the material to slide or flow downslope.

Did You Know?

The cost of damages by all types of mass wasting in the United States is conservatively estimated to exceed $2 billion in an average year.

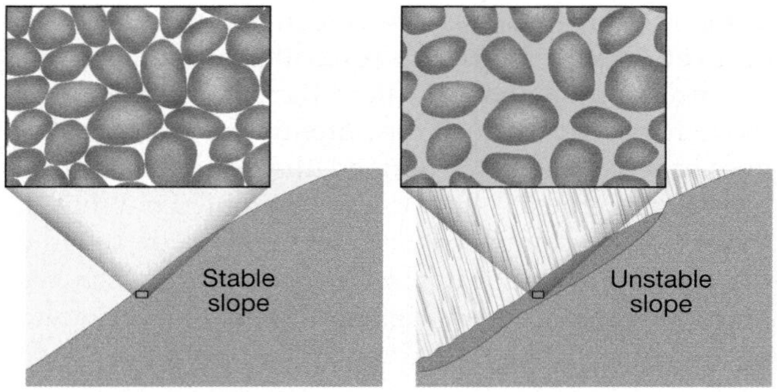

▲ **Figure 8.3** The effect of water on mass wasting can be great. **A.** When little or no water is present, friction among the closely packed soil particles on the slope holds them in place. **B.** When the soil is saturated, the grains are forced apart and friction is reduced, allowing the soil to move downslope.

Oversteepened Slopes

Oversteepening of slopes is another trigger of many mass movements. There are many situations in nature where oversteepening takes place. A stream undercutting a valley wall and waves pounding against the base of a cliff are but two familiar examples. Furthermore, through their activities, people often create oversteepened and unstable slopes that become prime sites for mass wasting.

Unconsolidated, granular particles (sand-sized or coarser) assume a stable slope called the **angle of repose.** This is the steepest angle at which material remains stable. Depending on the size and shape of the particles, the angle varies from 25 to 40 degrees. The larger, more angular particles maintain the steepest slopes. If the angle is increased, the rock debris will adjust by moving downslope.

Oversteepening is not just important because it triggers movements of unconsolidated granular materials. Oversteepening also produces unstable slopes and mass movements in cohesive soils, regolith, and bedrock. The response will not be immediate, as with loose, granular material, but sooner or later one or more mass-wasting processes will eliminate the oversteepening and restore stability to the slope.

Removal of Vegetation

Plants protect against erosion and contribute to the stability of slopes because their root systems bind soil and regolith together. In addition, plants shield the soil surface from the erosional effects of raindrop impact (see Figure 5.15, p. 132). Where plants are lacking, mass wasting is enhanced, especially if slopes are steep and water is plentiful. When anchoring vegetation is removed by forest fires or by people (for timber, farming, or development), surface materials frequently move downslope.

In July 1994 a severe wildfire swept Storm King Mountain west of Glenwood Springs, Colorado, denuding the slopes of vegetation. Two months later heavy rains resulted in numerous debris flows, one of which blocked Interstate 70 and threatened to dam the Colorado River. A 5-kilometer (3-mile) length of the highway was inundated with tons of rock, mud, and burned trees. The closure of Interstate 70 imposed costly delays on this major highway. Following extensive wildfires that occurred in the summer of 2000, similar types of mass wasting threaten highways and other development near fire-ravaged hillsides throughout the West (Figure 8.4).

In addition to eliminating plants that anchor the soil, fire can promote mass wasting in other ways. Following a wildfire, the upper part of the soil may become dry and loose. As a result, even in dry weather, the soil tends to move down steep slopes. Moreover, fire can also "bake" the ground, creating a water-repellant layer at a shallow depth. This nearly impermeable barrier prevents or slows the infiltration of water, resulting in increased surface runoff during rains. The consequence can be dangerous torrents of viscous mud and rock debris.

Earthquakes as Triggers

Conditions that favor mass wasting may exist in an area for a long time without movement occurring. An additional factor is sometimes necessary to trigger the movement. Among the more important and dramatic triggers are earthquakes. An earthquake and its aftershocks can dislodge enormous volumes of rock and unconsolidated material. The event in the Peruvian Andes described at the beginning of this chapter is one tragic example.

▼ **Figure 8.4** During the summer of 2000, wildfires raged in parts of many western states. Millions of acres were burned. The loss of anchoring vegetation sets the stage for accelerated mass wasting. This scene shows an area between Beavertail Hill and Bearmouth, Montana. *(Photo by Derek Pruitt/The Montana Standard/AP/Wide World Photos)*

The angle of repose here is about 30°. (G. Leavens/Photo Researchers)

Landslides Triggered by the Northridge Earthquake. In January 1994 an earthquake struck the Los Angeles region of southern California. Named for its epicenter in the town of Northridge, the 6.7-magnitude event produced estimated losses of $20 billion. Some of these losses were the result of thousands of landslides over an area of 10,000 square kilometers (3900 square miles) that were set in motion by the quake. Most were shallow rock falls and slides, but some were much larger and filled canyon bottoms with jumbles of soil, rock, and plant debris. The debris in canyon bottoms created a secondary threat because it can be mobilized during rainstorms to produce debris flows. Such flows are common and often disastrous in southern California.

The mass-wasting processes triggered by the Northridge earthquake destroyed dozens of homes and caused extensive damage to roads, pipelines, and well machinery in oil fields. In some places more than 75 percent of slope areas were denuded by landslides, making them vulnerable to subsequent mass wasting triggered by heavy rains.

Home destroyed by a landslide that was triggered by the 1994 Northridge earthquake. (Chromo Sohm/Stock Market)

Liquefaction. Intense ground shaking during earthquakes can cause water-saturated surface materials to lose their strength and behave as fluidlike masses that flow. This process, called *liquefaction,* was a major cause of property damage in Anchorage, Alaska, during the massive 1964 Good Friday earthquake described in Chapter 14.

Landslides Without Triggers?

Do rapid mass-wasting events always require some sort of trigger such as heavy rains or earthquakes? The answer is no. Such events sometimes occur without being triggered. For example, on the afternoon of May 9, 1999, a landslide killed 10 hikers and injured many others at Sacred Falls State Park near Hauula on the north shore of Oahu, Hawaii. The tragic event occurred when a mass of rock from a canyon wall plunged 150 meters (500 feet) down a nearly vertical slope to the valley floor. Because of safety concerns, the park was closed so that landslide specialists from the U.S. Geological Survey could investigate the site. Their study concluded that the landslide occurred *without being triggered* by any discernible external conditions.

Many rapid mass-wasting events occur without a discernible trigger. Slope materials gradually weaken over time under the influence of long-term weathering,

Warning sign near Arthur's Pass, New Zealand.
(R. Essel/Stock Market)

infiltration of water, and other physical processes. Eventually, if the strength falls below what is necessary to maintain slope stability, a landslide will occur. The timing of such events is random, and thus accurate prediction is impossible.

Classification of Mass-Wasting Processes

There is a broad array of different processes that geologists call mass wasting. Four processes are illustrated in Figure 8.5. Generally, the different types are classified based on the type of material involved, the kind of motion displayed, and the velocity of the movement.

Did You Know?

The U.S. Geological Survey estimates that between 25 and 50 people are killed by landslides annually in the United States. The worldwide death toll is much higher, more than 600 per year.

Type of Material

The classification of mass-wasting processes on the basis of the material involved in the movement depends upon whether the descending mass began as unconsolidated material or as bedrock. If soil and regolith dominate, terms such as debris, mud, or earth are used in the description. In contrast, when a mass of bedrock breaks loose and moves downslope, the term rock may be part of the description.

Type of Motion

In addition to characterizing the type of material involved in a mass-wasting event, the way in which the material moves may also be important. Generally, the kind of motion is described as either a fall, a slide, or a flow.

Fall. When the movement involves the freefall of detached individual pieces of any size, it is termed a **fall.** Fall is a common form of movement on slopes that are so steep that loose material cannot remain on the surface. The rock may fall directly to the base of the slope or move in a series of leaps and bounds over other rocks along the way. Many falls result when freeze and thaw cycles and/or the action of plant roots loosen rock to the point that gravity takes over. Although signs along bedrock cuts on highways warn of falling rock, few of us have actually witnessed such an event. However, as Figure 8.6 illustrates, they do indeed occur. In fact, this is the primary way in which **talus slopes** are built and maintained (see Figure 5.3, p. 118). Sometimes falls may trigger other forms of downslope movement. For

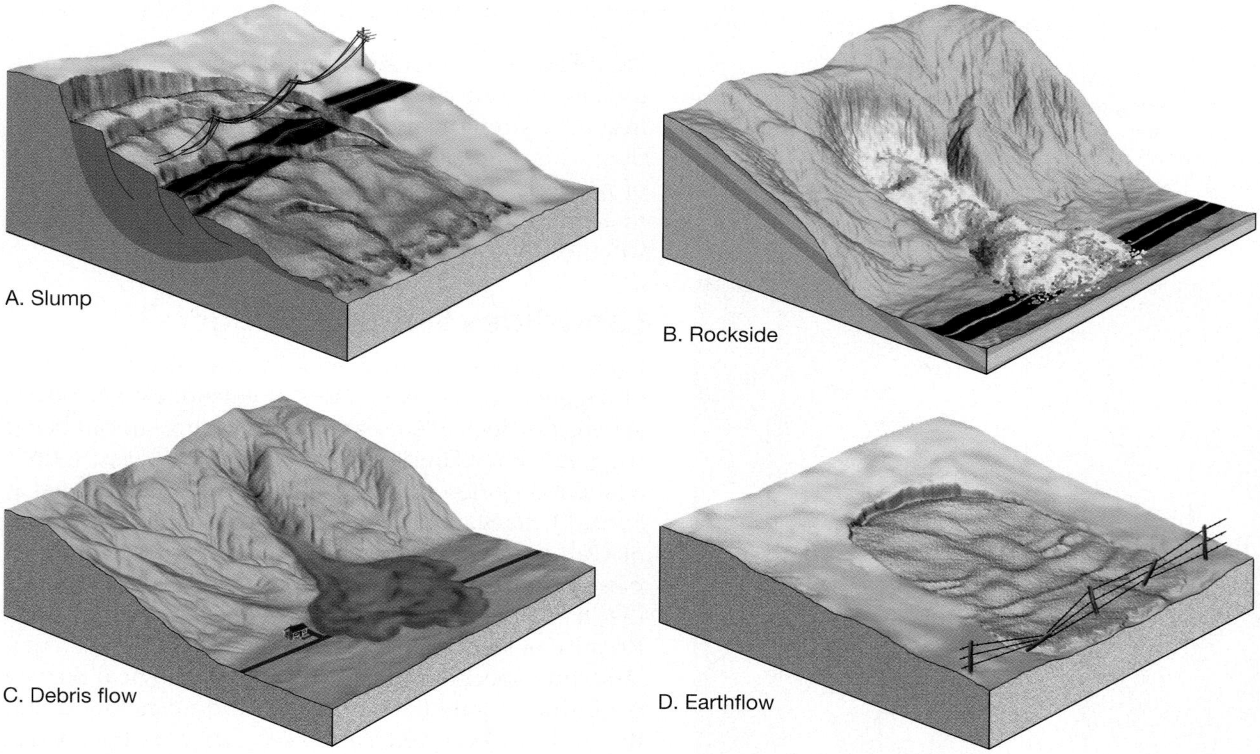

▲ **Figure 8.5** The four processes illustrated here are all considered to be relatively rapid forms of mass wasting. Because material in slumps **A.** and rockslides **B.** move along well-defined surfaces, they are said to move by sliding. By contrast, when material moves downslope as a viscous fluid, the movement is described as a flow. Debris flow **C.** and earthflow **D.** advance downslope in this manner.

▲ **Figure 8.6** Rockfall blocking Montana Highway 2 east of Pipestone Pass, May 1998. *(AP/Wide World Photo)*

example, recall that the Yungay disaster described at the beginning of the chapter was initiated by a mass of freefalling material that broke from the nearly vertical summit of Nevado Huascaran.

Slide. Many mass-wasting processes are described as **slides.** Slides occur whenever material remains fairly coherent and moves along a well-defined surface. Sometimes the surface is a joint, a fault, or a bedding plane that is approximately parallel to the slope. However, in the case of the movement called *slump,* the descending material moves en masse along a curved surface of rupture.

Flow. The third type of movement common to mass-wasting processes is termed **flow.** Flow occurs when material moves downslope as a viscous fluid. Most flows are saturated with water and typically move as lobes or tongues.

Talus slope in Banff National Park, Alberta, Canada. (M. Miller)

Rate of Movement

The event described at the beginning of this chapter clearly involved rapid movement. The rock and debris moved downslope at speeds well in excess of 200 kilometers (125 miles) per hour. This most rapid type of mass movement is termed a **rock avalanche.** Many researchers believe that rock avalanches, such as the one that produced the scene in Figure 8.7, must literally float on air as they move downslope. That is, high velocities result when air becomes trapped and compressed beneath the falling mass of debris, allowing it to move as a buoyant, flexible sheet across the surface.

Most mass movements, however, do not move with the speed of a rock avalanche. In fact, a great deal of mass wasting is imperceptibly slow. One process that we will examine later, termed *creep,* results in particle movements that are usually measured in millimeters or centimeters per year. Thus, as you can see, rates of movement can be spectacularly sudden or exceptionally gradual. Although various types of mass wasting are often classified as either rapid or slow, such a distinction is highly subjective because there is a wide range of rates between the two extremes. Even the velocity of a single process at a particular site can vary considerably.

Mudflow on the Isle of Wight, UK. (Tony Waltham)

◀ **Figure 8.7** This 4-kilometer-long tongue of rubble was deposited atop Alaska's Sherman Glacier by a rock avalanche. The event was triggered by a tremendous earthquake in March 1964. *(Photo by Austin Post, U.S. Geological Survey)*

Snow avalanches are a form of mass wasting.
(T. Murphy/Peter Arnold, Inc.)

Slump

GEODe — Mass Wasting ▼ Types of Mass Wasting

Slump refers to the downward sliding of a mass of rock or unconsolidated material moving as a unit along a curved surface (Figure 8.5A). Usually the slumped material does not travel spectacularly fast nor very far. This is a common form of mass wasting, especially in thick accumulations of cohesive materials such as clay. The rupture surface is characteristically spoon-shaped and concave upward or outward. As the movement occurs, a crescent-shaped scarp is created at the head, and the block's upper surface is sometimes tilted backward. Although slump may involve a single mass, it often consists of multiple blocks. Sometimes water collects between the base of the scarp and the top of the tilted block. As this water percolates downward along the surface of rupture, it may promote further instability and additional movement.

Slump commonly occurs because a slope has been oversteepened. The material on the upper portion of a slope is held in place by the material at the bottom of the slope. As this anchoring material at the base is removed, the material above is made unstable and reacts to the pull of gravity. One relatively common example is a valley wall that becomes oversteepened by a meandering river. Another is a coastal cliff that has been undercut by wave action at its base.

Slumping may also occur when a slope is overloaded, causing internal stress on the material below. This type of slump often occurs where weak, clay-rich material underlies layers of stronger, more resistant rock such as sandstone. The seepage of water through the upper layers reduces the strength of the clay below, causing slope failure.

Did You Know?

Snow avalanches are considered a type of mass wasting because these thundering downslope movements also move large quantities of rock and soil in addition to snow and ice. About 10,000 snow avalanches occur each year in the mountainous western United States.

Rockslide

GEODe — Mass Wasting ▼ Types of Mass Wasting

Rockslides frequently occur in high mountain areas such as the Andes, Alps, and Canadian Rockies. They are sudden and rapid movements that happen when detached segments of bedrock break loose and slide downslope (Figure 8.5B). As the moving mass thunders along the surface, it breaks into many smaller pieces. Such events are among the fastest and most destructive mass movements.

Rockslides usually take place where there is an inclined surface of weakness. Such surfaces tend to form where strata are tilted or where joints and fractures exist parallel to the slope. When rock in such a setting is undercut at the base of the slope, it loses support and eventually gives way. Sometimes the rockslide is triggered when rain or melting snow lubricates the underlying surface to the point that friction is no longer sufficient to hold the rock unit in place. As a result, rockslides tend to be more common during the spring, when heavy rains and melting snow are most prevalent.

As was noted earlier, earthquakes can trigger rockslides and other mass movements. The 1811 earthquake at New Madrid, Missouri, for example, caused slides in an area of more than 13,000 square kilometers (5000 square miles) along the Mississippi River valley. A more recent example occurred on August 17, 1959, when a severe earthquake west of Yellowstone National Park triggered a massive slide in the canyon of the Madison River in southwestern Montana (Figure 8.8). In a matter of moments an estimated 27 million cubic meters of rock, soil, and trees slid into the canyon. The debris dammed the river and buried a campground and highway. More than 20 unsuspecting campers perished.

Not far from the site of the Madison Canyon slide, the classic Gros Ventre rockslide occurred 34 years earlier. The Gros Ventre River flows west from the northernmost part of the Wind River Range in northwestern Wyoming, through the Grand Teton National Park, and eventually empties into the Snake River. On June 23, 1925, a massive rockslide took place in its valley, just east of the small town of Kelly. In the span of just a few minutes a great mass of sandstone, shale, and soil crashed down the south side of the valley, carrying with it a dense pine forest. The volume of debris, estimated at 38 million cubic meters (50 million cubic yards), created a 70-meter-high dam on the Gros Ventre River. Because the river was completely blocked, a lake was created. It filled so quickly that a house that had been 18 meters (60 feet) above the river was floated off its foundation 18 hours after the slide. In 1927, the lake overflowed the dam, partially draining the lake and resulting in a devastating flood downstream.

Why did the Gros Ventre rockslide take place? Figure 8.9 shows a diagrammatic cross-sectional view of the geology of the valley. Notice the following points: (1) The sedimentary strata in this area dip (tilt) 15 to 21 degrees; (2) underlying the bed of sandstone is a relatively thin layer of clay; and (3) at the bottom of the valley, the river had cut through much of the sandstone layer. During the spring of 1925, water from heavy rains and melting snow seeped through the sandstone, saturating the clay below. Because much of the sandstone layer had been cut through by the Gros Ventre River, the layer had virtually no support at the bottom of the slope. Eventually the sandstone could no longer hold its position on the wetted clay, and gravity pulled the mass down the side of the valley. The circumstances at this location were such that the event was inevitable.

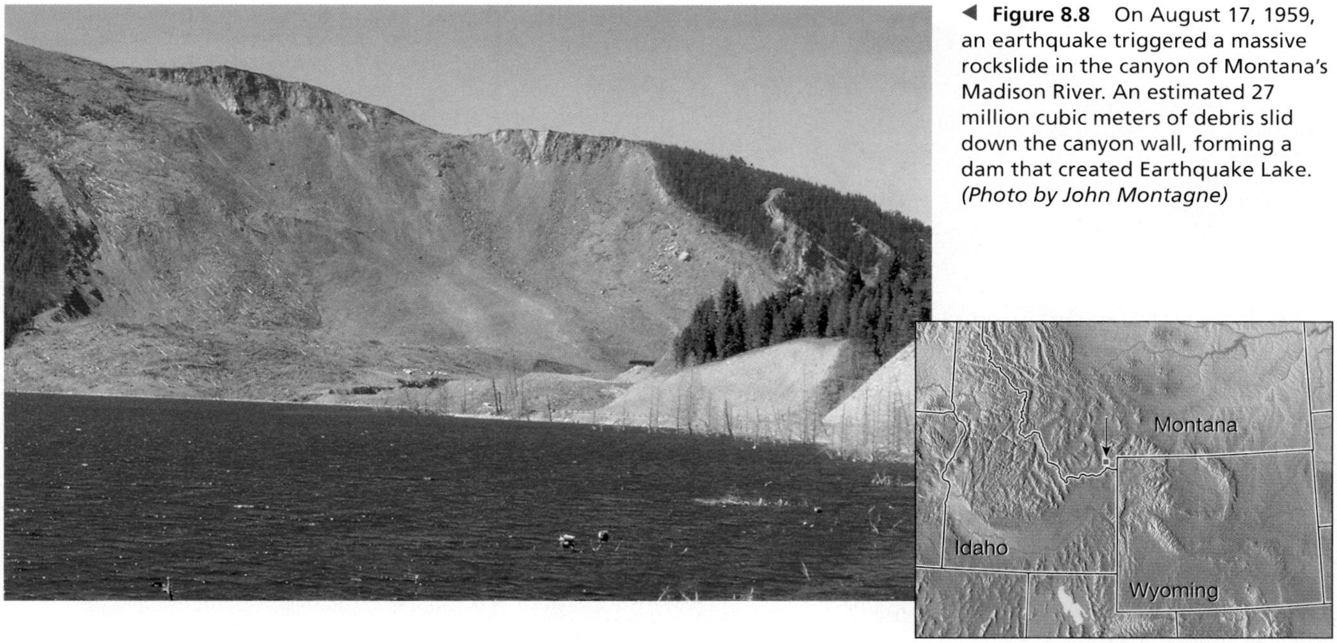

◀ **Figure 8.8** On August 17, 1959, an earthquake triggered a massive rockslide in the canyon of Montana's Madison River. An estimated 27 million cubic meters of debris slid down the canyon wall, forming a dam that created Earthquake Lake. *(Photo by John Montagne)*

Debris Flow

GEODe
Mass Wasting
▼ Types of Mass Wasting

Debris flow is a relatively rapid type of mass wasting that involves a flow of soil and regolith containing a large amount of water (see Figure 8.5C). Debris flows, which are also called **mudflows,** are most characteristic of semiarid mountainous regions and are also common on the slopes of some volcanoes. Because of their fluid properties, debris flows frequently follow canyons and stream channels. In populated areas, debris flows can pose a significant hazard to life and property.

Debris Flows in Semiarid Regions

When a cloudburst or rapidly melting mountain snows create a sudden flood in a semiarid region, large quantities of soil and regolith are washed into nearby stream channels because there is usually little vegetation to anchor the surface material. The end product is a flowing tongue of well-mixed mud, soil, rock, and water. Its consistency can range from that of wet concrete to a soupy mixture not much thicker than muddy water. The rate of flow therefore depends not only on the slope but also on the water content. When dense, debris flows are capable of carrying or pushing large boulders, trees, and even houses with relative ease.

Debris flows pose a serious hazard to development in relatively dry mountainous areas such as southern California. Here construction of homes on canyon hillsides and the removal of native vegetation by brush fires and other means have increased the frequency of these destructive events (Figure 8.10). Moreover, when a debris flow reaches the end of a steep, narrow canyon, it spreads out, covering the area beyond the mouth of the canyon with a mixture of wet debris. This material contributes to the buildup of fanlike deposits at canyon mouths.* The fans are relatively easy to build on, often have nice views, and are close to the mountains; in fact, like the nearby canyons, many have become preferred sites for development. Because debris flows occur only sporadically, the public is often unaware of the potential hazard of such sites (see Box 8.1).

Slump along California coast at Point Fermin. (John S. Shelton)

Lahars

Debris flows composed mostly of volcanic materials on the flanks of volcanoes are called **lahars.** The word originated in Indonesia, a volcanic region that has experienced many of these often destructive events. Historically, lahars have been one of the deadliest volcano hazards. They can occur either during an eruption or when a volcano is quiet. They take place when highly unstable layers of ash and debris become saturated with water and flow down steep volcanic slopes, generally following existing stream channels. Heavy rain-

*These structures are called *alluvial fans* and will be discussed in greater detail in Chapters 9 and 12.

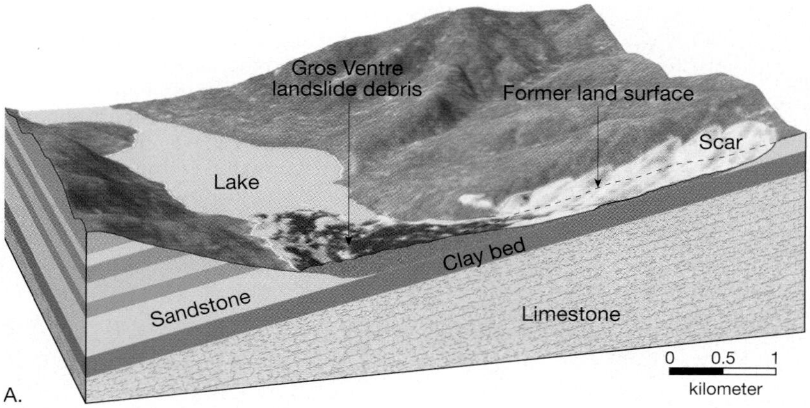

A.

B.

◀ **Figure 8.9** Part A shows a cross-sectional view of the Gros Ventre rockslide. The slide occurred when the tilted and undercut sandstone bed could no longer maintain its position atop the saturated bed of clay. As the photo in part B illustrates, even though the Gros Ventre rockslide occurred in 1925, the scar left on the side of Sheep Mountain is still a prominent feature. *(Part A after W. C. Alden, "Landslide and Flood at Gros Ventre, Wyoming," Transactions (AIME) 76 (1928); 348. Part B photo by Stephen Trimble)*

Did You Know?

Even after traveling many miles from the volcano and mixing with cold waters, some of the lahars associated with the May 1980 eruption of Mount St. Helens maintained temperatures between 184° and 196°F (remember, 212°F is the boiling point of water!).

falls often trigger these flows. Others are initiated when large volumes of ice and snow are suddenly melted by heat flowing to the surface from within the volcano or by the hot gases and near-molten debris emitted during a violent eruption.

When Mount St. Helens erupted in May 1980, several lahars were created. The flows and accompanying floods raced down the valleys of the north and south forks of the Toutle River at speeds that were often in excess of 30 kilometers (20 miles) per hour. Fortunately, the affected area was not densely settled. Nevertheless, more than 200 homes were destroyed or severely damaged. Most bridges met a similar fate.

In November 1985, lahars were produced during the eruption of Nevado del Ruiz, a 5300-meter (17,400-foot) volcano in the Andes Mountains of Colombia. The eruption melted much of the snow and ice that capped the uppermost 600 meters (2000 feet) of the peak, producing torrents of hot, thick mud, ash, and debris. The lahars moved outward from the volcano, following the valleys of three rain-swollen rivers that radiate from the peak. The flow that moved down the valley of the Lagunilla River was the most destructive. It devastated the town of Armero, 48 kilometers (30 miles) from the mountain. Most of the more than 25,000 deaths caused by the event occurred in this once thriving agricultural community.

Death and property damage due to the lahars also occurred in 13 other villages within the 180-square-kilometer (70-square-mile) disaster area. Although a great deal of pyroclastic material was explosively ejected from Nevado del Ruiz, it was the lahars triggered by this eruption that made this such a devastating natural disaster. In fact, it was the worst

House buried by a lahar created by the May 1980 eruption of Mount St. Helens. (D.R. Crandell/USGS)

▼ **Figure 8.10** In February 1998, a mudflow literally buried this pickup truck in Rio Nido, California. Heavy rains triggered the event. *(Photo by Michael Collier)*

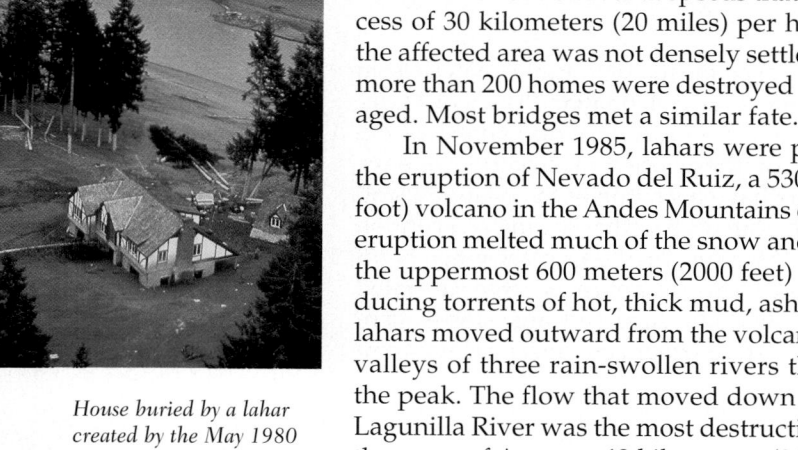

BOX 8.1

Debris Flows on Alluvial Fans: A Case Study from Venezuela*

In December 1999, heavy rains triggered thousands of landslides along the coast of Venezuela (Figure 8.A). Debris flows and flash floods caused severe property damage and the tragic loss of an estimated 19,000 lives. The sites of most of the death and destruction were *alluvial fans.* These landforms are gently sloping, cone- to fan-shaped accumulations of sediment that are commonly found where high-gradient streams leave narrow valleys in mountainous areas and abruptly meet flat terrain.

Several hundred thousand people live in the narrow coastal zone north of Caracas, Venezuela. They occupy alluvial fans located at the base of steep mountains that rise to elevations of more than 2000 meters (6600 feet) because these sites are the only areas that are not too sleep to build on (Figure 8.B). Such settings are highly vulnerable to rainfall-induced landslides.

An unusually wet period in December 1999 included rains of 20 centimeters (8 inches) on December 2 and 3, followed by an additional 91 centimeters (36 inches) between December 14 and 16. The heavy rains triggered thousands of debris flows and other types of mass wasting. Once created, these moving masses of mud and rock coalesced to form giant debris flows that moved rapidly through steep, narrow canyons before exiting onto the alluvial fans.

On virtually every alluvial fan in the area, debris flows and flash floods brought massive amounts of sediment, including boulders as large as 10 meters (33 feet) in diameter. Hundreds of houses and other structures were damaged or destroyed (Figure 8.C). Total damages approached $2 billion.

Figure 8.A Area of Venezuela affected by disastrous debris flows and flash floods in 1999.

Figure 8.B Aerial view of the highly developed alluvial fan at Caraballeda, Venezuela. (Kimberly White/REUTERS/CORBIS/Bettmann)

This example from Venezuela shows the potential for extreme loss of life and property damage where large numbers of people occupy alluvial fans. The possibility for similar events of comparable magnitude exists in other parts of the world.

Building communities on alluvial fans can transform natural processes into major lethal events. Kofi Annan, Secretary General of the United Nations, put it this way: "The term 'natural disaster' has become an increasingly anachronistic misnomer. In reality, human behavior transforms natural hazards into what should really be called unnatural disasters."**

*Based on material prepared by the U.S. Geological Survey.

**Matthew C. Larsen, et al. *Natural Hazards on Alluvial Fans. The Venezuela Debris Flow and Flash Flood Disaster,* U.S. Geological Survey Fact Sheet FS 103, p. 4.

Figure 8.C Debris-flow damage. Huge boulders (in excess of 300 tons) were transported by some flows. (AP/Wide World Photo)

volcanic disaster since 28,000 people died following the 1902 eruption of Mount Pelée on the Caribbean island of Martinique.*

Earthflow

GEODe Mass Wasting
▼ Types of Mass Wasting

We have seen that debris flows are frequently confined to channels in semiarid regions. In contrast, **earthflows** most often form on hillsides in humid areas during times of heavy precipitation or snowmelt (see Figure 8.5D, p. 190). When water saturates the soil and regolith on a hillside, the material may break away, leaving a scar on the slope and forming a tongue- or teardrop-shaped mass that flows downslope (Figure 8.11).

The materials most commonly involved are rich in clay and silt and contain only small proportions of sand and coarser particles. Earthflows range in size from bodies a few meters long, a few meters wide, and less than a meter deep to masses more than a kilometer long, several hundred meters wide, and more than 10 meters deep.

Did You Know?

Losses from the 1983 Thistle landslide in central Utah exceeded $400 million, making it the most costly single landslide in U.S. history. The slide dammed the Spanish Fork River and obliterated U.S. Highway 6 and the mainline of the Denver and Rio Grande Western Railroad. The town of Thistle was inundated by the lake created behind the landslide dam.

Because earthflows are quite viscous, they generally move at slower rates than the more fluid debris flows described in the preceding section. They are characterized by a slow and persistent movement and may remain active for periods ranging from days to years. Depending on the steepness of the slope and the material's consistency, measured velocities range from less than 1 millimeter per day up to several meters per day. Over the time span that earthflows are active, movement is typically faster during wet periods than during drier times. In addition to occurring as isolated hillside phenomena, earthflows commonly take place in association with large slumps. In this situation, they may be seen as tonguelike flows at the base of the slump block.

▲ **Figure 8.11** This small, tongue-shaped earthflow occurred on a newly formed slope along a recently constructed highway. It formed in clay-rich material following a period of heavy rain. Notice the small slump at the head of the earthflow. *(Photo by E. J. Tarbuck)*

*A discussion of the Mount Pelée eruption, as well as additional material on lahars, can be found in Chapter 4.

Slow Movements

GEODe Mass Wasting
▼ Types of Mass Wasting

Movements such as rockslides, rock avalanches, and lahars are certainly the most spectacular and catastrophic forms of mass wasting. These dangerous events deserve intensive study to enable more effective prediction, timely warnings, and better controls to save lives. However, because of their large size and spectacular nature, they give us a false impression of their importance as a mass-wasting process. Indeed, sudden movements are responsible for moving less material than the slower and far more subtle action of creep. Whereas rapid types of mass wasting are characteristic of mountains and steep hillsides, creep takes place on both steep and gentle slopes and is thus much more widespread.

Creep

Creep is a type of mass wasting that involves the gradual downhill movement of soil and regolith. One factor that contributes to creep is the alternate expansion and contraction of surface material caused by freezing and thawing or wetting and drying. As shown in Figure 8.12, freezing or wetting lifts particles at right angles to the slope, and thawing or drying allows the particles to fall back to a slightly lower level. Each cycle therefore

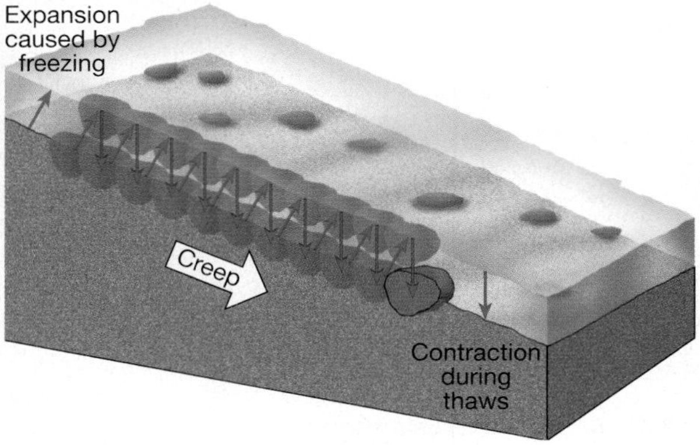

▲ **Figure 8.12** The repeated expansion and contraction of surface material causes a net downslope migration of soil and rock particles—a process called *creep.*

BOX 8.2

The Sensitive Permafrost Landscape

Many of the mass-wasting events described in this chapter had sudden and disastrous impacts on people. When the activities of people cause ice contained in permanently frozen ground to melt, the impact is more gradual and less deadly. Nevertheless, because permafrost regions are sensitive and fragile landscapes, the scars resulting from poorly planned actions can remain for generations.

Permanently frozen ground, known as *permafrost*, occurs where summers are too cool to melt more than a shallow surface layer. Deeper ground remains frozen year-round. When people disturb the surface by removing the insulating vegetation mat or by constructing roads and buildings, the delicate thermal balance is disturbed, and the permafrost can thaw (Figure 8.D). Thawing produces unstable ground that may slide, slump, subside, and undergo severe frost heaving.

Figure 8.D When a rail line was built across this permafrost landscape in Alaska, the ground subsided. (Photo by Lynn A. Yehle/O. J. Ferrians, Jr./U.S. Geological Survey)

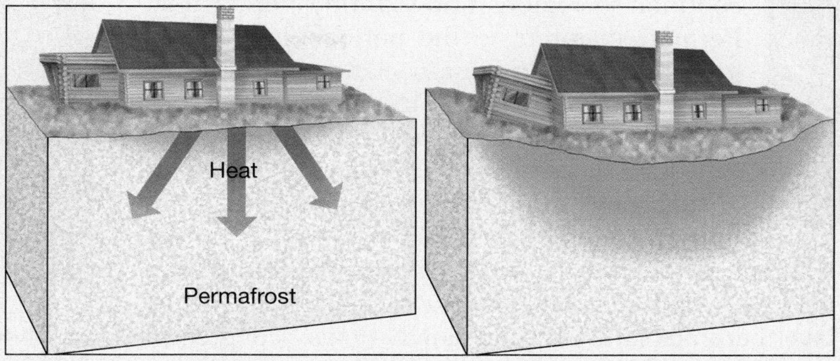

Figure 8.E This building, located south of Fairbanks, Alaska, subsided because of thawing permafrost. Notice that the right side, which was heated, settled much more than the unheated porch on the left.

As Figure 8.E illustrates, when a heated structure is built directly on permafrost that contains a high proportion of ice, thawing creates soggy material into which a building can sink. One solution is to place buildings and other structures on piles, like stilts. Such piles allow subfreezing air to circulate between the floor of the building and the soil and thereby keep the ground frozen.

When oil was discovered on Alaska's North Slope, many people were concerned about the building of a pipeline linking the oil fields of Prudhoe Bay to the ice-free port of Valdez 1300 kilometers (800 miles) to the south. There was serious concern that such a massive project might damage the sensitive permafrost environment. Many also worried about possible oil spills.

Because oil must be heated to about 60°C (140°F) to flow properly, special engineering procedures had to be developed to isolate this heat from the permafrost. Methods included insulating the pipe, elevating portions of the pipeline above ground level, and even placing cooling devices in the ground to keep it frozen. The Alaska pipeline is clearly one of the most complex and costly projects ever built in the Arctic tundra. Detailed studies and careful engineering helped minimize adverse effects resulting from the disturbance of frozen ground.

moves the material a tiny distance downslope. Creep is aided by anything that disturbs the soil. For example, raindrop impact and disturbance by plant roots and burrowing animals may contribute.

Creep is also promoted when the ground becomes saturated with water. Following a heavy rain or snowmelt, a waterlogged soil may lose its internal cohesion, allowing gravity to pull the material downslope. Because creep is imperceptibly slow, the process cannot be observed in action. What can be observed, however, are the effects of creep. Creep causes fences and utility poles to tilt and retaining walls to be displaced.

Solifluction

When soil is saturated with water, the soggy mass may flow downslope at a rate of a few millimeters or a few centimeters per day or per year. Such a process is called **solifluction** (literally, "soil flow"). It is a type of mass wasting that is common wherever water cannot escape from the saturated surface layer by infiltrating to deeper levels. A dense clay hardpan in

Did You Know?

Permafrost underlies nearly one-quarter of Earth's land area. In North America, more than 80 percent of Alaska and about 50 percent of Canada are underlain by permafrost.

In places in Alaska, a pipeline is suspended above ground to prevent melting of delicate permafrost. (Tom & Pat Leeson/Photo Researchers)

soil or an impermeable bedrock layer can promote solifluction.

Solifluction is a form of mass wasting that is common in regions underlain by **permafrost.** Permafrost refers to the permanently frozen ground that occurs in association with Earth's harsh tundra and ice-cap climates (see Box 8.2). Solifluction occurs in a zone above the permafrost called the *active layer,* which thaws to a depth of about a meter during the brief high-latitude summer and then refreezes in winter. During the summer season, water is unable to percolate into the impervious permafrost layer below. As a result, the active layer becomes saturated and slowly flows. The process can occur on slopes as gentle as 2 to 3 degrees. Where there is a well-developed mat of vegetation, a solifluction sheet may move in a series of well-defined lobes or as a series of partially overriding folds (Figure 8.13).

▲ **Figure 8.13** Solifluction lobes northeast of Fairbanks, Alaska. Solifluction occurs in permafrost regions when the active layer thaws in summer. *(Photo by James E. Patterson)*

The Chapter in Review

- *Mass wasting* refers to the downslope movement of rock, regolith, and soil under the direct influence of gravity. In the evolution of most landforms, mass wasting is the step that follows weathering. The combined effects of mass wasting and erosion by running water produce stream valleys.
- *Gravity is the controlling force of mass wasting.* Other factors that influence or trigger downslope movements are saturation of the material with water, oversteepening of slopes beyond the *angle of repose,* removal of vegetation, and ground shaking by earthquakes.
- The various processes included under the name of mass wasting are divided and described on the basis of (1) the type of material involved (debris, mud, earth, or rock); (2) the type of motion (fall, slide, or flow); and (3) the rate of movement (rapid or slow).
- The more rapid forms of mass wasting include *slump,* the downward sliding of a mass of rock or unconsolidated material moving as a unit along a curved surface; *rockslide,* blocks of bedrock breaking loose and sliding downslope; *debris flow,* a relatively rapid flow of soil and regolith containing a large amount of water; and *earthflow,* an unconfined flow of saturated, clay-rich soil that most often occurs on a hillside in a humid area following heavy precipitation or snowmelt.
- The slowest forms of mass wasting include *creep,* the gradual downhill movement of soil and regolith, and *solifluction,* the gradual flow of a saturated surface layer that is underlain by an impermeable zone. Common sites for solifluction include regions underlain by *permafrost* (permanently frozen ground associated with tundra and ice-cap climates).

Key Terms

angle of repose (p. 188)
creep (p. 196)
debris flow (p. 193)
earthflow (p. 196)
fall (p. 190)
flow (p. 191)
lahar (p. 193)
mass wasting (p. 187)
mudflow (p. 193)
permafrost (p. 198)
rock avalanche (p. 191)
rockslide (p. 192)
slide (p. 191)
slump (p. 192)
solifluction (p. 197)
talus slope (p. 190)

Questions for Review

1. Describe how mass-wasting processes contribute to the development of stream valleys.
2. What is the controlling force of mass wasting?
3. How does water affect mass wasting?
4. Describe the significance of the angle of repose.
5. How might the removal of vegetation by fire or logging promote mass wasting?
6. How are earthquakes linked to landslides?
7. Distinguish among *fall, slide,* and *flow.*
8. Why can rock avalanches move at such great speeds?

9. Slump and rockslide both move by sliding. In what ways do these processes differ?
10. What factors led to the massive rockslide at Gros Ventre, Wyoming?
11. Explain why building a home on an alluvial fan might not be a good idea. (See Box 8.1)
12. Compare and contrast *debris flow* and *earthflow.*
13. Describe the mass wasting that occurred at Mount St. Helens during its active period in 1980 and at Nevado del Ruiz in 1985.
14. Describe the mechanisms that contribute to the slow downslope movement called *creep.*
15. During what season does solifluction occur in permafrost regions?

Online Study Guide

The *Essentials of Geology* Web site uses the resources and flexibility of the Internet to aid in your study of the topics in this chapter. Written and developed by geology instructors, this site will help improve your understanding of geology. Visit **www.prenhall.com/lutgens** and click on the cover of *Essentials of Geology 9e* to find:

- Online review quizzes.
- Critical thinking exercises.
- Links to chapter-specific Web resources.
- Internet-wide key-term searches.

http://www.prenhall.com/lutgens

Hudson River, in New York State's Adirondack State Park. *(Photo by Carr Clifton Photography)*

CHAPTER 9

Running Water

Focus on Learning

To assist you in learning the important concepts in this chapter, you will find it helpful to focus on the following questions:

- What is the hydrologic cycle? What is the source of energy that powers the cycle?
- What are the factors that determine the velocity of water in a stream?
- In what way is base level related to a stream's ability to erode?
- The work of a stream includes what three processes?
- What are the two general types of stream valleys and some features associated with each?
- What are the common drainage patterns produced by streams?
- Why do floods occur? What are some basic flood-control strategies?

Rivers are very important to people. We use them as highways for moving goods, as sources of water for irrigation, and as an energy source. Their fertile floodplains have been cultivated since the dawn of civilization. When viewed as part of the Earth system, rivers and streams represent a basic link in the constant cycling of the planet's water. Moreover, running water is the dominant agent of landscape alteration, eroding more terrain and transporting more sediment than any other process. Because so many people live near rivers, floods are among the most destructive of all geologic hazards. Despite huge investments in levees and dams, rivers cannot always be controlled.

Rivers are a basic part of the hydrologic cycle.

Earth as a System: The Hydrologic Cycle

GEODe ESSENTIALS OF GEOLOGY

Running Water
▼ Hydrologic Cycle

All the rivers run into the sea; yet the sea is not full; unto the place from whence the rivers come, thither they return again.
(Ecclesiastes 1:7)

As the perceptive writer of Ecclesiastes indicated, water is continually on the move, from the ocean to the land and back again in an endless cycle. Water is everywhere on Earth—in the oceans, glaciers, rivers, lakes, air, soil, and in living tissue. All of these reservoirs constitute Earth's *hydrosphere.* In all, the water content of the hydrosphere is about 1.36 billion cubic kilometers (326 million cubic miles).

The vast bulk of it, about 97.2 percent, is stored in the global oceans. Ice sheets and glaciers account for another 2.15 percent, leaving only 0.65 percent to be divided among lakes, streams, subsurface water, and the atmosphere (Figure 9.1). Although the percentages of Earth's total water found in each of the latter sources is but a small fraction of the total inventory, the absolute quantities are great.

The water found in each of the reservoirs depicted in Figure 9.1 does not remain in these places indefinitely. Water can readily change from one state of matter (solid, liquid, or gas) to another at the temperatures and pressures that occur at Earth's surface. Therefore, water is constantly moving among the hydrosphere, the atmosphere, the geosphere, and the biosphere. This unending circulation of Earth's water supply is called the **hydrologic cycle.** The cycle shows us many critical interrelationships among different parts of the Earth system.

Did You Know?

Each year a field of crops may transpire the equivalent of a water layer 60 cm (2 ft) deep over the entire field. The same area of trees may pump twice this amount into the atmosphere.

Did You Know?

The 380,000 km^3 of water that annually cycles through the atmosphere is enough to cover Earth's surface to a depth of about 1 m (39 in.).

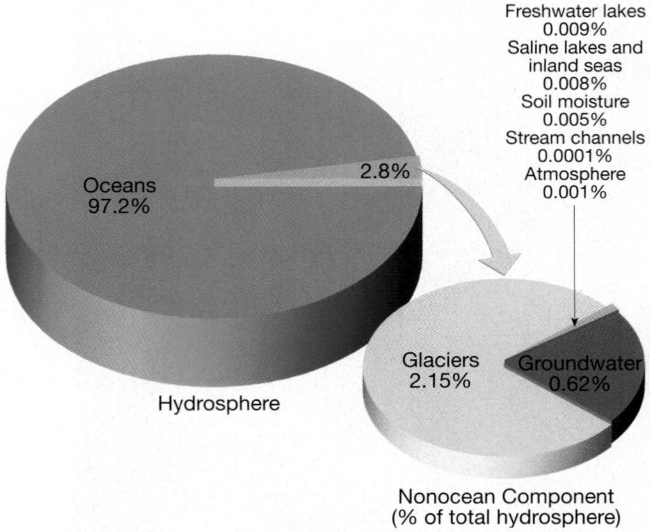

▲ **Figure 9.1** Distribution of Earth's water.

The hydrologic cycle is a gigantic worldwide system powered by energy from the Sun in which the atmosphere provides the vital link between the oceans and continents (Figure 9.2). Water evaporates into the atmosphere from the ocean and to a much lesser extent from the continents. Winds transport this moisture-laden air often great distances until conditions cause the moisture to condense into clouds and precipitation to fall. The precipitation that falls into the ocean has completed its cycle and is ready to begin another. The water that falls on land, however, must make its way back to the ocean.

What happens to precipitation once it has fallen on land? A portion of the water soaks into the ground (called **infiltration**), slowly moving downward, then laterally, finally seeping into lakes, streams, or directly into the ocean. When the rate of rainfall exceeds Earth's ability to absorb it, the surplus water flows over the surface into lakes and streams, a process called **runoff.** Much of the water that infiltrates or runs off eventually returns to the atmosphere because of evaporation from the soil, lakes, and streams. Also, some of the water that infiltrates the ground surface is absorbed by plants, which then release it into the atmosphere. This process is called **transpiration.** Because we cannot clearly distinguish between the amount of water that is evaporated and the amount that is transpired by plants, the term **evapotranspiration** is often used for the combined effect.

When precipitation falls in very cold areas—at high elevations or high latitudes—the water may not immediately soak in, run off, or evaporate. Instead, it may become part of a snowfield or a glacier. In this way, glaciers store large quantities of water on land. If present-

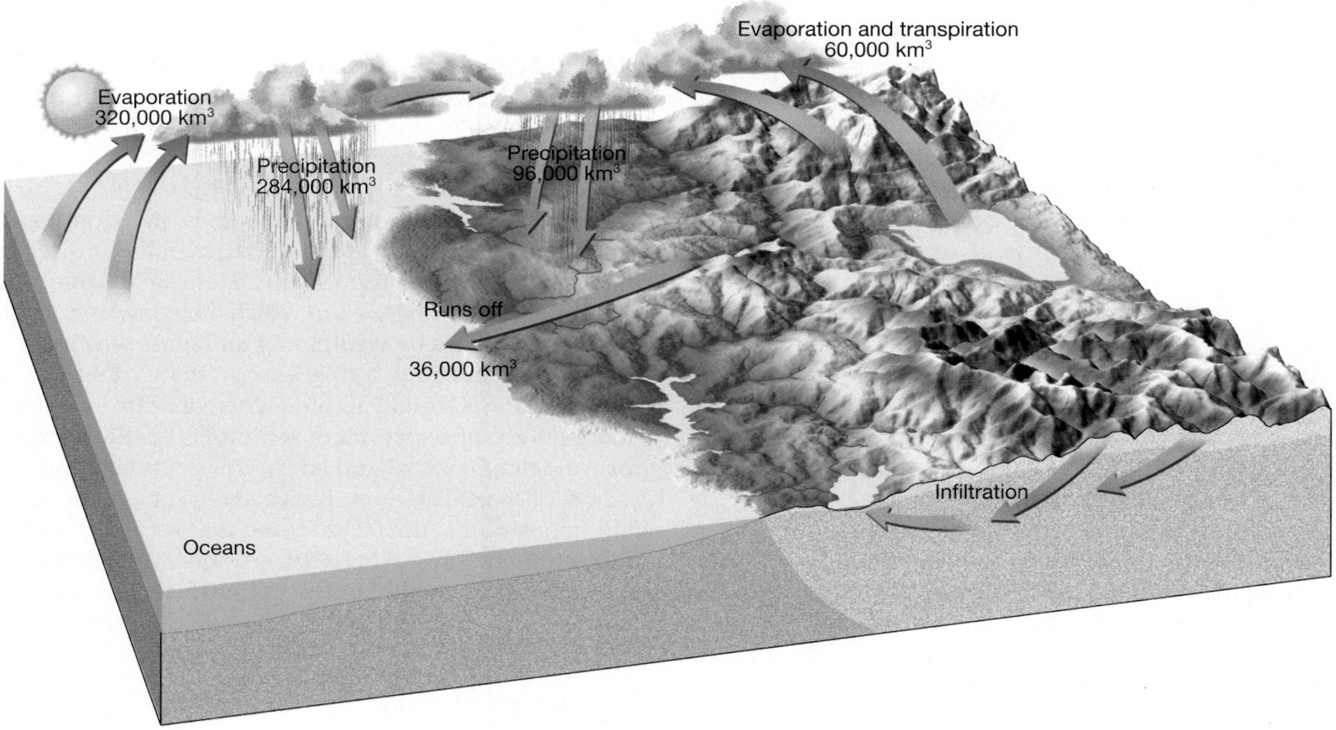

▲ **Figure 9.2** Earth's water balance. Each year, solar energy evaporates about 320,000 cubic kilometers of water from the oceans, while evaporation from the land (including lakes and streams) contributes 60,000 cubic kilometers of water. Of this total of 380,000 cubic kilometers of water, about 284,000 cubic kilometers fall back to the ocean, and the remaining 96,000 cubic kilometers fall on the land surface. Of that 96,000 cubic kilometers, only 60,000 cubic kilometers of water evaporate from the land, leaving 36,000 cubic kilometers of water to erode the land during the journey back to the oceans.

day glaciers were to melt and release all their water, sea level would rise by several dozen meters. This would submerge many heavily populated coastal areas. As you will see in Chapter 11, over the past 2 million years huge ice sheets have formed and melted on several occasions, each time changing the balance of the hydrologic cycle.

Figure 9.2 also shows Earth's overall *water balance,* or the volume of water that passes through each part of the cycle annually. The amount of water vapor in the air at any one time is just a tiny fraction of Earth's total water supply. But the *absolute* quantities that are cycled through the atmosphere over a one-year period are immense—some 380,000 cubic kilometers. Estimates show that over North America almost six times more water is carried by moving currents of air than is transported by all the continent's rivers.

It is important to know that the hydrologic cycle is *balanced.* Because the total amount of water vapor in the atmosphere remains about the same, the average annual precipitation over Earth must be equal to the quantity of water evaporated. However, for all of the continents taken together, precipitation exceeds evaporation. Conversely, over the oceans, evaporation exceeds precipitation. Since the level of the world ocean is not dropping, the system must be in balance. In Figure 9.2, the 36,000 cubic kilometers of water that annually runs off from the land to the ocean causes enormous erosion. In fact, this immense volume of moving water is *the single most important agent sculpturing Earth's land surface.*

To summarize, the hydrologic cycle represents the continuous movement of water from the oceans to the atmosphere, from the atmosphere to the land, and from the land back to the sea. The wearing down of Earth's land surface is largely attributable to the last of these steps and is the primary focus of the remainder of this chapter.

Transpiration by plants is one way that water enters the atmosphere.

Running Water

GEODe: Running Water
▼ Stream Characteristics

Although we have always depended on running water, its source eluded us for centuries. Not until the 1500s did we realize that streams were supplied by surface runoff and the migration of underground water.

Drainage Basins

All water in rivers ultimately has its source as rain and snow, although there can be considerable lag time before this water actually enters a river system. Upon reaching the surface, some precipitation soaks into the ground, while the remaining water flows over the surface. This *overland flow* is routed across the landscape by hillslopes into rills, gullies, streams, and finally large rivers that usually flow to the ocean.

Did You Know?

North America's largest river is the Mississippi. South of Cairo, Illinois, where the Ohio River joins it, the Mississippi is more than 1 mi (1.6 km) wide. Each year the "Mighty Mississippi" carries about half a billion tons of sediment to the Gulf of Mexico.

The land area that contributes water to a river system is called a **drainage basin** (Figure 9.3). The drainage basin of one stream is separated from the drainage basin of another by an imaginary line called a **divide** (Figure 9.3). Divides range in scale from a ridge separating two small gullies on a hillside to a *continental divide* which splits whole continents into enormous drainage basins. The Mississippi River has the largest drainage basin in North America. Extending between the Rocky Mountains in the West and the Appalachian Mountains in the East, the Mississippi River and its tributaries collect water from more than 3.2 million square kilometers (1.2 million square miles) of the continent.

Mountain streams have steep gradients.

River Systems

Rivers and streams can be simply defined as water flowing in a channel. They have three important roles in the formation of a landscape: They erode the channels in which they flow, they transport sediments provided by weathering and slope processes, and they produce a wide variety of erosional and depositional landforms. In fact, in most areas, including many arid regions, river systems have shaped the varied landscape that we humans inhabit.

A river system consists of three main parts in which different processes dominate: a zone of erosion, a zone of sediment transport, and a zone of sediment deposition. It is important to realize that some erosion, transport, and deposition occur in all three zones; however, within each zone, one of these processes is usually dominant. In addition, the parts of a river system are interdependent, so that the processes occurring in one part influence the others.

In large river systems erosion is the dominant process in the upstream area, which generally consists of mountainous or hilly topography. Here small tributary streams erode the channels in which they flow and carry material provided by weathering and mass wasting.

The region within a river system that is dominated by deposition is usually located where the stream enters a large body of water. Here sediments accumulate to form a delta, or are reworked by wave action to form a variety of coastal features. Between the zones of erosion and deposition is the *trunk stream* that serves to transport sediments. Taken together, erosion, transportation, and deposition are the processes by which rivers move Earth's surface materials and sculpt landscapes.

Streamflow

GEODe ESSENTIALS OF GEOLOGY Running Water
▼ Stream Characteristics

Water may flow in one of two ways, either as **laminar flow** or **turbulent flow.** In very slow moving streams the flow is often laminar and the water particles move in roughly straight-line paths that parallel the stream channel. However, streamflow is usually turbulent, with the water moving in an erratic fashion that can be characterized as a swirling motion. Strong turbulent flow may exhibit whirlpools and eddies, as well as rolling whitewater rapids. Even streams that appear smooth

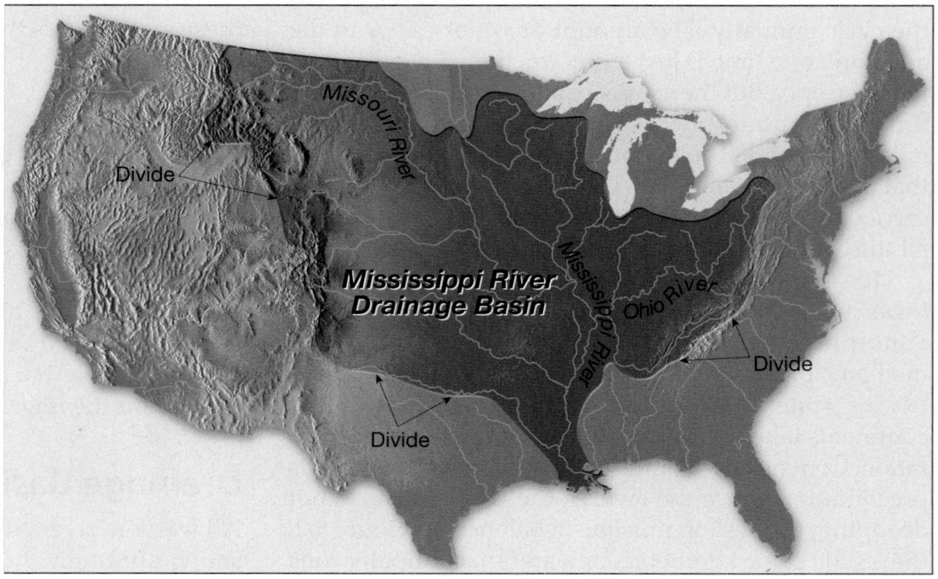

▶ **Figure 9.3** A drainage basin is the land area drained by a stream and its tributaries. The drainage basin of the Mississippi River, North America's largest river, covers about 3 million square kilometers. Divides are the boundaries that separate drainage basins from each other. Drainage basins and divides exist for all streams, regardless of size.

on the surface often exhibit turbulent flow near the bottom and sides of the channel. Turbulence contributes to the stream's ability to erode its channel because it acts to lift sediment from the streambed.

Water makes its way to the sea under the influence of gravity. The time required for the journey depends on the velocity of the stream. Velocity is the distance that water travels in a unit of time. Some sluggish streams flow at less than 1 kilometer per hour, whereas a few rapid ones may exceed 30 kilometers per hour. Velocities are measured at gaging stations (Figure 9.4A). Along straight stretches, the highest velocities are near the center of the channel just below the surface, where friction is lowest (Figure 9.4B). But when a stream curves, its zone of maximum speed shifts toward its outer bank (Figure 9.4C).

The ability of a stream to erode and transport materials depends on its velocity. Even slight changes in velocity can lead to significant changes in the load of sediment that water can transport. Several factors determine the velocity of a stream, including (1) gradient; (2) shape, size, and roughness of the channel; and (3) discharge.

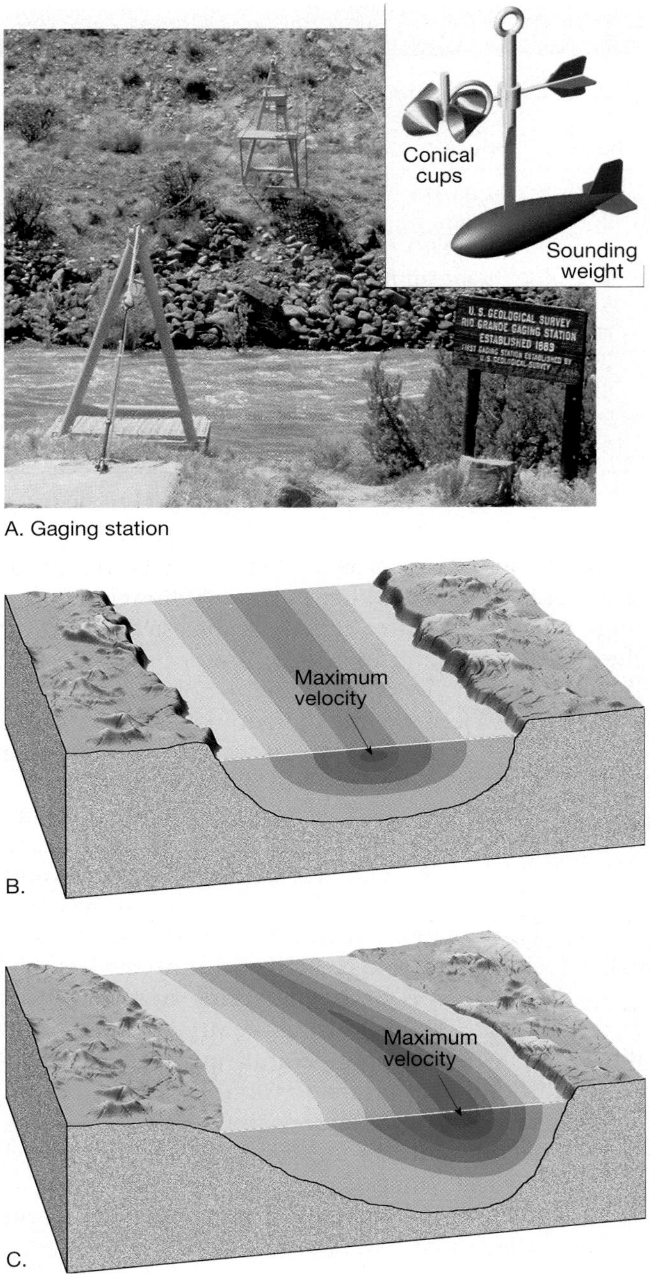

▲ **Figure 9.4** **A.** Continuous records of stage and discharge are collected by the U.S. Geological Survey at more than 7000 gaging stations in the United States. Average velocities are determined by using measurements from several spots across the stream. This station is on the Rio Grande south of Taos, New Mexico. *(Photo by E. J. Tarbuck)* **B.** Along straight stretches, stream velocity is highest at the center of the channel. **C.** When a stream curves, its zone of maximum speed shifts toward the outer bank.

Gradient and Channel Characteristics

The slope of a stream channel expressed as the vertical drop of a stream over a specified distance is **gradient.** Portions of the lower Mississippi River, for example, have very low gradients of 10 centimeters per kilometer or less. By contrast, some mountain stream channels decrease in elevation at a rate of more than 40 meters per kilometer, or a gradient 400 times steeper than the lower Mississippi (Figure 9.5). Gradient varies not only among different streams but also over a particular stream's length. The steeper the gradient, the more energy available for streamflow. If two streams were identical in every respect except gradient, the stream with the higher gradient would obviously have the greater velocity.

A stream's channel is a conduit that guides the flow of water, but the water encounters friction as it flows. The shape, size, and roughness of the channel affect the amount of friction. Larger channels have more efficient flow because a smaller proportion of water is in contact with the channel. A smooth channel promotes a more uniform flow, whereas an irregular channel filled with boulders creates enough turbulence to slow the stream significantly.

Discharge

The **discharge** of a stream is the volume of water flowing past a certain point in a given unit of time. This is usually measured in cubic meters per second or cubic feet per second. Discharge is determined by multiplying a stream's cross-sectional area by its velocity:

$$\text{discharge (m}^3\text{/second)} = \text{channel width (meters)} \times \text{channel depth (meters)} \times \text{velocity (meters/second)}$$

Table 9.1 lists the world's largest rivers in terms of discharge. The largest river in North America, the

Table 9.1 World's largest rivers ranked by discharge

			Drainage Area		Average Discharge	
Rank	River	Country	Square kilometers	Square miles	Cubic meters	Cubic feet
1	Amazon	Brazil	5,778,000	2,231,000	212,400	7,500,000
2	Congo	Zaire	4,014,500	1,550,000	39,650	1,400,000
3	Yangtze	China	1,942,500	750,000	21,800	770,000
4	Brahmaputra	Bangladesh	935,000	361,000	19,800	700,000
5	Ganges	India	1,059,300	409,000	18,700	660,000
6	Yenisei	Russia	2,590,000	1,000,000	17,400	614,000
7	Mississippi	United States	3,222,000	1,244,000	17,300	611,000
8	Orinoco	Venezuela	880,600	340,000	17,000	600,000
9	Lena	Russia	2,424,000	936,000	15,500	547,000
10	Parana	Argentina	2,305,000	890,000	14,900	526,000

Mississippi, discharges an average of 17,300 cubic meters (611,000 cubic feet) per second. Although this is a huge quantity of water, it is nevertheless dwarfed by the mighty Amazon in South America, the world's largest river. Fed by a vast rainy region that is nearly three-fourths the size of the conterminous United States, the Amazon discharges 12 times more water than the Mississippi.

Did You Know?

The Amazon River is responsible for about 20 percent of all the water reaching the ocean via rivers. Its nearest rival, Africa's Congo River, delivers about 4 percent of the total.

The discharges of most rivers are far from constant. This is true because of such variables as rainfall and snowmelt. In areas with seasonal variations in precipitation, streamflow will tend to be highest during the wet season, or during spring snowmelt, and lowest during the dry season or during periods when high temperature increases the water losses through evapotranspiration. However, not all channels maintain a continuous flow of water. Streams that exhibit flow only during "wet" periods are referred to as *intermittent streams.* In arid climates many streams carry water only occasionally after a heavy rainstorm and are called *ephemeral streams.*

Changes from Upstream to Downstream

One useful way of studying a stream is to examine its *profile.* A profile is simply a cross-sectional view of a stream from its source area (called the *head* or *headwaters*) to its *mouth,* the point downstream where the river empties into another water body. By examining Figure 9.6, you can see that the most obvious feature of a typical profile is a constantly decreasing gradient from the head to the mouth. Although many local irregularities may exist, the overall profile is a smooth concave upward curve.

The profile shows that the gradient decreases downstream. To see how other factors change in a downstream direction, observations and measurements must be made. When data are collected from several gaging stations along a river, they show that in a humid region discharge increases from the head toward the mouth. This should come as no surprise because, as we move downstream, more and more tributaries contribute water to the main channel (Figure 9.6). Furthermore, in most humid regions, additional water is added from the groundwater supply. Thus, as you move downstream, the stream's width, depth, and velocity change in response to the increased volume of water carried by the stream.

Streams that begin in mountainous areas where precipitation is abundant and then flow through arid regions may experience the opposite situation. Here discharge may actually decrease downstream because of water loss due to evaporation, infiltration into the streambed, and removal by irrigation. The Colorado River in the southwestern United States is such an example.

The Work of Running Water

Streams are Earth's most important erosional agent. Not only do they have the ability to downcut and widen their channels but streams also have the capacity to

▼ **Figure 9.5** Rapids are common in mountain streams where the gradient is steep and the channel is rough and irregular. *(Photo by Bruce Gaylord/Visuals Unlimited)*

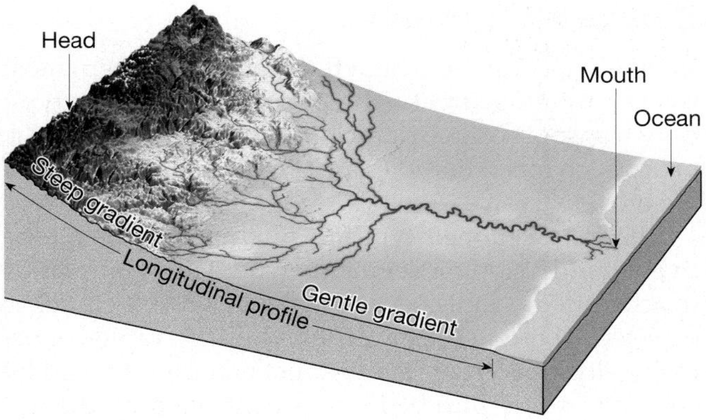

▲ **Figure 9.6** A longitudinal profile is a cross-section along the length of a stream. Note the concave-upward curve of the profile, with a steeper gradient upstream and a gentler gradient downstream. Moving downstream from the head, the discharge of most streams increases because tributaries and groundwater contribute water to the main channel.

transport the enormous quantities of sediment that are delivered to the stream by sheet flow, mass wasting, and groundwater. Eventually much of this material is dropped by the water to create a variety of depositional features.

Erosion

A stream's ability to accumulate and transport soil and weathered rock is aided by the work of raindrops, which knock sediment particles loose (see Figure 5.15 on p. 132). When the ground is saturated, rainwater begins to flow downslope, transporting some of the material it has dislodged. On barren slopes the flow of muddy water, called *sheet flow,* will often erode small channels, or *rills* which in time may evolve into larger *gullies* (see Figure 5.16 on p. 132).

Once the surface flow reaches a stream, its ability to erode is greatly enhanced by the increase in water volume. When the flow of water is sufficiently strong, it can dislodge particles from the channel and lift them into the moving water. In this manner, the force of running water swiftly erodes poorly consolidated materials on the bed and sides of a stream channel. On occasion, the banks of the channel may be undercut, dumping even more loose debris into the water to be carried downstream.

In addition to eroding unconsolidated materials, the hydraulic force of streamflow can also cut a channel into solid bedrock. A stream's ability to erode bedrock is greatly enhanced by the particles it carries. These particles can be any size, from large boulders in very fast-flowing waters to sand and gravel-size particles in somewhat slower flow. Just as the particles of grit on sandpaper can wear away a piece of wood, so too can the sand and gravel carried by a stream abrade a bedrock channel. Moreover, pebbles caught in swirling eddies can act like "drills" and bore circular *potholes* into the channel floor.

Transportation

All streams regardless of size transport some rock material. Streams also sort the solid sediment they transport because finer, lighter material is carried more rapidly than coarser, heavier rock debris. Depending on the nature of the rock material, the stream load consists of material (1) in solution (**dissolved load**), (2) in suspension (**suspended load**), and (3) sliding or rolling along the bottom (**bed load**).

> **Did You Know?**
>
> In the United States, hydroelectric power plants contribute about 5 percent of the country's energy needs. Most of this energy is produced at large dams.

Dissolved Load. Most of the *dissolved load* is brought to a stream by groundwater and is dispersed throughout the flow. The quantity of material carried in solution is highly variable and is most abundant in humid areas where limestone and other soluble rock forms the bedrock. Usually the amount of dissolved load is small and therefore is expressed as parts of dissolved material per million parts of water (parts per million, or ppm). Although some rivers may have a dissolved load of 1000 ppm or more, the average figure for the world's rivers is estimated at 115 to 120 ppm.

Suspended Load. Most large rivers carry the largest part of their load in *suspension.* Indeed, the visible cloud of sediment suspended in the water is the most obvious portion of a stream's load. Usually only fine particles consisting of silt and clay can be carried this way, but during a flood, sand and even gravel-size particles are transported as well. Also, during a flood, the total quantity of material carried in suspension increases dramatically, as can be verified by anyone whose home has been a site for the deposition of this material.

Potholes in the bed of a small Indiana stream. (Tom Till)

Bed Load. A portion of a stream's load consists of sand, gravel, and occasionally large boulders. These coarser particles, which are too large to be carried in suspension, move along the bottom (bed) of the stream channel and constitute the *bed load.* Unlike the suspended and dissolved loads, which are constantly in motion, the bed load is in motion only intermittently, when the force of the water is sufficient to move the larger particles. The smaller particles, mainly sand and gravel, are moved along the stream by *saltation,* which resembles a series of jumps or skips. Larger particles either roll or slide along the bottom, depending on their shape.

> **Did You Know?**
>
> When does a rill become a gully? As a rule of thumb, we know that *rills* have grown large enough to be called *gullies* when normal farm cultivation cannot eliminate the channels.

Competence and Capacity. Streams vary in their ability to carry a load. Their ability is determined by two

The greatest erosion and sediment transport occurs during floods. (T. Davis/Stone Allstock)

criteria. First, the **competence** of a stream measures the maximum size of particles it is capable of transporting. The stream's velocity determines its competence. If the velocity of a stream doubles, its competence increases four times; if the velocity triples, its competence increases nine times; and so forth. This explains how large boulders that seem immovable can be transported during a flood, which greatly increases a stream's velocity.

Second, the **capacity** of a stream is the maximum load it can carry. The capacity of a stream is directly related to its discharge. The greater the volume of water flowing in a stream, the greater is its capacity for hauling sediment.

By now it should be clear why the greatest erosion and transportation of sediment occur during floods. The increase in discharge results in a greater capacity, and the increase in velocity results in greater competence. With rising velocity the water becomes more turbulent, and larger and larger particles are set in motion. In just a few days or perhaps a few hours a stream in flood stage can erode and transport more sediment than it does during months of normal flow.

Deposition

Whenever a stream slows down, the situation reverses. As its velocity decreases, its competence is reduced and sediment begins to drop out, largest particles first. Each particle size has a *critical settling velocity.* As streamflow drops below the critical setting velocity of a certain particle size, sediment in that category begins to settle out. Thus, stream transport provides a mechanism by which solid particles of various sizes are separated. This process, called **sorting,** explains why particles of similar size are deposited together.

The material deposited by a stream is called **alluvium,** the general term for any stream-deposited sediment. Many different depositional features are composed of alluvium. Some occur within stream channels, some occur on the valley floor adjacent to the channel, and some exist at the mouth of the stream. We will consider the nature of these features later.

Stream Channels

A basic characteristic of streamflow that distinguishes it from overland flow is that it is usually confined to a channel. A stream channel can be thought of as an open conduit that consists of the streambed and banks that act to confine the flow except during floods.

Although somewhat oversimplified, we can divide stream channels into two types. *Bedrock channels* are those in which the streams are actively cutting into solid rock. In contrast, when the bed and banks are composed mainly of unconsolidated sediment, the channel is called an *alluvial channel.*

Bedrock Channels

In their headwaters, where the gradient is steep, most rivers cut into bedrock. These mountain streams typically transport coarse particles that actively abrade the bedrock channel. Potholes are often visible evidence of the erosional forces at work.

Bedrock channels typically alternate between relatively gently sloping segments where alluvium tends to accumulate, and steeper segments where bedrock is exposed. These steeper areas may contain rapids or occasionally a waterfall. The channel pattern exhibited by streams cutting into bedrock is controlled by the underlying geologic structure. Even when flowing over rather uniform bedrock, streams tend to exhibit a winding or irregular pattern rather than flowing in a straight channel. Anyone who has gone on a white-water rafting trip has observed the steep, winding nature of a stream flowing in a bedrock channel.

Alluvial Channels

Many stream channels are composed of loosely consolidated sediment (alluvium) and therefore can undergo major changes in shape because the sediments are continually being eroded, transported, and redeposited. The major factors affecting the shapes of these channels is the average size of the sediment being transported, the channel gradient, and the discharge.

Alluvial channel patterns reflect a stream's ability to transport its load at a uniform rate, while expending the least amount of energy. Thus, the size and type of sediment being carried help determine the nature of the stream channel. Two common types of alluvial channels are *meandering channels* and *braided channels.*

Meandering Streams. Streams that transport much of their load in suspension generally move in sweeping bends called **meanders.** These streams flow in relatively deep, smooth channels and transport mainly mud (silt and clay). The lower Mississippi River exhibits a channel of this type.

Because of the cohesiveness of consolidated mud, the banks of stream channels carrying fine particles tend to resist erosion. As a consequence, most of the erosion in such channels occurs on the outside of the meander, where velocity and turbulence are greatest. In time, the outside bank is undermined, especially during periods of high water. Because the outside of a meander is a zone of active erosion, it is often referred to as the **cut bank** (Figure 9.7). The debris acquired by the stream at the cut bank moves downstream with the coarser material generally being deposited as **point bars** in zones of decreased velocity on the insides of meanders. In this manner, meanders migrate laterally by eroding the outside of the bends and depositing on the inside.

In addition to migrating laterally, the bends in a channel also migrate down the valley. This occurs

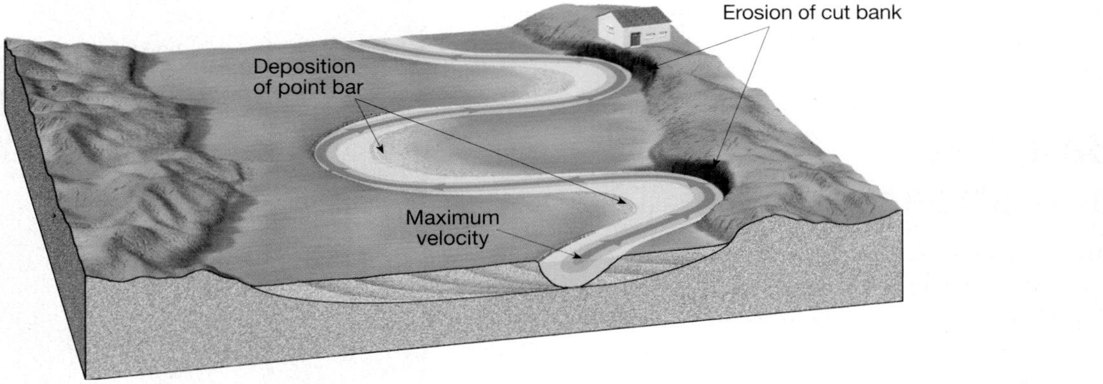

▲ **Figure 9.7** When a stream meanders, its zone of maximum speed shifts toward the outer bank. Because the outside of a meander is a zone of active erosion, it is called the *cut bank.* A point bar is deposited when water on the inside of a meander slows.

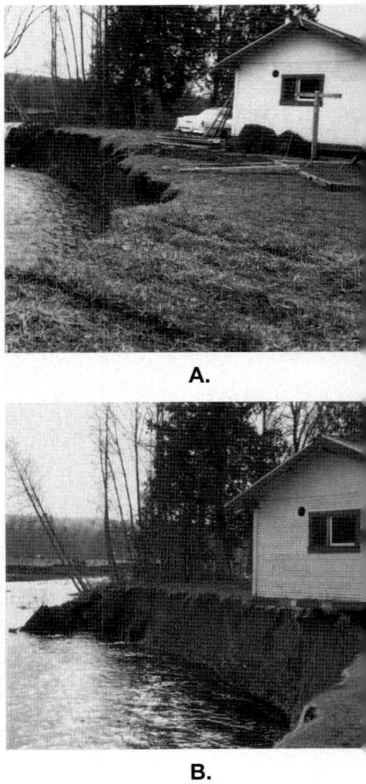

These two photos show erosion of a cut bank. (P. A. Glancy/USGS)

because erosion is more effective on the downstream (downslope) side of the meander. Sometimes the downstream migration of a meander is slowed when it reaches a more resistant material. This allows the next meander upstream to "catch up" and overtake it as shown in Figure 9.8. Gradually the neck of land between the meanders is narrowed. Eventually the river may erode through the narrow neck of land to the next loop (Figure 9.8). The new, shorter channel segment is called a **cutoff** and, because of its shape, the abandoned bend is called an **oxbow lake** (Figure 9.9).

Braided Streams. Some streams consist of a complex network of converging and diverging channels that thread their way among numerous islands or gravel bars (Figure 9.10). Because these channels have an interwoven appearance, these streams are said to be **braided.** Braided channels form where a large proportion of the stream's load consists of coarse material (sand and gravel) and the stream has a highly variable discharge. Because the bank material is readily erodable, braided channels are wide and shallow.

One circumstance in which braided streams form is at the end of a glacier where there is a large seasonal variation in discharge. Here, large amounts of ice-eroded sediment are dumped into the meltwater streams flowing away from the glacier. When flow is sluggish, the stream is unable to move all of the sediment and therefore deposits the coarsest material as bars that force the flow to split and follow several paths. Usually the laterally shifting channels completely rework most of the surface sediments each year, thereby transforming the entire streambed. In some braided streams, however, the bars have built up to form islands that are anchored by vegetation.

In summary, meandering channels develop where the load consists largely of fine-grained particles that are transported as suspended load in a deep, smooth channel. By contrast, wide, shallow braided channels

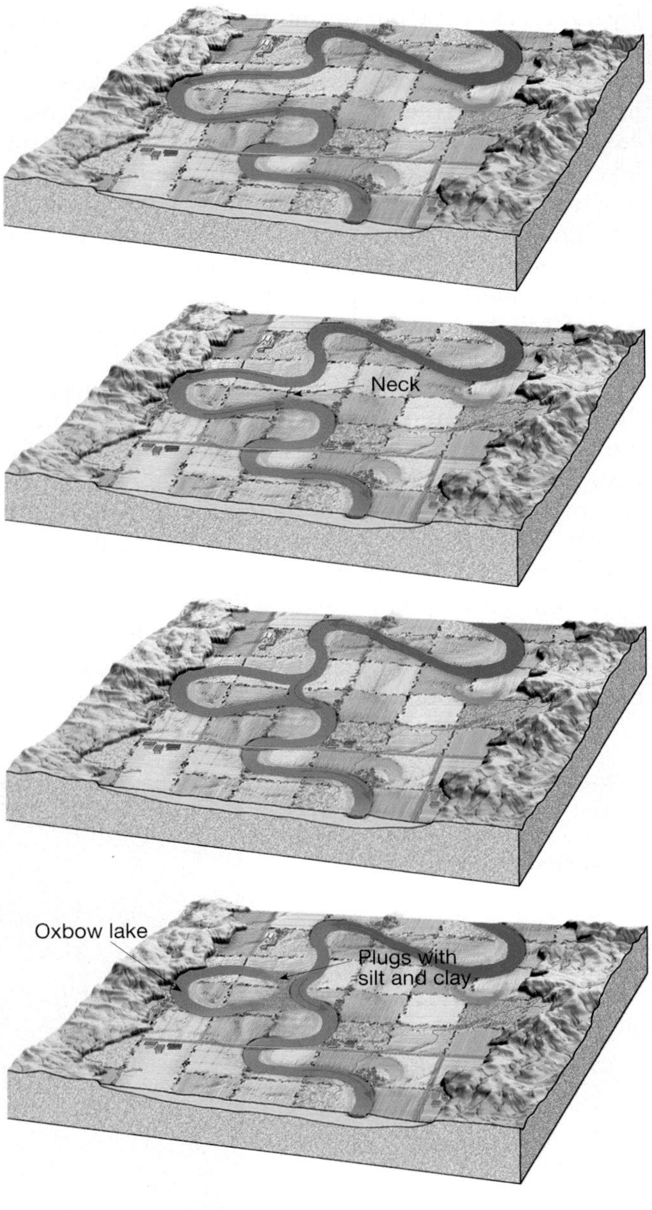

▲ **Figure 9.8** Formation of a cutoff and oxbow lake.

develop where coarse-grained alluvium is transported as bedload.

Base Level and Stream Erosion

Building a dam raises base level upstream. (Michael Collier)

Streams cannot endlessly erode their channels deeper and deeper. There is a lower limit to how deep a stream can erode, and that limit is called **base level.** Most often a stream's base level occurs where a stream enters the ocean, a lake, or another stream.

Two general types of base level are recognized. Sea level is considered the *ultimate base level,* because it is the lowest level to which stream erosion could lower the land. *Temporary,* or *local, base levels* include lakes, resistant layers of rock, and main streams that act as base level for their tributaries. For example, when a stream enters a lake, its velocity quickly approaches zero and its ability to erode ceases. Thus, the lake prevents the stream from eroding below its level at any point upstream from the lake. However, because the outlet of the lake can cut downward and drain the lake, the lake is only a temporary hindrance to the stream's ability to downcut its channel. In a similar manner, the layer of resistant rock at the lip of the waterfall in Figure 9.11 acts as a temporary base level. Until the ledge of hard rock is eliminated, it will limit the amount of downcutting upstream.

Any change in base level will cause a corresponding readjustment of stream activities. When a dam is built along a stream, the reservoir that forms behind it raises the base level of the stream (Figure 9.12). Upstream from the dam the gradient is reduced, lowering the stream's velocity and, hence, its sediment-transporting ability. The stream, now having too little energy to transport its entire load, will deposit sediment. This builds up its channel. Deposition will be the dominant process until the stream's gradient increases sufficiently to transport its load.

▲ **Figure 9.9** Oxbow lakes occupy abandoned meanders. As they fill with sediment, oxbow lakes gradually become swampy meander scars. Aerial view of an oxbow lake created by the meandering Green River near Bronx, Wyoming. *(Photo by Michael Collier)*

Shaping Stream Valleys

GEODe: Essentials of Geology
Running Water
▼ Reviewing Valleys and Stream-related Features

Streams, with the aid of weathering and mass wasting, shape the landscape through which they flow. As a result, streams continuously modify the valleys that they occupy.

A **stream valley** consists not only of the channel but also the surrounding terrain that directly con-

▶ **Figure 9.10** Braided stream choked with sediment near the terminus of a melting glacier. *(Photo by Bradford Washburn)*

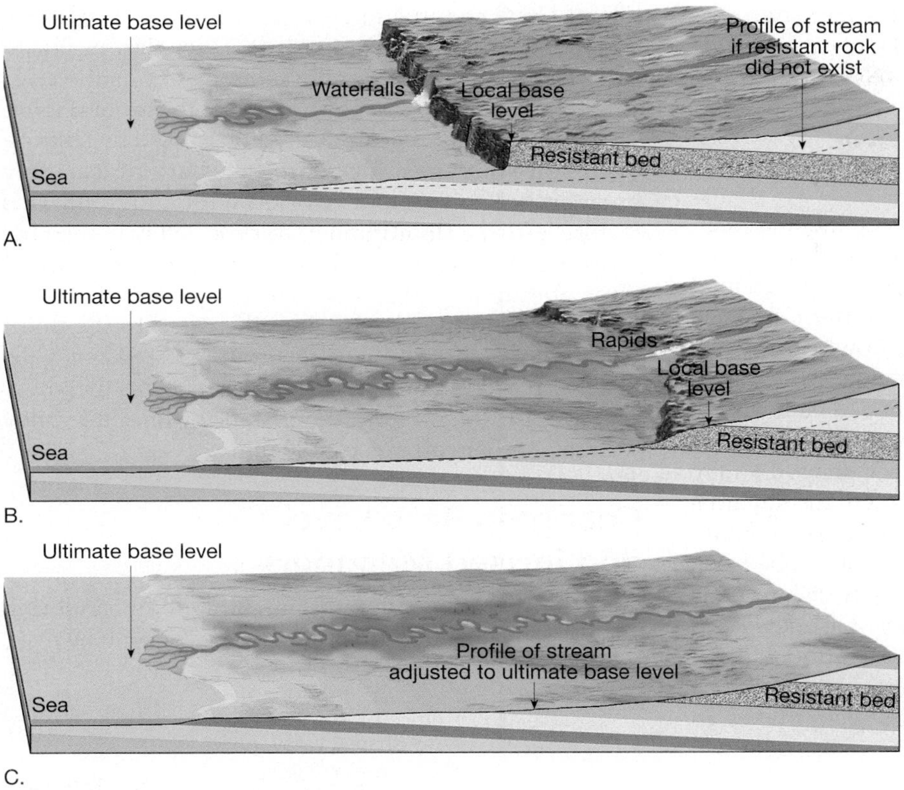

◀ **Figure 9.11** A resistant layer of rock can act as a local (temporary) base level. Because the durable layer is eroded more slowly, it limits the amount of downcutting upstream.

Narrow gorge cut in resistant bedrock. (Tom and Susan Bean)

tributes water to the stream. Thus it includes the valley bottom, which is the lower, flatter area that is partially or totally occupied by the stream channel, and the sloping valley walls that rise above the valley bottom on both sides. Most stream valleys are much broader at the top than is the width of their channel at the bottom. This would not be the case if the only agent responsible for eroding valleys were the streams flowing through them. The sides of most valleys are shaped by a combination of weathering, overland flow, and mass wasting. In some arid regions, where weathering is slow and where rock is particularly resistant, narrow valleys having nearly vertical walls are common.

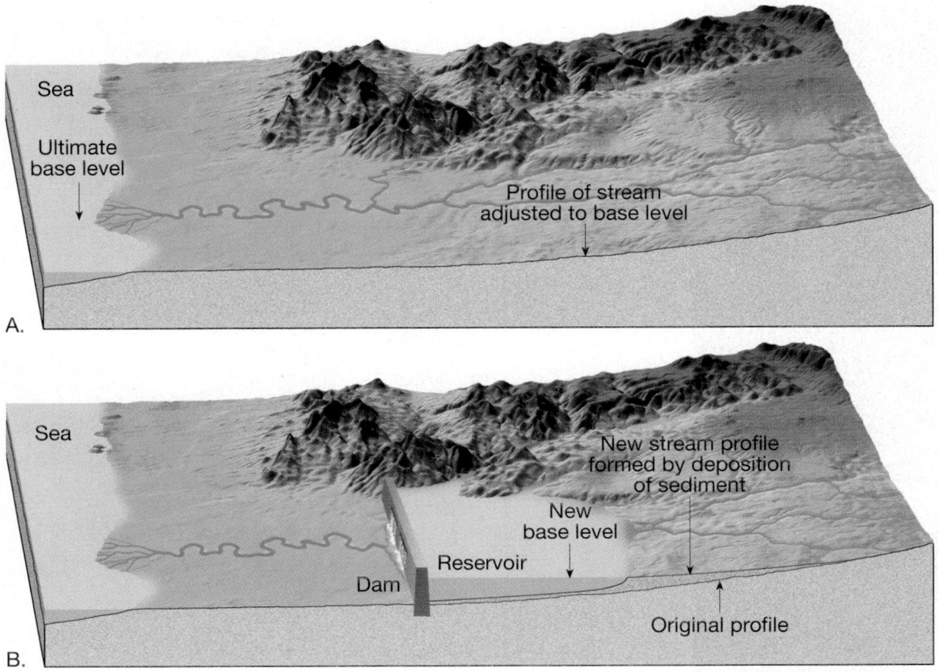

◀ **Figure 9.12** When a dam is built and a reservoir forms, the stream's base level is raised. This reduces the stream's velocity and leads to deposition and a reduction of the gradient upstream from the reservoir.

Floodplain of the meandering Big Sioux River. (Tom and Susan Bean)

Stream valleys can be divided into two general types—narrow V-shaped valleys and wide valleys with flat floors—with many gradations between.

Valley Deepening

When a stream's gradient is steep and the channel is well above base level, downcutting is the dominant activity. Abrasion caused by bed load sliding and rolling along the bottom, and the hydraulic power of fast-moving water, slowly lowers the streambed. The result is usually a V-shaped valley with steep sides. A classic example of a V-shaped valley is the section of the Yellowstone River shown in Figure 9.13.

The most prominent features of a V-shaped valley are *rapids* and *waterfalls*. Both occur where the stream's gradient increases significantly, a situation usually caused by variations in the erodability of the bedrock into which a stream channel is cutting. Resistant beds create rapids by acting as a temporary base level upstream while allowing downcutting to continue downstream. In time erosion usually eliminates the resistant rock. Waterfalls are places where the stream makes an abrupt vertical drop.

Valley Widening

Once a stream has cut its channel closer to base level, downward erosion becomes less dominant. At this point the stream's channel takes on a meandering pattern, and more of the stream's energy is directed from side to side. The result is a widening of the valley as the river cuts away first at one bank and then at the other (Figure 9.14). The continuous lateral erosion caused by shifting of the stream's meanders produces an increasingly broader, flat valley floor covered with alluvium. This feature, called a **floodplain,** is appropriately named because when a river overflows its banks during flood stage, it inundates the floodplain.

Over time the floodplain will widen to the point that the stream is only actively eroding the valley walls in a few places. In fact, in large rivers such as the lower Mississippi River valley, the distance from one valley wall to another can exceed 100 miles.

Changing Base Level and Incised Meanders

We usually expect a stream with a highly meandering course to be on a floodplain in a wide valley. However,

▼ **Figure 9.13** V-shaped valley of the Yellowstone River. The rapids and waterfalls indicate that the river is vigorously downcutting. *(Photo by Art Wolfe)*

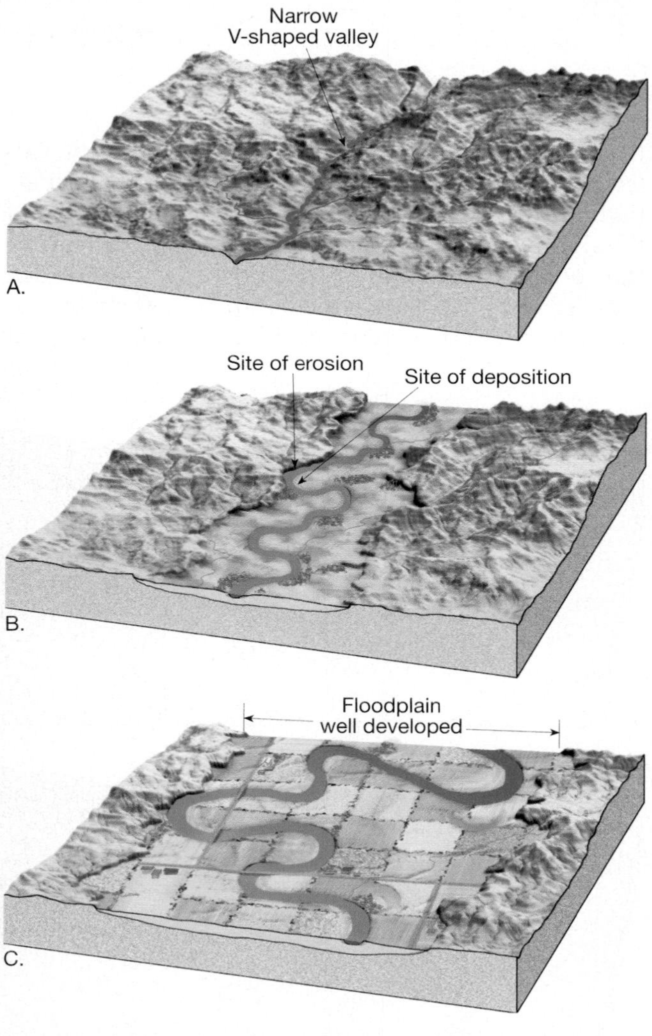

▲ **Figure 9.14** Stream eroding its floodplain.

▲ **Figure 9.15** Incised meanders of the Colorado River in Canyonlands National Park, Utah. Here, as the Colorado Plateau was gradually uplifted, the meandering river adjusted to being higher above base level by downcutting. *(Photo by Michael Collier)*

certain rivers exhibit meandering channels that flow in steep, narrow valleys. Such meanders are called **incised meanders** (Figure 9.15). How do such features form?

Originally, the meanders probably developed on the floodplain of a stream that was relatively near base level. Then, a change in base level caused the stream to begin downcutting. One of two events could have occurred. Either base level dropped or the land upon which the river was flowing was uplifted.

An example of the first circumstance happened during the Ice Age when large quantities of water were withdrawn from the ocean and locked up in glaciers on land. The result was that sea level (ultimate base level) dropped, causing rivers flowing into the ocean to begin to downcut.

Regional uplift of the land, the second cause for incised meanders, is exemplified by the Colorado Plateau in the southwestern United States. Here, as the plateau was gradually uplifted, numerous meandering rivers adjusted to being higher above base level by downcutting (Figure 9.15).

Depositional Landforms

As indicated earlier, whenever a stream's velocity slows, it begins to deposit some of the sediment it is carrying. Also recall that streams continually pick up sediment in one part of their channel and redeposit it downstream. These channel deposits are most often composed of sand and gravel and are commonly referred to as **bars.** Such features, however, are only temporary, for the material will be picked up again and eventually carried to the ocean. In addition to sand and gravel bars, streams also create other depositional features that have a somewhat longer life span. These include deltas, natural levees, and alluvial fans.

Deltas

When a stream enters the relatively still waters of an ocean or lake, its velocity drops abruptly, and the resulting deposits form a **delta** (Figure 9.16). As the delta grows outward, the stream's gradient continually lessens. This circumstance eventually causes the channel to become choked with sediment deposited from the slowing water. As a consequence, the river seeks a shorter, higher-gradient route to base level, as illustrated in Figure 9.16B. This illustration shows the main channel dividing into several smaller ones, called **distributaries.** Most deltas are characterized by these shifting channels that act in an opposite way to that of tributaries.

Did You Know?

The world's highest uninterrupted waterfall is Angel Falls on Venezuela's Churun River. Named for American aviator Jimmie Angel, who first sighted the falls from the air in 1933, the river plunges 979 m (3212 ft).

Rather than carrying water into the main channel, distributaries carry water away from the main channel. After numerous shifts of the channel, a delta may grow into a rough triangular shape like the Greek letter delta (Δ), for which it is named. Note, however, that many deltas do not exhibit the idealized shape. Differences in the configurations of shorelines and variations in the nature and strength of wave activity result in many shapes. Many large rivers have deltas extending over thousands of square kilometers. The delta of the Mississippi River is one example (see Box 9.1). It resulted from the accumulation of huge quantities of sediment derived from the vast region drained by the river and its tributaries. Today, New Orleans rests where there was ocean less than 5000 years ago. Figure 9.17 shows that portion of the Mississippi delta that has been built over the past 5000 to 6000 years. As you can see, the delta is actually a series of seven coalescing subdeltas. Each formed when the river left its existing channel in favor of a shorter, more direct path to the Gulf of Mexico. The individual subdeltas interfinger and partially cover one another to produce a very complex structure. The present subdelta, called a *bird-foot* delta because of the configuration of its distributaries, has been built by the Mississippi in the last 500 years.

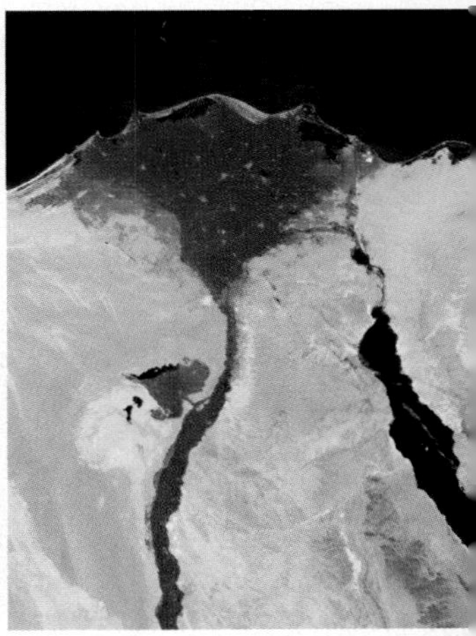

Egypt's Nile delta. (NASA)

Natural Levees

Some rivers occupy valleys with broad floodplains and build **natural levees** that parallel their channels on both banks (Figure 9.18). Natural levees are built by successive floods over many years. When a stream overflows its banks, its velocity immediately diminishes, leaving

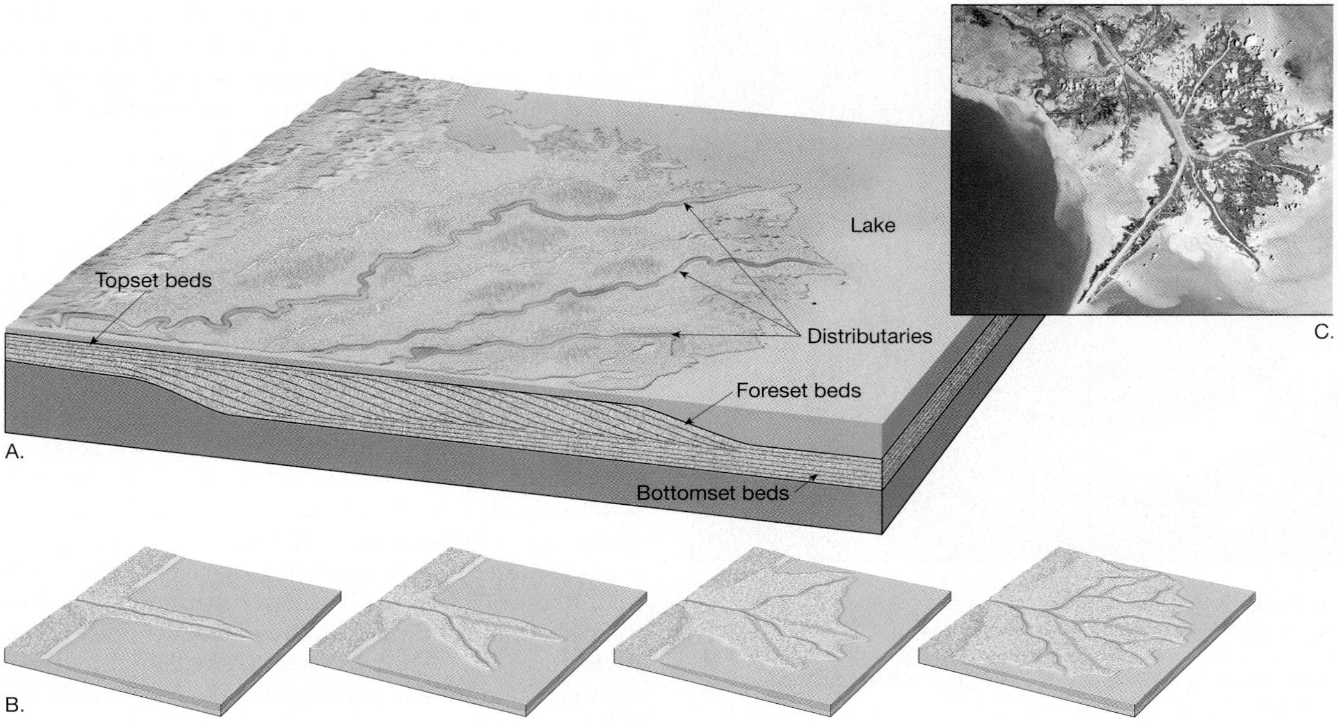

▲ **Figure 9.16** **A.** Structure of a simple delta that forms in the relatively quiet waters of a lake. **B.** Growth of a simple delta. As a stream extends its channel, the gradient is reduced. Frequently, during flood stage the river is diverted to a higher-gradient route, forming a new distributary. Old, abandoned distributaries are gradually invaded by aquatic vegetation and thus filled with sediment. *(After Ward's Natural Science Establishment, Inc., Rochester, N.Y.)* **C.** Satellite image of a portion of the delta of the Mississippi River in May 2001. For the past 500 years or so, the main flow of the river has been along its present course, extending southeast from New Orleans. During that span, the delta advanced into the Gulf of Mexico at a rate of about 10 kilometers (6 miles) per century. Notice the numerous distributaries. *(Image courtesy of JPL/Cal Tech/NASA)*

Alluvial fan in Death Valley. (Michael Collier)

coarse sediment deposited in strips bordering the channel. As the water spreads out over the valley, a lesser amount of fine sediment is deposited over the valley floor. This uneven distribution of material produces the very gentle slope of the natural levee.

The natural levees of the lower Mississippi rise 6 meters (20 feet) above the floodplain. The area behind the levee is characteristically poorly drained for the obvious reason that water cannot flow up the levee and into the river. Marshes called **backswamps** result. A tributary stream that cannot enter a river because levees block the way often has to flow parallel to the river until it can breach the levee. Such streams are called **yazoo tributaries** after the Yazoo River, which parallels the Mississippi for over 300 kilometers (about 190 miles).

Alluvial Fans

Alluvial fans typically develop where a high-gradient stream leaves a narrow valley in mountainous terrain and comes out suddenly onto a broad, flat plain or valley floor. Alluvial fans form in response to the abrupt drop in gradient combined with the change from a narrow channel of a mountain stream to less confined channels at the base of the mountains. The sudden drop in velocity causes the stream to dump its load of sediment quickly in a distinctive cone- or fan-shaped accumulation. The surface of the fan slopes outward in a broad arc from an apex at the mouth of the steep valley. Usually, coarse material is dropped near the apex of the fan, while fine material is carried toward the base of the deposit.

Drainage Patterns

Running Water
▼ Stream Characteristics

Drainage systems are networks of streams that together form distinctive patterns. The nature of a drainage pattern can vary greatly from one type of terrain to another, primarily in response to the kinds of rock on which the streams developed or the structural pattern of faults and folds.

The most commonly encountered drainage pattern is the **dendritic pattern** (Figure 9.19A). This pattern of

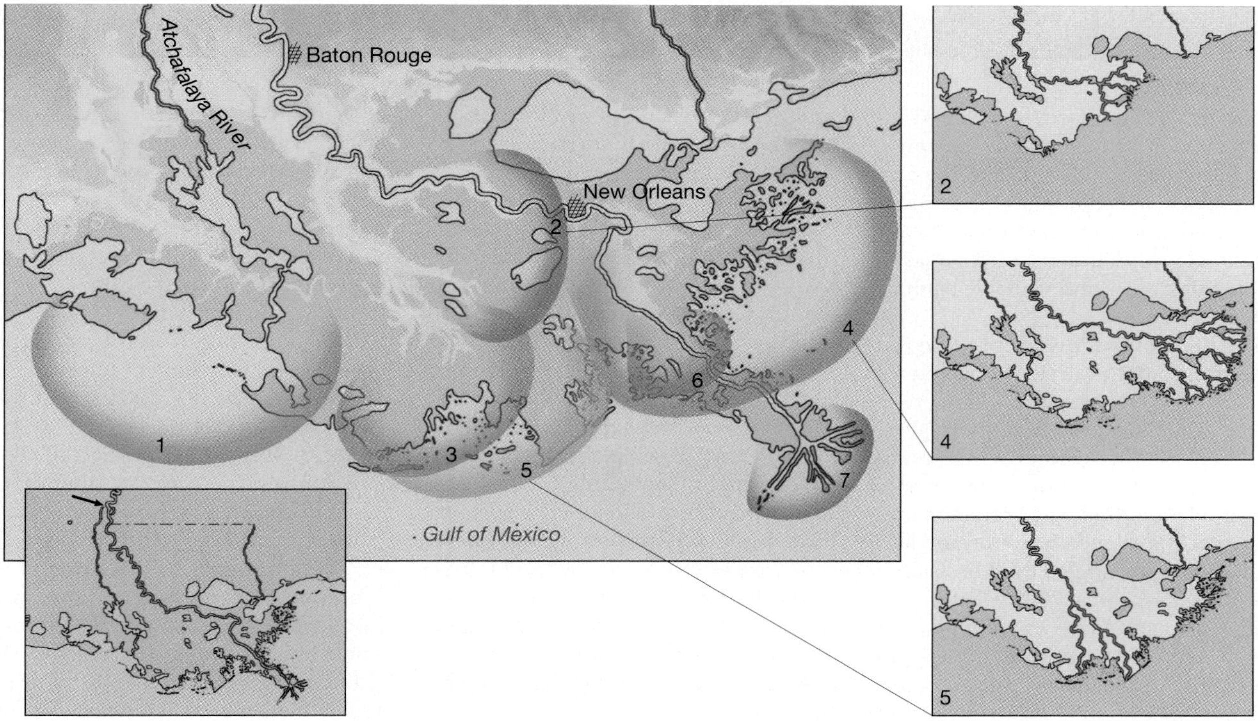

▲ **Figure 9.17** During the past 5000 to 6000 years, the Mississippi River has built a series of seven coalescing subdeltas. The numbers indicate the order in which the subdeltas were deposited. The present bird-foot delta (number 7) represents the activity of the past 500 years. Without ongoing human efforts, the present course will shift and follow the path of the Atchafalaya River. The inset on left shows the point where the Mississippi may someday break through (arrow) and the shorter path it would take to the Gulf of Mexico. *(After C. R. Kolb and J. R. Van Lopik)*

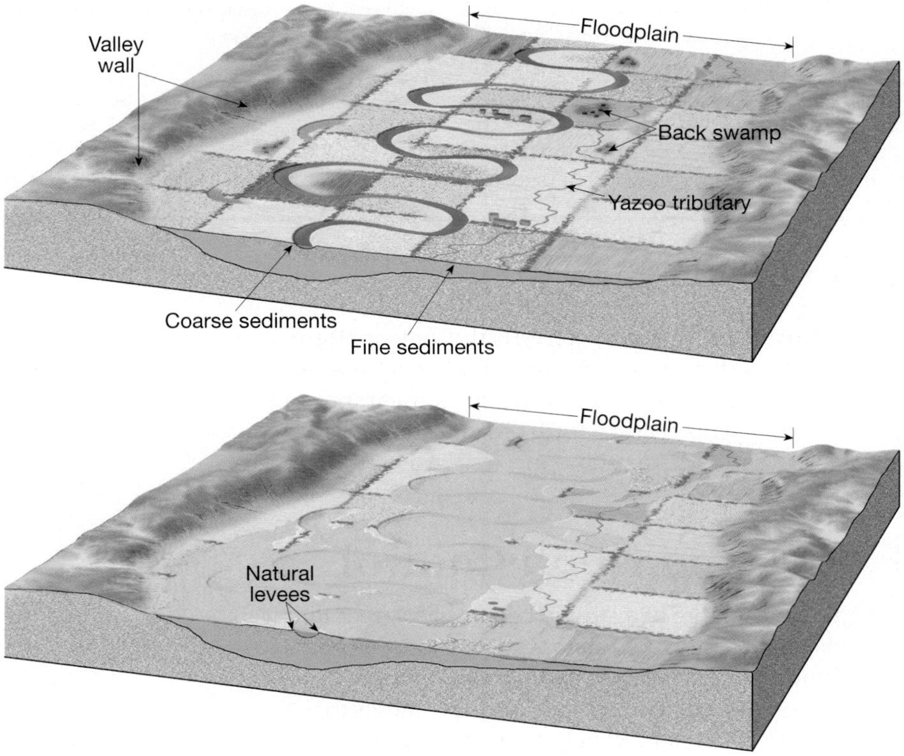

◀ **Figure 9.18** Natural levees are gently sloping deposits that are created by repeated floods. Because the ground next to the stream channel is higher than the adjacent floodplain, back swamps and yazoo tributaries may develop.

BOX 9.1

Coastal Wetlands Are Vanishing on the Mississippi Delta

Coastal wetlands form in sheltered environments that include swamps, tidal flats, coastal marshes, and bayous. They are rich in wildlife and provide nesting grounds and important stopovers for waterfowl and migratory birds, as well as spawning areas and valuable habitats for fish.

The delta of the Mississippi River in Louisiana contains about 40 percent of all coastal wetlands in the lower 48 states. Louisiana's wetlands are sheltered from the wave action of hurricanes and winter storms by low-lying offshore barrier islands. Both the wetlands and the protecting islands have formed as a result of the shifting of the Mississippi River during the past 7000 years.

The dependence of Louisiana's coastal wetlands on the Mississippi River and its distributaries as a direct source of sediment and fresh water leaves them vulnerable to changes in the river system. Moreover, the reliance on barrier islands for protection from storm waves leaves coastal wetlands vulnerable when these narrow offshore islands are eroded.

Today, the coastal wetlands of Louisiana are disappearing at an alarming rate. Although Louisiana contains 40 percent of the wetlands in the lower 48 states, it accounts for 80 percent of the wetland loss. According to the U.S. Geological Survey, Louisiana lost nearly 5000 square kilometers (1900 square miles) of coastal land between 1932 and 2000. The state continues to lose between 65 and 91 square kilometers (25 to 35 square miles) each year. At this rate another 1800 to 4500 square kilometers (700 to 1750 square miles) will vanish under the Gulf of Mexico by the year 2050.* Global climate change could increase the severity of the problem because rising sea level and stronger tropical storms would accelerate rates of coastal erosion.**

Why are Louisiana's wetlands shrinking? The answer is twofold: *natural change* and *human activity*. First, the Mississippi delta and its wetlands are continually changing naturally. As sediment accumulates and builds the delta in one area, erosion and subsidence cause losses elsewhere. When the river shifts, the zones of delta growth and destruction also shift. Second, ever since people arrived, the rate at which the delta and its wetlands are destroyed has accelerated.

Before Europeans settled the delta, the Mississippi River regularly overflowed its banks in seasonal floods. The huge quantities of sediment that were deposited renewed the soil and kept the delta from sinking below sea level. However, with settlement came flood-control efforts and the desire to maintain and improve navigation on the river. Artificial levees were constructed to contain the rising river during flood stage. Over time the levees were extended all the way to the mouth of the Mississippi to keep the channel open for navigation.

The effects have been straightforward. The levees prevent sediment and fresh water from being dispersed into the wetlands. Instead, the river is forced to carry its load to the deep waters at the mouth. Meanwhile, the processes of compaction, subsidence, and wave erosion continue. Because not enough sediment is added to offset these forces, the size of the delta and the extent of its wetlands gradually shrink.

The problem has been aggravated by a decline in the sediment transported by the Mississippi, decreasing by approximately 50 percent over the past 100 years. A substantial portion of the reduction results from trapping of sediment in large reservoirs created by dams built on tributaries to the Mississippi.

Another factor contributing to wetland decline is the fact that the delta is laced with 13,000 kilometers (8000 miles) of navigation channels and canals. These artificial openings to the sea allow salty Gulf waters to flow far inland. The invasion of saltwater and tidal action causes massive "brownouts" or marsh die-offs.

Understanding and modifying the impact of people is a necessary basis for any plan to reduce the loss of wetlands in the Mississippi delta. The U.S. Geological Survey estimates that restoring Louisiana's coasts will require about $14 billion over the next 40 years. What if nothing is done? State and federal officials estimate that costs of inaction could exceed $100 billion.

*See "Louisiana's Vanishing Wetlands, Going, Going . . ." In *Science*, Vol. 219, 15 September 2000, pp. 1860-63.
**For more on this possibility, see Box 13.2 "Coastal Vulnerability to Sea-Level Rise."

irregularly branching tributary streams resembles the branching pattern of a deciduous tree. In fact, the word *dendritic* means "treelike." The dendritic pattern forms where the underlying material is relatively uniform. Because the surface material is essentially uniform in its resistance to erosion, it does not control the pattern of streamflow. Rather, the pattern is determined chiefly by the direction of slope of the land.

When streams diverge from a central area like spokes from the hub of a wheel, the pattern is said to be **radial** (Figure 9.19B). This pattern typically develops on isolated volcanic cones and domal uplifts.

Figure 9.19C illustrates a **rectangular** pattern, in which many right-angle bends can be seen. This pattern develops when the bedrock is crisscrossed by a series of joints and/or faults. Because these structures are eroded more easily than unbroken rock, their geometric pattern guides the directions of valleys.

Figure 9.19D illustrates a **trellis drainage pattern,** a rectangular pattern in which tributary streams are nearly parallel to one another and have the appearance of a garden trellis. This pattern forms in areas underlain by alternating bands of resistant and less-resistant rock.

Floods and Flood Control

When the discharge of a stream becomes so great that it exceeds the capacity of its channel, it overflows its banks as a **flood.** Floods are the most common and most destructive of all geologic hazards. They are, nevertheless, simply part of the *natural* behavior of streams.

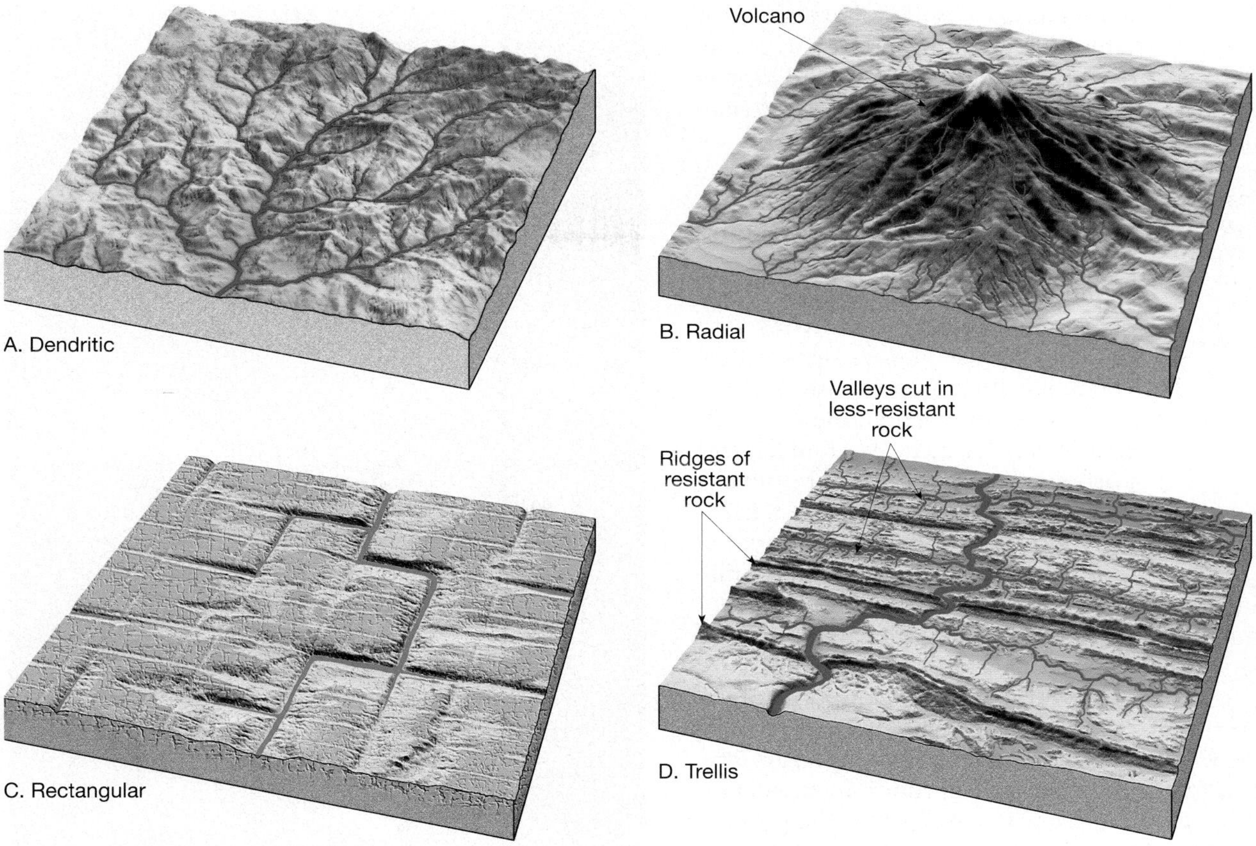

▲ **Figure 9.19** Drainage patterns. **A.** Dendritic. **B.** Radial. **C.** Rectangular. **D.** Trellis.

Most floods have a meteorological origin caused by atmospheric processes that can vary greatly in both time and space. Just an hour or less of intense thunderstorm rainfall can trigger floods in small valleys. By contrast, major floods in large river valleys are often the result of an extraordinary series of precipitation events over a broad region for an extended time span.

Land-use planning in river basins requires an understanding of the frequency and magnitude of floods. Probably the greatest immediate practical use of the data collected at stream-gaging stations is estimating the probability of various flood magnitudes.

Causes and Types of Floods

Floods can be the result of several naturally-occurring and human-induced factors. Among the common types of floods are regional floods, flash floods, ice-jam floods, and dam-failure floods.

Regional Floods. Some regional floods are seasonal. Rapid melting of snow in spring and/or heavy spring rains often overwhelm a river. The extensive 1997 flood along the Red River of the North is a notable example of an event triggered by rapid snowmelt. The flood was preceded by an especially snowy winter. As April began, the snow was melting and flooding seemed imminent, but on April 5 and 6, a blizzard rebuilt the shrinking snowdrifts to heights of 6 meters (20 feet) in some places. Then rapidly rising temperatures melted the snow in a matter of days, causing a record-breaking 500-year flood. Roughly 4.5 million acres were underwater, and the losses in the Grand Forks, North Dakota region exceeded $3.5 billion. Early spring floods are sometimes made worse if the ground is frozen. This reduces infiltration into the soil, thereby increasing runoff.

Extended wet periods any time of the year can create saturated soils, after which any additional rain runs off into streams until capacities are exceeded. Regional floods are often caused by slow-moving storm systems, including decaying hurricanes. The extensive and costly floods in eastern North Carolina in September 1999 were the result of torrential rains on already waterlogged soils from decaying Hurricane Floyd. Persistent wet weather patterns led to the exceptional rains and devastating floods in the upper Mississippi River Valley during the summer of 1993 (Figure 9.20).

Flash flooding in Las Vegas in August 2003. (John Locher)

Did You Know?

Flash floods are responsible for more deaths than tornadoes and hurricanes combined. For the nine-year period 1992–2001, the number of storm-related deaths in the United States from flooding averaged 127 per year. By contrast, tornado fatalities averaged 71 annually and hurricanes 16.

Flash Floods. A flash flood can occur with little warning and can be deadly because it produces a rapid rise in water levels and can have a devastating flow velocity. Several factors influence flash flooding. Among them are rainfall intensity and duration, surface conditions, and the topography. Mountainous areas are especially susceptible because steep slopes can funnel runoff into narrow canyons with disastrous consequences. This is illustrated by the Big Thompson River flood of July 31, 1976, in Colorado. During a four-hour period, more than 30 centimeters (12 inches) of rain fell on a portion of the river's small drainage basin. The flash flood in the narrow canyon lasted only a few hours but took 139 lives and caused tens of millions of dollars in damages.

Did You Know?

Just 15 cm (6 in) of fast-moving water can knock a person down. Many cars will float and be swept away in only 0.6 m (2 ft) of water. More than half of all U.S. flash-flood fatalities are auto-related.

Urban areas are susceptible to flash floods because a high percentage of the surface area is composed of impervious roofs, streets, and parking lots, where runoff is very rapid. Thus, less water infiltrates and the rate and amount of runoff increase. Further, because much less water soaks into the ground, the low-water (dry-season) flow in many urban streams, which is maintained by the movement of groundwater into the channel, is greatly reduced. As one might expect, the magnitude of these effects is a function of the percentage of land that is covered by impermeable surfaces.

Did You Know?

Humans have covered an amazing amount of land with buildings, parking lots, and roads. A recent study indicated that the area of such impervious surfaces in the United States (excluding Alaska and Hawaii) amounts to more than 112,600 km^2 (nearly 44,000 mi^2), which is slightly less than the area of the state of Ohio.

Ice-Jam Floods. Frozen rivers are susceptible to ice-jam floods. As the level of a stream rises, it will break up the ice and create ice flows that can pile up on channel obstructions. Such an ice jam creates a dam across the channel. Water upstream from the ice dam can rise rapidly and overflow the channel banks. When the ice dam fails, the water stored behind the dam is released, causing a flash flood downstream.

Dam-Failure Floods. Human interference with a stream system can cause floods. A prime example is the failure of a dam or an artificial levee. Dams and artificial levees are built for flood protection. They are designed to contain floods of a certain magnitude. If a larger flood occurs, the dam or levee is overtopped. If the dam or levee fails or is washed out, the water behind it is released to become a flash flood. The bursting of a dam in 1889 on the Little Conemaugh River caused the devastating Johnstown, Pennsylvania, flood that took 3000 lives.

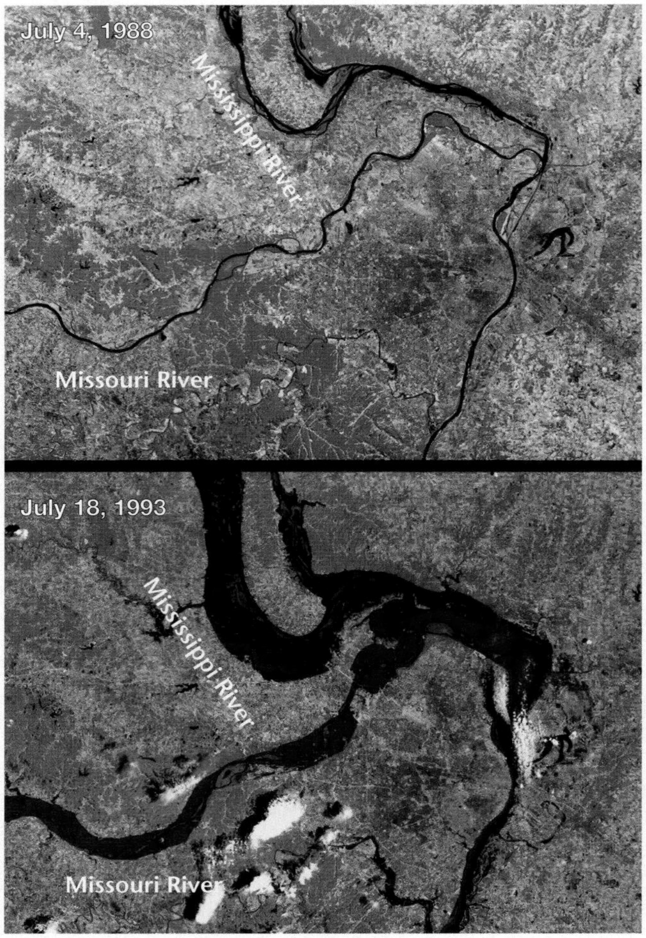

▲ **Figure 9.20** Satellite views of the Missouri River flowing into the Mississippi River. St. Louis is just south of their confluence. The upper image shows the rivers during a drought that occurred in summer 1988. The lower image depicts the peak of the record-breaking 1993 flood. Exceptional rains produced the wettest spring and early summer of the twentieth century in the upper Mississippi River basin. In all, nearly 14 million acres were inundated, displacing at least 50,000 people. *(Courtesy of Spaceimaging.com)*

Flood Control

Several strategies have been devised to eliminate or lessen the catastrophic effects of floods. Engineering efforts include the construction of artificial levees, the building of flood-control dams, and river channelization.

Artificial Levees. *Artificial levees* are earthen mounds built on the banks of a river to increase the volume of water the channel can hold. These most common of stream-containment structures have been used since ancient times and continue to be used today.

Artificial levees are usually easy to distinguish from natural levees because their slopes are much steeper. When a river is confined by levees during periods of high water, it frequently deposits material in its channel as the discharge diminishes. This is sediment that otherwise would have been dropped on the floodplain. Thus, each time there is a high flow, deposits are left on the river bed, and the bottom of the channel is built up. With the buildup of the bed, less water is required to overflow the original levee. As a result, the height of the levee may have to be raised periodically to protect the floodplain. Moreover, many artificial levees are not built to withstand periods of extreme flooding. For example, levee failures were numerous in the Midwest during the summer of 1993, when the upper Mississippi and many of its tributaries experienced record floods (Figure 9.21).

Flood-Control Dams. *Flood-control dams* are built to store floodwater and then let it out slowly. This lowers the flood crest by spreading it out over a longer time span. Ever since the 1920s, thousands of dams have been built on nearly every major river in the United States. Many dams have significant nonflood-related functions, such as providing water for irrigated agriculture and for hydroelectric power generation. Many reservoirs are also major regional recreational facilities.

Although dams may reduce flooding and provide other benefits, building these structures also has significant costs and consequences. For example, reservoirs created by dams may cover fertile farmland, useful forests, historic sites, and scenic valleys. Of course, dams trap sediment. Therefore, deltas and floodplains downstream erode because they are no longer replenished with silt during floods. Large dams can also cause significant ecological damage to river environments that took thousands of years to establish.

Building a dam is not a permanent solution to flooding. Sedimentation behind a dam means that the volume of its reservoir will gradually diminish, reducing the effectiveness of this flood-control measure.

Did You Know?

Artificial cutoffs are sometimes used for flood control because they shorten the length of a stream and thus increase its gradient. Artificial cutoffs created by the Army Corps of Engineers have shortened the Mississippi River by more than 240 km (150 mi).

Channelization. *Channelization* involves altering a stream channel to speed the flow of water to prevent it from reaching flood height. This may simply involve clearing a channel of obstructions or dredging a channel to make it wider and deeper.

A more radical alteration involves straightening a channel by creating *artificial cutoffs*. The idea is that by shortening the stream, the gradient and hence the velocity are increased. By increasing velocity, the larger discharge associated with flooding can be dispersed more rapidly.

Since the early 1930s, the U.S. Army Corps of Engineers has created many artificial cutoffs on the Mississippi for the purpose of increasing the efficiency of the channel and reducing the threat of flooding. The program has been somewhat successful in reducing the height of the river in flood stage. However, because the river's tendency toward meandering still exists, preventing the river from returning to its previous course has been difficult.

A Nonstructural Approach. All of the flood-control measures described so far have involved structural solutions aimed at "controlling" a river. These solutions are expensive and often give people residing on the floodplain a false sense of security.

Major reservoirs such as Arizona's Lake Powell provide important recreational facilities. (CORBIS)

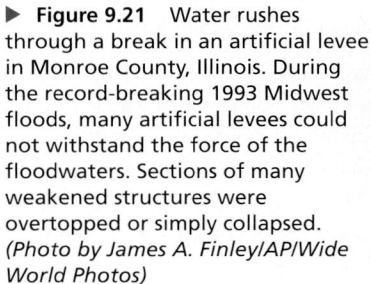

Figure 9.21 Water rushes through a break in an artificial levee in Monroe County, Illinois. During the record-breaking 1993 Midwest floods, many artificial levees could not withstand the force of the floodwaters. Sections of many weakened structures were overtopped or simply collapsed. *(Photo by James A. Finley/AP/Wide World Photos)*

Today many scientists and engineers advocate a nonstructural approach to flood control. They suggest that an alternative to artificial levees, dams, and channelization is sound floodplain management. By identifying high-risk areas, appropriate zoning regulations can be implemented to minimize development and promote more appropriate land use.

The Chapter in Review

- The *hydrologic cycle* describes the continuous interchange of water among the oceans, atmosphere, and continents. Powered by energy from the Sun, it is a global system in which the atmosphere provides the link between the oceans and continents. The processes involved in the hydrologic cycle include *precipitation, evaporation, infiltration* (the movement of water into rocks or soil through cracks and pore spaces), *runoff* (water that flows over the land), and *transpiration* (the release of water vapor to the atmosphere by plants). *Running water is the single most important agent sculpturing Earth's land surface.*
- The land area that contributes water to a stream is its *drainage basin.* Drainage basins are separated by imaginary lines called *divides.*
- River systems consist of three main parts: the zones of erosion, transportation, and deposition.
- The factors that determine a stream's *velocity* are *gradient* (slope of the stream channel), *shape, size,* and *roughness* of the channel and the stream's *discharge* (amount of water passing a given point per unit of time frequently measured in cubic feet per second). Most often, the gradient and roughness of a stream decrease downstream, while width, depth, discharge, and velocity increase.
- Streams transport their load of sediment in solution (*dissolved load*), in suspension (*suspended load*), and along the bottom of the channel (*bed load*). Much of the dissolved load is contributed by groundwater. Most streams carry the greatest part of their load in suspension. The bed load moves only intermittently and is usually the smallest portion of a stream's load.
- A stream's ability to transport solid particles is described using two criteria: *capacity* (the maximum load of solid particles a stream can carry) and *competence* (the maximum particle size a stream can transport). Competence increases as the square of stream velocity, so if velocity doubles, water's force increases fourfold.
- Streams deposit sediment when velocity slows and competence is reduced. This results in *sorting,* the process by which like-sized particles are deposited together. Stream deposits are called *alluvium* and may occur as channel deposits called *bars,* as floodplain deposits, which include *natural levees,* and as *deltas* or *alluvial fans* at the mouths of streams.
- Stream channels are of two basic types: *bedrock channels* and *alluvial channels.* Bedrock channels are most common in headwaters regions where gradients are steep. Rapids and waterfalls are common features. Two types of alluvial channels are *meandering channels* and *braided channels.*
- The two general types of *base level* (the lowest point to which a stream may erode its channel) are (1) *ultimate base level* and (2) *temporary,* or *local base level.* Any change in base level will cause a stream to adjust and establish a new balance. Lowering base level will cause a stream to downcut, whereas raising base level results in deposition of material in the channel.
- When a stream has cut its channel closer to base level, its energy is directed from side to side, and erosion produces a flat valley floor, or *floodplain.* Streams that flow upon flood plains often move in sweeping bends called *meanders.* Widespread meandering may result in shorter channel segments, called *cutoffs,* and/or abandoned bends, called *oxbow lakes.*
- Common *drainage patterns* (the form of a network of streams) produced by a main channel and its tributaries include (1) *dendritic,* (2) *radial,* (3) *rectangular,* and (4) *trellis.*
- *Floods* are triggered by heavy rains and/or snowmelt. Sometimes human interference can worsen or even cause floods. Flood-control measures include the building of *artificial levees* and dams, as well as *channelization,* which could involve creating *artificial cutoffs.* Many scientists and engineers advocate a nonstructural approach to flood control that involves more appropriate land use.

Key Terms

alluvial fan (p. 214)
alluvium (p. 208)
backswamp (p. 214)
bar (p. 213)
base level (p. 210)
bed load (p. 207)
braided stream (p. 209)
capacity (p. 208)
competence (p. 208)
cut bank (p. 208)
cutoff (p. 209)
delta (p. 213)
dendritic pattern (p. 214)
discharge (p. 205)
dissolved load (p. 207)
distributary (p. 213)
divide (p. 204)
drainage basin (p. 204)
evapotranspiration (p. 202)
flood (p. 215)
floodplain (p. 212)
gradient (p. 205)
hydrologic cycle (p. 202)
incised meander (p. 213)

infiltration (p. 202)
laminar flow (p. 204)
meander (p. 208)
natural levee (p. 213)
oxbow lake (p. 209)
point bar (p. 208)
radial pattern (p. 215)
rectangular pattern (p. 215)
runoff (p. 202)
sorting (p. 208)
stream valley (p. 210)
suspended load (p. 207)
transpiration (p. 202)
trellis drainage pattern (p. 215)
turbulent flow (p. 204)
yazoo tributary (p. 214)

Questions for Review

1. Describe the movement of water through the hydrologic cycle. Once precipitation has fallen on land, what paths are available to it?
2. Over the oceans, evaporation exceeds precipitation, yet sea level does not drop. Can you explain why?
3. What are the three main parts (zones) of a river system?
4. A stream starts out 2000 meters above sea level and travels 250 kilometers to the ocean. What is its average gradient in meters per kilometer?
5. Suppose that the stream mentioned in Question 4 developed extensive meanders so that its course was lengthened to 500 kilometers. Calculate its new gradient. How does meandering affect gradient?
6. When the discharge of a stream increases, what happens to the stream's velocity?
7. In what three ways does a stream transport its load?
8. If you collect a jar of water from a stream, what part of its load will settle to the bottom of the jar? What portion will remain in the water? What part of a stream's load would probably not be present in your sample?
9. Distinguish between *competence* and *capacity.*
10. Are bedrock channels more likely to be found near the head or near the mouth of a stream?
11. Describe a situation that might cause a stream channel to become braided.
12. Define *base level.* Name the main river in your area. For what streams does it act as base level? What is base level for the Mississippi River? The Missouri River?
13. Describe two situations that would trigger the formation of incised meanders.
14. Briefly describe the formation of a natural levee. How is this feature related to back swamps and yazoo tributaries?
15. List two major depositional features, other than natural levees, that are associated with streams. Under what circumstances does each form?
16. Each of the following statements refers to a particular drainage pattern. Identify the pattern.
    a. Streams diverging from a central high area, such as a dome.
    b. Branching treelike pattern.
    c. A pattern that develops when bedrock is crisscrossed by joints and faults.
17. Contrast regional floods and flash floods. Which type is deadliest?
18. List and briefly describe three basic flood-control strategies. What are some drawbacks of each?

Online Study Guide

WWW

The *Essentials of Geology* Web site uses the resources and flexibility of the Internet to aid in your study of the topics in this chapter. Written and developed by geology instructors, this site will help improve your understanding of geology. Visit **www.prenhall.com/lutgens** and click on the cover of *Essentials of Geology 9e* to find:

- Online review quizzes.
- Critical thinking exercises.
- Links to chapter-specific Web resources.
- Internet-wide key-term searches.

http://www.prenhall.com/lutgens

Groundwater provides over 190 billion liters (50 billion gallons) per day in support of the agricultural economy of the United States. *(Photo by Michael Collier)*

CHAPTER 10

Groundwater

Focus on Learning

To assist you in learning the important concepts in this chapter, you will find it helpful to focus on the following questions:

- What is the importance of groundwater as a resource and as a geological agent?
- What is groundwater, and what factors affect it's movement?
- How do springs, geysers, wells, and artesian wells form?
- What are some environmental problems associated with groundwater?
- What features are produced by the geologic work of groundwater?

Worldwide, wells and springs provide water for cities, crops, livestock, and industry. In some areas, however, overuse of this basic resource has resulted in water shortages, streamflow depletion, land subsidence, contamination by saltwater, increased pumping costs, and groundwater pollution.

Importance of Underground Water

GEODe: Essentials of Geology
Groundwater
▼ Importance and Distribution of Groundwater

Groundwater is one of our most important and widely available resources, yet people's perceptions of the subsurface environment from which it comes are often unclear and incorrect. The reason is that the groundwater environment is largely hidden from view except in caves and mines, and the impressions people gain from these subsurface openings are misleading. Observations on the land surface give an impression that Earth is solid. This view remains when we enter a cave and see water flowing in a channel that appears to have been cut into solid rock.

Did You Know?

In the United States, groundwater is the source of about 40 percent of the water used for all purposes (except hydroelectric power generation and power plant cooling). Groundwater is the drinking water for more than 140 million people nationwide. It supplies 40 percent of the water used for irrigation and provides more than 25 percent of industry's needs.

Because of such observations, many people believe that groundwater occurs only in underground rivers. In reality, most of the subsurface environment is not solid at all. It includes countless tiny *pore spaces* between grains of soil and sediment, plus narrow joints and fractures in bedrock. Together, these spaces add up to an immense volume. It is in these small openings that groundwater collects and moves.

Considering the entire hydrosphere, or all of Earth's water, only about six-tenths of 1 percent occurs underground. Nevertheless, this small percentage, stored in the rocks and sediments beneath Earth's surface, is a vast quantity. When the oceans are excluded and only sources of freshwater are considered, the significance of groundwater becomes more apparent.

Table 10.1 contains estimates of the distribution of fresh water in the hydrosphere. Clearly, the largest volume occurs as glacial ice. Second in rank is groundwater, with slightly more than 14 percent of the total. However, when ice is excluded and just liquid water is considered, more than 94 percent of all fresh water is groundwater. Without question, *groundwater represents the largest reservoir of fresh water that is readily available to humans.* Its value in terms of economics and human well-being is incalculable.

Geologically, groundwater is important as an erosional agent. The dissolving action of groundwater slowly removes rock, allowing surface depressions known as sinkholes to form as well as creating subterranean caverns (Figure 10.1). Groundwater is also an equalizer of streamflow. Much of the water that flows in rivers is not direct runoff from rain and snowmelt. Rather, a large percentage of precipitation soaks in and then moves slowly underground to stream channels. Groundwater is thus a form of storage that sustains streams during periods when rain does not fall. When we see water flowing in a river during a dry period, it is water from rain that fell at some earlier time and was stored underground.

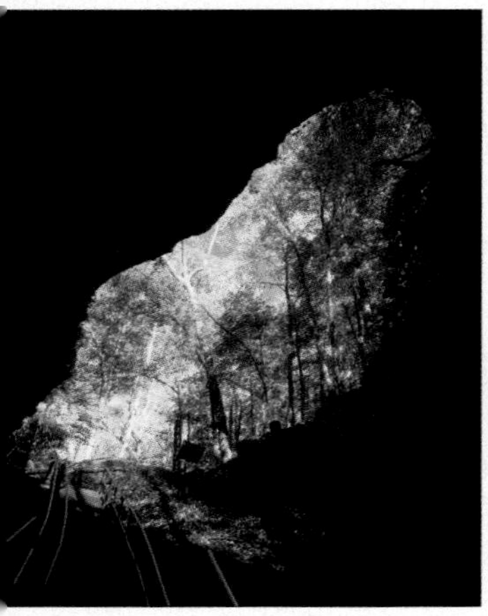

Cave entrance at Mammoth Cave, Kentucky. (Mark Muench)

Distribution of Underground Water

GEODe: Essentials of Geology
Groundwater
▼ Importance and Distribution of Groundwater

When rain falls, some of the water runs off, some evaporates, and the remainder soaks into the ground. This last path is the primary source of practically all under-

Table 10.1 Fresh water of the hydrosphere

Parts of the Hydrosphere	Volume of Fresh Water (km^3)	Share of Total Volume of Fresh Water (percent)	Rate of Water Exchange
Ice sheets and glaciers	24,000,000	84.945	8000 years
Groundwater	4,000,000	14.158	280 years
Lakes and reservoirs	155,000	0.549	7 years
Soil moisture	83,000	0.294	1 year
Water vapor in the atmosphere	14,000	0.049	9.9 days
River water	1,200	0.004	11.3 days
Total	28,253,200	100.00	

Source: U.S. Geological Survey Water Supply Paper 2220, 1987.

▲ **Figure 10.1** A view of the interior of Lehman Caves in Great Basin National Park, Nevada. The dissolving action of acidic groundwater created the caverns. Later, groundwater deposited the limestone decorations. *(Photo by David Muench/David Muench Photography, Inc.)*

ground water. The amount of water that takes each of these paths, however, varies greatly both in time and space. Influential factors include steepness of slope, nature of surface material, intensity of rainfall, and type and amount of vegetation. Heavy rains falling on steep slopes underlain by impervious materials will obviously result in a high percentage of the water running off. Conversely, if rain falls steadily and gently on more gradual slopes composed of materials that are easily penetrated by the water, a much larger percentage of water soaks into the ground.

Did You Know?

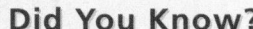

Although most caves and sinkholes are associated with regions underlain by limestone, these features can also form in gypsum and rock salt (halite) because these rocks are highly soluble and readily dissolved.

Some of the water that soaks in does not travel far, because it is held by molecular attraction as a surface film on soil particles. This near-surface zone is called the *belt of soil moisture.* It is crisscrossed by roots, voids left by decayed roots, and animal and worm burrows that enhance the infiltration of rainwater into the soil. Soil water is used by plants in life functions and transpiration. Some water also evaporates directly back into the atmosphere.

Rainfall intensity affects the amount of water that soaks in or runs off. (M. Fulton/Getty-Liaison)

Water that is not held as soil moisture will percolate downward until it reaches a zone where all of the open spaces in sediment and rock are completely filled with water. This is the **zone of saturation.** Water within it is called **groundwater.** The upper limit of this zone is known as the **water table.** The area above the water table where the soil, sediment, and rock are not saturated is called the **zone of aeration** (Figure 10.2). Although a considerable amount of water can be present in the zone of aeration, this water cannot be pumped by wells because it clings too tightly to rock and soil particles. By contrast, below the water table, the water pressure is great enough to allow water to enter wells, thus permitting groundwater to be withdrawn for use. We will examine wells more closely later in the chapter.

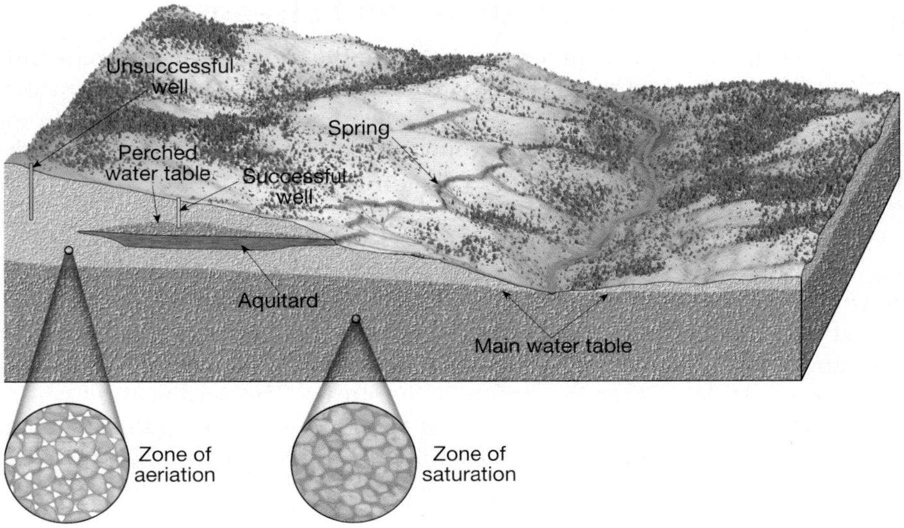

◀ **Figure 10.2** This diagram illustrates the relative positions of many features associated with subsurface water.

The Water Table

Groundwater
▼ Importance and Distribution of Groundwater

The water table, the upper limit of the zone of saturation, is a very significant feature of the groundwater system. The water-table level is important in predicting the productivity of wells, explaining the changes in the flow of springs and streams, and accounting for fluctuations in the levels of lakes.

Variations in the Water Table

The depth of the water table is highly variable and can range from zero, when it is at the surface, to hundreds of meters in some places. An important characteristic of the water table is that its configuration varies seasonally and from year to year because the addition of water to the groundwater system is closely related to the quantity, distribution, and timing of precipitation. Except where the water table is at the surface, we cannot observe it directly. Nevertheless, its elevation can be mapped and studied in detail where wells are numerous, because the water level in wells coincides with the water table (Figure 10.3). Such maps reveal that the water table is rarely level, as we might expect a table to be. Instead, its shape is usually a subdued replica of the surface topography, reaching its highest elevations beneath hills and then descending toward valleys. Where a wetland (swamp) is encountered, the water table is right at the surface. Lakes and streams generally occupy areas low enough that the water table is above the land surface.

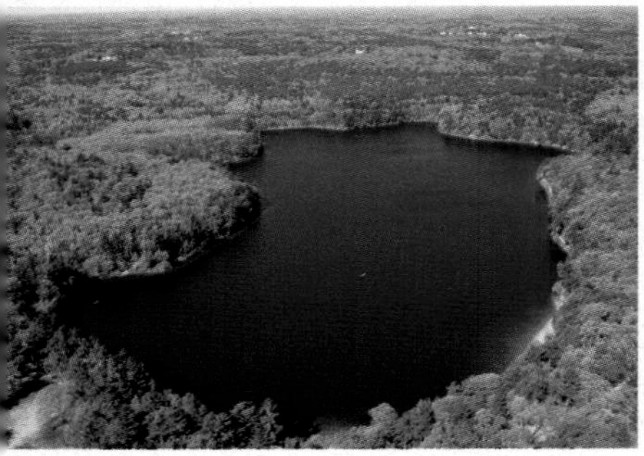

The surface of this pond represents the level of the water table. (S. Dunwell/ Getty)

Several factors contribute to the irregular surface of the water table. One important influence is the fact that groundwater moves very slowly and at varying rates under different conditions. Because of this, water tends to "pile up" beneath high areas between stream valleys. If rainfall were to cease completely, these water table "hills" would slowly subside and gradually approach the level of the valleys. However, new supplies of rainwater are usually added frequently enough to prevent this. Nevertheless, in times of extended drought, the water table may drop enough to dry up shallow wells. Other causes for the uneven water table are variations in rainfall and permeability from place to place.

Did You Know?

Groundwater pumping to support rapid population growth in south-central Arizona (including the Tucson and Phoenix areas) has resulted in water-level declines of between 90 and 150 m (300 and 500 ft).

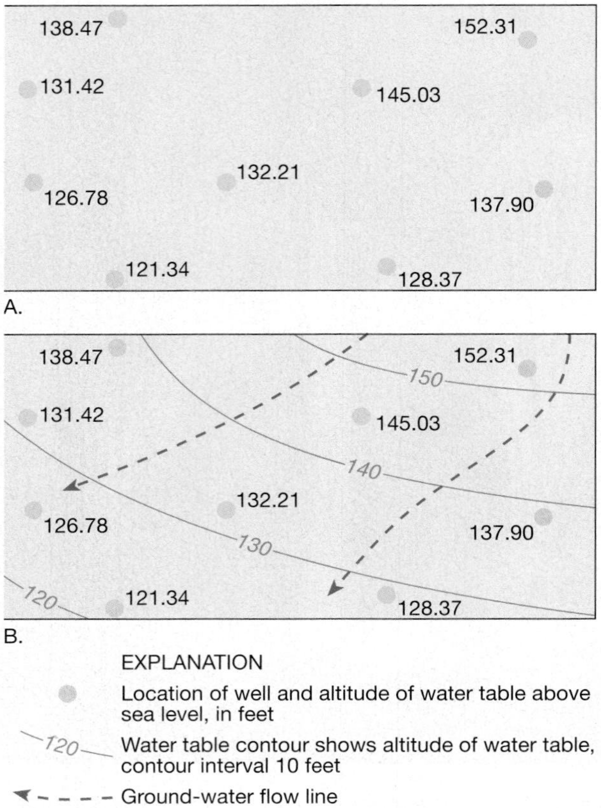

▲ **Figure 10.3** Preparing a map of the water table. The water level in wells coincides with the water table. **A.** First, the locations of wells and the elevation of the water table above sea level are plotted on a map. **B.** These data points are used to guide the drawing of water-table contour lines at regular intervals. On this sample map the interval is 10 feet. Groundwater flow lines can be added to show water movement in the upper portion of the zone of saturation. Groundwater tends to move approximately perpendicular to the contours and down the slope of the water table. *(After U.S Geological Survey)*

Interaction Between Groundwater and Streams

The interaction between the groundwater system and streams is a basic link in the hydrologic cycle. It can take place in one of three ways. Streams may gain water from the inflow of groundwater through the streambed. Such streams are called **gaining streams** (Figure 10.4A). For this to occur, the elevation of the water table must be higher than the level of the surface of the stream. Streams may lose water to the groundwater system by outflow through the streambed. The term **losing stream** is applied to this situation (Figure 10.4B). When this happens, the elevation of the water table must be lower than the surface of the stream. The third possibility is a combination of the first two—a stream gains in some sections and loses in others.

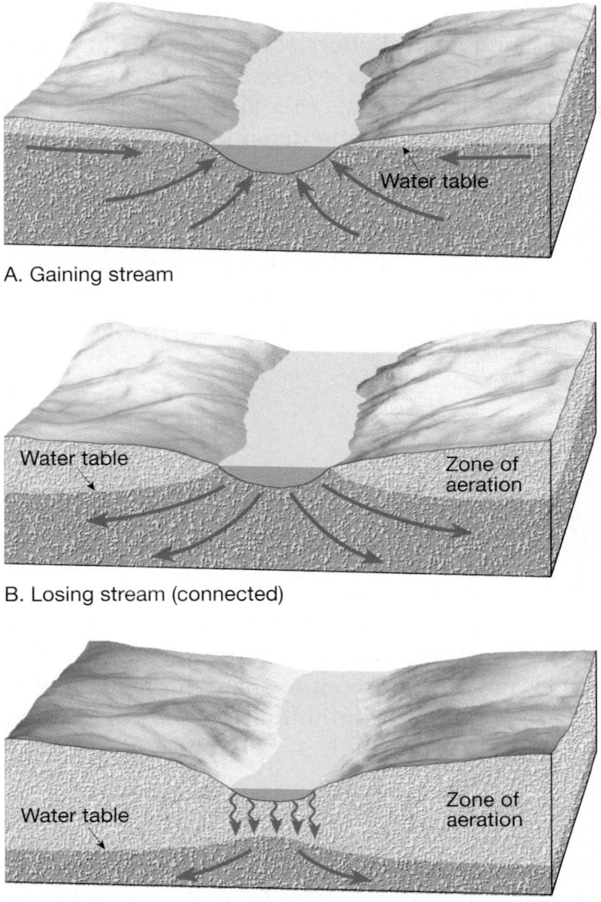

▲ **Figure 10.4** Interaction between the groundwater system and streams. **A.** Gaining streams receive water from the groundwater system. **B.** Losing streams lose water to the groundwater system. **C.** When losing streams are separated from the groundwater system by the zone of aeration, a bulge may form in the water table. *(After U.S. Geological Survey)*

Losing streams can be connected to the groundwater system by a continuous saturated zone, or they can be disconnected from the groundwater system by an unsaturated zone. Compare parts B and C in Figure 10.4. When the stream is disconnected, the water table may have a discernible bulge beneath the stream if the rate of water movement through the streambed and zone of aeration is greater than the rate of groundwater movement away from the bulge.

In some settings, a stream might always be a gaining stream or always be a losing stream. However, in many situations flow direction can vary a great deal along a stream; some sections receive groundwater and other sections lose water to the groundwater system. Moreover, the direction of flow can change over a short time span as the result of storms adding water near the streambank or when temporary flood peaks move down the channel.

Groundwater contributes to streams in most geologic and climatic settings. Even where streams are primarily losing water to the groundwater system, certain sections may receive groundwater inflow during some seasons. In one study of 54 streams in all parts of the United States, the analysis indicated that 52 percent of the streamflow was contributed by groundwater. The groundwater contribution ranged from a low of 14 percent to a maximum of 90 percent.

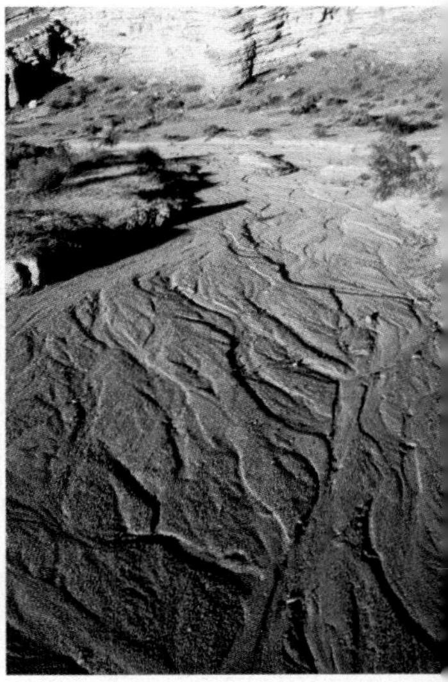

The water table here is disconnected from the stream.

Factors Influencing the Storage and Movement of Groundwater

The nature of subsurface materials strongly influences the rate of groundwater movement and the amount of groundwater that can be stored. Two factors are especially important—porosity and permeability.

Porosity

Water soaks into the ground because bedrock, sediment, and soil contain countless voids, or openings. These openings are similar to those of a sponge and are often called *pore spaces*. The quantity of groundwater that can be stored depends on the **porosity** of the material, which is the percentage of the total volume of rock or sediment that consists of pore spaces. Voids most often are spaces between sedimentary particles, but also common are joints, faults, cavities formed by the dissolving of soluble rocks such as limestone, and vesicles (voids left by gases escaping from lava).

Did You Know?

Because of its high porosity, excellent permeability, and great size, the Ogallala Formation, the largest aquifer in the United States, accumulated huge amounts of groundwater—enough freshwater to fill Lake Huron.

Variations in porosity can be great. Sediment is commonly quite porous, and open spaces may occupy 10 to 50 percent of the sediment's total volume. Pore space depends on the size and shape of the grains, how they are packed together, the degree of sorting, and in sedimentary rocks, the amount of cementing material. For example, clay may have a porosity as high as 50 percent, whereas some gravels may have only 20 percent voids.

Where sediments of various sizes are mixed, the porosity is reduced because the finer particles tend to fill the openings among the larger grains. Most igneous and metamorphic rocks, as well as some sedimentary rocks, are composed of tightly interlocking crystals, so the voids between the grains may be negligible. In these rocks, fractures must provide the voids.

Permeability, Aquitards, and Aquifers

Porosity alone cannot measure a material's capacity to yield groundwater. Rock or sediment might be very porous yet still not allow water to move through it. The pores must be *connected* to allow water flow, and they must be *large enough* to allow flow. Thus, the **permeability** of a material, its ability to *transmit* a fluid, is also very important.

Groundwater moves by twisting and turning through interconnected small openings. The smaller the pore spaces, the slower the water moves. For example, clay's ability to store water is great, owing to its high porosity, but its pore spaces are so small that water is unable to move through it. Thus, clay's porosity is high but its permeability is poor.

Impermeable layers that hinder or prevent water movement are termed **aquitards.** Clay is a good example. In contrast, larger particles, such as sand or gravel, have larger pore spaces. Therefore, the water moves with relative ease. Permeable rock strata or sediments that transmit groundwater freely are called **aquifers.** Sands and gravels are common examples.

In summary, you have seen that porosity is not always a reliable guide to the amount of groundwater that can be produced, and permeability is significant in determining the rate of groundwater movement and the quantity of water that might be pumped from a well.

Did You Know?

The rate of groundwater movement is highly variable. One method of measuring this movement involves introducing dye into a well. The time is measured until the coloring agent appears in another well at a known distance from the first. A typical rate for many aquifers is about 15 m per year (about 4 cm per day).

How Groundwater Moves

We noted the common misconception that groundwater occurs in underground rivers that resemble surface streams. Although subsurface streams do exist, they are not common. Rather, as you learned in the preceding sections, groundwater exists in the pore spaces and fractures in rock and sediment. Thus, contrary to any impressions of rapid flow that an underground river might evoke, the movement of most groundwater is exceedingly slow, from pore to pore. By exceedingly slow, we mean anywhere from millimeters per year to perhaps a kilometer per year, depending on conditions.

Groundwater contributes significantly to the flow of streams.

The energy that makes groundwater move is provided by the force of gravity. In response to gravity, water moves from areas where the water table is high to zones where the water table is lower (see Box 10.1). This means that water gravitates toward a stream channel, lake, or spring. Although some water takes the most direct path down the slope of the water table, much of the water follows long, curving paths toward the zone of discharge.

Figure 10.5 shows how water percolates into a stream from all possible directions. Some paths clearly turn upward, apparently against the force of gravity, and enter through the bottom of the channel. This is easily explained: The deeper you go into the zone of saturation, the greater the water pressure. Thus, the looping curves followed by water in the saturated zone may be thought of as a compromise between the downward pull of gravity and the tendency of water to move toward areas of reduced pressure. As a result, water at any given height is under greater pressure beneath a hill than beneath a stream channel, and the water tends to migrate toward points of lower pressure.

Springs

GEODe: Essentials of Geology — Groundwater ▼ Springs and Wells

Springs have aroused the curiosity and wonder of people for thousands of years. The fact that springs were, and to some people still are, rather mysterious phenomena is not difficult to understand, for here is water flowing freely from the ground in all kinds of weather in seemingly inexhaustible supply but with no obvious source.

Not until the middle of the 1600s did the French physicist Pierre Perrault invalidate the age-old assumption that precipitation could not adequately account for the amount of water emanating from springs and flowing in rivers. Over several years Perrault computed the quantity of water that fell on France's Seine River basin. He then calculated the mean annual runoff by measuring the river's discharge. After allowing for the loss of water by evaporation, he showed that there *was* sufficient water remaining to feed the springs. Thanks to Perrault's pioneering efforts and the measurements by many afterward, we now know that the source of springs is water from the zone of saturation and that the ultimate source of this water is precipitation.

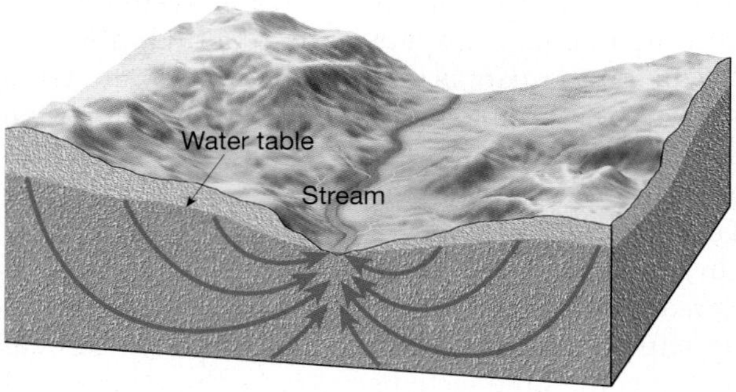

▲ **Figure 10.5** Arrows indicate groundwater movement through uniformly permeable material. The looping curves may be thought of as a compromise between the downward pull of gravity and the tendency of water to move toward areas of reduced pressure.

BOX 10.1

Measuring Groundwater Movement

The foundations of our modern understanding of groundwater movement began in the mid-nineteenth century with the work of the French scientist-engineer Henri Darcy. Among the experiments carried out by Darcy was one that showed that the velocity of groundwater flow is proportional to the slope of the water table—the steeper the slope, the faster the water moves (because the steeper the slope, the greater the pressure difference between two points). The water-table slope is known as the hydraulic gradient and can be expressed as follows:

$$\text{hydraulic gradient} = \frac{h_1 - h_2}{d}.$$

Where h_1 is the elevation of one point on the water table, h_2 is the elevation of a second point, and d is the horizontal distance between the two points (Figure 10.A)

Darcy also discovered that the flow velocity varied with the permeability of the sediment—groundwater flows more rapidly through sediments having greater permeability than through materials having lower permeability. This factor is known as *hydraulic conductivity* and is a coefficient that takes into account the permeability of the aquifer and the viscosity of the fluid.

To determine discharge (Q)—that is, the actual volume of water that flows through an aquifer in a specified time—the following equation is used:

$$Q = \frac{K\,A(h_1 - h_2)}{d}$$

Where $\frac{h_1 - h_2}{d}$ is the hydraulic gradient, K is the coefficient that represents hydraulic conductivity and A is the cross-sectional area of the aquifer. This expression has come to be called *Darcy's law.*

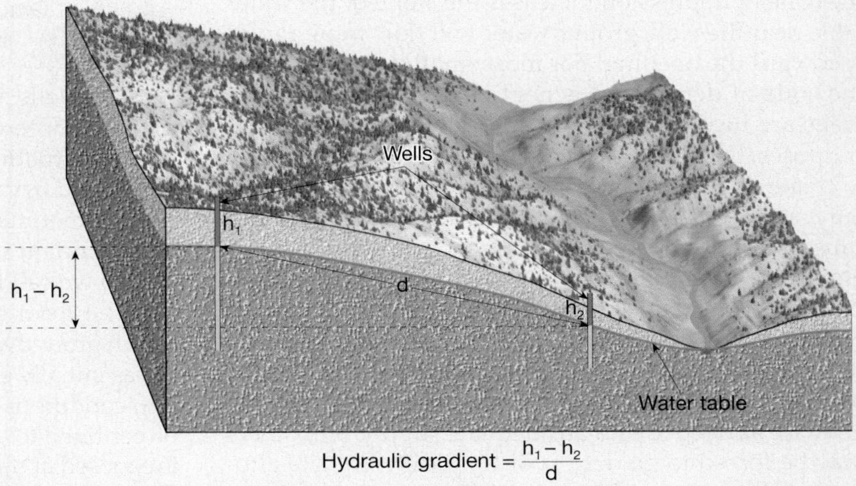

Figure 10.A The hydraulic gradient is determined by measuring the difference in elevation between two points on the water table (h_1–h_2) divided by the distance between them, d. Wells are used to determine the height of the water table.

Whenever the water table intersects Earth's surface, a natural outflow of groundwater results, which we call a **spring.** Springs such as the one pictured in Figure 10.6 form when an aquitard blocks the downward movement of groundwater and forces it to move laterally. Where the permeable bed outcrops, a spring results. Another situation leading to the formation of a spring is illustrated in Figure 10.2. Here an aquitard is situated above the main water table. As water percolates downward, a portion of it is intercepted by the aquitard, thereby creating a localized zone of saturation called a **perched water table.**

Springs, however, are not confined to places where a perched water table creates a flow at the surface. Many geological situations lead to the formation of springs because subsurface conditions vary greatly from place to place. Even in areas underlain by impermeable crystalline rocks, permeable zones may exist in the form of fractures or solution channels. If these openings fill with water and intersect the ground surface along a slope, a spring will result.

Wells

GEODe ESSENTIALS OF GEOLOGY
Groundwater
▼ Springs and Wells

The most common device used by people for removing groundwater is the **well,** a hole bored into the zone of saturation (see Figure 10.2). Wells serve as small reservoirs into which groundwater migrates and from which it can be pumped to the surface. The use of wells dates back many centuries and continues to be an important method of obtaining water today. By far the single greatest use of this water in the United States is irrigation for agriculture. More than 65 percent of the groundwater used each year is for this purpose. Industrial uses rank a distant second, followed by the amount used in city water systems and rural homes.

The water-table level may fluctuate considerably during the course of a year, dropping during dry seasons and rising following periods of rain. Therefore, to ensure a continuous supply of water, a well must penetrate below the water table. Often when water is withdrawn from a well, the water table around the well is lowered. This effect, termed **drawdown,** decreases with increasing distance from the well. The result is a depression in the water table, roughly conical in shape, known as a **cone of depression** (Figure 10.7). Because the cone of depression increases the slope of the water table near the well, groundwater will flow more rapidly toward the opening. For most small domestic wells, the cone of depression is negligible. However, when wells are heavily pumped for irrigation or industrial purposes, the withdrawal of water can be great enough to create a very wide and steep cone of depression. This may substantially lower the water table in an area and cause nearby shallow wells to become "high and dry." Figure 10.7 illustrates this situation.

Well drillers at work.
(B. Mogen/Visuals Unlimited)

Digging a successful well is a familiar problem for people in areas where groundwater is the primary source of supply. One well may be successful at a depth of 10 meters (33 feet), whereas a neighbor may have to go twice as deep to find an adequate supply. Still others may be forced to go deeper or try a different site altogether. When subsurface materials are heterogeneous, the amount of water a well is capable of providing may vary a great deal over short distances. For example, when two nearby wells are drilled to the same level and only one is successful, it may be caused by the presence of a perched water table beneath one of them. Such a case is shown in Figure 10.2. Massive igneous and metamorphic rocks provide a second example. These crystalline rocks are usually not very permeable except where they are cut by many intersecting joints and fractures. Therefore, when a well drilled into such rock does not intersect an adequate network of fractures, it is likely to be unproductive.

▼ **Figure 10.6** Spring in Arizona's Marble Canyon. *(Photo by Michael Collier)*

Artesian Wells

GEODe: Groundwater
▼ Springs and Wells

In most wells, water cannot rise on its own. If water is first encountered at 30 meters depth, it remains at that level, fluctuating perhaps a meter or two with seasonal wet and dry periods. However, in some wells, water rises, sometimes overflowing at the surface. Such wells are abundant in the *Artois* region of northern France, and so we call these self-rising wells *artesian.*

The term **artesian** is applied to any situation in which groundwater rises in a well above the level where it was initially encountered. For such a situation to occur, two conditions must exist (Figure 10.8): (1) Water must be confined to an aquifer that is inclined so that one end is exposed at the surface, where it can receive water; and (2) impermeable layers (aquitards), both above and below the aquifer, must be present to prevent the water from escaping. When such a layer is tapped, the pressure created by the weight of the water above will force the water to rise. If there were no friction, the water in the well would rise to the level of the water at the top of the aquifer. However, friction reduces the height of this pressure surface. The greater the distance from the recharge area (area where water enters the inclined aquifer), the greater the friction and the less the rise of water.

In Figure 10.8, Well 1 is a *nonflowing artesian well,* because at this location the pressure surface is below ground level. When the pressure surface is above the ground and a well is drilled into the aquifer, a *flowing artesian well* is created (Well 2, Figure 10.8). Not all artesian systems are wells. *Artesian springs* also exist. Here groundwater reaches the surface by rising through a natural fracture rather than through an artificially produced hole.

Artesian systems act as conduits, transmitting water from remote areas of recharge great distances to the points of discharge. In this manner, water that fell in central Wisconsin years ago is now taken from the ground and used by communities many kilometers away in Illinois. In South Dakota, such a system brings water from the Black Hills in the west, eastward across the state.

On a different scale, city water systems can be considered as examples of artificial artesian systems (Figure 10.9). The water tower, into which water is pumped, can be considered the area of recharge, the

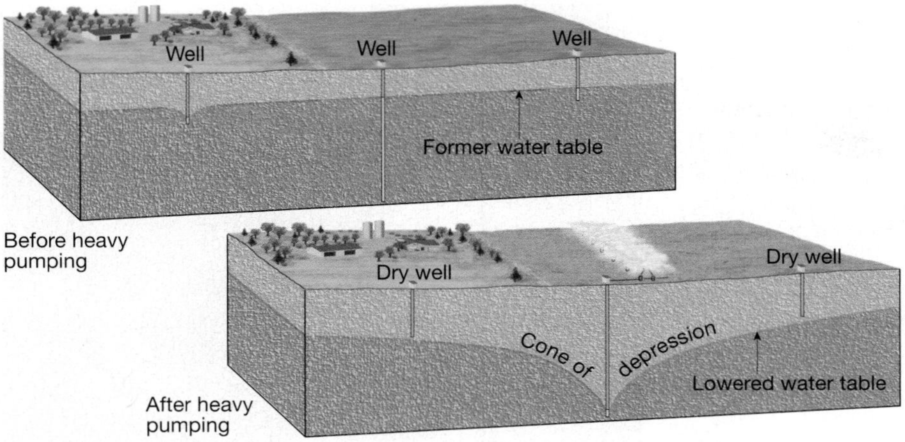

◀ **Figure 10.7** A cone of depression in the water table often forms around a pumping well. If heavy pumping lowers the water table, the shallow wells may be left dry.

pipes the confined aquifer, and the faucets in homes the flowing artesian wells.

Environmental Problems Associated with Groundwater

As with many of our valuable natural resources, groundwater is being exploited at an increasing rate. In some areas, overuse threatens the groundwater supply. In other places, groundwater withdrawal has caused the ground and everything resting upon it to sink. Still other localities are concerned with the possible contamination of their groundwater supply.

Treating Groundwater as a Nonrenewable Resource

Many natural systems tend to establish a condition of equilibrium. The groundwater system is no exception. The water table's height reflects a balance between the rate of water added by precipitation and the rate of water removed by discharge and withdrawal. Any imbalance will either raise or lower the water table. A long-term drop in the water table can occur if there is either a decrease in recharge due to a prolonged drought, or an increase in groundwater discharge or withdrawal.

A flowing artesian well. (James Patterson)

For many, groundwater appears to be an endlessly renewable resource, because it is continually replenished by rainfall and melting snow. But in some regions groundwater has been and continues to be treated as a *nonrenewable resource.* Where this occurs, the water available to recharge the aquifer falls significantly short of the amount being withdrawn.

The High Plains, a relatively dry region that extends from South Dakota to western Texas, provides one example (Figure 10.10A). Here an extensive agricultural economy is largely dependent on irrigation.

Did You Know?

Several different processes can cause ground subsidence, but according to the U.S. Geological Survey, more than 80 percent of the subsidence in the United States is related to the withdrawal of groundwater.

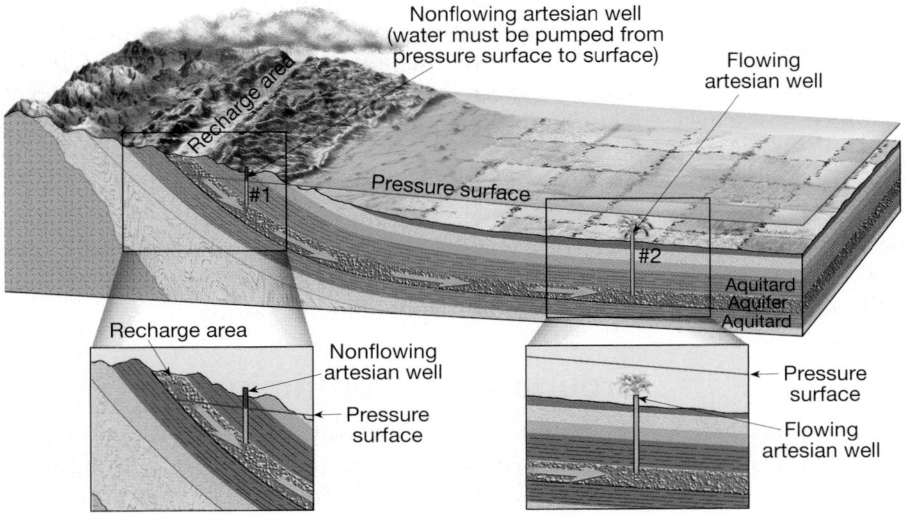

◀ **Figure 10.8** Artesian systems occur when an inclined aquifer is surrounded by impermeable beds.

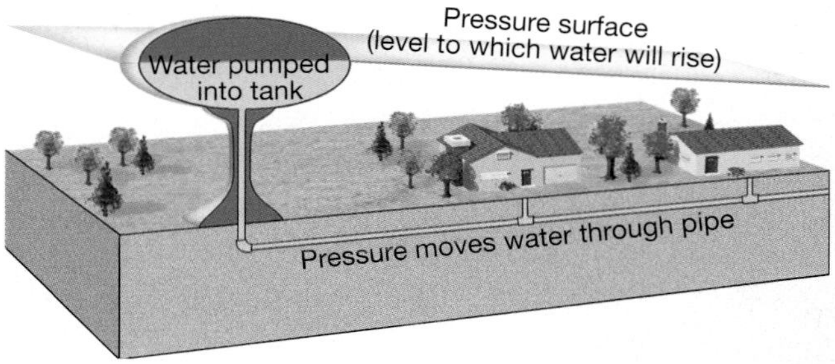

▲ **Figure 10.9** City water systems can be considered to be artificial artesian systems.

A.

B.

▲ **Figure 10.10** **A.** The High Plains extend from the western Dakotas south to Texas. Despite being a land of little rain, this is an important agricultural region. The reason is a vast endowment of groundwater that makes irrigation possible through most of the region. The source of most of this water is the Ogallala formation, the largest aquifer in the United States. **B.** In some agricultural regions, water is pumped from the ground faster than it is replenished. In such instances, groundwater is being treated as a nonrenewable resource. This aerial views shows circular crop fields irrigated by center pivot irrigation systems in semiarid eastern Colorado. *(Photo by James L. Amos/CORBIS/Bettmann)*

A "gusherlike" flowing artesian well in South Dakota early in the twentieth century. (N.H. Darton/USGS)

Today nearly 170,000 wells are being used to irrigate more than 65,000 square kilometers (16 million acres) of land (Figure 10.10B). In the southern part of this region, which includes the Texas panhandle, the natural recharge of the aquifer is very slow and the problem of declining groundwater levels is acute. In fact, in years of average or below-average precipitation, recharge is negligible because nearly all of the meager rainfall is returned to the atmosphere by evaporation and transpiration.

Therefore, where intense irrigation has been practiced for an extended period, depletion of groundwater can be severe. Declines in the water table as great as 1 meter per year have led to an overall drop of between 15 and 60 meters (50 and 200 feet) in some areas. Under these circumstances, it can be said that the groundwater is literally being "mined." Even if pumping were to cease immediately, it would take thousands of years for the groundwater to be fully replenished.

Groundwater depletion has been a concern in the High Plains and other areas of the West for many years, but it is worth pointing out that the problem is not confined to this part of the country. Increased demands on groundwater resources have overstressed aquifers in many areas, not just in arid and semiarid regions.

When removal of groundwater exceeds recharge, the water table falls. (B. Kamin/Visuals Unlimited)

Land Subsidence Caused by Groundwater Withdrawal

As you will see later in this chapter, surface subsidence can result from natural processes related to groundwater. However, the ground may also sink when water is pumped from wells faster than natural recharge processes can replace it. This effect is particularly pronounced in areas underlain by thick layers of loose sediments. As water is withdrawn, the weight of the overburden packs the sediment grains more tightly together and the ground subsides.

Many areas can be used to illustrate such land subsidence caused by excessive pumping of groundwater from relatively loose sediment. A classic example in the United States occurred in the San Joaquin Valley of California. This important agricultural region relies heavily on irrigation. Land subsidence due to groundwater withdrawal began in the valley in the mid-1920s and locally exceeded 8 meters (28 feet) by 1970. Then, because of the importation of surface water and a decrease in groundwater pumping, water levels in the aquifer recovered and subsidence ceased.

However, during a drought from 1976 to 1977, heavy groundwater pumping led to renewed subsidence. This time, water levels dropped at a much faster rate than during the previous period because of the reduced storage capacity caused by earlier compaction of material in the aquifer. In all, more than 13,400 square kilometers (5200 square miles) of irrigable land, half the entire valley, were affected by subsidence. Damage to structures, including highways, bridges, water lines, and wells, was extensive. Many other cases of land subsidence due to groundwater pumping exist in the United States and elsewhere in the world.

Groundwater Contamination

The pollution of groundwater is a serious matter, particularly in areas where aquifers provide a large part of the water supply. One common source of groundwater pollution is sewage. Its sources include an ever increasing number of septic tanks, as well as inadequate or broken sewer systems and farm wastes.

If sewage water that is contaminated with bacteria enters the groundwater system, it may become purified through natural processes. The harmful bacteria may be mechanically filtered by the sediment through which the water percolates, destroyed by chemical oxidation, and/or assimilated by other organisms. For purification to occur, however, the aquifer must be of the correct composition. For example, extremely permeable aquifers (such as highly fractured crystalline rock, coarse gravel, or cavernous limestone) have such large openings that contaminated groundwater might travel long distances without being cleansed. In this case, the water flows too rapidly and is not in contact with the surrounding material long enough for purification to occur. This is the problem at Well 1 in Figure 10.11.

In contrast, when the aquifer is composed of sand or permeable sandstone, the water can sometimes be purified after traveling only a few dozen meters through it. The openings between sand grains are large enough to permit water movement, yet the movement of the water is slow enough to allow ample time for its purification (Well 2, Figure 10.11).

Other sources and types of contamination also threaten groundwater supplies. These include widely used substances such as highway salt, fertilizers that are spread across the land surface, and pesticides. In addition, a wide array of chemicals and industrial materials may leak from pipelines, storage tanks, landfills, and holding ponds. Some of these pollutants are classified as *hazardous,* meaning that they are either flammable, corrosive, explosive, or toxic. As rainwater oozes through the refuse, it may dissolve a variety of potential contaminants. If the leached material reaches the water table, it will mix with the groundwater and contaminate the supply. Similar problems may result from leakage of shallow excavations called holding ponds into which a variety of liquid wastes are disposed.

Because groundwater movement is usually slow, polluted water may go undetected for a long time. In fact, most contamination is discovered only after drinking water has been affected and people become ill. By this time, the volume of polluted water may be very large, and even if the source of contamination is removed immediately, the problem is not solved. Although the sources of groundwater contamination are numerous, there are relatively few solutions.

Once the source of the problem has been identified and eliminated, the most common practice is simply to abandon the water supply and allow the pollutants to be flushed away gradually. This is the least costly and easiest solution, but the aquifer must remain unused for many years. To accelerate this process, polluted water is sometimes pumped out and treated. Following removal of the tainted water, the aquifer is allowed to recharge naturally, or in some cases, the treated water or other fresh water is pumped back in. This process is costly, time-consuming, and it may be risky because there is no way to be certain that all of the contamination has been removed. Clearly, the most effective solution to groundwater contamination is prevention.

The marks on this utility pole indicate the level of the adjacent land in preceding years in the San Joaquin Valley. (USGS)

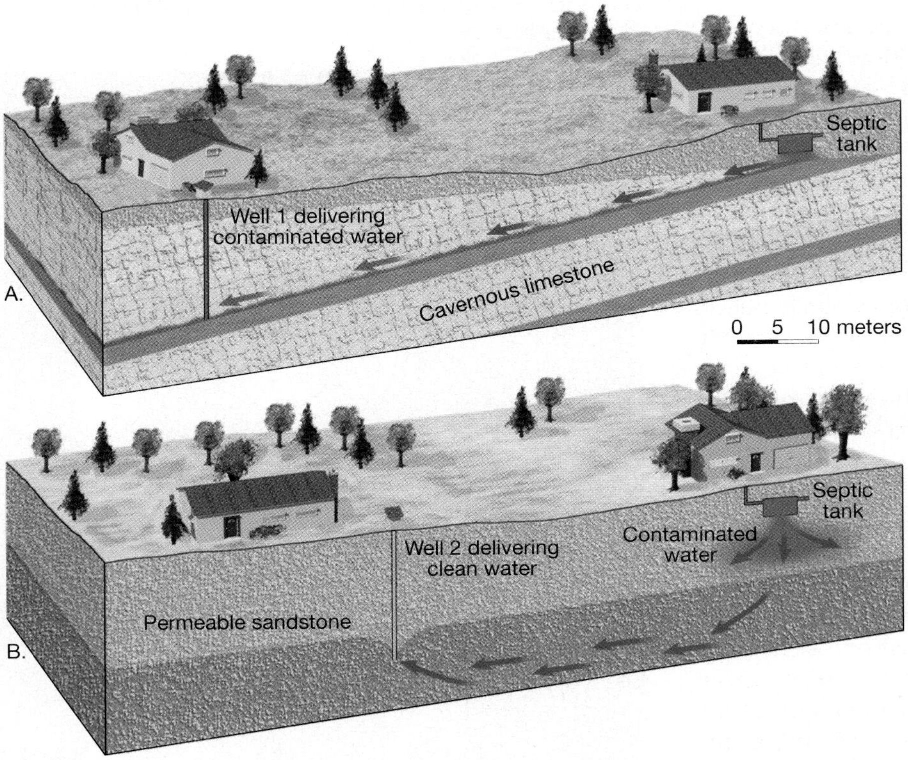

◀ **Figure 10.11** **A.** Although the contaminated water has traveled more than 100 meters before reaching Well 1, the water moves too rapidly through the cavernous limestone to be purified. **B.** As the discharge from the septic tank percolates through the permeable sandstone, it is purified in a relatively short distance.

Hot Springs and Geysers

By definition, the water in **hot springs** is 6–9°C (11–16°F) warmer than the mean annual air temperature for the localities where they occur. In the United States alone, there are well over 1000 such springs.

Temperatures in deep mines and oil wells usually rise with increasing depth, an average of about 2°C per 100 meters (1°F per 100 feet). Therefore, when groundwater circulates at great depths, it becomes heated. If it rises to the surface, the water may emerge as a hot spring. The water of some hot springs in the eastern United States is heated in this manner. The great majority (over 95 percent) of the hot springs (and geysers) in the United States are found in the West. The reason for such a distribution is that the source of heat for most hot springs is cooling igneous rock, and it is in the West that igneous activity has occurred most recently.

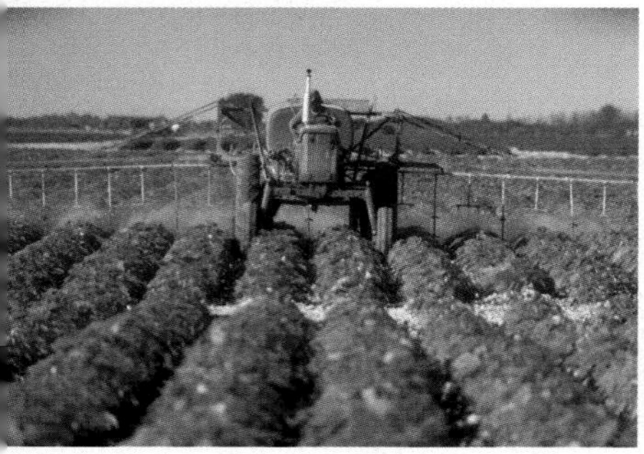

Agricultural chemicals are one source of groundwater contamination.
(R. Morsch/CORBIS)

Geysers are intermittent hot springs or fountains where columns of water are ejected with great force at various intervals, often rising 30 to 60 meters (100 to 200 feet) into the air. After the jet of water ceases, a column of steam rushes out, usually with a thunderous roar. Perhaps the most famous geyser in the world is Old Faithful in Yellowstone National Park, which erupts about once each hour (Figure 10.12). The great abundance, diversity, and spectacular nature of Yellowstone's geysers and other thermal features undoubtedly were the primary reason for its becoming the first national park in the United States. Geysers are also found in other parts of the world, notably New Zealand and Iceland. In fact, the Icelandic word *geysa,* "to gush," gives us the name *geyser*.

Geysers occur where extensive underground chambers exist within hot igneous rocks. How they operate is shown in Figure 10.13. As relatively cool groundwater enters the chambers, it is heated by the surrounding rock. At the bottom of the chambers, the water is under great pressure because of the weight of the overlying water. This great pressure prevents the water from boiling at the normal surface temperature of 100°C (212°F).

Leaking landfills can pollute groundwater.
(F. Rossotto/CORBIS)

For example, water at the bottom of a 300-meter (1000-foot) water-filled chamber must reach nearly 230°C (450°F) to boil. The heating causes the water to expand, with the result that some is forced out at the surface. This loss of water reduces the pressure on the remaining water in the chamber, which lowers the boiling point. A portion of the water deep within the chamber quickly turns to steam, and the geyser erupts (Figure 10.13). Following the eruption, cool groundwater again seeps into the chamber, and the cycle begins anew.

When groundwater from hot springs and geysers flows out at the surface, material in solution is often precipitated, producing an accumulation of chemical sedimentary rock. The material deposited at any given place commonly reflects the chemical makeup of the rock through which the water circulated. When the water contains dissolved silica, a material called *siliceous sinter* or *geyserite* is deposited around the spring. When the water contains dissolved calcium carbonate, a form of limestone called *travertine* or *calcareous tufa* is deposited. The latter term is used if the material is spongy and porous.

The deposits at Mammoth Hot Springs in Yellowstone National Park are more spectacular than most. As the hot water flows upward through a series of channels and then out at the surface, the reduced pressure allows carbon dioxide to separate and escape from the water. The loss of carbon dioxide causes the water to become supersaturated with calcium carbonate, which then precipitates. In addition to containing dissolved silica and calcium carbonate, some hot springs contain sulfur, which gives water a poor taste and unpleasant odor. Undoubtedly, Rotten Egg Spring, Nevada, is such a situation.

Geothermal Energy

Geothermal energy is harnessed by tapping natural underground reservoirs of steam and hot water. These occur where subsurface temperatures are high, owing to relatively recent volcanic activity. Geothermal energy is put to use in two ways: The steam and hot water are used for heating and to generate electricity.

Iceland is a large volcanic island with many active volcanoes (Figure 10.14). In Iceland's capital, Reykjavik,

▼ **Figure 10.12** A wintertime eruption of Old Faithful, one of the world's most famous geysers. It emits as much as 45,000 liters (almost 12,000 gallons) of hot water and steam about once each hour. *(Photo by Marc Muench, David Muench Photography, Inc.)*

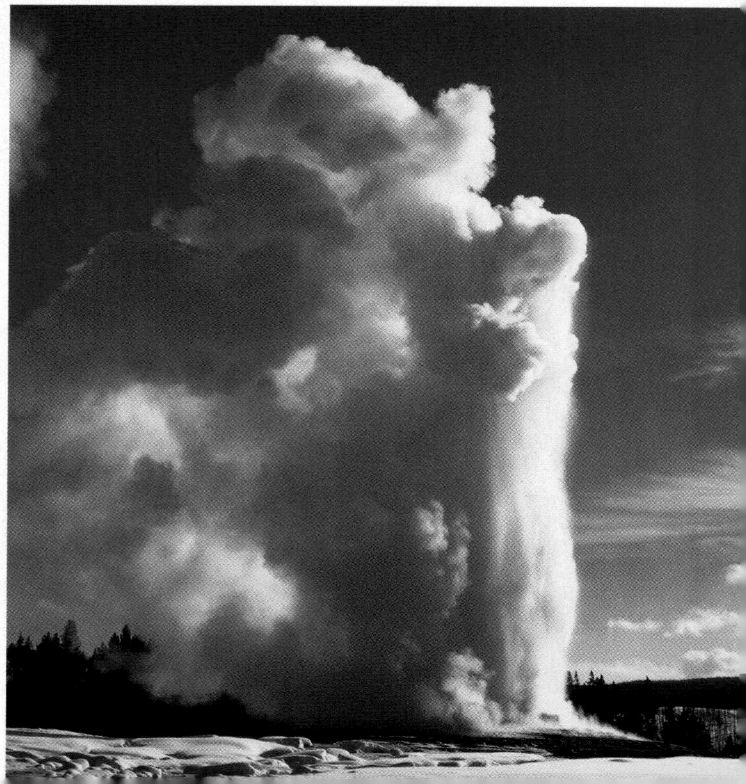

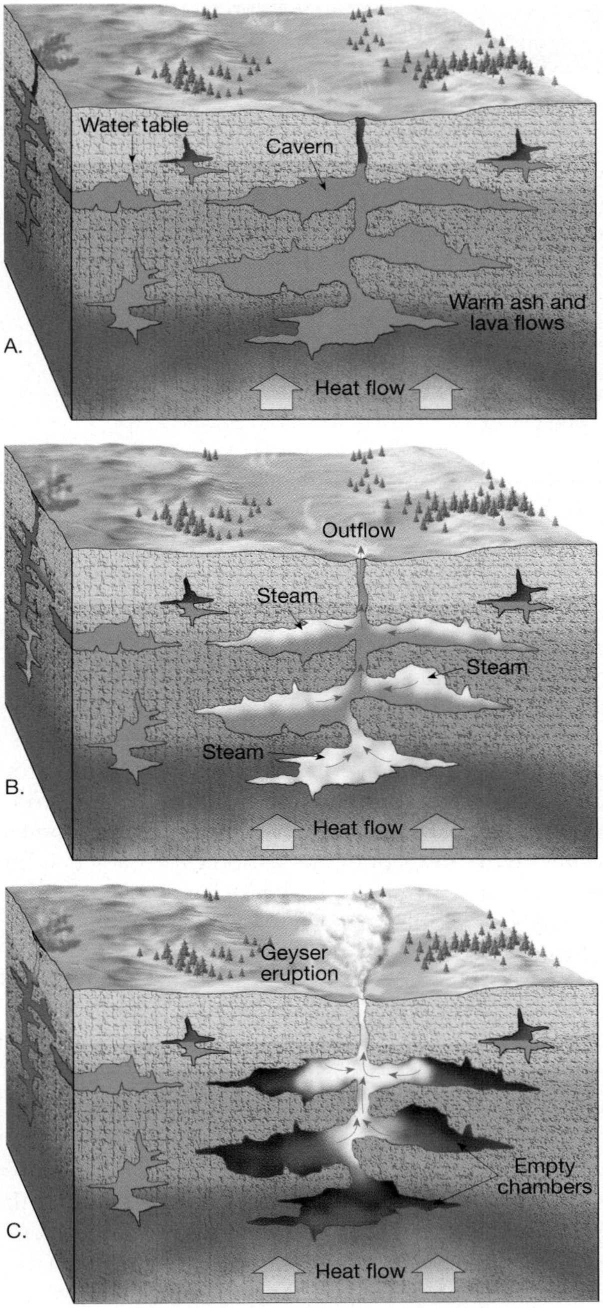

▲ **Figure 10.13** Idealized diagrams of a geyser. A geyser can form if the heat is not distributed by convection. **A.** In this figure, the water near the bottom is heated to near its boiling point. The boiling point is higher there than at the surface because the weight of the water above increases the pressure. **B.** The water higher in the geyser system is also heated; therefore, it expands and flows out at the top, reducing the pressure on the water at the bottom. **C.** At the reduced pressure on the bottom, boiling occurs. Some of the bottom water flashes into steam, and the expanding steam causes an eruption.

hot water is pumped into buildings throughout the city for space heating. (Reykjavik literally means "bay of steam.") It also warms greenhouses, where fruits and vegetables are grown year-round. In the United States, localities in several western states use hot water from geothermal sources for space heating.

As for generating electricity geothermally, the Italians were first to do so in 1904, so the idea is not new. By the turn of the twenty-first century, more than 250 geothermal power plants in 22 countries were producing more than 8000 megawatts (million watts). These plants provide power to more than 60 million people. The leading producers of geothermal power are listed in Table 10.2.

The first commercial geothermal power plant in the United States was built in 1960 at The Geysers, north of San Francisco. The Geysers remains the world's largest geothermal power plant, generating about 1700 megawatts or nearly 60 percent of U.S. geothermal power. In addition to The Geysers, geothermal development is occurring elsewhere in the western United States, including Nevada, Utah, and the Imperial Valley in southern California.

Did You Know?

A spectacular example of subsidence occurred in Mexico City. Heavy pumping from thousands of wells drilled into water-saturated sediments caused portions of the city to subside 6 m (20 ft) or more. Some buildings sunk so much that access from the street now occurs at what used to be the second floor.

What geologic factors favor a geothermal reservoir of commercial value?

1. *A potent source of heat,* such as a large magma chamber deep enough to ensure adequate pressure and slow cooling, yet not so deep that the natural water circulation is inhibited. Such magma chambers are most likely in regions of recent volcanic activity.
2. *Large and porous reservoirs with channels connected to the heat source,* near which water can circulate and then be stored in the reservoir.
3. *A cap of low permeability rocks* that inhibits the flow of water and heat to the surface. A deep, well-insulated reservoir contains much more stored energy than does a similar but uninsulated reservoir.

This hot spring in Iceland is surrounded by a lava field. (Markus Diouhy/ Das Fotoarchiv)

We must recognize that geothermal power is not inexhaustible. When hot fluids are pumped from volcanically heated reservoirs, water cannot be replaced and then heated sufficiently to recharge the reservoir. Experience shows that steam and hot water from individual wells usually lasts no more than 10 to 15 years, so more wells must be drilled to maintain power production. Eventually the field is depleted.

As with other alternative methods of power production, geothermal sources are not expected to provide a high percentage of the world's growing energy needs. Nevertheless, in regions where its potential can be developed, its use will no doubt continue to grow.

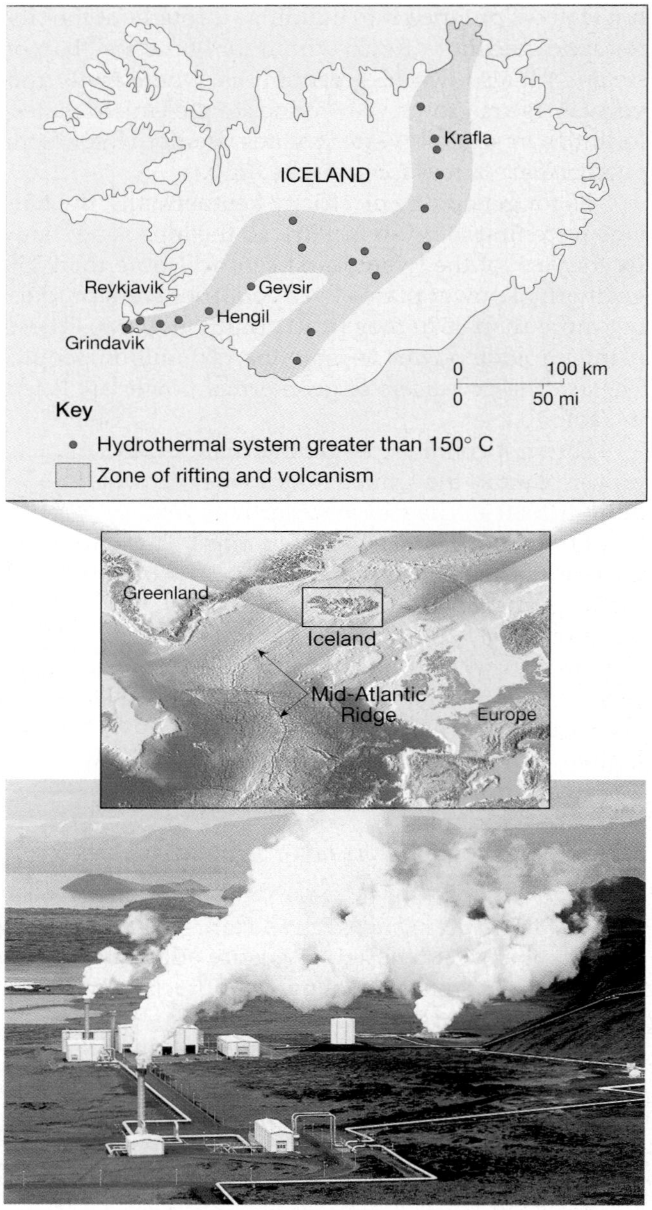

▲ **Figure 10.14** Iceland straddles the Mid-Atlantic Ridge. This divergent plate boundary is the site of numerous active volcanoes and geothermal systems. Because the entire country consists of geologically young volcanic rocks, warm water can be encountered in holes drilled almost anywhere. More than 45 percent of Iceland's energy comes from geothermal sources. The photo shows a power station in southwestern Iceland. The steam is used to generate electricity. Hot (83°C) water from the plant is sent via an insulated pipeline to Reykjavik for space heating. *(Photo by Simon Fraser/Science Photo Library/Photo Researchers, Inc.)*

The Geysers geothermal power plant. (Pacific Gas and Electric)

Table 10.2 Worldwide Geothermal Power Production, 2000

Producing Country	Megawatts
United States	2850
Philippines	1848
Italy	768.5
Mexico	743
Indonesia	589.5
Japan	530
New Zealand	345
Iceland	140
Costa Rica	120
El Salvador	105
All others	178
Total	8217

Source: Geothermal Education Office.

The Geologic Work of Groundwater

Groundwater dissolves rock. This fact is the key to understanding how caverns and sinkholes form. Because soluble rocks, especially limestone, underlie millions of square kilometers of Earth's surface, it is here that groundwater carries on its important role as an erosional agent. Limestone is nearly insoluble in pure water, but it is quite easily dissolved by water containing small quantities of carbonic acid, and most groundwater contains this acid. It forms because rainwater readily dissolves carbon dioxide from the air and from decaying plants. Therefore, when groundwater comes in contact with limestone, the carbonic acid reacts with calcite (calcium carbonate) in the rocks to form calcium bicarbonate, a soluble material that is then carried away in solution.

Caverns

The most spectacular results of groundwater's erosional handiwork are limestone **caverns.** In the United States alone, about 17,000 caves have been discovered, and new ones are being found every year. Although most are modest, some have spectacular dimensions. Mammoth Cave in Kentucky and Carlsbad Caverns in southeastern New Mexico are two famous examples that are both national parks. The Mammoth Cave system is the most extensive in the world, with more than 557 kilometers (345 miles) of interconnected passages. The dimensions at Carlsbad Caverns are impressive in a different way. Here we find the largest and perhaps most spectacular single chamber. The Big Room at Carlsbad Caverns has an area equivalent to 14 football fields and enough height to accommodate the U.S. Capitol Building.

Most caverns are created at or just below the water table in the zone of saturation. Here acidic groundwater follows lines of weakness in the rock, such as joints and bedding planes. As time passes, the dissolving process slowly creates cavities and gradually enlarges them into caverns. The material dissolved by groundwater is carried away and discharged into streams and transported to the ocean.

In many caves, development has occurred at several levels, with the current cavern-forming activity occurring at the lowest elevation. This situation reflects the close relationship between the formation of major subterranean passages and the river valleys into which they drain. As streams cut their valleys deeper, the water table drops as the elevation of the river drops.

Certainly the features that arouse the greatest curiosity for most cavern visitors are the stone formations that give some caverns a wonderland appearance. These are not *erosional* features like the cavern itself, but are depositional features created by the seemingly endless dripping of water over great spans of time. The calcium carbonate that is left behind produces the limestone we call *travertine.* These cave deposits, however, are also commonly called *dripstone,* an obvious reference to their mode of origin. Although the formation of caverns takes place in the zone of saturation, the deposition of dripstone is not possible until the caverns are above the water table in the zone of aeration. As soon as the chamber is filled with air, the stage is set for the decoration phase of cavern building to begin.

The various dripstone features found in caverns are collectively called **speleothems,** no two of which are exactly alike (Figure 10.15). Perhaps the most familiar speleothems are **stalactites.** These icicle-like pendants hang from the ceiling of the cavern and form where water seeps through cracks above. When the water reaches the air in the cave, some of the carbon dioxide in solution escapes from the drop and calcium carbonate precipitates. Deposition occurs as a ring around the edge of the water drop. As drop after drop follows, each leaves an infinitesimal trace of calcite behind, and a hollow limestone tube is created. Water then moves through the tube, remains suspended momentarily at the end, contributes a tiny ring of calcite, and falls to the cavern floor.

The stalactite just described is appropriately called a *soda straw* (Figure 10.16). Often the hollow tube of the soda straw becomes plugged or its supply of water increases. In either case, the water is forced to flow, and hence deposit, along the outside of the tube. As deposition continues, the stalactite takes on the more common conical shape.

Speleothems that form on the floor of a cavern and reach upward toward the ceiling are called **stalagmites.** The water supplying the calcite for stalagmite growth falls from the ceiling and splatters over the surface. As a result, stalagmites do not have a central tube, and they are usually more massive in appearance and more rounded on their upper ends than stalactites. Given enough time, a downward-growing stalactite and an upward-growing stalagmite may join to form a *column.*

Mammoth Hot Springs at Yellowstone National Park. (Stephen Trimble)

Karst Topography

Many areas of the world have landscapes that to a large extent have been shaped by the dissolving power of groundwater. Such areas are said to exhibit **karst topography,** named for the Krs Plateau, located along the northeastern shore of the Adriatic Sea in the border area between Slovenia (formerly a part of Yugoslavia) and Italy where such topography is strikingly developed. In the United States, karst landscapes occur in many areas that are underlain by limestone, including portions of Kentucky, Tennessee, Alabama, southern Indiana, and central and northern Florida (Figure 10.17). Generally, arid and semiarid areas are too dry to develop karst topography. When solution features exist in such regions, they are likely to be remnants of a time when rainier conditions prevailed.

Did You Know?

The largest geyser, if we use the word "large" to mean "tall," is Yellowstone's Steamboat Geyser. During a major eruption, it can spew plumes of water 90 m (300 ft) high for up to 40 minutes. This is followed by a steam phase that features bursts of hot mist that can extend 150 m (500 ft) into the sky.

◀ **Figure 10.15** Speleothems are of many types, including stalactites, stalagmites, and columns. Big Room, Carlsbad Caverns National Park. The large stalagmite is called the Totem Pole. *(Photo by David Muench/David Muench Photography)*

▲ **Figure 10.16** "Live" soda-straw stalactites. Lehman Caves, Great Basin National Park, Nevada. *(Photo by Tom & Susan Bean, Inc.)*

Did You Know?

The U.S. geothermal generating capacity of more than 2800 megawatts is enough to power 3.5 million American homes. It would take about 63 million barrels of oil to produce an equivalent amount of electricity.

Karst areas typically have irregular terrain punctuated with many depressions, called **sinkholes** or **sinks** (Figure 10.18). In the limestone areas of Florida, Kentucky, and southern Indiana, there are literally tens of thousands of these depressions varying in depth from just a meter or two to a maximum of more than 50 meters

Sinkholes commonly form in two ways. Some develop gradually over many years without any physical disturbance to the rock. In these situations, the limestone immediately below the soil is dissolved by downward-seeping rainwater that is freshly charged with carbon dioxide. With time, the bedrock surface is lowered and the fractures into which the water seeps are enlarged. As the fractures grow in size, soil subsides into the widening voids, from which it is removed by groundwater flowing in the passages below. These depressions are usually shallow and have gentle slopes.

By contrast, sinkholes can also form abruptly and without warning when the roof of a cavern collapses under its own weight. Typically, the depressions created in this manner are steep-sided and deep. When they form in populous areas, they may represent a serious geologic hazard. Such a situation is clearly the case in Figure 10.18B and Box 10.2.

In addition to a surface pockmarked by sinkholes, karst regions characteristically show a striking lack of surface drainage (streams). Following a rainfall, the runoff is quickly funneled below ground through the sinks. It then flows through caverns until it finally reaches the water table. Where streams do exist at the surface, their paths are usually short. The names of such streams often give a clue to their fate. In the Mammoth Cave area of Kentucky, for example, there is Sinking Creek, Little Sinking Creek, and Sinking Branch. Other sinkholes become plugged with clay and debris to create small lakes or ponds.

Some regions of karst development exhibit landscapes that look very different from the sinkhole-studded terrain depicted in Figure 10.17. One striking example is an extensive region in southern China that is described as exhibiting *tower karst.* As the image in Figure 10.19 shows, the term *tower* is appropriate because the landscape consists of a maze of isolated steep-sided hills that rise abruptly from the ground. Each is

▶ **Figure 10.17** This diagram shows a hypothetical region with well-developed karst features. Sinkholes are plentiful, and surface streams are funneled below ground. With the passage of time, caverns grow larger and the number and size of sinkholes increase. Collapse of caverns and coalescence of sinkholes may form large, flat-floored depressions. Eventually, the dissolving power of acidic groundwater may remove most of the limestone from the area, leaving only isolated remnants.

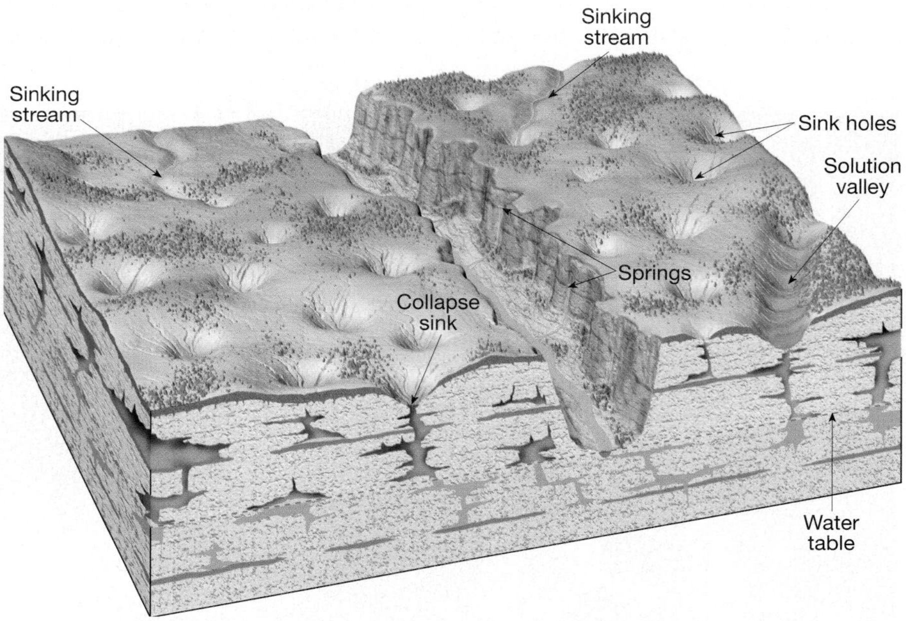

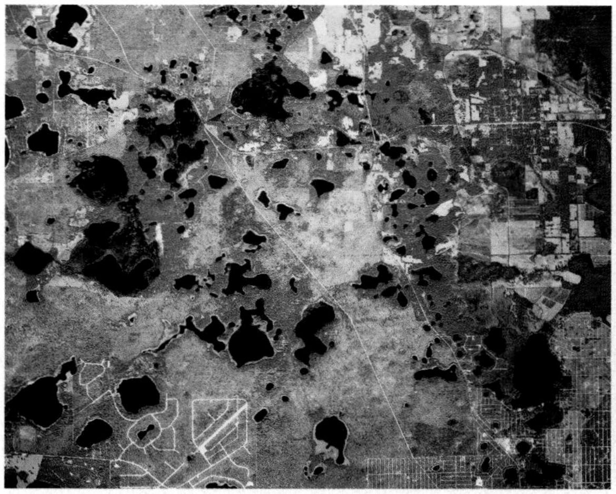

A.

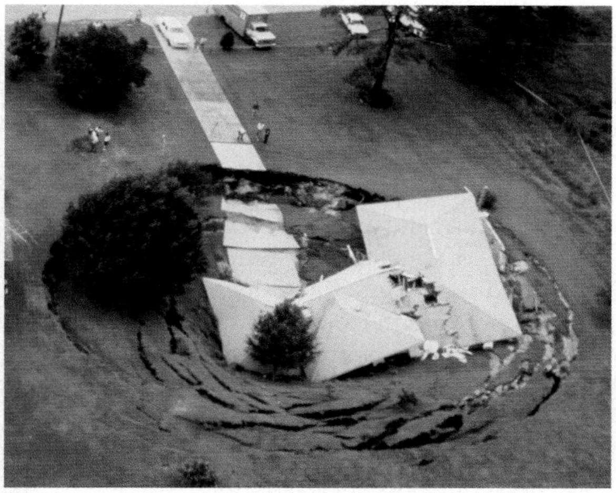

B.

◀ **Figure 10.18** **A.** This high-altitude infrared image shows an area of karst topography in central Florida. The numerous lakes occupy sinkholes. *(Courtesy of USDA-ASCS)* **B.** This small sinkhole formed suddenly in 1991 when the roof of a cavern collapsed, destroying this home in Frostproof, Florida. *(Photo by* St. Petersburg Times/*Getty Images, Inc./Liaison.)*

BOX 10.2

The Case of the Disappearing Lake

Lake Chesterfield was a pleasant 9.3-hectare (23-acre) man-made lake in a quiet suburb of St. Louis, where people living along its shore could fish from their small paddleboats—until it disappeared! In June 2004 residents witnessed the entire lake drain in less than three days (Figure 10.B).

"It was like someone pulled the plug," said Donna Ripp, who lives across the street from the lake, which is now a giant mud hole. Ripp said she began to notice the water level sinking and that by the second day, the lake was half empty. A day later it was completely gone.

What happened? The culprit is clear. At the north end of the lake there is a gaping sinkhole estimated to be about 20 meters (70 feet) in diameter. What geologists are now investigating is what the larger subterranean network looks like. This part of Missouri has many caves, including many that are large enough for humans to explore.

Geologist David Taylor, who inspected the lake shortly after the water drained into the ground, said the sinkhole itself "is really not that big." But it doesn't take a very large sinkhole to knock out an entire lake. Taylor said a hole 0.3 meter (1 foot) in diameter can drain at least 3800 liters (about 1000 gallons) a minute. Taylor is the head of a St. Charles–based company called Strata Services, Inc., that specializes in repairing lakes that are draining into subterranean cavities. "In my business I have fixed hundreds of leaky lakes," he said.

But before Taylor can consider repairing Lake Chesterfield, he and his colleagues first must get a sense of the network of cavities under the lake—a task that he said is exceedingly difficult. "There's all kinds of crazy stuff going on down there," he said. "This is all subsurface work. It's very unpredictable and very difficult."

Taylor found that the subsurface cavity responsible for the sinkhole under Lake Chesterfield runs laterally underground for several kilometers. A tracing dye placed near the sinkhole reemerges in a spring about 5.5 kilometers (3.5 miles) from the lake. In order to develop a better picture of the subterranean cavities, Taylor drilled five test holes at 12-meter (40-foot) intervals, finding that two revealed empty cavities below. But he estimated that it would require 600 holes in a 12-meter grid to even begin to understand the region completely.

Once that picture emerges, Taylor's company then fills the cavities with a cementlike substance so that other sinkholes don't open and create a similar problem. "If we just put a Band-Aid over the hole and fill the lake back up, the same thing will happen again," he said.

In the meantime, nearby residents are getting a crash course on karst topography. "I didn't even know there were underground caves here until all this happened." Donna Ripp said.

Figure 10.B **Residents examine emptied Lake Chesterfield, a 23-acre reservoir that drained in three days when a sinkhole opened beneath it.** (Photo by Hillary Levin/*St. Louis Post-Dispatch*)

riddled with interconnected caves and passageways. This type of karst topography forms in wet tropical and subtropical regions having thick beds of highly jointed limestone. Here groundwater has dissolved large volumes of limestone, leaving only these residual towers. Karst development is more rapid in tropical climates due to the abundant rainfall and the greater availability of carbon dioxide from the decay of lush tropical vegetation. The extra carbon dioxide in the soil means there is more carbonic acid for dissolving limestone. Other tropical areas of advanced karst development include portions of Puerto Rico, western Cuba, and northern Vietnam.

Did You Know?

America's largest bat colonies are found in caves. For example, Braken Cave in central Texas is the summer home of 20 million Mexican free-tail bats. They spend their days in total darkness more than 3 km inside the cave. Each night they leave the cave to feed, consuming more than 200,000 kg (220 tons) of insects.

▲ **Figure 10.19** One of the best-known and most distinctive regions of tower karst development is the Guilin District of southeastern China. *(Photo by Jose Fuste Raga/CORBIS/The Stock Market)*

The Chapter in Review

- As a resource, *groundwater* represents the largest reservoir of fresh water that is readily available to humans. Geologically, the dissolving action of groundwater produces *caves* and *sinkholes.* Groundwater is also an equalizer of streamflow.
- Groundwater is water that completely fills the pore spaces in sediment and rock in the subsurface *zone of saturation.* The upper limit of this zone is the *water table.* The *zone of aeration* is above the water table where the soil, sediment, and rock are not saturated.
- The interaction between streams and groundwater takes place in one of three ways: Streams gain water from the inflow of groundwater (*gaining stream*); they lose water through the streambed to the groundwater system (*losing stream*); or they do both, gaining in some sections and losing in others.
- The quantity of water that can be stored in a material depends upon its *porosity* (the volume of open spaces). The *permeability* (the ability to transmit a fluid through interconnected pore spaces) of a material is a very important factor controlling the movement of groundwater.
- Materials with very small pore spaces (such as clay) hinder or prevent groundwater movement and are called *aquitards. Aquifers* consist of materials with larger pore spaces (such as sand) that are permeable and transmit groundwater freely.
- Groundwater moves in looping curves that are a compromise between the downward pull of gravity and the tendency of water to move toward areas of reduced pressure.
- *Springs* occur whenever the water table intersects the land surface and a natural flow of groundwater results. *Wells,* openings bored into the zone of saturation, withdraw groundwater and create roughly conical depressions in the water table known as *cones of depression. Artesian wells* occur when water rises above the level at which it was initially encountered.
- When groundwater circulates at great depths, it becomes heated. If it rises, the water may emerge as a *hot spring.* Geysers occur when groundwater is heated in underground chambers, expands, and some water quickly changes to steam, causing the geyser to erupt. The source of heat for most hot springs and geysers is hot igneous rock. *Geothermal energy* is harnessed by tapping natural underground reservoirs of steam and hot water.
- Some of the current environmental problems involving groundwater include (1) *overuse* by intense irrigation, (2) *land subsidence* caused by groundwater withdrawal, and (3) contamination by pollutants.
- Most *caverns* form in limestone at or below the water table when acidic groundwater dissolves rock along lines of

weakness, such as joints and bedding planes. The various *dripstone* features found in caverns are collectively called *speleothems.* Landscapes that to a large extent have been shaped by the dissolving power of groundwater exhibit *karst topography,* an irregular terrain punctuated with many depressions, called *sinkholes* or *sinks.*

Key Terms

aquifer (p. 228)
aquitard (p. 228)
artesian (p. 230)
cavern (p. 236)
cone of depression (p. 229)
drawdown (p. 229)
gaining stream (p. 226)
geothermal energy (p. 234)
geyser (p. 234)
groundwater (p. 225)
hot spring (p. 234)
karst topography (p. 237)
losing stream (p. 226)
perched water table (p. 228)
permeability (p. 228)
porosity (p. 227)
sinkhole (sink) (p. 238)
speleothem (p. 237)
spring (p. 228)
stalactite (p. 237)
stalagmite (p. 237)
water table (p. 225)
well (p. 229)
zone of aeration (p. 225)
zone of saturation (p. 225)

Questions for Review

1. What percentage of fresh water is groundwater? If glacial ice is excluded and only liquid fresh water is considered, about what percentage is groundwater?
2. Geologically, groundwater is important as an erosional agent. Name another significant geological role of groundwater.
3. Define *groundwater,* and relate it to the water table.
4. Contrast a *gaining stream* and a *losing stream.*
5. Distinguish between *porosity* and *permeability.*
6. What is the difference between an *aquitard* and an *aquifer*?
7. Under what circumstances can a material have a high porosity but not be a good aquifer?
8. When an aquitard is situated above the main water table, a localized saturated zone may be created. What term is applied to such a situation?
9. What is meant by the term *artesian*? List two conditions that must be present for artesian wells to exist.
10. What problem is associated with the pumping of groundwater for irrigation in the southern part of the High Plains?
11. Briefly explain what happened in the San Joaquin Valley of California as the result of excessive groundwater withdrawal.
12. Which aquifer would be most effective in purifying polluted groundwater: coarse gravel, sand, or cavernous limestone?
13. What is meant when a groundwater pollutant is classified as hazardous?
14. What is the source of heat for most hot springs and geysers? How is this reflected in the distribution of these features?
15. Is geothermal power considered an inexhaustible energy source?
16. Name two common speleothems and distinguish between them.
17. Areas whose landscapes are largely a reflection of the erosional work of groundwater are said to exhibit what kind of topography?
18. Describe two ways in which sinkholes are created.

Online Study Guide

The *Essentials of Geology* Web site uses the resources and flexibility of the Internet to aid in your study of the topics in this chapter. Written and developed by geology instructors, this site will help improve your understanding of geology. Visit **www.prenhall.com/lutgens** and click on the cover of *Essentials of Geology 9e* to find:

- Online review quizzes.
- Critical thinking exercises.
- Links to chapter-specific Web resources.
- Internet-wide key-term searches.

http://www.prenhall.com/lutgens

Alsek Glacier, St. Elias Mountains, Glacier Bay National Park, Alaska. *(Copyright © by Carr Clifton. All rights reserved.)*

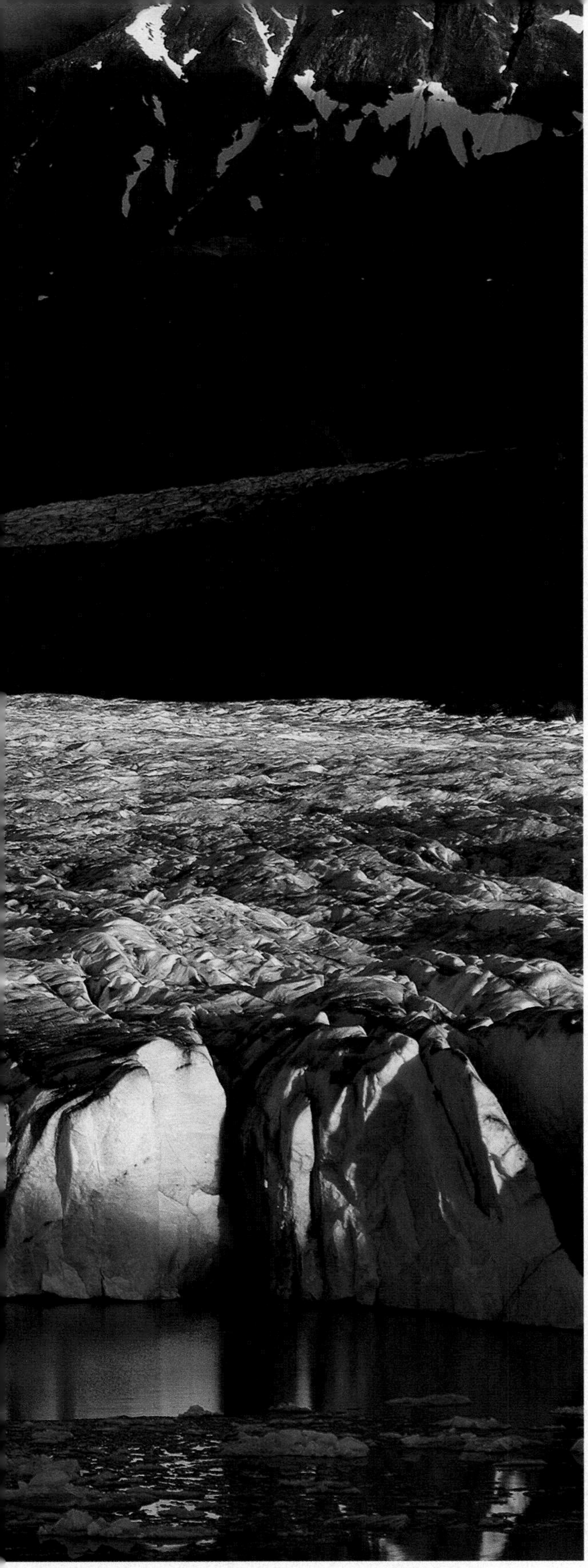

CHAPTER 11

Glaciers and Glaciation

Focus on Learning

To assist you in learning the important concepts in this chapter, you will find it helpful to focus on the following questions:

- What is a glacier? What are the different types of glaciers? Where are glaciers located?
- How do glaciers move, and what are the various processes of glacial erosion?
- What are the features created by glacial erosion and deposition? What materials make up the depositional features?
- What is the evidence for the Ice Age? What are some indirect effects of Ice Age glaciers?
- What are some proposals that attempt to explain the causes of glacial ages?

Cape Cod consists largely of glacial deposits. (NASA)

Climate has a strong influence on the nature and intensity of Earth's external processes. This fact is dramatically illustrated in this chapter because the existence and extent of glaciers is largely controlled by Earth's changing climate.

Like the running water and groundwater that were the focus of the preceding two chapters, glaciers represent a significant erosional process. These moving masses of ice are responsible for creating many unique landforms and are part of an important link in the rock cycle in which the products of weathering are transported and deposited as sediment.

Today, glaciers cover nearly 10 percent of Earth's land surface. However, in the recent geologic past, ice sheets were three times more extensive, covering vast areas with ice thousands of meters thick. Many regions still bear the mark of these glaciers. The basic character of such diverse places as the Alps, Cape Cod, and Yosemite Valley was fashioned by now vanished masses of glacial ice. Moreover, Long Island, the Great Lakes, and the fiords of Norway and Alaska all owe their existence to glaciers. Glaciers, of course, are not just a phenomenon of the geologic past. As you will see, they are still sculpting and depositing debris in many regions today (Figure 11.1).

Did You Know?

Approximately 160,000 glaciers presently occupy Earth's polar regions and high mountain environments. Today's glaciers cover only about one-third the area that was covered at the height of the most recent ice age.

Glaciers: A Part of Two Basic Cycles

GEODe Essentials of Geology — Glaciers and Glaciation ▼ Introduction

Glaciers are a part of two fundamental cycles in the Earth system—the hydrologic cycle and the rock cycle. Earlier you learned that the water of the hydrosphere is constantly cycled through the atmosphere, biosphere, and geosphere. Time and time again water is evaporated from the oceans into the atmosphere, precipitated upon the land, and carried by rivers and underground flow back to the sea. However, when precipitation falls at high elevations or high latitudes, the water may not immediately make its way toward the sea. Instead, it may become part of a glacier. Although the ice will eventually melt, allowing the water to continue its path to the sea, water can be stored as glacial ice for many tens, hundreds, or even thousands of years.

A **glacier** is a thick ice mass that forms over hundreds or thousands of years. It originates on land from the accumulation, compaction, and recrystallization of snow. A glacier appears to be motionless, but it is not—glaciers move very slowly. Like running water, groundwater, wind, and waves, glaciers are dynamic erosional agents that accumulate, transport, and deposit sediment. As such, glaciers are among the processes that perform a very basic function in the rock cycle. Although glaciers are found in many parts of the world

▼ **Figure 11.1** Aerial view of Biafo Glacier, a valley glacier in the mountains of Pakistan. Northern Pakistan contains the greatest concentration of glacial ice in Asia. *(Photo by Galen Rowell/Mountain Light Photography, Inc.)*

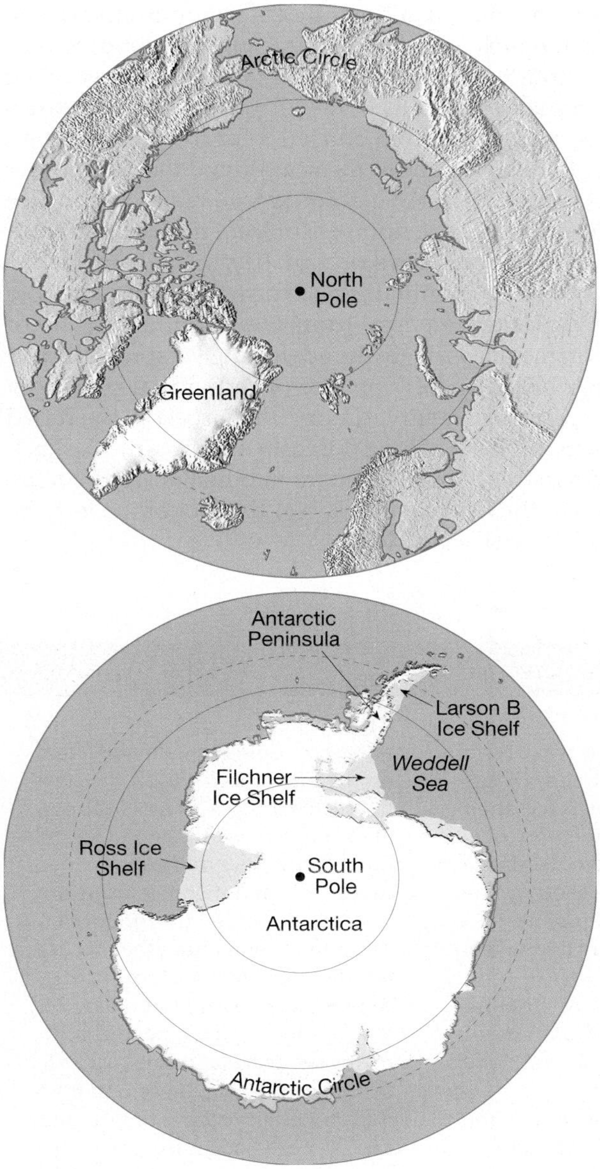

▲ **Figure 11.2** The only present-day continental ice sheets are those covering Greenland and Antarctica. Their combined areas represent almost 10 percent of Earth's land area. Greenland's ice sheet occupies 1.7 million square kilometers, or about 80 percent of the island. The area of the Antarctic Ice Sheet is almost 14 million square kilometers. Ice shelves occupy an additional 1.4 million square kilometers adjacent to the Antarctic Ice Sheet.

today, most are located in remote areas, either near Earth's poles or in high mountains.

Valley (Alpine) Glaciers

Literally thousands of relatively small glaciers exist in lofty mountain areas, where they usually follow valleys that were originally occupied by streams. Unlike the rivers that previously flowed in these valleys, the glaciers advance slowly, perhaps only a few centimeters per day. Because of their setting, these moving ice masses are termed **valley glaciers** or **alpine glaciers** (Figure 11.1). Each glacier is a stream of ice, bounded by precipitous rock walls, that flows downvalley from a snow accumulation center near its head. Like rivers, valley glaciers can be long or short, wide or narrow, single or with branching tributaries. Generally, the widths of alpine glaciers are small compared to their lengths. Some extend for just a fraction of a kilometer, whereas others go on for many tens of kilometers. The west branch of the Hubbard Glacier, for example, runs through 112 kilometers (nearly 70 miles) of mountainous terrain in Alaska and the Yukon Territory.

Ice Sheets

In contrast to valley glaciers, **ice sheets** exist on a much larger scale. Although many ice sheets have existed in the past, just two achieve this status at present (Figure 11.2). In the Northern Hemisphere, Greenland is covered by an imposing ice sheet that occupies 1.7 million square kilometers (0.7 million square miles), or about 80 percent of this large island. Averaging nearly 1500 meters (5000 feet) thick, in places the ice extends 3000 meters (10,000 feet) above the island's bedrock floor.

Most U.S. glaciers are located in Alaska. (Michael Collier)

In the Southern Hemisphere, the huge Antarctic Ice Sheet attains a maximum thickness of nearly 4300 meters (14,000 feet) and covers an area of more than 13.6 million square kilometers (5.3 million square miles). Because of the proportions of these huge features, they are often called *continental ice sheets*. Indeed, the combined areas of present-day continental ice sheets represent almost 10 percent of Earth's land area.

These enormous masses flow out in all directions from one or more snow-accumulation centers and completely obscure all but the highest areas of underlying terrain. Even sharp variations in the topography beneath the glacier usually appear as relatively subdued undulations on the surface of the ice. Such topographic differences, however, do affect the behavior of the ice sheets, especially near their margins, by guiding flow in certain directions and creating zones of faster and slower movement.

Along portions of the Antarctic coast, glacial ice flows into the adjacent ocean, creating features called **ice shelves.** They are large, relatively flat masses of floating ice that extend seaward from the coast but remain attached to the land along one or more sides. The shelves are thickest on their landward sides, and they become thinner seaward. They are sustained by ice from the adjacent ice sheet as well as being nourished by snowfall and the freezing of seawater to their bases. Antarctica's ice shelves extend over approximately 1.4 million square kilo-

Did You Know?

The total volume of all valley glaciers is estimated to be about 210,000 km^3 (50,400 mi^3). This is comparable to the combined volume of all the world's fresh water and saline lakes.

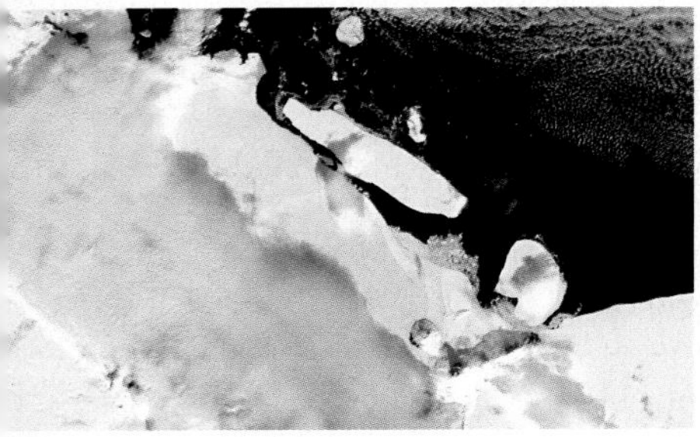

This huge iceberg broke from an Antarctic ice shelf. It covers an area of 3000 km^2 (1200 mi^2). (NASA)

meters (0.6 million square miles). The Ross and Filchner ice shelves are the largest, with the Ross Ice Shelf alone covering an area approximately the size of Texas (Figure 11.2). In recent years, satellite monitoring has shown that some ice shelves are breaking apart. Box 11.1 explores this topic.

Other Types of Glaciers

In addition to valley glaciers and ice sheets, other types of glaciers are also identified. Covering some uplands and plateaus are masses of glacial ice called **ice caps.** Like ice sheets, ice caps completely bury the underlying landscape but are much smaller than the continental-scale features. Ice caps occur in many places, including Iceland and several of the large islands in the Arctic Ocean.

Often ice caps and ice sheets feed **outlet glaciers.** These tongues of ice flow down valleys extending outward from the margins of these larger ice masses. The tongues are essentially valley glaciers that are avenues for ice movement from an ice cap or ice sheet through mountainous terrain to the sea. Where they encounter the ocean, some outlet glaciers spread out as floating ice shelves. Often large numbers of icebergs are produced.

Piedmont glaciers occupy broad lowlands at the bases of steep mountains and form when one or more valley glaciers emerge from the confining walls of mountain valleys. Here the advancing ice spreads out to form a broad sheet. The size of individual piedmont glaciers varies greatly. Among the largest is the broad Malaspina Glacier along the coast of southern Alaska. It covers more than 5000 square kilometers (2000 square miles) of the flat coastal plain at the foot of the lofty St. Elias Range (Figure 11.3).

BOX 11.1

The Collapse of Antarctic Ice Shelves

Studies using recent satellite imagery show that portions of some ice shelves are breaking apart. For example, during a 35-day span in February and March 2002 an ice shelf on the eastern side of the Antarctic Peninsula, known as the Larsen B ice shelf, broke apart and separated from the continent (Figure 11.A). The event sent thousands of icebergs adrift in the adjacent Weddell Sea (see Figure 11.2). In all, about 3250 square kilometers (1250 square miles) of the ice shelf broke apart. (For reference, the entire state of Rhode Island covers 2717 square kilometers.) This was not an isolated event but part of a trend. Over a five-year span, the Larsen B ice shelf shrunk by 5700 square kilometers (more than 2200 square miles). Moreover, since 1974 the extent of seven ice shelves surrounding the Antarctic Peninsula declined by about 13,500 square kilometers (nearly 5300 square miles).

Figure 11.A This satellite image shows the Larsen B ice shelf during its collapse in early 2002. (Image courtesy of NASA)

Why did these masses of floating ice break apart? What if the trend continues? Could there be any serious consequences?

Scientists attribute the breakup of the ice shelves to a strong regional climate warming. Since about 1950, Antarctic temperatures have risen by 2.5°C (4.3°F). The rate of warming has been about 0.5°C (nearly 1°F) per decade. If temperatures continue to rise, an ice shelf adjacent to Larsen B may start to recede in coming decades. Moreover, regional warming of just a few degrees Celsius may be sufficient to cause portions of the huge Ross Ice Shelf to become unstable and begin to break apart (See Figure 11.2).

What might the consequences be? Scientists at the National Snow and Ice Data Center (NSIDC) suggest the following:

> While the breakup of the ice shelves in the Peninsula has little consequence for sea level rise, the breakup of other shelves in the Antarctic could have a major effect on the rate of ice flow off the continent. Ice shelves act as a buttress, or braking system, for glaciers. Further, the shelves keep warmer marine air at a distance from the glaciers; therefore, they moderate the amount of melting that occurs on the glaciers' surfaces. Once their ice shelves are removed, the glaciers increase in speed due to meltwater percolation and/or a reduction of braking forces, and they may begin to dump more ice into the ocean. Glacier ice speed increases are already observed in Peninsula areas where ice shelves disintegrated in prior years.*

The addition of large quantities of glacial ice to the ocean could indeed cause a significant rise in sea level.

Remember that what is being suggested here is still speculative because our knowledge of the dynamics of Antarctica's ice shelves and glaciers is incomplete. Additional satellite monitoring and field studies will be necessary if we are to more accurately predict potential rises in global sea level triggered by the mechanism described here.

*National Snow and Ice Data Center, "Antarctic Ice-Shelf Collapses," 21 March 2002. **http://nsidc.org/iceshelves/larsenb2002.**

▲ **Figure 11.3** Malaspina glacier in southeastern Alaska is a classic piedmont glacier. Piedmont glaciers occur where valley glaciers exit a mountain range onto broad lowlands, are no longer laterally confined, and spread to become wide lobes. Malaspina Glacier is actually a compound glacier, formed by the merger of several valley glaciers, the most prominent of which seen here are Agassiz Glacier (left) and Seward Glacier (right). In total, Malaspina Glacier is up to 65 km (40 mi) wide and extends up to 45 km (28 mi) from the mountain front nearly to the sea. *(Image from NASA/JPL)*

How Glaciers Move

GEODe: Glaciers and Glaciation
▼ Budget of a Glacier

The movement of glacial ice is generally referred to as *flow*. The fact that glacial movement is described in this way seems paradoxical—how can a solid flow? The way in which ice flows is complex and is of two basic types. The first of these, *plastic flow*, involves movement *within* the ice. Ice behaves as a brittle solid until the pressure upon it is equivalent to the weight of about 50 meters (165 feet) of ice. Once that load is surpassed, ice behaves as a plastic material, and flow begins. A second and often equally important mechanism of glacial movement consists of the entire ice mass slipping along the ground. The lowest portions of most glaciers are thought to move by this sliding process called *basal slip* (Figure 11.4).

The upper 50 meters or so of a glacier is not under sufficient pressure to exhibit plastic flow. Rather, the ice in this uppermost zone is brittle and is appropriately referred to as the *zone of fracture*. The ice in this zone is carried along "piggyback" style by the ice below. When the glacier moves over irregular terrain, the zone of fracture is subjected to tension, resulting in cracks called **crevasses** (Figure 11.5). These gaping cracks, which often make travel across glaciers dangerous, may extend to depths of 50 meters (165 feet). Beyond this depth, plastic flow seals them off.

Rates of Glacial Movement

Unlike streamflow, glacial movement is not obvious. If we could watch an alpine glacier move, we would see that like the water in a river, all of the ice in the valley does not move downvalley at an equal rate. Just as friction with the bedrock floor slows the movement of the ice at the bottom of the glacier, the drag created by the valley walls leads to the flow being greatest in the center of the glacier. This was first demonstrated by experiments during the nineteenth century, in which markers were carefully placed in a straight line across the top of a valley glacier. Periodically, the positions of the stakes were recorded, revealing the type of movement just described. More about these experiments may be found in Box 1.1 on page 8.

How rapidly does glacial ice move? Average velocities vary considerably from one glacier to another. Some move so slowly that trees and other vegetation may become well established in the debris that has accumulated on the glacier's surface, whereas others may move at rates of up to several meters per day. For example, Byrd Glacier, an outlet glacier in Antarctica that was the subject of a 10-year study using satellite images, moved at an average rate of 750 to 800 meters per year (about 2 meters per day). Other glaciers in the study advanced at one-fourth that rate.

The advance of some glaciers is characterized by periods of extremely rapid movements called *surges*. Glaciers that exhibit such movement may flow along in an apparently normal manner, then speed up for a relatively short time before returning to the normal rate again. The flow rates during surges are as much as 100 times the normal rate.

Satellite image of Iceland's Vantnajükull ice cap. (NASA)

Did You Know?

The Antarctic ice sheet represents 80 percent of the world's ice and nearly two-thirds of Earth's fresh water.

Budget of a Glacier

Snow is the raw material from which glacial ice originates; therefore, glaciers form in areas where more snow falls in winter than melts during the summer. Glaciers are constantly gaining and losing ice. Snow accumulation and ice formation occur in the **zone of accumulation.** Its outer limits are defined by the *snowline.* The elevation of the snowline varies greatly. In polar regions, it may be sea level, whereas in tropical areas, the snowline exists only high in mountain areas, often at altitudes exceeding 4500 meters (15,000 feet). Above the snowline, in the zone of accumulation, the addition of snow thickens the glacier and promotes movement. Below the snowline is the **zone of wastage.** Here there is a net loss to the glacier as all of the snow from the previous winter melts, as does some of the glacial ice (Figure 11.6).

In addition to melting, glaciers also waste as large pieces of ice break off the front of the glacier in a process called **calving.** Calving creates *icebergs* in places where the glacier has reached the sea or a lake (Figure 11.7). Because icebergs are just slightly less dense than seawater, they float very low in the water, with about 90 percent of their mass submerged. Along the margins of Antarctica's ice shelves, calving is the primary means by which these masses lose ice. The relatively flat icebergs produced here can be several kilometers across and 600 meters thick. By comparison, thousands of irregularly shaped icebergs are produced by outlet glaciers flowing from the margins of the Greenland Ice Sheet. Many drift southward and find their way into the North Atlantic, where they can pose a hazard to navigation.

Whether the margin of a glacier is advancing, retreating, or remaining stationary depends on the budget of the glacier. The **glacial budget** is the balance, or lack of balance, between accumulation at the upper end of the glacier and loss at the lower end. This loss is termed **ablation.** If ice accumulation exceeds ablation, the glacial front advances until the two factors balance. When this happens, the terminus of the glacier is stationary.

If a warming trend increases ablation and/or if a drop in snowfall decreases accumulation, the ice front will retreat. As the terminus of the glacier retreats, the extent of the zone of wastage diminishes. Therefore, in time a new balance will be reached between accumulation and wastage, and the ice front will again become stationary.

Whether the margin of a glacier is advancing, retreating, or stationary, the ice within the glacier continues to flow forward. In the case of a receding

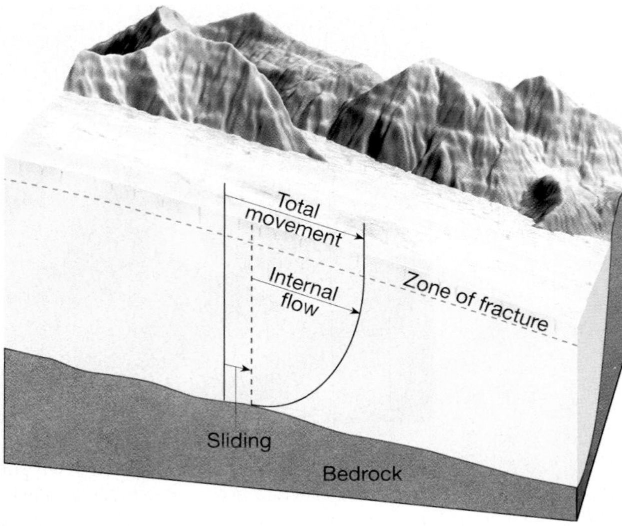

▲ **Figure 11.4** Vertical cross-section through a glacier to illustrate ice movement. Glacial movement is divided into two components. Below about 50 m (160 ft), ice behaves plastically and flows. In addition, the entire mass of ice may slide along the ground. The ice in the zone of fracture is carried along "piggyback" style. Notice that the rate of movement is slowest at the base of the glacier, where frictional drag is greatest.

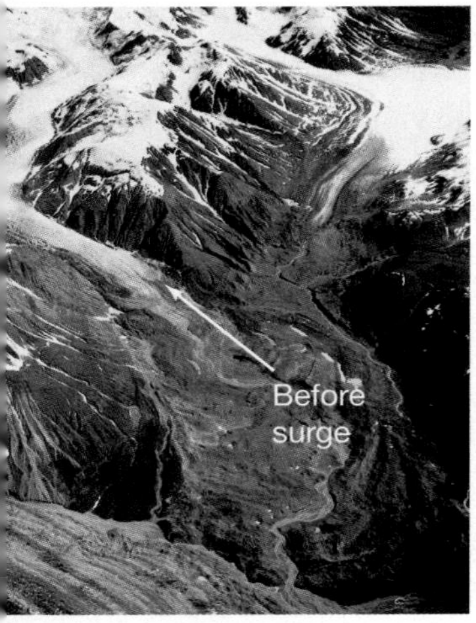

August 1964

Surge of Variegated Glacier—"before." (Austin Post/USGS)

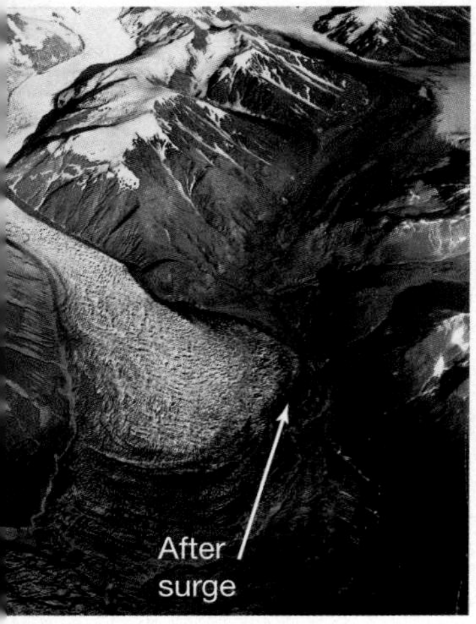

August 1965

Surge of Variegated Glacier—"after." (Austin Post/USGS)

▶ **Figure 11.5** Hikers challenged by a large crevasse on Peru's Mt. Huascaran. Crevasses form in the brittle ice of the zone of fracture. They do not continue down into the zone of flow. *(Photo by Galen Rowell/Mountain Light Photography, Inc.)*

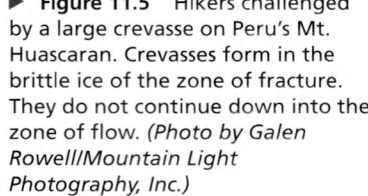

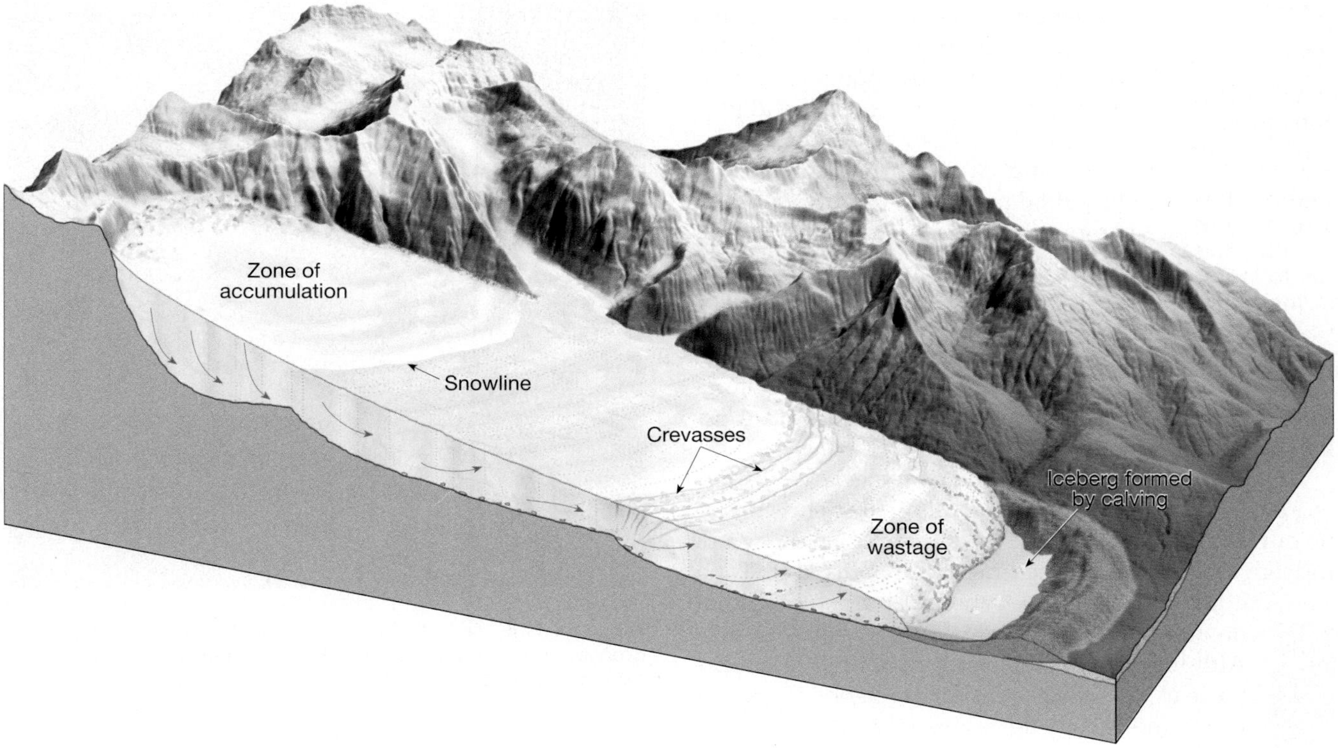

▲ **Figure 11.6** The snowline separates the zone of accumulation and the zone of wastage. Above the snowline, more snow falls each winter than melts each summer. Below the snowline, the snow from the previous winter completely melts as does some of the underlying ice. Whether the margin of a glacier advances, retreats, or remains stationary depends on the balance between accumulation and wastage (ablation). When a glacier moves across irregular terrain, *crevasses* form in the brittle upper portion.

glacier, the ice still flows forward, but not rapidly enough to offset ablation. This point is illustrated well in Figure 1.A on page 8. As the line of stakes within the Rhone Glacier continued to move downvalley, the terminus of the glacier slowly retreated upvalley.

Glacial Erosion

Glaciers are capable of great erosion. For anyone who has observed the terminus of an alpine glacier, the evidence of its erosive force is clear. You can witness

◀ **Figure 11.7** Icebergs are created when large pieces calve from the front of a glacier after it reaches a water body. Here ice is calving from the terminus of Alaska's Hubbard Glacier in Wrangell–St. Elias National Park, Alaska. Only about 10 percent of an iceberg protrudes above the water line. *(Photo by Tom & Susan Bean, Inc.)*

Did You Know?

Because Iceland has active volcanoes but is also a place with abundant glaciers, it is often called the "land of fire and ice." When volcanoes buried beneath the island's ice caps erupt, they trigger rapid melting that can cause major floods. Because the problem is so common here, there is an Icelandic word, *jükulhlaup* ("glacier flood"), to refer to the phenomenon.

firsthand the release of rock fragments of various sizes from the ice as it melts. All signs lead to the conclusion that the ice has scraped, scoured, and torn rock from the floor and walls of the valley and carried it downvalley. It should be pointed out, however, that in mountainous regions mass-wasting processes also make substantial contributions to the sediment load of a glacier. A glance back at Figure 8.7 on page 191 provides a striking example. The image shows a 4-kilometer-long tongue of rock avalanche debris deposited atop Alaska's Sherman Glacier.

Once rock debris is acquired by a glacier, the enormous competency of ice will not allow the debris to settle out like the load carried by a stream or by the wind. Indeed, as a medium of sediment transport, ice has no equal. Glaciers can transport huge blocks that no other erosional agent could possibly budge. Although today's glaciers are of limited importance as erosional agents, many landscapes that were modified by the widespread glaciers of the most recent Ice Age still reflect to a high degree the work of ice.

The red boat is next to an iceberg in Canada's Labrador Sea. (Yva Momatiuk/John Eastcott)

Glaciers erode the land primarily in two ways—*plucking* and *abrasion.* First, as a glacier flows over a fractured bedrock surface, it loosens and lifts blocks of rock and incorporates them into the ice. This process, known as **plucking,** occurs when meltwater penetrates the cracks and joints of bedrock beneath a glacier and freezes. As the water expands, it exerts tremendous leverage that pries the rock loose. In this manner sediment of all sizes, ranging from particles as fine as flour to blocks as big as houses, becomes part of the glacier's load.

The second major erosional process is **abrasion.** As the ice and its load of rock fragments slide over bedrock, they function like sandpaper to smooth and polish the surface below. The pulverized rock produced by the glacial gristmill is appropriately called **rock flour.** So much rock flour may be produced that meltwater streams flowing out of a glacier often have the grayish appearance of skim milk and offer visible evidence of the grinding power of ice.

Glacially transported boulder.

When the ice at the bottom of a glacier contains large fragments of rock, long scratches and grooves called **glacial striations** may be gouged into the bedrock (Figure 11.8). These linear grooves provide clues to the direction of ice flow. By mapping the striations over large areas, patterns of glacial flow can often be reconstructed.

In contrast, not all abrasive action produces striations. The rock surfaces over which the glacier moves may also become highly polished by the ice and its load of finer particles. The broad expanses of smoothly polished granite in Yosemite National Park provide an excellent example.

▲ **Figure 11.8** Glacial abrasion created the scratches and grooves in this bedrock. Glacier Bay National Park, Alaska. *(Photo by Carr Clifton)*

As is the case with other agents of erosion, the rate of glacial erosion is highly variable. This differential erosion by ice is largely controlled by four factors: (1) rate of glacial movement; (2) thickness of the ice; (3) shape, abundance, and hardness of the rock fragments contained in the ice at the base of the glacier; and (4) the erodibility of the surface beneath the glacier. Variations in any or all of these factors from time to time and/or from place to place mean that the features, effects, and degree of landscape modification in glaciated regions can vary greatly.

Landforms Created by Glacial Erosion

GEODe ESSENTIALS OF GEOLOGY Glaciers and Glaciation
▼ Reviewing Glacial Features

The erosional effects of valley glaciers and ice sheets are quite different. A visitor to a glaciated mountain region is likely to see a sharp and angular topography. The reason is that as alpine glaciers move downvalley, they tend to accentuate the irregularities of the mountain landscape by creating steeper canyon walls and making

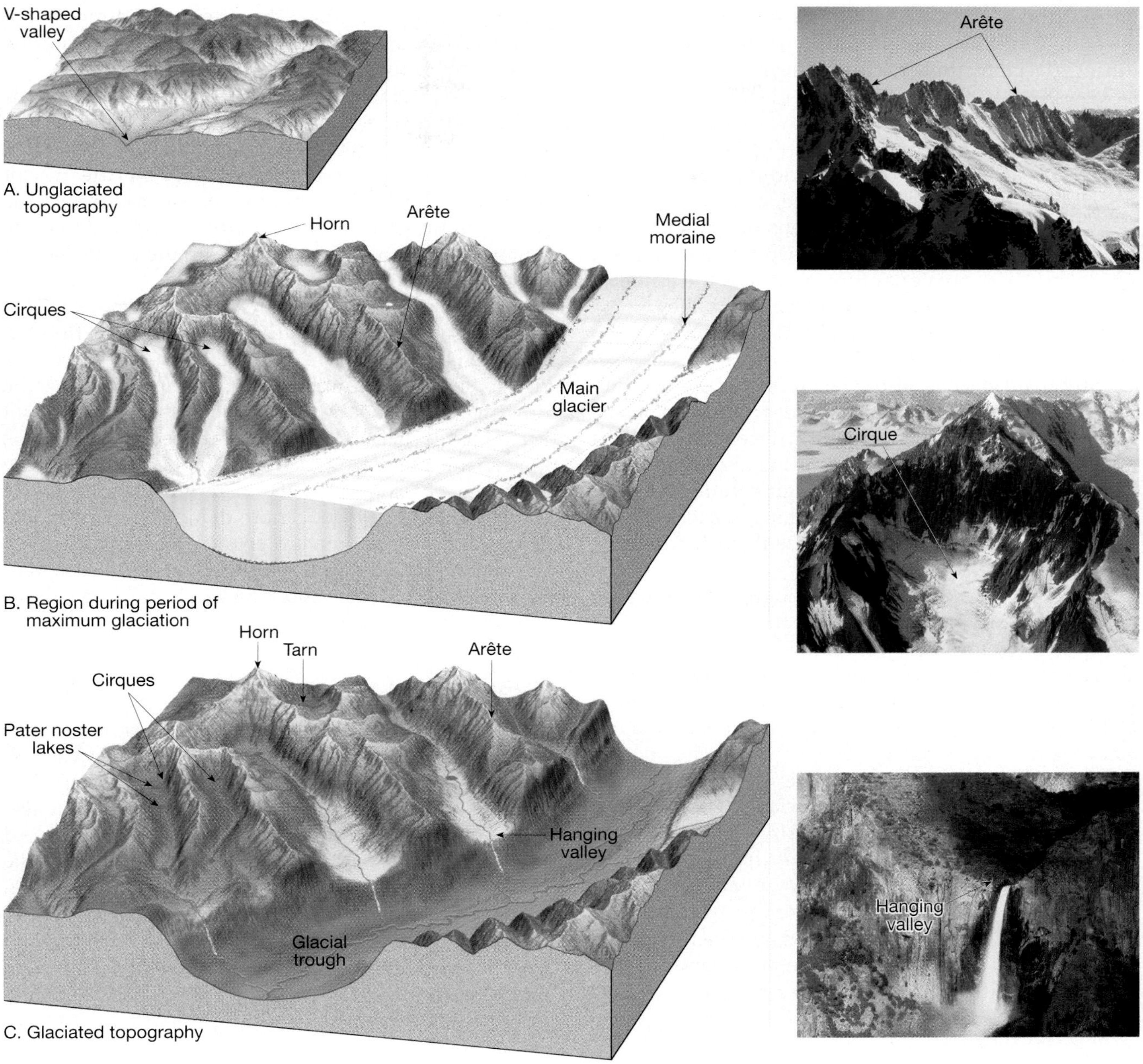

▲ **Figure 11.9** Erosional landforms created by alpine glaciers. The unglaciated landscape in part **A** is modified by valley glaciers in part **B.** After the ice recedes, in part **C,** the terrain looks very different than before glaciation. *(Arête photo from James E. Patterson Collection, Cirque photo by Marli Miller, Hanging Valley photo by Mark Muench/Muench Photography, Inc.)*

bold peaks even more jagged. By contrast, continental ice sheets generally override the terrain and hence subdue rather than accentuate the irregularities they encounter. Although the erosional potential of ice sheets is enormous, landforms carved by these huge ice masses usually do not inspire the same wonderment and awe as do the erosional features created by valley glaciers. Much of the rugged mountain scenery so celebrated for its majestic beauty is the product of erosion by alpine glaciers. Take a moment to study Figure 11.9 which shows a hypothetical mountain area before, during, and after glaciation. You will refer to this often in the following discussion.

Glaciated Valleys

A hike up a glaciated valley reveals a number of striking ice-created features. The valley itself is often a dramatic sight. Unlike streams, which create their own valleys, glaciers take the path of least resistance by following the course of existing stream valleys. Prior to glaciation, mountain valleys are characteristically narrow and V-shaped because streams are well above base level and are therefore downcutting. However, during glaciation these narrow valleys undergo a transformation as the glacier widens and deepens them, creating a U-shaped **glacial trough** (Figure 11.9C and Figure 11.10). In addition to producing a broader and deeper valley, the glacier also straightens the valley. As ice flows around sharp curves, its great erosional force removes the spurs of land that extend into the valley.

Did You Know?

The Kutiah Glacier in Pakistan holds the record for the fastest glacial surge ever recorded. In 1953, it raced more than 12 km (7.4 mi) in three months, averaging about 130 m (430 ft) per day.

The intensity of glacial erosion depends in part upon the thickness of the ice. Consequently, main glaciers, also called *trunk glaciers,* cut their valleys deeper than do their smaller tributary glaciers. Thus, when the glaciers eventually recede, the valleys of tributary glaciers are left standing above the main glacial trough and are termed **hanging valleys** (Figure 11.9C). Rivers flowing through hanging valleys can produce spectacular waterfalls, such as those in Yosemite National Park.

Glacially polished granite in Yosemite National Park.

As hikers walk up a glacial trough, they may pass a series of bedrock depressions on the valley floor that were probably formed by plucking and scoured by the abrasive force of the ice. If these depressions are filled with water, they are called **pater noster lakes** (Figure 11.10). The Latin name means "our Father" and is a reference to a string of rosary beads.

At the head of a glacial valley is a very characteristic and often imposing feature associated with an alpine glacier called a **cirque.** As Figure 11.9 illustrates, these bowl-shaped depressions have precipitous walls on three sides but are open on the downvalley side. The cirque is the focal point of the glacier's growth, because it is the area of snow accumulation and ice formation. They begin as irregularities in the mountainside that are subsequently enlarged by the frost wedging and plucking that occur along the sides and bottom of the glacier. After the glacier has melted away, the cirque basin is often occupied by a small lake.

Before leaving the topic of glacial troughs and their associated features, one more rather well-known feature should be discussed—fiords. **Fiords** are deep, often spectacular, steep-sided inlets of the sea that are present at high latitudes where mountains are adjacent to the ocean (Figure 11.11). They are drowned glacial troughs that became submerged as the ice left the valley and sea level rose following the Ice Age.

The depths of fiords can exceed 1000 meters (3300 feet). However, the great depths of these flooded troughs is only partly explained by the post–Ice Age rise in sea level. Unlike the situation governing the downward erosional work of rivers, sea level does not act as base level for glaciers. As a consequence, glaciers are capable of eroding their beds far below the surface of the sea. For example, a 300-meter-thick alpine glacier can carve its valley floor more than 250 meters below sea level before downward erosion ceases and the ice begins to float. Norway, British Columbia, Greenland, New Zealand, Chile, and Alaska all have coastlines characterized by fiords.

Arêtes and Horns

A visit to the Alps, the Northern Rockies, or many other scenic mountain landscapes carved by valley glaciers reveals not only glacial troughs, cirques, pater noster lakes, and the other related features just discussed. You also are likely to see sinuous, sharp-edged ridges called **arêtes** (French for *knife-edge*) and sharp, pyramid-like peaks called **horns** projecting above the surroundings. Both features can originate from the same basic process, the enlargement of cirques produced by plucking and frost action (Figure 11.9B, C). In the case of the spires of rock called horns, cirques around a single high mountain

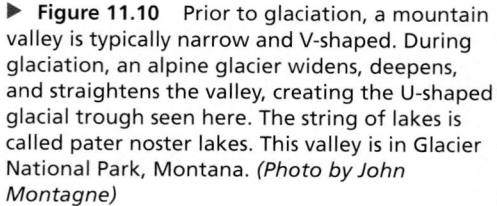

▶ **Figure 11.10** Prior to glaciation, a mountain valley is typically narrow and V-shaped. During glaciation, an alpine glacier widens, deepens, and straightens the valley, creating the U-shaped glacial trough seen here. The string of lakes is called pater noster lakes. This valley is in Glacier National Park, Montana. *(Photo by John Montagne)*

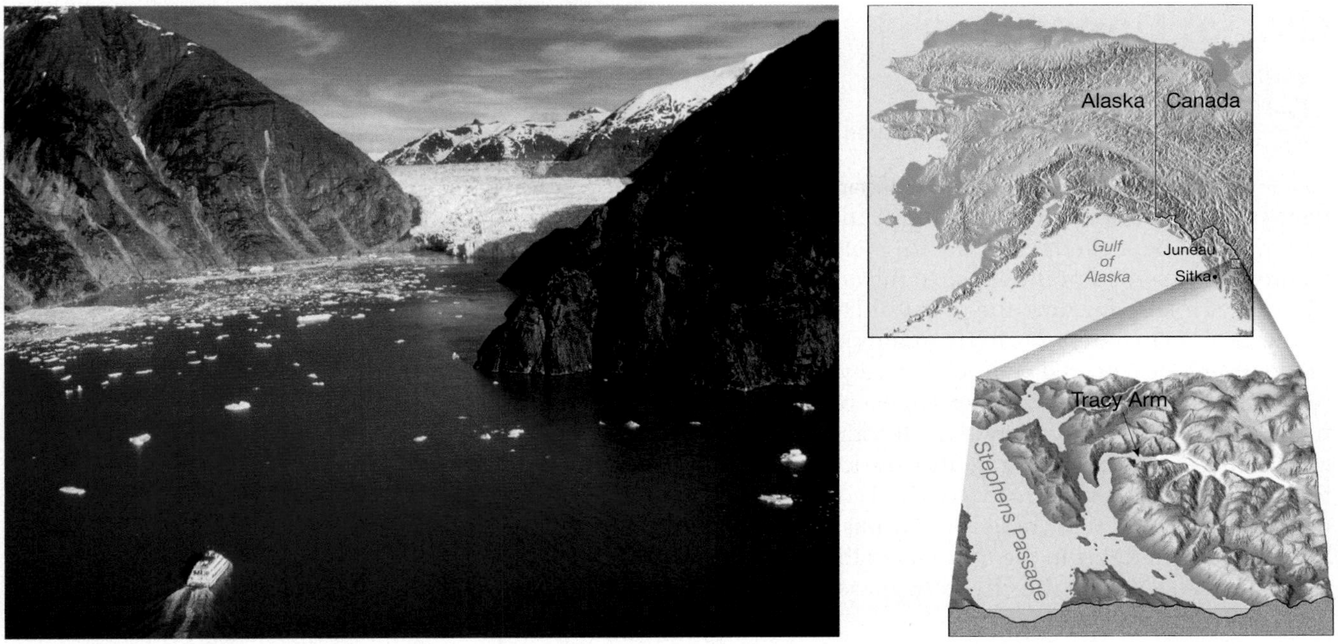

▲ **Figure 11.11** Like other fiords, this one at Tracy Arm, Alaska, is a drowned glacial trough. *(Photo by Tom & Susan Bean, Inc.)*

are responsible. As the cirques enlarge and converge, an isolated horn is produced. The most famous example is the Matterhorn in the Swiss Alps (Figure 11.12).

Arêtes can be formed in a similar manner, except that the cirques are not clustered around a point but rather exist on opposite sides of a divide. As the cirques grow, the divide separating them is reduced to a narrow knifelike partition. An arête, however, may also be created in another way. When two glaciers occupy parallel valleys, an arête can form when the divide separating the moving tongues of ice is progressively narrowed as the glaciers scour and widen their adjacent valleys.

Roches Moutonnées

In many glaciated landscapes, but most frequently where continental ice sheets have modified the terrain, the ice carves small streamlined hills from protruding bedrock knobs. Such an asymmetrical knob of bedrock is called a **roche moutonnée** (French for *sheep rock*). They are formed when glacial abrasion smoothes the gentle slope facing the oncoming ice sheet and plucking steepens the opposite side as the ice rides over the knob (Figure 11.13). Roches moutonnées indicate the direction of glacial flow, because the gentler slope is generally on the side from which the ice advanced.

▼ **Figure 11.12** Horns are sharp, pyramid-like peaks that are fashioned by alpine glaciers. This example is the famous Matterhorn in the Swiss Alps. *(Photo by E. J. Tarbuck)*

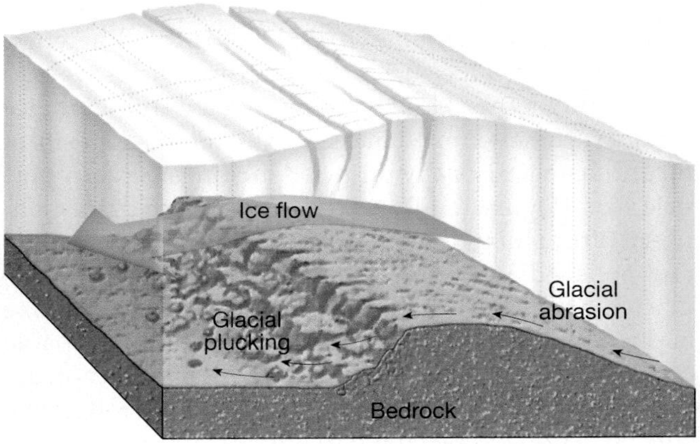

▲ **Figure 11.13** Roche moutonnée. The gentle slope was abraded and the steep side was plucked. The ice moved from right to left.

Glacial Deposits

GEODe Essentials of Geology
Glaciers and Glaciation
▼ Reviewing Glacial Features

Glaciers pick up and transport a huge load of debris as they slowly advance across the land. Ultimately these materials are deposited when the ice melts. In regions where glacial sediment is deposited, it can play a truly significant role in forming the physical landscape. For example, in many areas once covered by the ice sheets of the recent Ice Age, the bedrock is rarely exposed because glacial deposits that are tens or even hundreds of meters thick completely mantle the terrain. The general effect of these deposits is to reduce the local relief and thus level the topography. Indeed, rural country scenes that are familiar to many of us—rocky pastures in New England, wheat fields in the Dakotas, rolling farmland in the Midwest—result directly from glacial deposition.

A close look at glacial till often reveals cobbles that were scratched as they were dragged by the glacier.

Types of Glacial Drift

Long before the theory of an extensive Ice Age was proposed, much of the soil and rock debris covering portions of Europe was recognized as coming from elsewhere. At the time, these foreign materials were believed to have been "drifted" into their present positions by floating ice during an ancient flood. As a consequence, the term *drift* was applied to this sediment. Although rooted in a concept that was not correct, this term was so well established by the time the true glacial origin of the debris became widely recognized that it remained in the basic glacial vocabulary. Today, **glacial drift** is an all-embracing term for sediments of glacial origin, no matter how, where, or in what form they were deposited.

Did You Know?

If the Antarctic ice sheet were to melt at a suitable uniform rate, it could feed the Mississippi River for more than 50,000 years or it could maintain the flow of the Amazon River for approximately 5000 years.

Glacial drift is divided into two distinct types: (1) materials deposited directly by the glacier, which are known as *till*, and (2) sediments laid down by glacial meltwater, called *stratified drift*. Here is the difference: **Till** is deposited as glacial ice melts and drops its load of rock fragments. Unlike moving water and wind, ice cannot sort the sediment it carries; therefore, deposits of till are characteristically unsorted mixtures of many particle sizes (Figure 11.14). A close examination of this sediment often shows that many of the pieces are scratched and polished as a result of being dragged along by the glacier. **Stratified drift** is sorted according to the size and weight of the fragments. Because ice is not capable of such sorting activity, these sediments are not deposited directly by the glacier. Rather, they reflect the sorting action of glacial meltwater.

Some deposits of stratified drift are made by streams issuing directly from the glacier. Other stratified deposits involve sediment that was originally laid down as till and later picked up, transported, and redeposited by meltwater beyond the margin of the ice. Accumulations of stratified drift often consist largely of sand and gravel, because the meltwater is not capable of moving larger material and because the finer rock flour remains suspended and is commonly carried far from the glacier. An indication that stratified drift consists primarily of sand and gravel can be seen in many areas where these deposits are actively mined as aggregate for road work and other construction projects.

When boulders are found in the till or lying free on the surface, they are called **glacial erratics** if they are different from the bedrock below (Figure 11.15). Of course, this means that they must have been derived from a source outside the area where they are found.

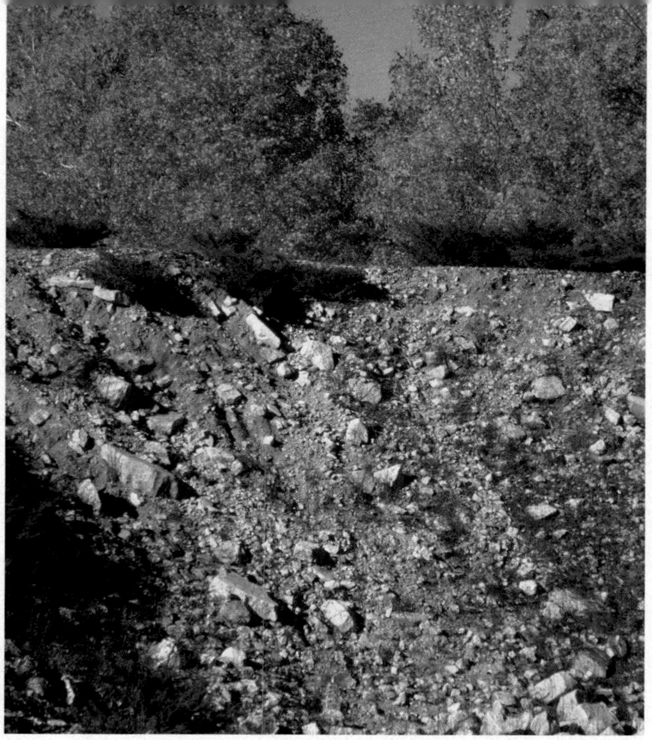

▲ **Figure 11.14** Glacial till is an unsorted mixture of many different sediment sizes. *(Photo by E. J. Tarbuck)*

▲ **Figure 11.15** A large glacial erratic, central Wyoming. *(Photo by Yva Momatiuk and John Eastcott/Photo Researchers, Inc.)*

Although the locality of origin for most erratics is unknown, the origin of some can be determined. In many cases, boulders were transported as far as 500 kilometers from their source area and, in a few instances, more than 1000 kilometers. Therefore, by studying glacial erratics as well as the mineral composition of the till, geologists can sometimes trace the path of a lobe of ice. In portions of New England, as well as other areas, erratics can be seen dotting pastures and farm fields. In some places, these rocks were cleared from fields and piled to make fences and walls.

Moraines, Outwash Plains, and Kettles

Perhaps the most widespread features created by glacial deposition are *moraines,* which are simply layers or ridges of till. Several types of moraines are identified; some are common only to mountain valleys, and others are associated with areas affected by either ice sheets or valley glaciers. Lateral and medial moraines fall in the first category, whereas end moraines and ground moraines are in the second.

Lateral and Medial Moraines. The sides of a valley glacier accumulate large quantities of debris from the valley walls. When the glacier wastes away, these materials are left as ridges, called **lateral moraines,** along the sides of the valley (Figure 11.16A). **Medial moraines** are formed when two valley glaciers coalesce to form a single ice stream. The till that was once carried along the edges of each glacier joins to form a single dark stripe of debris within the newly enlarged glacier. The creation of these dark stripes within the ice stream is one obvious proof that glacial ice moves, because the medial moraine could not form if the ice did not flow downvalley (Figure 11.16B). It is common to see several medial moraines within a large alpine glacier, because a streak will form whenever a tributary glacier joins the main valley.

Did You Know?

Glaciers are found on all continents except Australia. Surprisingly, tropical Africa has a small area of glacial ice atop its highest mountain, Mount Kilimanjaro. However, studies show that the mountain has lost more than 80 percent of its ice since 1912 and that all of it may be gone by 2020.

End and Ground Moraines. An **end moraine** is a ridge of till that forms at the terminus of a glacier. These relatively common landforms are deposited when a state of equilibrium is attained between ablation and ice accumulation. That is, the end moraine forms when the ice is melting and evaporating near the end of the glacier at a rate equal to the forward advance of the glacier from its region of nourishment. Although the terminus of the glacier is now stationary, the ice continues to flow forward, delivering a continuous supply of sediment in the same manner a conveyor belt delivers goods to the end of a production line. As the ice melts, the till is dropped and the end moraine grows. The longer the ice front remains stable, the larger the ridge of till will become.

Roche moutonnée in Yosemite National Park.

Eventually the time comes when ablation exceeds nourishment. At this point, the front of the glacier begins to recede in the direction from which it originally advanced. However, as the ice front retreats, the conveyor-belt action of the glacier continues to provide fresh supplies of sediment to the terminus. In this manner a

A.

B.

▲ **Figure 11.16** **A.** Well-developed lateral moraines deposited by the shrinking Athabaska Glacier in Canada's Jasper National Park. *(Photo by David Barnes/CORBIS/Stock Market)* **B.** Medial moraines form when the lateral moraines of merging glaciers join. Saint Elias National Park, Alaska. *(Photo by Tom and Susan Bean, Inc.)*

large quantity of till is deposited as the ice melts away, creating a rock-strewn, undulating plain. This gently rolling layer of till deposited as the ice front recedes is termed **ground moraine.** Ground moraine has a leveling effect, filling in low spots and clogging old stream channels, often leading to a derangement of the existing drainage system. In areas where this layer of till is still relatively fresh, such as the northern Great Lakes region, poorly drained swampy lands are quite common.

Did You Know?

There are many areas where scientists report that glaciers are wasting away. For example, 150 years ago there were 147 glaciers in Montana's Glacier National Park. Today only 37 remain, and these may vanish by 2030. Similarly, glaciers all across the Alps are retreating and disappearing every year.

Periodically, a glacier will retreat to a point where ablation and nourishment once again balance. When this happens, the ice front stabilizes and a new end moraine forms.

The pattern of end moraine formation and ground moraine deposition may be repeated many times before the glacier has completely vanished. Such a pattern is illustrated by Figure 11.17. The very first end moraine to form signifies the farthest advance of the glacier and is called the *terminal end moraine.* Those end moraines that form as the ice front occasionally stabilizes during retreat are termed *recessional end moraines.* Terminal and recessional moraines are essentially alike; the only difference between them is their relative positions.

Did You Know?

If all of Earth's glaciers melted, sea level would rise approximately 70 m (230 ft) worldwide.

End moraines deposited by the most recent stage of Ice Age glaciation are prominent features in many parts of the Midwest and Northeast. In Wisconsin, the wooded, hilly terrain of the Kettle Moraine near Milwaukee is a particularly picturesque example. A well-known example in the Northeast is Long Island. This linear strip of glacial sediment that extends northeastward from New York City is part of an end moraine complex that stretches from eastern Pennsylvania to Cape Cod, Massachusetts (Figure 11.18).

Figure 11.19 represents a hypothetical area during glaciation and following the retreat of ice sheets. It shows the end moraines that were just described as well as the depositional features that are discussed in the sections that follow. This figure depicts landscape features similar to what might be encountered if you were traveling in the upper Midwest or New England. As you read upcoming sections dealing with other glacial deposits, you will be referred to this figure several times.

Outwash Plains and Valley Trains. At the same time that an end moraine is forming, water from the melting glacier cascades over the till, sweeping some of it out in front of the growing ridge of unsorted debris. Meltwater generally emerges from the ice in rapidly moving streams that are often choked with suspended material and carry a substantial bed load as well. As the water leaves the glacier, it moves onto the relatively flat surface beyond and rapidly loses velocity. As a consequence, much of its bed load is dropped and the meltwater begins weaving a complex pattern of braided channels (Figure 11.19). In this way, a broad, ramplike surface composed of stratified drift is built adjacent to the downstream edge of most end moraines. When the feature is formed in association with an ice sheet, it is termed an **outwash plain,** and when largely confined to a mountain valley, it is usually referred to as a **valley train.**

Kettles. Often end moraines, outwash plains and valley trains are pockmarked with basins or depressions known as **kettles** (Figure 11.19). Kettles form when

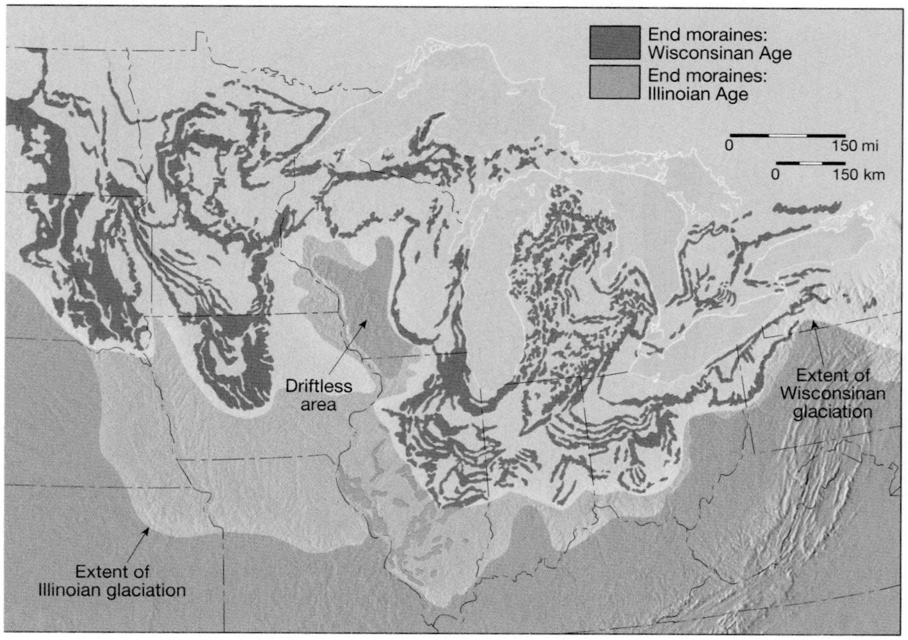

◀ **Figure 11.17** End moraines of the Great Lakes region. Those deposited during the most recent (Wisconsinan) stage are most prominent.

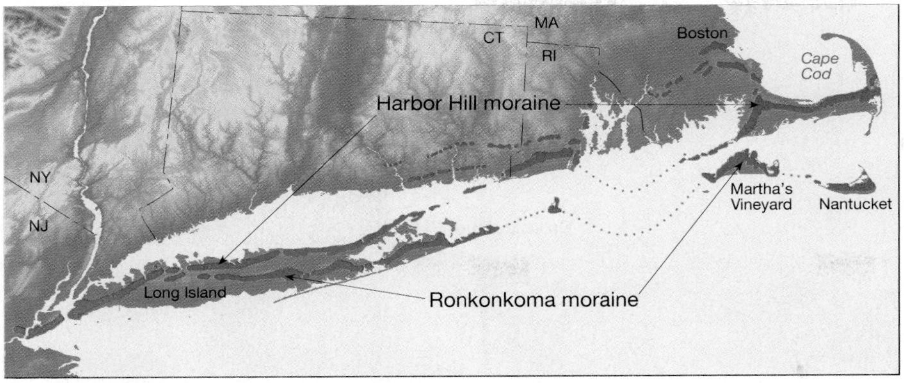

◀ **Figure 11.18** End moraines make up substantial parts of Long Island, Cape Cod, Martha's Vineyard, and Nantucket. Although portions are submerged, the Ronkonkoma moraine extends through central Long Island, Martha's Vineyard, and Nantucket. It was deposited about 20,000 years ago. The Harbor Hill moraine, which formed about 14,000 years ago, extends along the north shore of Long Island, through southern Rhode Island and Cape Cod.

blocks of stagnant ice become wholly or partly buried in drift and eventually melt, leaving pits in the glacial sediment. Although most kettles do not exceed 2 kilometers in diameter, some with diameters exceeding 10 kilometers occur in Minnesota. Likewise, the typical depth of most kettles is less than 10 meters, although the vertical dimensions of some approach 50 meters. In many cases water eventually fills the depression and forms a pond or lake. One well-known example is Walden Pond near Concord, Massachusetts. It is here that Henry David Thoreau lived alone for two years in the 1840s and about which he wrote his famous book *Walden, or Life in the Woods.*

Drumlins, Eskers, and Kames

Moraines are not the only landforms deposited by glaciers. Some landscapes are characterized by numerous elongate parallel hills made of till. Other areas exhibit conical hills and relatively narrow winding ridges composed largely of stratified drift.

Drumlins. **Drumlins** are streamlined asymmetrical hills composed of till (Figure 11.19). They range in height from 15 to 60 meters (50 to 200 feet) and average 0.4 to 0.8 kilometer (0.25 to 0.50 mile) in length. The steep side of the hill faces the direction from which the ice advanced, whereas the gentler slope points in the direction the ice moved. Drumlins are not found singly, but rather occur in clusters, called *drumlin fields.* One such cluster, east of Rochester, New York, is estimated to contain about 10,000 drumlins. Their streamlined shape indicates that they were molded in the zone of flow within an active glacier. It is thought that drumlins originate when glaciers advance over previously deposited drift and reshape the material.

Eskers and Kames. In some areas that were once occupied by glaciers, sinuous ridges composed largely of sand and gravel can be found. Known as **eskers,** these ridges are deposited by meltwater rivers flowing within, on top of, and beneath a mass of motionless, stagnant glacial ice (Figure 11.19). Many sediment sizes are carried by the torrents of meltwater in the ice-banked channels, but only the coarser material can settle out of the turbulent stream. In some areas they are mined for sand and gravel, and for this reason, eskers are disappearing in some localities.

Kames are steep-sided hills that, like eskers, are composed of sand and gravel (Figure 11.19). Kames originate when glacial meltwater washes sediment into openings and depressions in the stagnant wasting terminus of a glacier. When the ice eventually melts away, the stratified drift is left behind as mounds or hills.

Glaciers of the Ice Age

At various points in the preceding pages we mentioned the Ice Age, a time when ice sheets and alpine glaciers were far more extensive than they are today. There was a time when the most popular explanation for what we now know to be glacial deposits was that the material had been drifted in by means of icebergs or perhaps simply swept across the landscape by a catastrophic flood. However, during the nineteenth century, field investigations by many scientists provided convincing proof that an extensive Ice Age was responsible for these deposits and for many other features.

Stone wall in Wisconsin made of glacial erratics. (Tom Bean)

By the beginning of the twentieth century, geologists had largely determined the extent of Ice Age glaciation. Further, they discovered that many glaciated regions had not one layer of drift but several. Close examination of these older deposits showed well-developed zones of chemical weathering and soil formation as well as the remains of plants that require warm temperatures. The evidence was clear: There had not been just one glacial advance but many, each separated by extended periods when climates were as warm or warmer than at present. The Ice Age had not simply been a time when the ice advanced over the land,

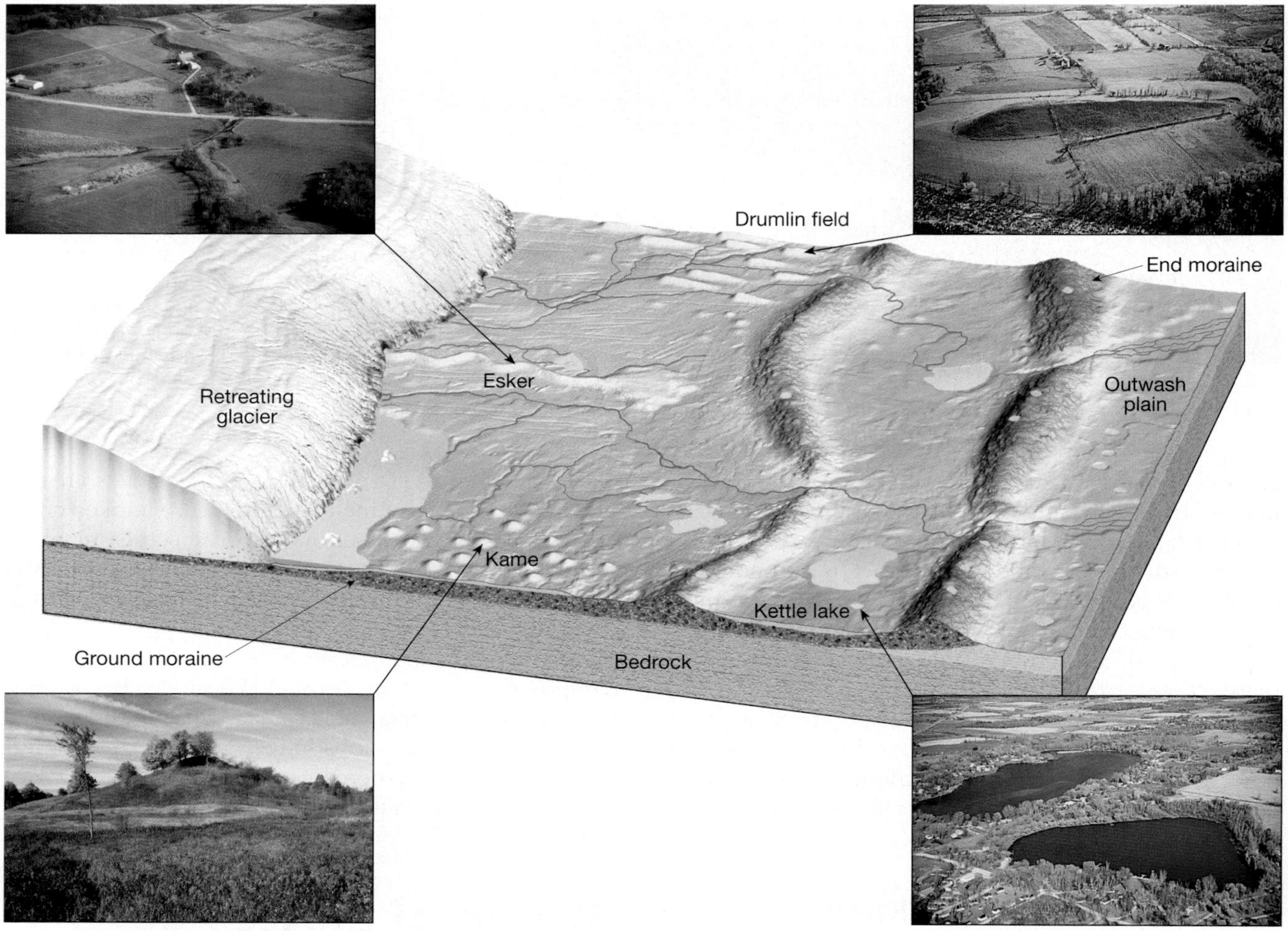

▲ **Figure 11.19** This hypothetical area illustrates many common depositional landforms. The outermost end moraine marks the limit of glacial advance and is called the *terminal end moraine*. End moraines that form as the ice front occasionally becomes stationary during retreat are called *recessional end moraines. (Drumlin photo courtesy of Ward's Natural Science Establishment; Kame, Esker, and Kettle photos by Richard P. Jacobs/JLM Visuals)*

Walden Pond, made famous by Thoreau, occupies a kettle. (David Muench)

lingered for a while, and then receded. Rather, the period was a very complex event characterized by a number of advances and withdrawals of glacial ice.

The glacial record on land is punctuated by many erosional gaps. This makes it difficult to reconstruct the episodes of the Ice Age clearly. But sediment on the ocean floor provides an uninterrupted record of climate cycles for this period. Studies of cores drilled from these seafloor sediments show that glacial/interglacial cycles have occurred about every 100,000 years. About 20 such cycles of cooling and warming were identified for the span we call the Ice Age.

During the glacial age, ice left its imprint on almost 30 percent of Earth's land area, including about 10 million square kilometers of North America, 5 million square kilometers of Europe, and 4 million square kilometers of Siberia (Figure 11.20). The amount of glacial ice in the Northern Hemisphere was roughly twice that of the Southern Hemisphere. The primary reason is that the Southern Hemisphere has little land in the middle latitudes, and therefore the southern polar ice could not spread far beyond the margins of Antarctica. By contrast, North America and Eurasia provided great expanses of land for the spread of ice sheets.

Today we know that the Ice Age began between 2 million and 3 million years ago. This means that most of the major glacial episodes occurred during a division of the geologic time scale called the **Pleistocene epoch.** Although the Pleistocene is commonly used as a synonym for the Ice Age, this epoch does not encompass it all. The Antarctic Ice Sheet, for example, formed at least 14 million years ago and, in fact, might be much older.

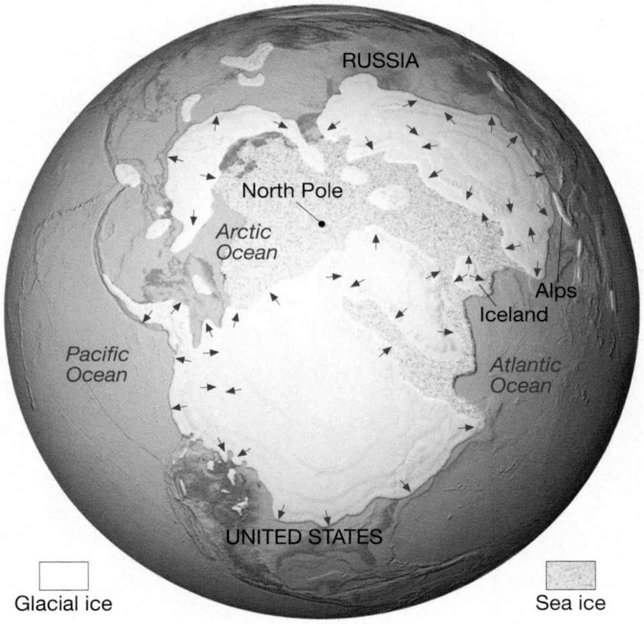

▲ **Figure 11.20** Maximum extent of glaciation in the Northern Hemisphere during the Ice Age.

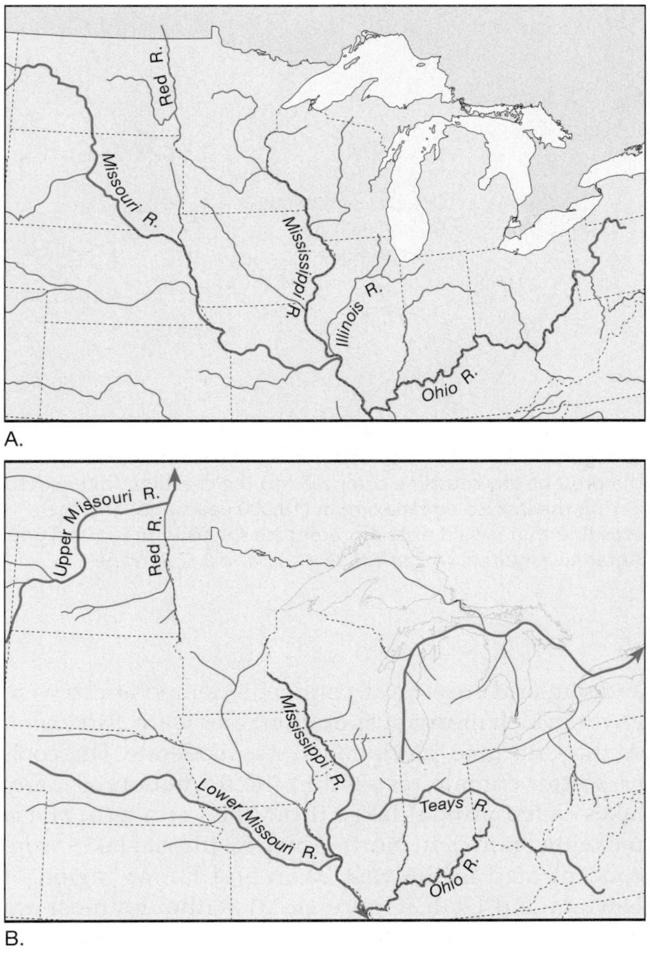

▲ **Figure 11.21** **A.** This map shows the Great Lakes and the familiar present-day pattern of rivers in the central United States. Pleistocene ice sheets played a major role in creating this pattern. **B.** Reconstruction of drainage systems in the central United States prior to the Ice Age. The pattern was very different from today, and there were no Great Lakes.

Some Indirect Effects of Ice Age Glaciers

In addition to the massive erosional and depositional work carried on by Pleistocene glaciers, the ice sheets had other, sometimes profound, effects on the landscape. For example, as the ice advanced and retreated, animals and plants were forced to migrate. This led to stresses that some organisms could not tolerate. Furthermore, many present-day stream courses bear little resemblance to their preglacial routes. The Missouri River once flowed northward toward Canada's Hudson Bay. The Mississippi River followed a path through central Illinois, and the head of the Ohio River reached only as far as Indiana (Figure 11.21). Other rivers that today carry only a trickle of water but nevertheless occupy broad channels are testimony to the fact that they once carried torrents of glacial meltwater.

In areas that were centers of ice accumulation, such as Scandinavia and northern Canada, the land has been slowly rising for the past several thousand years. The land had downwarped under the tremendous weight of 3-kilometer-thick masses of ice. Following the removal of this immense load, the crust has been adjusting by gradually rebounding upward ever since.

A far-reaching effect of the Ice Age was the worldwide change in sea level that accompanied each advance and retreat of the ice sheets. The snow that nourishes glaciers ultimately comes from moisture evaporated from the oceans. Therefore, when the ice sheets increased in size, sea level fell and the shoreline shifted seaward (Figure 11.22). Estimates suggest that sea level was as much as 100 meters (330 feet) lower than today. Thus, land that is presently flooded by the oceans was dry. The Atlantic Coast of the United States lay more than 100 kilometers (60 miles) to the east of New York City. Moreover, France and Britain were joined where the English Channel is today, Alaska and Siberia were connected across the Bering Strait, and Southeast Asia was tied by dry land to the islands of Indonesia.

Did You Know?

The Arctic Ocean surrounds the North Pole, so the ice in this region is sea ice (frozen seawater) *not* glacial ice. Remember, glaciers form on land from the accumulation of snow.

The formation and growth of ice sheets was an obvious response to significant changes in climate. But the existence of the glaciers themselves triggered climatic changes in the regions beyond their margins. In arid

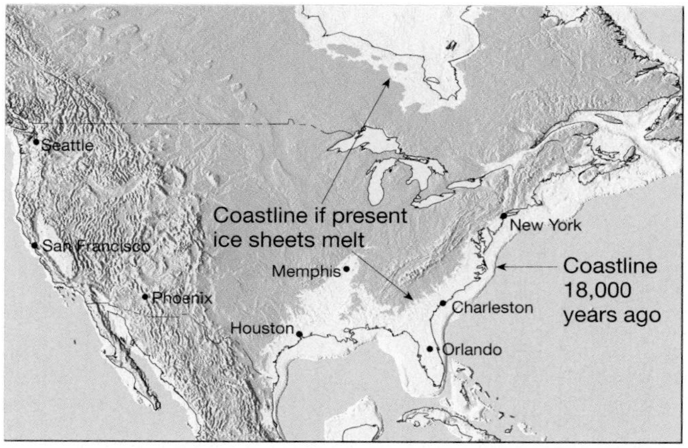

▲ **Figure 11.22** This map of a portion of North America shows the present-day coastline compared to the coastline that existed during the last ice-age maximum (18,000 years ago) and the coastline that would exist if present ice sheets in Greenland and Antarctica melted. *(After R.H. Dott, Jr. and R.L. Battan).*

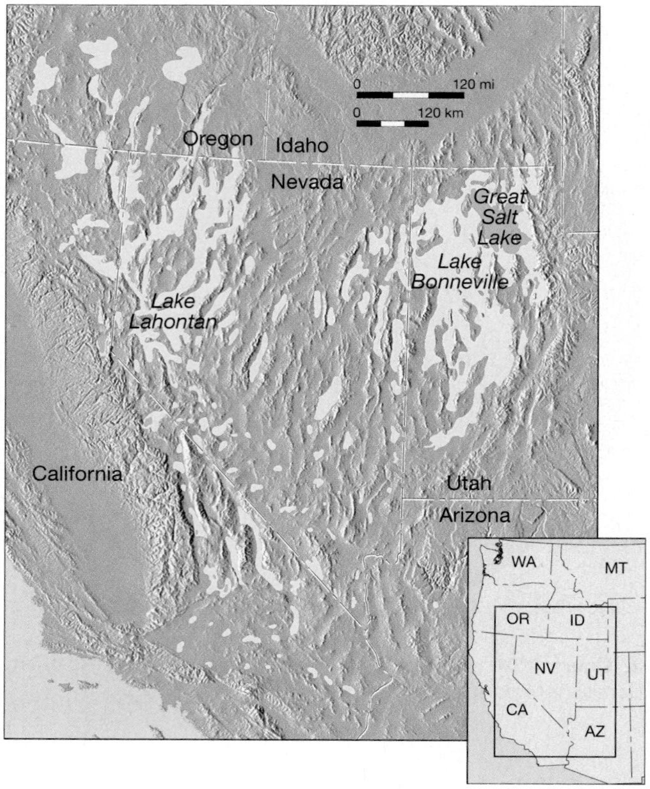

▲ **Figure 11.23** Pluvial lakes of the Western United States. By far the largest of the pluvial lakes in the vast Basin and Range region of Nevada and Utah was Lake Bonneville. With maximum depths exceeding 300 meters and an area of 50,000 square kilometers, Lake Bonneville was nearly the same size as present-day Lake Michigan. The Great Salt Lake is a remnant of this huge pluvial lake. *(After R. F. Flint)*

and semiarid areas on all continents, temperatures were lower, which meant evaporation rates were also lower. At the same time, precipitation was moderate. This cooler, wetter climate resulted in the formation of many lakes called **pluvial lakes** (from the Latin term *pluvia* meaning "rain"). In North America, pluvial lakes were concentrated in the vast Basin and Range region of Nevada and Utah (Figure 11.23). Although most are now gone, a few remnants remain, the largest being Utah's Great Salt Lake.

Causes of Glaciation

A great deal is known about glaciers and glaciation. Much has been learned about glacier formation and movement, the extent of glaciers past and present, and the features created by glaciers, both erosional and depositional. However, the causes of glacial ages are not completely understood.

Although widespread glaciation has been rare in Earth's history, the Pleistocene Ice Age is not the only glacial period for which a record exists. Two Precambrian glacial episodes have been identified in the geologic record, the first approximately 2 billion years ago and the second about 600 million years ago. Further, a well-documented record of an earlier glacial age is found in late Paleozoic rocks that are about 250 million years old and which exist on several landmasses.

Any theory that attempts to explain the causes of glacial ages must successfully answer two basic questions: (1) *What causes the onset of glacial conditions?* For continental ice sheets to have formed, average temperatures must have been somewhat lower than at present and perhaps substantially lower than throughout much of geologic time. Thus, a successful theory would have to account for the cooling that finally leads to glacial conditions. (2) *What caused the alternation of glacial and interglacial stages that have been documented for the Pleistocene epoch?* The first question deals with long-term trends in temperature on a scale of millions of years, but this second question relates to much shorter-term changes.

Although the literature of science contains many hypotheses relating to the possible causes of glacial periods, we will discuss only a few major ideas to summarize current thought.

Did You Know?

Studies have shown that the retreat of glaciers may reduce the stability of faults and hasten earthquake activity. When the weight of the glacier is removed, the ground rebounds. In tectonically active areas experiencing post-glacial rebound, earthquakes may occur sooner and/or be stronger than if the ice were present.

Plate Tectonics

Probably the most attractive proposal for explaining the fact that extensive glaciations have occurred only a few times in the geologic past comes from the theory of plate tectonics.* Because glaciers can form only on land, we know that landmasses must exist somewhere in the higher latitudes before an ice age can commence. Many scientists suggest that ice ages have occurred only when Earth's shifting crustal plates have carried the continents from tropical latitudes to more poleward positions.

Glacial features in present-day Africa, Australia, South America, and India indicate that these regions, which are now tropical or subtropical, experienced an ice age near the end of the Paleozoic era, about 250 million years ago. However, there is no evidence that ice sheets existed during this same period in what are today the higher latitudes of North America and Eurasia. For many years this puzzled scientists. Was the climate in these relatively tropical latitudes once like it is today in Greenland and Antarctica? Why did glaciers not form in North America and Eurasia? Until the plate tectonics theory was formulated, there had been no reasonable explanation.

Today scientists understand that the areas containing these ancient glacial features were joined together as a single supercontinent (Pangaea) located at latitudes far to the south of their present positions. Later this landmass broke apart, and its pieces, each moving on a different plate, migrated toward their present locations (Figure 11.24). Now we know that during the geologic past, plate movements accounted for many dramatic climatic changes as landmasses shifted in relation to one another and moved to different latitudinal positions.

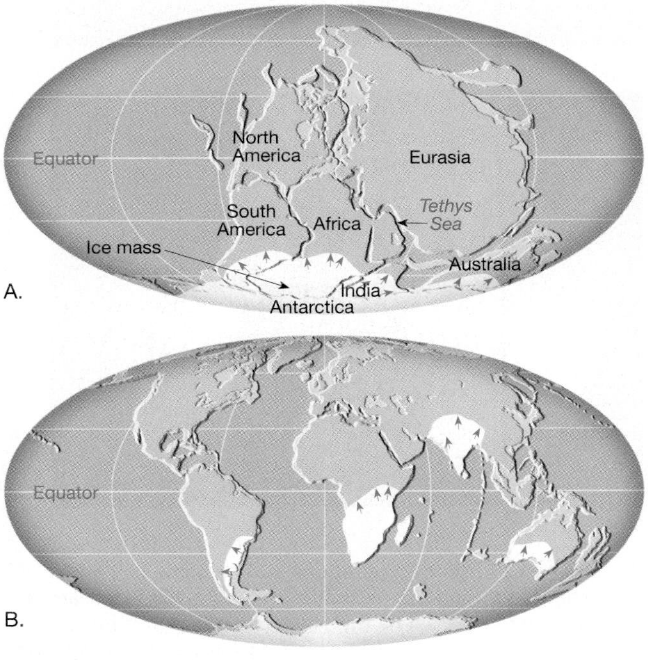

▲ **Figure 11.24** **A.** The supercontinent Pangaea showing the area covered by glacial ice about 300 million years ago. **B.** The continents as they are today. The white areas indicate where evidence of the old ice sheets exists.

*A brief overview of the theory appears in Chapter 1, and a more extensive discussion is presented in Chapter 15.

Changes in oceanic circulation also must have occurred, altering the transport of heat and moisture and consequently the climate as well. Because the rate of plate movement is very slow—a few centimeters per year—appreciable changes in the positions of the continents occur only over great spans of geologic time. Thus, climate changes brought about by shifting plates are extremely gradual and happen on a scale of millions of years.

Swiss-born Louis Agassiz (1807–1873) was instrumental in formulating modern ideas about the Ice Age. (Harvard University Archives)

Variations in Earth's Orbit

Because climatic changes brought about by moving plates are extremely gradual, the plate tectonics theory cannot be used to explain the alternation between glacial and interglacial climates that occurred during the Pleistocene epoch. Therefore, we must look to some other triggering mechanism that may cause climatic change on a scale of thousands rather than millions of years. Today many scientists strongly suspect that the climatic oscillations that characterized the Pleistocene may be linked to variations in Earth's orbit. This hypothesis was first developed and strongly advocated by the Yugoslavian scientist Milutin Milankovitch and is based on the premise that variations in incoming solar radiation are a principal factor controlling Earth's climate.

> **Did You Know?**
>
> Antarctica's ice sheet weighs so much that it depresses Earth's crust by an estimated 900 m (3000 ft) or more.

Milankovitch formulated a comprehensive mathematical model based on the following elements (Figure 11.25):

1. Variations in the shape (*eccentricity*) of Earth's orbit about the Sun;
2. Changes in *obliquity;* that is, changes in the angle that the axis makes with the plane of Earth's orbit; and
3. The wobbling of Earth's axis, called *precession.*

Using these factors, Milankovitch calculated variations in the receipt of solar energy and the corresponding surface temperature of Earth back into time in an attempt to correlate these changes with the climate fluctuations of the Pleistocene. In explaining climatic changes that result from these three variables, note that they cause little or no variation in the total solar energy reaching the ground. Instead, their impact is felt because they change the degree of contrast between the seasons. Somewhat milder winters in the middle to high latitudes means greater snowfall totals, whereas cooler summers would bring a reduction in snowmelt.

Utah's Great Salt Lake is a remnant of a much larger pluvial lake. (S. Smith/CORBIS)

Among the studies that have added credibility to the astronomical hypothesis of Milankovitch is one in

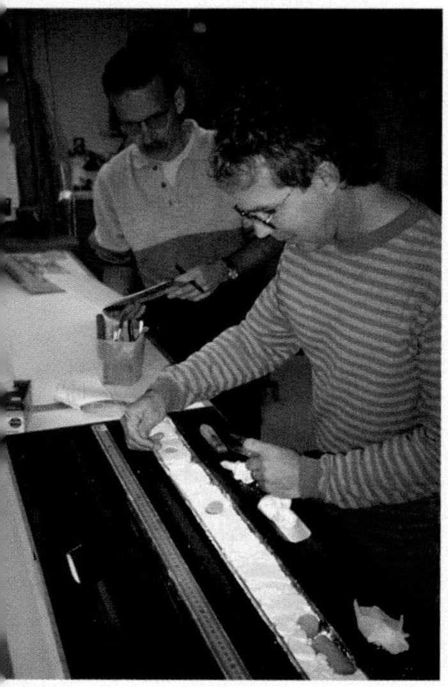

Ocean-floor sediments provide important data about Ice Age history and climate change. (A. Trujillo)

which deep-sea sediments containing certain climatically sensitive microorganisms were analyzed to establish a chronology of temperature changes going back nearly 500,000 years.* This time scale of climatic change was then compared to astronomical calculations of eccentricity, obliquity, and precession to determine whether a correlation did indeed exist.

Although the study was very involved and mathematically complex, the conclusions were straightforward. The researchers found that major variations in climate over the past several hundred thousand years were closely associated with changes in the geometry of Earth's orbit; that is, cycles of climatic change were shown to correspond closely with the periods of obliquity, precession, and orbital eccentricity. More specifically, the authors stated: "It is concluded that changes in the earth's orbital geometry are the fundamental cause of the succession of Quaternary ice ages."†

Let us briefly summarize the ideas that were just described. The theory of plate tectonics provides us with an explanation for the widely spaced and nonperiodic onset of glacial conditions at various times in the geologic past, whereas the astronomical model proposed by Milankovitch and supported by the work of J. D. Hays and his colleagues furnishes an explanation for the alternating glacial and interglacial episodes of the Pleistocene.

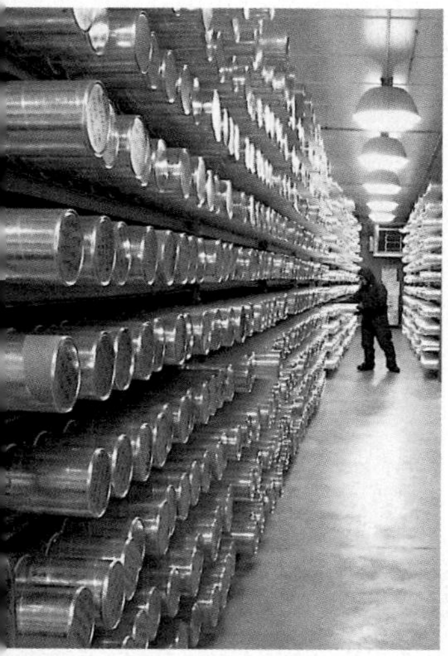

·age area for ice cores ·he National Ice Core ·oratory. (USGS/National ·ore Laboratory)

Other Factors

Variations in Earth's orbit correlate closely with the timing of glacial-interglacial cycles. However, the variations in solar energy reaching Earth's surface caused by these orbital changes do not adequately explain the magnitude of the temperature changes that occurred during the most recent ice age. Other factors must also have contributed. One factor involves variations in the chemical composition of the atmosphere. Other influences involve changes in the reflectivity of Earth's surface and in ocean circulation. Let's take a brief look at these factors.

Chemical analyses of air bubbles that become trapped in glacial ice at the time of ice formation indicate that the ice-age atmosphere contained less of the gases carbon dioxide and methane than the post ice-age atmosphere (see Box 11.2). Carbon dioxide and methane are important "greenhouse" gases, which means that they trap radiation emitted by Earth and contribute to the heating of the atmosphere. When the amount of carbon dioxide and methane in the atmosphere increase,

*J. D. Hays, John Imbrie, and N. J. Shackelton, Variations in the Earth's Orbit: Pacemaker of the Ice Ages, *Science* 194 (1976): 1121–32.

†J. D. Hays et al., ibid., p. 1131. The term *quaternary* refers to the period on the geologic time scale that encompasses the last 1.6 million years.

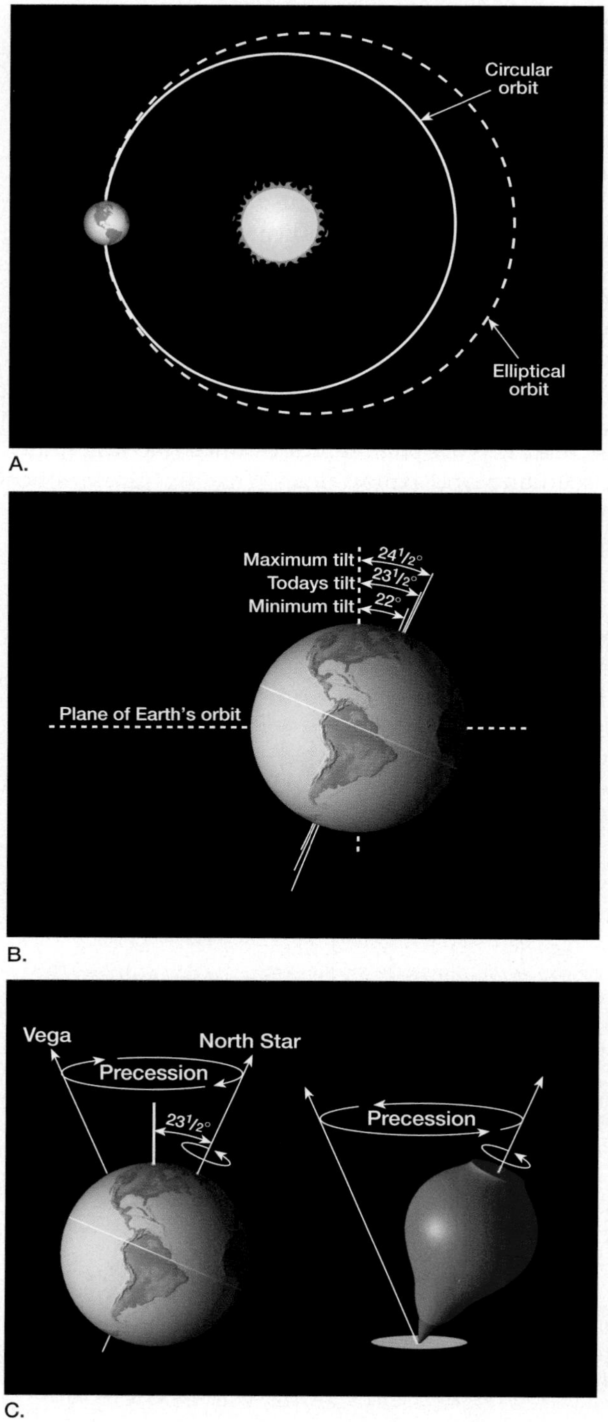

▲ **Figure 11.25** Orbital variations. **A.** The shape of Earth's orbit changes during a cycle that spans about 100,000 years. It gradually changes from nearly circular to one that is more elliptical and then back again. This diagram greatly exaggerates the amount of change. **B.** Today the axis of rotation is tilted about 23.5° to the plane of Earth's orbit. During a cycle of 41,000 years, this angle varies from 21.5° to 24.5°. **C.** Precession. Earth's axis wobbles like that of a spinning top. Consequently, the axis points to different spots in the sky during a cycle of about 26,000 years.

BOX 11.2

Glacial Ice—A Storehouse of Climate Data

High-technology and precision instrumentation are now available to study the composition and dynamics of the atmosphere. But such tools are recent inventions and therefore have been providing data for only a short time span. To understand fully the behavior of the atmosphere and to anticipate future climate change, we must somehow discover how climate has changed over broad expanses of time.

Ice cores are an indispensable source of data for reconstructing past climates. Research based on vertical cores taken from the Greenland and Antarctic ice sheets, as well as many alpine glaciers, has changed our basic understanding of how the climate system works.

Scientists collect samples with a drilling rig, like a small version of an oil drill. A hollow shaft follows the drill head into the ice, and an ice core is extracted. In this way, cores that sometimes exceed 2000 meters (6500 feet) in length and may represent more than 200,000 years of climate history are acquired for study.

The ice provides a detailed record of changing air temperatures and snowfall. Air bubbles trapped in the ice record variations in atmospheric composition. Changes in carbon dioxide and methane are linked to fluctuating temperatures. The cores also include atmospheric fallout such as windblown dust, volcanic ash, pollen, and modern-day pollution.

Past temperatures are determined by *oxygen isotope analysis*. This technique is based on precise measurement of the ratio between two isotopes of oxygen: O^{16}, which is the most common, and the heavier O^{18}. More O^{18} is evaporated from the oceans when temperatures are high, and less is evaporated when temperatures are low. Therefore, the heavier isotope is more abundant in the precipitation of warm eras and less abundant during colder periods. Using this principle, scientists are able to produce a record of past temperature changes. A portion of such a record is shown in Figure 11.B.

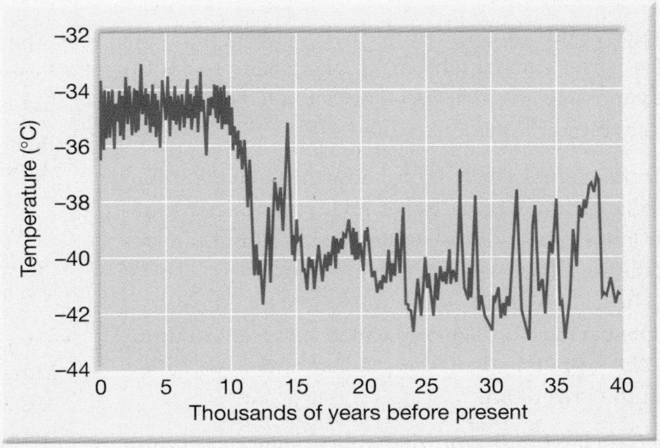

Figure 11.B This graph showing temperature variations over the past 40,000 years is derived from oxygen isotope analysis of ice cores recovered from the Greenland ice sheet. (After U.S. Geological Survey)

global temperatures rise, and when there is a reduction in these gases, as occurred during the Ice Age, temperatures fall. Therefore, reductions in the concentrations of greenhouse gases help explain the magnitude of the temperature drop that occurred during glacial times. Although scientists know that concentrations of carbon dioxide and methane dropped, they do not know what caused the drop. As often occurs in science, observations gathered during one investigation yield information and raise questions that require further analysis and explanation.

Obviously, whenever Earth enters an ice age, extensive areas of land that were once ice free are covered with ice and snow. In addition, a colder climate causes the area covered by sea ice (frozen surface sea water) to expand as well. Ice and snow reflect a large portion of incoming solar energy back to space. Thus, energy that would have warmed Earth's surface and the air above is lost and global cooling is reinforced.*

Yet another factor that influences climate during glacial times relates to ocean currents. Research has shown that ocean circulation changes during ice ages. For example, studies suggest that the warm current that transports large amounts of heat from the topics toward higher latitudes in the North Atlantic was significantly weaker during the ice age. This would lead to a colder climate in Europe, amplifying the cooling attributable to orbital variations.

Sea ice (frozen surface seawater) in the Arctic Ocean. (Wayne Lynch/DRK)

In conclusion, we emphasize that the ideas just discussed do not represent the only possible explanations for glacial ages. Although interesting and attractive, these proposals are certainly not without critics, nor are they the only possibilities currently under study. Other factors may be, and probably are, involved.

*Recall from Chapter 1 that something that reinforces (adds to) the initial change is called a *positive feedback mechanism*. To review this idea, see the discussion on feedback mechanisms in the section on "Earth as a System" in Chapter 1.

The Chapter in Review

- A *glacier* is a thick mass of ice originating on the land from the compaction and recrystallization of snow, and it shows evidence of past or present flow. Today, *valley* or *alpine glaciers* are found in mountain areas where they usually follow valleys that were originally occupied by streams. *Ice sheets* exist on a much larger scale, covering most of Greenland and Antarctica.
- Near the surface of a glacier, in the *zone of fracture*, ice is brittle. However, below about 50 meters, pressure is great, causing ice to *flow* like a *plastic material*. A second important mechanism of glacial movement consists of the entire ice mass *slipping* along the ground.
- The average velocity of glacial movement is generally quite slow, but it varies considerably from one glacier to another. The advance of some glaciers is characterized by periods of extremely rapid movements called *surges*.
- Glaciers form in areas where more snow falls in winter than melts during summer. Snow accumulation and ice formation occur in the *zone of accumulation*. Its outer limits are defined by the *snowline*. Beyond the snowline is the *zone of wastage*, where there is a net loss to the glacier. The *glacial budget* is the balance, or lack of balance, between accumulation at the upper end of the glacier, and loss, called *ablation*, at the lower end.
- Glaciers erode land and acquire debris by *plucking* (lifting pieces of bedrock out of place) and *abrasion* (grinding and scraping of a rock surface). Mass-wasting processes also make significant contributions to the load of many alpine glaciers. Erosional features produced by valley glaciers include *glacial troughs, hanging valleys, pater noster lakes, fiords, cirques, arêtes, horns,* and *roches moutonnées*.
- Any sediment of glacial origin is called *drift*. The two distinct types of glacial drift are (1) *till*, which is unsorted sediment deposited directly by the ice; and (2) *stratified drift*, which is relatively well-sorted sediment laid down by glacial meltwater.
- The most widespread features created by glacial deposition are layers or ridges of till, called *moraines*. Associated with valley glaciers are *lateral moraines*, formed along the sides of the valley, and *medial moraines*, formed between two valley glaciers that have joined. *End moraines*, which mark the former position of the front of a glacier, and *ground moraine*, an undulating layer of till deposited as the ice front retreats, are common to both valley glaciers and ice sheets. An *outwash plain* is often associated with the end moraine of an ice sheet. A *valley train* may form when the glacier is confined to a valley. Other depositional features include *drumlins* (streamlined asymmetrical hills composed of till), *eskers* (sinuous ridges composed largely of sand and gravel deposited by streams flowing in tunnels beneath the ice, near the terminus of a glacier), and *kames* (steep-sided hills composed of sand and gravel).
- The *Ice Age*, which began 2 million to 3 million years ago, was a very complex period characterized by a number of advances and withdrawals of glacial ice. Most of the major glacial episodes occurred during a division of the geologic time scale called the *Pleistocene epoch*. Perhaps the most convincing evidence for the occurrence of several glacial advances during the Ice Age is the widespread existence of *multiple layers of drift* and an uninterrupted record of climate cycles preserved in *seafloor sediments*.
- In addition to massive erosional and depositional work, other effects of Ice Age glaciers included the *migration* of organisms, *changes in stream courses, adjustment of the crust* by rebounding after the removal of the immense load of ice, and *climate changes* caused by the existence of the glaciers themselves. In the sea, the most far-reaching effect of the Ice Age was the *worldwide change* in *sea level* that accompanied each advance and retreat of the ice sheets.
- Any theory that attempts to explain the causes of glacial ages must answer two basic questions: (1) What causes the onset of glacial conditions? and (2) What caused the alternating glacial and interglacial stages that have been documented for the Pleistocene epoch? Two of the many hypotheses for the cause of glacial ages involve (1) plate tectonics and (2) variations in Earth's orbit. Other factors that are related to climate change during glacial ages include changes in atmospheric composition, variations in the amount of sunlight reflected by Earth's surface, and changes in ocean circulation.

Key Terms

ablation (p. 248)
abrasion (p. 250)
alpine glacier (p. 245)
arête (p. 252)
calving (p. 248)
cirque (p. 252)
crevasse (p. 247)
drumlin (p. 257)
end moraine (p. 255)
esker (p. 257)
fiord (p. 252)
glacial budget (p. 248)
glacial drift (p. 254)
glacial erratic (p. 254)
glacial striations (p. 250)
glacial trough (p. 252)
glacier (p. 244)
ground moraine (p. 256)
hanging valley (p. 252)
horn (p. 252)
ice cap (p. 246)
ice sheet (p. 245)
ice shelf (p. 245)
kame (p. 257)
kettle (p. 256)
lateral moraine (p. 255)
medial moraine (p. 255)
outlet glacier (p. 246)
outwash plain (p. 256)
pater noster lakes (p. 252)
piedmont glacier (p. 246)
Pleistocene epoch (p. 258)
plucking (p. 250)
pluvial lake (p. 260)
roche moutonnée (p. 253)
rock flour (p. 250)
stratified drift (p. 254)
till (p. 254)
valley glacier (p. 245)
valley train (p. 256)
zone of accumulation (p. 248)
zone of wastage (p. 248)

Questions for Review

1. Where are glaciers found today? What percentage of Earth's land area do glaciers cover?
2. Describe how glaciers fit into the hydrologic cycle. What role do they play in the rock cycle?
3. Each of the following statements refers to a particular type of glacier. Name the type of glacier.
   a. The term *continental* is often used to describe this type of glacier.
   b. This type of glacier is also called an *alpine glacier.*
   c. This is a glacier formed when one or more valley glaciers spreads out at the base of a steep mountain front.
   d. Greenland is the only example in the Northern Hemisphere.
   e. This is a stream of ice leading from the margin of an ice sheet through the mountains to the sea.
4. Describe the two components of glacial flow. At what rates do glaciers move? In a valley glacier, does all of the ice move at the same rate? Explain.
5. Why do crevasses form in the upper portion of a glacier but not below a depth of about 50 meters (160 feet)?
6. Under what circumstances will the front of a glacier advance? Retreat? Remain stationary?
7. How do glaciers accumulate their load of sediment?
8. How does a glaciated mountain valley differ in appearance from a mountain valley that was not glaciated?
9. List and describe the erosional features you might expect to see in an area where valley glaciers exist or have recently existed.
10. What is glacial drift? What is the difference between till and stratified drift? What general effect do glacial deposits have on the landscape?
11. List the four basic moraine types. What do all moraines have in common? What is the significance of *terminal* and *recessional moraines*?
12. List and briefly describe four depositional features other than moraines.
13. Examine the photo of the drumlin in Figure 11.19. From what direction (right or left) did the ice sheet advance in this area?
14. How does a kettle form?
15. About what percentage of Earth's land surface was covered at some time by Pleistocene glaciers? How does this compare to the area presently covered by ice sheets and glaciers? (Check your answer with Question 1.)
16. During the Pleistocene epoch the extent of glacial ice in the Northern Hemisphere was about twice as great as in the Southern Hemisphere. Briefly explain why this was the case.
17. List three indirect effects of Ice Age glaciers.
18. How might plate tectonics help us understand the cause of ice ages? Can plate tectonics explain the alternation between glacial and interglacial climates during the Pleistocene?

Online Study Guide

WWW

The *Essentials of Geology* Web site uses the resources and flexibility of the Internet to aid in your study of the topics in this chapter. Written and developed by geology instructors, this site will help improve your understanding of geology. Visit **www.prenhall.com/lutgens** and click on the cover of *Essentials of Geology 9e* to find:

- On-line review quizzes.
- Web-based critical thinking and writing exercises.
- Links to chapter-specific Web resources.
- Internet-wide key term searches.

http://www.prenhall.com/lutgens

Arizona's Organ Pipe Cactus National Monument is in the Sonoran Desert. Aho Mountains in the background. *(Photo by Jeff Lepore/Photo Researchers, Inc.)*

CHAPTER 12

Deserts and Wind

Focus on Learning

To assist you in learning the important concepts in this chapter, you will find it helpful to focus on the following questions:

- What are the causes of deserts in the lower and middle latitudes?
- What are the roles of weathering, water, and wind in arid and semiarid climates?
- How have many of the landscapes in the dry Basin and Range region of the United States evolved?
- How does wind erode?
- What are some depositional features produced by wind?

Climate has a strong influence on the nature and intensity of Earth's external processes. This was clearly demonstrated in the preceding chapter on glaciers. Another excellent example of the strong link between climate and geology is seen when we examine the development of arid landscapes. The word *desert* literally means *deserted* or *unoccupied.* For many dry regions this is a very appropriate description, although where water is available in deserts, plants and animals thrive. Nevertheless, the world's dry regions are among the least familiar land areas on Earth outside the polar realm.

Desert landscapes frequently appear stark. Their profiles are not softened by a carpet of soil and abundant plant life. Instead, barren rocky outcrops with steep, angular slopes are common. At some places the rocks are tinted orange and red. At others they are gray and brown and streaked with black. For many visitors, desert scenery exhibits a striking beauty; to others the terrain seems bleak. No matter which feeling is elicited, it is clear that deserts are very different from the more humid places where most people live.

Did You Know?

The dry regions of the world encompass about 42 million km² (16.4 mi²), a surprising 30 percent of Earth's land surface. No other climate group covers so large a land area.

As you will see, arid regions are not dominated by a single geologic process. Rather, the effects of tectonic forces, running water, and wind are all apparent. Because these processes combine in different ways from place to place, the appearance of desert landscapes varies a great deal as well (Figure 12.1).

Distribution and Causes of Dry Lands

GEODe Essentials of Geology
Deserts and Winds
▼ Distribution and Causes of Dry Lands

We all recognize that deserts are dry places, but just what is meant by the term *dry?* That is, how much rain defines the boundary between humid and dry regions? Sometimes it is arbitrarily defined by a single rainfall figure—for example, 25 centimeters per year of precipitation. However, the concept of *dryness* is very relative; it refers to *any situation in which water deficiency exists.* Hence, climatologists define **dry climate** as one in which yearly precipitation is less than the potential loss of water by evaporation.

Dryness, then, is related not only to annual rainfall totals but is also a function of evaporation, which in turn closely depends upon temperature. As temperatures climb, potential evaporation also increases. Twenty-five centimeters of rain may support only a sparse vegetative cover in Nevada, whereas the same amount of precipitation falling in northern Scandinavia is sufficient to support forests.

Within these water-deficient regions, two climatic types are commonly recognized: **desert,** which is *arid,* and **steppe,** which is *semiarid.* The two share many features; their differences are primarily a matter of degree. The steppe is a marginal and more humid variant of the desert and is a transition zone that surrounds the desert and separates it from bordering humid climates. The world map showing the distribution of desert and

▼ **Figure 12.1** A scene in the Great Basin Desert. The appearance of desert landscapes varies a great deal from place to place. *(Photo by Charlie Ott/Photo Researchers, Inc.)*

steppe regions reveals that dry lands are concentrated in the subtropics and in the midlatitudes (Figure 12.2).

Low-Latitude Deserts

The heart of the low-latitude dry climates lies in the vicinities of the Tropics of Cancer and Capricorn. Figure 12.2 shows a virtually unbroken desert environment stretching for more than 9300 kilometers (nearly 5800 miles) from the Atlantic coast of North Africa to the dry lands of northwestern India. In addition to this single great expanse, the Northern Hemisphere contains another, much smaller area of subtropical desert and steppe in northern Mexico and the southwestern United States.

In the Southern Hemisphere, dry climates dominate Australia. Almost 40 percent of the continent is desert, and much of the remainder is steppe. In addition, arid and semiarid areas occur in southern Africa and make a limited appearance in coastal Chile and Peru.

What causes these bands of low-latitude desert? The answer is the global distribution of air pressure and winds. Figure 12.3, an idealized diagram of Earth's general circulation, helps visualize the relationship. Heated air in the pressure belt known as the *equatorial low* rises to great heights (usually between 15 and 20 kilometers) and then spreads out. As the upper-level flow reaches 20° to 30° latitude, north or south, it sinks toward the surface. Air that rises through the atmosphere expands and cools, a process that leads to the development of clouds and precipitation. For this reason, the areas under the influence of the equatorial low are among the rainiest on Earth. Just the opposite is true for the regions in the vicinity of 30° north and south latitude, where high pressure predominates. Here, in the zones known as the *subtropical highs,* air is subsiding. When air sinks, it is compressed and warmed. Such conditions are just the opposite of what is needed to produce clouds and precipitation. Consequently, these regions are known for their clear skies, sunshine, and ongoing drought.

Middle-Latitude Deserts

Unlike their low-latitude counterparts, middle-latitude deserts and steppes are not controlled by the subsiding air masses associated with high pressure. Instead, these dry lands exist principally because they are sheltered in the deep interiors of large landmasses. They are far removed from the ocean, which is the ultimate source of moisture for cloud formation and precipitation. One well-known example is the Gobi Desert of central Asia, shown on the map north of India (Figure 12.2).

The presence of high mountains across the paths of prevailing winds further separates these areas from water-bearing maritime air masses. Also, the mountains force the air to lose much of its water. The mechanism is simple: As prevailing winds meet mountain barriers, the air is forced to ascend. When air rises, it expands and cools, a process that can produce

The cloudless deserts of Africa and Arabia are clearly visible in this image. (NASA/Photo Researchers)

Did You Know?

The Sahara of North Africa is the world's largest desert. Extending from the Atlantic Ocean to the Red Sea, it covers about 9 million km^2 (3.5 million mi^2)—an area about the size of the U.S. By comparison, the largest desert in the U.S., Nevada's Great Basin Desert, has an area that is less than 5 percent as large as the Sahara.

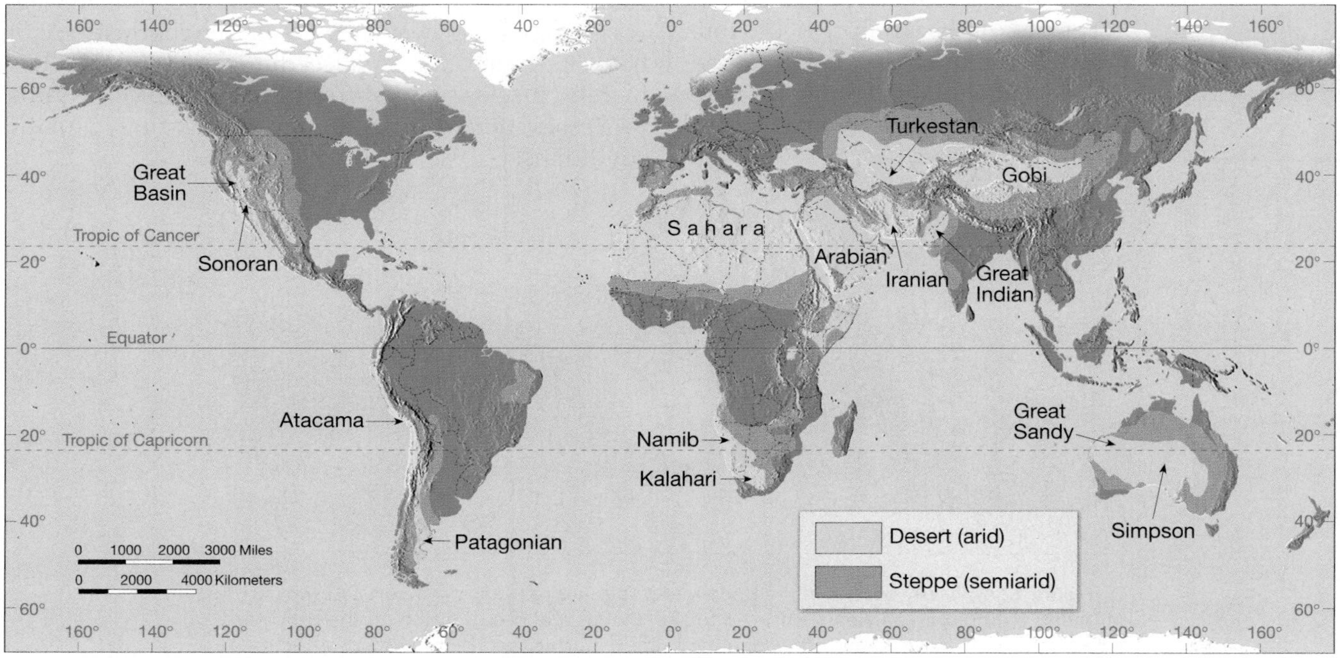

▲ **Figure 12.2** Arid and semiarid climates cover about 30 percent of Earth's land surface. No other climate group covers so large an area.

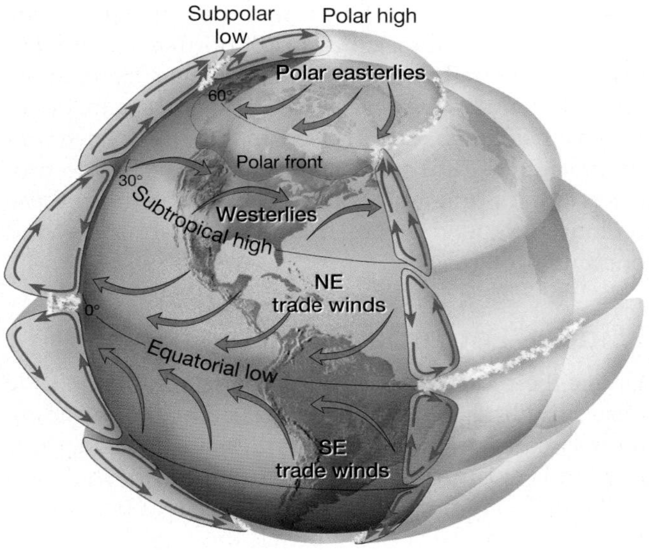

▲ **Figure 12.3** Idealized diagram of Earth's general circulation. The deserts and steppes that are centered in the latitude belt between 20° and 30° north and south coincide with the subtropical high-pressure belts. Here dry, subsiding air inhibits cloud formation and precipitation. By contrast, the pressure belt known as the equatorial low is associated with areas that are among the rainiest on Earth.

Did You Know?

Some of the highest temperatures ever recorded were in deserts. For instance, the highest authentically recorded temperature in the U.S.—as well as the entire Western Hemisphere—is 57°C (134°F), measured at Death Valley, California, on July 10, 1913. The world-record high temperature of nearly 59°C (137°F) was recorded in the Sahara at Azizia, Libya, on September 13, 1922.

clouds and precipitation. The windward sides of mountains, therefore, often have high precipitation.

By contrast, the leeward sides of mountains are usually much drier (Figure 12.4). This situation exists because air reaching the leeward side has lost much of its moisture, and if the air descends, it is compressed and warmed, making cloud formation even less likely. The dry region that results is often referred to as a **rainshadow desert.**

Because most midlatitude deserts occupy sites on the leeward sides of mountains, they can also be classified as rainshadow deserts. In North America, the Coast Ranges, Sierra Nevada, and Cascades are the foremost mountain barriers to moisture from the Pacific. In Asia, the great Himalayan chain prevents the summertime monsoon flow of moist Indian Ocean air from reaching the interior.

Because the Southern Hemisphere lacks extensive land areas in the middle latitudes, only a small area of desert and steppe occurs in this latitude range, existing primarily near the southern tip of South America in the rainshadow of the towering Andes.

Middle-latitude deserts provide an example of how tectonic processes affect climate. Rainshadow deserts exist by virtue of the mountains produced when plates collide. Without such mountain-building episodes, wetter climates would prevail where many dry regions exist today.

Geologic Processes in Arid Climates

GEODe Essentials of Geology — Deserts and Winds
▼ Common Misconceptions About Deserts

The angular hills, the sheer canyon walls, and the desert surface of pebbles or sand contrast sharply with the rounded hills and curving slopes of more humid places. Indeed, to a visitor from a humid region, a desert landscape may seem to have been shaped by forces altogether different from those operating in well-watered areas. However, although the contrasts might be striking, they do not reflect different processes. They merely disclose the differing effects of the same processes that operate under contrasting climatic conditions.

▲ **Figure 12.4** Many deserts in the middle latitudes are rainshadow deserts. As moving air meets a mountain barrier, it is forced to rise. Clouds and precipitation on the windward side often result. Air descending the leeward side is much drier. The mountains effectively cut the leeward side off from the sources of moisture, producing a rainshadow desert. The Great Basin desert is a rainshadow desert that covers nearly all of Nevada and portions of adjacent states.

Weathering

In humid regions, relatively well-developed soils support an almost continuous cover of vegetation. Here the slopes and rock edges are rounded, reflecting the strong influence of chemical weathering in a humid climate. By contrast, much of the weathered debris in deserts consists of unaltered rock and mineral fragments—the results of mechanical weathering processes. In dry lands, rock weathering of any type is greatly reduced because of the lack of moisture and the scarcity of organic acids from decaying plants. However, chemical weathering is not completely lacking in deserts. Over long spans of time, clays and thin soils do form, and many iron-bearing silicate minerals oxidize, producing the rust-colored stain found tinting some desert landscapes.

The Role of Water

Permanent streams are normal in humid regions, but practically all desert stream beds are dry most of the time (Figure 12.5A). Deserts have **ephemeral streams,** which means that they carry water only in response to specific episodes of rainfall. A typical ephemeral stream might flow only a few days or perhaps just a few hours during the year. In some years, the channel might carry no water at all.

This fact is obvious even to the casual traveler who notices numerous bridges with no streams beneath them or numerous dips in the road where dry channels cross. However, when the rare heavy showers do come, so much rain falls in such a short time that all of it cannot soak in. Because desert vegetative cover is sparse, runoff is largely unhindered and consequently rapid, often creating flash floods along valley floors (Figure 12.5B). These floods are quite unlike floods in humid regions. A flood on a river like the Mississippi may take several days to reach its crest and then subside. But desert floods arrive suddenly and subside quickly. Because much of the surface material in a desert is not anchored by vegetation, the amount of erosional work that occurs during a single short-lived rain event is impressive.

Humid regions are notable for their integrated drainage systems. But in arid regions streams usually lack an extensive system of tributaries. In fact, a basic characteristic of desert streams is that they are small and die out before reaching the sea. Because the water table is usually far below the surface, few desert streams can draw upon it as streams do in humid regions (see Figure 10.4, p. 227). Without a steady supply of water, the combination of evaporation and infiltration soon depletes the stream.

The few permanent streams that do cross arid regions, such as the Colorado and Nile rivers, originate *outside* the desert, often in well-watered mountains. Here the water supply must be great to compensate for the losses occurring as the stream crosses the desert (see Box 12.1). For example, after the Nile leaves its headwaters in the lakes and mountains of central Africa, it traverses almost 3000 kilometers (1900 miles) of the Sahara *without a single tributary.* By contrast, in humid regions the discharge of a river increases as it flows downstream because tributaries and groundwater contribute additional water along the way.

It should be emphasized that *running water, although infrequent, nevertheless does most of the erosional work in deserts.* This is contrary to a common belief that wind is the most important erosional agent sculpturing desert landscapes. Although wind erosion is indeed more significant in dry areas than elsewhere, most desert landforms are carved by running water. As

California's Death Valley is part of a rainshadow desert. (James E. Paterson)

Did You Know?

All deserts aren't hot. Cold temperatures are experienced in middle-latitude deserts. For example, at Ulan Bator in Mongolia's Gobi Desert, the average *high* temperature on January days is only −19°C (−2°F)!

A.

B.

▲ **Figure 12.5** **A.** Most of the time, desert stream channels are dry. **B.** An ephemeral stream shortly after a heavy shower. Although such floods are short-lived, large amounts of erosion occur. *(Photos by E. J. Tarbuck)*

BOX 12.1

The Disappearing Aral Sea

The Aral Sea lies on the border between Uzbekistan and Kazakhstan in central Asia (Figure 12.A). The setting is the Turkestan desert, a middle-latitude desert in the rainshadow of Afghanistan's high mountains. In this region of interior drainage, two large rivers, the Amu Darya and the Syr Darya, carry water from the mountains of northern Afghanistan across the desert to the Aral Sea. Water leaves the sea by evaporation. Thus, the size of the water body depends upon the balance between river inflow and evaporation.

In 1960 the Aral Sea was one of the world's largest inland water bodies, with an area of about 67,000 square kilometers (26,000 square miles). Only the Caspian Sea, Lake Superior, and Lake Victoria were larger. By the year 2000 the area of the Aral Sea was less than 50 percent of its 1960 size, and its volume was reduced by 80 percent. The shrinking of this water body is depicted in Figure 12.B. By about 2010 all that will remain will be three shallow remnants.

What caused the Aral Sea to dry up over the past 40 years? The answer is that the flow of water from the mountains that supplied the sea was significantly reduced and then all but eliminated. As recently as 1965, the Aral Sea received about 50 cubic kilometers (12 cubic miles) of fresh water per year. By the early 1980s this number fell to nearly zero. The reason was that the waters of the Amu Darya and Syr Darya were diverted to supply a major expansion of irrigated agriculture in this dry realm.

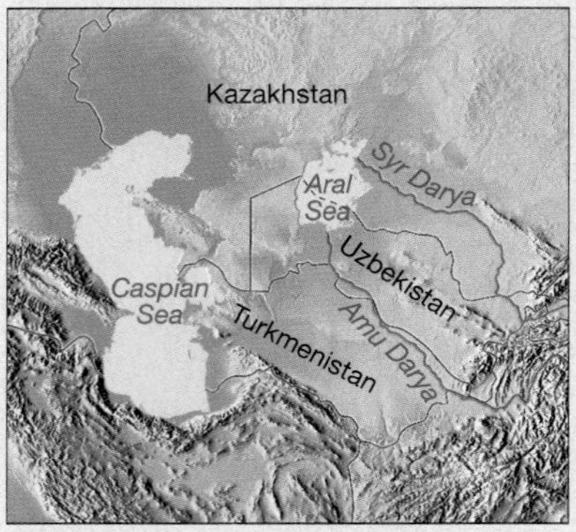

Figure 12.A The Aral Sea lies east of the Caspian Sea in the Turkestan desert. Two rivers, the Amu Darya and Syr Darya, deliver water to the sea from the mountains to the south.

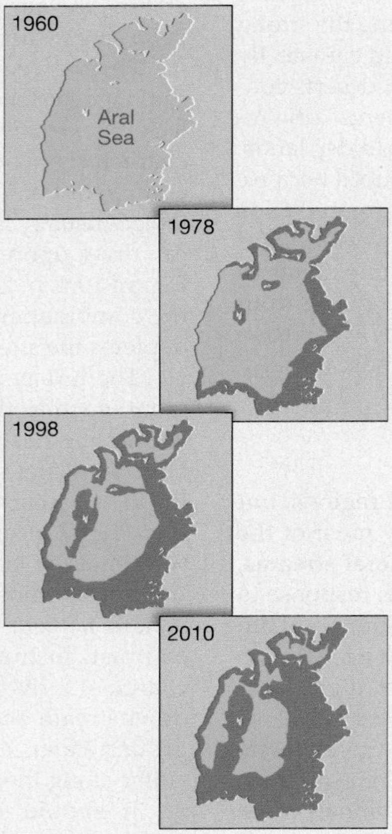

Figure 12.B The shrinking Aral Sea. By the year 2010 all that will remain are three small remnants.

The intensive irrigation greatly increased agricultural productivity, but not without significant costs. The deltas of the two major rivers have lost their wetlands, and wildlife has disappeared. The once thriving fishing industry is dead, and the 24 species of fish that once lived in the Aral Sea are no longer there. The shoreline is now tens of kilometers from the towns that were once fishing centers (Figure 12.C).

The shrinking sea has exposed millions of acres of former seabed to sun and wind. The surface is encrusted with salt and with agrochemicals brought by the rivers. Strong winds routinely pick up and deposit thousands of tons of newly exposed material every year. This process has not only contributed to a significant reduction in air quality for people living in the region but has also appreciably affected crop yields due to the deposition of salt-rich sediments on arable land.

Figure 12.C In the town of Jamboul, Kazakhstan, boats now lie in the sand because the Aral Sea has dried up. (Photo by Ergun Cagatay/Getty Images, Inc. - Liaison)

The shrinking Aral Sea has had a noticeable impact on the region's climate. Without the moderating effect of a large water body, there are greater extremes of temperature, a shorter growing season, and reduced local precipitation. These changes have caused many farms to switch from growing cotton to growing rice, which demands even more diverted water.

Environmental experts agree that the current situation cannot be sustained. Could this crisis be reversed if enough fresh water were to once again flow into the Aral Sea?

Prospects appear grim. Experts estimate that restoring the Aral Sea to about twice its present size would require stopping all irrigation from the two major rivers for 50 years. This could not be done without ruining the economies of the countries that rely on that water.

The decline of the Aral Sea is a major environmental disaster that sadly is of human making.

you will see shortly, the main role of wind is in the transportation and deposition of sediment, which creates and shapes the ridges and mounds we call *dunes.*

Basin and Range: The Evolution of a Mountainous Desert Landscape

GEODe Essentials of Geology — Deserts and Winds ▼ Reviewing Landforms and Landscapes

Because arid regions typically lack permanent streams, they are characterized as having **interior drainage.** This means that they have a discontinuous pattern of intermittent streams that do not flow out of the desert to the ocean. In the United States, the dry Basin and Range region provides an excellent example. The region includes southern Oregon, all of Nevada, western Utah, southeastern California, southern Arizona, and southern New Mexico. The name Basin and Range is an apt description for this almost 800,000-square-kilometer (more than 300,000-square-mile) area, as it is characterized by more than 200 relatively small mountain ranges that rise 900 to 1500 meters (3000 to 5000 feet) above the basins that separate them.

In this region, as in others like it around the world, most erosion occurs without reference to the ocean (ultimate base level), because the interior drainage never reaches the sea. Even where permanent streams flow to the ocean, few tributaries exist, and thus only a narrow strip of land adjacent to the stream has sea level as its ultimate level of land reduction.

The block models in Figure 12.6 depict how the landscape has evolved in the Basin and Range region. During and following uplift of the mountains, running water begins carving the elevated mass and depositing

Did You Know?

Not all low-latitude deserts are hot, sunny places with low humidity and cloudless skies. In West Coast subtropical deserts, such as the Atacama and Namib, cold ocean currents are responsible for cool temperatures, frequent fogs, and gloomy periods of dense (but rainless) low clouds.

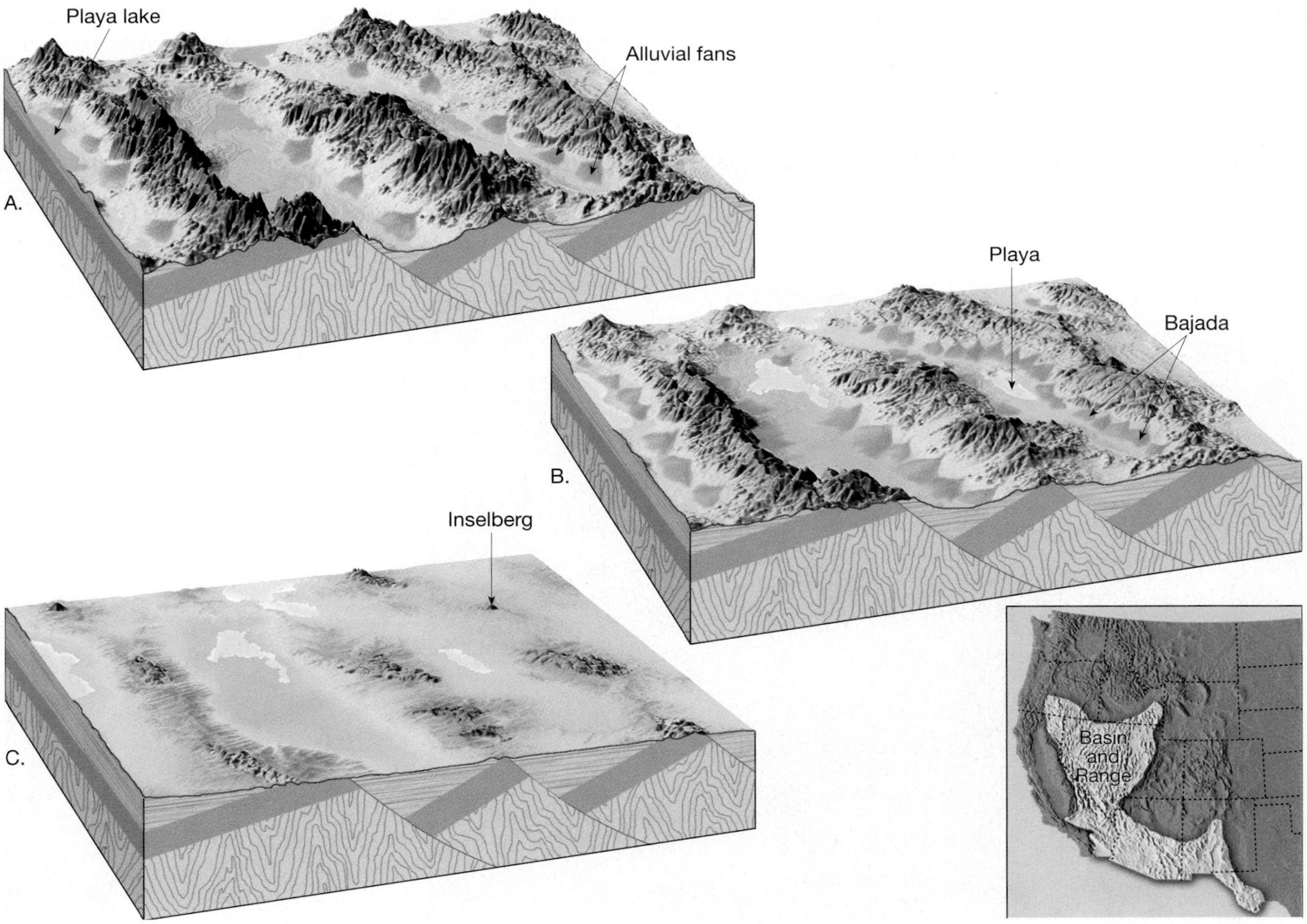

▲ **Figure 12.6** Stages of landscape evolution in a mountainous desert such as the Basin and Range region of the American West. As erosion of the mountains and deposition in the basins continue, relief diminishes. **A.** Early stage. **B.** Middle stage. **C.** Late stage.

Did You Know?

In the dry western United States, different names are used for ephemeral streams, including *wash* and *arroyo*. In other parts of the world, a dry desert stream might be called a *wadi* (Arabia and North Africa), a *donga* (South America), or a *nullah* (India).

large quantities of debris in the basin. During this early stage, relief is greatest, and as erosion lowers the mountains and sediment fills the basins, elevation differences diminish.

When the occasional torrents of water produced by sporadic rains move down the mountain canyons, they are heavily loaded with sediment. Emerging from the confines of the canyon, the runoff spreads over the gentler slopes at the base of the mountains and quickly loses velocity. Consequently, most of its load is dumped within a short distance. The result is a cone of debris known as an **alluvial fan** at the mouth of a canyon (Figure 12.7). Over the years, a fan enlarges, eventually coalescing with fans from adjacent canyons to produce an apron of sediment (*bajada*) along the mountain front (Figure 12.6B).

Desert rains are often in the form of thunderstorms. (Warren Faidley/DRK)

On the rare occasions of abundant rainfall or snowmelt in the mountains, streams may flow across the alluvial fans to the center of the basin, converting the basin floor into a shallow **playa lake.** Playa lakes last only a few days or weeks, before evaporation and infiltration remove the water. The dry, flat lake bed that remains is termed a *playa*.

Playas are typically composed of fine silts and clays and are occasionally encrusted with salts left behind by evaporation. These precipitated salts may be unusual. A case in point is the sodium borate (better known as borax) mined from ancient playa lake deposits in Death Valley, California.

With the ongoing erosion of the mountain mass and the accompanying sedimentation, the local relief continues to diminish. Eventually nearly the entire mountain mass is gone. Thus, by the late stages of erosion, the mountain areas are reduced to a few large bedrock knobs (called *inselbergs*) projecting above the sediment-filled basin (Figure 12.6C).

Each of the stages of landscape evolution in an arid climate depicted in Figure 12.6 can be observed in the Basin and Range region. Recently uplifted mountains in an early stage of erosion are found in southern Oregon and northern Nevada. Death Valley, California, and southern Nevada fit into the more advanced middle stage, whereas the late stage, with its inselbergs, can be seen in southern Arizona.

Transportation of Sediment by Wind

Moving air, like moving water, is turbulent and able to pick up loose debris and transport it to other locations. Just as in a stream, the velocity of wind increases with height above the surface. Also like a stream, wind transports fine particles in suspension while heavier ones are carried as bed load. However, the transport of sediment by wind differs from that by running water in two significant ways. First, wind's lower density compared to

▲ **Figure 12.7** Aerial view of alluvial fans in Death Valley, California. The size of the fan depends on the size of the drainage basin. As the fans grow, they eventually coalesce to form a bajada. *(Photo by Michael Collier)*

water renders it less capable of picking up and transporting coarse materials. Second, because wind is not confined to channels, it can spread sediment over large areas, as well as high into the atmosphere.

Bed Load

The **bed load** carried by wind consists of sand grains. Observations in the field and experiments using wind tunnels indicate that windblown sand moves by skipping and bouncing along the surface—a process termed **saltation.** The term is not a reference to salt, but instead derives from the Latin word meaning "to jump."

The movement of sand grains begins when wind reaches a velocity sufficient to overcome the inertia of the resting particles. At first the sand rolls along the surface. When a moving sand grain strikes another grain, one or both of them may jump into the air. Once in the air, the grains are carried forward by the wind until gravity pulls them back toward the surface. When the sand hits the surface, it either bounces back into the air or dislodges other grains, which then jump upward. In this manner, a chain reaction is established, filling the air near the ground with saltating sand grains in a short period of time (Figure 12.8).

Bouncing sand grains never travel far from the surface. Even when winds are very strong, the height of the saltating sand seldom exceeds a meter and usually is no greater than a half meter.

Suspended Load

Unlike sand, finer dust particles can be swept high into the atmosphere by the wind. Because dust is often composed of rather flat particles that have large surface areas compared to their weight, it is relatively easy for turbulent air to counterbalance the pull of gravity and keep these fine particles airborne for hours or even days. Although both silt and clay can be carried in suspension, silt commonly makes up the bulk of the **suspended load** because the reduced level of chemical weathering in deserts produces only small amounts of clay.

▼ **Figure 12.8** A cloud of saltating sand grains moving up the gentle slope of a dune. *(Photo by Stephen Trimble)*

Fine particles are easily carried by the wind, but they are not so easily picked up to begin with. The reason is that the wind velocity is practically zero within a very thin layer close to the ground. Thus, the wind cannot lift the sediment by itself. Instead, the dust must be ejected or spattered into the moving air currents by bouncing sand grains or other disturbances. This idea is illustrated nicely by a dry unpaved country road on a windy day. Left undisturbed, little dust is raised by the wind. However, as a car or truck moves over the road, the layer of silt is kicked up, creating a thick cloud of dust.

Salt flats are sometimes present in desert basins. (Marli Miller)

Although the suspended load is usually deposited relatively near its source, high winds are capable of carrying large quantities of dust great distances. In the 1930s, silt that was picked up in Kansas was transported to New England and beyond into the North Atlantic. Similarly, dust blown from the Sahara has been traced as far as the West Indies.

Wind Erosion

Compared to running water and glaciers, wind is a relatively insignificant erosional agent. Recall that even in deserts, most erosion is performed by intermittent running water, not by the wind. Wind erosion is more effective in arid lands than in humid areas because in humid regions moisture binds particles together and vegetation anchors the soil. For wind to be effective, dryness and scanty vegetation are important prerequisites.

When such circumstances exist, wind may pick up, transport, and deposit great quantities of fine sediment. During the 1930s, parts of the Great Plains experienced great dust storms. The plowing under of the natural vegetative cover for farming, followed by severe drought, made the land ripe for wind erosion and led to the area being labeled the Dust Bowl (see Box 12.2).

Did You Know?

Deserts don't necessarily consist of mile after mile of drifting sand dunes. Surprisingly, sand accumulations represent only a small percentage of the total desert area. In the Sahara, dunes cover only one-tenth of its area. The sandiest of all deserts, the Arabian, is one-third sand-covered.

Deflation, Blowouts, and Desert Pavement

One way that wind erodes is by **deflation,** the lifting and removal of loose material. Although the effects of deflation are sometimes difficult to notice because the entire surface is being lowered at the same time, they can be significant. In portions of the 1930s Dust Bowl, vast areas of land were lowered by as much as a meter in only a few years.

The most noticeable results of deflation in some places are shallow depressions called **blowouts** (Figure

BOX 12.2

Dust Bowl—Soil Erosion in the Great Plains

During a span of dry years in the 1930s, large dust storms plagued the Great Plains. Because of the size and severity of these storms, the region came to be called the Dust Bowl, and the time period the Dirty Thirties. The heart of the Dust Bowl was nearly 100 million acres in the panhandles of Texas and Oklahoma and adjacent parts of Colorado, New Mexico, and Kansas. To a lesser extent, dust storms were also a problem over much of the Great Plains, from North Dakota to west-central Texas.

At times dust storms were so severe that they were called "black blizzards" and "black rollers" because visibility was reduced to only a few feet. Numerous storms lasted for hours and stripped huge volumes of topsoil from the land (Figure 12.D).

In the spring of 1934, a windstorm that lasted for a day and a half created a dust cloud 2000 kilometers (1200 miles) long. As the sediment moved east, New York had "muddy rains" and Vermont "black snows." Another storm carried dust more than 3 kilometers (2 miles) into the atmosphere and transported it 3000 kilometers from its source in Colorado to create "midday twilight!" in New England and New York.

What caused the Dust Bowl? Clearly, the fact that portions of the Great Plains experienced some of North America's strongest winds is important. However, it was the expansion of agriculture that set the stage for the disastrous period of soil erosion. Mechanization allowed the rapid transformation of the grass-covered prairies of this semiarid region into farms. Between the 1870s and 1930, cultivation expanded nearly tenfold, from about 10 million acres to more than 100 million acres.

Figure 12.D Dust blackens the sky on May 21, 1937, near Elkhart, Kansas. It was because of storms like this that portions of the Great Plains were called the "Dust Bowl" in the 1930s. (Photo courtesy of the Library of Congress)

As long as precipitation was adequate, the soil remained in place. However, when a prolonged drought struck in the 1930s, the unprotected fields were vulnerable to the wind. The result was severe soil loss, crop failure, and economic hardship.

Beginning in 1939, a return to rainier conditions brought relief. New farming practices that reduced soil loss by wind were instituted. Although dust storms are less numerous and not as severe as in the Dirty Thirties, soil erosion by strong winds still occurs periodically whenever the combination of drought and unprotected soil exists.

Did You Know?

The driest place on Earth is in Chile's Atacama Desert, a narrow belt of dry land that extends along the Pacific Coast of South America. The town of Arica has an average rainfall of just 0.03 in. per year. Over a span of 59 years, this region received a *total* of less than 2 in. of rain.

12.9). In the Great Plains region, from Texas north to Montana, thousands of blowouts are visible on the landscape. They range from small dimples less than a meter deep and 3 meters wide to depressions that approach 50 meters in depth and several kilometers across. The factor that controls the depths of these basins (that is, acts as base level) is the local water table. When blowouts are lowered to the water table, damp ground and vegetation prevent further deflation.

▲ **Figure 12.9** This photo was taken north of Granville, North Dakota, in July 1936, during a prolonged drought. Strong winds removed the soil that was not anchored by vegetation. The mounds are 1.2 meters (4 feet) high and show the level of the land prior to deflation. *(Photo courtesy of the State Historical Society of North Dakota)*

In portions of many deserts, the surface is a layer of coarse pebbles and gravels too large to be moved by the wind. Such a layer, called **desert pavement,** is created as the wind lowers the surface by removing fine material until eventually only a continuous cover of coarse sediment remains (Figure 12.10). Once desert pavement becomes established, a process that may take hundreds of years, the surface is protected from further deflation if left undisturbed. However, as the layer is only one or two stones thick, disruption by vehicles or animals can dislodge the pavement and expose the fine-grained material below. If this happens, the surface is once again subject to deflation.

Wind Abrasion

Like glaciers and streams, wind also erodes by **abrasion.** In dry regions and along some beaches, windblown sand cuts and polishes exposed rock surfaces. It sometimes creates interestingly shaped stones called **ventifacts** (Figure 12.11). The side of the stone exposed to the prevailing wind is abraded, leaving it polished,

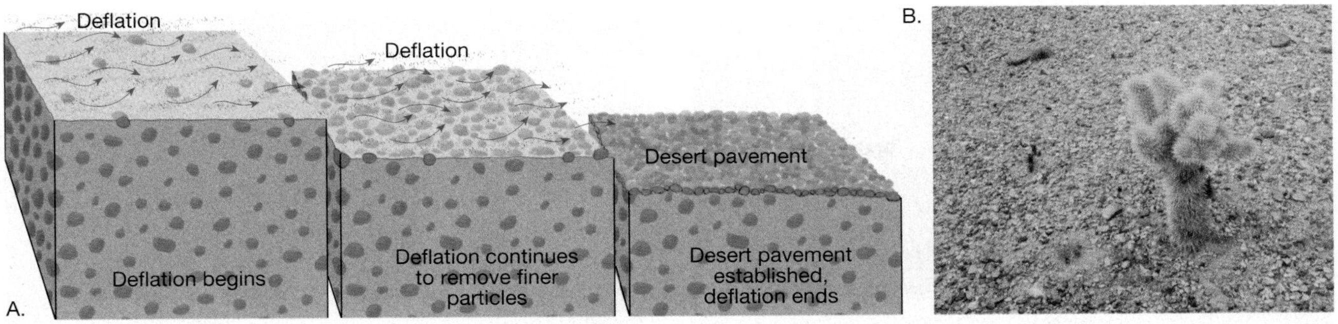

▲ **Figure 12.10** **A.** Formation of desert pavement. As these cross-sections illustrate, coarse particles gradually become concentrated into a tightly packed layer as deflation lowers the surface by removing sand and silt. **B.** If left undisturbed, desert pavement such as this in Arizona's Sonoran Desert will protect the surface from further deflation. *(Photo by David Muench)*

pitted, and with sharp edges. If the wind is not consistently from one direction or if the pebble becomes reoriented, it may have several faceted surfaces.

Abrasion is often credited for accomplishments beyond its actual capabilities. Such features as balanced rocks that stand high atop narrow pedestals and intricate detailing on tall pinnacles are not the results of abrasion. Because sand seldom travels more than a meter above the surface, the wind's sandblasting effect is obviously quite limited in vertical extent.

Wind Deposits

GEODe: ESSENTIALS OF GEOLOGY
Deserts and Winds
▼ Reviewing Landforms and Landscapes

Although wind is relatively unimportant in producing *erosional* landforms, significant *depositional* landforms are created by the wind in some regions. Accumulations of windblown sediment are particularly conspicuous in the world's dry lands and along many sandy coasts.

▼ **Figure 12.11** Ventifacts are rocks that are polished and shaped by sandblasting. *(Photo by Stephen Trimble)*

Wind deposits are of two distinctive types: (1) mounds and ridges of sand from the wind's bed load, which we call *dunes,* and (2) extensive blankets of silt, called *loess,* that once were carried in suspension.

Sand Deposits

As is the case with running water, wind drops its load of sediment when its velocity falls and the energy available for transport diminishes. Thus, sand begins to accumulate wherever an obstruction across the path of the wind slows its movement. Unlike many deposits of silt, which form blanketlike layers over large areas, winds commonly deposit sand in mounds or ridges called **dunes** (Figure 12.12).

As moving air encounters an object, such as a clump of vegetation or a rock, the wind sweeps around and over it, leaving a shadow of slower moving air behind the obstacle as well as a smaller zone of quieter air just in front of the obstacle. Some of the saltating sand grains moving with the wind come to rest in these wind shadows. As the accumulation of sand continues, it becomes a more imposing barrier to the wind and thus a more efficient trap for even more sand. If there is a sufficient supply of sand and the wind blows steadily for a long enough time, the mound of sand grows into a dune.

Thick plumes of dust blowing from the Sahara toward southern Europe. (NASA)

Many dunes have an asymmetrical profile with the leeward (sheltered) slope being steep and the windward slope more gently inclined. Sand moves up the gentle slope on the windward side by saltation. Just beyond the crest of the dune, where the wind velocity is reduced, the sand accumulates. As more sand collects, the slope steepens, and eventually some of it slides or slumps under the pull of gravity. In this way, the leeward slope of the dune, called the **slip face,** maintains an angle of about 34 degrees, the angle of repose for loose dry sand (Figure 12.12B). (Recall from Chapter 8

A.

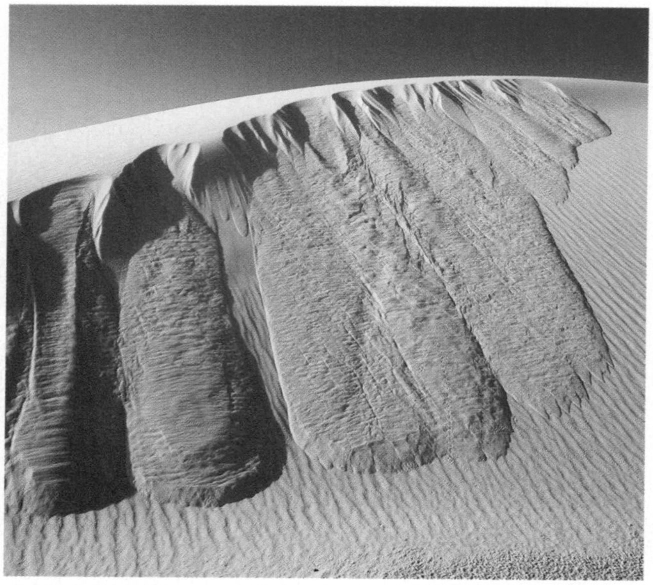

B.

▲ **Figure 12.12** **A.** Dunes composed of gypsum sand at White Sands National Monument in southeastern New Mexico. These dunes are gradually migrating from the background (top) toward the foreground (bottom) of the photo. Strong winds move sand up the more gentle windward slopes. As sand accumulates near the dune crest, the slope steepens and some of the sand slides down the steeper *slip face*. **B.** Sand sliding down the steep slip face of a dune in White Sands National Monument, New Mexico. *(Photos by Michael Collier)*

This 1937 photo shows an abandoned Dust Bowl era farm in Oklahoma. (U.S. Dept. of Agriculture)

that the angle of repose is the steepest angle at which loose material remains stable.) Continued sand accumulation, coupled with periodic slides down the slip face, results in the slow migration of the dune in the direction of air movement.

As sand is deposited on the slip face, layers form that are inclined in the direction the wind is blowing. These sloping layers are called **cross beds.** When the dunes are eventually buried under other layers of sediment and become part of the sedimentary rock record, their asymmetrical shape is destroyed, but the cross beds remain as testimony to their origin. Nowhere is cross-bedding more prominent than in the sandstone walls of Zion Canyon in southern Utah (Figure 12.13).

Did You Know?

Migrating dunes can lead to costly problems. For example, to keep a portion of Highway 95 near Winnemucca, Nevada, open to traffic, sand must be removed about three times a year. Each time between 1500 and 4000 m^3 of sand are removed. Attempts at stabilizing the dunes with vegetation have failed.

Types of Sand Dunes

Dunes are not just random heaps of windblown sediment. Rather, they are accumulations that usually assume surprisingly consistent patterns. A broad assortment of dune forms exists, generally simplified to a few major types for discussion. Of course, gradations exist among different forms as well as irregularly shaped dunes that do not fit easily into any category. Several factors influence the form and size that dunes ultimately assume. These include wind direction and velocity, availability of sand, and the amount of vegetation present. Six basic dune types are shown in Figure 12.14, with arrows indicating wind directions.

Barchan Dunes. Solitary sand dunes shaped like crescents and with their tips pointing downwind are called **barchan dunes** (Figure 12.14A). These dunes form where supplies of sand are limited and the surface is relatively flat, hard, and lacking vegetation. They migrate slowly with the wind at a rate of up to 15 meters per year. Their size is usually modest, with the largest barchans reaching heights of about 30 meters while the maximum spread between their horns approaches 300 meters. When the wind direction is nearly constant, the crescent form of these dunes is nearly symmetrical. However, when the wind direction is not perfectly fixed, one tip becomes larger than the other.

Transverse Dunes. In regions where the prevailing winds are steady, sand is plentiful, and vegetation is sparse or absent, the dunes form a series of long ridges that are separated by troughs and oriented at right angles to the prevailing wind. Because of this orientation, they are termed **transverse dunes** (Figure 12.14B). Typically, many coastal dunes are of this type. In addition, transverse dunes are common in many arid regions where the extensive surface of wavy sand is sometimes

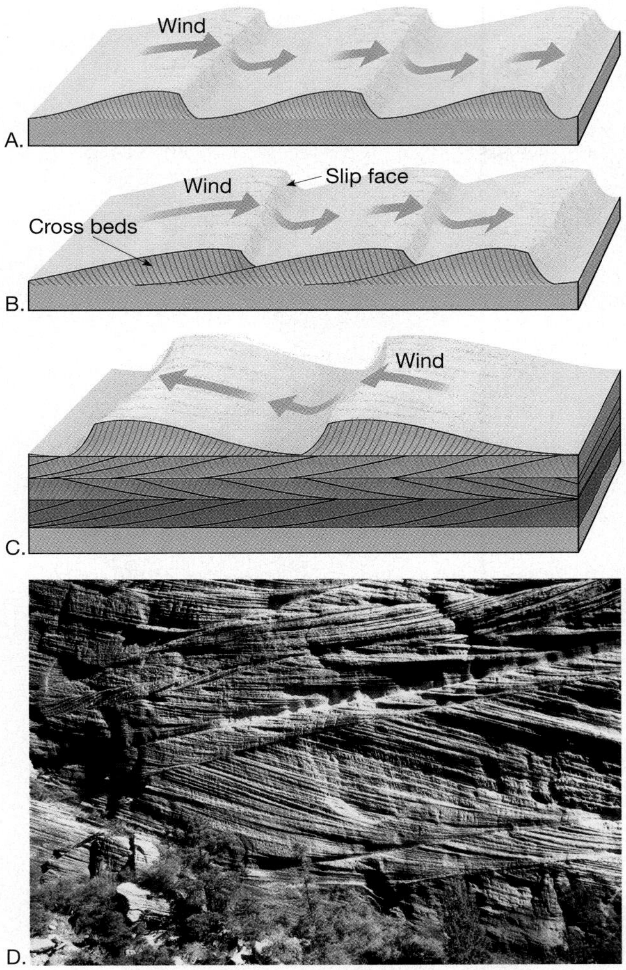

▲ **Figure 12.13** As parts **A** and **B** illustrate, dunes commonly have an asymmetrical shape. The steeper leeward side is called the slip face. Sand grains deposited on the slip face at the angle of repose create the cross-bedding of the dunes. **C.** A complex pattern develops in response to changes in wind direction. Also notice that when dunes are buried and become part of the sedimentary record, the cross-bedded structure is preserved. **D.** Cross beds are an obvious characteristic of the Navajo Sandstone in Zion National Park, Utah. *(Photo by Marli Miller)*

called a *sand sea.* In some parts of the Sahara and Arabian deserts, transverse dunes reach heights of 200 meters, are 1 to 3 kilometers across, and can extend for distances of 100 kilometers or more.

There is a relatively common dune form that is intermediate between isolated barchans and extensive waves of transverse dunes. Such dunes, called **barchanoid dunes,** form scalloped rows of sand oriented at right angles to the wind (Figure 12.14C). The rows resemble a series of barchans that have been positioned side by side. Visitors exploring the gypsum dunes at White Sands National Monument, New Mexico, will recognize this form (Figure 12.12A).

Aerial view of a solitary barchan. (John S. Shelton)

Longitudinal Dunes. **Longitudinal dunes** are long ridges of sand that form more or less parallel to the prevailing wind and where sand supplies are moderate (Figure 12.14D). Apparently the prevailing wind direction must vary somewhat but still remain in the same quadrant of the compass. Although the smaller types are only 3 or 4 meters high and several dozens of meters long, in some large deserts longitudinal dunes can reach great size. For example, in portions of North Africa, Arabia, and central Australia, these dunes can approach a height of 100 meters and extend for distances of more than 100 kilometers (62 miles).

Parabolic Dunes. Unlike the other dunes that have been described thus far, **parabolic dunes** form where vegetation partially covers the sand. The shape of these dunes resembles the shape of barchans except that their tips point into the wind rather than downwind (Figure 12.14E). Parabolic dunes often form along coasts where there are strong onshore winds and abundant sand. If the sand's sparse vegetative cover is disturbed at some spot, deflation creates a blowout. Sand is then transported out of the depression and deposited as a curved rim that grows higher as deflation enlarges the blowout.

Did You Know?

The highest dunes in the world are located along the southwest coast of Africa in the Namib Desert. In places, these huge dunes reach heights of 300 to 350 m (1000 to 1167 ft). The dunes at Great Sand Dunes National Park in southern Colorado are the highest in North America, rising over 210 m (700 ft) above the surrounding terrain.

Star Dunes. Confined largely to parts of the Sahara and Arabian deserts, **star dunes** are isolated hills of sand that exhibit a complex form (Figure 12.14F). Their name is derived from the fact that the bases of these dunes resemble multipointed stars. Usually three or four sharp-crested ridges diverge from a central high point that in some cases may approach a height of 90 meters. As their form suggests, star dunes develop where wind directions are variable.

Dunes encroaching upon irrigated fields in Egypt. (Georg Gerster/Photo Researchers)

Loess (Silt) Deposits

In some parts of the world the surface topography is mantled with deposits of windblown silt, called **loess.** Over thousands of years dust storms deposited this material. When loess is breached by streams or road cuts, it tends to maintain vertical cliffs and lacks any visible layers, as you can see in Figure 12.15.

The distribution of loess worldwide indicates that there are two principal sources for this sediment: deserts and glacial deposits. The thickest and most extensive

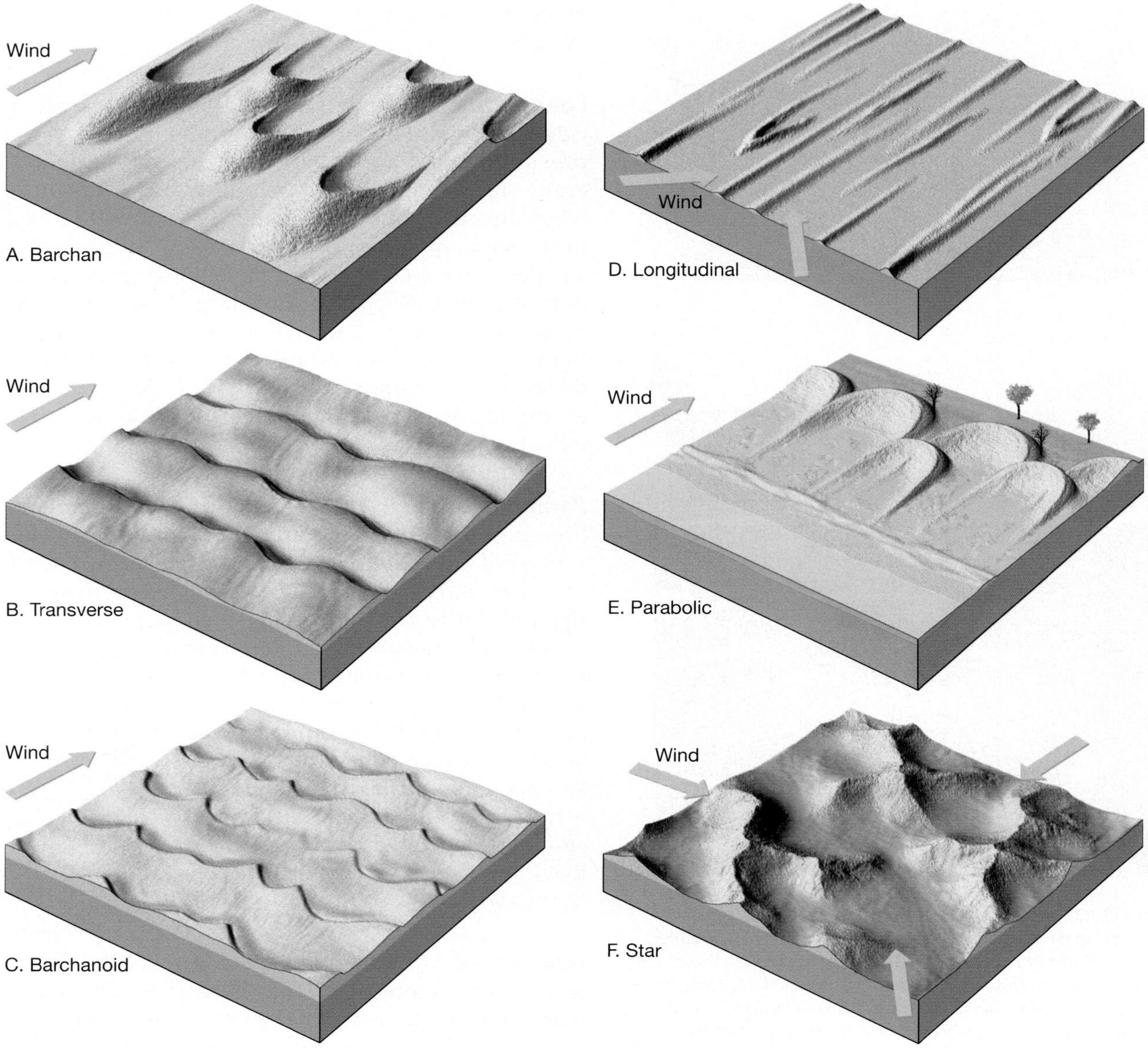

▲ **Figure 12.14** Sand dune types. **A.** Barchan dunes. **B.** Transverse dunes. **C.** Barchanoid dunes. **D.** Longitudinal dunes. **E.** Parabolic dunes. **F.** Star dunes.

This cutaway section of a sand dune shows cross-bedding. (John S. Shelton)

deposits of loess on Earth occur in western and northern China. They were blown there from the extensive desert basins of Central Asia. Accumulations of 30 meters are common, and thicknesses of more than 100 meters have been measured. It is this fine, buff colored sediment that gives the Yellow River (Hwang Ho) its name.

In the United States, deposits of loess are significant in many areas, including South Dakota, Nebraska, Iowa, Missouri, and Illinois, as well as portions of the Columbia Plateau in the Pacific Northwest. The correlation between the distribution of loess and important farming regions in the Midwest and eastern Washington State is not just a coincidence, because soils derived from this wind-deposited sediment are among the most fertile in the world.

Unlike the deposits in China, which originated in deserts, the loess in the United States and Europe is an indirect product of glaciation. Its source is deposits of

▲ **Figure 12.15** This vertical loess bluff near the Mississippi River in southern Illinois is about 3 meters high. *(Photo by James E. Patterson)*

stratified drift. During the retreat of the ice sheets, many river valleys were choked with sediment deposited by meltwater. Strong westerly winds sweeping across the barren floodplains picked up the finer sediment and dropped it as a blanket on the eastern sides of the valleys. Such an origin is confirmed by the fact that loess deposits are thickest and coarsest on the lee side of such major glacial drainage outlets as the Mississippi and Illinois rivers and rapidly thin with increasing distance from the valleys.

Star dune in the Namib Desert. (G. Gerster/Photo Researchers)

The Chapter in Review

- The *concept of dryness is relative;* it refers to any situation in which a water deficiency exists. Dry regions encompass about 30 percent of Earth's land surface. Two climatic types are commonly recognized: *desert,* which is arid, and *steppe* (a marginal and more humid variant of desert), which is semiarid. *Low-latitude deserts* coincide with the zones of subtropical highs in lower latitudes. In contrast, *middle-latitude deserts* exist principally because of their positions in the deep interiors of large landmasses far removed from the ocean.

- The same geologic processes that operate in humid regions also operate in deserts, but under contrasting climatic conditions. In dry lands *rock weathering of any type is greatly reduced* because of the lack of moisture and the scarcity of organic acids from decaying plants. Much of the weathered debris in deserts is the result of *mechanical weathering.* Practically all desert streams are dry most of the time and are said to be *ephemeral.* Stream courses in deserts are seldom well integrated and lack an extensive system of tributaries. Nevertheless, *running water is responsible for most of the erosional work in a desert.* Although wind erosion is more significant in dry areas than elsewhere, the main role of wind in a desert is in the transportation and deposition of sediment.

- Because arid regions typically lack permanent streams, they are characterized as having *interior drainage.* Many of the landscapes of the Basin and Range region of the western and southwestern United States are the result of streams eroding uplifted mountain blocks and depositing the sediment in interior basins. *Alluvial fans, playas,* and *playa lakes* are features often associated with these landscapes. In the late stages of erosion, the mountain areas are reduced to a few large bedrock knobs, called *inselbergs,* projecting above the sediment-filled basin.

- The transport of sediment by wind differs from that by running water in two ways. First, wind has a low density compared to water; thus, it is not capable of picking up and transporting coarse materials. Second, because wind is not confined to channels, it can spread sediment over large areas. The *bed load* of wind consists of sand grains skipping and bouncing along the surface in a process termed *saltation.* Fine dust particles are capable of being carried by the wind great distances as *suspended load.*

- Compared to running water and glaciers, wind is a relatively insignificant erosional agent. *Deflation,* the lifting and removal of loose material, often produces shallow depressions called *blowouts.* In portions of many deserts, the surface is a layer of coarse pebbles and gravels, called *desert pavement,* too large to be moved by the wind. Wind also erodes by *abrasion,* sometimes creating interestingly shaped stones termed *ventifacts.* Because sand seldom travels more than a meter above the surface, the effect of abrasion is obviously limited in vertical extent.

- Wind deposits are of two distinct types: (1) *mounds and ridges of sand,* called *dunes,* that are formed from sediment carried as part of the wind's bed load; and (2) extensive *blankets of silt,* called *loess,* that once were carried by wind in *suspension.* The profiles of many dunes show an asymmetrical shape, with the leeward (sheltered) slope being steep and the windward slope more gently inclined. The *types of sand dunes* include (1) *barchan dunes,* (2) *transverse dunes,* (3) *barchanoid dunes,* (4) *longitudinal dunes,* (5) *parabolic dunes,* and (6) *star dunes.* The thickest and most extensive deposits of loess occur in western and northern China. Unlike the deposits in China, which originated in deserts, the loess in the United States and Europe is an indirect product of glaciation.

Key Terms

abrasion (p. 276)
alluvial fan (p. 274)
barchan dune (p. 278)
barchanoid dune (p. 279)
bed load (p. 275)
blowout (p. 275)
cross beds (p. 278)
deflation (p. 275)
desert (p. 268)
desert pavement (p. 276)
dry climate (p. 268)
dune (p. 277)
ephemeral stream (p. 271)
interior drainage (p. 273)
loess (p. 279)
longitudinal dune (p. 279)
parabolic dune (p. 279)
playa lake (p. 274)
rainshadow desert (p. 270)
saltation (p. 275)
slip face (p. 277)
star dune (p. 279)
steppe (p. 268)
suspended load (p. 275)
transverse dune (p. 278)
ventifact (p. 276)

Questions for Review

1. How extensive are the desert and steppe regions of Earth?
2. What is the primary cause of subtropical deserts? Of middle-latitude deserts?
3. In which hemisphere (Northern or Southern) are middle-latitude deserts most common?
4. Why is the amount of precipitation that is used to define whether a place has a dry or humid climate a variable figure?
5. Why is rock weathering reduced in deserts as compared to humid climates?
6. When a permanent stream such as the Nile River crosses a desert, does discharge increase or decrease? How does this compare to a river in a humid area?
7. What is the most important erosional agent in deserts?
8. Describe the features and characteristics associated with each of the stages in the evolution of a mountainous desert. Where in the United States can these stages be observed?
9. Why is sea level (ultimate base level) not a significant factor influencing erosion in desert regions?
10. Why is the Aral Sea shrinking? (See Box 12.1.)
11. Describe the way in which wind transports sand. During very strong winds, how high above the surface can sand be carried?
12. Why is wind erosion relatively more important in arid regions than in humid areas?
13. What factor limits the depths of blowouts?
14. How do sand dunes migrate?
15. Indicate which type of dune is associated with each of the statements below.
    a. Dunes whose tips point into the wind.
    b. Long sand ridges oriented at right angles to the wind.
    c. Dunes that often form along coasts where strong winds create a blowout.
    d. Solitary dunes whose tips point downward.
    e. Long sand ridges that are oriented more or less parallel to the prevailing wind.
    f. An isolated dune consisting of three or four sharp-crested ridges diverging from a central high point.
    g. Scalloped rows of sand oriented at right angles to the wind.
16. What is *loess?* Where are deposits of loess found? What are the origins of this sediment?

Online Study Guide

The *Essentials of Geology* Web site uses the resources and flexibility of the Internet to aid in your study of the topics in this chapter. Written and developed by geology instructors, this site will help improve your understanding of geology. Visit **www.prenhall.com/lutgens** and click on the cover of *Essentials of Geology 9e* to find:

- Online review quizzes.
- Critical thinking exercises.
- Links to chapter-specific Web resources.
- Internet-wide key-term searches.

http://www.prenhall.com/lutgens

Wind-generated waves provide most of the energy that shapes and modifies shorelines. (Photo by Bob Barbour/Minden Pictures)

CHAPTER 13

Shorelines

Focus on Learning

To assist you in learning the important concepts in this chapter, you will find it helpful to focus on the following questions:

- Why is the shoreline considered a dynamic interface?
- What are the various parts of the coastal zone?
- What factors influence the height, length, and period of a wave?
- What is the motion of water particles within a wave?
- How do waves erode?
- What are some typical features produced by wave erosion and from sediment deposited by beach drift and longshore currents?
- What are the local factors that influence shoreline erosion, and what are some basic responses to shoreline erosion problems?
- How do emergent and submergent coasts differ in their formation and characteristic features?
- How are tides produced?

The restless waters of the ocean are constantly in motion. Winds generate surface currents, the gravity of the Moon and Sun produces tides, and density differences create deep-ocean circulation. Further, waves carry the energy from storms to distant shores, where their impact erodes the land.

Shorelines are dynamic environments. Their topography, geologic makeup, and climate vary greatly from place to place. Continental and oceanic processes converge along coasts to create landscapes that frequently undergo rapid change. When it comes to the deposition of sediment, they are transition zones between marine and continental environments.

Today, the coastal zone is experiencing intensive human activity. Unfortunately, people often treat the shoreline as if it were a stable platform on which structures can safely be built. This attitude inevitably leads to conflicts between people and nature. As you will see, many coastal landforms, especially beaches and barrier islands, are relatively fragile, short-lived geological features that are inappropriate sites for development.

The Shoreline: A Dynamic Interface

Nowhere is the restless nature of the ocean's water more noticeable than along the shore—the dynamic interface among air, land, and sea. An *interface* is a common boundary where different parts of a system interact. This is certainly an appropriate designation for the coastal zone. Here we can see the rhythmic rise and fall of tides and observe waves constantly rolling in and breaking. Sometimes the waves are low and gentle. At other times, they pound the shore with awesome fury.

Cape Cod consists mainly of glacial materials.

Although it may not be readily apparent, the shoreline is constantly being shaped and modified by waves. For example, along Cape Cod, Massachusetts, wave activity is eroding cliffs of poorly consolidated glacial sediments so aggressively that the cliffs are retreating inland at up to 1 meter per year (Figure 13.1A). By contrast, at Point Reyes, California, the far more durable bedrock cliffs are less susceptible to wave attack and are therefore retreating much more slowly (Figure 13.1B). Along both coasts, wave activity moves sediment toward and away from the shore as well as along it. Such activity sometimes produces narrow sandbars that frequently change size and shape as storm waves come and go.

Rocky cliffs of Point Reyes. (David Muench)

The nature of present-day shorelines is not just the result of the relentless attack on the land by the sea. Indeed, the shore has a complex character that results from multiple geologic processes. For example, practically all coastal areas were affected by the worldwide rise in sea level that accompanied the melting of ice sheets at the close of the Pleistocene epoch. As the sea encroached landward, the shoreline retreated, becoming superimposed upon existing landscapes that had resulted from such diverse processes as stream erosion, glaciation, volcanic activity, and the forces of mountain building.

The Coastal Zone

In general conversation a number of terms are used when referring to the boundary between land and sea. In the preceding section, the terms *shore, shoreline, coastal zone,* and *coast* were all used. Moreover, when many think of the land-sea interface, the word *beach* comes to mind. Let's take a moment to clarify these terms and introduce some other terminology used by those who study the land-sea boundary zone. You will find it helpful to refer to Figure 13.2, which is an idealized profile of the coastal zone.

The **shoreline** is the line that marks the contact between land and sea. Each day, as tides rise and fall, the position of the shoreline migrates. Over longer time spans, the average position of the shoreline gradually shifts as sea level rises or falls.

The **shore** is the area that extends between the lowest tide level and the highest elevation on land that is affected by storm waves. By contrast, the **coast** extends inland from the shore as far as ocean-related features can be found. The **coastline** marks the coast's seaward edge, whereas the inland boundary is not always obvious or easy to determine.

As Figure 13.2 illustrates, the shore is divided into the *foreshore* and the *backshore.* The **foreshore** is the area exposed when the tide is out (low tide) and submerged when the tide is in (high tide). The **backshore** is landward of the high-tide shoreline. It is usually dry, being affected by waves only during storms. Two other zones are commonly identified. The **nearshore zone** lies between the low-tide shoreline and the line where waves break at low tide. Seaward of the nearshore zone is the **offshore zone.**

For many a beach is the sandy area where people lie in the sun and walk along the water's edge. Technically, a **beach** is an accumulation of sediment found along the landward margin of the ocean or a lake. Along straight coasts, beaches may extend for tens or hundreds of kilometers. Where coasts are irregular, beach formation may be confined to the relatively quiet waters of bays.

Beaches consist of one or more **berms,** which are relatively flat platforms often composed of sand that are adjacent to coastal dunes or cliffs and marked by a change in slope at the seaward edge. Another part of the beach is the **beach face,** which is the wet sloping

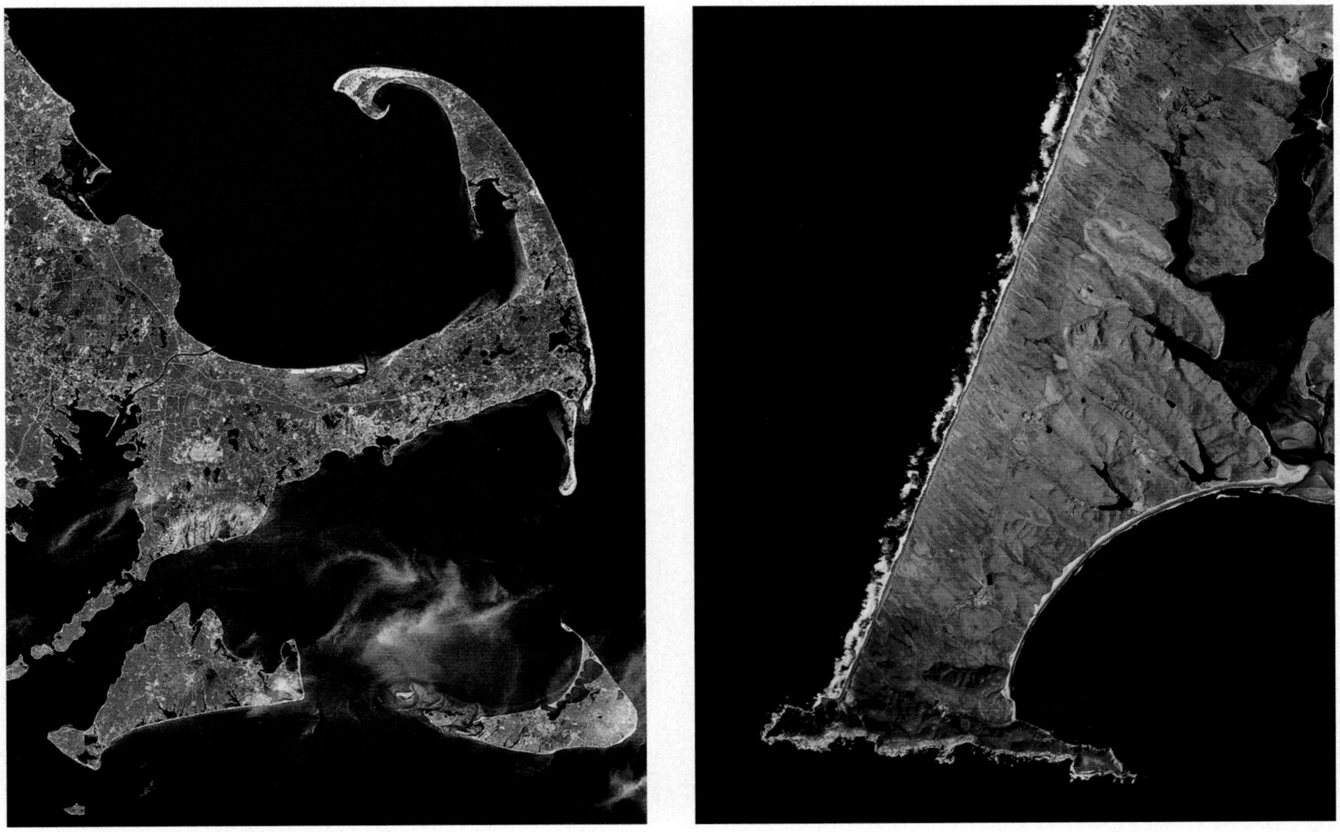

A.

B.

▲ **Figure 13.1** **A.** This satellite image includes the familiar outline of Cape Cod. Boston is to the upper left. The two large islands off the south shore of Cape Cod are Martha's Vineyard (left) and Nantucket (right). Although the work of waves constantly modifies this coastal landscape, shoreline processes are not responsible for creating it. Rather, the present size and shape of Cape Cod result from the positioning of moraines and other glacial materials deposited during the Pleistocene epoch. *(Satellite Image courtesy of Earth Satellite Corporation/Science Photo Library/Photo Researchers, Inc.)* **B.** High-altitude image of the Point Reyes area north of San Francisco, California. The 5.5-kilometer-long south-facing cliffs at Point Reyes (bottom of photo) are exposed to the full force of the waves from the Pacific Ocean. Nevertheless, this promontory retreats slowly because the bedrock from which it formed is very resistant. *(High-altitude image courtesy of USDA-ASCS.)*

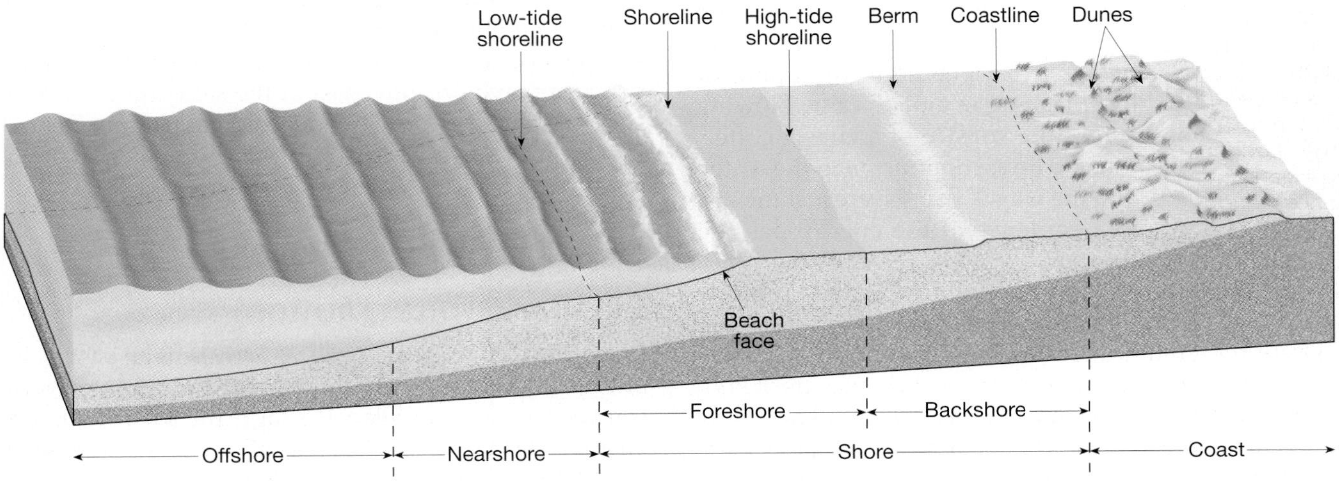

▲ **Figure 13.2** The coastal zone consists of several parts. The beach is an accumulation of sediment on the landward margin of the ocean or a lake. It can be thought of as material in transit along the shore.

This beach in Australia is heavily developed (Peter Hendrie)

surface that extends from the berm to the shoreline. Where beaches are sandy, sunbathers usually prefer the berm, whereas joggers prefer the wet, hard-packed sand of the beach face.

Beaches are composed of whatever material is locally abundant. The sediment for some beaches is derived from the erosion of adjacent cliffs or nearby coastal mountains. Other beaches are built from sediment delivered to the coast by rivers.

Although the mineral makeup of many beaches is dominated by durable quartz grains, other minerals may be dominant. For example, in areas such as southern Florida, where there are no mountains or other sources of rock-forming minerals nearby, most beaches are composed of shell fragments and the remains of organisms that live in coastal waters. Some beaches on volcanic islands in the open ocean are composed of weathered grains of the basaltic lava that comprise the islands, or of coarse debris eroded from coral reefs that develop around islands in low latitudes.

Regardless of the composition, the material that comprises the beach does not stay in one place. Instead, crashing waves are constantly moving it. Thus, beaches can be thought of as material in transit along the shore.

Waves

GEODe Shorelines
▼ Waves and Beaches

Ocean waves are energy traveling along the interface between ocean and atmosphere, often transferring energy from a storm far out at sea over distances of several thousand kilometers. That's why even on calm days the ocean still has waves that travel across its surface. When observing waves, always remember that you are watching *energy* travel though a medium (water). If you make waves by tossing a pebble into a pond, or by splashing in a pool, or by blowing across the surface of a cup of coffee, you are imparting *energy* to the water, and the waves you see are just the visible evidence of the energy passing through.

Wind-generated waves provide most of the energy that shapes and modifies shorelines. Where the land and sea meet, waves that may have traveled unimpeded for hundreds or thousands of kilometers suddenly encounter a barrier that will not allow them to advance further. Stated another way, the shore is where a practically irresistible force confronts an almost immovable object. The conflict that results is never-ending and sometimes dramatic (Box 13.1).

Did You Know?

About half of the world's human population lives on or within 100 km (60 mi) of a coast. The proportion of the U.S. population residing within 75 km (45 mi) of a coast in 2010 is projected to be well in excess of 50 percent. The concentration of such large numbers of people so near the shore means that hurricanes and tsunamis place millions at risk.

Wave Characteristics

Most ocean waves derive their energy and motion from the wind. When a breeze is less than 3 kilometers (2 miles) per hour, only small wavelets appear. At greater wind speeds, more stable waves gradually form and advance with the wind.

Characteristics of ocean waves are illustrated in Figure 13.3, which shows a simple, nonbreaking waveform. The tops of the waves are the *crests*, which are separated by *troughs*. Halfway between the crests and troughs is the *still water level*, which is the level the water would occupy if there were no waves. The vertical distance between trough and crest is called the **wave height,** and the horizontal distance between successive crests (or troughs) is the **wavelength.** The time it takes one full wave—one wavelength—to pass a fixed position is the **wave period.**

The height, length, and period that are eventually achieved by a wave depend upon three factors: (1) wind speed; (2) length of time the wind has blown; and (3) *fetch*, the distance that the wind has traveled across the open water. As the quantity of energy transferred from the wind to the water increases, the height and steepness of the waves increase as well. Eventually a critical point is reached where waves grow so tall that they topple over, forming ocean breakers called *whitecaps*.

For a particular wind speed, there is a maximum fetch and duration of wind beyond which waves will no longer increase in size. When the maximum fetch and duration are reached for a given wind velocity, the waves are said to be fully developed. The reason that waves can grow no further is that they are losing as much energy through the breaking of whitecaps as they are receiving from the wind.

When the wind stops or changes direction, or the waves leave the stormy area where they were created, they continue on without relation to local winds. The waves also undergo a gradual change to *swells* that are lower in height and longer in length and may carry the storm's energy to distant shores. Because many independent wave systems exist at the same time, the sea surface acquires a complex and irregular pattern. Hence, the sea waves we watch from the shore are usually a mixture of swells from faraway storms and waves created by local winds.

Circular Orbital Motion

Waves can travel great distances across ocean basins. In one study, waves generated near Antarctica were tracked as they traveled through the Pacific Ocean basin. After more than 10,000 kilometers (over 6000 miles), the waves finally expended their energy a week later along the shoreline of the Aleutian Islands of Alaska. The water itself doesn't travel the entire distance,

BOX 13.1

The Move of the Century—Relocating the Cape Hatteras Lighthouse

In spite of efforts to protect structures that are too close to the shore, they can still be in danger of being destroyed by receding shorelines and the destructive power of waves. Such was the case for one of the nation's most prominent landmarks, the candy-striped lighthouse at Cape Hatteras, North Carolina, which is 21 stories tall—the nation's tallest lighthouse.

The lighthouse was built in 1870 on the Cape Hatteras barrier island 457 meters (1500 feet) from the shoreline to guide mariners through the dangerous offshore shoals known as the "Graveyard of the Atlantic." As the barrier island began migrating toward land, its beach narrowed. When the waves began to lap just 37 meters (120 feet) from its brick and granite base, there was concern that even a moderate-strength hurricane could trigger beach erosion sufficient to topple the lighthouse.

In 1970 the U.S. Navy built three groins in front of the lighthouse in an effort to protect the lighthouse from further erosion. The groins initially slowed erosion but disrupted sand flow in the surf zone, which caused the flattening of nearby dunes and the formation of a bay south of the lighthouse. Attempts to increase the width of the beach in front of the lighthouse included beach nourishment and artificial offshore beds of seaweed, both of which failed to widen the beach substantially. In the 1980s the Army Corps of Engineers proposed building a massive stone seawall around the lighthouse but decided the eroding coast would eventually move out from under the structure, leaving it stranded at sea on its own island. In 1988 the National Academy of Sciences determined that the shoreline in front of the lighthouse would retreat so far as to destroy the lighthouse and recommended relocation of the tower as had been done with smaller lighthouses. In 1999 the National Park Service, which owns the lighthouse, finally authorized moving the structure to a safer location.

Moving the lighthouse, which weighs 4395 metric tons (4830 short tons), was accomplished by severing it from its foundation and carefully hoisting it onto a platform of steel beams fitted with roller dollies. Once on the platform, it was slowly rolled along a specially designed steel track using a series of hydraulic jacks. A strip of vegetation was cleared to make a runway along which the lighthouse traveled 1.5 meters (5 feet) at a time, with the track picked up from behind and reconstructed in front of the tower as it moved. In less than a month, the lighthouse was gingerly transported 884 meters (2900 feet) from its original location, making it one of the largest structures ever successfully moved.

After its $12 million move, the lighthouse now resides in a scrub oak and pine woodland (Figure 13.A). Although it now stands farther inland, the light's slightly higher elevation makes it visible just as far out to sea, where it continues to warn mariners of the hazardous shoals. At the current rate of shoreline retreat, the lighthouse should be safe from the threat of waves for at least another century.

Figure 13.A When North Carolina's Cape Hatteras Lighthouse was threatened by shoreline erosion in 1999, it was relocated 4889 meters (1600 feet) from the shore. (Photo by Drew Wilson © 1999, *Virginian-Pilot*)

but the wave form does. As the wave travels, the water passes the energy along by moving in a circle. This movement is called *circular orbital motion.*

Observation of an object floating in waves reveals that it moves not only up and down but also slightly forward and backward with each successive wave. Figure 13.4 shows that a floating object moves up and backward as the crest approaches, up and forward as the crest passes, down and forward after the crest, down and backward as the trough approaches, and rises and moves backward again as the next crest advances. When the movement of the toy boat shown in Figure 13.4 is traced as a wave passes, it can be seen that the boat moves in a circle and it returns to essentially the same place. Circular orbital motion allows a waveform (the wave's shape) to move forward *through the water* while the individual water particles that transmit the wave move around in a circle. Wind moving across a field of wheat causes a similar phenomenon:

Did You Know?

Ocean waves may someday be a significant source of electrical power. In November 2000 the world's first commercial wave-power station began operating off the Scottish coast. The 500-kilowatt station uses technology in which incoming waves push air up and down inside a partially submerged tube. The air rushing into and out of the top of the tube is used to drive a turbine.

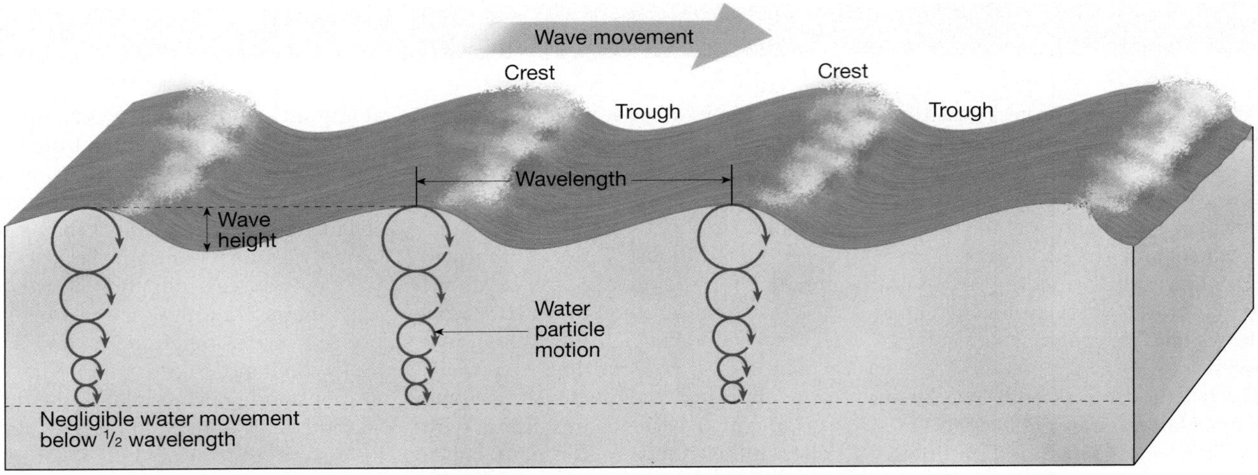

▲ **Figure 13.3** This diagram illustrates the basic parts of a wave as well as the movement of water particles with the passage of the wave. Negligible water movement occurs below a depth equal to one half the wavelength (the level of the dashed line).

The black sands at this beach are derived from dark volcanic rock.

Most ocean waves are wind-generated.

The wheat itself doesn't travel across the field, but the waves do.

The energy contributed by the wind to the water is transmitted not only along the surface of the sea but also downward. However, beneath the surface the circular motion rapidly diminishes until, at a depth equal to one half the wavelength measured from still water level, the movement of water particles becomes negligible. This depth is known as the *wave base*. The dramatic decrease of wave energy with depth is shown by the rapidly diminishing diameters of water-particle orbits in Figure 13.3.

Waves in the Surf Zone

As long as a wave is in deep water, it is unaffected by water depth (Figure 13.5 left). However, when a wave approaches the shore, the water becomes shallower and influences wave behavior. The wave begins to "feel bottom" at a water depth equal to its wave base. Such depths interfere with water movement at the base of the wave and slow its advance (Figure 13.5, center).

As a wave advances toward the shore, the slightly faster waves farther out to sea catch up, decreasing the wavelength. As the speed and length of the wave diminish, the wave steadily grows higher. Finally, a critical point is reached when the wave is too steep to support itself and the wave front collapses, or *breaks* (Figure 13.5, right), causing water to advance up the shore.

The turbulent water created by breaking waves is called **surf.** On the landward margin of the surf zone the turbulent sheet of water from collapsing breakers, called *swash*, moves up the slope of the beach. When the energy of the swash has been expended, the water flows back down the beach toward the surf zone as *backwash*.

Wave Erosion

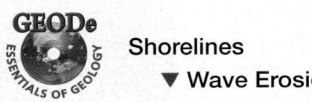

GEODe: Essentials of Geology
Shorelines
▼ Wave Erosion

During calm weather wave action is minimal. However, just as streams do most of their work during floods, so too do waves perform most of their work during storms. The impact of high, storm-induced waves against the shore can be awesome (Figure 13.6). Each breaking wave may hurl thousands of tons of water against the land, sometimes making the ground tremble. The pressures exerted by Atlantic waves in wintertime, for example, average nearly 10,000 kilograms per square meter (more than 2000 pounds per square foot). The force during storms is even greater.

It is no wonder that cracks and crevices are quickly opened in cliffs, coastal structures, and anything else that is subjected to these enormous shocks. Water is forced into every opening, greatly compressing air trapped in the cracks. When the wave subsides, the air expands rapidly, dislodging rock fragments and enlarging and extending fractures.

In addition to the erosion caused by wave impact and pressure, **abrasion,** the sawing and grinding action of water armed with rock fragments, is also important. In fact, abrasion is probably more intense in the surf zone than in any other environment. Smooth, rounded stones and pebbles along the shore are obvious reminders of the grinding action of rock against rock in the surf zone (Figure 13.7A). Further, such fragments

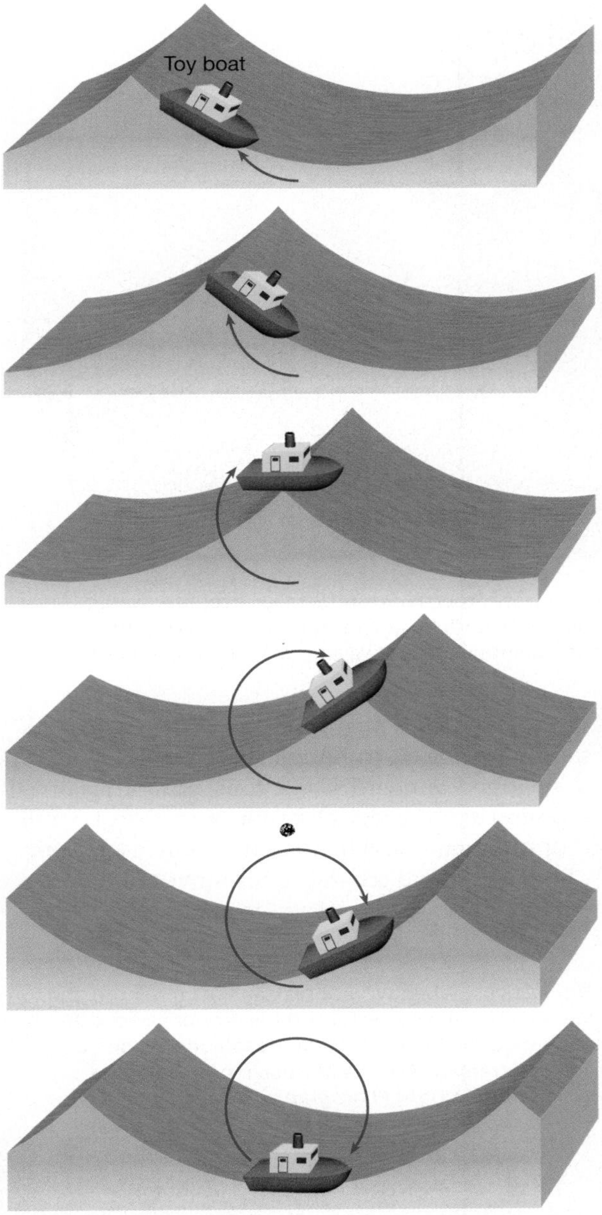

▲ **Figure 13.4** The movements of the toy boat show that the wave form advances but the water does not advance appreciably from its original position. In this sequence, the wave moves from left to right as the toy boat (and the water in which it is floating) rotates in an imaginary circle.

are used as tools by the waves as they cut horizontally into the land (Figure 13.7B).

Sand Movement on the Beach

Beaches are sometimes called "rivers of sand." The reason is that the energy from breaking waves often causes large quantities of sand to move along the beach face and in the surf zone roughly parallel to the shoreline. Wave energy also causes sand to move perpendicular to (toward and away from) the shoreline.

Movement Perpendicular to the Shoreline

If you stand ankle-deep in water at the beach, you will see that swash and backwash move sand toward and away from the shoreline. Whether there is a net loss or addition of sand depends on the level of wave activity. When wave activity is relatively light (less energetic waves), much of the swash soaks into the beach, which reduces the backwash. Consquently, the swash dominates and causes a net movement of sand up the beach face toward the berm.

When high-energy waves prevail, the beach is saturated from previous waves, so much less of the swash soaks in. As a result, the berm erodes because backwash is strong and causes a net movement of sand down the beach face.

Along many beaches, light wave activity is the rule during the summer. Therefore, a wide sand berm gradually develops. During winter, when storms are frequent and more powerful, strong wave activity erodes and narrows the berm. A wide berm that may have taken months to build can be dramatically narrowed in just a few hours by the high-energy waves created by a strong winter storm.

Breaking waves in the surf zone release a great deal of energy. (Victor Cavataio)

Wave Refraction

The bending of waves, called **wave refraction,** plays an important part in shoreline processes. It affects the distribution of energy along the shore and thus strongly influences where and to what degree erosion, sediment transport, and deposition will take place.

Waves seldom approach the shore straight on. Rather, most waves move toward the shore at an angle. However, when they reach the shallow water of a smoothly sloping bottom, they are bent and tend to become parallel to the shore. Such bending occurs because the part of the wave nearest the shore reaches shallow water and slows first, whereas the end that is still in deep water continues forward at its full speed. The net result is a wave front that may approach nearly parallel to the shore regardless of the original direction of the wave.

Because of refraction, wave impact is concentrated against the sides and ends of headlands that project into the water, whereas wave attack is weakened in bays. This differential wave attack along irregular coastlines is illustrated in Figure 13.8. Because the waves reach the shallow water in front of the headland sooner than they do in adjacent bays, they are bent more nearly parallel to the protruding land and strike it from all three sides. By contrast, refraction in the bays causes waves to diverge and expend less energy. In these zones of

Cliff undercut by wave erosion along the Oregon coast.

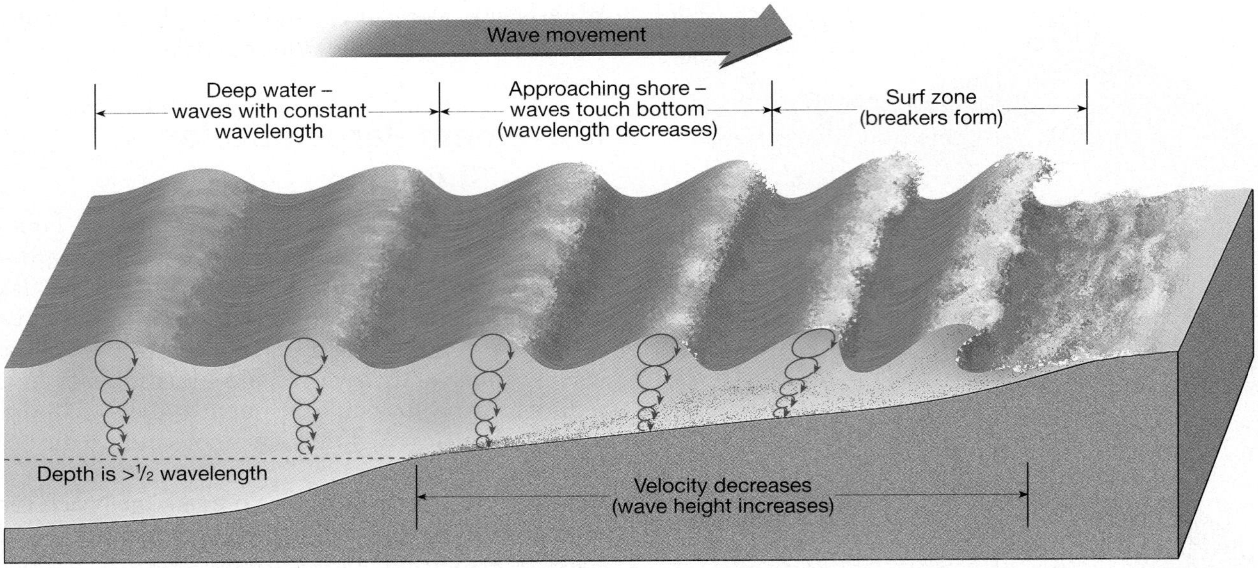

▲ **Figure 13.5** Changes that occur when a wave moves onto shore. The waves touch bottom as they encounter water depths less than half a wavelength. The wave speed decreases, and the waves stack up against the shore, causing the wavelength to decrease. This results in an increase in wave height to the point where the waves pitch forward and break in the surf zone.

weakened wave activity, sediments can accumulate and form sheltered sandy beaches. Over a long period, erosion of the headlands and deposition in the bays will straighten an irregular shoreline.

Beach Drift and Longshore Currents

Wave bending around the end of a beach. (James Patterson)

Although waves are refracted, most still reach the shore at an angle, however slight. Consequently, the uprush of water from each breaking wave (the swash) is not head-on, but oblique. However, the backwash is straight down the slope of the beach. The effect of this pattern of water movement is to transport particles of sediment in a zigzag pattern along the beach face (Figure 13.9). This movement is called **beach drift,** and it can transport sand and pebbles hundreds or even thousands of meters each day. However, a more typical rate is 5 to 10 meters per day.

Did You Know?

Rip currents are strong narrow-surface or near-surface currents flowing seaward through the surf zone at nearly right angles to the shore. They represent the return of water that has been piled up on shore by incoming waves and can attain speeds of 7–8 km (4–5 mi) per hour. Because they move faster than most people can swim, rip currents pose a hazard to swimmers.

Oblique waves also produce currents within the surf zone that flow parallel to the shore and move substantially more sediment than beach drift. Because the water here is turbulent, these **longshore currents** easily move the fine suspended sand and roll larger sand and gravel along the bottom. When the sediment transported by longshore currents is added to the quantity moved by beach drift, the total can be very large. At Sandy Hook, New Jersey, for example, the quantity of sand transported along the shore over a 48-year period averaged almost 750,000 tons annually. For a 10-year period at Oxnard, California, more than 1.5 million tons of sediment moved along the shore each year.

Both rivers and coastal zones move water and sediment from one area *(upstream)* to another *(downstream).*

▼ **Figure 13.6** When waves break against the shore, the force of the water can be powerful and the erosional work that is accomplished can be great. Marin Headlands, Golden Gate National Recreation Area, California. *(Photo by Galen Rowell/Mountain Light Photography, Inc.)*

A.

B.

▲ **Figure 13.7** **A.** Abrasion can be intense in the surf zone. Smooth, rounded stones along the shore are an obvious reminder of this fact. Garrapata State Park, California. *(Photo by Carr Clifton)* **B.** Sandstone cliff undercut by wave erosion at Gabriola Island, British Columbia, Canada. *(Photo by Fletcher and Baylis/Photo Researchers, Inc.)*

Beach drift and longshore currents, however, move in a zigzag pattern, whereas rivers flow mostly in a turbulent, swirling fashion. Additionally, the direction of flow of longshore currents along a shoreline can change, whereas rivers flow in the same direction (downhill). Longshore currents change direction because the direction that waves approach the beach changes seasonally. Nevertheless, longshore currents generally flow southward along both the Atlantic and Pacific shores of the United States.

Shoreline Features

A fascinating assortment of shoreline features can be observed along the world's coastal regions. These shoreline features vary depending on the type of rocks

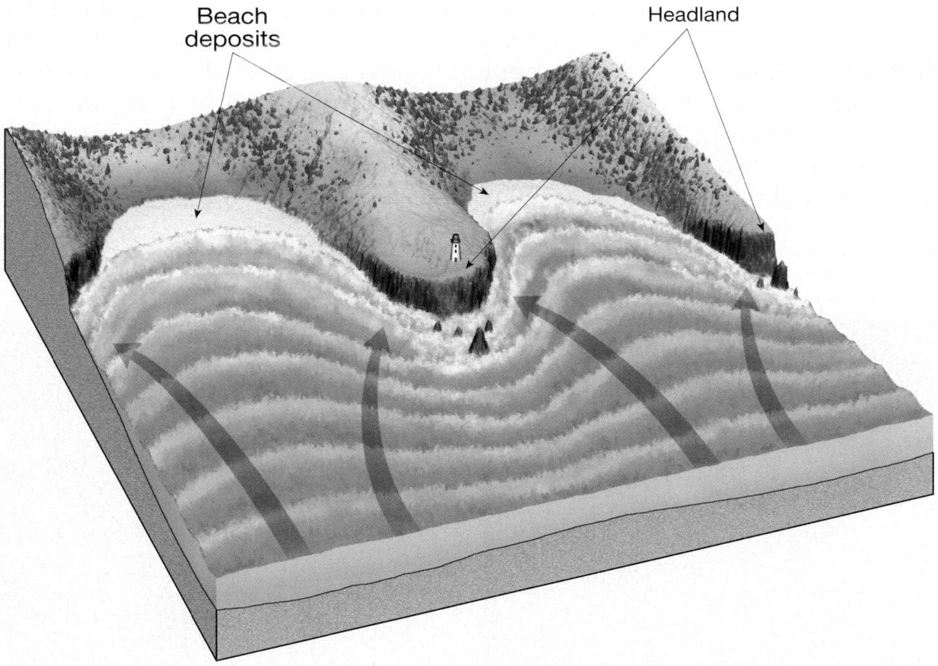

◀ **Figure 13.8** Wave refraction along an irregular coastline. As waves first touch bottom in the shallows off the headlands, they are slowed, causing the waves to refract (bend) and align nearly parallel to the shoreline. This causes wave energy to be concentrated at headlands (resulting in erosion) and dispersed in bays (resulting in deposition).

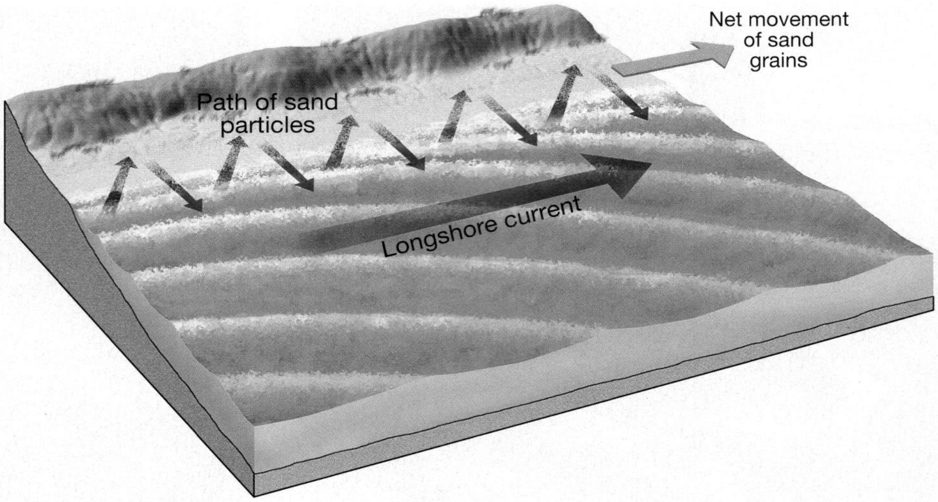

▶ **Figure 13.9** Beach drift and longshore currents are created by obliquely breaking waves. Beach drift occurs as incoming waves carry sand obliquely up the beach, while the water from spent waves carries it directly down the slope of the beach. Similar movements occur offshore in the surf zone to create the longshore current. These processes transport large quantities of material along the beach and in the surf zone.

Did You Know?

During a storm that struck the coast of Scotland, a 1350-ton portion of a steel-and-concrete breakwater was torn from the rest of the structure and moved toward shore. Five years later the 2600-ton unit that replaced the first met a similar fate.

exposed along the shore, the intensity of waves, the nature of coastal currents, and whether the coast is stable, sinking, or rising. Features that owe their origin primarily to the work of erosion are called *erosional features,* while deposits of sediment produce *depositional features.*

Erosional Features

Many coastal landforms owe their origin to erosional processes. Such erosional features are common along the rugged and irregular New England coast and along the steep shorelines of the West Coast of the United States.

Wave-Cut Cliffs, Wave-Cut Platforms, and Marine Terraces. **Wave-cut cliffs,** as the name implies, originate by the cutting action of the surf against the base of coastal land. As erosion progresses, rocks overhanging the notch at the base of the cliff crumble into the surf and the cliff retreats. A relatively flat, benchlike surface, called a **wave-cut platform,** is left behind by the receding cliff (Figure 13.10, left). The platform broadens as wave attack continues. Some debris produced by the breaking waves remains along the water's edge as sediment on the beach, while the remainder is transported farther seaward. If a wave-cut platform is uplifted above sea level by tectonic forces, it becomes a **marine terrace** (Figure 13.10, right). Marine terraces are easily recognized by their gentle seaward-sloping shape and are often desirable sites for coastal roads, buildings, or agriculture.

Sea Arches and Sea Stacks. Headlands that extend into the sea are vigorously attacked by waves because of refraction. The surf erodes the rock selectively, wearing away the softer or more highly fractured rock at the fastest rate. At first, sea caves may form. When two caves on opposite sides of a headland unite, a **sea arch** results (Figure 13.11). Eventually, the arch falls in, leaving an isolated remnant, or **sea stack,** on the wave-cut platform (Figure 13.11). In time, it too will be consumed by the action of the waves.

▶ **Figure 13.10** Wave-cut platform and marine terrace. A wave-cut platform is exposed at low tide along the California coast at Bolinas Point near San Francisco. An elevated wave-cut platform has been uplifted to create the marine terrace. *(Photo by John S. Shelton)*

◀ **Figure 13.11** Sea arch and sea stack at the tip of Mexico's Baja Peninsula. *(Photo by Mark A. Johnson/The Stock Market)*

Depositional Features

Sediment eroded from the beach is transported along the shore and deposited in areas where wave energy is low. Such processes produce a variety of depositional features.

Spits, Bars, and Tombolos. Where beach drift and longshore currents are active, several features related to the movement of sediment along the shore may develop. A **spit** is an elongated ridge of sand that projects from the land into the mouth of an adjacent bay. Often the end in the water hooks landward in response to the dominant direction of the longshore current (Figure 13.12). The term **baymouth bar** is applied to a sandbar that completely crosses a bay, sealing it off from the open ocean (Figure 13.12). Such a feature tends to form across bays where currents are weak, allowing a spit to extend to the other side. A **tombolo,** a ridge of sand that connects an island to the mainland or to another island, forms in much the same manner as a spit.

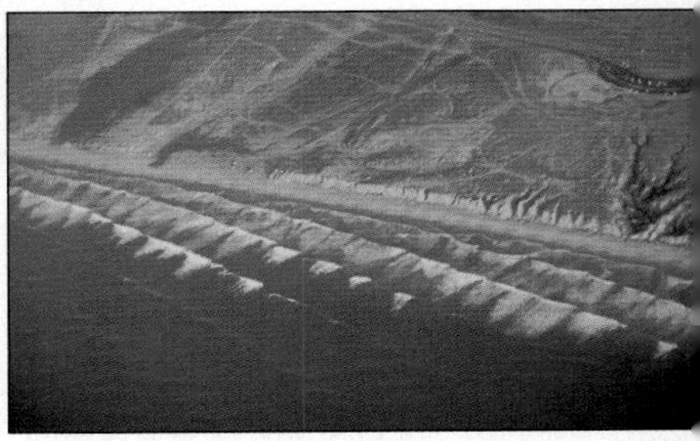

Waves approaching the beach at a slight angle produce a longshore current moving from left to right. (John S. Shelton)

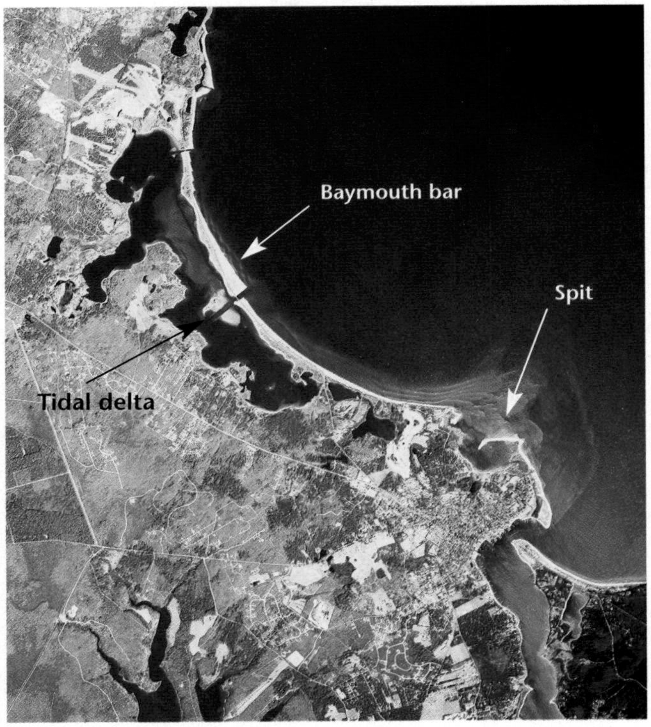

▲ **Figure 13.12** High-altitude image of a well-developed spit and baymouth bar along the coast of Martha's Vineyard, Massachusetts. Also notice the tidal delta in the lagoon adjacent to the inlet through the baymouth bar. *(Image courtesy of USDA-ASCS)*

Barrier Islands. The Atlantic and Gulf Coastal Plains are relatively flat and slope gently seaward. The shore zone is characterized by **barrier islands.** These low ridges of sand parallel the coast at distances from 3 to 30 kilometers offshore. From Cape Cod, Massachusetts, to Padre Island, Texas, nearly 300 barrier islands rim the coast (Figure 13.13).

Most barrier islands are from 1 to 5 kilometers wide and between 15 and 30 kilometers long. The tallest features are sand dunes, which usually reach heights of 5 to 10 meters. The lagoons that separate these narrow islands from the shore are zones of relatively quiet water that allow small craft traveling between New York and northern Florida to avoid the rough waters of the North Atlantic.

Barrier islands probably form in several ways. Some originate as spits that were subsequently severed from the mainland by wave erosion or by the general rise in sea level following the last episode of glaciation. Others are created when turbulent waters in the line of breakers heap up sand that has been scoured from the bottom. Finally, some barrier islands may be former sand-dune ridges that originated along the shore during

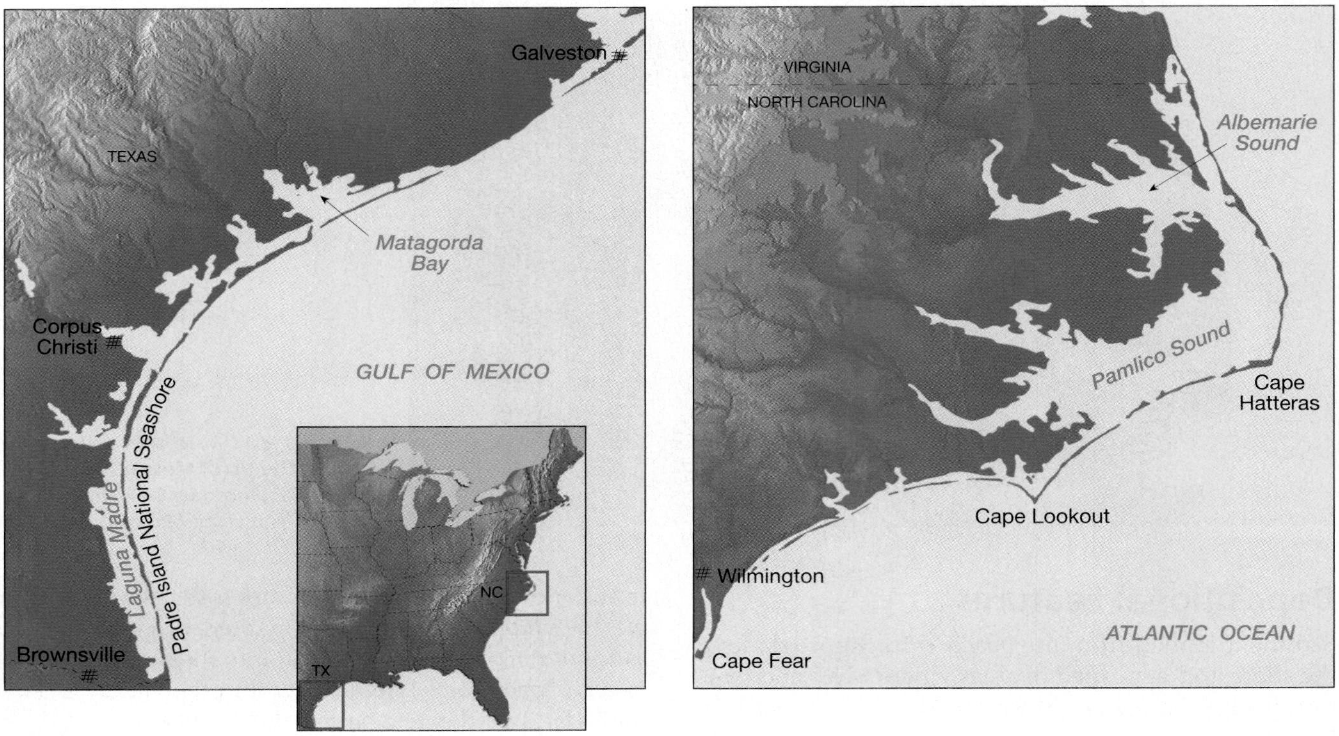

▲ **Figure 13.13** Nearly 300 barrier islands rim the Gulf and Atlantic coasts. The islands along the south Texas coast and along the coast of North Carolina are excellent examples.

Did You Know?

Along shorelines composed of unconsolidated material rather than hard rock, the rate of erosion by breaking waves can be extraordinary. In parts of Britain, where waves have the easy task of eroding glacial deposits of sand, gravel, and clay, the coast has been worn back 3 to 5 km (2–3 mi) since Roman times (2000 years ago), sweeping away many villages and ancient landmarks.

Wave-cut cliff along California coast.

the last glacial period, when sea level was lower. As the ice sheets melted, sea level rose and flooded the area behind the beach-dune complex.

The Evolving Shore

A shoreline continually undergoes modification regardless of its initial configuration. At first most coastlines are irregular, although the degree of, and reason for, the irregularity may differ considerably from place to place. Along a coastline of varied geology, the pounding surf may initially increase its irregularity because the waves will erode the weaker rocks more easily than the stronger ones. However, if a shoreline remains stable, marine erosion and deposition will eventually produce a straighter, more regular coast.

Figure 13.14 illustrates the evolution of an initially irregular coast. As waves erode the headlands, creating cliffs and a wave-cut platform, sediment is carried along the shore. Some material is deposited in the bays, whereas other debris is formed into spits and baymouth bars. At the same time, rivers fill the bays with sediment. Ultimately, a generally straight, smooth coast results.

Stabilizing the Shore

GEODe: Shorelines ▼ Waves and Beaches

Today the coastal zone teems with human activity. Unfortunately, people often treat the shoreline as if it were a stable platform on which structures can be built safely. This approach jeopardizes both people and the shoreline because many coastal landforms are relatively fragile, short-lived features that are easily damaged by development. And as anyone who has endured a tropical storm or tsunami knows, the shoreline is not always a safe place to live.

Compared with natural hazards such as earthquakes, volcanic eruptions, and landslides, shoreline erosion appears to be a more continuous and predictable process that causes relatively modest damage to limited areas. In reality, the shoreline is a dynamic place that can change rapidly in response to natural forces. Exceptional storms are capable of eroding beaches and cliffs at rates that far exceed the long-term average. Such bursts of accelerated erosion not only have a significant impact on the natural evolution of a coast but can also have a profound impact on people who reside in the coastal zone. Erosion along our coasts causes significant property damage. Huge sums are spent annually not only to repair damage but also

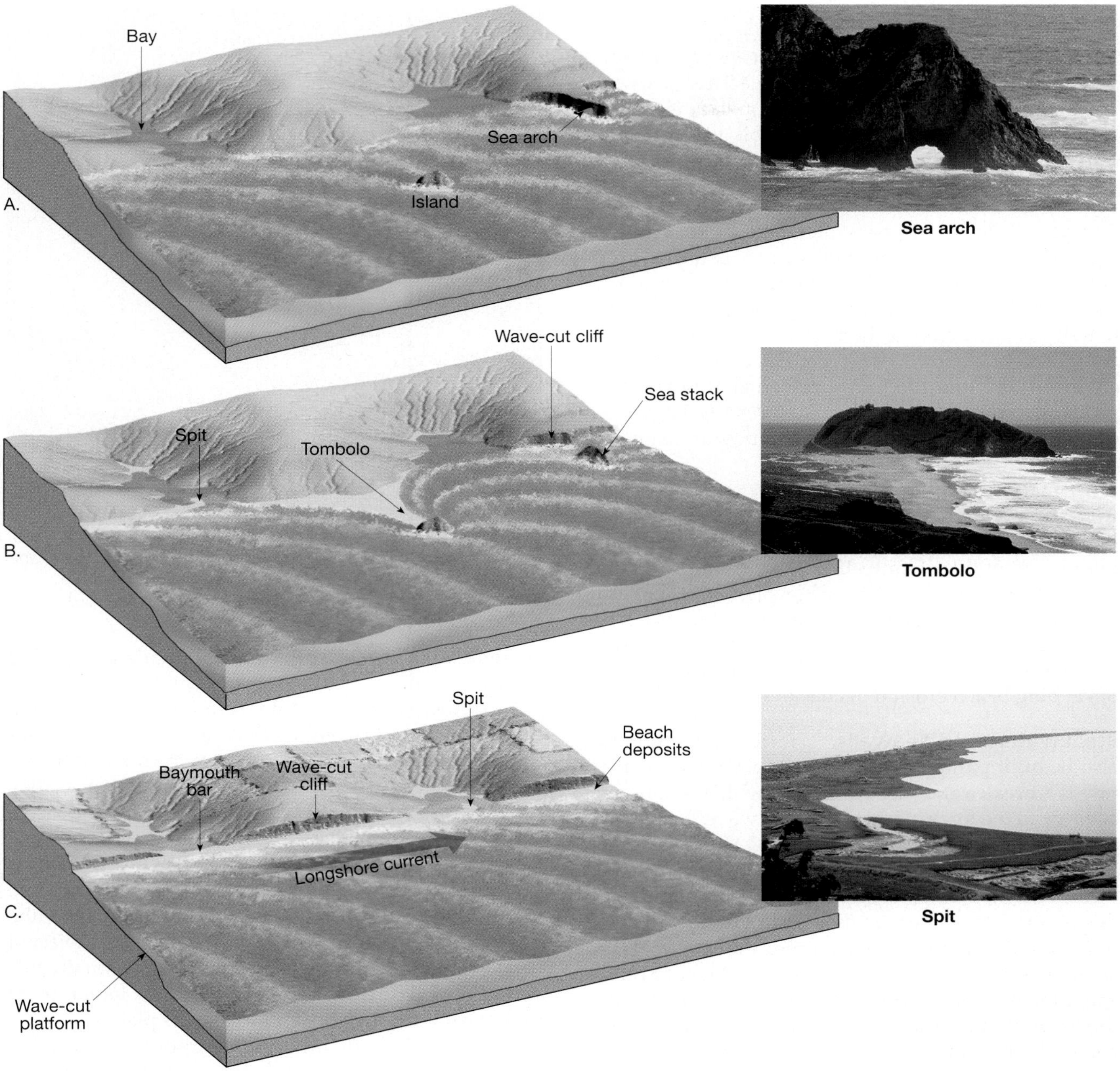

▲ **Figure 13.14** These diagrams illustrate the changes that can take place through time along an initially irregular coastline that remains relatively stable. The coastline shown in part **A** gradually evolves to **B,** and then **C.** The diagrams also serve to illustrate many of the features described in the section on shoreline features. *(Photos by E. J. Tarbuck)*

to prevent or control erosion. Already a problem at many sites, shoreline erosion is certain to become increasingly serious as extensive coastal development continues.

Although the same processes cause change along every coast, not all coasts respond in the same way. Interactions among different processes and the relative importance of each process depend on local factors, which include (1) the proximity of a coast to sediment-laden rivers, (2) the degree of tectonic activity, (3) the topography and composition of the land, (4) prevailing winds and weather patterns, and (5) the configuration of the coastline and nearshore areas.

During the past 100 years, growing affluence and increasing demands for recreation have brought unprecedented development to many coastal areas. As both the number and the value of structures have increased, so too have efforts to protect property from storm waves by stabilizing the shore. Also, controlling the natural migration of sand is an ongoing struggle in many coastal areas. Such interference can result in unwanted changes that are difficult and expensive to correct.

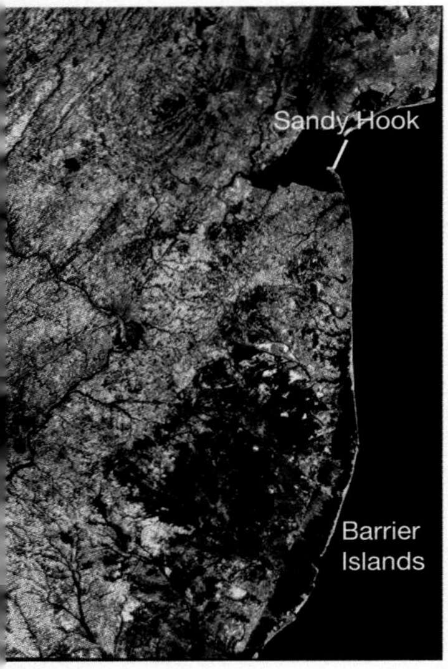

Satellite image of New Jersey showing barrier islands and Sandy Hook spit. (Phillips Petroleum Co.)

Hard Stabilization

Structures built to protect a coast from erosion or to prevent the movement of sand along a beach are known as **hard stabilization.** Hard stabilization can take many forms and often results in predictable yet unwanted outcomes. Hard stabilization includes groins, breakwaters, and seawalls.

Groins. To maintain or widen beaches that are losing sand, groins are sometimes constructed. A **groin** is a barrier built at a right angle to the beach to trap sand that is moving parallel to the shore. Groins are usually constructed of large rocks but may also be composed of wood. These structures often do their job so effectively that the longshore current beyond the groin becomes sand-starved. As a result, the current erodes sand from the beach on the downstream side of the groin.

To offset this effect, property owners downstream from the structure may erect a groin on their property. In this manner, the number of groins multiplies, resulting in a *groin field* (Figure 13.15). An example of such proliferation is the shoreline of New Jersey, where hundreds of these structures have been built. Because it has been shown that groins often do not provide a satisfactory solution, they are no longer the preferred method of keeping beach erosion in check.

Breakwaters and Seawalls. Hard stabilization can also be built parallel to the shoreline. One such structure is a **breakwater,** the purpose of which is to protect boats from the force of large breaking waves by creating a quiet water zone near the shoreline. However, when this is done, the reduced wave activity along the shore behind the structure may allow sand to accumulate. If this happens, the marina will eventually fill with sand while the downstream beach erodes and retreats. At Santa Monica, California, where the building of a breakwater created such a problem, the city had to install a dredge to remove sand from the protected quiet water zone and deposit it down the beach where longshore currents and beach drift could recirculate the sand (Figure 13.16).

▼ **Figure 13.15** Groins along the New Jersey shore at Cape May. *(Photo by John S. Shelton)*

▲ **Figure 13.16** Aerial view of a breakwater at Santa Monica, California. The breakwater appears as a faint line in the water behind which many boats are anchored. The construction of the breakwater disrupted longshore transport and caused the seaward growth of the beach. *(Photo by John S. Shelton)*

Another type of hard stabilization built parallel to the shoreline is a **seawall,** which is designed to armor the coast and defend property from the force of breaking waves. Waves expend much of their energy as they move across an open beach. Seawalls cut this process short by reflecting the force of unspent waves seaward. As a consequence, the beach to the seaward side of the seawall experiences significant erosion and may, in some instances, be eliminated entirely (Figure 13.17). Once the width of the beach is reduced, the seawall is subjected to even greater pounding by the waves. Eventually this battering will cause the wall to fail, and a larger, more expensive wall must be built to take its place.

The wisdom of building temporary protective structures along shorelines is increasingly questioned. The feelings of many coastal scientists and engineers are expressed in the following excerpt from a position paper that grew out of a conference on America's eroding shoreline:

> It is now clear that halting the receding shoreline with protective structures benefits only a few and seriously degrades or destroys the natural beach and the value it holds for the majority. Protective structures divert the ocean's energy temporarily from private properties, but usually refocus the energy on the adjacent natural beaches. Many interrupt the natural sand flow in coastal currents, robbing many beaches of vital sand replacement.*

*Strategy for Beach Preservation Proposed, *Geotimes 30* (No. 12, December 1985): 15.

◀ **Figure 13.17** Seabright in northern New Jersey once had a broad, sandy beach. A seawall 5 to 6 meters (16 to 20 feet) high and 8 kilometers (5 miles) long was built to protect the town and the railroad that brought tourists to the beach. As you can see, after the wall was built, the beach narrowed dramatically. *(Photo by Rafael Macia/Photo Researchers, Inc.)*

Alternatives to Hard Stabilization

Armoring the coast with hard stabilization has several potential drawbacks, including the cost of the structure and the loss of sand on the beach. Alternatives to hard stabilization include beach nourishment and relocation.

Beach Nourishment. **Beach nourishment** represents an approach to stabilizing shoreline sands without hard stabilization. As the term implies, this practice simply involves the addition of large quantities of sand to the beach system (Figure 13.18). By building the beaches seaward, beach quality and storm protection are both improved. Beach nourishment, however, is not a permanent solution to the problem of shrinking beaches. The same processes that removed the sand in the first place will eventually remove the replacement sand as well. In addition, beach nourishment is very expensive because huge volumes of sand must be transported to the beach from offshore areas, nearby rivers, or other source areas.

In some instances, beach nourishment can lead to unwanted environmental effects. For example, beach replenishment at Waikiki Beach, Hawaii, involved replacing coarse calcareous sand with softer, muddier calcareous sand. Destruction of the soft beach sand by breaking waves increased the water's turbidity and killed offshore coral reefs. At Miami Beach, increased turbidity also damaged local coral communities.

Beach nourishment appears to be an economically viable long-range solution to the beach preservation problem only in areas where there exists dense development, large supplies of sand, relatively low wave energy, and reconcilable environmental issues. Unfortunately, few areas possess all these attributes.

Relocation. Instead of building structures such as groins and seawalls to hold the beach in place or adding sand to replenish eroding beaches, another option is available. Many coastal scientists and planners are calling for a policy shift from defending and rebuilding beaches and coastal property in high hazard areas to removing or relocating storm-damaged buildings in those places and letting nature reclaim the beach. This approach is similar to that adopted by the federal government for river floodplains following the devastating 1993 Mississippi River floods in which vulnerable structures are abandoned and relocated on higher, safer ground.

Such proposals, of course, are controversial. People with significant nearshore investments shudder at the thought of not rebuilding and not defending coastal developments from the erosional wrath of the sea. Others, however, argue that with sea level rising, the impact of coastal storms will only get worse in the decades to come (see Box 13.2). This group advocates that oft-damaged structures be abandoned or relocated to improve personal safety and to reduce costs.

Erosion Problems Along U.S. Coasts

The shoreline along the Pacific Coast of the United States is strikingly different from that characterizing the Atlantic and Gulf Coast regions. Some of the differences are related to plate tectonics. The West Coast represents the leading edge of the North American plate, and because of this, it experiences active uplift and deformation. By contrast, the East Coast is a tectonically quiet region that is far from any active plate margin. Because of this basic geological difference, there are differences in the nature of shoreline erosion problems along America's opposite coasts.

Atlantic and Gulf Coasts. Much of the development along the Atlantic and Gulf coasts has occurred on barrier islands. Typically, barrier islands consist of a wide beach that is backed by dunes and separated from

Did You Know?

The worst natural disaster in U.S. history came as a result of a hurricane that struck an unprepared Galveston, Texas, on September 8, 1900. The strength of the storm and the lack of adequate warning cost the lives of 6000 people in the city and at least 2000 more elsewhere.

A. B.

▲ **Figure 13.18** Miami Beach. **A.** Before beach nourishment and **B.** after beach nourishment. *(Courtesy of the U.S. Army Corps of Engineers, Vicksburg District)*

Hurricanes are the ultimate coastal hazard. Hurricane Ivan on September 17, 2004. (NASA)

the mainland by marshy lagoons. The broad expanses of sand and exposure to the ocean have made barrier islands exceedingly attractive sites for development. Unfortunately, development has taken place more rapidly than our understanding of barrier island dynamics.

Because barrier islands face the open ocean, they receive the full force of major storms that strike the coast. When a storm occurs, the barriers absorb the energy of the waves primarily through the movement of sand. This process and the dilemma that results have been described as follows:

> Waves may move sand from the beach to offshore areas or, conversely, into the dunes; they may erode the dunes, depositing sand onto the beach or carrying it out to sea; or they may carry sand from the beach and the dunes into the marshes behind the barrier, a process known as overwash. The common factor is movement. Just as a flexible reed may survive a wind that destroys an oak tree, so the barriers survive hurricanes and nor'easters not through unyielding strength but by giving before the storm.
>
> This picture changes when a barrier is developed for homes or a resort. Storm waves that previously rushed harmlessly through gaps between the dunes now encounter buildings and roadways. Moreover, because the dynamic nature of the barrier is readily perceived only during storms, homeowners tend to attribute damage to a particular storm, rather than to the basic mobility of coastal barriers. With their homes or investments at stake, local residents are more likely to seek to hold the sand in place and the waves at bay than to admit that development was improperly placed to begin with.*

Did You Know?

The average cost of sand used to replenish beaches is between \$5–\$10 per cubic yard. A typical dump truck holds about 60 cubic yards. A small project requiring several hundred cubic yards of sand could exceed \$10,000 per year. Large projects run into the millions of dollars.

*Frank Lowenstein, Beaches or Bedrooms—The Choice as Sea Level Rises, *Oceanus* 28 (No. 3, Fall 1985): 22.

Pacific Coast. In contrast to the broad, gently sloping coastal plains of the Atlantic and Gulf coasts, much of the Pacific Coast is characterized by relatively narrow beaches that are backed by steep cliffs and mountain ranges. Recall that America's western margin is a more rugged and tectonically active region than the eastern margin. Because uplift continues, a rise in sea level in the West is not so readily apparent. Nevertheless, like the shoreline erosion problems facing the East's barrier islands, West Coast difficulties also stem largely from the alteration of a natural system by people.

A major problem facing the Pacific shoreline, and especially portions of southern California, is a significant narrowing of many beaches. The bulk of the sand on many of these beaches is supplied by rivers that transport it from the mountainous regions to the coast. Over the years this natural flow of material to the coast has been interrupted by dams built for irrigation and flood control. The reservoirs effectively trap the sand that would otherwise nourish the beach environment. When the beaches were wider, they served to protect the cliffs behind them from the force of storm waves. Now, however, the waves move across the narrowed beaches without losing much energy and cause more rapid erosion of the sea cliffs.

Although the retreat of the cliffs provides material to replace some of the sand impounded behind dams, it also endangers homes and roads built on the bluffs. In addition, development atop the cliffs aggravates the problem. Urbanization increases runoff, which, if not carefully controlled, can result in serious bluff erosion. Watering lawns and gardens adds significant quantities of water to the slope. This water percolates downward toward the base of the cliff, where it may emerge in small seeps. This action reduces the slope's stability and facilitates mass wasting.

BOX 13.2

Coastal Vulnerability to Sea-Level Rise

Human activities, especially the combustion of fossil fuels, have been adding vast amounts of carbon dioxide and other trace gases to the atmosphere for 200 years or more. The prospect is that emissions of these gases will continue to increase during the twenty-first century. One consequence of this change in the composition of the atmosphere is an enhancement of Earth's greenhouse effect with a resulting increase in global temperatures. During the twentieth century, average global temperatures increased by about 0.6°C. During the twenty-first century, the increase is projected to be considerably greater.

One probable impact of a human-induced global warming is a rise in sea level. How is a warmer atmosphere related to a global rise in sea level? The most obvious connection—the melting of glaciers—is important but *not* the most significant factor. More significant is that a warmer atmosphere causes an increase in ocean volume due to thermal expansion. Higher air temperatures warm the adjacent upper layers of the ocean, which in turn causes the water to expand and sea level to rise.

Research indicates that sea level has risen between 10 and 25 centimeters (4 and 8 inches) over the past century and that the trend will continue at an accelerated rate. Some models indicate that the rise may approach or even exceed 50 centimeters (20 inches) by the year 2100. Such a change may seem modest, but scientists realize that any rise in sea level along a *gently* sloping shoreline, such as the Atlantic and Gulf coasts of the United States, will lead to significant erosion and severe permanent inland flooding (Figure 13.B). If this happens, many beaches and wetlands will be eliminated, and coastal civilization would be severely disrupted.

Because rising sea level is a gradual phenomenon, it may be overlooked by coastal residents as an important contributor to shoreline erosion problems. Rather, the blame may be assigned to other forces, especially storm activity. Although a given storm may be the immediate cause, the magnitude of its destruction may result from the relatively small sea-level rise that allowed the storm's power to cross a much greater land area.

One of the most challenging problems for coastal scientists today is determining the physical response of the coastline to sea-level rise. Predicting shoreline retreat and land-loss rates is critical to formulating coastal management strategies. To date, long-term planning for our shorelines has been piecemeal, if at all. Consequently, development continues without adequate consideration of the potential costs of erosion, flooding, and storm damage.

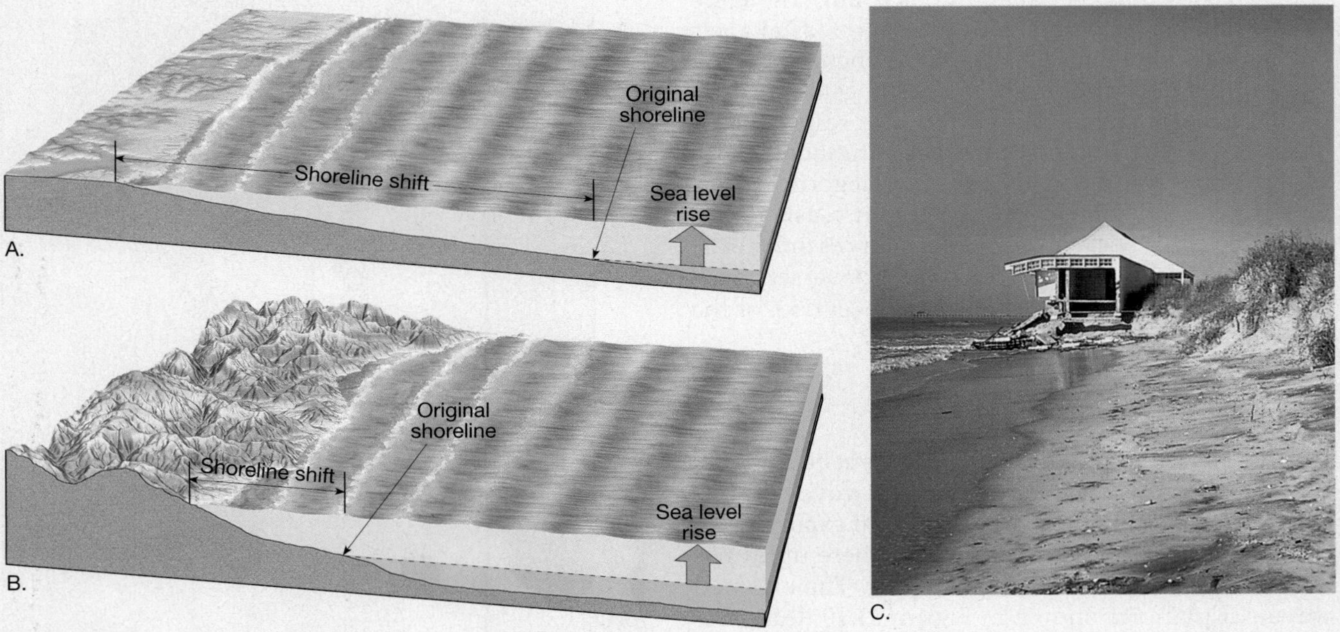

Figure 13.B The slope of a shoreline is critical to determining the degree to which sea-level changes will affect it. A. When the slope is gentle, small changes in sea level cause a substantial shift. B. The same sea-level rise along a steep coast results in only a small shoreline shift. C. As sea level gradually rises, the shoreline retreats, and structures that were once thought to be safe from wave attack are exposed to the force of the sea. (Photo by Kenneth Hasson)

Shoreline erosion along the Pacific Coast varies considerably from one year to the next, largely because of the sporadic occurrence of storms. As a consequence, when the infrequent but serious episodes of erosion occur, the damage is often blamed on the unusual storms and not on coastal development or the sediment-trapping dams that may be great distances away. If, as predicted, sea level rises at an increasing rate in the years to come, increased shoreline erosion and sea-cliff retreat should be expected along many parts of the Pacific Coast.

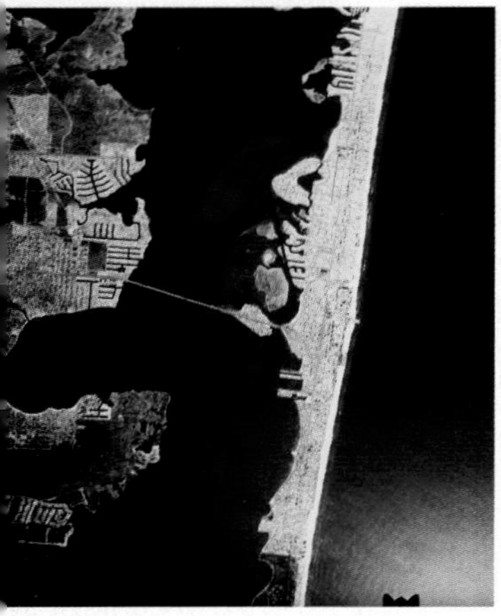

This high-altitude infrared image shows a portion of an urbanized barrier island off the New Jersey coast. (USDA-ASCS)

Coastal Classification

The great variety of shorelines demonstrates their complexity. Indeed, to understand any particular coastal area, many factors must be considered, including rock types, size and direction of waves, frequency of storms, range between high and low tides, and offshore topography. In addition, practically all coastal areas were affected by the worldwide rise in sea level that accompanied the melting of Ice Age glaciers at the close of the Pleistocene epoch. Finally, tectonic events that elevate or drop the land or change the volume of ocean basins must be taken into account. The large number of factors that influence coastal areas make shoreline classification difficult.

Many geologists classify coasts based on changes that have occurred with respect to sea level. This commonly used classification divides coasts into two very general categories: emergent and submergent. **Emergent coasts** develop either because an area experiences uplift or as a result of a drop in sea level. Conversely, **submergent coasts** are created when sea level rises or the land adjacent to the sea subsides.

Emergent coast of California's San Clemente Island. (John S. Shelton)

Emergent Coasts

In some areas the coast is clearly emergent because rising land or a falling water level exposes wave-cut cliffs and platforms above sea level. Excellent examples include portions of coastal California where uplift has occurred in the recent geological past. The elevated wave-cut platform shown in Figure 13.10 illustrates this. In the case of the Palos Verdes Hills, south of Los Angeles, seven different terrace levels exist, indicating seven episodes of uplift. The ever persistent sea is now cutting a new platform at the base of the cliff. If uplift follows, it too will become an elevated marine terrace.

Other examples of emergent coasts include regions that were once buried beneath great ice sheets. When glaciers were present, their weight depressed the crust, and when the ice melted, the crust began gradually to spring back. Consequently, prehistoric shoreline features today are found high above sea level. The Hudson Bay region of Canada is one such area, portions of which are still rising at a rate of more than 1 centimeter per year.

The rugged coast of Maine. (Roy Rainford)

Submergent Coasts

In contrast to the preceding examples, other coastal areas show definite signs of submergence. Shorelines that have been submerged in the relatively recent past are often highly irregular because the sea typically floods the lower reaches of river valleys flowing into the ocean. The ridges separating the valleys, however, remain above sea level and project into the sea as

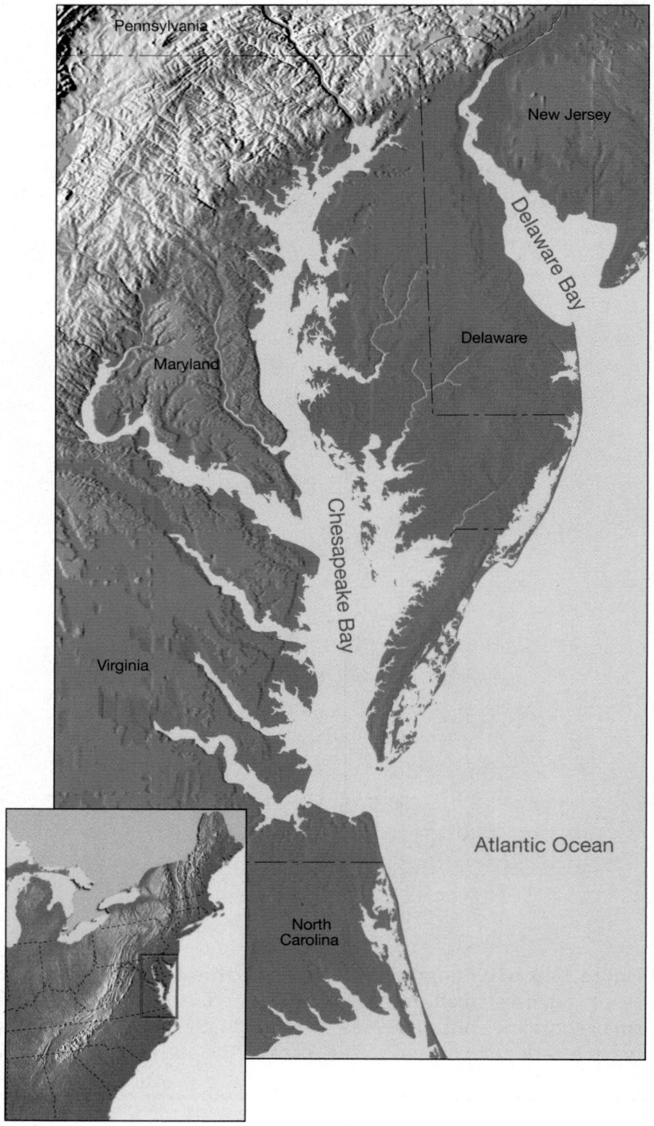

▲ **Figure 13.19** Estuaries along the East Coast of the United States. The lower portions of many river valleys were submerged by the rise in sea level that followed the end of the Ice Age. Chesapeake Bay and Delaware Bay are especially prominent examples.

headlands. These drowned river mouths, which are called **estuaries,** characterize many coasts today. Along the Atlantic coastline, the Chesapeake and Delaware bays are examples of estuaries created by submergence (Figure 13.19). The picturesque coast of Maine, particularly in the vicinity of Acadia National Park, is another excellent example of an area that was flooded by the postglacial rise in sea level and transformed into a highly irregular submerged coastline.

Keep in mind that most coasts have a complicated geologic history. With respect to sea level, many have at various times emerged and then submerged again. Each time, they retain some of the features created during the previous situation.

Tides

Tides are daily changes in the elevation of the ocean surface. Their rhythmic rise and fall along coastlines have been known since antiquity. Other than waves, they are the easiest ocean movements to observe. An exceptional example of extreme daily tides is shown in Figure 13.20.

Although known for centuries, tides were not explained satisfactorily until Sir Isaac Newton applied the law of gravitation to them. Newton showed that there is a mutual attractive force between two bodies, as between Earth and the Moon. Because the atmosphere and the ocean both are fluids and thus free to move, both are deformed by this force. Hence, ocean tides result from the gravitational attraction exerted upon Earth by the Moon and, to a lesser extent, by the Sun.

Causes of Tides

To illustrate how tides are produced, we will assume Earth is a rotating sphere covered to a uniform depth with water (Figure 13.21). It is easy to see how the Moon's gravitational force can cause the water to bulge on the side of Earth nearest the Moon. In addition, however, an equally large tidal bulge is produced on the side of Earth directly opposite the Moon.

Both tidal bulges are caused, as Newton discovered, by the pull of gravity. Gravity is inversely proportional to the square of the distance between two objects, meaning simply that it quickly weakens with distance. In this case, the two objects are the Moon and Earth. Because the force of gravity decreases with distance, the Moon's gravitational pull on Earth is slightly greater on the near side of Earth than on the far side. The result of this differential pulling is to stretch (elongate) the "solid" Earth very slightly. In contrast, the world ocean, which is mobile, is deformed quite dramatically by this effect to produce the two opposing tidal bulges.

Because the position of the Moon changes only moderately in a single day, the tidal bulges remain in place while Earth rotates through them. For this reason, if you stand on the seashore for 24 hours, Earth will rotate you through alternating areas of deeper and shallower water. As you are carried into each tidal bulge, the tide rises, and as you are carried into the intervening troughs between the tidal bulges, the tide falls. Therefore, most places on Earth experience two high tides and two low tides each tidal day.

Further, the tidal bulges migrate as the Moon revolves around Earth every 29 days. As a result, the tides,

Did You Know?

Tides can be used to generate electrical power by constructing a dam across the mouth of a bay or estuary in a coastal area having a large tidal range. The narrow opening between the bay and the open ocean magnifies the variations in water level as tides rise and fall and the strong in-and-out flow is used to drive turbines. The largest plant yet constructed is along the coast of France.

Did You Know?

Tidal waves are more accurately known as *tsunami* and have *nothing* to do with the tides. They are long-wavelength, fast-moving, often large, and sometimes destructive waves that originate from sudden changes in the topography of the seafloor. Since a mechanism that often triggers these dangerous waves is an earthquake, tsunami are appropriately termed *seismic sea waves.* For more about tsunami, see Chapter 14.

▲ **Figure 13.20** High tide and low tide on Nova Scotia's Minas Basin in the Bay of Fundy. Tidal flats are exposed during low tide. *(Photos courtesy of Nova Scotia Department of Tourism)*

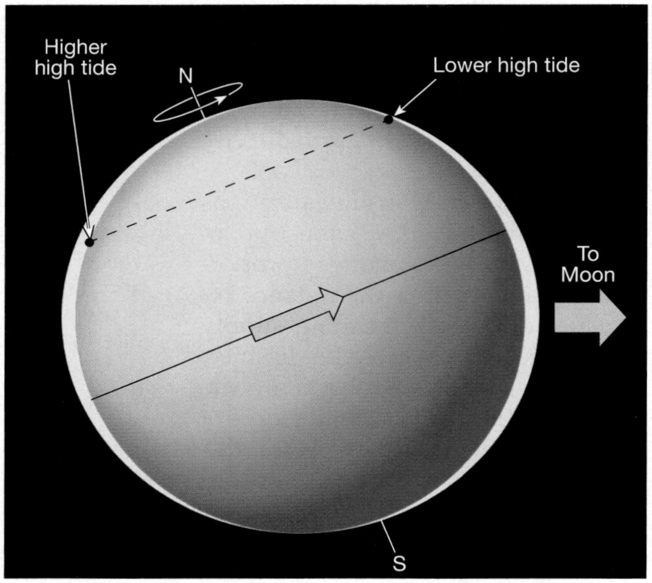

▲ **Figure 13.21** Idealized tidal bulges on Earth caused by the Moon. If Earth were covered to a uniform depth with water, there would be two tidal bulges: one on the side of Earth facing the Moon (right) and the other on the opposite side of Earth (left). Depending on the Moon's position, tidal bulges may be inclined relative to Earth's equator. In this situation, Earth's rotation causes an observer to experience two unequal high tides during a day.

like the time of moonrise, shift about 50 minutes later each day. After 29 days the cycle is complete and a new one begins.

In many locations, there may be an inequality between the high tides during a given day. Depending on the Moon's position, the tidal bulges may be inclined to the equator as in Figure 13.21. This figure illustrates that one high tide experienced by an observer in the Northern Hemisphere is considerably higher than the high tide half a day later. In contrast, a Southern Hemisphere observer would experience the opposite effect.

Tidal power plant at Annapolis Royal, Nova Scotia, on the Bay of Fundy. (James P. Blair)

Monthly Tidal Cycle

The primary body that influences the tides is the Moon, which makes one complete revolution around Earth every 29 and a half days. The Sun, however, also influences the tides. It is far larger than the Moon, but because it is much farther away, its effect is considerably less. In fact, the Sun's tide-generating effect is only about 46 percent that of the Moon's.

Near the times of new and full moons, the Sun and Moon are aligned and their forces are added together (Figure 13.22A). Accordingly, the combined gravity of these two tide-producing bodies causes larger tidal bulges (higher high tides) and deeper tidal troughs (lower low tides), producing a large tidal range. These are called the **spring tides,** which have no connection with the spring season but occur twice a month during the time when the Earth–Moon–Sun system is aligned. Conversely, at about the time of the first and third quarters of the Moon, the gravitational forces of the Moon and Sun act on Earth at right angles, and each partially offsets the influence of the other (Figure 13.22B). As a result, the daily tidal range is less. These are called **neap tides,** which also occur twice each month. Each month, then, there are two spring tides and two neap tides, each about one week apart.

Tidal Currents

Tidal current is the term used to describe the *horizontal* flow of water accompanying the rise and fall of the tide. These water movements induced by tidal forces can be important in some coastal areas. Tidal currents that advance into the coastal zone as the tide rises are called *flood currents.* As the tide falls, seaward-moving water generates *ebb currents.* Periods of little or no current,

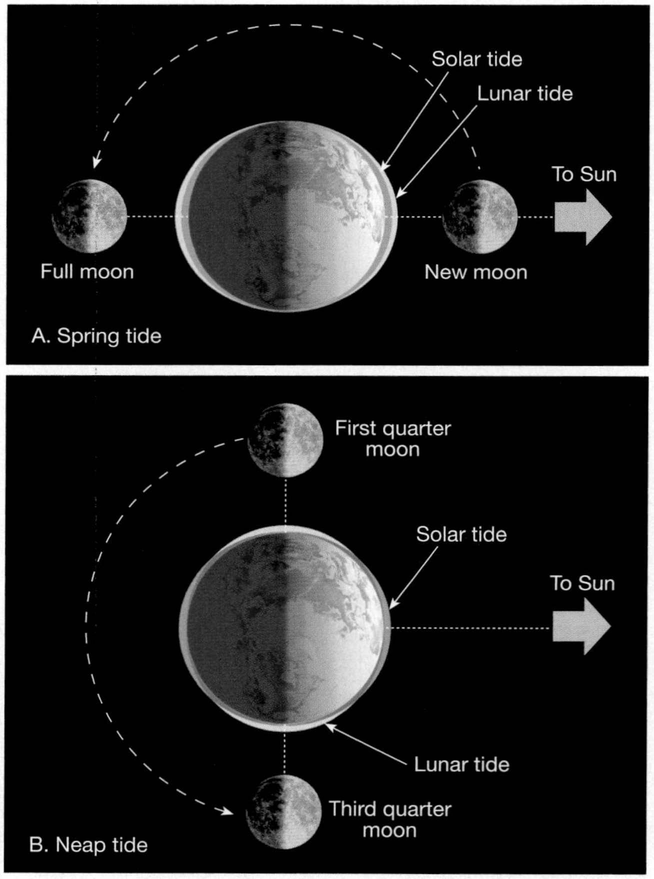

▲ **Figure 13.22** Earth–Moon–Sun positions and the tides. **A.** When the Moon is in the full or new position, the tidal bulges created by the Sun and Moon are aligned, there is a large tidal range on Earth, and spring tides are experienced. **B.** When the Moon is in the first-or third-quarter position, the tidal bulges produced by the Moon are at right angles to the bulges created by the Sun. Tidal ranges are smaller, and neap tides are experienced.

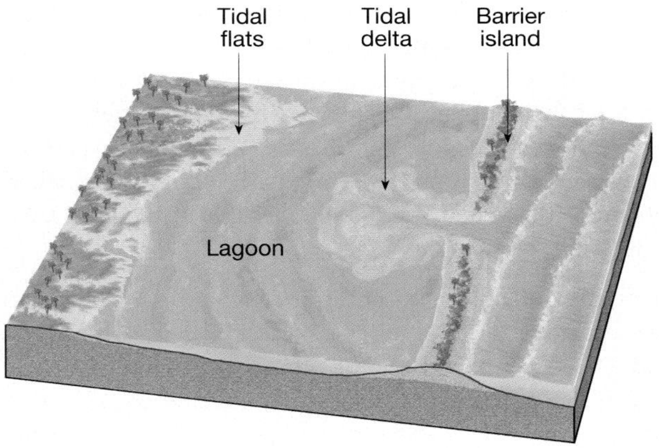

▲ **Figure 13.23** As a rapidly moving tidal current moves through a barrier island's inlet into the quiet waters of the lagoon, the current slows and deposits sediment, creating a tidal delta. Because this tidal delta has developed on the landward side of the inlet, it is called a *flood delta.*

called *slack water,* separate flood and ebb. The areas affected by these alternating tidal currents are called **tidal flats** (see Figure 13.20, right). Depending on the nature of the coastal zone, tidal flats vary from narrow strips seaward of the beach to zones that may extend for several kilometers.

Although tidal currents are not important in the open sea, they can be rapid in bays, river estuaries, straits, and other narrow places. Off the coast of Brittany in France, for example, tidal currents that accompany a high tide of 12 meters (40 feet) may attain a speed of 20 kilometers (12 miles) per hour. Although tidal currents are not generally considered to be major agents of erosion and sediment transport, notable exceptions occur where tides move through narrow inlets. Here they scour the small entrances to many good harbors that would otherwise be blocked.

Extensive tidal flats exposed at low tide. (Nova Scotia Department of Tourism)

Sometimes deposits called **tidal deltas** are created by tidal currents (Figure 13.23). They may develop either as *flood deltas* landward of an inlet or as *ebb deltas* on the seaward side of an inlet. Because wave activity and longshore currents are reduced on the sheltered landward side, flood deltas are more common and actually more prominent (see Figure 13.12). They form after the tidal current moves rapidly through an inlet. As the current emerges into more open waters from the narrow passage, it slows and deposits its load of sediment.

Did You Know?

The world's largest tidal range (difference between successive high and low tides) is found in the northern end of Nova Scotia's Bay of Fundy. Here the maximum spring tidal range is about 17 m (56 ft). This leaves boats "high and dry" during low tide (see Figure 13.20).

The Chapter in Review

- The *shore* is the area extending between the lowest tide level and the highest elevation on land that is affected by storm waves. The *coast* extends inland from the shore as far as ocean-related features can be found. The shore is divided into the *foreshore* and *backshore.* Seaward of the foreshore are the *nearshore* and *offshore* zones.

- A *beach* is an accumulation of sediment found along the landward margin of the ocean or a lake. Among its parts are one or more *berms* and the *beach face.* Beaches are composed of whatever material is locally abundant and should be thought of as material in transit along the shore.

- *Waves are moving energy* and *most ocean waves are initiated by the wind.* The three factors that influence the *height, wavelength,* and *period* of a wave are (1) *wind speed,* (2) *length of time the wind has blown,* and (3) *fetch,* the distance the wind has traveled across open water. Once waves leave a storm area, they are termed *swells,* which are symmetrical, longer-wavelength waves.

- As waves travel, *water particles transmit energy by circular orbital motion,* which extends to a depth equal to one half the wavelength. When a wave travels into shallow water, it experiences physical changes that can cause the wave to collapse, or *break,* and form *surf.*

- The three factors that influence the *height, wavelength,* and *period* of a wave are (1) *wind speed,* (2) *length of time the wind has blown,* and (3) *fetch,* the distance that the wind has traveled across the open water.

- Wave erosion is caused by *wave impact pressure* and *abrasion* (the sawing and grinding action of water armed with rock fragments). The bending of waves is called *wave refraction.* Owing to refraction, wave impact is concentrated against the sides and ends of headlands.

- Most waves reach the shore at an angle. The uprush (swash) and backwash of water from each breaking wave moves the sediment in a zigzag pattern along the beach. This movement, called *beach drift,* can transport sand hundreds or even thousands of meters each day. Oblique waves also produce *longshore currents* within the surf zone that flow parallel to the shore.

- Features produced by *shoreline erosion* include *wave-cut cliffs* (which originate from the cutting action of the surf against the base of coastal land), *wave-cut platforms* (relatively flat, benchlike surfaces left behind by receding cliffs), *sea arches* (formed when a headland is eroded and two caves from opposite sides unite), and *sea stacks* (formed when the roof of a sea arch collapses).

- Some of the depositional features formed when sediment is moved by beach drift and longshore currents are *spits* (elongated ridges of sand that project from the land into the mouth of an adjacent bay), *baymouth bars* (sandbars that completely cross a bay), and *tombolos* (ridges of sand that connect an island to the mainland or to another island). Along the Atlantic and Gulf coastal plains, the shore zone is characterized by *barrier islands,* low ridges of sand that parallel the coast at distances from 3 to 30 kilometers offshore.
- Local factors that influence shoreline erosion are (1) the proximity of a coast to sediment-laden rivers, (2) the degree of tectonic activity, (3) the topography and composition of the land, (4) prevailing winds and weather patterns, and (5) the configuration of the coastline and nearshore areas.
- *Hard stabilization* involves building solid, massive structures in an attempt to protect a coast from erosion or prevent the movement of sand along the beach. Hard stabilization includes *groins* (short walls constructed at a right angle to the shore to trap moving sand), *breakwaters* (structures built parallel to the shore to protect it from the force of large breaking waves), and *seawalls* (armoring the coast to prevent waves from reaching the area behind the wall). *Alternatives to hard stabilization* include *beach nourishment,* which involves the addition of sand to replenish eroding beaches, and *relocation* of damaged or threatened buildings.
- One common classification of coasts is based upon changes that have occurred with respect to sea level. *Emergent coasts,* often with wave-cut cliffs and wave-cut platforms above sea level, develop either because an area experiences uplift or as a result of a drop in sea level. Conversely, *submergent coasts,* with their drowned river mouths, called *estuaries,* are created when sea level rises or the land adjacent to the sea subsides.
- *Tides,* the daily rise and fall in the elevation of the ocean surface, are caused by the *gravitational attraction* of the Moon and, to a lesser extent, by the Sun. Near the times of new and full moons, the Sun and Moon are aligned, and their gravitational forces are added together to produce especially high and low tides. These are called the *spring tides.* Conversely, at about the times of the first and third quarters of the Moon, when the gravitational forces of the Moon and Sun are at right angles, the daily tidal range is less. These are called *neap tides.*
- *Tidal currents* are horizontal movements of water that accompany the rise and fall of tides. *Tidal flats* are the areas that are affected by the advancing and retreating tidal currents. When tidal currents slow after emerging from narrow inlets, they deposit sediment that may eventually create *tidal deltas.*

Key Terms

abrasion (p. 290)
backshore (p. 286)
barrier island (p. 295)
baymouth bar (p. 295)
beach (p. 286)
beach drift (p. 292)
beach face (p. 286)
beach nourishment (p. 299)
berm (p. 286)
breakwater (p. 298)
coast (p. 296)
coastline (p. 286)
emergent coast (p. 302)
estuary (p. 303)
foreshore (p. 286)
groin (p. 298)
hard stabilization (p. 298)
longshore current (p. 292)
marine terrace (p. 294)
neap tide (p. 304)
nearshore (p. 286)
offshore (p. 286)
sea arch (p. 294)
sea stack (p. 294)
seawall (p. 298)
shore (p. 286)
shoreline (p. 286)
spit (p. 295)
spring tide (p. 304)
submergent coast (p. 302)
surf (p. 290)
tidal current (p. 304)
tidal deltas (p. 305)
tidal flats (p. 305)
tide (p. 303)
tombolo (p. 295)
wave-cut cliff (p. 294)
wave-cut platform (p. 294)
wave height (p. 288)
wavelength (p. 288)
wave period (p. 288)
wave refraction (p. 291)

Questions for Review

1. Distinguish among shore, shoreline, coast, and coastline.
2. What is a beach? Briefly distinguish between *beach face* and *berm.* What are the sources of beach sediment?
3. List three factors that determine the height, length, and period of a wave.
4. Describe the motion of a floating object as a wave passes.
5. Describe the physical changes that occur to a wave's speed, wavelength, and height as it moves into shallow water and breaks.
6. Describe two ways in which waves cause erosion.
7. What is *wave refraction?* What is the effect of this process along irregular coastlines?
8. Why are beaches often called "rivers of sand"?
9. Describe the formation of the following features: *wave-cut cliff, wave-cut platform, marine terrace, sea stack, spit, baymouth bar,* and *tombolo.*
10. List three ways that barrier islands may form.
11. In what direction are beach drift and longshore currents moving sand in Figure 13.15, p. 298 Is it moving toward the top or toward the bottom of the photo?
12. List the types of hard stabilization and describe what each is intended to do. What effect does each one have on the distribution of sand on the beach?

13. List two alternatives to hard stabilization, indicating potential problems with each one.

14. How is a warmer atmosphere related to a global rise in sea level? (See Box 13.2.)

15. Relate the damming of rivers to the shrinking of beaches at some locations along the West Coast of the United States. Why do narrower beaches lead to accelerated sea-cliff retreat?

16. What observable features would lead you to classify a coastal area as emergent?

17. Are estuaries associated with submergent or emergent coasts? Explain.

18. Discuss the origin of ocean tides. Explain why the Sun's influence on Earth's tides is only about half that of the Moon's, even though the Sun is so much more massive than the Moon.

19. Explain why an observer can experience two unequal high tides during one day.

20. Distinguish between flood current and ebb current.

Online Study Guide

The *Essentials of Geology* Web site uses the resources and flexibility of the Internet to aid in your study of the topics in this chapter. Written and developed by geology instructors, this site will help improve your understanding of geology. Visit **www.prenhall.com/lutgens** and click on the cover of *Essentials of Geology 9e* to find:

- Online review quizzes.
- Critical thinking exercises.
- Links to chapter-specific Web resources.
- Internet-wide key-term searches.

http://www.prenhall.com/lutgens

Destruction in the coastal town of Aceh, on the Indonesian island of Sumatra, from a massive tsunami on December 26, 2004.
(Photo by Michael L Bak/Department of Defense/Getty Images)

CHAPTER 14

Earthquakes and Earth's Interior

Focus on Learning

To assist you in learning the important concepts in this chapter, you will find it helpful to focus on the following questions:

- What is an earthquake?
- What are the types of earthquake waves?
- How is the epicenter of an earthquake determined?
- Where are the principal earthquake zones on Earth?
- How is earthquake strength expressed?
- What are the major zones of Earth's interior?
- How do continental crust and oceanic crust differ?

On October 17, 1989, at 5:04 p.m. Pacific Daylight Time, millions of television viewers around the world were settling in to watch the third game of the World Series. Instead, they saw their television sets go black as tremors hit San Francisco's Candlestick Park. Although the earthquake was centered in a remote section of the Santa Cruz Mountains, 100 kilometers to the south, major damage occurred in the Marina District of San Francisco (Figure 14.1).

The most tragic result of the violent shaking was the collapse of some double-decked sections of Interstate 880, also known as the Nimitz Freeway. The ground motions caused the upper deck to sway, shattering the concrete support columns along a mile-long section of the freeway. The upper deck then collapsed onto the lower roadway, flattening cars as if they were aluminum beverage cans. This earthquake, named the Loma Prieta quake for its point of origin, claimed 67 lives.

In mid-January 1994, less than five years after the Loma Prieta earthquake devastated portions of the San Francisco Bay area, a major earthquake struck the Northridge area of Los Angeles. Although not the fabled "Big One," this moderate 6.7-magnitude earthquake left 57 dead, over 5000 injured, and tens of thousands of households without water and electricity. The damage exceeded $40 billion and was attributed to an apparently unknown fault that ruptured 18 kilometers (11 miles) beneath Northridge.

The Northridge earthquake began at 4:31 a.m. and lasted roughly 40 seconds. During this brief period, the quake terrorized the entire Los Angeles area. In the three-story Northridge Meadows apartment complex, 16 people died when sections of the upper floors collapsed onto the first-floor units. Nearly 300 schools were seriously damaged, and a dozen major roadways buckled. Among these were two of California's major arteries—the Golden State Freeway (Interstate 5), where an overpass collapsed completely and blocked the roadway, and the Santa Monica Freeway. Fortunately, these roadways had practically no traffic at this early morning hour.

In nearby Granada Hills, broken gas lines were set ablaze while the streets flooded from broken water mains. Seventy homes burned in the Sylmar area. A 64-car freight train derailed, including some cars carrying hazardous cargo. But it is remarkable that the destruction was not greater. Unquestionably, the upgrading of structures to meet the requirements of building codes developed for this earthquake-prone area helped minimize what could have been a much greater human tragedy.

Did You Know?

Literally thousands of earthquakes occur daily! Fortunately, the majority of them are too small to be felt by people, and the majority of larger ones occur in remote regions. Their existence is known only because of sensitive seismographs.

Fire triggered when a gas line ruptured during an earthquake in southern California in 1994.
(AFP/Getty Images)

▼ **Figure 14.1** Damage in San Francisco's Marina District following the 1989 Loma Prieta earthquake. *(Photo by David Weintraub/Photo Researchers, Inc.)*

What Is an Earthquake?

GEODe

Earthquakes and Earth's Interior (Part A)
▼ What is an Earthquake?

An **earthquake** is the vibration of Earth produced by the rapid release of energy. Most often, earthquakes are caused by slippage along a fault in Earth's crust. The energy released radiates in all directions from its source, the **focus,** in the form of waves (Figure 14.2). These waves are analogous to those produced when a stone is dropped into a calm pond. Just as the impact of the stone sets water waves in motion, an earthquake generates seismic waves that radiate throughout Earth. Even though the energy dissipates rapidly with increasing distance from the focus, sensitive instruments located throughout the world record the event.

Over 30,000 earthquakes that are strong enough to be felt occur worldwide annually. Fortunately, most are minor tremors and do very little damage. Generally, only about 75 significant earthquakes take place each year, and many of these occur in remote regions. However, occasionally a large earthquake occurs near a large population center. Under these conditions, an earthquake is among the most destructive natural forces on Earth.

The shaking of the ground, coupled with the liquefaction of some soils, wreaks havoc on buildings and other structures. In addition, when a quake occurs in a populated area, power and gas lines are often ruptured, causing numerous fires. In the famous 1906 San Francisco earthquake, much of the damage was caused by fires (Figure 14.3), which quickly became uncontrollable when broken water mains left firefighters with only trickles of water.

Earthquakes and Faults

The tremendous energy released by atomic explosions or by volcanic eruptions can produce an earthquake, but these events are relatively weak and infrequent.

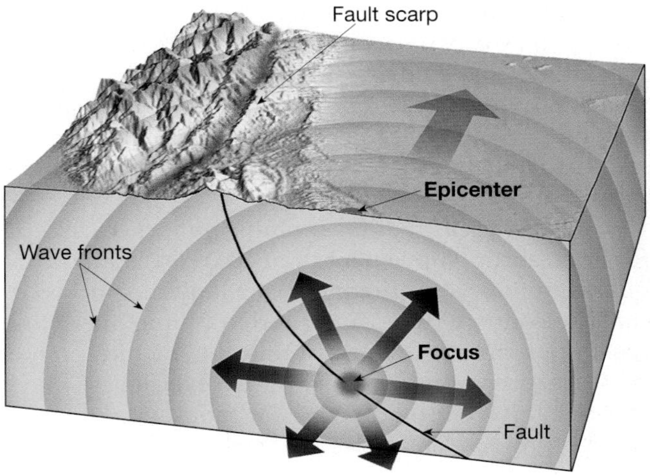

▲ **Figure 14.2** The focus of all earthquakes is located at depth. The surface location directly above it is called the epicenter.

What mechanism produces a destructive earthquake? Ample evidence exists that Earth is not a static planet. We know that Earth's crust has been uplifted at times, because we have found numerous ancient wave-cut benches many meters above the level of the highest tides. Other regions exhibit evidence of extensive subsidence. In addition to these vertical displacements, offsets in fence lines, roads, and other structures indicate that horizontal movement is common. These movements are associated with large fractures in Earth's crust called **faults.**

Typically, earthquakes occur along preexisting faults that formed in the distant past along zones of weakness in Earth's crust. Some are very large and can generate major earthquakes. One example is the San Andreas Fault, which is a transform fault boundary that separates two great sections of Earth's lithosphere: the North American plate and the Pacific plate. Other faults are small and produce only minor earthquakes.

Slippage along a fault produced an offset in this orange grove. (John S. Shelton)

◀ **Figure 14.3** San Francisco in flames after the 1906 earthquake. *(Reproduced from the collection of the Library of Congress)*

Aerial view of the San Andreas Fault. (Michael Collier)

Most faults are not perfectly straight or continuous; instead, they consist of numerous branches and smaller fractures that display kinks and offsets. Such a pattern is displayed in Figure 17.B, (p. 400) which shows that the San Andreas Fault is actually a system that consists of several large faults and innumerable small fractures (not shown).

It is also clear that most faults are locked, except for brief, abrupt movements that accompany an earthquake rupture. The primary reason faults are locked is that the confining pressure exerted by the overlying crust is enormous. Because of this, the fractures in the crust are essentially squeezed shut. Nevertheless, even faults that have been inactive for thousands of years can rupture again if the stresses acting on the region increase sufficiently.

Discovering the Cause of Earthquakes

The actual mechanism of earthquake generation eluded geologists until H. F. Reid of Johns Hopkins University conducted a study following the great 1906 San Francisco earthquake. The earthquake was accompanied by horizontal surface displacements of several meters along the northern portion of the San Andreas Fault. Field investigations determined that during this single earthquake, the Pacific plate lurched as much as 4.7 meters (15 feet) northward past the adjacent North American plate.

The mechanism for earthquake formation that Reid deduced from this information is illustrated in Figure 14.4. In part A of the figure, you see an existing fault, or break in the rock. In part B, tectonic forces ever so slowly deform the crustal rocks on both sides of the fault, as demonstrated by the bent features. Under these conditions, rocks are bending and storing elastic energy, much like a wooden stick does if bent. Eventually, the frictional resistance holding the rocks in place is overcome. As slippage occurs at the weakest point (the focus), displacement will exert stress farther along the fault, where additional slippage will occur, releasing the built-up strain (Figure 14.4C). This slippage allows the deformed rock to "snap back." The vibrations we know as an earthquake occur as the rock elastically returns to its original shape. The "springing back" of the rock was termed **elastic rebound** by Reid, because the rock behaves elastically, much like a stretched rubber band does when it is released.

Most of the motion along faults can be satisfactorily explained by the plate tectonics theory, which states that large slabs of Earth's lithosphere are in continual slow motion. These mobile plates interact with neighboring plates, straining and deforming the rocks at their margins. In fact, it is along faults associated with plate boundaries that most earthquakes occur. Furthermore, earthquakes are repetitive: As soon as one is over, the continuous motion of the plates adds strain to the rocks until they eventually fail again.

In summary, most earthquakes are produced by the rapid release of elastic energy stored in rock that has been subjected to great stress. Once the strength of the rock is exceeded, it suddenly ruptures, causing the vibrations of an earthquake. Earthquakes most often occur along existing faults whenever the frictional forces on the fault surfaces are overcome.

Did You Know?

Humans have inadvertently triggered earthquakes. In 1962 Denver began experiencing frequent tremors. The earthquakes were located near an army waste-disposal well used to inject waste into the ground. Investigators concluded that the pressurized fluids made their way along a buried fault surface, which reduced friction and triggered fault slippage and earthquakes. Sure enough, when the pumping halted, so did the tremors.

Foreshocks and Aftershocks

The intense vibrations of the 1906 San Francisco earthquake lasted about 40 seconds. Although most of the displacement along the fault occurred in this rather short period, additional movements along this and other nearby faults lasted for several days following the main quake. The adjustments that follow a major earthquake often generate smaller earthquakes called **aftershocks.** Although these aftershocks are usually much weaker than the main earthquake, they can sometimes destroy already badly weakened structures. This occurred, for example, during a 1988 earthquake in Armenia. A large aftershock of magnitude 5.8 collapsed many structures that had been weakened by the main tremor.

In addition, small earthquakes called **foreshocks** often precede a major earthquake by days or, in some cases, by as much as several years. Monitoring of these foreshocks has been used as a means of predicting forthcoming major earthquakes, with mixed success. We will consider the topic of earthquake prediction in a later section of this chapter.

San Andreas Fault: An Active Earthquake Zone

The San Andreas is undoubtedly the most studied fault system in the world. Over the years, investigations have shown that displacement occurs along discrete segments that are 100 to 200 kilometers long. Further, each fault segment behaves somewhat differently from the others. Some portions of the San Andreas exhibit a slow, gradual displacement known as **fault creep,** which occurs relatively smoothly and therefore with little noticeable seismic activity. Other segments regularly slip, producing small earthquakes.

Still other segments remain locked and store elastic energy for hundreds of years before rupturing in great earthquakes. The latter process is described as *stick-slip motion,* because the fault exhibits alternating

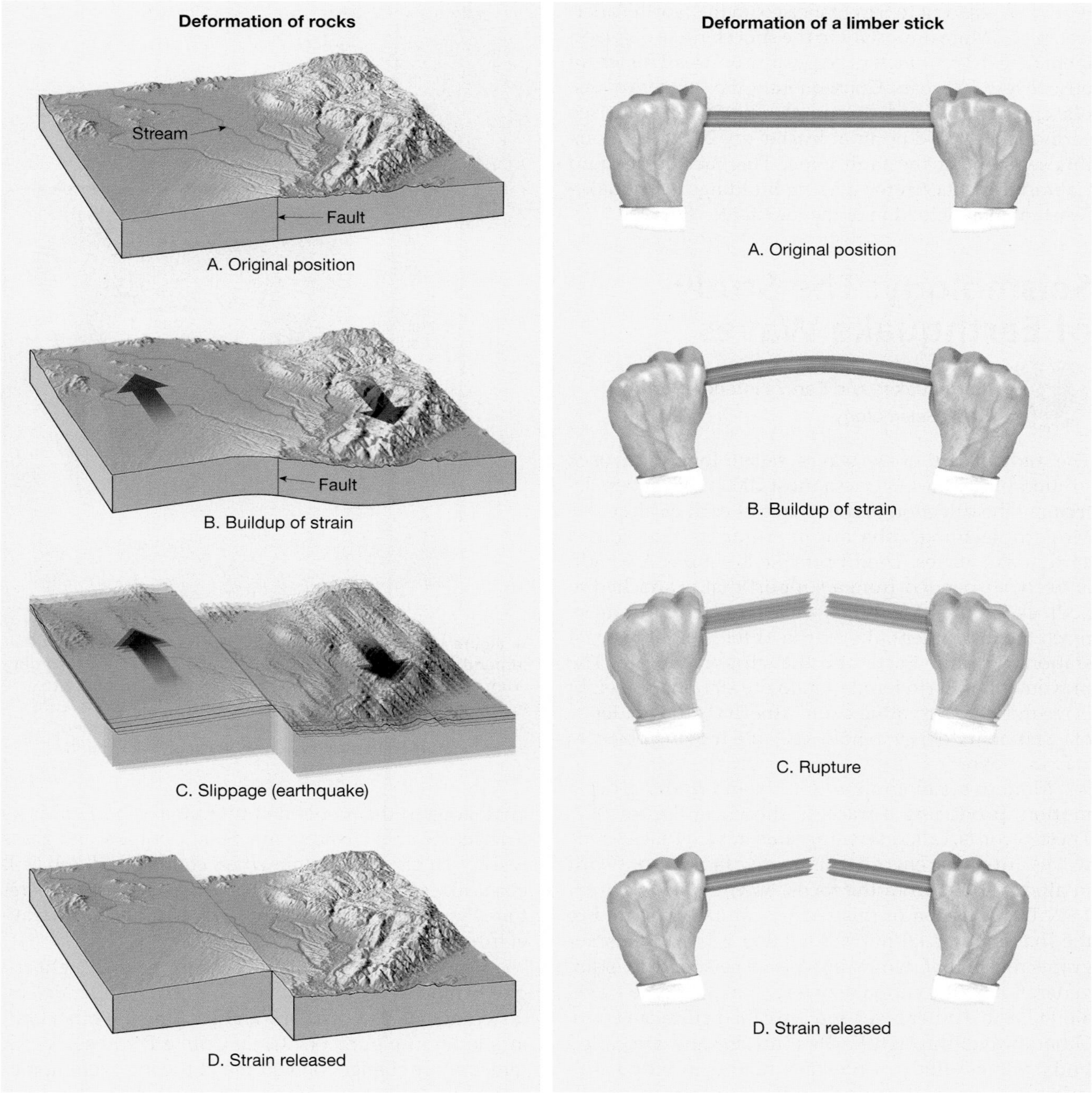

▲ **Figure 14.4** Elastic rebound. As rock is deformed it bends, storing elastic energy. Once the rock is strained beyond its breaking point it ruptures, releasing the stored-up energy in the form of earthquake waves.

periods of locked behavior followed by sudden slippage. It is estimated that great earthquakes should occur about every 50 to 200 years along those sections of the San Andreas Fault that exhibit stick-slip motion. This knowledge is useful when assigning a potential earthquake risk to a given segment of the fault zone.

The tectonic forces along the San Andreas fault zone that were responsible for the 1906 San Francisco earthquake are still active. Currently, laser beams are used to measure the relative motion between the opposite sides of this fault. These measurements reveal a displacement of 2 to 5 centimeters (1 to 2 inches) per year. Although this seems slow, it produces substantial movement over millions of years.

To illustrate, in 30 million years, this rate of displacement would slide the western portion of California northward so that Los Angeles, on the Pacific plate, would be adjacent to San Francisco on the North American plate! More important in the short term, a displacement of just 2 centimeters per year produces 2 meters of offset every 100 years. Consequently, the 4 meters of displacement produced during the 1906 San Francisco earthquake should occur at least every 200 years along this segment of the fault zone. This fact lies behind California's concern for making buildings earthquake-resistant in anticipation of the inevitable "Big One."

Seismology: The Study of Earthquake Waves

GEODe ESSENTIALS OF GEOLOGY Earthquakes and Earth's Interior (Part A)
▼ Seismology

The study of earthquake waves, **seismology,** dates back to attempts by the Chinese almost 2000 years ago to determine the direction to the source of each earthquake. Modern **seismographs** are instruments that record earthquake waves. Their principle is simple. A weight is freely suspended from a support that is attached to bedrock (Figure 14.5). When waves from an earthquake reach the instrument, the inertia of the weight keeps it stationary, while Earth and the support vibrate. The movement of Earth in relation to the stationary weight is recorded on a rotating drum. (Inertia is the tendency of a stationary object to hold still, or a moving object to stay in motion.)

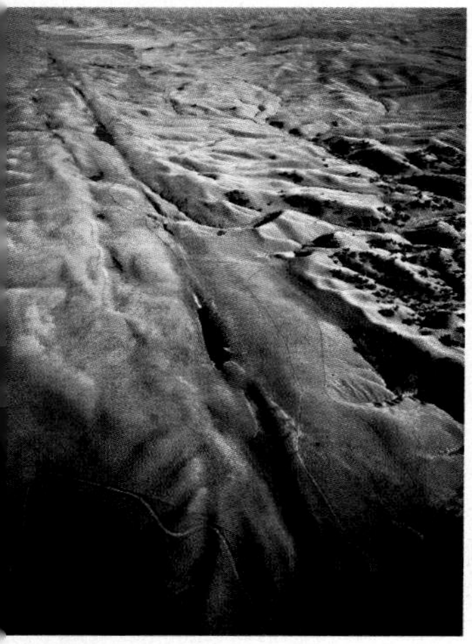

Sag ponds. Slippage along the San Andreas Fault produced this low area that collected water. (Michael Collier)

Modern seismographs amplify and record ground motion, producing a trace as shown in Figure 14.6. These records, called **seismograms,** reveal that seismic waves are elastic energy. This energy radiates outward in all directions from the focus, as you saw in Figure 14.2. Transmission of this energy can be compared to the shaking of gelatin in a bowl that is jarred. Seismograms reveal that two main types of seismic waves are generated by the slippage of a rock mass. Some travel along Earth's outer layer and are called **surface waves.** Others travel through Earth's interior and are called **body waves.** Body waves are further divided into **primary waves (P waves)** and **secondary waves (S waves).**

Did You Know?

The first instrument to detect earthquakes was developed in about 132 AD in China by Chang Heng. (China has a long history of devastating earthquakes.) Chang Heng's instrument is thought to have detected unfelt earthquakes and estimated the direction to the epicenters.

Body waves are divided into P and S waves by their mode of travel through intervening materials. The P waves are push-pull waves—they push (compress) and pull (expand) rocks in the direction the wave is traveling (Figure 14.7A, B). Imagine holding someone by the shoulders and shaking that person. This push-pull movement is how P waves move through Earth. This wave motion is analogous to that generated by human vocal cords as they move air to create sound. Solids, liquids, and gases resist a change in volume when compressed and will elastically spring back once the force is removed. Therefore, P waves, which are compressional waves, can travel through all these materials.

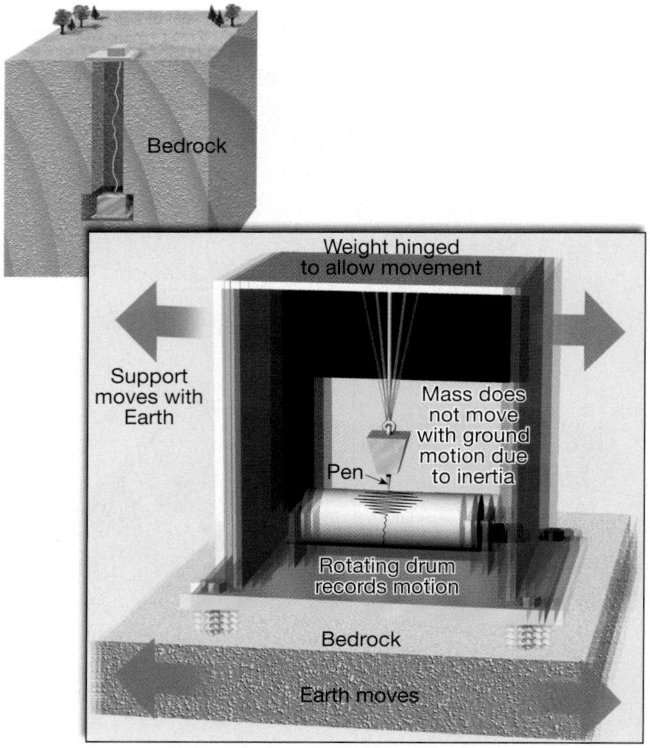

▲ **Figure 14.5** Principle of the seismograph. The inertia of the suspended mass tends to keep it motionless, while the recording drum, which is anchored to bedrock, vibrates in response to seismic waves. Thus, the stationary mass provides a reference point from which to measure the amount of displacement occurring as the seismic wave passes through the ground.

Conversely, S waves shake the particles at right angles to their direction of travel. This can be illustrated by fastening one end of a rope and shaking the other end, as shown in Figure 14.7 C, D. Unlike P waves, which temporarily change the volume of the intervening material by alternately compressing and expanding it, S waves temporarily change the shape of the material that transmits them. Because fluids (gases and liquids) do not respond elastically to changes in shape, they will not transmit S waves.

The motion of surface waves is somewhat more complex. As surface waves travel along the ground, they cause the ground and anything resting upon it to move, much like ocean swells toss a ship. In addition to their up-and-down motion, surface waves have a side-to-side motion similar to an S wave oriented in a horizontal plane. This latter motion is particularly damaging to the foundations of structures.

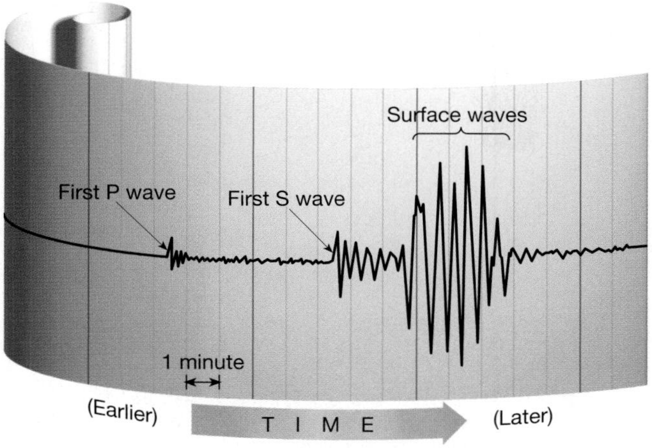

▲ **Figure 14.6** Typical seismic record. Note the time interval (about 5 minutes) between the arrival of the first P wave and the arrival of the first S wave.

By observing a typical seismic record, as shown in Figure 14.6, you can see a major difference among these seismic waves: P waves arrive at the recording station first, then S waves, and then surface waves. This is a consequence of their speeds. To illustrate, the velocity of P waves through granite within the crust is about 6 kilometers per second. Under the same conditions, S waves travel at 3.5 kilometers per second. Differences in density and elastic properties of the rock greatly influence the velocities of these waves. Generally, in any solid material, P waves travel about 1.7 times faster than S waves, and surface waves can be expected to travel at 90 percent of the velocity of the S waves.

As you shall see, seismic waves allow us to determine the location and magnitude of earthquakes. In addition, seismic waves provide us with a tool for probing Earth's interior.

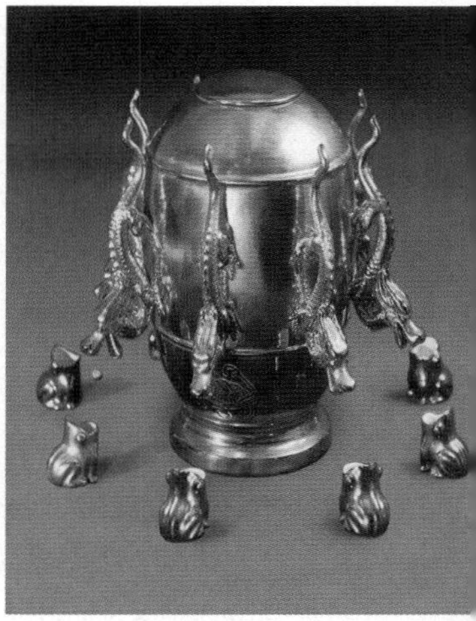

Ancient Chinese seismograph. During an Earth tremor, the dragons located in the direction of the main vibrations would drop a ball into the mouths of the frogs below.

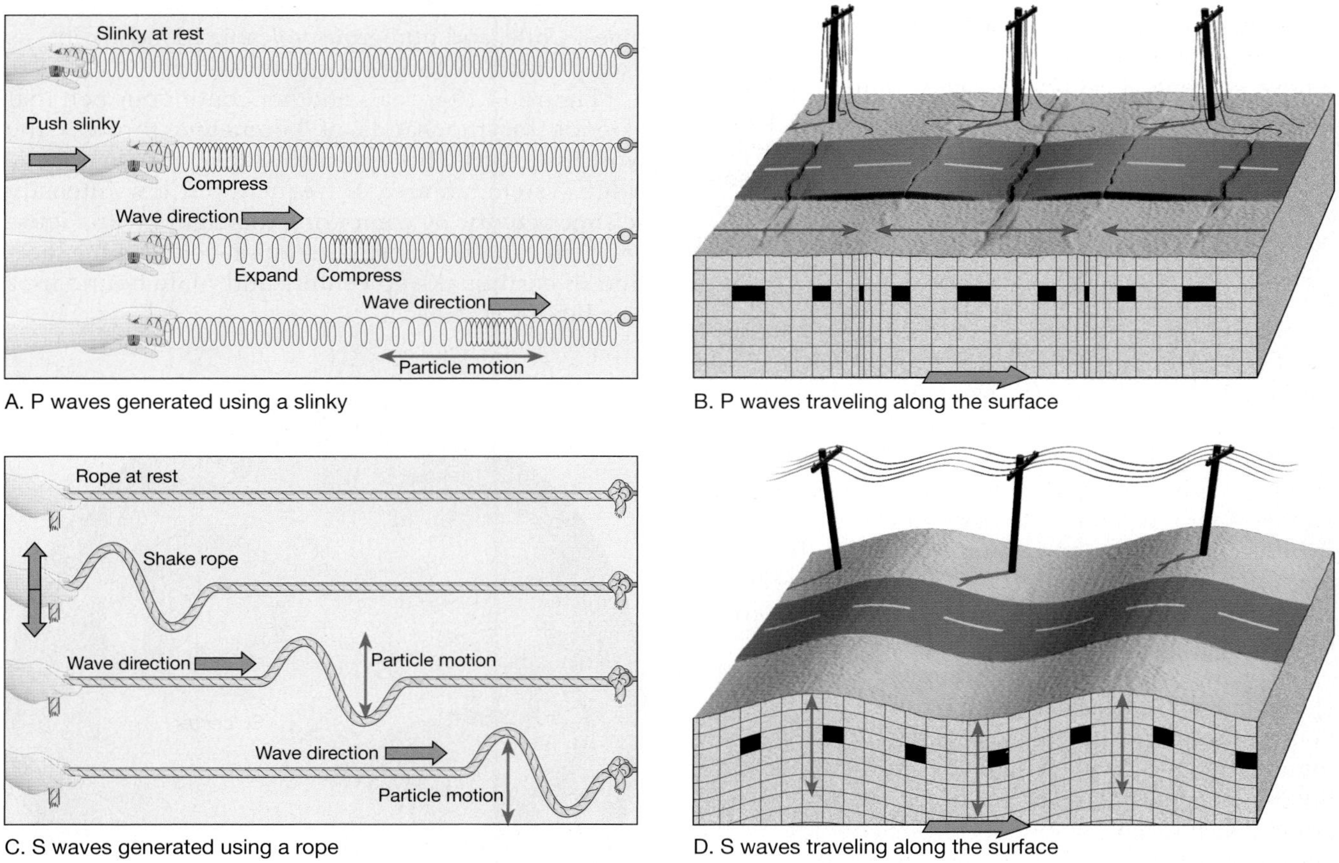

▲ **Figure 14.7** Types of seismic waves and their characteristic motion. (Note that during a strong earthquake, ground shaking consists of a combination of various kinds of seismic waves.) **A.** As illustrated by a slinky, P waves are compressional waves that alternately compress and expand the material through which they pass. **B.** The back-and-forth motion produced as P waves travel along the surface can cause the ground to buckle and fracture, and may cause power lines to break. **C.** S waves cause material to oscillate at right angles to the direction of wave motion. **D.** Because S waves can travel in any plane, they produce up-and-down and sideways shaking of the ground.

Locating an Earthquake

GEODe ESSENTIALS OF GEOLOGY
Earthquakes and Earth's Interior (Part A)
▼ Locating the Source of an Earthquake

Recall that the *focus* is the place within Earth where earthquake waves originate. The **epicenter** is the location on the surface directly above the focus (see Figure 14.2).

The difference in velocities of P and S waves provides a method for locating the epicenter. The principle used is analogous to a race between two autos, one faster than the other. The P wave always wins the race, arriving ahead of the S wave. But the greater the length of the race, the greater the difference in the arrival times at the finish line (the seismic station). Therefore, the greater the interval measured on a seismogram between the arrival of the first P wave and the first S wave, the greater the distance to the earthquake source.

Did You Know?

Most people associate seismology with earthquakes, but seismographs are really just sensitive instruments that record vibrations from any source, including underground nuclear tests, volcanic eruptions, or simply waves beating on a distant shore.

A system for locating earthquake epicenters was developed by using seismograms from earthquakes whose epicenters could be easily pinpointed from physical evidence. From these seismograms, travel-time graphs were constructed (Figure 14.8). The first travel-time graphs were greatly improved when seismograms became available from nuclear explosions, because the precise location and time of detonation were known.

Using the sample seismogram in Figure 14.6 and the travel-time curve in Figure 14.8, we can determine the distance separating the recording station from the earthquake in two steps: (1) using the seismogram, determine the time interval between the arrival of the first P wave and the first S wave, and (2) using the travel-time graph, find the P-S interval on the vertical axis and use that information to determine the distance to the epicenter on the horizontal axis. From this information, we can determine that this earthquake occurred 3400 kilometers (2100 miles) from the recording instrument.

Now we know the *distance*, but what *direction*? The epicenter could be in any direction from the seismic station. As shown in Figure 14.9, the precise location can be found when the distance is known from three or more different seismic stations. On a globe, we draw a circle around each seismic station. Each circle represents the epicenter distance for each station. The point where the three circles intersect is the epicenter of the quake. This method is called *triangulation*.

About 95 percent of the energy released by earthquakes originates in a few relatively narrow zones (Figure 14.10). The greatest energy is released along a path around the outer edge of the Pacific Ocean known as the *circum-Pacific belt*. Included in this zone are regions of great seismic activity, such as Japan, the Philippines, Chile, and numerous volcanic island chains, as exemplified by Alaska's Aleutian Islands.

Figure 14.10 reveals another continuous belt that extends for thousands of kilometers through the world's oceans. This zone coincides with the oceanic ridge system, an area of frequent but low-intensity seismic activity. By comparing this figure with Figure 15.9 you can see a close correlation between the location of earthquake epicenters and plate boundaries (see Box 14.1).

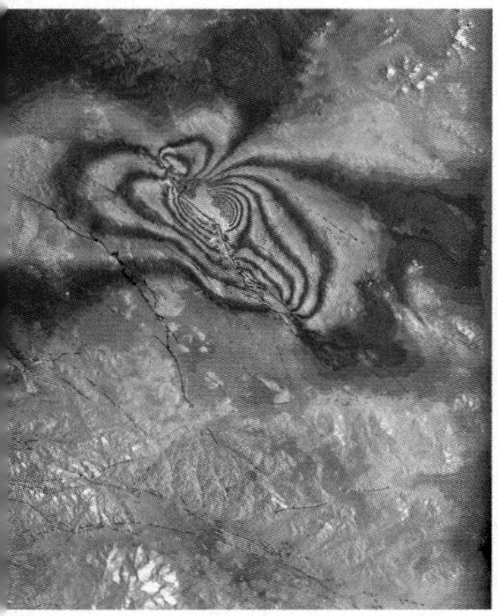

Color satellite map of ground displacement caused by the 1999 Hector Mine earthquake in California. (NASA)

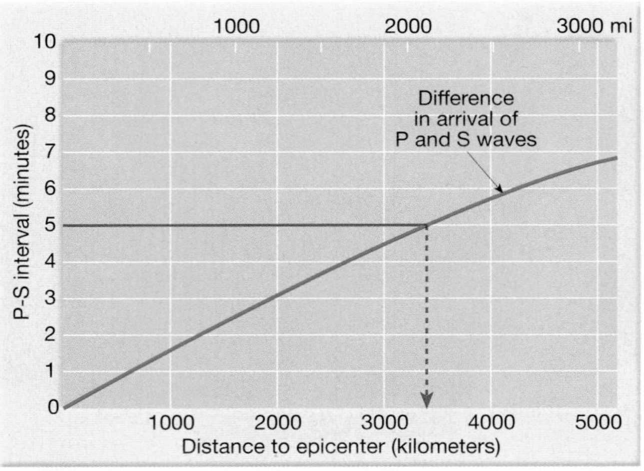

▲ **Figure 14.8** A travel-time graph is used to determine distance to the epicenter. The difference in arrival time of the first P wave and the first S wave in the example is 5 minutes. Thus, the epicenter is roughly 3400 kilometers (2100 miles) away.

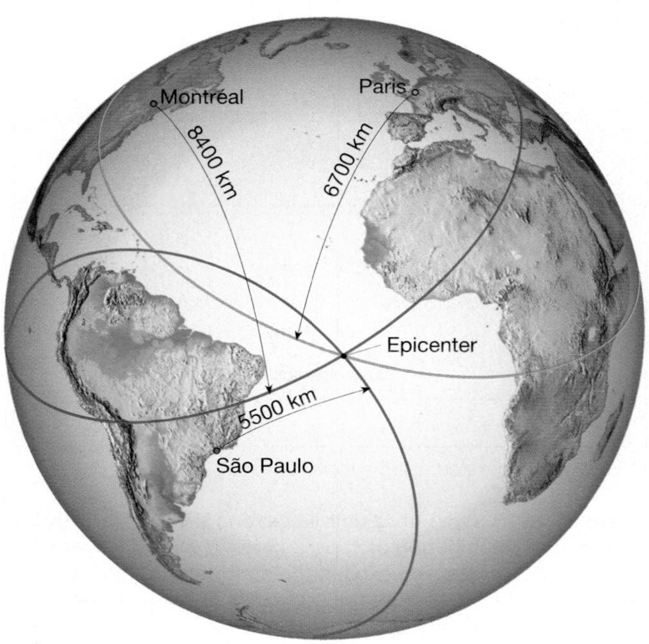

▲ **Figure 14.9** An earthquake epicenter is located using the distance obtained from three seismic stations.

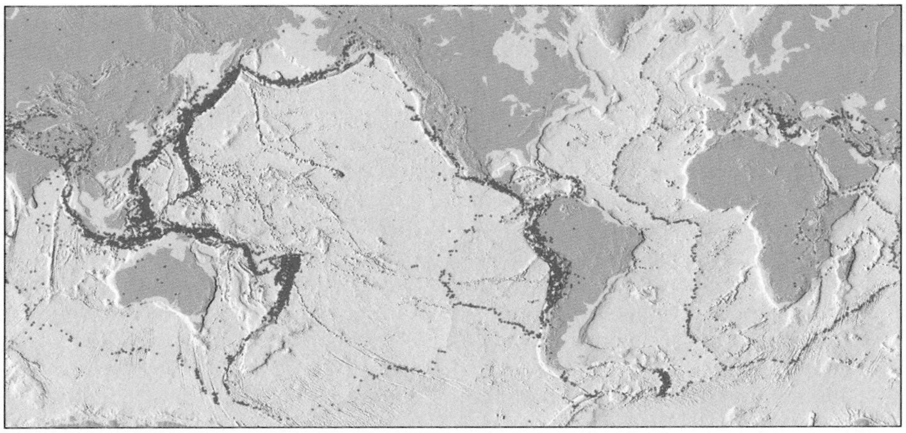

◀ **Figure 14.10** Distribution of the 14,229 earthquakes with magnitudes equal to or greater than 5 for a ten year period. *(Data from National Geophysical Data Center/NOAA)*

Measuring the Size of Earthquakes

Historically, seismologists have employed a variety of methods to obtain two fundamentally different measures that describe the size of an earthquake—intensity and magnitude. The first of these to be used was **intensity**—a measure of the degree of earthquake shaking at a given locale based on the amount of damage. With the development of seismographs, it became clear that a quantitative measure of an earthquake based on seismic records rather than uncertain personal estimates of damage was desirable. The measurement that was developed, called **magnitude,** relies on calculations that use data provided by seismic records (and other techniques) to estimate the amount of energy released at the source of the earthquake.

As it turns out, both intensity and magnitude provide useful, although quite different, information about earthquake strength. Consequently, both measures are still used to describe the relative sizes of earthquakes.

Intensity Scales

Until a little over a century ago, historical records provided the only accounts of the severity of earthquake shaking and destruction. Using these descriptions—which were compiled without any established standards for reporting—made accurate comparisons of earthquake sizes difficult, at best.

In order to standardize the study of earthquake severity, workers developed various intensity scales that considered damage done to buildings, as well as individual descriptions of the event, and secondary effects—landslides and the extent of ground rupture. By 1902, Guiseppe Mercalli had developed a relatively reliable intensity scale, which in a modified form is still used today. The **Modified Mercalli Intensity Scale** shown in Table 14.1 was developed using California buildings as its standard, but it is appropriate for use throughout most of the United States and Canada to estimate the strength of an earthquake. For example, if some well-built wood structures and most masonry buildings are destroyed by an earthquake, a region would be assigned an intensity of X on the Mercalli scale (Table 14.1).

Despite their usefulness in providing seismologists with a tool to compare earthquake severity, particularly in regions where there are no seismographs, intensity scales have severe drawbacks. In particular, intensity scales are based on effects (largely destruction) of earthquakes that depend not only on the severity of ground shaking but also on factors, such as population density, building design, and the nature of surface materials. The modest 6.9-magnitude earthquake in Armenia in 1988 was extremely destructive, mainly because of inferior building construction, whereas the 1985 Mexico City quake was deadly because of the soft sediment upon which the city rests. Thus, the destruction wrought by earthquakes may not be a true measure of the earthquake's actual size.

Earthquake damage near Ismit, Turkey, 1999. (CORBIS/SYGMA)

Magnitude Scales

In order to compare earthquakes across the globe, a measure was needed that does not rely on parameters that vary considerably from one part of the world to another, such as construction practices. As a consequence, a number of magnitude scales were developed.

Richter Magnitude. In 1935 Charles Richter of the California Institute of Technology developed the first magnitude scale using seismic records to estimate the relative sizes of earthquakes. As shown in Figure 14.11 (top), the **Richter scale** is based on the amplitude of the largest seismic wave (P, S, or surface wave) recorded on a seismogram. Because seismic waves weaken as the distance between the earthquake focus and the seismograph increases (in a manner similar to

Did You Know?

Some of the most interesting uses of seismographs involve the reconstruction of unfortunate events, such as airline crashes, pipeline explosions, and mining disasters. For example, a seismologist helped with the investigation of Pan Am Flight 103, which was brought down in 1988 over Lockerbie, Scotland, by a terrorist's bomb. Nearby seismographs recorded six separate impacts, indicating that the plane had broken into that many large pieces before it crashed.

BOX 14.1

Damaging Earthquakes East of the Rockies

When you think "earthquake," you probably think of California and Japan. However, six major earthquakes have occurred in the central and eastern United States since colonial times. Three of these had estimated Richter magnitudes of 7.5, 7.3, and 7.8, and they were centered near the Mississippi River Valley in southeastern Missouri. Occurring on December 16, 1811; January 23, 1812; and February 7, 1812, these earthquakes, plus numerous smaller tremors, destroyed the town of New Madrid, Missouri, triggered massive landslides, and caused damage over a six-state area. The course of the Mississippi River was altered, and Tennessee's Reelfoot Lake was enlarged. The distances over which these earthquakes were felt are truly remarkable. Chimneys were reported downed in Cincinnati, Ohio, and Richmond, Virginia, while Boston residents, located 1770 kilometers (1100 miles) to the northeast, felt the tremor.

Despite the history of the New Madrid earthquake, Memphis, Tennessee, the largest population center in the area, does not have adequate earthquake provisions in its building code. Further, because Memphis is located on unconsolidated floodplain deposits, buildings are more susceptible to damage than are similar structures built on bedrock. It has been estimated that if an earthquake the size of the 1811–1812 New Madrid event were to strike in the next decade, it would result in casualties in the thousands and damages in tens of billions of dollars.

Damaging earthquakes that occurred in Aurora, Illinois (1909), and Valentine, Texas (1931), remind us that other areas in the central United States are vulnerable.

The greatest historical earthquake in the eastern states occurred August 31, 1886, in Charleston, South Carolina. The event, which spanned one minute, caused 60 deaths, numerous injuries, and great economic loss within a radius of 200 kilometers (120 miles) of Charleston. Within eight minutes, effects were felt as far away as Chicago and St. Louis, where strong vibrations shook the upper floors of buildings, causing people to rush outdoors. In Charleston alone, over 100 buildings were destroyed and 90 percent of the remaining structures damaged. It was difficult to find a chimney still standing (Figure 14.A).

Numerous other strong earthquakes have been recorded in the eastern United States. New England and adjacent areas have experienced sizable shocks ever since colonial times. The first reported earthquake in the Northeast took place in Plymouth, Massachusetts, in 1683 and was followed in 1755 by the destructive Cambridge, Massachusetts, quake. Moreover, from the time that records have been kept, New York state alone has experienced over 300 earthquakes large enough to be felt.

Earthquakes in the central and eastern United States occur far less frequently than in California, yet history indicates that the East is vulnerable. Further, these shocks east of the Rockies have generally produced structural damage over a larger area than counterparts of similar magnitude in California. The reason is that the underlying bedrock in the central and eastern United States is older and more rigid. As a result, seismic waves are able to travel greater distances with less attenuation than in the western United States. It is estimated that for earthquakes of similar magnitude, the region of maximum ground motion in the East may be up to 10 times larger than in the West. Consequently, the higher rate of earthquake occurrence in the western United States is balanced somewhat by the fact that central and eastern U.S. quakes can damage larger areas.

Figure 14.A Damage to Charleston, South Carolina, caused by the August 31, 1886, earthquake. Damage ranged from toppled chimneys and broken plaster to total collapse. (Photo courtesy of U.S. Geological Survey)

Table 14.1 Modified Mercalli Intensity Scale

I	Not felt except by a very few under especially favorable circumstances.
II	Felt only by a few persons at rest, especially on upper floors of buildings.
III	Felt quite noticeably indoors, especially on upper floors of buildings, but many people do not recognize it as an earthquake.
IV	During the day felt indoors by many, outdoors by few. Sensation like heavy truck striking building.
V	Felt by nearly everyone, many awakened. Disturbances of trees, poles, and other tall objects sometimes noticed.
VI	Felt by all; many frightened and run outdoors. Some heavy furniture moved; few instances of fallen plaster or damaged chimneys. Damage slight.
VII	Everybody runs outdoors. Damage negligible in buildings of good design and construction; slight-to-moderate in well-built ordinary structures; considerable in poorly built or badly designed structures.
VIII	Damage slight in specially designed structures; considerable in ordinary substantial buildings with partial collapse; great in poorly built structures. (Fall of chimneys, factory stacks, columns, monuments, walls.)
IX	Damage considerable in specially designed structures. Buildings shifted off foundations. Ground cracked conspicuously.
X	Some well-built wooden structures destroyed. Most masonry and frame structures destroyed. Ground badly cracked.
XI	Few, if any, (masonry) structures remain standing. Bridges destroyed. Broad fissures in ground.
XII	Damage total. Waves seen on ground surfaces. Objects thrown upward into air.

light), Richter developed a method that accounted for the decrease in wave amplitude with increased distance. Theoretically, as long as equivalent instruments were used, monitoring stations at various locations would obtain the same Richter magnitude for each recorded earthquake. (Richter selected the Wood-Anderson seismograph as the standard recording device.)

Although the Richter scale has no upper limit, the largest magnitude recorded on a Wood-Anderson seismograph was 8.9. These great shocks released approximately 10^{26} ergs of energy—roughly equivalent to the detonation of 1 billion tons of TNT. Conversely, earthquakes with a Richter magnitude of less than 2.0 are not felt by humans. With the development of more sensitive instruments, tremors of a magnitude of minus 2 were recorded. Table 14.2 shows how Richter magnitudes and their effects are related.

Earthquakes vary enormously in strength, and great earthquakes produce wave amplitudes that are thousands of times larger than those generated by weak tremors. To accommodate this wide variation, Richter used a *logarithmic scale* to express magnitude, where a *tenfold* increase in wave amplitude corresponds to an increase of 1 on the magnitude scale. Thus, the amount of ground shaking for a 5-magnitude earthquake is 10 times greater than that produced by an earthquake having a Richter magnitude of 4.

In addition, each unit of Richter magnitude equates to roughly a *32-fold energy increase.* Thus, an earthquake with a magnitude of 6.5 releases 32 times more energy than one with a magnitude of 5.5, and roughly 1000 times more energy than a 4.5-magnitude quake. A major earthquake with a magnitude of 8.5 releases millions of times more energy than the smallest earthquakes felt by humans.

Other Magnitude Scales. Richter's original goal was modest in that he only attempted to rank the earthquakes of southern California (shallow-focus earthquakes) into groups of large, medium, and small magnitude. Hence, Richter magnitude was designed to study nearby (or local) earthquakes and is denoted by the symbol M_L—where *M* is for *magnitude* and *L* is for *local.*

The convenience of describing the size of an earthquake by a single number that could be calculated quickly from seismographs makes the Richter scale a powerful tool. Further, unlike intensity scales that could only be applied to populated areas of the globe, Richter magnitudes could be assigned to earthquakes in more remote regions and even to events that occurred in the ocean basins. As a result, the method devised by Richter was adapted to a number of different seismographs located throughout the world. In time, seismologists modified Richter's work and developed new magnitude scales.

However, despite their usefulness, none of these "Richter-like" magnitude scales are adequate for describing very large earthquakes. For example, the 1906 San Francisco earthquake and the 1964 Alaskan earthquake have roughly the same Richter magnitudes of about 8.3. However, the Alaskan earthquake released considerably more energy than the San Francisco quake, based on the size of the fault zone and the amount of displacement observed. Thus, the Richter scale (as well as the other related magnitude scales) are said to be *saturated* for large earthquakes because they cannot distinguish between the size of very large events.

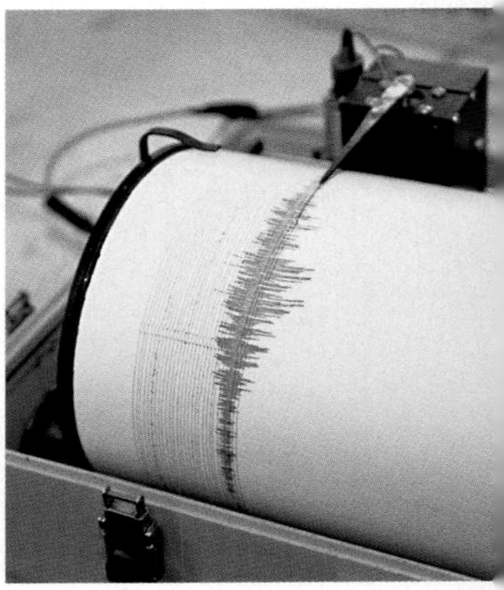

Seismograph recording seismic activity. (Zephyr/Photo Researchers, Inc.)

Did You Know?

During the 1811–1812 New Madrid earthquake, the ground subsided as much as 15 feet and created Lake St. Francis west of the Mississippi and enlarged Reelfoot Lake to the east. Other regions rose, creating temporary waterfalls in the bed of the Mississippi River.

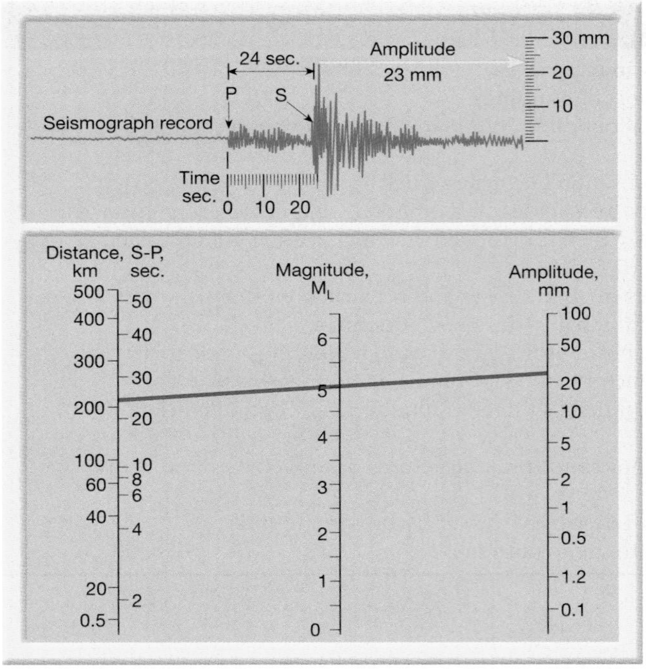

▲ **Figure 14.11** Illustration showing how the Richter magnitude of an earthquake can be determined graphically using a seismograph record from a Wood-Anderson instrument. First, measure the height (amplitude) of the largest wave on the seismogram (23 mm) and then the distance to the focus using the time interval between S and P waves (24 seconds). Next, draw a line between the distance scale (left) and the wave amplitude scale (right). By doing this, you should obtain the Richter magnitude (M_L) of 5. *(Data from California Institute of Technology)*

Moment Magnitude. In recent years seismologists have been employing a more precise measure called **moment magnitude** (M_W), which can be calculated using several techniques. In one method the moment magnitude is calculated from field studies using a combination of factors that include the average amount of displacement along the fault, the area of the rupture surface, and the shear strength of the faulted rock—a measure of how much energy a rock can store before it suddenly slips and releases this energy in the form of an earthquake (and heat).

The moment magnitude can also be readily calculated from seismograms by examining very long period seismic waves. The values obtained have been calibrated so that small- and moderate-sized earthquakes have moment magnitudes that are roughly equivalent to Richter magnitudes. However, moment magnitudes are much better for describing very large earthquakes. For example, on the moment magnitude scale, the 1906 San Francisco earthquake, which had a Richter magnitude of 8.3, would be demoted to 7.9 on the moment magnitude scale, whereas the 1964 Alaskan earthquake with an 8.3 Richter magnitude would be increased to 9.2. The strongest earthquake on record is the 1960 Chilean earthquake with a moment magnitude of 9.5.

Moment magnitude has gained wide acceptance among seismologists and engineers because: (1) it is the only magnitude scale that estimates adequately the size of very large earthquakes; (2) it is a measure that can be derived mathematically from the size of the rupture surface and the amount of displacement, thus it better reflects the total energy released during an earthquake; and (3) it can be verified by two independent methods—field studies that are based on measurements of fault displacement and by seismographic methods using long-period waves.

Did You Know?

The greatest loss of life attributed to an earthquake occurred in the Hwang River of northern China. Here the Chinese made their homes by carving numerous structures into a powdery, windblown material called *loess.* Early on the morning of January 23, 1556, an earthquake struck the region, collapsing the cavelike structures and killing about 830,000 people. Sadly, in 1920 a similar event collapsed loess structures in much the same way, taking about 200,000 lives.

Table 14.2 Earthquake magnitudes and expected world incidence

Richter Magnitudes	Effects Near Epicenter	Estimated Number per Year
< 2.0	Generally not felt, but recorded.	600,000
2.0–2.9	Potentially perceptible.	300,000
3.0–3.9	Felt by some.	49,000
4.0–4.9	Felt by most.	6200
5.0–5.9	Damaging shocks.	800
6.0–6.9	Destructive in populous regions.	266
7.0–7.9	Major earthquakes. Inflict serious damage.	18
≥ 8.0	Great earthquakes. Destroy communities near epicenter.	1.4

Source: Earthquake Information Bulletin and others.

Destruction from Earthquakes

The most violent earthquake to jar North America this century—the Good Friday Alaskan Earthquake—occurred in 1964. Felt throughout that state, the earthquake had a Richter magnitude of 8.3 and reportedly lasted 3 to 4 minutes. This event left 131 people dead, thousands homeless, and the economy of the state badly disrupted because it occurred near major towns and seaports (Figure 14.12). Had the schools and business districts been open on this holiday, the toll surely would have been higher. Within 24 hours of the initial shock, 28 aftershocks were recorded, 10 of which exceeded a Richter magnitude of 6.

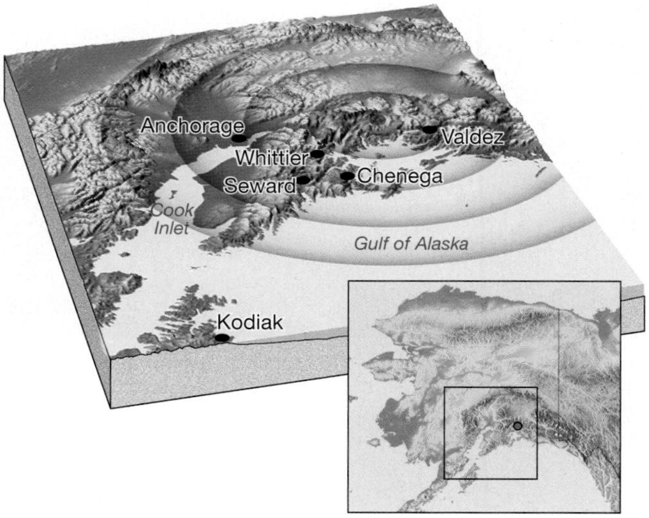

▲ **Figure 14.12** Region most affected by the Good Friday earthquake of 1964. Note the epicenter (red dot). *(After U.S. Geological Survey)*

Destruction from Seismic Vibrations

The 1964 Alaskan earthquake provided geologists with new insights into the role of *ground shaking* as a destructive force. As the energy released by an earthquake travels along Earth's surface, it causes the ground to vibrate in a complex manner by moving up and down as well as from side to side. The amount of structural damage attributable to the vibrations depends on several factors, including (1) the intensity and (2) the duration of vibrations, (3) the nature of the material upon which the structure rests, and (4) the design of the structure.

All multistory structures in Anchorage were damaged by the vibrations, but the more flexible wood-frame residential buildings fared best. Figure 14.13 offers a striking example of how construction variations affect earthquake damage. You can see that the steel-frame structure on the left withstood the vibrations, whereas the poorly designed J.C. Penney building was badly damaged. Engineers have learned that unreinforced masonry buildings are the most serious safety threats in earthquakes.

▼ **Figure 14.13** Damage to the five-story JC Penney Co. building, Anchorage, Alaska. Very little structural damage was incurred by the adjacent building. *(Courtesy of NOAA)*

Collapse of a street in Anchorage, Alaska, caused by the 1964 earthquake.

Most large structures in Anchorage were damaged, even though they were built according to the earthquake provisions of the Uniform Building Code. Perhaps some of that destruction can be attributed to the unusually long duration of this earthquake. Most quakes consist of tremors that last less than a minute. For example, the 1994 Northridge earthquake was felt for about 40 seconds, and the strong vibrations of the 1989 Loma Prieta earthquake lasted less than 15 seconds. But the Alaska quake reverberated for 3 to 4 minutes.

Amplification of Seismic Waves. Although the region within 20 to 50 kilometers of the epicenter will experience about the same intensity of ground shaking, destruction varies considerably within this area. Differences are mainly attributable to the nature of the ground on which structures are built. Soft sediments, for example, generally amplify the vibrations more than solid bedrock. Thus, the buildings located in Anchorage, which were situated on unconsolidated sediments, experienced heavy structural damage. By contrast, most of the town of Whittier, although much nearer the epicenter, rests on a firm foundation of granite and hence suffered much less damage. However, Whittier was damaged by a tsunami (described in the next section).

Did You Know?

It is a commonly held belief that moderate earthquakes decrease the chances of a major earthquake in the same region, but this is not the case. When you compare the amount of energy released by earthquakes of different magnitudes, it turns out that thousands of moderate tremors would be needed to release the huge amount of energy released during one "great" earthquake.

The 1985 Mexican earthquake gave seismologists and engineers a vivid reminder of what had been learned from the 1964 Alaskan earthquake. The Mexican coast, where the earthquake was centered, experienced unusually mild tremors despite the strength of the quake. As expected, the seismic waves became progressively weaker with increasing distance from the epicenter. However, in the central section of Mexico City, nearly 400 kilometers from the source, the vibrations intensified to five times that experienced in outlying districts (Figure 14.14). Much of this amplified ground motion can be attributed to soft sediments, remnants of an ancient lakebed, that underlie portions of the city.

Liquefaction. In areas where unconsolidated materials are saturated with water, earthquake vibrations can generate a phenomenon known as **liquefaction.** Under these conditions, what had been a stable soil turns into a mobile fluid that is not capable of supporting buildings or other structures (Figure 14.15). As a result, underground objects such as storage tanks and

These "mud volcanoes" formed when geysers of mud and water shot from the ground, an indication of liquefaction. (Richard Hilton)

sewer lines may literally float toward the surface of their newly liquefied environment. Buildings and other structures may settle and collapse. During the 1989 Loma Prieta earthquake, in San Francisco's Marina District, foundations failed and geysers of sand and water shot from the ground, indicating that liquefaction had occurred.

▲ **Figure 14.14** During the 1985 Mexican earthquake, multistory buildings swayed back and forth as much as 1 meter. Many, including the hotel shown here, collapsed or were seriously damaged. *(Photo by James L. Beck)*

Tsunami

Large undersea earthquakes occasionally set in motion massive waves of water called **seismic sea waves,** or **tsunami.** (The name *tsunami* is a Japanese word meaning "harbor wave," for Japanese harbors have suffered from them many times.) These destructive waves often are called tidal waves by the media. However, this name is inappropriate, for these waves are most often generated by earthquakes and to a lesser extent by submarine landslides, volcanic eruptions, and meteorite impacts. They are *not* produced by the tidal effect of the Moon or Sun.

Students experiencing the nature of liquefaction. (M. Miller)

Tsunami triggered by an earthquake occur where a slab of oceanic crust is displaced vertically along a fault (Figure 14.16), or where vibrations from a quake trigger an underwater landslide.

Once formed, a tsunami resembles the ripples created when a pebble is dropped into a pond. In contrast to ripples, tsunami advance across the ocean at speeds between 500 and 950 kilometers (300 and 600 miles) per hour. Despite this, a tsunami in the open ocean can pass undetected because its height is usually less than 1 meter and the distance between wave crests is great, ranging from 100 to 700 kilometers. However, upon entering shallower coastal water, these destructive waves are slowed and the water begins to pile up to heights that occasionally exceed 30 meters (100 feet), as shown in Figure 14.16. As the crest of a tsunami approaches shore, it appears as a rapid rise in sea level with a turbulent and chaotic surface.

Damage on Sumatra from the December 26, 2004 tsunami. (James R. McGury)

Usually the first warning of an approaching tsunami is a rather rapid withdrawal of water from beaches. Some residents living near the Pacific basin have learned to heed this warning and move to higher ground, because about 5 to 30 minutes following the retreat of water a surge capable of extending hundreds of meters inland may occur. In a successive fashion, each surge is followed by a rapid oceanward retreat of the water. These waves are separated by intervals of between 10 and 60 minutes. They are able to traverse thousands of kilometers of the ocean before their energy is dissipated.

Tsunami Warning System. In 1946, a large tsunami struck the Hawaiian Islands without warning. A wave more than 15 meters (50 feet) high left several coastal villages in shambles. This destruction motivated the U.S. Coast and Geodetic Survey to establish a tsunami warning system for coastal areas of the Pacific. From seismic observatories throughout the region, large earthquakes are reported to the Tsunami Warning Center in Honolulu. Scientists at the Center use tidal gauges to determine whether a tsunami has formed. Within an hour a warning is issued. Although tsunamis travel very rapidly, there is sufficient time to evacuate all but the region nearest the epicenter. For example, a tsunami generated near the Aleutian Islands would take five hours to reach Hawaii, and one generated near the coast of Chile would travel 15 hours before reaching Hawaii (Figure 14.17).

Tsunami Damage from the 2004 Indonesian Earthquake. A massive undersea earthquake of moment magnitude 9.0 occurred near the island of Sumatra on December 26, 2004, and sent waves of water racing across the Indian Ocean and Bay of Bengal. This tsunami was one of the deadliest natural disasters of any kind in modern times, claiming more than 200,000 lives. As water surged several kilometers inland, cars

▼ **Figure 14.15** Effects of liquefaction. This tilted building rests on unconsolidated sediment that imitated quicksand during the 1985 Mexican earthquake. *(Photo by James L. Beck)*

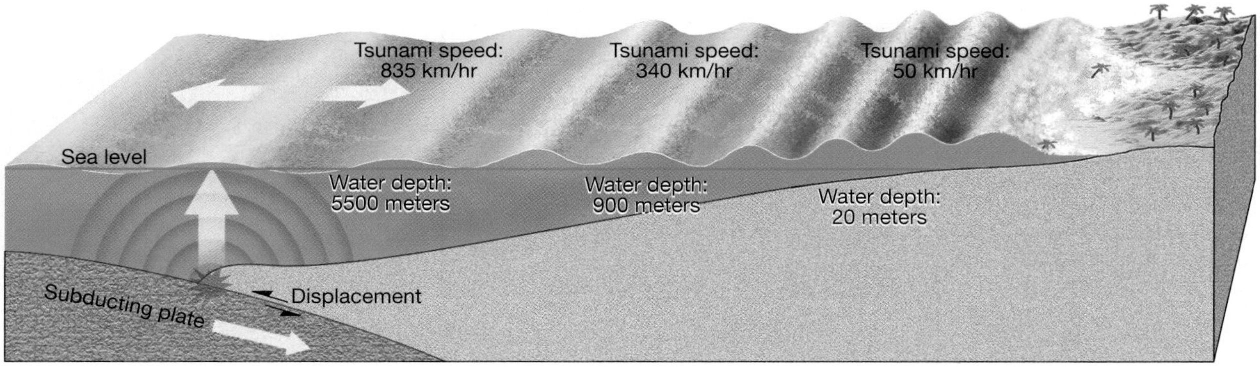

▲ **Figure 14.16** Schematic drawing of a tsunami generated by displacement of the ocean floor. The speed of a wave moving across the ocean correlates with ocean depth. As shown, waves moving in deep water advance at speeds in excess of 800 kilometers per hour. Speed gradually slows to 50 kilometers per hour at depths of 20 meters. Decreasing depth slows the movement of the wave column. As waves slow in shallow water, they grow in height until they topple and rush onto shore with tremendous force. The size and spacing of these swells are not to scale.

and trucks were flung around like toys in a bathtub, and fishing boats were rammed into homes. In some locations, the backwash of water dragged bodies and huge amounts of debris out to sea.

The destruction was indiscriminate, destroying luxury resorts and poor fishing hamlets on the Indian Ocean coast (Figure 14.18). Devastation was most severe along the southeast coast of Sri Lanka, in the Indonesian province of Aceh, in the Indian state of Tamil Nada, and on Thailand's resort island of Phuket. Damages were reported as far away as the Somalia coast of Africa, 4100 kilometers (2500 miles) west of the earthquake epicenter.

The killer waves generated by this massive quake achieved heights as great as 10 meters (33 feet) and struck many unprepared areas within three hours of the event. Although the Pacific basin contains deep-sea buoys and tide gauges that can spot tsunami waves at sea, the Indian Ocean does not. (The deep-sea buoys have pressure sensors that detect changes in pressure as the earthquake's energy travels through the ocean, and tide gauges measure the rise and fall in sea level.) The rarity of tsunami in the Indian Ocean also contributed to the lack of preparedness for such an event. It should come as no surprise that the countries of India, Indonesia, and Thailand have announced plans to establish a tsunami warning system for the Indian Ocean.

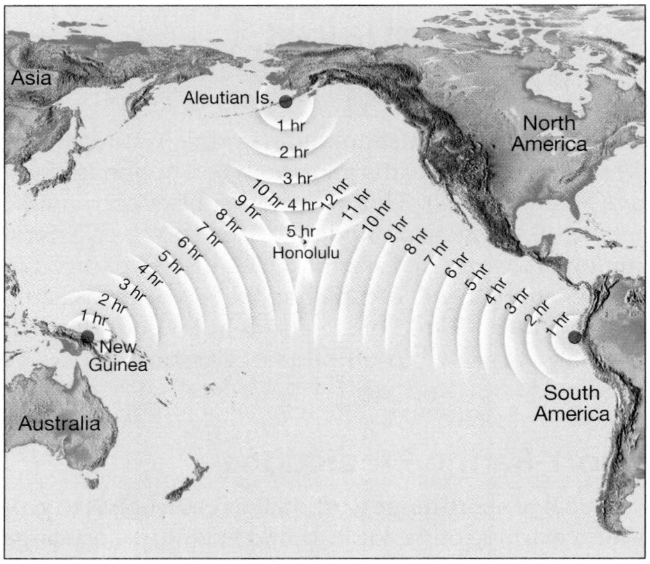

▲ **Figure 14.17** Tsunami travel times to Honolulu, Hawaii, from selected locations throughout the Pacific. *(Data from NOAA)*

Did You Know?

In Japanese folklore, earthquakes were caused by a giant catfish that lived beneath the ground. Whenever the catfish flopped about, the ground shook. Meanwhile, people in Russia's Kamchatka Peninsula believed that earthquakes occurred when a mighty dog, Kozei, shook freshly fallen snow from its coat.

Landslides and Ground Subsidence

In the 1964 Alaskan earthquake, the greatest damage to structures was from landslides and ground subsidence triggered by the vibrations. At Valdez and Seward, the violent shaking caused river-delta materials to experience liquefaction; the subsequent slumping carried both waterfronts away. Because the disaster could happen again, the entire town of Valdez was relocated about 7 kilometers away on more stable ground. In Valdez, 31 persons on a dock perished when it slid into the sea.

Most of the damage in Anchorage was attributed to landslides. Many homes were destroyed in Turnagain Heights when a layer of clay lost its strength and over 200 acres of land slid toward the ocean (Figure 14.19). A portion of this landslide has been left in its natural condition as a reminder of this destructive event. The site was named Earthquake Park. Downtown Anchorage was also disrupted as sections of the main business district dropped by as much as 3 meters (10 feet).

Tsunami damage at Crescent City, California, caused by the 1964 Alaskan earthquake.

▶ **Figure 14.18** A massive earthquake of magnitude 9.0 off the Indonesian island of Sumatra sent a tsunami racing across the Indian Ocean and Bay of Bengal on December 26, 2004. Here, unsuspecting foreign tourists, who at first walked out on the sand after the water receded, now rush toward shore as the first of six tsunami start to roll toward Hat Rai Lay Beach near Krabi in southern Thailand. *(AFP/Getty Images Inc.)*

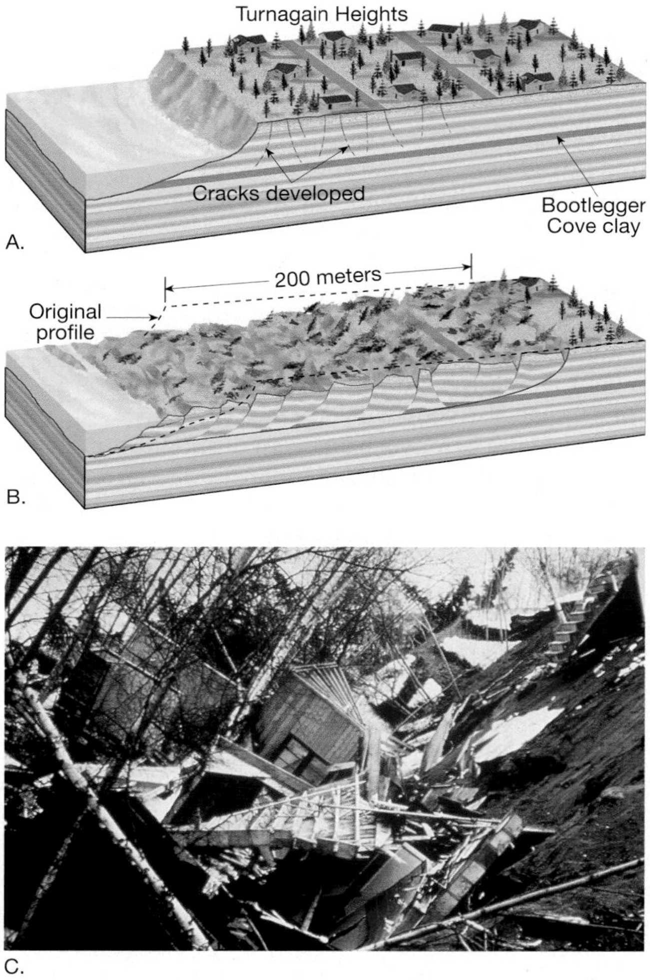

▲ **Figure 14.19** Turnagain Heights slide caused by the 1964 Alaskan earthquake. **A.** Vibrations from the earthquake caused cracks to appear near the edge of the bluff. **B.** Within seconds blocks of land began to slide toward the sea on a weak layer of clay. In less than 5 minutes, as much as 200 meters of the Turnagain Heights bluff area had been destroyed. **C.** Photo of a small portion of the Turnagain Heights slide. *(Photo courtesy of U.S. Geological Survey)*

Fire

The 1906 San Francisco earthquake reminds us of the formidable threat of fire. The central city contained mostly large, older wooden structures and brick buildings. The greatest destruction was caused by fires that started when gas and electrical lines were severed. The fires raged uncontrolled for three days and devastated over 500 city blocks (see Figure 14.3, p. 311). The problem was compounded by the initial ground shaking, which broke the city's water lines into hundreds of unconnected pieces.

The fire was finally contained when buildings were dynamited along a wide boulevard to create a fire break, the same strategy used in fighting a forest fire. Although only a few deaths were attributed to the fires, such is not always the case. A 1923 earthquake in Japan (their worst quake prior to the 1995 Kobe tremor) triggered an estimated 250 fires, which devastated the city of Yokohama and destroyed more than half the homes in Tokyo. Over 100,000 deaths were attributed to the fires, which were driven by unusually high winds.

Can Earthquakes Be Predicted?

The vibrations that shook Northridge, California, in 1994 inflicted 57 deaths and about $40 billion in damage (Figure 14.20). This was from a brief earthquake (about 40 seconds) of moderate rating (M_W 6.7). Seismologists warn that earthquakes of comparable or greater strength will occur along the San Andreas Fault, which cuts a 1300-kilometer (800-mile) path through the state. The obvious question is: Can earthquakes be predicted?

Short-Range Predictions

The goal of short-range earthquake prediction is to provide a warning of the location and magnitude of a large earthquake within a narrow time frame. Substantial efforts to achieve this objective are being advanced in Japan, the United States, China, and Russia—countries

▲ **Figure 14.20** Damage to Interstate 5 during the January 17, 1994, Northridge earthquake. *(Photo by Tom McHugh/Photo Researchers, Inc.)*

where earthquake risks are high (Table 14.3). This research has concentrated on monitoring possible *precursors*—phenomena that precede, and thus provide a warning of, a forthcoming earthquake. In California, for example, seismologists are measuring uplift, subsidence, and strain in the rocks near active faults. Some Japanese scientists are studying peculiar anomalous behavior that may precede a quake.

One claim of a successful short-range prediction was made by Chinese seismologists after the February 4, 1975, earthquake in Liaoning Province. According to reports, very few people were killed, although more than 1 million lived near the epicenter, because the earthquake was predicted and the population was evacuated. Recently, some Western seismologists have questioned this claim and suggest instead that an intense swarm of foreshocks that began 24 hours before the main earthquake may have caused many people to evacuate spontaneously. Further, an official Chinese government report issued 10 years later stated that 1328 people died and 16,980 injuries resulted from this quake.

One year after the Liaoning earthquake, at least 240,000 people died in the Tangshan, China, earthquake, which was not predicted. The Chinese have also issued false alarms. In a province near Hong Kong, people reportedly left their dwellings for over a month, but no earthquake followed. Clearly, whatever method the Chinese employ for short-range predictions, it is *not* reliable.

For a short-range prediction scheme to warrant general acceptance, it must be both accurate and reliable. Thus, *it must have a small range of uncertainty as to location and timing, and it must produce few failures, or false alarms.* Can you imagine the debate that would precede an order to evacuate a large city in the United States, such as Los Angeles or San Francisco? The cost of evacuating millions of people, arranging for living accommodations, and providing for their lost work time and wages would be staggering.

Long-Range Forecasts

In contrast to short-range predictions, which aim to forecast earthquakes within a time frame of hours or, at most, days, long-range forecasts give the probability of a certain magnitude earthquake occurring on a time scale of 30 to 100 years or more. Stated another way, these forecasts give statistical estimates of the expected intensity of ground motion for a given area over

Did You Know?

During an earthquake near Port Royal, Jamaica, the water-saturated sand on which the city was built vigorously shook. As a result, the sand particles lost contact with one another, giving the mixture the consistency of a thick milk shake. Anything supported by the ground, such as buildings and people, either floated or sank. One eyewitness stated: "[W]hole streets with inhabitants were swallowed up. . . . Some were swallowed quite down, and cast up again by great quantities of water; others went down and were never more seen."

Table 14.3 Some notable earthquakes

Year	Location	Deaths (est.)	Magnitude†	Comments
1556	Shensi, China	830,000		Possibly the greatest natural disaster.
1755	Lisbon, Portugal	70,000		Tsunami damage extensive.
*1811–1812	New Madrid, Missouri	Few	7.9	Three major earthquakes.
*1886	Charleston, South Carolina	60		Greatest historical earthquake in the eastern United States.
*1906	San Francisco, California	1500	7.8	Fires caused extensive damage.
1908	Messina, Italy	120,000		
1923	Tokyo, Japan	143,000	7.9	Fire caused extensive destruction.
1960	Southern Chile	5700	9.5	Possibly the largest-magnitude earthquake ever recorded.
*1964	Alaska	131	9.2	Greatest North American earthquake ever recorded.
1970	Peru	66,000	7.8	Great rock avalanche.
*1971	San Fernando, California	65	6.5	Damage exceeded $1 billion.
1975	Liaoning Province, China	1328	7.5	First major earthquake to be predicted.
1976	Tangshan, China	240,000	7.6	Not predicted.
1985	Mexico City	9500	8.1	Major damage occurred 400 km from epicenter.
1988	Armenia	25,000	6.9	Poor construction practices.
*1989	Loma Prieta	62	6.9	Damages exceeded $6 billion.
1990	Iran	50,000	7.3	Landslides and poor construction practices caused great damage.
1993	Latur, India	10,000	6.4	Located in stable continental interior.
*1994	Northridge, California	57	6.7	Damages in excess of $40 billion.
1995	Kobe, Japan	5472	6.9	Damage estimated to exceed $100 billion.
1999	Izmit, Turkey	17,127	7.4	Nearly 44,000 injured and more than 250,000 displaced.
1999	Chi-Chi, Taiwan	2300	7.6	Severe destruction; 8700 injuries.
2001	El Salvador	1000	7.6	Triggered many landslides.
2001	Bhuj, India	20,000	7.9	1 million or more homeless.
2003	Bam, Iran	41,000	6.6	Ancient city with poor construction.
2004	Indian Ocean	200,000	9.0	Devastating tsunami damage.

**U.S. earthquakes.*
†*Widely differing magnitudes have been estimated for some of these earthquakes. When available, moment magnitudes are used.*
Source: U.S. Geological Survey.

a specified time frame. Although long-range forecasts may not be as informative as we would prefer, the data are important for updating the Uniform Building Code, which contains nationwide standards for designing earthquake-resistant structures.

Long-range predictions are based on the premise that earthquakes are repetitive or cyclical, like the weather. In other words, as soon as one earthquake is over, the continuing motions of Earth's plates begin to build strain in the rocks again until they fail once more. This has led seismologists to study historical records of earthquakes to see if there are any discernable patterns so that the probability of recurrence might be established.

Did You Know?

The tsunami generated in the 1964 Alaskan earthquake heavily damaged communities along the Gulf of Alaska and killed 107 people. By contrast, only nine people died in Anchorage as a direct result of the vibrations. Furthermore, despite a one-hour warning, 12 persons perished in Crescent City, California, from the fifth and most destructive tsunami wave generated by this quake.

One study conducted by the U.S. Geological Survey gives the probability of a rupture occurring along various segments of the San Andreas Fault for the 30 years between 1988 and 2018 (Figure 14.21). From this investigation, the area around the Santa Cruz Mountains was given a 30 percent probability of producing a 6.5-magnitude earthquake during this time period. In fact, it produced the Loma Prieta quake in 1989, of 7.1 magnitude.

The region along the San Andreas Fault given the highest probability (90 percent) of generating a quake is the Parkfield section. This area has been called the "Old Faithful" of earthquake zones because activity here has been very regular since record-keeping began in 1857. In late September 2004, a 6.0-magnitude earthquake again struck this area. Although the event was more than a decade overdue, it did demonstrate the potential usefulness of long-range forecasts. Another section between Parkfield and the Santa Cruz Mountains is given a very low probability of generating an earthquake. This area has experienced very little seismic activity in historical times; rather, it exhibits a slow,

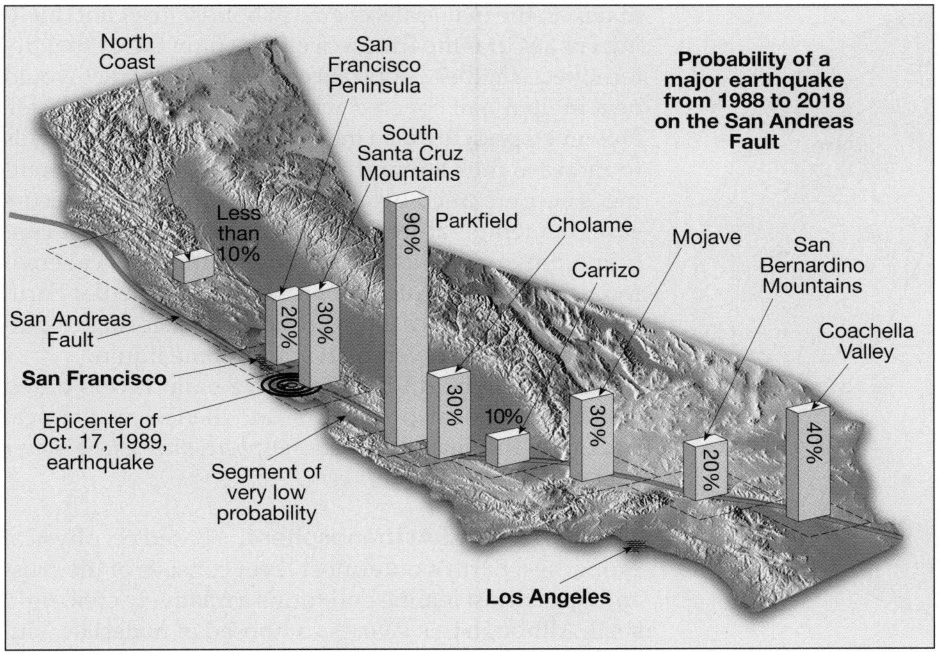

◀ **Figure 14.21** Probability of a major earthquake from 1988 to 2018 on the San Andreas Fault.

continual movement known as *fault creep*. Such movement is beneficial because it prevents strain from building to high levels in the rocks.

In summary, it appears that the best prospects for making useful earthquake predictions involve forecasting magnitudes and locations on time scales of years or perhaps even decades. These forecasts are important because they provide information used to develop the Uniform Building Code and assist in land-use planning.

Earthquakes and Earth's Interior

GEODe Essentials of Geology — Earthquakes and Earth's Interior (Part B) ▼ Earth's Layered Structure

Earth's interior lies just below us; however, direct access is very limited. The deepest well yet drilled has penetrated Earth's crust only 12 kilometers (7.5 miles), less than 0.2 percent of the distance to the planet's center. Consequently, most knowledge of our planet's interior comes from the study of earthquake waves that travel through Earth and vibrate the surface at some distant point.

If Earth were a perfectly homogeneous body, seismic waves would spread through it in a straight line at a constant speed. However, this is not the case for Earth. It so happens that the seismic waves reaching seismographs located farther from an earthquake travel at faster average speeds than those recorded at locations closer to the event. This general increase in speed with depth is a consequence of increased pressure, which enhances the elastic properties of deeply buried rock. As a result, the paths of seismic rays through Earth are refracted (bent) as they travel (Figure 14.22).

As more sensitive seismographs were developed, it became apparent that in addition to gradual changes in seismic-wave velocities, rather abrupt velocity changes also occur at particular depths. Because these changes were detected worldwide, seismologists concluded that Earth must be composed of distinct shells having varying compositions and/or mechanical properties (Figure 14.22).

Lasers used to measure movement along the San Andreas Fault. (John Nakata/USGS)

Layers Defined by Composition

Compositional layering in Earth's interior likely resulted from density sorting that took place during an early period of partial melting. During this period the heavier elements, principally iron and nickel, sank as the lighter rocky components floated upward. This segregation of material is still occurring, but at a much reduced rate. Because of this chemical differentiation, Earth's interior is not homogeneous. Rather, it consists of three major regions that have markedly different chemical compositions (Figure 14.23).

The principal compositional layers of Earth include:

- the **crust,** Earth's comparatively thin outer skin that ranges in thickness from 3 kilometers (2 miles) at the oceanic ridges to over 70 kilometers (40 miles) in some mountain belts, such as the Andes and Himalayas;

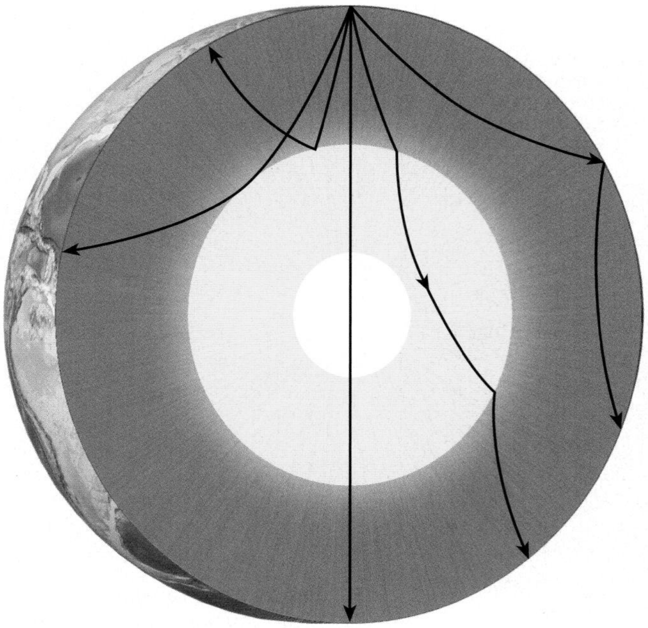

▲ **Figure 14.22** A few of the many possible paths that seismic rays take through Earth.

- the **mantle,** a solid rocky (silica-rich) shell that extends to a depth of about 2900 kilometers (1800 miles);
- the **core,** an iron-rich sphere having a radius of 3486 kilometers (2161 miles).

We will consider the composition and structure of these major divisions of Earth's interior later in the chapter.

Layers Defined by Physical Properties

Earth's interior is characterized by a gradual increase in temperature, pressure, and density with depth. Estimates put the temperature at a depth of 100 kilometers at between 1200°C and 1400°C, whereas the temperature at Earth's center may exceed 6700°C. Clearly, Earth's interior has retained much of the energy acquired during its formative years, despite the fact that heat is continuously flowing toward the surface, where it is lost to space. The increase in pressure with depth causes a corresponding increase in rock density.

The gradual increase in temperature and pressure with depth affects the physical properties and hence the mechanical behavior of Earth materials. When a substance is heated, its chemical bonds weaken and its mechanical strength (resistance to deformation) is reduced. If the temperature exceeds the melting point of an Earth material, the material's chemical bonds break and melting ensues. If temperature were the only factor that determined whether a substance melted, our planet would be a molten ball covered with a thin, solid outer shell. However, pressure also increases with depth and tends to increase rock strength. Furthermore, because melting is accompanied by an increase in volume, it occurs at higher temperatures at depth because of greater confining pressure. Thus, depending on the physical environment (temperature and pressure), a particular Earth material may behave like a brittle solid, deform in a puttylike manner, or even melt and become liquid.

> **Did You Know?**
>
> The best advice for your safety during an earthquake is to *duck, cover,* and *hold.* When you feel an earthquake, *duck* under a desk or sturdy table. Stay away from windows, bookcases, file cabinets, heavy mirrors, hanging plants, and other heavy objects that could fall. Stay under *cover* until the shaking stops. And, *hold* on to the desk or table: If it moves, move with it.

Earth can be divided into five main layers based on their physical properties and hence mechanical strength—the *lithosphere, asthenosphere, mesosphere (lower mantle), outer core,* and *inner core.*

Lithosphere and Asthenosphere. Based on physical properties, Earth's outermost layer consists of the crust and uppermost mantle and forms a relatively cool, rigid shell. Although this layer is composed of materials with markedly different chemical compositions, it tends to act as a unit that exhibits rigid behavior—mainly because it is cool and thus strong. This layer, called the **lithosphere** (*sphere of rock*), averages about 100 kilometers in thickness but may be 250 kilometers thick or more below the older portions of the continents (Figure 14.23). Within the ocean basins, the lithosphere is only a few kilometers thick along the oceanic ridges but increases to perhaps 100 kilometers in regions of older and cooler oceanic crust.

Beneath the lithosphere in the upper mantle lies a soft, comparatively weak layer known as the **asthenosphere** (weak sphere). The top portion of the asthenosphere has a temperature/pressure regime that results in a small amount of melting. Within this very weak zone, the lithosphere is mechanically detached from the layer below. The result is that the lithosphere is able to move independently of the asthenosphere, a topic we will consider in the next chapter.

It is important to emphasize that the strength of various Earth materials is a function of both their composition and of the temperature and pressure of their environment. You should not get the idea that the entire lithosphere behaves like a brittle solid similar to rocks found on the surface. Rather, the rocks of the lithosphere get progressively hotter and weaker (more easily deformed) with increasing depth. At the depth of the uppermost asthenosphere, the rocks are close enough to their melting temperatures (some melting may actually occur) that they are very easily deformed. Thus, the uppermost asthenosphere is weak because it is near its melting point, just as hot wax is weaker than cold wax.

Mesosphere or Lower Mantle. Below the zone of weakness in the uppermost asthenosphere, increased pressure counteracts the effects of higher temperature

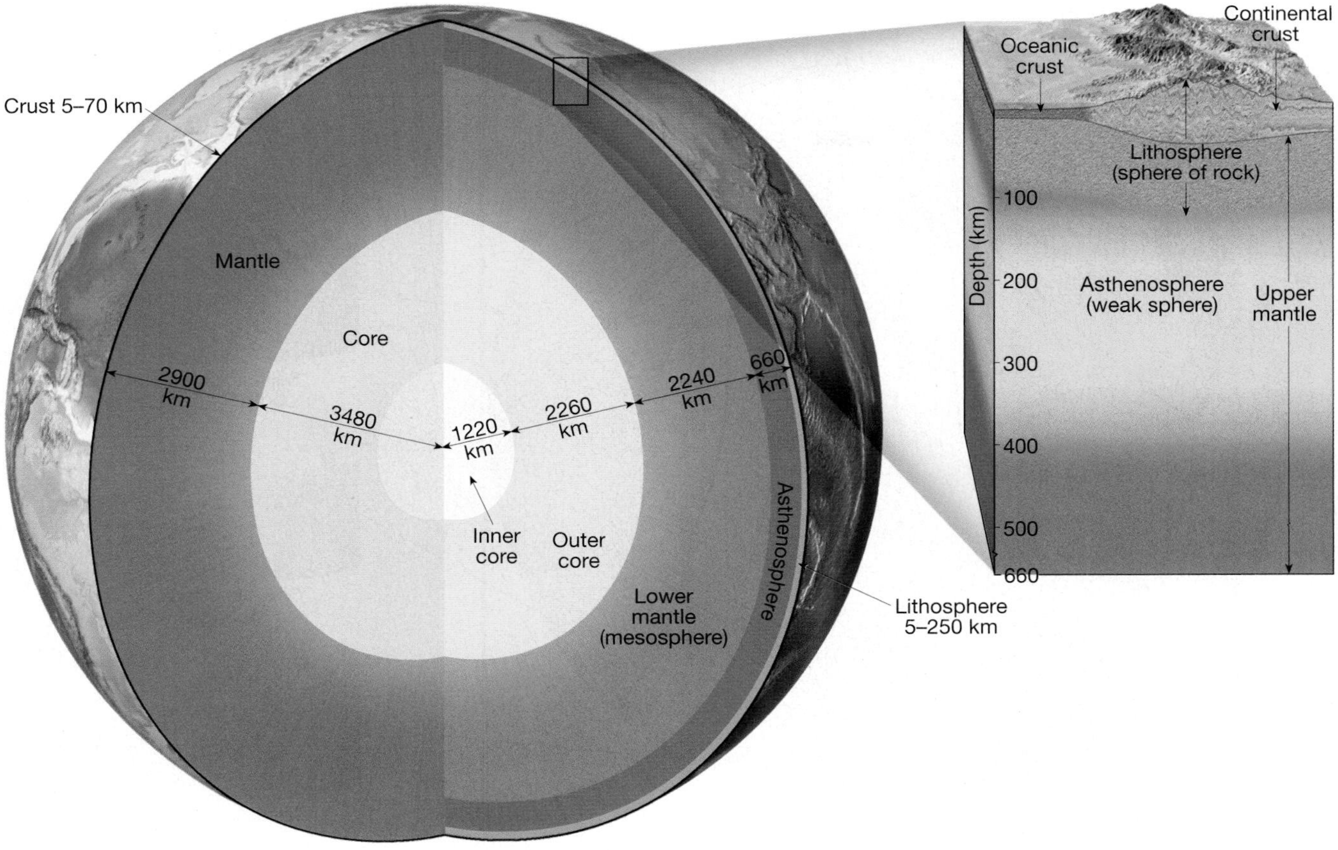

▲ **Figure 14.23** Views of Earth's layered structure. The left side of the large cross section shows that Earth's interior is divided into three different layers based on compositional differences—the crust, mantle, and core. The right side of the large cross section depicts the five main layers of Earth's interior based on physical properties and hence mechanical strength—the lithosphere, asthenosphere, lower mantle (mesosphere), outer core, and inner core. The block diagram to the right of the large cross section shows an enlarged view of the upper portion of Earth's interior.

and the rocks gradually strengthen with depth. Between the depths of 660 kilometers and 2900 kilometers a more rigid layer called the **mesosphere** (*middle sphere*) or **lower mantle** is found (Figure 14.23). Despite their strength, the rocks of the mesosphere are still very hot and capable of very gradual flow.

Inner and Outer Core. The core, which is composed mostly of an iron-nickel alloy, is divided into two regions that exhibit very different mechanical strengths (Figure 14.23). The **outer core** is a *liquid layer* 2270 kilometers (1410 miles) thick. It is the convective flow of metallic iron within this zone that generates Earth's magnetic field. The **inner core** is a sphere having a radius of 1216 kilometers (754 miles). Despite its higher temperature, the material in the inner core is stronger (because of immense pressure) than the outer core and behaves like a *solid.*

Discovering Earth's Major Layers

The story of how seismologists discovered Earth's core and layers is interesting. In 1909, a pioneering Yugoslavian seismologist, Andrija Mohorovii, presented the first convincing evidence for layering within Earth. By studying seismic records, he found that the velocity of seismic waves increases abruptly below about 50 kilometers of depth. This boundary separates the crust from the underlying mantle and is known as the **Mohorovičić discontinuity** in his honor. For reasons that are obvious, the name for this boundary was quickly shortened to **Moho.**

A few years later another boundary was discovered, by the German seismologist Beno Gutenberg. Generally, seismic waves from even small earthquakes are strong enough to travel around the world. This is why a seismograph in Antarctica can record earthquakes in California or Italy. But Gutenberg observed that P waves diminish and eventually die out about 105 degrees around the globe from an earthquake. Then, about 140 degrees away, the P waves reappear, but about 2 minutes later than would be expected, based on the distance traveled (Figure 14.24). This belt, where direct P waves are absent, is about 35 degrees wide and has been named the **shadow zone.**

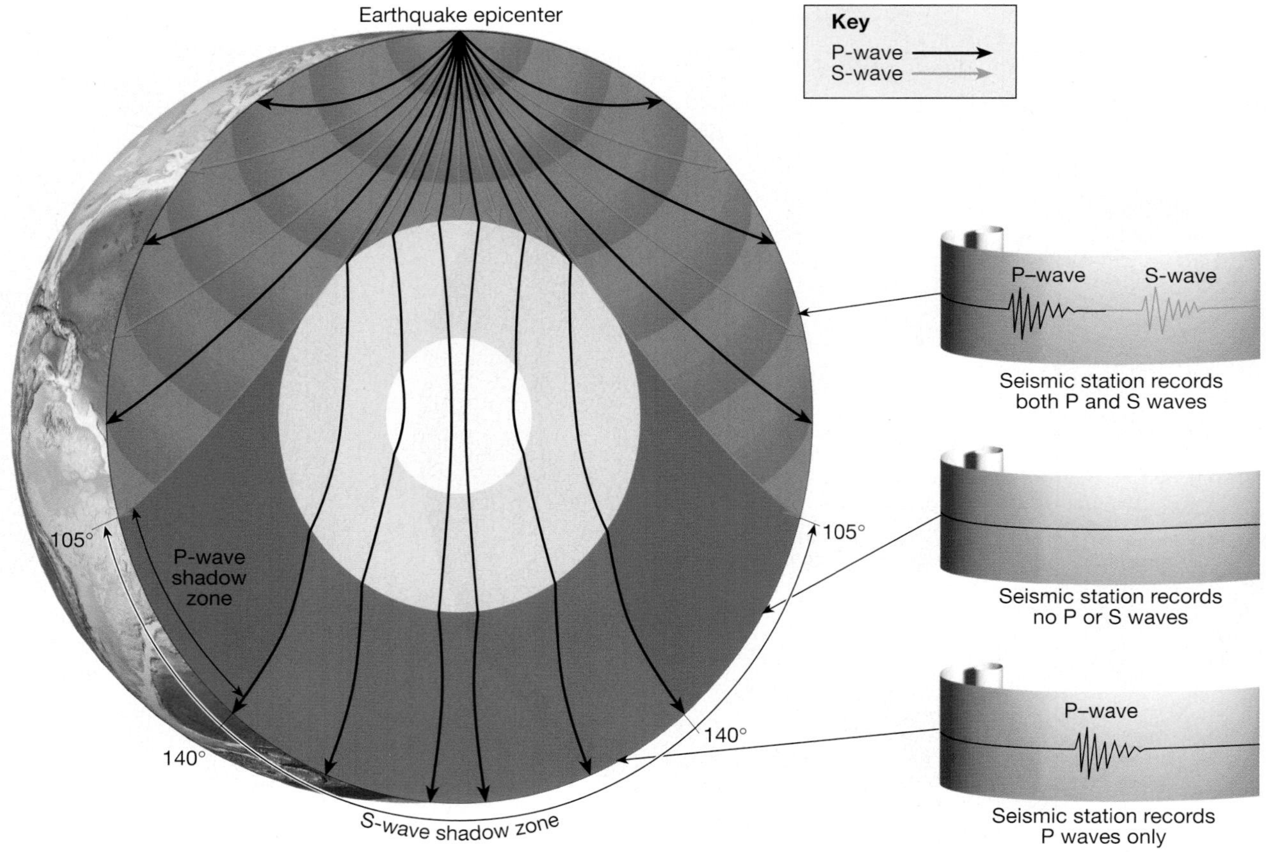

▲ **Figure 14.24** Paths of P and S waves traveling through Earth's interior. The abrupt change in physical properties at the mantle–core boundary causes the wave paths to bend sharply. This abrupt change in wave direction results in a shadow zone for P waves between about 105 and 140 degrees. Further, any location more than 105 degrees from the earthquake epicenter will not receive direct S waves since the outer core will not transmit them.

Inge Lehmann discovered Earth's inner core in 1936. (Susan M. Landon)

Gutenberg realized that the shadow zone could be explained if Earth contained a core composed of material unlike the overlying mantle. The core must somehow hinder the transmission of P waves in a manner similar to the light rays being blocked by an opaque object that casts a shadow. However, rather than actually stopping the P waves, the shadow zone bends them, as shown in Figure 14.24. It was further learned that S waves could not travel through the core. Therefore, geologists concluded that at least a portion of this region is liquid.

In 1936, the last major subdivision of Earth's interior was predicted by Inge Lehmann, a Danish seismologist. Lehmann discovered a new region of seismic reflection within the core. Hence, Earth has a core within a core. The size of the inner core was not accurately calculated until the early 1960s, when underground nuclear tests were conducted in Nevada. Because the precise locations and times of the explosions were known, echoes from seismic waves that bounced off the inner core provided an accurate means of determining its size.

Over the past few decades, advances in seismology have allowed for much refinement of the gross view of Earth's interior presented so far. Of major importance was the discovery of the lithosphere and asthenosphere.

Discovering Earth's Composition

We have examined Earth's structure, so let us now look at the composition of each layer. Composition tells us much about how our planet has developed over its estimated age of 4.5 billion years.

Earth's crust varies in thickness, exceeding 70 kilometers (40 miles) in some mountainous regions and being thinner than 3 kilometers (2 miles) in some oceanic areas (see Figure 14.23). Early seismic data indicated that the continental crust, which is mostly made of lighter, granitic rocks, is quite different in composition from denser oceanic crust. Until recently, however, scientists had only seismic evidence from which to determine the composition of oceanic crust, because it lies beneath an average of 3 kilometers of water as well as hundreds of meters of sediment. With the development of deep-sea drilling technology, the recovery of ocean floor samples was made possible. The samples were of basaltic composition—very different from the rocks that make up the continents.

Our knowledge of the rocks of the mantle and core is much more speculative. However, we do have some clues. Recall that some of the lava that reaches Earth's surface originates in the partially melted asthenosphere, within the mantle. In the laboratory, experiments have shown that partial melting of a rock called *peridotite* results in a melt that has a basaltic composition similar to lava that emerges during volcanic activity of oceanic islands. Denser rocks like peridotite are thought to make up the mantle and provide the lava for oceanic eruptions.

Surprisingly, meteorites or "shooting stars" that collide with Earth provide evidence of Earth's inner composition. Because meteorites are part of the solar system, they are assumed to be representative samples. Their composition ranges from metallic meteorites made of iron and nickel to stony meteorites composed of dense rock similar to peridotite.

Because Earth's crust contains a much smaller percentage of iron than do meteorites, geologists believe that the dense iron and other dense metals literally sank toward Earth's center during the planet's early history. By the same token, lighter substances may have floated to the surface, creating the less dense crust. Thus, Earth's core is thought to be mainly dense iron and nickel, similar to metallic meteorites, whereas the surrounding mantle is believed to be composed of rocks similar to stony meteorites.

The concept of a molten iron outer core is further supported by Earth's magnetic field. Our planet acts as a large magnet. The most widely accepted mechanism explaining why Earth has a magnetic field requires that the core be made of a material that conducts electricity, such as iron, and that it is mobile enough to allow circulation. Both of these conditions are met by the model of Earth's core that was established on the basis of seismic data.

Not only does an iron core explain Earth's magnetic field, but it also explains the high density of inner Earth—about 14 times that of water at Earth's center. Even under the extreme pressure at those depths, average crustal rocks with densities 2.8 times that of water would not have the density calculated for the core. But iron, which is three times more dense than crustal rocks, has the required density.

In summary, although earthquakes can be very destructive, much of our knowledge about Earth's interior comes from the study of these phenomena. As our knowledge of earthquakes and their causes improves, we learn more of our planet's internal workings and how to reduce the consequences of tremors. In the next chapter, you will see that most earthquakes originate at the boundaries of Earth's great lithospheric plates.

Did You Know?

Although there are no historical records of a tsunami produced by a meteorite impact in the oceans, such events have occurred. Geologic evidence indicates that the most recent occurrence was a megatsunami that devastated portions of the Australian coast about the year 1500. A meteorite impact 65 million years ago near Mexico's Yucatan Peninsula created one of the largest impact-induced tsunami ever. The giant wave swept hundreds of kilometers inland around the shore of the Gulf of Mexico.

The Chapter in Review

- *Earthquakes* are vibrations of Earth produced by the rapid release of energy from rocks that rupture because they have been subjected to stresses beyond their limit. This energy, which takes the form of waves, radiates in all directions from the earthquake's source, called the *focus.* The movements that produce most earthquakes occur along large fractures, called *faults,* that are usually associated with plate boundaries.

- Along a fault, rocks store energy as they are bent. As slippage occurs at the weakest point (the focus), displacement will exert stress farther along a fault, where additional slippage will occur until most of the built-up strain is released. An earthquake occurs as the rock elastically returns to its original shape. The springing back of the rock is termed *elastic rebound.* Small earthquakes, called *foreshocks,* often precede a major earthquake. The adjustments that follow a major earthquake often generate smaller earthquakes called *aftershocks.*

- Two main types of *seismic waves* are generated during an earthquake: (1) *surface waves,* which travel along the outer layer of Earth; and (2) *body waves,* which travel through Earth's interior. Body waves are further divided into *primary,* or *P, waves,* which push (compress) and pull (expand) rocks in the direction the wave is traveling, and *secondary,* or *S, waves,* which shake the particles in rock at right angles to their direction of travel. The P waves can travel through solids, liquids, and gases. Fluids (gases and liquids) will not transmit S waves. In any solid material, P waves travel about 1.7 times faster than do S waves.

- The location on Earth's surface directly above the focus of an earthquake is the *epicenter.* An epicenter is determined using the difference in velocities of P and S waves. Using the difference in arrival times between P and S waves, the distance separating a recording station from the earthquake can be determined. When the distances are known from three or more seismic stations, the epicenter can be located using a method called *triangulation.*

- *A close correlation exists between earthquake epicenters and plate boundaries.* The principal earthquake epicenter zones are along the outer margin of the Pacific Ocean, known as the *circum-Pacific belt,* and through the world's oceans along the *oceanic ridge system.*

- Seismologists use two fundamentally different measures to describe the size of an earthquake—intensity and magnitude. *Intensity* is a measure of the degree of ground shaking at a given locale based on the amount of damage. The *Modified Mercalli Intensity Scale* uses damages to buildings in California as the basis for estimating the intensity of ground shaking for a local earthquake. *Magnitude* is calculated from seismic records and estimates the amount of energy released at the source of an earthquake. Using the *Richter scale,* the magnitude of an earthquake is estimated by measuring the *amplitude* (maximum displacement) of the largest seismic wave recorded. A logarithmic scale is used to express magnitude, in which a tenfold increase in ground shaking corresponds to an increase of 1 on the magnitude scale. *Moment magnitude* is currently used to estimate the size of moderate and large earthquakes. It is calculated using the average displacement of the fault, the area of the fault surface, and the sheer strength of the faulted rock.

- The most obvious factors determining the amount of destruction accompanying an earthquake are the magnitude of the earthquake and the proximity of the quake to a populated area. Structural damage attributable to earthquake vibrations depends on several factors, including (1) wave amplitudes, (2) the duration of the vibrations, (3) the nature of the material upon which the structure rests, and (4) the design of the structure. Secondary effects of earthquakes include *tsunamis,* landslides, ground subsidence, and fire.

- Substantial research to predict earthquakes is under way in Japan, the United States, China, and Russia—countries where earthquake risk is high. No reliable method of short-range prediction has yet been devised. Long-range forecasts are based on the premise that earthquakes are repetitive or cyclical. Seismologists study the history of earthquakes for patterns so their occurrences might be predicted.

- Earth's internal structure is divided into layers based on differences in chemical composition and on the basis of changes in physical properties. Compositionally, Earth is divided into a thin outer *crust,* a soild rocky *mantle,* and a dense *core.* Based on physical properties, the layers of Earth are (1) the *lithosphere*—the cool, rigid, outermost layer that averages about 100 kilometers thick, (2) the *asthenosphere,* a relatively weak layer located in the mantle beneath the lithosphere, (3) the more rigid *mesosphere,* where rocks are very hot and capable of very gradual flow, (4) the liquid *outer core,* where Earth's magentic field is generated, and (5) the solid *inner core.*

- The *continental crust* is primarily made of *granitic* rock, whereas the *oceanic crust* has a *basaltic* composition. Rocks similar to *peridotite* make up the *mantle.* The *core* is made up mainly of *iron* and *nickel.* An iron core explains the high density of Earth's interior as well as Earth's magnetic field.

Key Terms

aftershock (p. 312)
asthenosphere (p. 328)
body wave (p. 314)
core (p. 328)
crust (p. 327)
earthquake (p. 311)
elastic rebound (p. 312)
epicenter (p. 316)
fault (p. 311)
fault creep (p. 312)
focus (p. 311)
foreshock (p. 312)
inner core (p. 329)
intensity (p. 317)
liquefaction (p. 321)
lithosphere (p. 328)
lower mantle (p. 329)
magnitude (p. 317)
mantle (p. 328)
mesosphere (p. 329)
Modified Mercalli Intensity Scale (p. 317)
Mohorovičić discontinuity (Moho) (p. 329)
moment magnitude (p. 320)
outer core (p. 329)
primary (P) wave (p. 314)
Richter magnitude (p. 317)
Richter scale (p. 317)
secondary (S) wave (p. 314)
seismic sea wave (tsunami) (p. 322)
seismogram (p. 314)
seismograph (p. 314)
seismology (p. 314)
shadow zone (p. 329)
surface wave (p. 314)

Questions for Review

1. What is an *earthquake*? Under what circumstances do earthquakes occur?
2. How are *faults, foci,* and *epicenters* related?
3. Who was first to explain the actual mechanism by which earthquakes are generated?
4. Explain what is meant by *elastic rebound.*
5. Faults that are experiencing no active creep may be considered safe. Rebut or defend this statement.
6. Describe the principle of a *seismograph.*
7. Using Figure 14.8, determine the distance between an earthquake and a seismic station if the first S wave arrives three minutes after the first P wave.
8. List the major differences between *P* and *S waves.*
9. Which type of seismic wave causes the greatest destruction to buildings?

10. Most strong earthquakes occur in a zone on the globe known as the _____.

11. What factor contributed most to the extensive damage that occurred in the central portion of Mexico City during the 1985 earthquake?

12. The 1988 Armenian earthquake had a Richter magnitude of 6.9, less than the 1994 Northridge California earthquake. Nevertheless, the loss of life was far greater in the Armenian event. Why?

13. An earthquake measuring 7 on the Richter scale releases about _____ times more energy than an earthquake with a magnitude of 6.

14. List three reasons why the *moment magnitude scale* has gained popularity among seismologists.

15. List four factors that affect the amount of destruction caused by seismic vibrations.

16. In addition to the destruction created directly by seismic vibrations, list three other types of destruction associated with earthquakes.

17. Distinguish between the *Mercalli scale* and the *Richter scale.*

18. What is a *tsunami*? How is one generated?

19. Cite some reasons why an earthquake with a moderate magnitude might cause more extensive damage than a quake with a high magnitude.

20. What evidence do we have that Earth's outer core is molten?

21. Contrast the physical makeup of the *asthenosphere* and the *lithosphere.*

22. Why are meteorites considered important clues to the composition of Earth's interior?

23. Describe the composition of the following:
a. continental crust
b. oceanic crust
c. mantle
d. core

Online Study Guide

The *Essentials of Geology* Web site uses the resources and flexibility of the Internet to aid in your study of the topics in this chapter. Written and developed by geology instructors, this site will help improve your understanding of geology. Visit **www.prenhall.com/lutgens** and click on the cover of *Essentials of Geology 9e* to find:

- Online review quizzes.
- Critical thinking exercises.
- Links to chapter-specific Web resources.
- Internet-wide key-term searches.

http://www.prenhall.com/lutgens

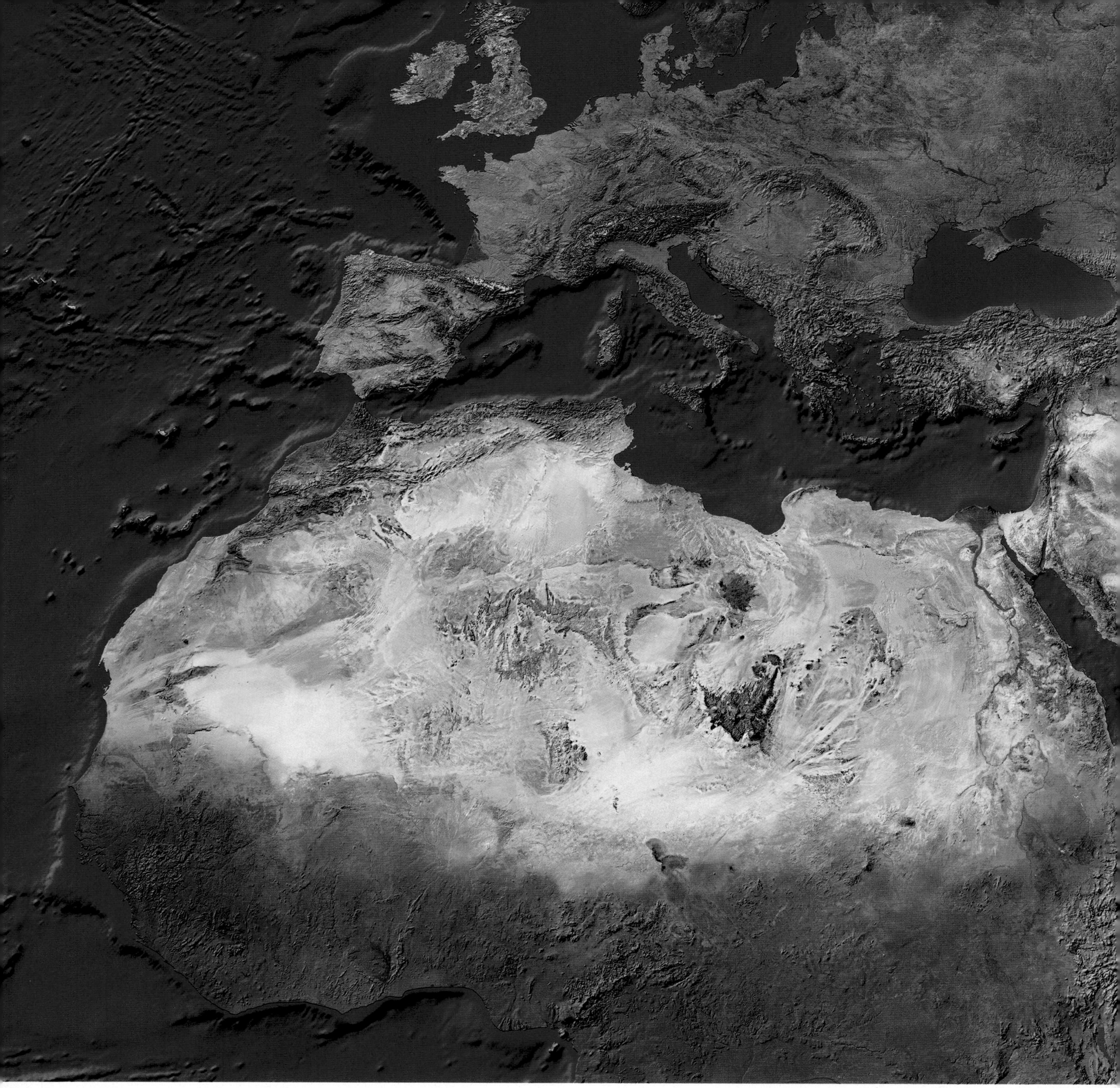

Composite satellite image of Europe, North Africa, and the Arabian Peninsula *(Image ® by Worldsat International, Inc., 2001. www.worldsat.ca. All rights reserved.)*

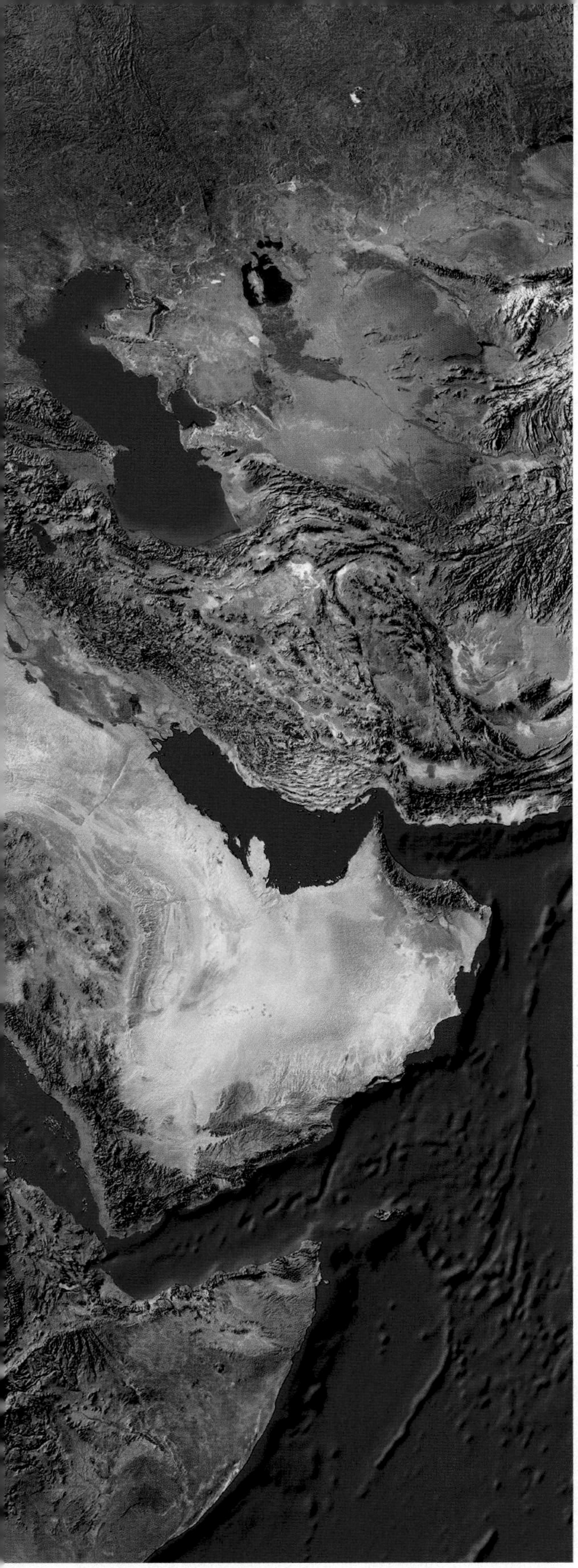

CHAPTER 15

Plate Tectonics: A Scientific Theory Unfolds

Focus on Learning

To assist you in learning the important concepts in this chapter, you will find it helpful to focus on the following questions:

- What evidence was used to support the continental drift hypothesis?
- What was one of the main objections to the continental drift hypothesis?
- What is the theory of plate tectonics?
- In what major way does the plate tectonics theory depart from the continental drift hypothesis?
- What are the three types of plate boundaries?
- What is the evidence used to support the plate tectonics theory?
- What models have been proposed to explain the driving mechanism for plate motion?

Early in the twentieth century, most geologists believed that the geographic positions of the ocean basins and continents were fixed. During the last few decades, however, vast amounts of new data have dramatically changed our understanding of the nature and workings of our planet. Earth scientists now realize that the continents gradually migrate across the globe. Where landmasses split apart, new ocean basins are created between the diverging blocks. Meanwhile, older portions of the seafloor are carried back into the mantle in regions where trenches occur in the deep ocean floor. Because of these movements, blocks of continental crust eventually collide and form Earth's great mountain ranges (Figure 15.1). In short, a revolutionary new model of Earth's tectonic* processes has emerged.

This profound reversal of scientific understanding has been appropriately described as a scientific *revolution.* The revolution began as a relatively straightforward proposal by Alfred Wegener, called *continental drift.* After many years of heated debate, Wegener's hypothesis of drifting continents was rejected by the vast majority of the scientific community. The concept of a mobile Earth was particularly distasteful to North American geologists, perhaps because much of the supporting evidence had been gathered from the southern continents, with which most North American geologists were unfamiliar.

During the 1950s and 1960s, new kinds of evidence began to rekindle interest in this nearly abandoned proposal. By 1968 these new developments led to the unfolding of a far more encompassing explanation, which incorporated aspects of continental drift and seafloor spreading—a theory known as *plate tectonics.*

In this chapter, we will examine the events that led to this dramatic reversal of scientific opinion in an attempt to provide some insight into how science works. We will also briefly trace the developments of the concept of continental drift, examine why it was first rejected, and consider the evidence that finally led to the acceptance of the theory of plate tectonics.

*Tectonics refers to the deformation of Earth's crust and results in the formation of structural features such as mountains.

Continental Drift: An Idea Before Its Time

The idea that continents, particularly South America and Africa, fit together like pieces of a jigsaw puzzle *originated* with improved world maps. However, little significance was given this idea until 1915, when Alfred Wegener, a German meteorologist and geophysicist, published *The Origin of Continents and Oceans.* In this book, Wegener set forth his radical hypothesis of **continental drift.**

▼ **Figure 15.1** Climbers camping on a shear rock face of a mountain known as K7 in Pakistan's Karakoram, a part of the Himalayas. These mountains formed as India collided with Eurasia. *(Photo by Jimmy Chin/National Geographic/Getty)*

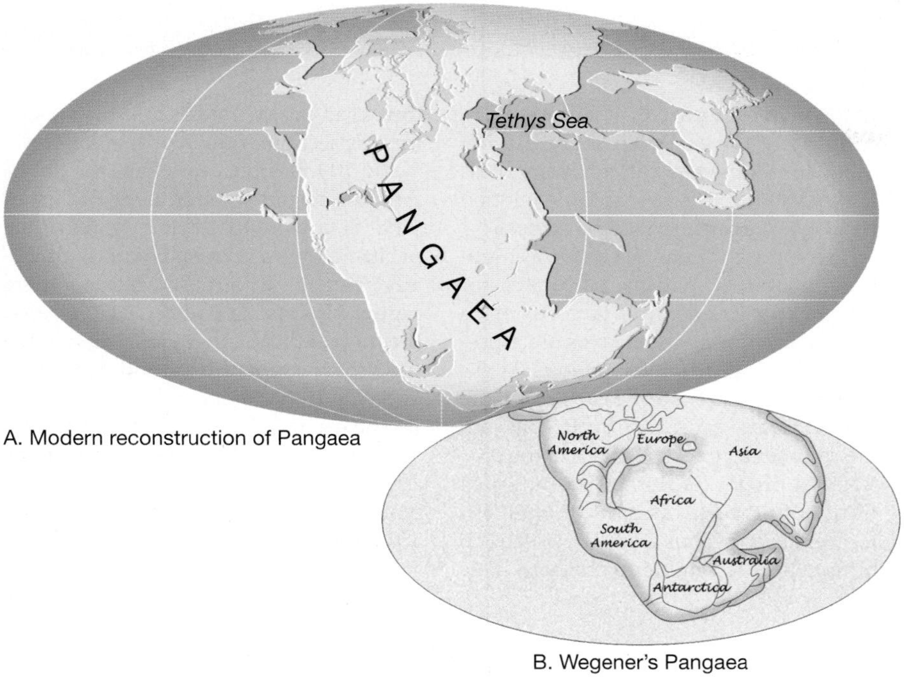

◀ **Figure 15.2** Reconstruction of Pangaea as it is thought to have appeared 200 million years ago. **A.** Modern reconstruction. **B.** Reconstruction done by Wegener in 1915.

Wegener suggested that a supercontinent he called **Pangaea** (meaning "all land") once existed (Figure 15.2). He further hypothesized that, about 200 million years ago, this supercontinent began breaking into smaller continents, which then "drifted" to their present positions. (see Box 15.1).

Wegener and others collected substantial evidence to support these claims. The fit of South America and Africa, and the geographic distribution of fossils, rock structures, and ancient climates all seemed to support the idea that these now separate landmasses were once joined. Let us examine their evidence.

Evidence: The Continental Jigsaw Puzzle

Like a few others before him, Wegener first suspected that the continents might have been joined when he noticed the remarkable similarity between the coastlines on opposite sides of the South Atlantic. However, his use of present-day shorelines to make a fit of the continents was challenged immediately by other Earth scientists. These opponents correctly argued that shorelines are continually modified by erosional processes, and even if continental displacement had taken place, a good fit today would be unlikely. Wegener appeared to be aware of this problem, and, in fact, his original jigsaw fit of the continents was only very crude.

Scientists have determined that a much better approximation of the true outer boundary of the continents is the seaward edge of the continental shelf, which lies submerged several hundred meters below sea level. In the early 1960s, scientists produced a map that attempted to fit the edges of the continental shelves at a depth of 900 meters (3000 feet). The remarkable fit that was obtained is shown in Figure 15.3. Although the continents overlap in a few places, these are regions where streams have deposited large quantities of sediment, thus enlarging the continental shelves. The overall fit was even better than researchers suspected it would be.

Did You Know?

Although Alfred Wegener is rightfully credited with formulating the continental drift hypothesis, he was not the first to suggest continental mobility. An American geologist, F. B. Taylor published the first paper to outline this important idea. However, Taylor's paper provided little supporting evidence, whereas Wegener spent much of his professional life trying to substantiate his views.

Evidence: Fossils Match Across the Seas

Although the seed for Wegener's hypothesis came from the remarkable similarities of the continental margins on opposite sides of the Atlantic, he thought the idea of a mobile Earth was improbable. Not until he learned that identical fossil organisms were known from rocks in both South America and Africa did he begin to seriously pursue this idea. Through a review of the literature, Wegener learned that most paleontologists (scientists who study fossils and the evolution of life) were in agreement that some type of land connection was needed to explain the existence of identical fossils of Mesozoic life forms on widely separated landmasses. (Just as modern life forms native to North America are quite different from those of Africa, one would expect that during the Mesozoic era, organisms on widely separated continents would be quite distinct).

BOX 15.1

The Breakup of Pangaea

Wegener used evidence from fossils, rock types, and ancient climates to create a jigsaw-puzzle fit of the continents—thereby creating his supercontinent of Pangaea. In a similar manner, but employing modern tools not available to Wegener, geologists have recreated the steps in the breakup of this supercontinent, an event that began nearly 200 million years ago. From this work, the dates when individual crustal fragments separated from one another and their relative motions have been well established (Figure 15.A).

An important consequence of the breakup of Pangaea was the creation of a "new" ocean basin: the Atlantic. As you can see in part B of Figure 15.A, splitting of the supercontinent did not occur simultaneously along the margins of the Atlantic. The first split developed between North America and Africa.

Here, the continental crust was highly fractured, providing pathways for huge quantities of fluid lavas to reach the surface. Today these lavas are represented by weathered igneous rocks found along the Eastern Seaboard of the United States—primarily buried beneath the sedimentary rocks that form the continental shelf. Radiometric dating of these solidified lavas indicate that rifting began in various stages between 180 million and 165 million years ago. This time span can be used as the "birth date" for this section of the North Atlantic.

By 130 million years ago, the South Atlantic began to open near the tip of what is now South Africa. As this zone of rifting migrated northward, it gradually opened the South Atlantic (compare Figure 15.A, parts B and C). Continued breakup of the southern landmass led to the separation of Africa and Antarctica and sent India on a northward journey. By the early Cenozoic, about 50 million years ago, Australia had separated from Antarctica, and the South Atlantic had emerged as a full-fledged ocean (Figure 15.A, part D).

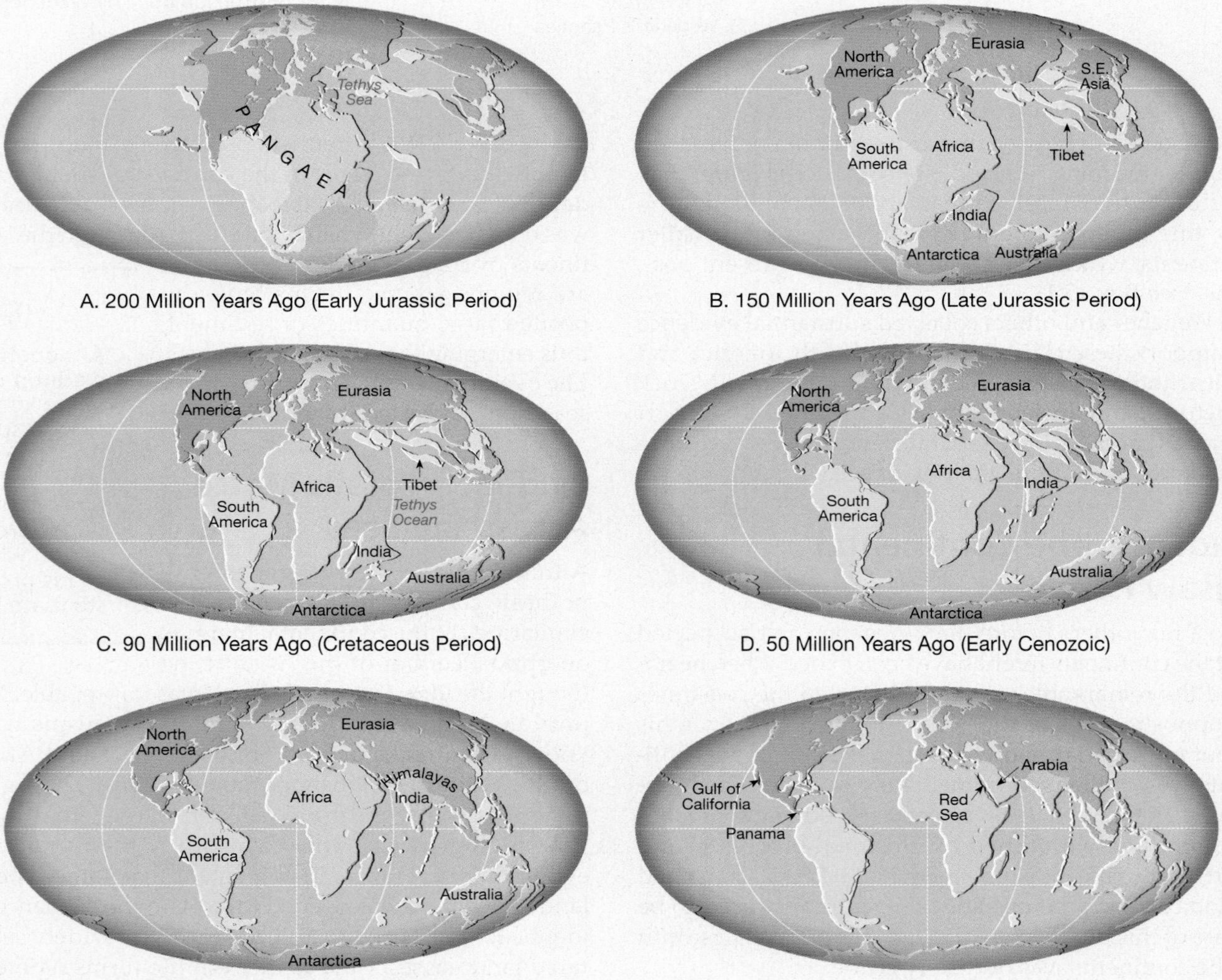

Figure 15.A Several views of the breakup of Pangaea over a period of 200 million years.

A modern map (Figure 15.A, part F) shows that India eventually collided with Asia, an event that began, about 45 million years ago and created the Himalayas as well as the Tibetan Highlands. About the same time, the separation of Greenland from Eurasia completed the breakup of the northern landmass. During the last 20 million years or so of Earth history, Arabia rifted from Africa to form the Red Sea, and Baja California separated from Mexico to form the Gulf of California (Figure 15.A, part E). Meanwhile, the Panama Arc joined North America and South America to produce our globe's familiar, modern appearance.

Mesosaurus. To add credibility to his argument for the existence of a supercontinent, Wegener cited documented cases of several fossil organisms that were found on different landmasses despite the unlikely possibility that their living forms could have crossed the vast ocean presently separating these continents. The classic example is *Mesosaurus,* an aquatic fishcatching reptile whose fossil remains are found only in black shales of Permian age (about 260 million years ago) in eastern South America and southern Africa (Figure 15.4). If *Mesosaurus* had been able to make the long journey across the vast South Atlantic Ocean, its remains should be more widely distributed. As this is not the case, Wegener argued that South America and Africa must have been joined during that period of Earth history.

How did scientists during Wegener's era explain the existence of identical fossil organisms in places separated by thousands of kilometers of open ocean? Transoceanic land bridges were the most widely accepted explanations of such migrations (Figure 15.5). We know, for example, that during the recent Ice Age, the lowering of sea level allowed animals to cross the narrow Bering Strait between Asia and North America. Was it possible that land bridges once connected Africa and South America but later subsided below sea level? Modern maps of the seafloor substantiate Wegener's contention that land bridges of this magnitude had not existed. If they had, remnants would still lie below sea level.

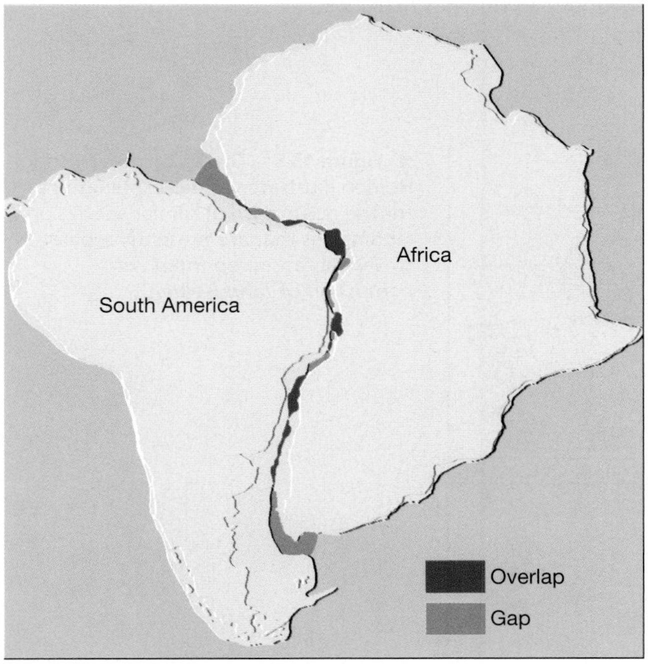

▲ **Figure 15.3** This shows the best fit of South America and Africa along the continental slope at a depth of 500 fathoms (about 900 meters). The areas where continental blocks overlap appear in brown. *(After A. G. Smith. "Continental Drift," in* Understanding the Earth, *edited by I. G. Gass)*

Alfred Wegener, during an expedition to Greenland. (Bildarchiv Preussischer Kulturbesitz, Berlin)

Present-Day Organisms. In a later edition of his book, Wegener also cited the distribution of present-day organisms as evidence to support the drifting of continents. For example, modern organisms with similar ancestries clearly had to evolve in isolation during the last few tens of millions of years. Most obvious of these are the Australian marsupials (such as kangaroos), which have a direct fossil link to the marsupial opossums found in the Americas. After the breakup of Pangaea, the Australian marsupials followed a different evolutionary path than related life forms in the Americas.

After the breakup of Pangaea, the Australian marsupials evolved differently than their relatives in the Americas. (Martin Harvey)

Evidence: Rock Types and Structures Match

Anyone who has worked a picture puzzle knows that in addition to the pieces fitting together, the picture must be continuous as well. The "picture" that must match in the "continental drift puzzle" is one of rock types and mountain belts found on the continents. If the continents were once together, the rocks found in a particular region on one continent should closely match in age and type those found in adjacent positions on the adjoining continent. Wegener found evidence of 2.2-billion-year-old igneous rocks in Brazil that closely resembled similarly aged rocks in Africa.

Similar evidence exists in the form of mountain belts that terminate at one coastline, only to reappear on landmasses across the ocean. For instance, the mountain belt that includes the Appalachians trends northeastward through the eastern United States and

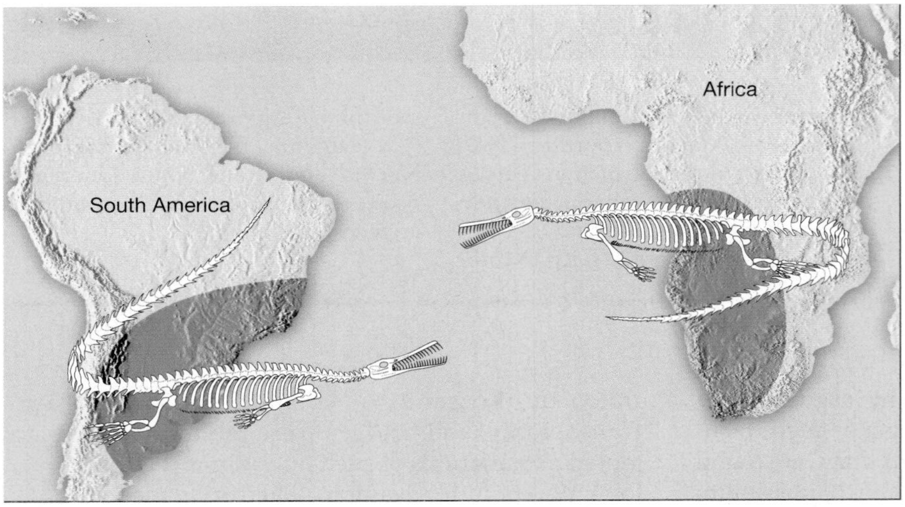

◀ **Figure 15.4** Fossils of *Mesosaurus* have been found on both sides of the South Atlantic and nowhere else in the world. Fossil remains of this and other organisms on the continents of Africa and South America appear to link these landmasses during the late Paleozoic and early Mesozoic eras.

The Appalachians are similar in age and structure to the mountains in northwestern Europe. (Tim Fitzharris)

disappears off the coast of Newfoundland. Mountains of comparable age and structure are found in Greenland, the British Isles, and Scandinavia. When these landmasses are reassembled, as in Figure 15.6, the mountain chains form a nearly continuous belt.

Wegener must have been convinced that the similarities in rock structure on both sides of the Atlantic linked these landmasses when he said, "It is just as if we were to refit the torn pieces of a newspaper by matching their edges and then check whether the lines of print run smoothly across. If they do, there is nothing left but to conclude that the pieces were in fact joined in this way.*

Evidence: Ancient Climates

Because Alfred Wegener was a meteorologist by profession, he was keenly interested in obtaining paleoclimatic data to support continental drift. His efforts

*Alfred Wegener, *The Origin of Continents and Oceans.* Translated from the 4th revised German edition of 1929 by J. Birman (London: Methuen, 1966).

◀ **Figure 15.5** These sketches by John Holden illustrate various explanations for the occurrence of similar species on landmasses that are presently separated by vast oceans. *(Reprinted with permission of John Holden)*

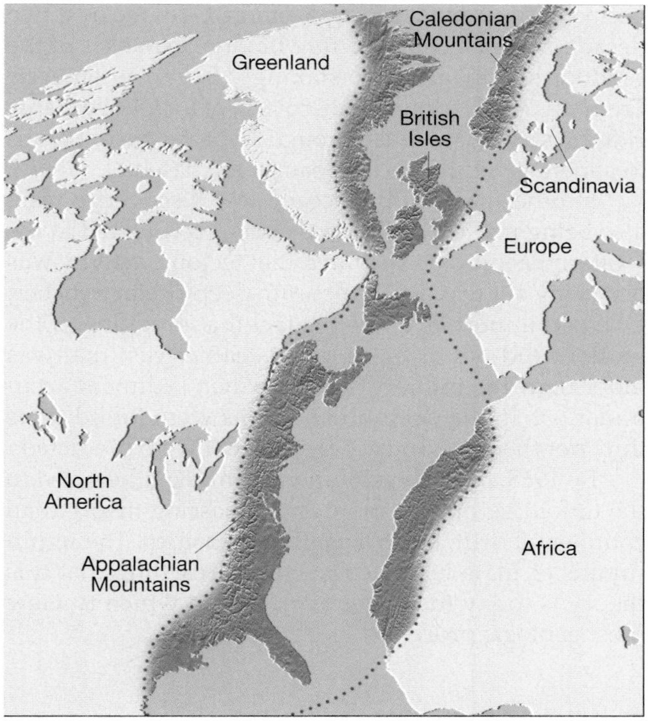

▲ **Figure 15.6** Matching mountain ranges across the North Atlantic. The Appalachian Mountains trend along the eastern flank of North America and disappear off the coast of Newfoundland. Mountains of comparable age and structure are found in Greenland, the British Isles, and Scandinavia. When these landmasses are placed in their predrift locations, these ancient mountain chains form a nearly continuous belt. These folded mountain belts formed roughly 300 million years ago as the landmasses collided during the formation of the supercontinent of Pangaea.

were rewarded when he found evidence for apparently dramatic global climatic changes during the geologic past. In particular, he learned of ancient glacial deposits that indicated that near the end of the Paleozoic era (about 300 million years ago), ice sheets covered extensive areas of the Southern Hemisphere and India (Figure 15.7A). Layers of glacially transported sediments of the same age were found in southern Africa and South America, as well as in India and Australia. Much of the land area containing evidence of this late Paleozoic glaciation presently lies within 30 degrees of the equator in subtropical or tropical climates.

Could Earth have gone through a period of sufficient cooling to have generated extensive ice sheets in areas that are presently tropical? Wegener rejected this explanation because during the late Paleozoic, large tropical swamps existed in the Northern Hemisphere. These swamps, with their lush vegetation, eventually became the major coal fields of the eastern United States, Europe, and Siberia.

Fossils from these coal fields indicate that the tree ferns that produced the coal deposits had large fronds, which are indicative of tropical settings. Furthermore, unlike trees in colder climates, these trees lacked growth rings, a characteristic of tropical plants that grow in regions having minimal fluctuations in temperature.

Wegener suggested that a more plausible explanation for the late Paleozoic glaciation was provided by the supercontinent of Pangaea. In this configuration the southern continents are joined together and located near the South Pole (Figure 15.7B). This would account for the conditions necessary to generate extensive expanses of glacial ice over much of the Southern Hemisphere. At the same time, this geography would place today's northern landmasses nearer the equator and account for their vast coal deposits. Wegener was so convinced that his explanation was correct that he wrote, "This evidence is so compelling that by comparison all other criteria must take a back seat."

How does a glacier develop in hot, arid central Australia? How do land animals migrate across wide

Did You Know?

Alfred Wegener, best known for his continental drift hypothesis, also wrote numerous scientific papers on weather and climate. Following his interest in meteorology, Wegener made four extended trips to the Greenland Ice Sheet in order to study its harsh winter weather. In November of 1930, while making a month-long trek across the ice sheet, Wegener and a companion perished.

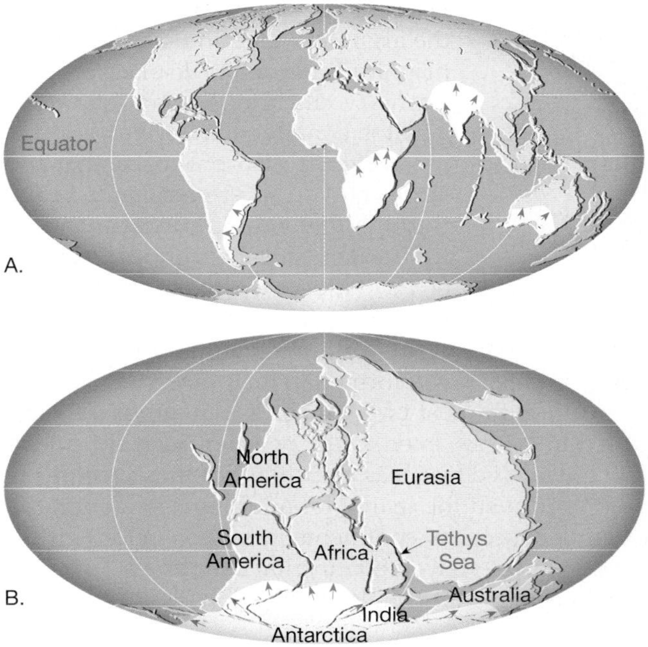

▲ **Figure 15.7** Paleoclimatic evidence for continental drift. **A.** Near the end of the Paleozoic era (about 300 million years ago) ice sheets covered extensive areas of the Southern Hemisphere and India. Arrows show the direction of Ice movement that can be inferred from glacial grooves in the bedrock. **B.** Shown are the continents restored to their former position with the South Pole located roughly between Antarctica and Africa. This configuration accounts for the conditions necessary to generate a vast ice sheet and also explains the directions of ice movement that radiated away from the South Pole.

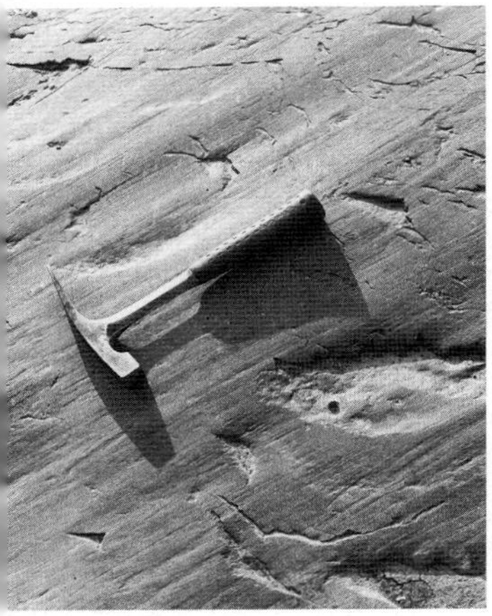

Glacial striations in bedrock of Australia indicate direction of ice movement. (W. B. Hamilton/USGS)

expanses of open water? As compelling as this evidence may have been, 50 years passed before most of the scientific community accepted the concept of continental drift and the logical conclusions to which it led.

The Great Debate

Wegener's proposal did not attract much open criticism until 1924, when his book was translated into English. From this time on, until his death in 1930, his drift hypothesis encountered a great deal of hostile criticism. To quote the respected American geologist T. C. Chamberlin, "Wegener's hypothesis . . . takes considerable liberty with our globe, and is less bound by restrictions or tied down by awkward, ugly facts than most of its rival theories. Its appeal seems to lie in the fact that it plays a game in which there are few restrictive rules and no sharply drawn code of conduct."

One of the main objections to Wegener's hypothesis stemmed from his inability to provide a mechanism that was capable of moving the continents across the globe. Wegener proposed that the tidal influence of the Moon was strong enough to give the continents a westward motion. However, the prominent physicist Harold Jeffreys quickly countered with the argument that tidal friction of the magnitude needed to displace the continents would bring Earth's rotation to a halt in a matter of a few years.

Wegener also proposed that the larger and sturdier continents broke through the oceanic crust, much like ice breakers cut through ice. However, no evidence existed to suggest that the ocean floor was weak enough to permit passage of the continents without themselves being appreciably deformed in the process.

Did You Know?

A group of scientists proposed an interesting although incorrect explanation for the cause of continental drift. This proposal suggested that early in Earth's history, our planet was only about half its current diameter and completely covered by continental crust. Through time, Earth expanded, causing the continents to split into their current configurations, while new seafloor "filled in" the spaces as they drifted apart.

Although most of Wegener's contemporaries opposed his views, even to the point of open ridicule, a few considered his ideas plausible. For those geologists who continued the search for additional evidence, the exciting concept of continents in motion held their interest. Others viewed continental drift as a solution to previously unexplainable observations.

Plate Tectonics: The New Paradigm

GEODe Plate Tectonics
▼ Introduction

Following World War II, oceanographers equipped with new marine tools and ample funding from the U.S. Office of Naval Research embarked on an unprecedented period of oceanographic exploration. Over the next two decades a much better picture of large expanses of the seafloor slowly and painstakingly began to emerge. From this work came the discovery of a global **oceanic ridge system** that winds through all of the major oceans in a manner similar to the seams on a baseball.

In other parts of the ocean, new discoveries were also being made. Earthquake studies conducted in the western Pacific demonstrated that tectonic activity was occurring at great depths beneath deep-ocean trenches. Of equal importance was the fact that dredging of the seafloor did not bring up any oceanic crust that was older than 180 million years. Further, sediment accumulations in the deep-ocean basins were found to be thin, not the thousands of meters that were predicted.

By 1968, these developments, among others, led to the unfolding of a far more encompassing theory than continental drift, known as **plate tectonics.** The implications of plate tectonics are so far-reaching that this theory is today the framework within which to view most geologic processes.

Earth's Major Plates

According to the plate tectonics model, the uppermost mantle, along with the overlying crust, behave as a strong, rigid layer, known as the **lithosphere,** which is broken into pieces called **plates** (Figure 15.8). Lithospheric plates are thinnest in the oceans where their thickness may vary from as little as a few kilometers at the oceanic ridges to 100 kilometers in the deep-ocean basins. By contrast, continental lithosphere is generally 100–150 kilometers thick but may be more than 250 kilometers thick below older portions of landmasses. The lithosphere overlies a weaker region in the mantle known as the **asthenosphere** (weak sphere). The temperature/pressure regime in the upper asthenosphere is such that the rocks there are near their melting temperatures. This results in a very weak zone that permits the lithosphere to be effectively detached from the layers below. Thus, the weak rock within the upper asthenosphere allows Earth's rigid outer shell to move.

As shown in Figure 15.9, seven major lithospheric plates are recognized. They are the North American, South American, Pacific, African, Eurasian, Australian-Indian, and Antarctic plates. The largest is the Pacific plate, which encompasses a significant portion of the Pacific Ocean basin. Notice from Figure 15.9 that most of the large plates include an entire continent plus a large area of ocean floor (for example, the South American plate). This is a major departure from Wegener's continental drift hypothesis, which proposed that the continents moved through the ocean floor, not with it. Note also that none of the plates are defined entirely by the margins of a continent.

Intermediate-sized plates include the Caribbean, Nazca, Philippine, Arabian, Cocos, Scotia, and Juan de Fuca plates. In addition, there are over a dozen smaller

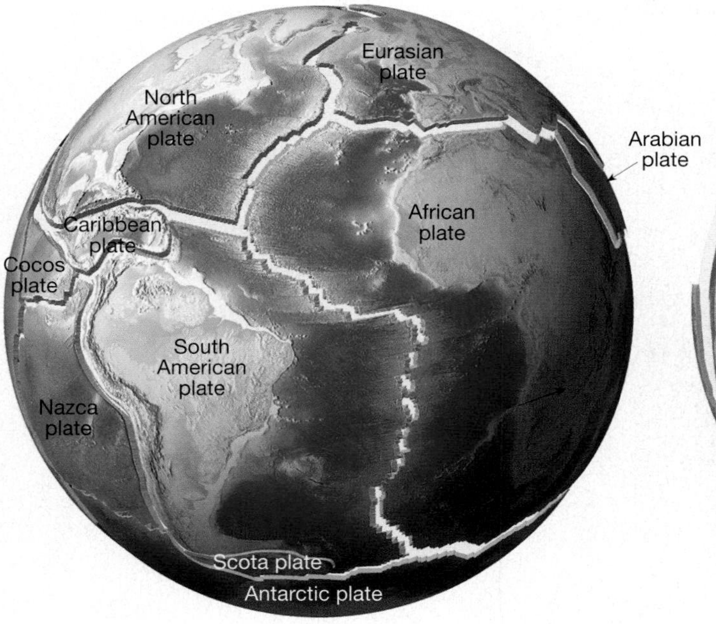

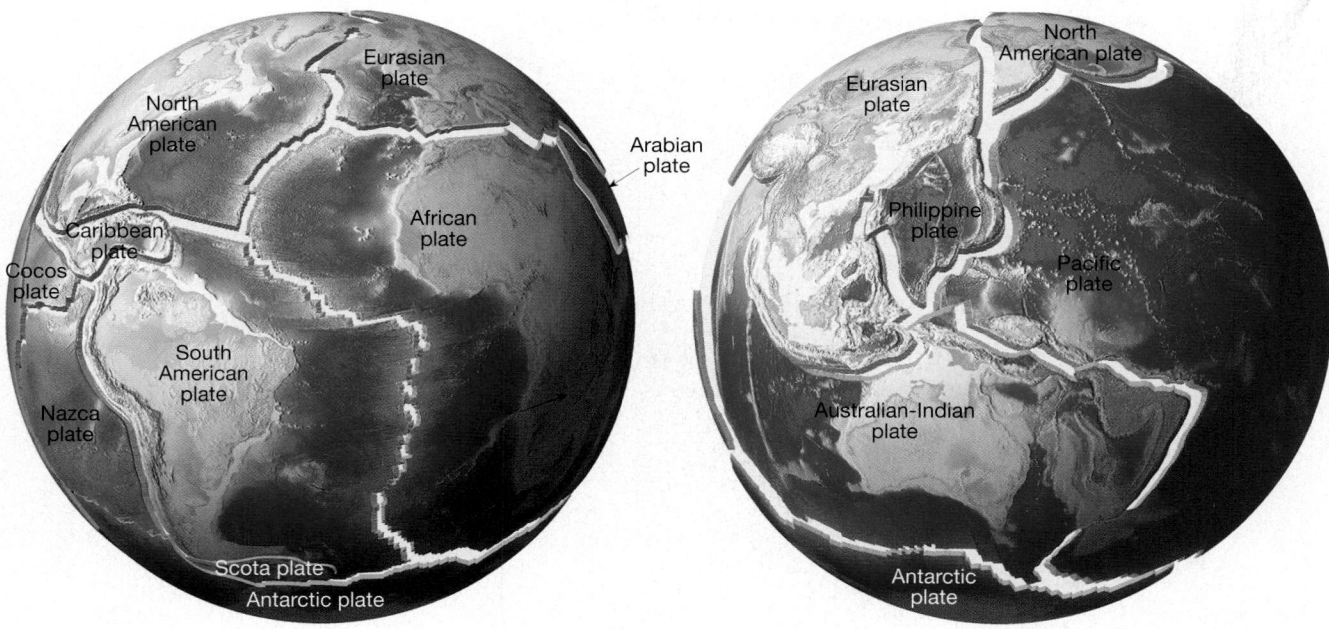

▲ **Figure 15.8** Illustration of some of Earth's lithospheric plates.

plates that have been identified but are not shown in Figure 15.9.

One of the main tenets of the plate tectonic theory is that plates move as coherent units relative to all other plates. As plates move, the distance between two locations on the same plate—New York and Denver, for example—remains relatively constant, whereas the distance between sites on different plates, such as New York and London, gradually changes. (Recently it has been shown that plates can suffer *some* internal deformation, particularly oceanic lithosphere.)

Lithospheric plates move relative to each other at a very slow but continuous rate that averages about 5 centimeters (2 inches) per year. This movement is ultimately driven by the unequal distribution of heat within Earth. Hot material found deep in the mantle moves slowly upward and serves as one part of our planet's internal convection system. Concurrently, cooler, denser slabs of oceanic lithosphere descend into the mantle, setting Earth's rigid outer shell into motion. Ultimately, the titanic, grinding movements of Earth's lithospheric plates generate earthquakes, create volcanoes, and deform large masses of rock into mountains.

Plate Boundaries

Lithospheric plates move as coherent units relative to all other plates. Although the interiors of plates may experience some deformation, all major interactions among individual plates (and therefore most deformation) occur along their *boundaries.* In fact, plate boundaries were first established by plotting the locations of earthquakes. Moreover, plates are bounded by three distinct types of boundaries, which are differentiated by the type of movement they exhibit. These boundaries are depicted at the bottom of Figure 15.9 and are briefly described here:

1. **Divergent boundaries** (*constructive margins*)—where two plates move apart, resulting in upwelling of material from the mantle to create new seafloor (Figure 15.9A).
2. **Convergent boundaries** (*destructive margins*)—where two plates move together, resulting in oceanic lithosphere descending beneath an overriding plate, eventually to be reabsorbed into the mantle, or possibly in the collision of two continental blocks to create a mountain system (Figure 15.9B).
3. **Transform fault boundaries** (*conservative margins*)—where two plates grind past each other without the production or destruction of lithosphere (Figure 15.9C).

The deep-diving submersible Alvin *is used to study the ocean floor.* (Rod Catanach)

Each plate is bounded by a combination of these three types of plate margins. For example, the Juan de Fuca plate has a divergent zone on the west, a convergent boundary on the east, and numerous transform faults, which offset segments of the oceanic ridge (Figure 15.9).

In the following sections we will briefly summarize the nature of the three types of plate boundaries.

Divergent Boundaries

GEODe ESSENTIALS OF GEOLOGY
Plate Tectonics
▼ Divergent Boundaries

Most **divergent boundaries** are located along the crests of oceanic ridges and can be thought of as *constructive plate margins* since this is where new oceanic lithosphere

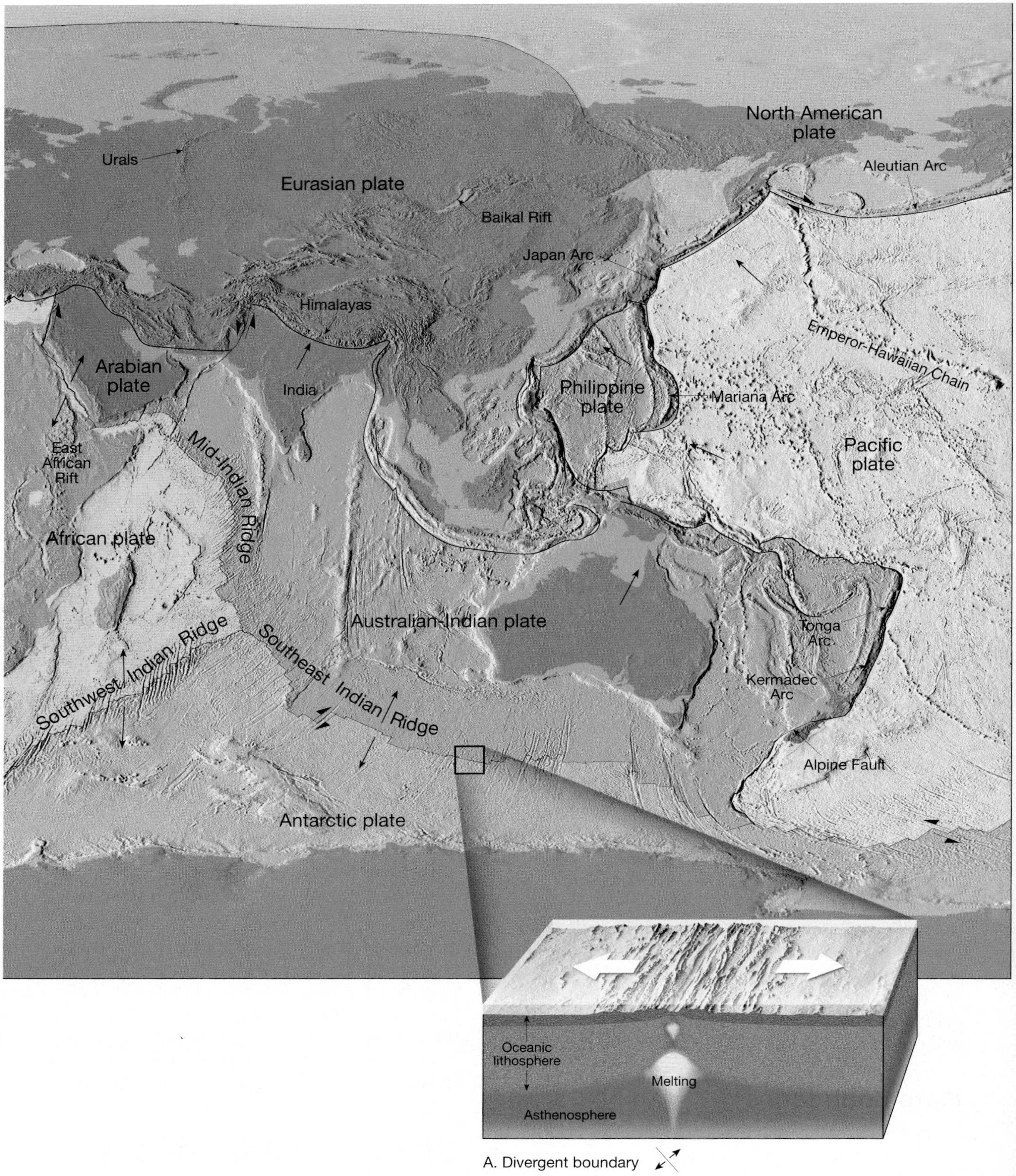

▲ **Figure 15.9** A mosaic of rigid plates constitutes Earth's outer shell. *(After W. B. Hamilton, U.S. Geological Survey)*

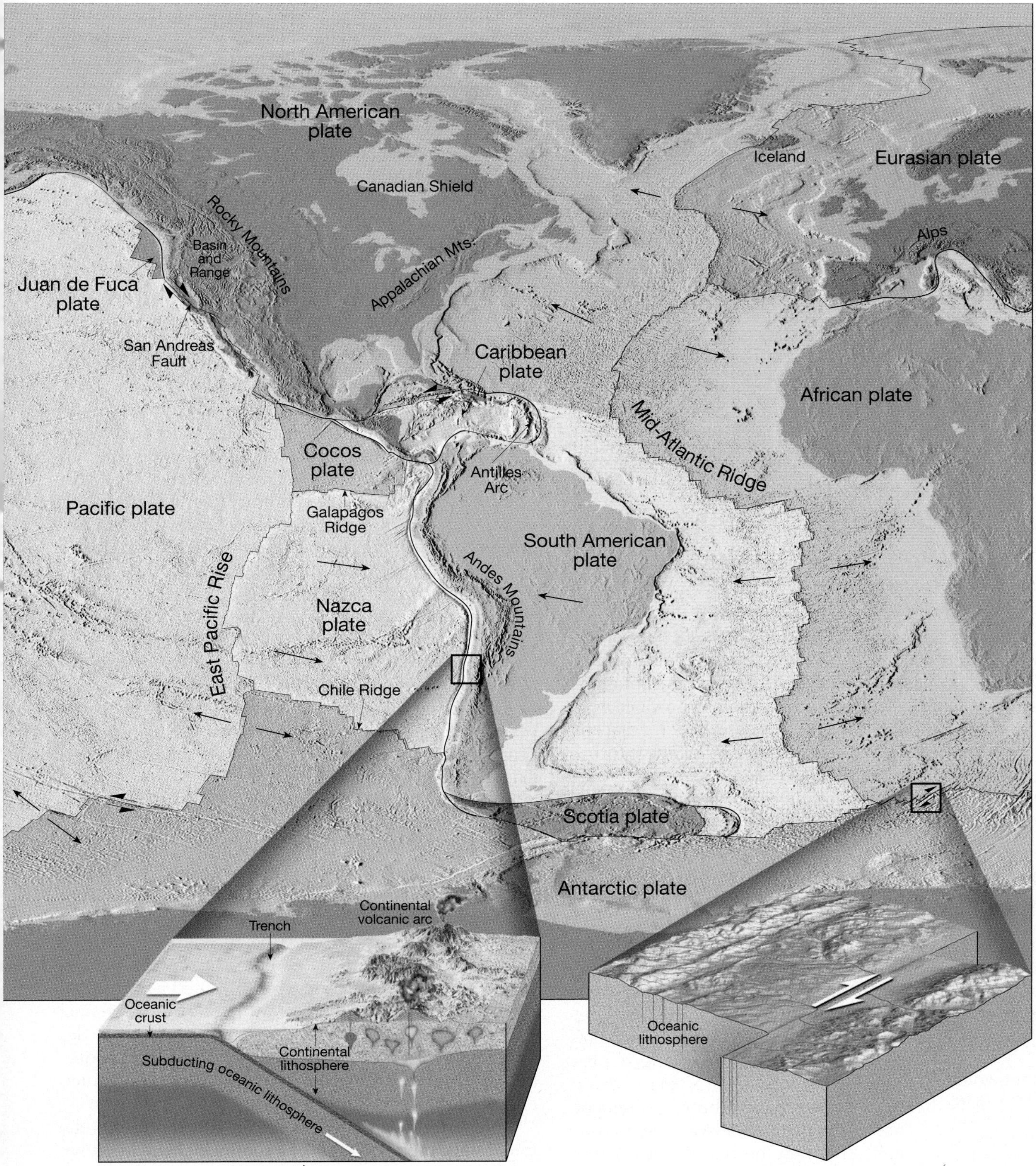

B. Convergent boundary

C. Transform fault boundary

▲ **Figure 15.9**

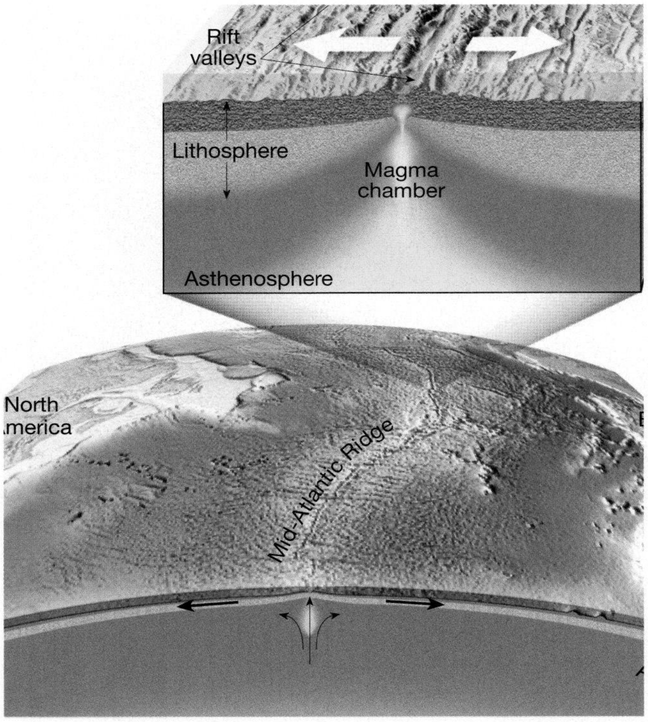

▲ **Figure 15.10** Most divergent plate boundaries are situated along the crests of oceanic ridges.

is generated (Figure 15.10). Here, as the plates move away from the ridge axis, the fractures that form are filled with molten rock that wells up from the hot mantle below. Gradually, this magma cools to produce new slivers of seafloor. In a continuous manner, adjacent plates spread apart and new oceanic lithosphere forms between them. As we shall see later, divergent boundaries are not confined to the ocean floor but can also form on the continents.

Oceanic Ridges and Seafloor Spreading

Along well-developed divergent plate boundaries, the seafloor is elevated, forming the *oceanic ridge*. The interconnected oceanic ridge system is the longest topographic feature on Earth's surface, exceeding 70,000 kilometers (43,000 miles) in length. Representing 20 percent of Earth's surface, the oceanic ridge system winds through all major ocean basins like the seam on a baseball. Although the crest of the oceanic ridge is commonly 2 to 3 kilometers higher than the adjacent ocean basins, the term "ridge" may be misleading as this feature is not narrow but has widths from 1000 to 4000 kilometers. Further, along the axis of some ridge segments is a deep downfaulted structure called a **rift valley.**

Did You Know?

An observer on another planet would notice, after only a few million years, that all the continents and ocean basins on Earth are indeed moving. The Moon, on the other hand, is tectonically dead, so it would look virtually unchanged millions of years into the future.

The mechanism that operates along the oceanic ridge system to create new seafloor is appropriately called **seafloor spreading.** Typical rates of spreading average around 5 centimeters (2 inches) per year. This is roughly the same rate at which human fingernails grow. Comparatively slow spreading rates of 2 centimeters per year are found along the Mid-Atlantic Ridge, whereas spreading rates exceeding 15 centimeters (6 inches) have been measured along sections of the East Pacific Rise. Although these rates of lithospheric production are slow on a human time scale, they are nevertheless rapid enough to have generated all of Earth's ocean basins within the last 200 million years. In fact, none of the ocean floor that has been dated exceeds 180 million years in age.

The primary reason for the elevated position of the oceanic ridge is that newly created oceanic crust is hot, and occupies more volume, which makes it less dense than cooler rocks. As new lithosphere is formed along the oceanic ridge, it is slowly yet continually displaced away from the zone of upwelling along the ridge axis. Thus, it begins to cool and contract, thereby increasing in density. This thermal contraction accounts for the greater ocean depths that exist away from the ridge crest.

It takes about 80 million years before the cooling and contracting cease completely. By this time, rock that was once part of the elevated oceanic ridge system is located in the deep-ocean basin, where it is buried by substantial accumulations of sediment. In addition, cooling causes the mantle rocks below the oceanic crust to strengthen, thereby adding to the plate's thickness. Stated another way, the thickness of oceanic lithosphere is age-dependent. The older (cooler) it is, the greater its thickness.

Continental Rifting

Divergent plate boundaries can also develop within a continent, in which case the landmass may split into two or more smaller segments, as Alfred Wegener had proposed for the breakup of Pangaea (Figure 15.11). The splitting of a continent is thought to begin with the formation of an elongated depression called a *continental rift*. A modern example of a continental rift is the East African Rift (Figure 15.12). Whether this rift will develop into a full-fledged spreading center and eventually split the continent of Africa is a matter of much speculation.

Nevertheless, the East African Rift represents the initial stage in the breakup of a continent. Here tensional forces have stretched and thinned the continental crust. As a result, molten rock ascends from the asthensophere and initiates volcanic activity at the surface (Figure 15.11A). Large volcanic mountains such as Mount Kilimanjaro and Mount Kenya exemplify the extensive volcanic activity that accompanies continental rifting. Research suggests that if tensional forces are main-

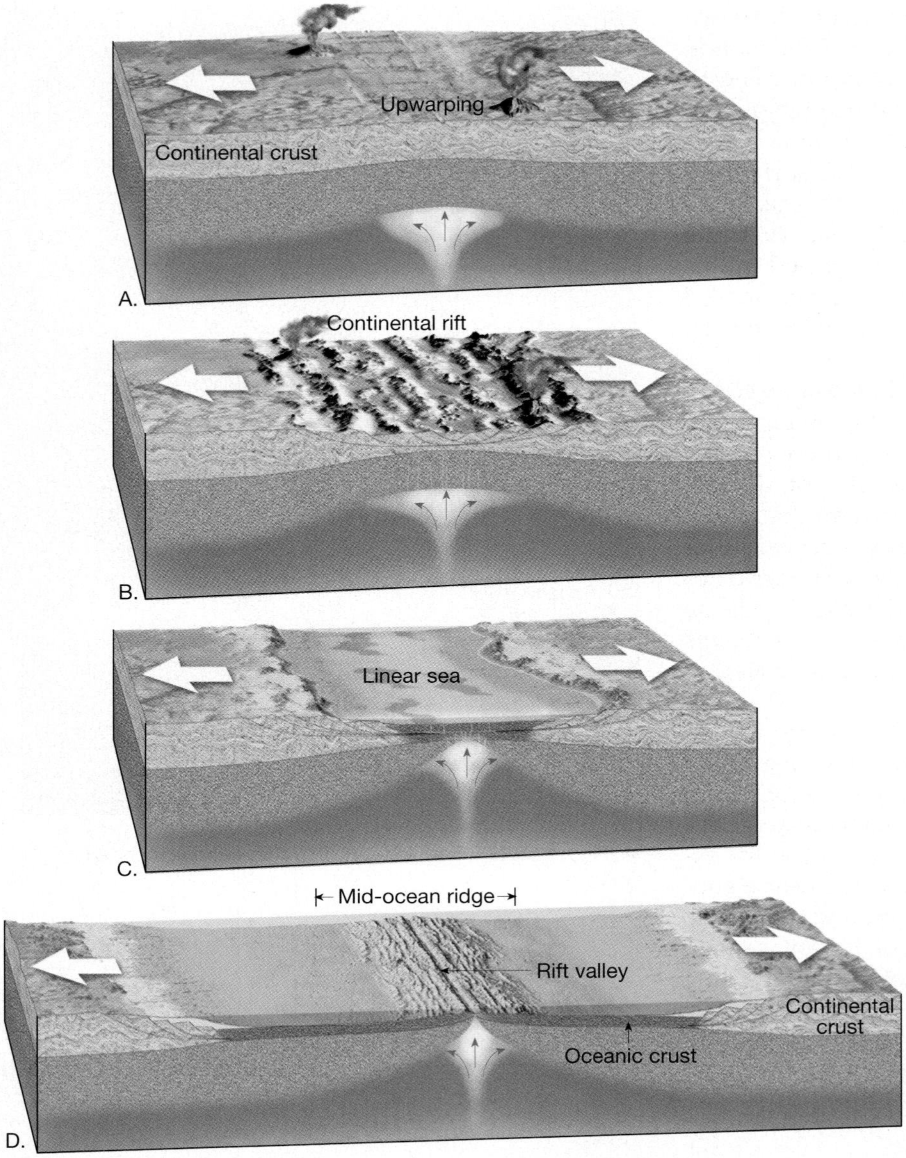

◀ **Figure 15.11** Continental rifting and the formation of a new ocean basin. **A.** Continental rifting is thought to occur where tensional forces stretch and thin the crust. As a result, molten rock ascends from the asthenosphere and initiates volcanic activity at the surface. **B.** As the crust is pulled apart, large slabs of rock sink, generating a rift valley. **C.** Further spreading generates a narrow sea. **D.** Eventually, an expansive ocean basin and ridge system are created.

tained, the rift valley will lengthen and deepen, eventually extending out to the margin of the continent, splitting it in two (Figure 15.11C). At this point, the rift becomes a narrow sea with an outlet to the ocean, similar to the Red Sea. The Red Sea formed when the Arabian Peninsula rifted from Africa, an event that began about 20 million years ago. If spreading continues, the Red Sea will grow wider and develop an elevated oceanic ridge similar to the Mid-Atlantic Ridge (Figure 15.11D).

Not all rift valleys develop into full-fledged spreading centers. Running through the central United States is an aborted rift zone that extends from Lake Superior into central Kansas. This once active rift valley is filled with rock that was extruded onto the crust more than a billion years ago. Why one rift valley develops into a full-fledged oceanic spreading center while others are abandoned is not yet known.

Convergent Boundaries

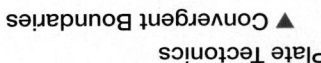

GEODe: Essentials of Geology — Plate Tectonics ▼ Convergent Boundaries

Although new lithosphere is constantly being produced at the oceanic ridges, our planet is not growing larger—its total surface area remains constant. To balance the addition of newly created lithosphere, older portions of oceanic lithosphere descend into the mantle along **convergent boundaries.** Because lithosphere is "destroyed" at convergent boundaries, they are also called *destructive plate margins.*

Did You Know?

The remains of some of the earliest humans, *Homo habilis* and *Homo erectus,* were discovered by Louis and Mary Leakey in the East African Rift. Scientists consider this region to be the "birthplace" of the human race.

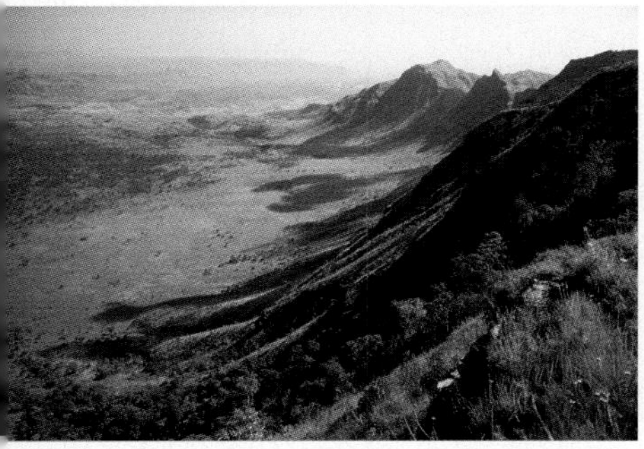

Steep cliffs along the East African Rift in Kenya. (David Keith Jones)

Convergent plate margins occur where two plates move toward each other and the motion is accommodated by one plate sliding beneath the other. As two plates slowly converge, the leading edge of one is bent downward, allowing it to slide beneath the other. The surface expression produced by the descending plate is a **deep-ocean trench,** such as the Peru–Chile trench (Figure 15.13). Trenches formed in this manner may be thousands of kilometers long, 8 to 12 kilometers deep, and between 50 and 100 kilometers wide.

Convergent boundaries are also called **subduction zones** because they are sites where lithosphere is descending (being subducted) into the asthenosphere. Subduction occurs because the density of the descending lithospheric plate is greater than that of the underlying asthenosphere. In general, oceanic lithosphere is more dense than the underlying asthenosphere, whereas continental lithosphere is less dense and resists subduction. As a consequence, it is always the lithosphere that is capped with oceanic crust that is subducted.

Slabs of oceanic lithosphere descend into the asthenosphere at angles that vary from a few degrees to nearly vertical (90 degrees), but average about 45 degrees. The angle at which oceanic lithosphere descends into the asthenosphere depends on its density. For example, when a spreading center is located near a subduction zone, the lithosphere is young and, therefore, warm and buoyant. Hence, the angle of descent is small. This is the situation along parts of the Peru–Chile

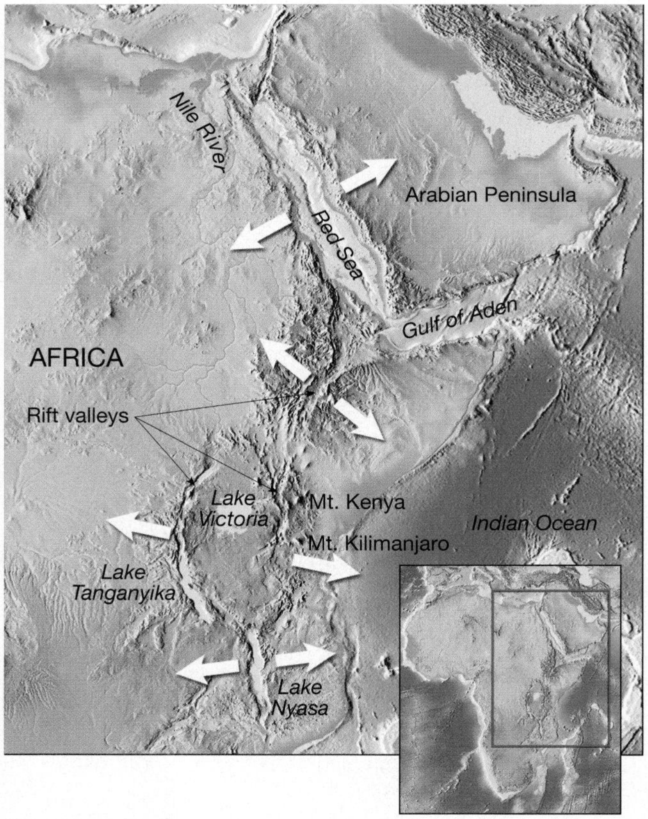

▲ **Figure 15.12** East African rift valleys and associated features.

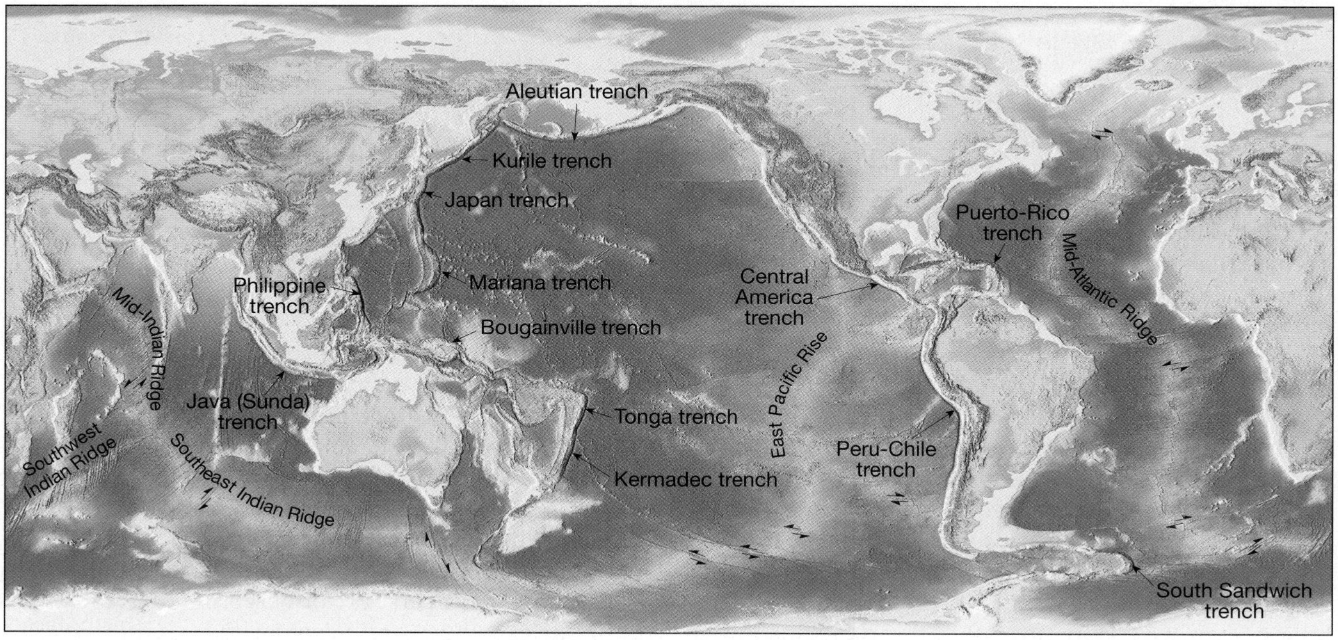

▲ **Figure 15.13** Distribution of the world's oceanic trenches, ridge system, and transform faults. Where transform faults offset ridge segments, they permit the ridge to change direction (curve) as can be seen in the Atlantic Ocean.

trench. Low dip angles usually result in considerable interaction between the descending slab and the overriding plate. Consequently, these regions experience great earthquakes.

As oceanic lithosphere ages (gets farther from the spreading center), it gradually cools, which causes it to thicken and increase in density. Once oceanic lithosphere is about 15 million years old, it becomes more dense than the supporting asthenosphere and will sink when given the opportunity. In parts of the western Pacific, some oceanic lithosphere is over 180 million years old. This is the thickest and most dense in today's oceans. The subducting slabs in this region typically descend at angles approaching 90 degrees. Examples can be found in the subduction zones associated with the Tonga, Mariana, and Kurile trenches (Figure 15.13).

Although all convergent zones have the same basic characteristics, they are highly variable features. Each is controlled by the type of crustal material involved and the tectonic setting. Convergent boundaries can form between two oceanic plates, one oceanic and one continental plate, or two continental plates. All three situations are illustrated in Figure 15.14.

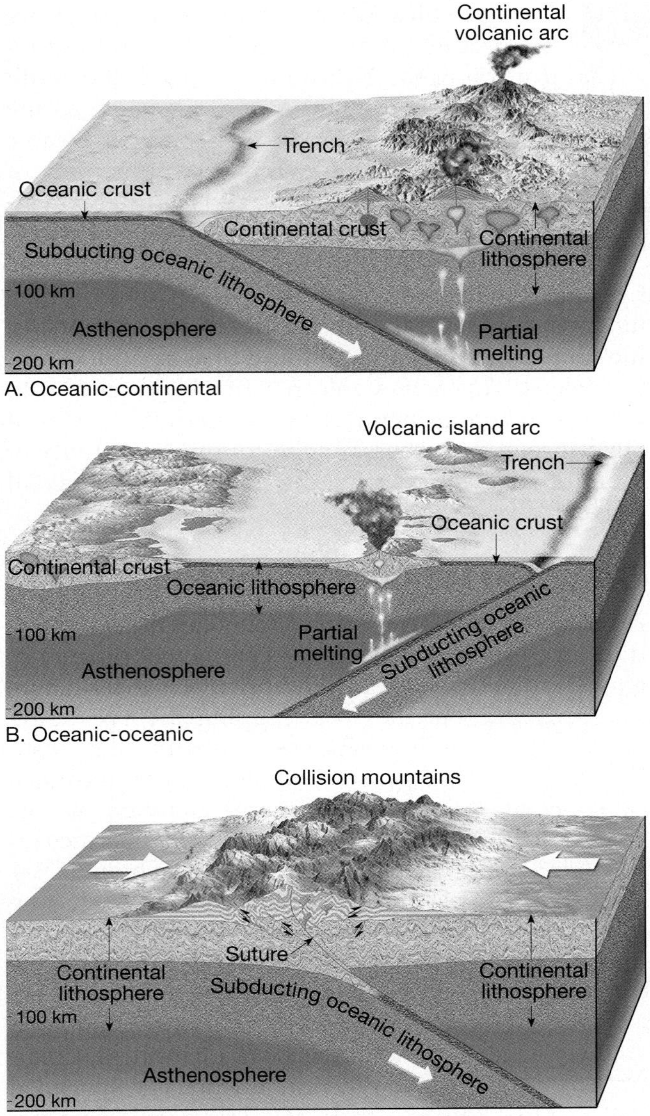

▲ **Figure 15.14** Three types of convergent plate boundaries. **A.** Oceanic–continental. **B.** Oceanic–oceanic. **C.** Continental–continental.

Oceanic–Continental Convergence

Whenever the leading edge of a plate capped with continental crust converges with a slab of oceanic lithosphere, the buoyant continental block remains "floating," while the denser oceanic slab sinks into the mantle (Figure 15.14A). When a descending oceanic slab reaches a depth of about 100 kilometers, melting is triggered within the wedge of hot asthenosphere that lies above it. But how does the subduction of a cool slab of oceanic lithosphere cause mantle rock to melt? The answer lies in the fact that volatiles (mainly water) act like salt does to melt ice. That is, "wet" rock, in a high-pressure environment, melts at substantially lower temperatures than "dry" rock of the same composition.

Sediments and oceanic crust contain a large amount of water which is carried to great depths by a subducting plate. As the plate plunges downward, water is "squeezed" from the pore spaces as confining pressure increases. At even greater depths, heat and pressure drive water from hydrated (water-rich) minerals such as the *amphiboles.* At a depth of roughly 100 kilometers and several kilometers from the upper boundary of the cool subducting oceanic slab, the mantle is sufficiently hot that the introduction of water leads to some melting. This process, called **partial melting,** generates as little as 10 percent molten material, which is intermixed with unmelted mantle rock. Being less dense than the surrounding mantle, this hot mobile mixture (magma) gradually rises toward the surface as a teardrop-shaped structure. Depending on the environment, these mantle-derived magmas may ascend through the crust and give rise to a volcanic eruption. However, much of this molten rock never reaches the surface; rather, it solidifies at depth where it acts to thicken the crust.

Partial melting of mantle rock generates molten rock that has a *basaltic composition* similar to what erupts on the island of Hawaii. In a continental setting, however, basaltic magma typically melts and assimilates some of the crustal rocks through which it ascends. The result is the formation of a silica-rich (SiO_2) magma having an *intermediate (andesitic) composition.* On occasions when andesitic magmas reach the surface, they often erupt explosively, generating large columns of volcanic ash and gases. A classic example of such an eruption is the 1980 eruption of Mount St. Helens.

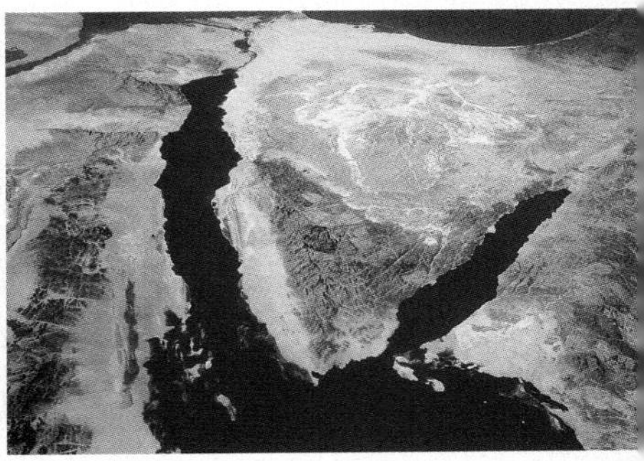

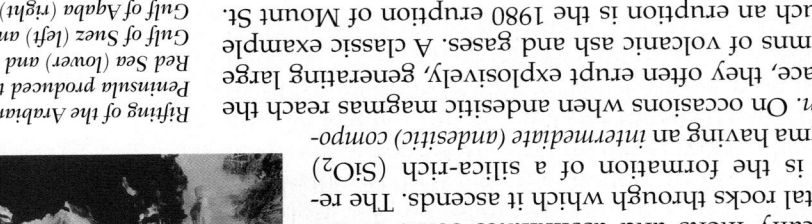

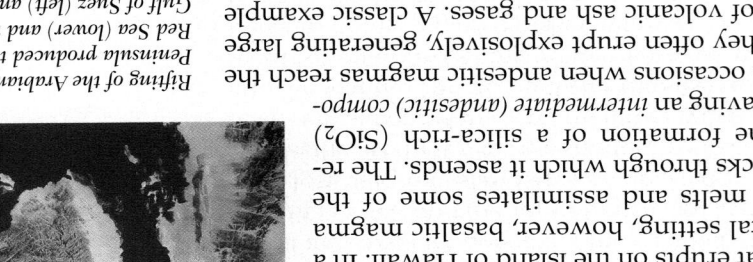

Rifting of the Arabian Peninsula produced the Red Sea (lower) and the Gulf of Suez (left) and Gulf of Aqaba (right). (NASA)

Ash from the eruption of Mount St. Helens in May 1980. (David Frasier)

The volcanoes of the towering Andes are the product of magma generated by the subduction of the Nazca plate beneath the South American continent. Mountains such as the Andes, which are produced in part by volcanic activity associated with the subduction of oceanic lithosphere, are called **continental volcanic arcs.** Another active continental volcanic arc is located in the western United States. The Cascade Range of Washington, Oregon, and California consists of several well-known volcanic mountains, including Mount Rainier, Mount Shasta, and Mount St. Helens. This active volcanic arc also extends into Canada, where it includes Mount Garibaldi, Mount Silverthrone and others. As the continuing activity of Mount St. Helens testifies, the Cascade Range is still active.

Oceanic–Oceanic Convergence

An oceanic–oceanic convergent boundary has many features in common with oceanic–continental plate margins. The differences are mainly attributable to the nature of the crust capping the overriding plate. Where two oceanic slabs converge, one descends beneath the other, initiating volcanic activity by the same mechanism that operates at oceanic–continental plate boundaries. Water "squeezed" from the subducting slab of oceanic lithosphere triggers melting in the hot wedge of mantle rock that lies above. In this setting, volcanoes grow up from the ocean floor, rather than upon a continental platform. When subduction is sustained, it will eventually build a chain of volcanic structures that emerge as islands. The volcanic islands are spaced about 80 kilometers apart and are built upon submerged ridges of volcanic material a few hundred kilometers wide. This newly formed land consisting of an arc-shaped chain of small volcanic islands is called a **volcanic island arc,** or simply an **island arc** (Figure 15.14B).

Did You Know?

Many of the tropical islands in the Caribbean, where Americans dream of taking winter holidays, are of volcanic origin. Located within this volcanic island arc is the island of Martinique, where Mount Pelée erupted in 1902, killing about 28,000 people, and the island of Montserrat, where recent volcanism required the entire island to be evacuated.

The Aleutian, Mariana, and Tonga islands are examples of volcanic island arcs. Island arcs such as these are generally located 100 to 300 kilometers (60 to 200 miles) from a deep-ocean trench. Located adjacent to the island arcs just mentioned are the Aleutian trench, the Mariana trench, and the Tonga trench (see Figure 15.13).

Most volcanic island arcs are located in the western Pacific. At these sites the subducting Pacific crust is relatively old and dense and therefore will readily sink into the mantle. This accounts for the steep angle of descent (often approaching 90 degrees) common in the trenches of this region. Further, many of these subduction zones lack the large earthquakes that are associated with some other convergent zones, such as the Peru–Chile trench.

Relatively young island arcs are fairly simple structures that are underlain by deformed oceanic crust that is generally less than 20 kilometers (12 miles) thick. Examples include the arcs of Tonga, the Aleutians, and the Lesser Antilles. By contrast, older island arcs are more complex and are underlain by crust that ranges in thickness from 20 to 35 kilometers. Examples include the Japanese and Indonesian arcs, which are built upon material generated by earlier episodes of subduction or sometimes on a small piece of continental crust.

Continental–Continental Convergence

As you saw earlier, when an oceanic plate is subducted beneath continental lithosphere, an Andean-type volcanic arc develops along the margin of the continent. However, if the subducting plate also contains continental lithosphere, continued subduction eventually brings the two continental blocks together (Figure 15.15A). Whereas oceanic lithosphere is relatively dense and sinks into the asthenosphere, continental lithosphere is buoyant, which prevents it from being subducted to any great depth. The result is a collision between the two continental fragments (Figure 15.15C).

Such a collision occurred when the subcontinent of India "rammed" into Asia and produced the Himalayas—the most spectacular mountain range on Earth (see Figure 15.1). During this collision, the continental crust buckled, fractured, and was generally shortened and thickened. In addition to the Himalayas, several other major mountain systems, including the Alps, Appalachians, and Urals, formed during continental collisions.

Prior to a continental collision, the landmasses involved are separated by an ocean basin. As the continental blocks converge, the intervening seafloor is subducted beneath one of the plates. Subduction initiates partial melting in the overlying mantle, which in turn results in the growth of a volcanic arc. Depending on the location of the subduction zone, the volcanic arc could develop on either of the converging landmasses, or if the subduction zone developed several hundred kilometers seaward from the coast, a volcanic island arc would form. Eventually, as the intervening seafloor is consumed, these continental masses collide. This folds and deforms the accumulation of sediments and sedimentary rocks along the continental margin as if they had been placed in a gigantic vise. The result is the formation of a new mountain range composed of deformed and metamorphosed sedimentary rocks, fragments of the volcanic arc, and often slivers of oceanic crust.

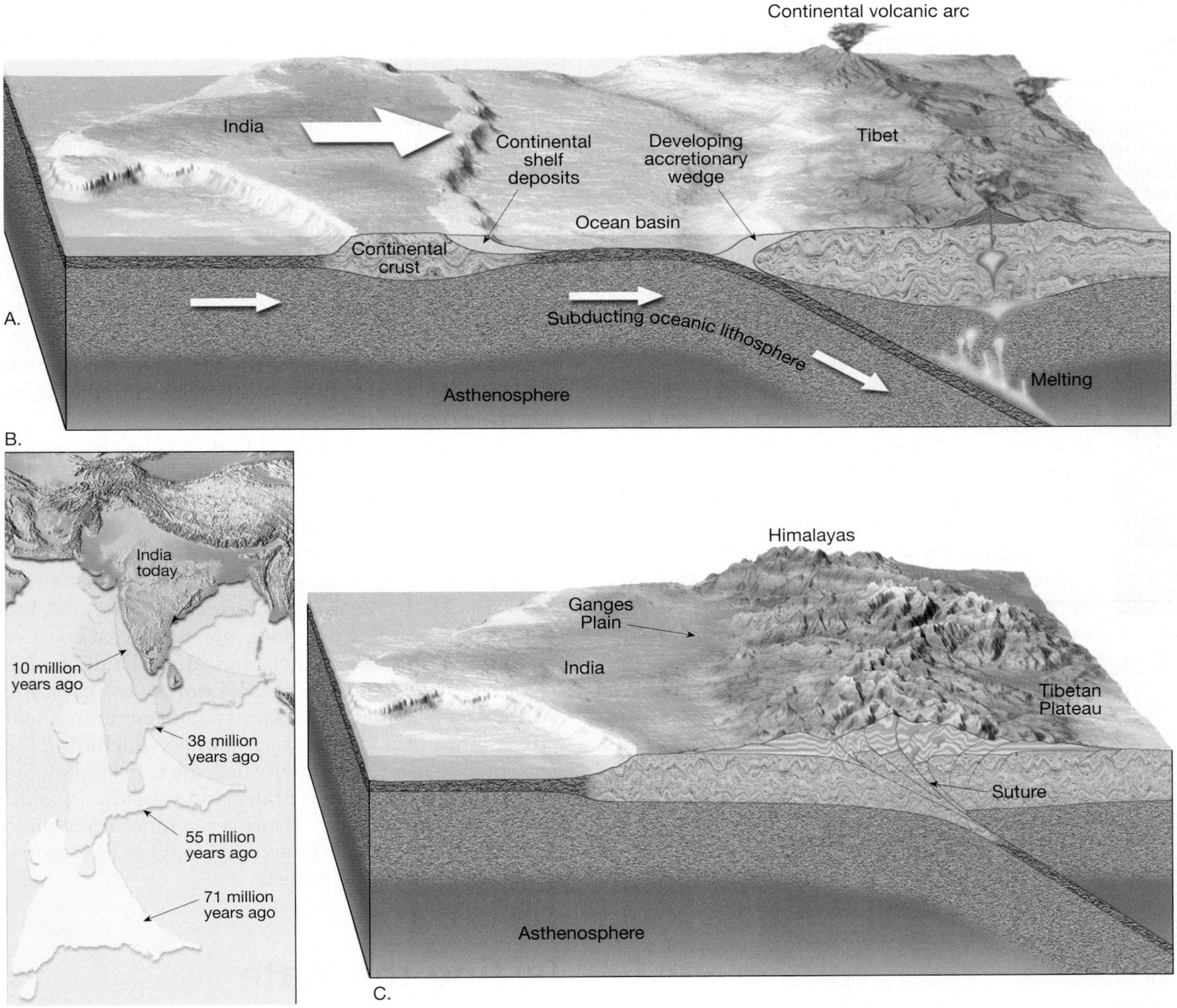

▲ **Figure 15.15** The ongoing collision of India and Asia, starting about 45 million years ago, produced the majestic Himalayas. **A.** Converging plates generated a subduction zone, while partial melting triggered by the subducting oceanic slab produced a continental volcanic arc. Sediments scraped from the subducting plate were added to the accretionary wedge. **B.** Position of India in relation to Eurasia at various times. *(Modified after Peter Molnar)* **C.** Eventually the two landmasses collided, deforming and elevating the sediments that had been deposited along their continental margins.

Transform Fault Boundaries

GEODe ESSENTIALS OF GEOLOGY

Plate Tectonics

▼ Transform Fault Boundaries

The third type of plate boundary is the **transform fault** where plates slide horizontally past one another without the production or destruction of lithosphere (*conservative plate margins*). Transform faults were first identified where they join offset segments of an oceanic ridge (Figure 15.16). At first it was erroneously assumed that the ridge system originally formed a long and continuous chain that was later offset by horizontal displacement along these large faults. However, the displacement along these faults was found to be in the exact opposite direction required to produce the offset ridge segments.

The true nature of transform faults was discovered in 1965 by J. Tuzo Wilson of the University of Toronto. Wilson suggested that these large faults connect the global active belts (convergent boundaries, divergent

Japan's Mount Fuji is a volcano that formed as a result of subduction along the Japanese Trench. (National Geographic)

Did You Know?

The Great Alpine Fault of New Zealand is a transform fault that runs down South Island, marking the line where two plates meet. The northwestern part of South Island sits on the Australian plate, while the rest of the island lies on the Pacific plate. Like its sister fault, California's San Andreas displacement has been measured in hundreds of miles.

The French Alps were created by the collision of the African and Eurasian plates. (Art Wolfe)

boundaries, and other transform faults) into a continuous network that divides Earth's outer shell into several rigid plates. Thus, Wilson became the first to suggest that Earth was made of individual plates, while at the same time identifying the faults along which relative motion between the plates is made possible.

Most transform faults join two segments of an oceanic ridge (Figure 15.16). Here, they are part of prominent linear breaks in the oceanic crust known as **fracture zones,** which include both the active transform faults as well as their inactive extentions into the plate interior. These fracture zones are present approximately every 100 kilometers along the trend of a ridge axis. As shown in Figure 15.16, active transform faults lie *only between* the two offset ridge segments. Here seafloor produced at one ridge axis moves in the opposite direction as seafloor produced at an opposing ridge segment. Thus, between the ridge segments these adjacent slabs of oceanic crust are grinding past each other along the fault. Beyond the ridge crests are the inactive zones, where the fractures are preserved as linear topographic scars. The trend of these fracture zones roughly parallels the direction of plate motion at the time of their formation. Thus, these structures can be used to map the direction of plate motion in the geologic past.

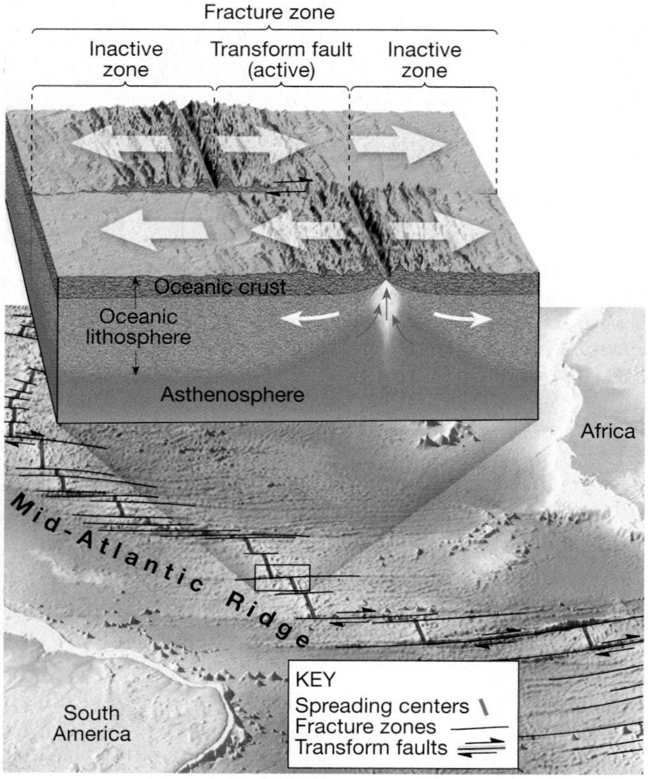

▲ **Figure 15.16** Illustration of a transform fault joining segments of the Mid-Atlantic Ridge.

In another role, transform faults provide the means by which the oceanic crust created at ridge crests can be transported to a site of destruction: the deep-ocean trenches. Figure 15.17 illustrates this situation. Notice that the Juan de Fuca plate moves in a southeasterly direction, eventually being subducted under the West Coast of the United States. The southern end of this plate is bounded by the Mendocino fault. This transform fault boundary connects the Juan de Fuca ridge to the Cascadia subduction zone (Figure 15.17). Therefore, it facilitates the movement of the crustal material created at the ridge crest to its destination beneath the North American continent.

Although most transform faults are located within the ocean basins, a few cut through continental crust. Two examples are the earthquake-prone San Andreas Fault of California and the Alpine Fault of New Zealand. Notice in Figure 15.17 that the San Andreas Fault connects a spreading center located in the Gulf of California to the Cascadia subduction zone and the Mendocino fault located along the northwest coast of the United States. Along the San Andreas Fault, the Pacific plate is moving toward the northwest, past the North American plate. If this movement continues, that part of California west of the fault zone, including the Baja Peninsula, will eventually become an island off the West Coast of the United States and Canada. It could eventually reach Alaska. However, a more immediate concern is the earthquake activity triggered by movements along this fault system.

Testing the Plate Tectonics Model

With the development of the theory of plate tectonics, researchers from all of the Earth sciences began testing this new model of how Earth works. Some of the evidence supporting continental drift and seafloor spreading has already been presented. In addition, some of the evidence that was instrumental in solidifying the support for this new idea follows. Note that much of this evidence was not new; rather, it was new interpretations of already existing data that swayed the tide of opinion.

Evidence: Ocean Drilling

Some of the most convincing evidence confirming seafloor spreading has come from drilling into the ocean floor. From 1968 until 1983, the source of these important data was the Deep Sea Drilling Project, an international program sponsored by several major oceanographic institutions and the National Science

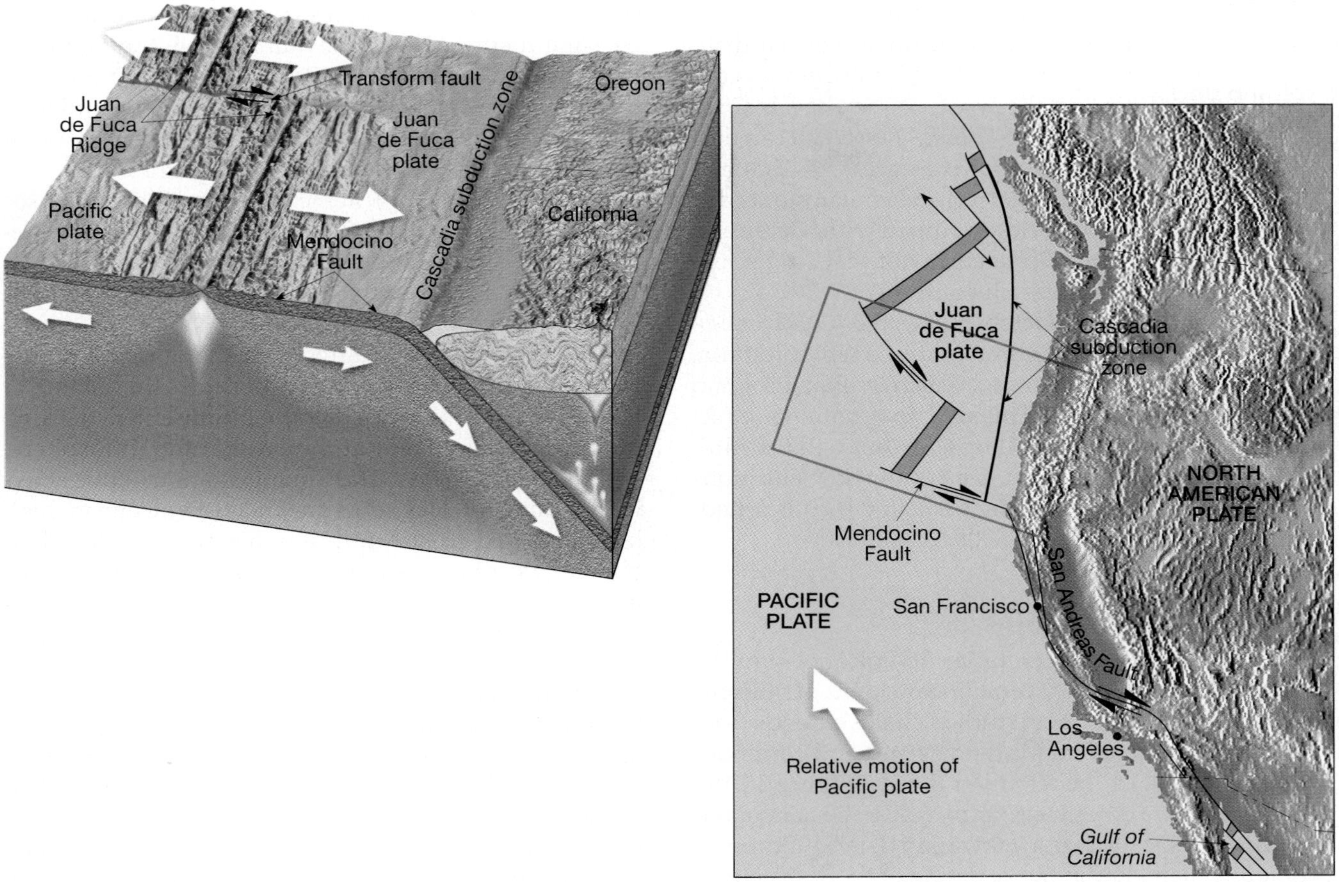

▲ **Figure 15.17** The Mendocino transform fault permits seafloor generated at the Juan de Fuca ridge to move southeastward past the Pacific plate and beneath the North American plate. Thus, this transform fault connects a divergent boundary to a subduction zone. Furthermore, the San Andreas Fault, also a transform fault, connects two spreading centers: the Juan de Fuca ridge and a divergent zone located in the Gulf of California.

Foundation. The primary goal was to gather first-hand information about the age of the ocean basins and the processes that formed them. To accomplish this, a new drilling ship, the *Glomar Challenger,* was built.

Operations began in August 1968, in the South Atlantic. At several sites holes were drilled through the entire thickness of sediments to the basaltic rock below. An important objective was to gather samples of sediment from just above the igneous crust as a means of dating the seafloor at each site.* Because sedimentation begins immediately after the oceanic crust forms, remains of microorganisms found in the oldest sediments—those resting directly on the crust—can be used to date the ocean floor at that site.

When the oldest sediment from each drill site was plotted against its distance from the ridge crest, the plot demonstrated that the age of the sediment increased with increasing distance from the ridge. This finding supported the seafloor-spreading hypothesis, which predicted the youngest oceanic crust would be found at the ridge crest and the oldest oceanic crust would be adjacent to the continental margins.

The data from the Deep Sea Drilling Project also reinforced the idea that the ocean basins are geologically youthful because no sediment with an age in excess of 180 million years was found. By comparison, continental crust that exceeds 4 billion years in age has been dated.

The thickness of ocean-floor sediments provided additional verification of seafloor spreading. Drill cores from the *Glomar Challenger* revealed that sediments are almost entirely absent on the ridge crest and the sediment thickens with increasing distance from the ridge. Because the ridge crest is younger than the areas farther away from it, this pattern of sediment distribution should be expected if the seafloor-spreading hypothesis is correct.

The Ocean Drilling Program succeeded the Deep Sea Drilling Project and, like its predecessor, was a

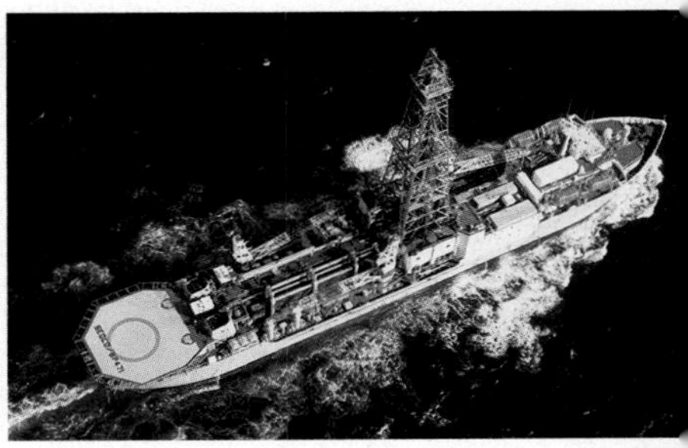

The JOIDES Resolution, *the drilling ship of the Ocean Drilling Program.* (Ocean Drilling Program)

*Radiometric dates of the ocean crust itself are unreliable because of the alteration of basalt by seawater.

Did You Know?

Olympus Mons is a huge volcano on Mars that strongly resembles the Hawaiian shield volcanoes. Rising 25 kilometers above the surrounding plains, Olympus Mons owes its massiveness to the fact that plate tectonics does not operate on Mars. Consequently, instead of being carried away from the hot spot by plate motion, as occurred with the Hawaiian volcanoes, Olympus Mons remained fixed and grew to a gigantic size.

major international program. The more technologically advanced drilling ship, the *JOIDES Resolution*, continued the work of the *Glomar Challenger.** The *JOIDES Resolution* can drill in water depths as great as 8200 meters (27,000 feet) and contains onboard laboratories equipped with a large and sophisticated array of seagoing scientific research equipment.

In October 2003, the *JOIDES Resolution* became part of a new program, the Integrated Ocean Drilling Program (IODP). This new international effort will not rely on just one drilling ship, but will use multiple vessels for exploration. One of the new additions is the massive 210-meter-long *Chikyu,* which is scheduled to begin operations in 2006.

Drilling rig on board the JOIDES Resolution.
(Ocean Drilling Program)

Evidence: Hot Spots

Mapping of seamounts (submarine volcanoes) in the Pacific Ocean revealed several linear chains of volcanic structures. One of the most studied chains extends from the Hawaiian Islands to Midway Island and continues northward toward the Aleutian trench (Figure 15.18). This nearly continuous string of volcanic islands and seamounts is called the Hawaiian Island–Emperor Seamount chain. Radiometric dating of these structures showed that the volcanoes increase in age with increasing distance from Hawaii. Hawaii, the youngest island in the chain, began rising from the seafloor less than a million years ago, whereas Midway Island is 27 million years old, and Suiko Seamount, near the Aleutian trench, is 65 million years old (Figure 15.18).

Taking a closer look at the Hawaiian Islands, we see a similar increase in age from the volcanically active island of Hawaii, at the southeastern end of the chain, to the inactive volcanoes that make up the island of Kauai in the northwest.

Researchers are in agreement that a rising plume of mantle material is located beneath the island of Hawaii. As the ascending **mantle plume** enters the low-pressure environment at the base of the lithosphere, melting occurs. The surface manifestation of this activity is a **hot spot.** Hot spots are areas of volcanism, high heat flow, and crustal uplifting that are a few hundred kilometers across. As the Pacific plate moved over this hot spot, successive volcanic structures were built. As shown in Figure 15.18, the age of each volcano indicates the time when it was situated over the relatively stationary mantle plume.

Kauai is the oldest of the large islands in the Hawaiian chain. Five million years ago, when it was positioned over the hot spot, Kauai was the only Hawaiian island. Visible evidence of the age of Kauai can be seen by examining its extinct volcanoes, which have been eroded into jagged peaks and vast canyons. By contrast, the relatively young island of Hawaii exhibits fresh lava flows, and two of Hawaii's volcanoes, Mauna Loa and Kilauea, remain active.

*JOIDES stands for Joint Oceanographic Institutions for Deep Earth Sampling.

▶ **Figure 15.18** The chain of islands and seamounts that extends from Hawaii to the Aleutian trench results from the movement of the Pacific plate over an apparently stationary hot spot. Radiometric dating of the Hawaiian Islands shows that the volcanic activity decreases in age toward the island of Hawaii.

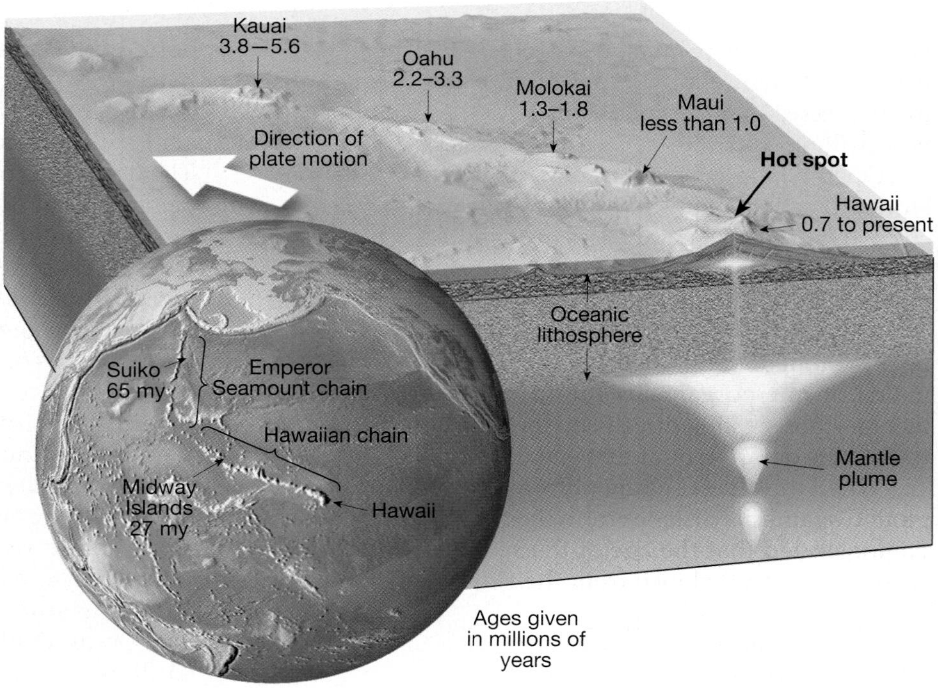

Two island groups parallel the Hawaiian Island-Emperor Seamount chain. One chain consists of the Tuamotu and Line islands, and the other includes the Austral, Gilbert, and Marshall islands. In each case, the most recent volcanic activity has occurred at the southeastern end of the chain, and the islands get progressively older to the northwest. Thus, like the Hawaiian Island–Emperor Seamount chain, these volcanic structures apparently formed by the same motion of the Pacific plate over fixed mantle plumes. Not only does this evidence support the fact that the plates do indeed move relative to Earth's interior but the hot spot "tracks" trace the direction of plate motion. For example, hot spots found on the floor of the Atlantic have increased our understanding of the migration of landmasses following the breakup of Pangaea.

Research suggests that at least some mantle plumes originate at great depth, perhaps at the mantle-core boundary. Others, however, may have a much shallower origin. Of the 40 or so hot spots that have been identified, over a dozen are located near divergent boundaries. For example, the mantle plume located beneath Iceland is responsible for the large accumulation of volcanic rocks found along the northern section of the Mid-Atlantic Ridge.

The existence of mantle plumes and their association with hot spots is well documented. Most mantle plumes are long-lived features that appear to maintain relatively fixed positions within the mantle. However, recent evidence has shown that some hot spots may slowly migrate. If this is the case, models of past plate motion that are based on a fixed hot-spot frame of reference will need to be reevaluated.

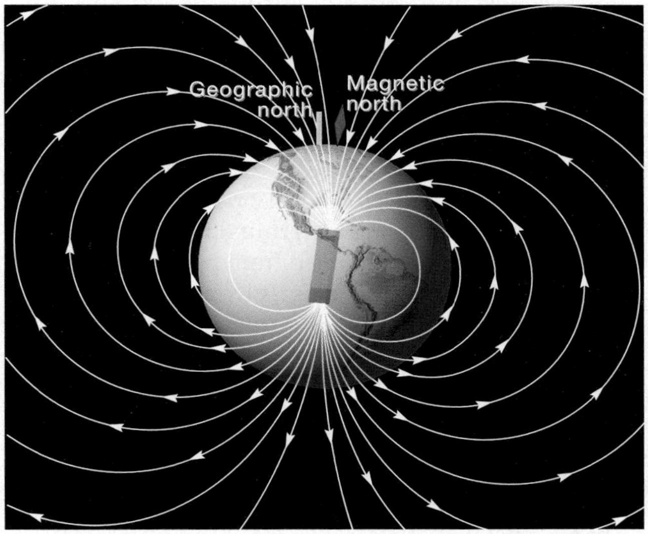

▲ **Figure 15.19** Earth's magnetic field consists of lines of force much like those a giant bar magnet would produce if placed at the center of Earth.

Hot-spot volcanism, Kilauea, Hawaii. (USGS)

Evidence: Paleomagnetism

Anyone who has used a compass to find direction knows that Earth's magnetic field has a north and a south magnetic pole. Today these magnetic poles align closely, but not exactly, with the geographic poles. (The geographic poles, or true north and south poles, are where Earth's rotational axis intersects the surface.) Earth's magnetic field is similar to that produced by a simple bar magnet. Invisible lines of force pass through the planet and extend from one magnetic pole to the other as shown in Figure 15.19. A compass needle, itself a small magnet free to rotate on an axis, becomes aligned with the magnetic lines of force and points to the magnetic poles.

Unlike the pull of gravity, we cannot feel Earth's magnetic field, yet its presence is revealed because it deflects a compass needle. In a similar manner, certain rocks contain minerals that serve as "fossil compasses." These iron-rich minerals, such as *magnetite,* are abundant in lava flows of basaltic composition.* When heated above a temperature known as the **Curie point,** these magnetic minerals lose their magnetism. However, when these iron-rich grains cool below their Curie point (about 585°C for magnetite), they gradually become magnetized in the direction of the existing magnetic lines of force. Once the minerals solidify, the magnetism they possess will usually remain "frozen" in this position. In this regard, they behave much like a compass needle; they "point" toward the position of the magnetic poles at the time of their formation. Then, if the rock is moved, the rock magnetism will retain its original alignment. Rocks that formed thousands or millions of years ago and contain a "record" of the direction of the magnetic poles at the time of their formation are said to possess **fossil magnetism,** or **paleomagnetism.**

Apparent Polar Wandering. A study of rock magnetism conducted during the 1950s in Europe led to an interesting discovery. The magnetic alignment in the iron-rich minerals in lava flows of different ages indicated that many different paleomagnetic poles once existed. A plot of the apparent positions of the magnetic north pole with respect to Europe revealed that during the past 500 million years, the location of the pole had gradually wandered from a location near Hawaii northward through eastern Siberia and finally to its present location (Figure 15.20A). This was strong evidence that either the magnetic poles had migrated through time, an idea known as *polar wandering,* or that the lava flows moved—in other words, Europe had drifted in relation to the poles.

Although the magnetic poles are known to move in an erratic path around the geographic poles, studies of paleomagnetism from numerous locations show that the positions of the magnetic poles, averaged over

*Some sediments and sedimentary rocks contain enough iron-bearing mineral grains to acquire a measurable amount of magnetization.

thousands of years, correspond closely to the positions of the geographic poles. Therefore, a more acceptable explanation for the apparent polar wandering paths was provided by Wegener's hypothesis. If the magnetic poles remain stationary, their *apparent movement* is produced by continental drift.

The latter idea was further supported by comparing the latitude of Europe as determined from fossil magnetism with evidence obtained from paleoclimatic studies. Recall that during the Pennsylvanian period (about 300 million years ago) coal-producing swamps covered much of Europe. During this same time period, paleomagnetic evidence places Europe near the equator—a fact consistent with the tropical environment indicated by these coal deposits.

Did You Know?

When all of the continents were joined to form Pangaea, the rest of Earth's surface was covered with a huge ocean called *Panthalassa* (*pan* = all, *thalassa* = sea). Today all that remains of Panthalassa is the Pacific Ocean, which has been decreasing in size since the breakup of Pangaea.

Further evidence for continental drift came a few years later when a polar-wandering path was constructed for North America (Figure 15.20A). It turned out that paths for North America and Europe had similar shapes but were separated by about 24° of longitude. At the time these rocks crystallized, could there have been two magnetic north poles that migrated parallel to each other? Investigators found no evidence to support this possibility. The differences in these migration paths, however, can be reconciled if the two presently separated continents are placed next to one another, as we now believe they were prior to the opening of the Atlantic Ocean. Notice in Figure 15.20B that these apparent wandering paths nearly coincided during the period from about 400 to 160 million years ago. This is evidence that North America and Europe were joined during this period and moved relative to the poles as part of the same continent.

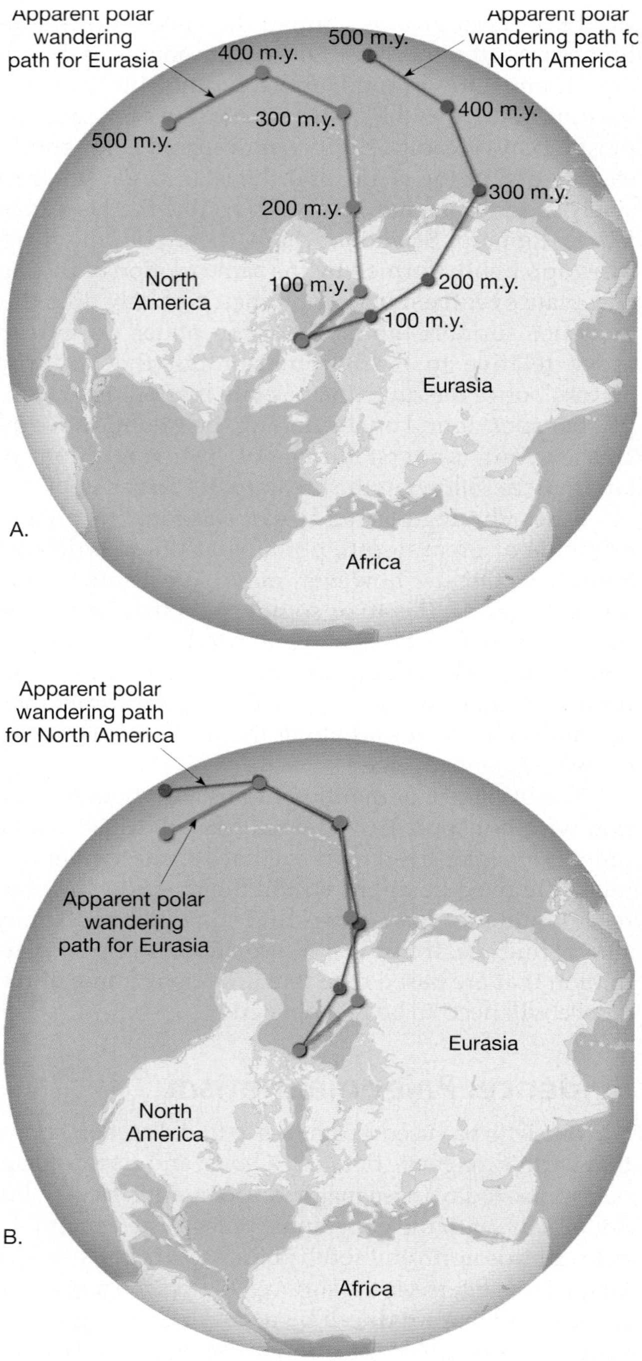

▲ **Figure 15.20** Simplified apparent polar wandering paths as established from North American and Eurasian paleomagnetic data. **A.** The more westerly path determined from North American data is thought to have been caused by the westward drift of North America by about 24 degrees from Eurasia. **B.** The positions of the wandering paths when the landmasses are reassembled in their predrift locations.

Magnetic Reversals and Seafloor Spreading. Another discovery came when geophysicists learned that Earth's magnetic field periodically reverses polarity; that is, the north magnetic pole becomes the south magnetic pole, and vice versa. A rock solidifying during one of the periods of reverse polarity will be magnetized with the polarity opposite that of rocks being formed today. When rocks exhibit the same magnetism as the present magnetic field, they are said to possess **normal polarity,** whereas rocks exhibiting the opposite magnetism are said to have **reverse polarity.**

Evidence for **magnetic reversals** was obtained when investigators measured the magnetism of lavas and sediments of various ages around the world. They found that normally and reversely magnetized rocks of a given age in one location matched the magnetism of rocks of the same age found in all other locations. This was convincing evidence that Earth's magnetic field had indeed reversed.

Once the concept of magnetic reversals was confirmed, researchers set out to establish a time scale for magnetic reversals. The task was to measure the magnetic polarity of hundreds of lava flows and use radiometric dating techniques to establish their ages.

Figure 15.21 shows the **magnetic time scale** established for the last few million years. The major divisions of the magnetic time scale are called *chrons* and last for roughly 1 million years. As more measurements became available, researchers realized that several, short-lived reversals (less than 200,000 years long) occur during any one chron.

Meanwhile, oceanographers had begun to do magnetic surveys of the ocean floor in conjunction with their efforts to construct detailed maps of seafloor topography. These magnetic surveys were accomplished by towing very sensitive instruments called **magnetometers** behind research vessels. The goal of these geophysical surveys was to map variations in the strength of Earth's magnetic field that arise from differences in the magnetic properties of the underlying crustal rocks.

The first comprehensive study of this type was carried out off the Pacific Coast of North America and had an unexpected outcome. Researchers discovered alternating stripes of high- and low-intensity magnetism as shown in Figure 15.22.

This relatively simple pattern of magnetic variation defied explanation until 1963, when Fred Vine and D. H. Matthews demonstrated that the high- and low-intensity stripes supported the concept of seafloor spreading. Vine and Matthews suggested that the stripes of high-intensity magnetism are regions where the paleomagnetism of the ocean crust exhibits normal polarity (Figure 15.22). Consequently, these rocks *enhance* (reinforce) Earth's magnetic field. Conversely, the low-intensity stripes are regions where the ocean crust is polarized in the reverse direction and therefore *weakens* the existing magnetic field. But how do parallel stripes of normally and reversely magnetized rock become distributed across the ocean floor?

Vine and Matthews reasoned that as magma solidifies along narrow rifts at the crest of an oceanic ridge, it is magnetized with the polarity of the existing magnetic field (Figure 15.23). Because of seafloor spreading, this stripe of magnetized crust would gradually increase in width. When Earth's magnetic field reverses polarity, any newly formed seafloor (having the opposite polarity) would form in the middle of the old stripe. Gradually, the two parts of the old stripe are carried in opposite directions away from the ridge crest. Subsequent reversals would build a pattern of normal and reverse stripes as shown in Figure 15.23. Because new rock is added in equal amounts to both trailing edges of the spreading ocean floor, we should expect that the pattern of stripes (size and polarity) found on one side of an oceanic ridge to be a mirror image of the other side. A few years later a survey across the Mid-Atlantic Ridge just south of Iceland revealed a pattern of magnetic

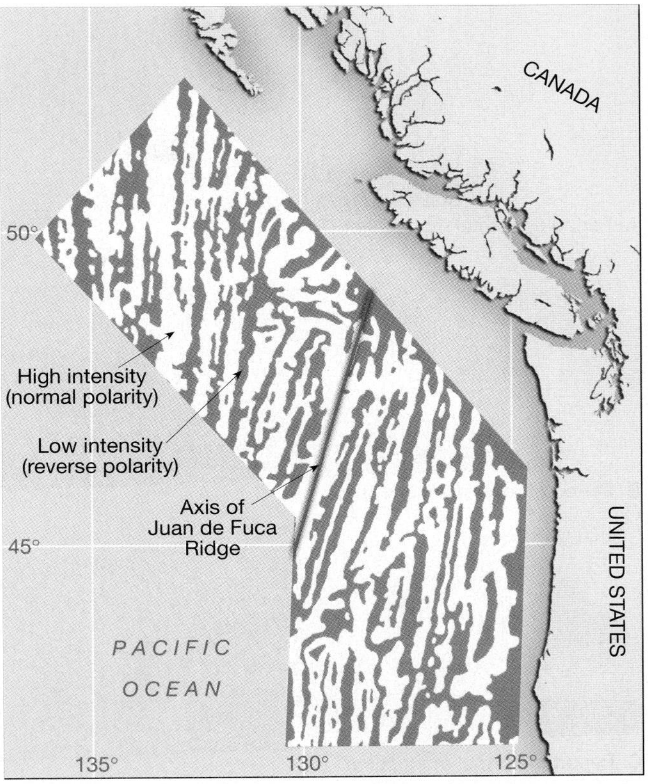

▲ **Figure 15.22** Pattern of alternating stripes of high- and low-intensity magnetism discovered off the Pacific Coast of North America.

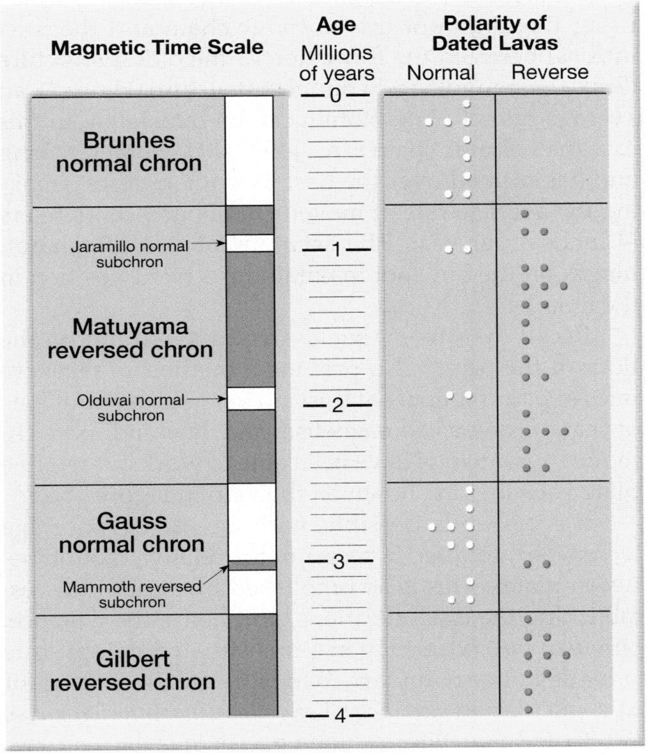

▲ **Figure 15.21** Time scale of Earth's magnetic field in the recent past. This time scale was developed by establishing the magnetic polarity for lava flows of known age. *(Data from Allen Cox and G. B. Dalrymple)*

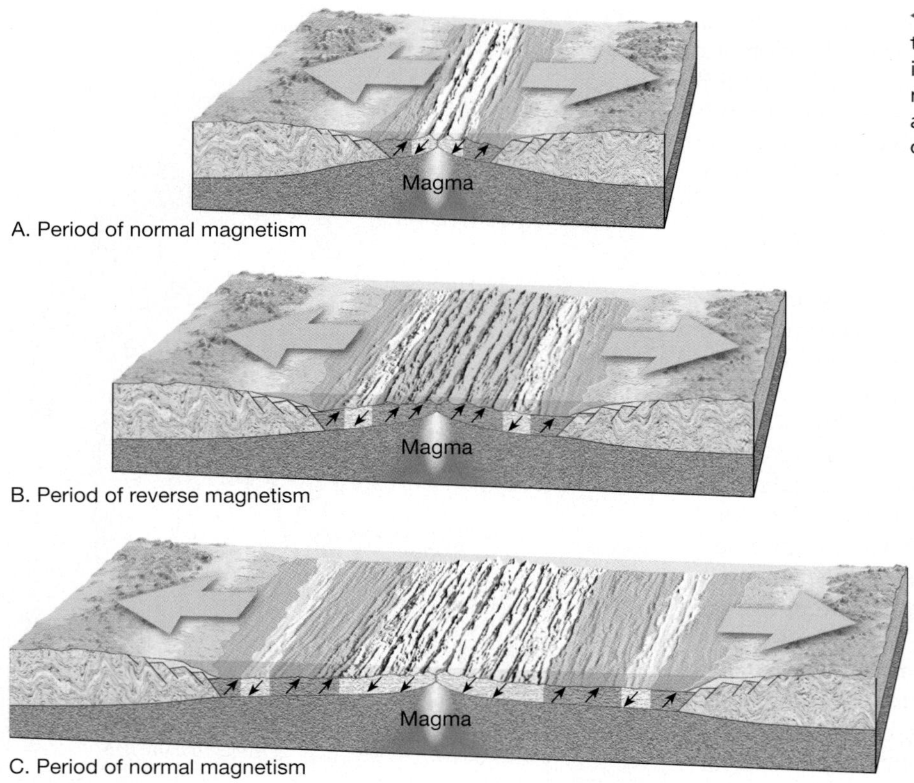

◀ **Figure 15.23** As new basalt is added to the ocean floor at the mid-ocean ridges, it is magnetized according to Earth's existing magnetic field. Hence, it behaves much like a tape recorder as it records each reversal of the planet's magnetic field.

stripes exhibiting a remarkable degree of symmetry to the ridge axis (Figure 15.24).

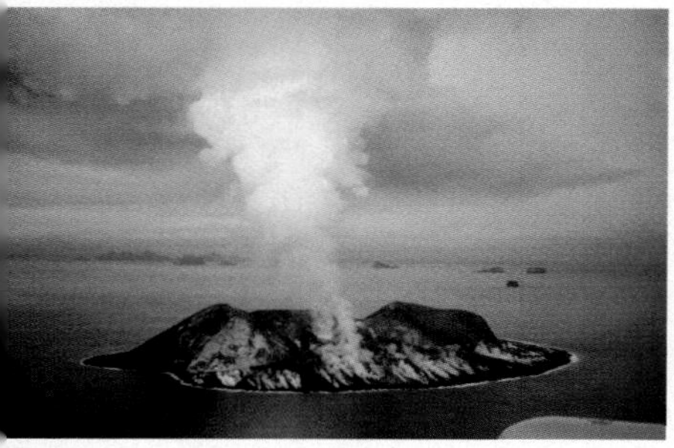

Surtsey, a volcanic island south of Iceland, formed along the Mid-Atlantic Ridge. (S. Jonasson/FLPA)

Because the dates of magnetic reversals going back nearly 200 million years have been established, the rate at which spreading occurs at the various ridges can be determined accurately. In the Pacific Ocean, for example, the magnetic stripes are much wider for corresponding time intervals than those of the Atlantic Ocean. Hence, we conclude that a faster spreading rate exists for the divergent boundaries of the Pacific as compared to the Atlantic. When we apply numerical dates to these magnetic events, we find that the spreading rate for the North Atlantic Ridge is only 2 centimeters (1 inch) per year. The rate is somewhat faster for the South Atlantic. The spreading rates for the East Pacific Rise generally range between 6 and 12 centimeters (2.5 to 5 inches) per year, with a maximum rate of about 20 centimeters (8 inches) a year. Thus, we have a magnetic tape recorder that records changes in Earth's magnetic field. This recorder also permits us to determine the rate of seafloor spreading.

Measuring Plate Motion

A number of methods have been employed to establish the direction and rate of plate motion. As noted earlier, hot spot "tracks" like those of the Hawaiian Island–Emperor Seamount chain trace the movement of the Pacific Plate relative to the mantle below. Further, by measuring the length of this volcanic chain and the time interval between the formation of the oldest structure (Suiko Seamount) and youngest structure (Hawaii), an average rate of plate motion can be calculated. In this case the volcanic chain is roughly 3000 kilometers long and has formed over the past 65 million years—making the average rate of movement about 9 centimeters (4 inches) per year. The accuracy of this calculation hinges on the hot spot maintaining a fixed position in the mantle.

Recall that the magnetic stripes measured on the floor of the ocean also provide a method to measure rates of plate motion—at least as averaged over millions of years. Using paleomagnetism and other indirect techniques, researchers have been able to work out relative plate velocities as shown on the map in Figure 15.25.

It is currently possible, with the use of space-age technology, to directly measure the relative motion between plates. This is accomplished by periodically establishing the exact locations, and hence the distance, between two observing stations situated on opposite sides of a plate boundary. Two of the methods used for this calculation are *Very Long Baseline Interferometry* (VLBI) and a satellite positioning technique that employs the *Global Positioning System* (GPS). The Very Long Baseline Interferometry system utilizes large radio telescopes to record signals from very distant quasars (quasi-stellar objects). Quasars lie billions of light-years from Earth,

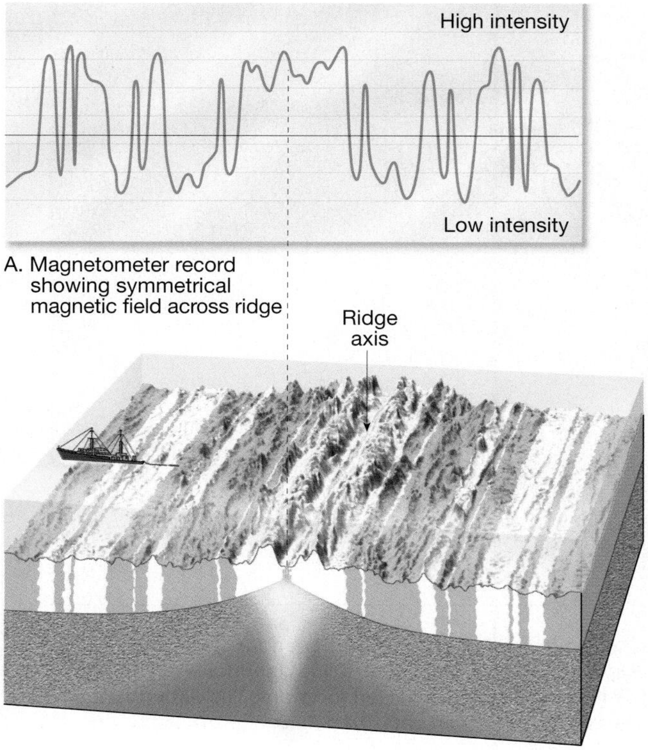

▲ **Figure 15.24** The ocean floor as a magnetic tape recorder. **A.** Schematic representation of magnetic intensities recorded as a magnetometer is towed across a segment of the oceanic ridge. **B.** Notice the symmetrical stripes of low- and high-intensity magnetism that parallel the ridge crest. Vine and Matthews suggested that the stripes of high-intensity magnetism occur where normally magnetized oceanic basalts enhance the existing magnetic field. Conversely, the low-intensity stripes are regions where the crust is polarized in the reverse direction, which weakens the existing magnetic field.

so they act as stationary reference points. The millisecond differences in the arrival times of the same signal at different Earth-bound observatories provide a means of establishing the precise distance between receivers. A typical survey may take a day to perform and involves two widely spaced radio telescopes observing perhaps a dozen quasars, 5 to 10 times each. This scheme provides an estimate of the distance between these observatories, which is accurate to about 2 centimeters. By repeating this experiment at a later date, researchers can establish the relative motion of these sites. This method has been particularly useful in establishing large-scale plate motions, such as the separation that is occurring between the United States and Europe.

You may be familiar with GPS, which uses 21 satellites to accurately locate any individual who is equipped with a handheld receiver. By using two spaced receivers, signals obtained by these instruments can be used to calculate their relative positions with considerable accuracy. Techniques using GPS receivers have been shown to be useful in establishing small-scale crustal movements such as those that occur along local faults in regions known to be tectonically active.

Confirming data obtained from these and other techniques leave little doubt that real plate motion has been detected (Figure 15.25). Calculations show that Hawaii is moving in a northwesterly direction and is approaching Japan at 8.3 centimeters per year. A site located in Maryland is retreating from one in England at a rate of about 1.7 centimeters per year—a rate that is close to the 2.3-centimeters-per-year spreading rate that was established from paleomagnetic evidence.

Radiotelescopes can be used to track plate motions. (NRAO)

What Drives Plate Motion?

The plate tectonics theory *describes* plate motion and the role that this motion plays in generating and/or modifying the major features of Earth's crust. Therefore, acceptance of plate tectonics does not rely on knowing exactly what drives plate motion. This is fortunate, because none of the models yet proposed can account for all major facets of plate tectonics. Nevertheless, researchers generally agree on the following:

1. Convective flow in the rocky 2900-kilometer-thick mantle—in which warm, buoyant rock rises and cooler, dense material sinks—is the underlying driving force for plate movement.
2. Mantle convection and plate tectonics are part of the same system. Subducting oceanic plates drive the cold downward-moving portion of convective flow while shallow upwelling of hot rock along the oceanic ridge and buoyant mantle plumes are the upward-flowing arm of the convective mechanism.
3. The slow movements of Earth's plates and mantle are ultimately driven by the unequal distribution of heat within Earth's interior.

What is not known with any high degree of certainty is the precise nature of this convective flow.

Forces that Drive Plate Motion

Several mechanisms contribute to plate motion; these include *slab pull, ridge push,* and *slab suction.* There is general agreement that the subduction of cold, dense slabs of oceanic lithosphere is the main driving force of plate motion (Figure 15.26). As these slabs sink into the asthenosphere, they "pull" the trailing plate along. This phenomenon, called **slab pull,** results because old slabs of oceanic lithosphere are more dense than the underlying asthenosphere and hence "sink like a rock."

Another important driving force is called **ridge push** (Figure 15.26). This gravity-driven mechanism results from the elevated position of the oceanic ridge,

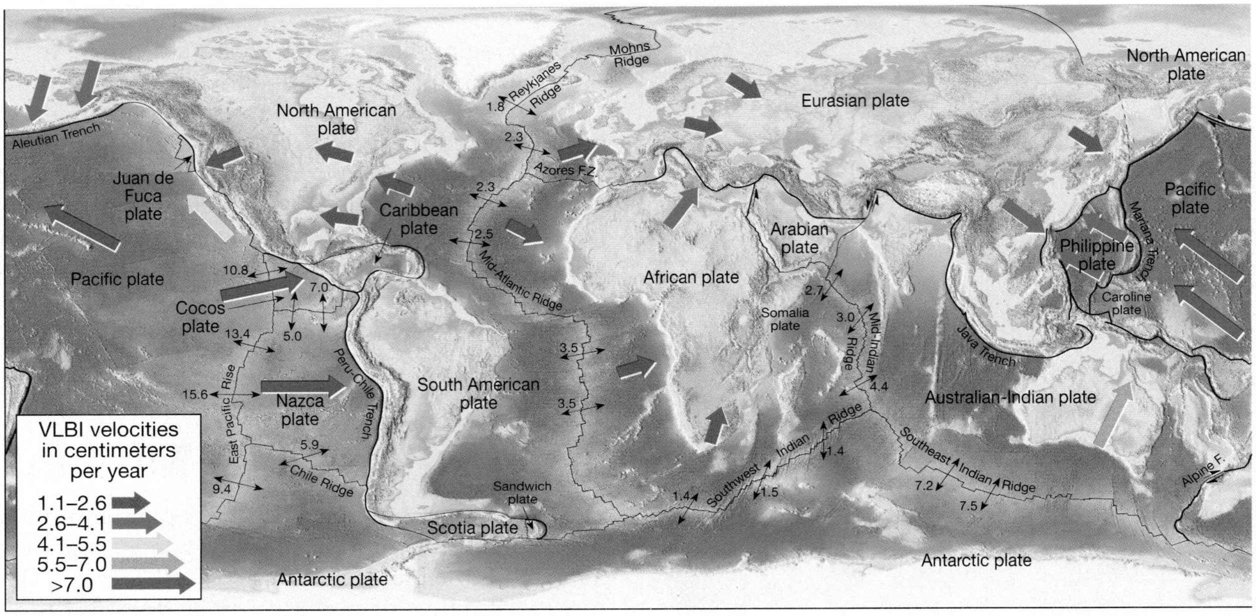

▲ **Figure 15.25** This map illustrates directions and rates of plate motion in centimeters per year. Seafloor-spreading velocities (as shown with black arrows and labels) are based on the spacing of dated magnetic stripes (anomalies). The colored arrows show Very Long Baseline Interferometry (VLBI) data of plate motion at selected locations. The data obtained by these methods are typically consistent. *(Seafloor data from DeMets and others, VLBI data from Ryan and others)*

which causes slabs of lithosphere to "slide" down the flanks of the ridge. Ridge push appears to contribute far less to plate motions than slab pull. Note that despite its greater average height above the seafloor, spreading rates along the Mid-Atlantic Ridge are considerably less than spreading rates along the less steep East Pacific Rise. Also supporting the notion that slab pull is more important than ridge push is the fact that when more than 20 percent of the perimeter of a plate consists of subduction zones, rates of plate movement are relatively rapid. Examples include the Pacific, Nazca, and Cocos plates, all of which have spreading rates that exceed 10 centimeters per year.

Yet another driving force arises from the drag of a subducting slab on the adjacent mantle. The result is an induced mantle circulation that pulls both the subducting and overriding plates toward the trench. Because this mantle flow tends to "suck" in nearby plates (similar to pulling the plug on a bathtub), it is called **slab suction** (Figure 15.26). Even if a subducting slab becomes detached from the overlying plate, its descent will continue to create flow in the mantle and hence will continue to drive plate motion.

▶ **Figure 15.26** Illustration of some of the forces that act on plates.

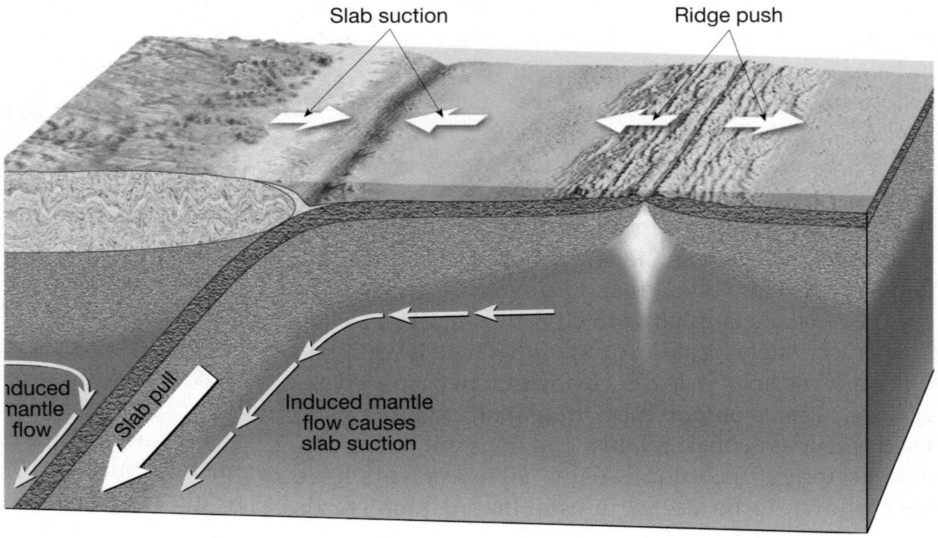

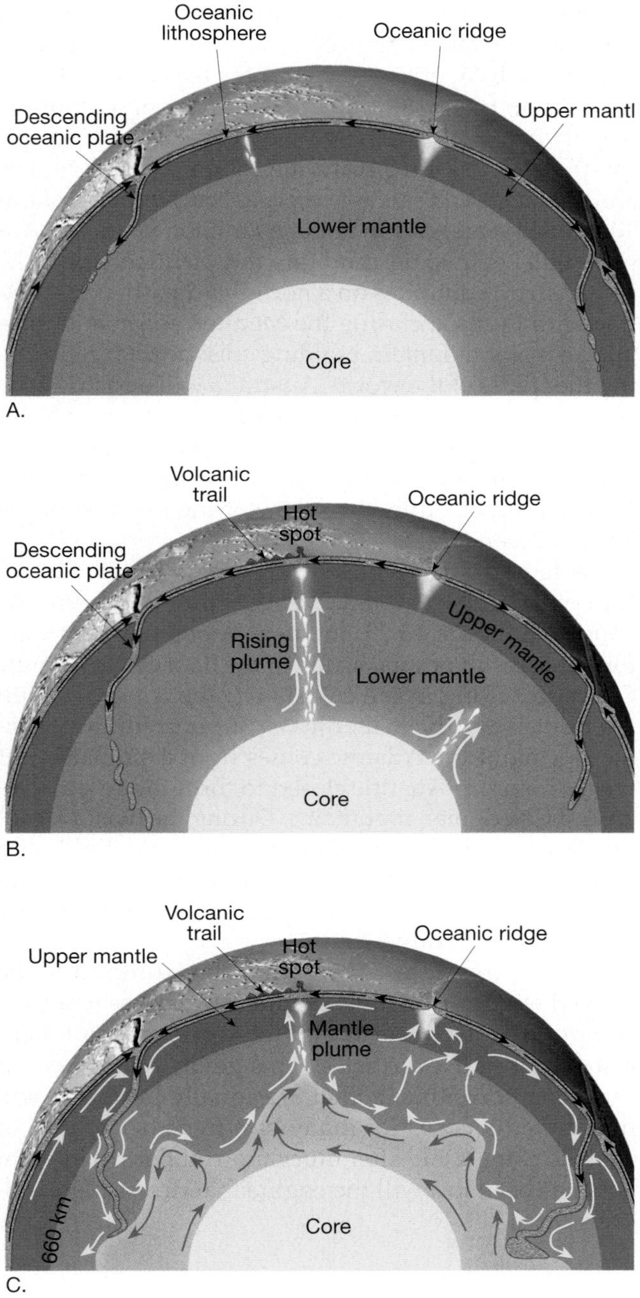

▲ **Figure 15.27** Proposed models for mantle convection. **A.** The model shown in this illustration consists of two convection layers—a thin, convective layer above 660 kilometers and a thick one below. **B.** In this whole-mantle convection model, cold oceanic lithosphere descends into the lowermost mantle, while hot mantle plumes transport heat toward the surface. **C.** This deep-layer model suggests that the mantle operates similar to a lava lamp on a low setting. Earth's heat causes these layers of convection to slowly swell and shrink in complex patterns without substantial mixing. Some material from the lower layer flows upward as mantle plumes.

Models of Plate-Mantle Convection

Any model of plate-mantle convection must be consistent with observed physical and chemical properties of the mantle. When seafloor spreading was first proposed, geologists suggested that convection in the mantle consisted of upcurrents coming from deep in the mantle beneath oceanic ridges. Upon reaching the base of the lithosphere, these currents were thought to spread laterally and pull the plates apart. Thus, plates were viewed as being carried passively by the flow in the mantle. However, based on physical evidence, it became clear that upwelling beneath oceanic ridges is shallow and not related to deep convection in the mantle. It is the horizontal movement of lithospheric plates away from the ridge that causes mantle upwelling, not the other way around. We have also learned that plate motion is the dominant source of convective flow in the mantle. As the plates move, they drag the adjacent material along, thereby inducing flow in the mantle. Thus, modern models have plates being an integral part of mantle convection and perhaps even its most active component.

Layering at 660 Kilometers. One of the earliest proposals regarding mantle convection came to be called the "layer cake" model. As shown in Figure 15.27A, this model has two zones of convection—a thin convective layer above 660 kilometers and a thick one located below. This model successfully explains why the basaltic lavas that erupt along oceanic ridges have a somewhat different composition than those that erupt in Hawaii as a result of hot-spot activity. The mid-ocean ridge basalts come from the upper convective layer, which is well mixed, whereas the mantle plume that feeds the Hawaiian volcanoes taps a deeper source of magma that resides in the lower convective layer.

Despite evidence that supports this model, other research suggests that subducting slabs of cold oceanic lithosphere penetrate the 660-kilometer boundary. If true, the subducting lithosphere should serve to mix the upper and lower layers. As a result, the layered mantle structure proposed in this model would not exist.

Did You Know?

Because plate tectonic processes are powered by heat from Earth's interior, the forces that drive plate motion will cease sometime in the distant future. The work of external forces (wind, water, and ice), however, will continue to erode Earth's surface. Eventually, landmasses will be nearly flat. What a different world it will be—an Earth with no earthquakes, no volcanoes, and no mountains.

Whole-Mantle Convection. Other researchers favor some type of *whole-mantle convection.* In a whole-mantle convection model, slabs of cold oceanic lithosphere descend to great depths and stir the entire mantle (Figure 15.27B). Simultaneously, hot mantle plumes originating near the mantle-core boundary transport heat and material toward the surface.

One whole-mantle model suggests that the subducting slabs of oceanic lithosphere accumulate at the mantle-core boundary. Over time this material is thought to melt and buoyantly rise toward the surface in the form of a mantle plume.

Deep-Layer Model. A remaining possibility is layering deeper in the mantle. One deep-layer model has been described as analogous to a lava lamp on a low setting. As shown in Figure 15.27C, the lower, perhaps one-third, of the mantle is like the colored fluid in the bottom layer of a lava lamp. Like a lava lamp on low, heat from Earth's interior causes the two layers to slowly swell and shrink in complex patterns without substantial mixing. A small amount of material from the lower layer flows upward as mantle plumes to generate hot-spot volcanism at the surface.

This model provides the two chemically different mantle sources for basalt that are required by observational data. Further, it is compatible with seismic images that show cold lithospheric plates sinking deep into the mantle. Despite its attractiveness, there is little seismic evidence to suggest that a deep mantle layer of this nature exists, except for the very thin layer located at the mantle-core boundary.

Although there is still much to be learned about the mechanisms that cause plates to move, some facts are clear. The unequal distribution of heat in Earth generates some type of thermal convection that ultimately drives plate-mantle motion. Furthermore, descending lithospheric plates are active components of downwelling, and serve to transport cold material into the mantle. Exactly how this convective flow operates is yet to be determined.

Did You Know?

Researchers have estimated that the continents join to form supercontinents roughly every 500 million years. Since it has been about 200 million years since Pangaea broke up, we have only 300 million years to wait before the next supercontinent is completed.

Plate Tectonics into the Future

Geologists have also extrapolated present-day plate movements into the future. Figure 15.28 illustrates where Earth's landmasses may be 50 million years from now if present plate movements persist during this time span.

In North America we see that the Baja Peninsula and the portion of southern California that lies west of the San Andreas Fault will have slid past the North American plate. If this northward migration takes place, Los Angeles and San Francisco will pass each other in about 10 million years, and in about 60 million years Los Angeles will begin to descend into the Aleutian Trench.

If Africa continues on a northward path, it will collide with Eurasia, closing the Mediterranean and initiating a major mountain-building episode (Figure 15.28). In other parts of the world, Australia will be astride the equator and, along with New Guinea, will be on a collision course with Asia. Meanwhile, North and South America will begin to separate, while the Atlantic and Indian oceans continue to grow at the expense of the Pacific Ocean.

A few geologists have even speculated on the nature of the globe 250 million years into the future. As shown in Figure 15.29, the next supercontinent may form as a result of subduction of the floor of the Atlantic Ocean, resulting in the collision of the Americas with the Eurasian–African landmass. Support for the possible closing of the Atlantic comes from a similar event when the proto-Atlantic closed to form the Appalachian and Caledonian mountains. During the next 250 million years, Australia is also projected to collide with Southeast Asia. If this scenario is accurate, the dispersal of Pangaea will end when the continents reorganize into the next supercontinent.

Such projections, although interesting, must be viewed with considerable skepticism because many assumptions must be correct for these events to unfold as just described. Nevertheless, changes in the shapes and positions of continents that are equally profound will undoubtedly occur for many hundreds of millions of years to come. Only after much more of Earth's internal heat has been lost will the engine that drives plate motions cease.

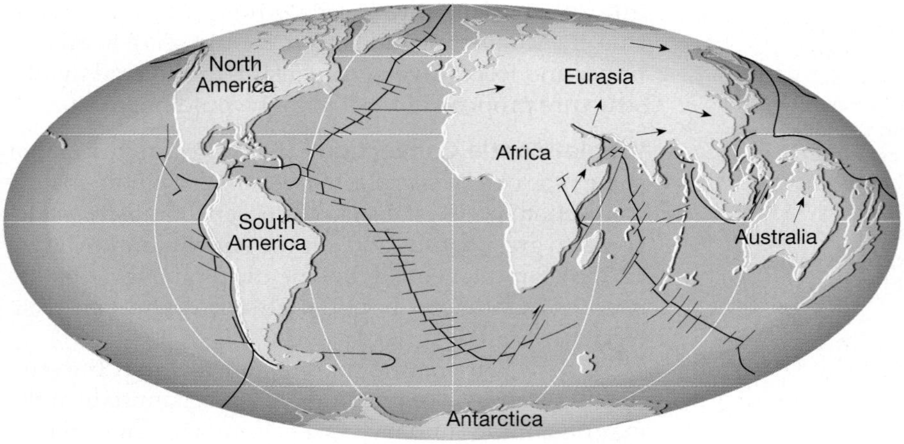

◀ **Figure 15.28** The world as it may look 50 million years from now. *(Modified after Robert S. Dietz, John C. Holden, C. Scotese, and others)*

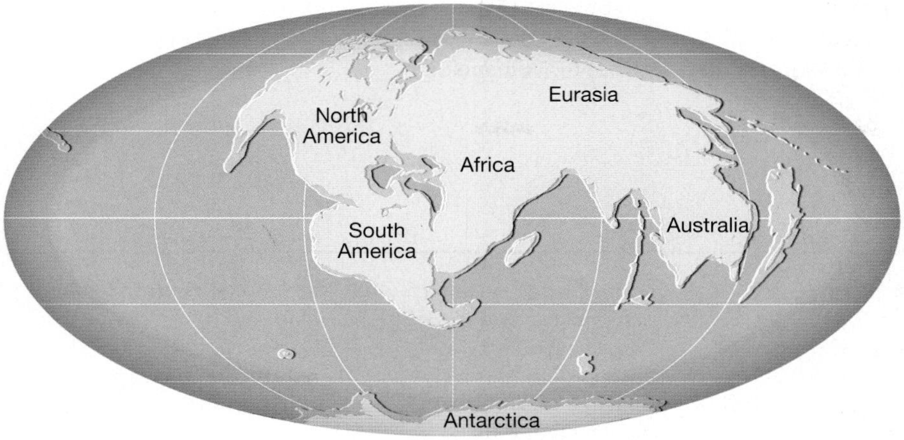

◀ **Figure 15.29** Reconstruction of Earth as it may appear 250 million years into the future. *(Modified after C. Scotese and others)*

The Chapter in Review

- In the early 1900s *Alfred Wegener* set forth his *continental drift* hypothesis. One of its major tenets was that a supercontinent called Pangaea began breaking apart into smaller continents about 200 million years ago. The smaller continental fragments then "drifted" to their present positions. To support the claim that the now-separate continents were once joined, Wegener and others used the *fit of South America and Africa, the distribution of ancient climates, fossil evidence, and rock structures.*
- One of the main objections to the continental drift hypothesis was the inability of its supporters to provide an acceptable mechanism for the movement of continents.
- The theory of *plate tectonics,* a far more encompassing theory than continental drift, holds that Earth's rigid outer shell, called the *lithosphere,* consists of seven large and numerous smaller segments called *plates* that are in motion relative to each other. Most of Earth's *seismic activity, volcanism,* and *mountain building* occur along the dynamic margins of these plates.
- A major departure of the plate tectonics theory from the continental drift hypothesis is that large plates contain both continental and ocean crust and the entire plate moves. By contrast, in continental drift, Wegener proposed that the sturdier continents "drifted" by breaking through the oceanic crust, much like ice breakers cut through ice.
- *Divergent plate boundaries* occur where plates move apart, resulting in upwelling of material from the mantle to create new seafloor. Most divergent boundaries occur along the axis of the oceanic ridge system and are associated with seafloor spreading, which occurs at rates between about 2 and 15 centimeters per year. New divergent boundaries may form within a continent (for example, the East African rift valleys), where they may fragment a landmass and develop a new ocean basin.
- *Convergent plate boundaries* occur where plates move together, resulting in the subduction of oceanic lithosphere into the mantle along a deep oceanic trench. Convergence between an oceanic and continental block results in subduction of the oceanic slab and the formation of a *continental volcanic arc* such as the Andes of South America. Oceanic–oceanic convergence results in an arc-shaped chain of volcanic islands called a *volcanic island arc.* When two plates carrying continental crust converge, both plates are too buoyant to be subducted. The result is a "collision" resulting in the formation of a mountain belt such as the Himalayas.
- *Transform fault boundaries* occur where plates grind past each other without the production or destruction of lithosphere. Most transform faults join two segments of an oceanic ridge. Others connect spreading centers to subduction zones and thus facilitate the transport of oceanic crust created at a ridge crest to its site of destruction, at a deep-ocean trench. Still others, like the San Andreas Fault, cut through continental crust.
- The theory of plate tectonics is supported by (1) *paleomagnetism,* the direction and intensity of Earth's magnetism in the geologic past; (2) the global distribution of *earthquakes* and their close association with plate boundaries; (3) the ages of *sediments* from the floors of the deep-ocean basins; and (4) the existence of island groups that formed over *hotspots* and that provide a frame of reference for tracing the direction of plate motion.
- Three basic models for mantle convection are currently being evaluated. Mechanisms that contribute to this convective flow are slab-pull, ridge-push, and mantle plumes. *Slab-pull* occurs where cold, dense oceanic lithosphere is subducted and pulls the trailing lithosphere along. *Ridge-push* results when gravity sets the elevated slabs astride oceanic ridges in motion. Hot, buoyant *mantle plumes* are considered the upward flowing arms of mantle convection.

One model suggests that mantle convection occurs in two layers separated at a depth of 660 kilometers. Another model proposes whole-mantle convection that stirs the entire 2900-kilometer-thick rocky mantle. Yet another model suggests that the bottom third of the mantle gradually bulges upward in some areas and sinks in others without appreciable mixing.

Key Terms

asthenosphere (p. 342)
continental drift (p. 336)
continental volcanic arc (p. 350)
convergent boundary (p. 347)
Curie point (p. 355)
deep-ocean trench (p. 348)
divergent boundary (p. 343)
fossil magnetism (p. 355)
fracture zone (p. 352)
hot spot (p. 354)
island arc (p. 350)
lithosphere (p. 342)
magnetic reversal (p. 356)
magnetic time scale (p. 357)
magnetometer (p. 357)
mantle plume (p. 354)
normal polarity (p. 356)
oceanic ridge system (p. 342)
paleomagnetism (p. 355)
Pangaea (p. 337)
partial melting (p. 349)
plate (p. 342)
plate tectonics (p. 342)
reverse polarity (p. 356)
ridge push (p. 359)
rift (rift valley) (p. 346)
seafloor spreading (p. 346)
slab pull (p. 359)
slab suction (p. 360)
subduction zone (p. 348)
transform fault boundary (p. 351)
volcanic island arc (p. 350)

Questions for Review

1. Who is credited with developing the continental drift hypothesis?
2. What was probably the first evidence that led some to suspect the continents were once connected?
3. What was Pangaea?
4. List the evidence that Wegener and his supporters gathered to substantiate the continental drift hypothesis.
5. Explain why the discovery of the fossil remains of *Mesosaurus* in both South America and Africa, but nowhere else, supports the continental drift hypothesis.
6. Early in this century, what was the prevailing view of how land animals migrated across vast expanses of ocean?
7. How did Wegener account for the existence of glaciers in the southern landmasses, while at the same time areas in North America, Europe, and Siberia supported lush tropical swamps?
8. On what basis were plate boundaries first established?
9. What are the three major types of plate boundaries? Describe the relative plate motion at each of these boundaries.
10. What is seafloor spreading? Where is active seafloor spreading occurring today?
11. What is a subduction zone? With what type of plate boundary is it associated?
12. Where is lithosphere being consumed? Why must the production and destruction of lithosphere be going on at approximately the same rate?
13. Briefly describe how the Himalaya Mountains formed.
14. Differentiate between transform faults and the other two types of plate boundaries.
15. Some predict that California will sink into the ocean. Is this idea consistent with the theory of plate tectonics?
16. Define the term *paleomagnetism*.
17. How does the continental drift hypothesis account for the apparent wandering of Earth's magnetic poles?
18. What is the age of the oldest sediments recovered by deep-ocean drilling? How do the ages of these sediments compare to the ages of the oldest continental rocks?
19. How do hot spots and the plate tectonics theory account for the fact that the Hawaiian Islands vary in age?
20. Briefly describe the three mechanisms that drive plate motion.
21. With what type of plate boundary are the following places or features associated: Himalayas, Aleutian Islands, Red Sea, Andes Mountains, San Andreas Fault, Iceland, Japan, Mount St. Helens?

Online Study Guide

The *Essentials of Geology* Web site uses the resources and flexibility of the Internet to aid in your study of the topics in this chapter. Written and developed by geology instructors, this site will help improve your understanding of geology. Visit **www.prenhall.com/lutgens** and click on the cover of *Essentials of Geology 9e* to find:

- On-line review quizzes.
- Critical thinking exercises.
- Links to chapter-specific Web resources.
- Internet-wide key term searches.

http://www.prenhall.com/lutgens

The *Thomas G. Thompson* is the University of Washington's oceanographic research vessel. *(Photo by Paul Sounders/CORBIS)*

CHAPTER 16

Origin and Evolution of the Ocean Floor

Focus on Learning

To assist you in learning the important concepts in this chapter, you will find it helpful to focus on the following questions:

- How is the ocean floor mapped?
- What are the three major topographic provinces of the ocean floor and the important features associated with each?
- What is the nature and origin of the oceanic ridge system?
- How does oceanic crust form, and how is it different from continental crust?
- What is the relationship between continental rifting and new ocean basins?
- What is the supercontinent cycle?

The ocean is the largest feature on Earth, covering more than 70 percent of our planet's surface. One of the main reasons that Wegener's continental drift hypothesis was not widely accepted when first proposed was that so little was known about the ocean floor. Until the twentieth century, investigators used weighted lines to measure water depth. In deep water these depth measurements, or soundings, took hours to perform and could be wildly inaccurate.

With the development of new marine tools following World War II, our knowledge of the diverse topography of the ocean floor grew rapidly. One of the most interesting discoveries was the global oceanic ridge system. This broad elevated landform, which stands 2 to 3 kilometers above the adjacent deep-ocean basins, is the longest topographic feature on Earth.

Today we know that oceanic ridges mark divergent, or constructive, plate margins where new oceanic lithosphere originates. We also know that deep-ocean trenches represent convergent plate boundaries, where oceanic lithosphere is subducted into the mantle. Because the processes of plate tectonics are creating oceanic crust at mid-ocean ridges and consuming it at subduction zones, the oceanic crust is continually being renewed and recycled.

In this chapter, we will examine the topography of the ocean floor and look at the processes that produced its varied features. You will also learn about the composition, structure, and origin of oceanic crust. In addition, you will examine those processes that recycle oceanic lithosphere and consider how this activity causes Earth's landmasses to migrate about the face of the globe.

An Emerging Picture of the Ocean Floor

GEODe ESSENTIALS OF GEOLOGY

Origin and Evolution of the Ocean Floor
▼ Mapping the Ocean Floor

If all water were drained from the ocean basins, a great variety of features would be seen, including broad volcanic peaks, deep trenches, extensive plains, linear mountain chains, and large plateaus. In fact, the scenery would be nearly as diverse as that on the continents.

An understanding of seafloor features came with the development of techniques that measure the depth of the oceans. **Bathymetry** is the measurement of ocean depths and the charting of the shape or topography of the ocean floor.

Mapping the Seafloor

The first understanding of the ocean floor's varied topography did not unfold until the historic three-and-a-half-year voyage of the HMS *Challenger* (Figure 16.1). From December 1872 to May 1876, the *Challenger* expe-

▼ **Figure 16.1** The first systematic bathymetric measurements of the ocean were made aboard the H.M.S. *Challenger* during its historic three-and-a-half-year voyage. Inset shows route of the H.M.S. *Challenger,* which departed England in December of 1872 and returned in May 1876. *(From C.W. Thompson and Sir John Murray, Report on the Scientific Results of the Voyage of the* H.M.S. Challenger, *Vol. 1, Great Britain: Challenger Office, 1895, Plate 1. Library of Congress)*

dition made the first—and perhaps still most comprehensive—study of the global ocean ever attempted by one agency. The 127,500-kilometer (79,000-mile) trip took the ship and its crew of scientists to every ocean except the Arctic. Throughout the voyage, they sampled a multitude of ocean properties, including water depth, which was accomplished by laboriously lowering a long weighted line overboard. Not many years later, the knowledge gained by the *Challenger* of the ocean's great depth and varied topography was further expanded with the laying of communication cables, especially in the North Atlantic Ocean. However, as long as a weighted line was the only way to measure ocean depths, knowledge of seafloor features remained extremely limited.

Bathymetric Techniques. Today sound energy is used to measure water depths. The basic approach employs some type of **sonar,** an acronym for *so*und *na*vigation and *r*anging. The first devices that used sound to measure water depth, called **echo sounders,** were developed early in the twentieth century. Echo sounders work by transmitting a sound wave (called a *ping*) into the water in order to produce an echo when it bounces off any object, such as a marine organism or the ocean floor (Figure 16.2A). A sensitive receiver intercepts the echo reflected from the bottom, and a clock precisely measures the travel time to fractions of a second. By knowing the speed of sound waves in water—about 1500 meters (4900 feet) per second—and the time required for the energy pulse to reach the ocean floor and return, depth can be calculated. The depths determined from continuous monitoring of these echoes are plotted so a profile of the ocean floor is obtained. By laboriously combining profiles from several adjacent traverses, a chart of the seafloor can be produced.

Following World War II, the U.S. Navy developed *sidescan sonar* to look for mines and other explosive devices. These torpedo-shaped instruments can be towed behind a ship where they send out a fan of sound extending to either side of the ship's track. By combining swaths of sidescan sonar data, researchers produced the first photograph-like images of the seafloor. Although sidescan sonar provides valuable views of the seafloor, it does not provide bathymetric (water depth) data.

This problem is not present in the *high-resolution multibeam* instruments developed during the 1990s. These systems use hull-mounted sound sources that send out a fan of sound, then record reflections from the seafloor through a set of narrowly focused receivers aimed at different angles. Thus, rather than obtaining the depth of a single point every few seconds, this technique makes it possible for a survey ship to map the features of the ocean floor along a strip tens of kilometers wide (Figure 16.2B). When a ship uses multibeam sonar to make a map of a section of seafloor, it travels through the area in a regularly spaced back-and-forth pattern known as, appropriately enough, "mowing the lawn." Furthermore, these systems can collect bathymetric data of such high

Did You Know?

The U.S. Navy utilizes the biological sonar of bottlenose dolphins to help defend its ships and facilities. Specifically, dolphins have been trained to detect mines intended to blow up ships. When navy ships enter mine-infested waters, the trained dolphins are released and use their sonar to locate the deadly mines.

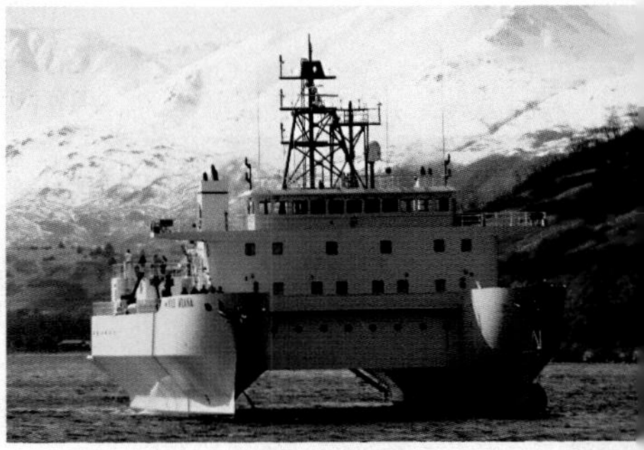

Ultramodern research vessel Kilo Moana. (Associated Press)

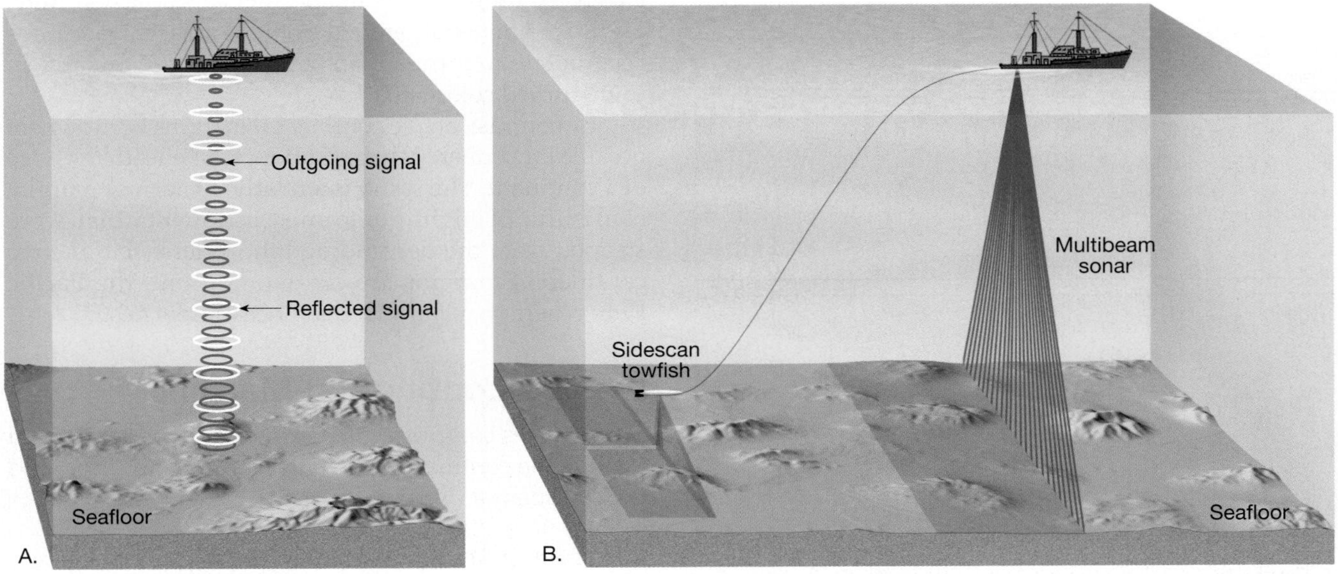

▲ **Figure 16.2** Various types of sonar. **A.** An echo sounder determines the water depth by measuring the time interval required for an acoustic wave to travel from a ship to the seafloor and back. The speed of sound in water is 1500 m/sec. Therefore, depth = ½ (1500 m/sec × echo travel time). **B.** Modern multibeam sonar and sidescan sonar obtain an "image" of a narrow swath of seafloor every few seconds.

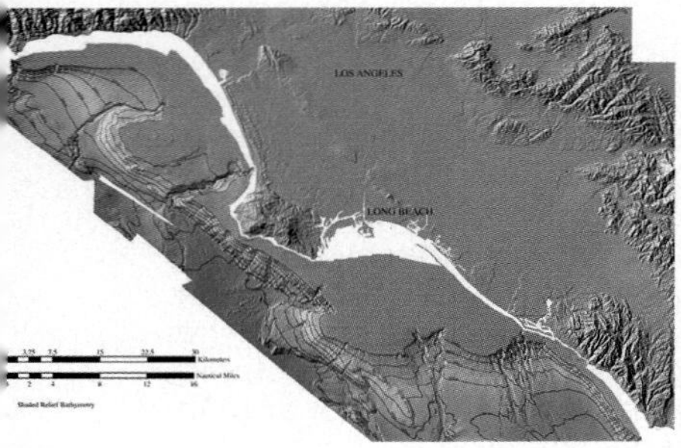

Sonar map of seafloor in the Los Angeles area of California. (USGS)

resolution that they can distinguish depths that differ by less than a meter.

Despite their greater efficiency and enhanced detail, research vessels equipped with multibeam sonar travel at a mere 10 to 20 kilometers (6 to 12 miles) per hour. It would take at least 100 vessels outfitted with this equipment hundreds of years to map the entire seafloor. This explains why only about 5 percent of the seafloor has been mapped in detail—and why large areas of the seafloor have not yet been mapped with sonar at all.

Seismic Reflection Profiles. Marine geologists are also interested in viewing the rock structure beneath the sediments that blanket much of the seafloor. This can be accomplished by making a **seismic reflection profile.** To construct such a profile, strong low-frequency sounds are produced by explosions (depth charges) or air guns. These sound waves penetrate beneath the seafloor and reflect off the contacts between rock layers, and fault zones, just like sonar reflects off the bottom of the sea. Figure 16.3 shows a seismic profile of a portion of the Madeira abyssal plain in the eastern Atlantic. Although the seafloor here is flat, notice the irregular ocean crust buried by a thick accumulation of sediments.

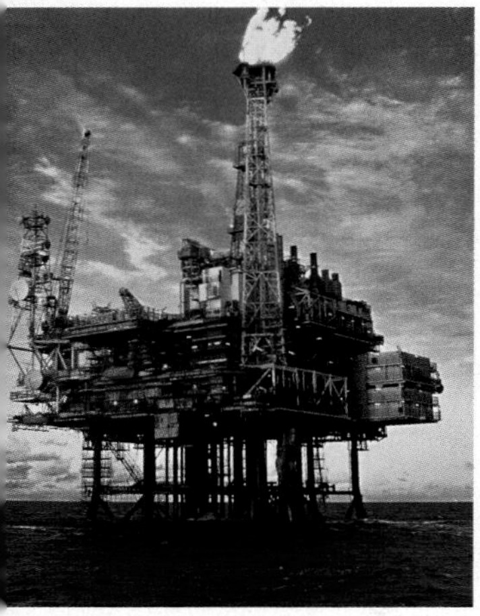

Offshore drilling rig extracting petroleum from the continental shelf. (Mark Lehman)

Viewing the Ocean Floor from Space

Another technological breakthrough that has led to an enhanced understanding of the seafloor involves measuring the shape of the surface of the global ocean from space. After compensating for waves, tides, currents, and atmospheric effects, it was discovered that the water's surface is not perfectly "flat." This is because gravity attracts water toward regions where massive seafloor features occur. Therefore, mountains and ridges produce elevated areas on the ocean surface, and, conversely, canyons and trenches cause slight depressions. Satellites equipped with *radar altimeters* are able to measure these subtle differences by bouncing microwaves off the sea surface. These devices can measure variations as small as 3 to 6 centimeters. Such data have added greatly to the knowledge of ocean-floor topography.

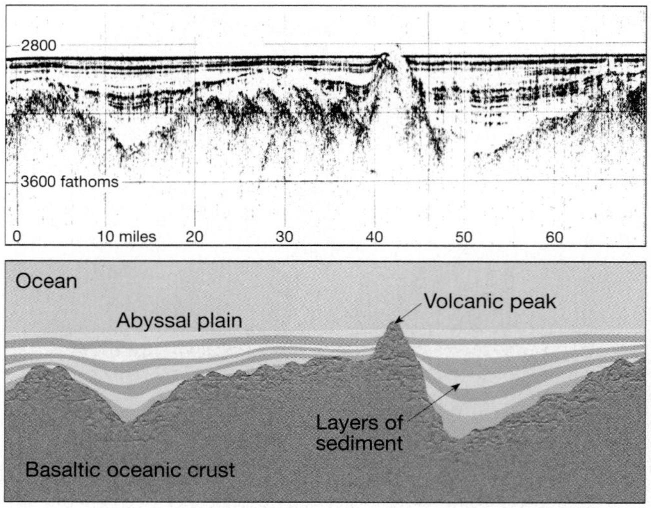

▲ **Figure 16.3** Seismic cross section and matching sketch across a portion of the Madeira abyssal plain in the eastern Atlantic Ocean, showing the irregular oceanic crust buried by sediments. *(Data courtesy of Charles Hollister, Woods Hole Oceanographic Institution)*

Provinces of the Ocean Floor

Oceanographers studying the topography of the ocean floor have delineated three major units: *continental margins, deep-ocean basins,* and *oceanic (mid-ocean) ridges.* The map in Figure 16.4 outlines these provinces for the North Atlantic Ocean, and the profile at the bottom of the illustration shows the varied topography. Such profiles usually have their vertical dimension exaggerated many times—40 times in this case—to make topographic features more conspicuous. Vertical exaggeration, however, makes slopes shown in seafloor profiles appear to be *much* steeper than they actually are.

Continental Margins

Two main types of **continental margins** have been identified—*passive* and *active.* Passive margins are found along most of the coastal areas that surround the Atlantic and Indian oceans, including the east coasts of North and South America, as well as the coastal areas of Europe and Africa. Passive margins are *not* situated along an active plate boundary and therefore experience very little volcanism and few earthquakes. Here, weathered materials eroded from the adjacent landmass accumulate to form a thick, broad wedge of relatively undisturbed sediments.

By contrast, active continental margins occur where oceanic lithosphere is being subducted beneath the edge of a continent. The result is a relatively narrow margin, consisting of highly deformed sediments that were scraped from the descending lithospheric slab. Active continental margins are common around the Pacific Rim, where they parallel deep-ocean trenches.

Passive Continental Margins

The features comprising a **passive continental margin** include the continental shelf, the continental slope, and the continental rise (Figure 16.5).

Continental Shelf. The **continental shelf** is a gently sloping submerged surface extending from the shoreline toward the deep-ocean basin. Because it is underlain by continental crust, it is clearly a flooded extension of the continents.

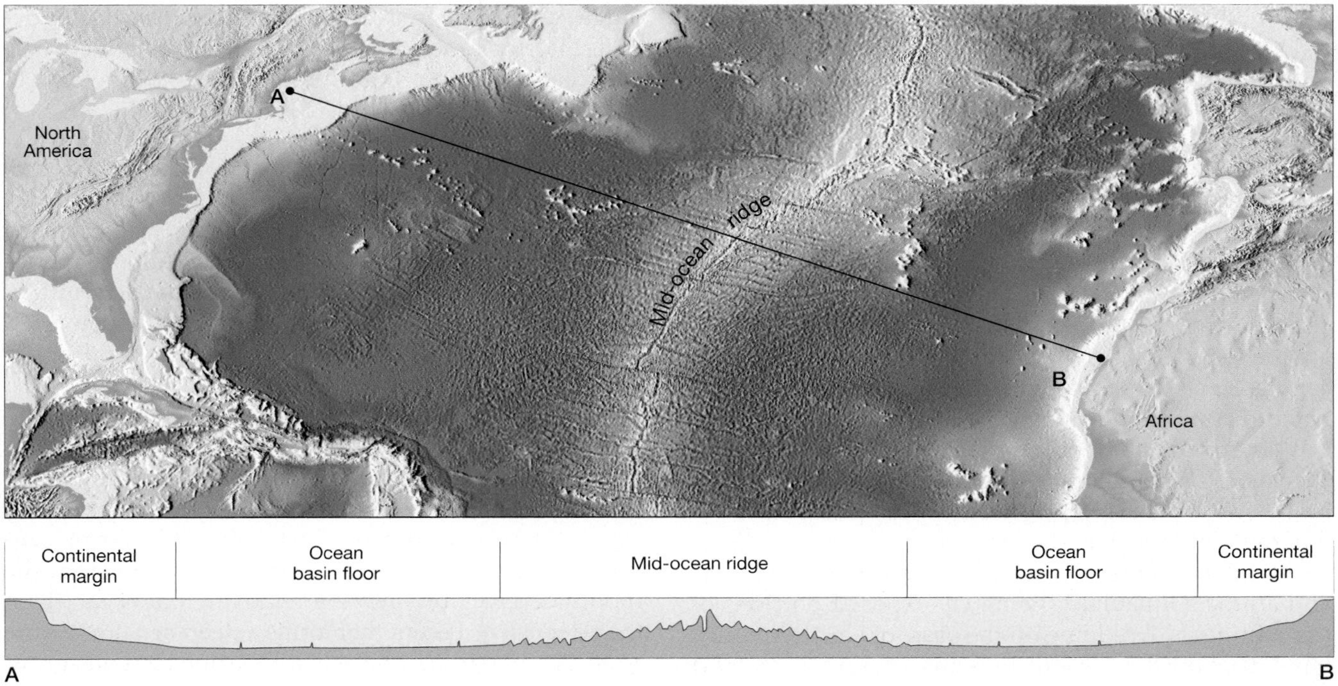

▲ **Figure 16.4** Major topographic divisions of the North Atlantic and a profile from New England to the coast of North Africa.

The continental shelf varies greatly in width. Although almost nonexistent along some continents, the shelf extends seaward more than 1500 kilometers (930 miles) along others. On average, the continental shelf is about 80 kilometers (50 miles) wide and 130 meters (425 feet) deep at its seaward edge. The average inclination of the continental shelf is only about one-tenth of 1 degree, a drop of only about 2 meters per kilometer (10 feet per mile). The slope is so slight that it would appear to an observer to be a horizontal surface.

Although continental shelves represent only 7.5 percent of the total ocean area, they have economic and political significance because they contain important mineral deposits, including large reservoirs of oil and natural gas, as well as huge sand and

Did You Know?

The salinity of water in the Dead Sea is nearly 10 times greater than the salinity of normal seawater. As a result, the water is so dense and buoyant that people and objects easily float.

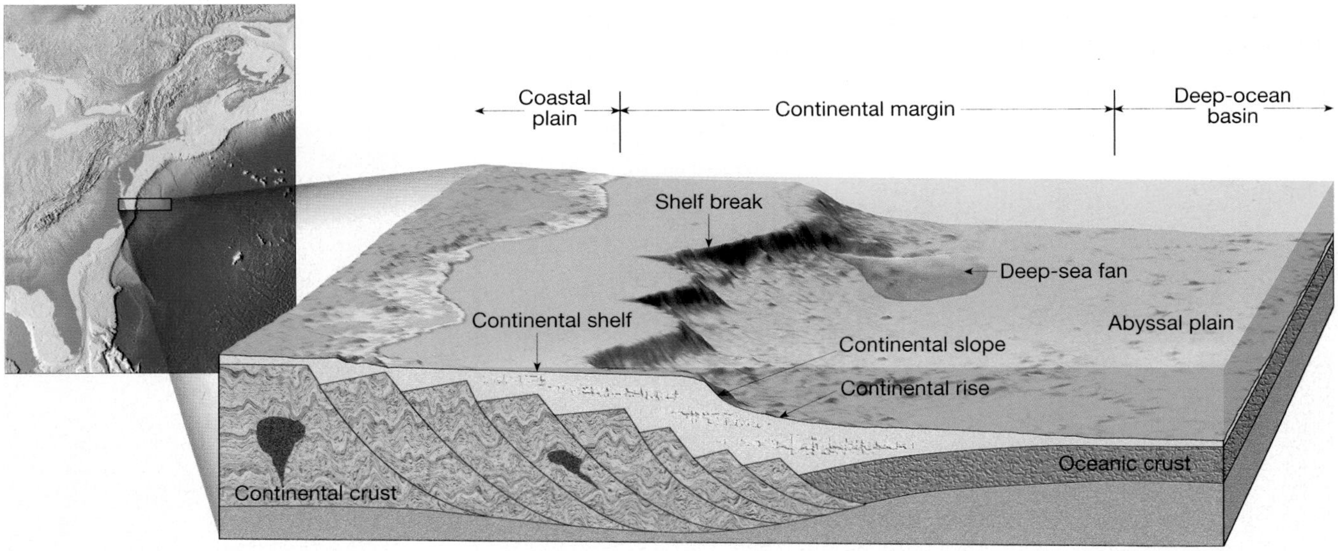

▲ **Figure 16.5** Schematic view showing the provinces of a passive continental margin. Note that the steepness of the slopes shown for the continental shelf and continental slope are greatly exaggerated. The continental shelf has an average slope of one-tenth of 1 degree, while the continental slope has an average slope of about 5 degrees.

Did You Know?

Faced with a shortage of land along its heavily populated coast, Japan has constructed artificial islands on the adjacent continental shelf. One of these islands is located about 3 miles off the coast of Osaka, where the Kansai Airport was built.

Upper portion of La Jolla Submarine Canyon. The erosional work of turbidity currents helped create this feature. (Francis P. Shepard)

gravel deposits. The waters of the continental shelf also contain many important fishing grounds, which are significant sources of food.

Even though the continental shelf is relatively featureless, some areas are mantled by extensive glacial deposits and thus are quite rugged. In addition, some continental shelves are dissected by large valleys running from the coastline into deeper waters. Many of these *shelf valleys* are the seaward extensions of river valleys on the adjacent landmass. Such valleys appear to have been excavated during the Pleistocene epoch (Ice Age). During this time great quantities of water were stored in vast ice sheets on the continents. This caused sea level to drop by 100 meters (330 feet) or more, exposing large areas of the continental shelves. Because of this drop in sea level, rivers extended their courses, and land-dwelling plants and animals inhabited the newly exposed portions of the continents. Dredging off the coast of North America has retrieved the ancient remains of numerous land dwellers, including mammoths, mastodons, and horses, adding to the evidence that portions of the continental shelves were once above sea level.

Most passive continental shelves, such as those along the East Coast of the United States, consist of shallow-water deposits that can reach several kilometers in thickness. Such deposits have led researchers to conclude that these thick accumulations of sediment are produced along a gradually subsiding continental margin.

Continental Slope. Marking the seaward edge of the continental shelf is the **continental slope,** a relatively steep structure (as compared with the shelf) that marks the boundary between continental crust and oceanic crust (Figure 16.5). Although the inclination of the continental slope varies greatly from place to place, it averages about 5 degrees and in places may exceed 25 degrees. Further, the continental slope is a relatively narrow feature, averaging only about 20 kilometers (12 miles) in width.

Continental Rise. In regions where trenches do not exist, the steep continental slope merges into a more gradual incline known as the **continental rise.** Here the slope drops to about one-third degree, or about 6 meters per kilometer (32 feet per mile). Whereas the width of the continental slope averages about 20 kilometers (12 miles), the continental rise may extend for hundreds of kilometers into the deep-ocean basin.

The continental rise consists of a thick accumulation of sediment that moved downslope from the continental shelf to the deep-ocean floor. The sediments are delivered to the base of the continental slope by *turbidity currents* that periodically flow down submarine canyons. When these muddy currents emerge from the mouth of a canyon onto the relatively flat ocean floor, they deposit sediment that forms a **deep-sea fan** (Figure 16.5). As fans from adjacent submarine canyons grow, they merge laterally with one another to produce a continuous covering of sediment at the base of the continental slope forming the continental rise.

Active Continental Margins

Along some coasts the continental slope descends abruptly into a deep-ocean trench. In this situation, the landward wall of the trench and the continental slope are essentially the same feature. In such locations, the continental shelf is very narrow, if it exists at all.

Active continental margins are located primarily around the Pacific Ocean in areas where oceanic lithosphere is being subducted beneath the leading edge of a continent (Figure 16.6). Here sediments from the ocean floor and pieces of oceanic crust are scraped from the

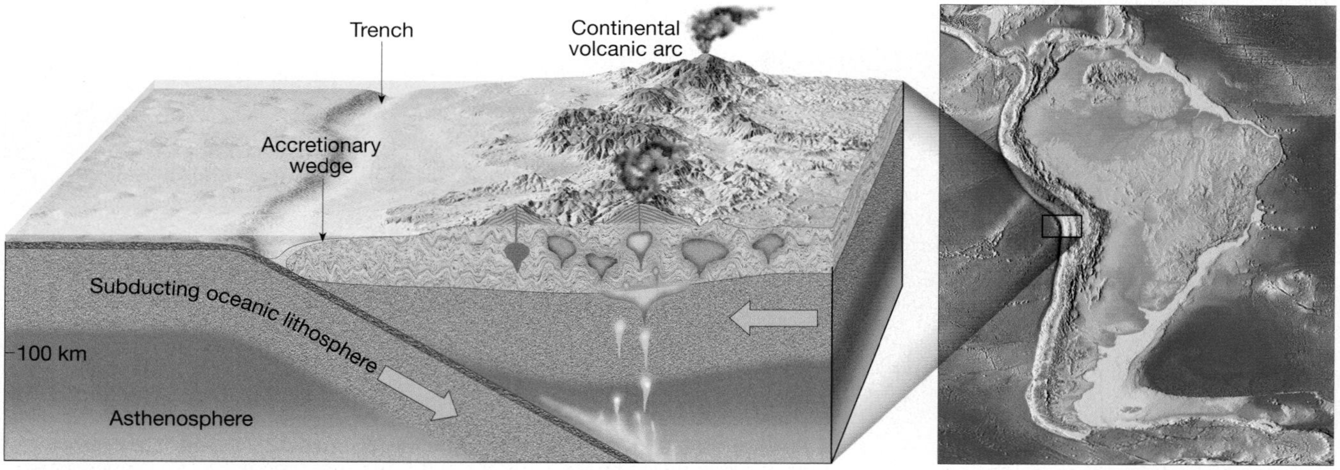

▲ **Figure 16.6** Active continental margin. Here sediments from the ocean floor are scraped from the descending plate and added to the continental crust as an accretionary wedge.

descending oceanic plate and plastered against the edge of the overriding continent. This chaotic accumulation of deformed sediment and scraps of oceanic crust is called an **accretionary wedge.** Prolonged plate subduction, along with the accretion of sediments on the landward side of the trench, can produce a large accumulation of sediments along a continental margin. A large accretionary wedge, for example, is found along the northern coast of Japan's Honshu Island.

Some subduction zones have little or no accumulation of sediments, indicating that ocean sediments are being carried into the mantle with the subducting plate. These tend to be regions where old oceanic lithosphere is subducting nearly vertically into the mantle. In these locations the continental margin is very narrow, as the trench may lie a mere 50 kilometers (31 miles) offshore.

Features of Deep-Ocean Basins

Between the continental margin and the oceanic ridge lies the **deep-ocean basin** (see Figure 16.4). The size of this region—almost 30 percent of Earth's surface—is roughly comparable to the percentage of land above sea level. This region includes remarkably flat areas known as *abyssal plains;* tall volcanic peaks called *seamounts* and *guyots; deep-ocean trenches,* which are extremely deep linear depressions in the ocean floor; and large flood basalt provinces called *oceanic plateaus.*

Deep-Ocean Trenches

Deep-ocean trenches are long, relatively narrow creases in the seafloor that form the deepest parts of the ocean (Table 16.1). Most trenches are located along the margins of the Pacific Ocean (Figure 16.7) where some exceed 10,000 meters (33,000 feet) in depth. A portion of one trench—the Challenger Deep in the Mariana Trench—has been measured at 11,022 meters (36,163 feet) below sea level, making it the deepest known part of the world ocean. Only two trenches are located in the Atlantic—the Puerto Rico Trench adjacent to the Lesser Antilles arc and the South Sandwich Trench.

Although deep-ocean trenches represent only a small portion of the area of the ocean floor, they are nevertheless significant geologic features. Trenches are sites of plate convergence where lithospheric plates subduct and plunge back into the mantle. In addition to earthquakes being created as one plate "scrapes" against another, volcanic activity is also associated with these regions. Recall that the release of volatiles—especially water—from a descending plate triggers melting in the wedge of asthenosphere above. This buoyant material slowly migrates upward and gives rise to volcanic activity at the surface. Thus, trenches are often paralleled by an arc-shaped row of active volcanoes called a *volcanic island arc.* Furthermore, *continental volcanic arcs,* such as those making up portions of the Andes and Cascades, are located parallel to trenches that lie adjacent to continental margins. The large number of trenches and associated volcanic activity along the margins of the Pacific Ocean explains why the region is known as the *Ring of Fire.*

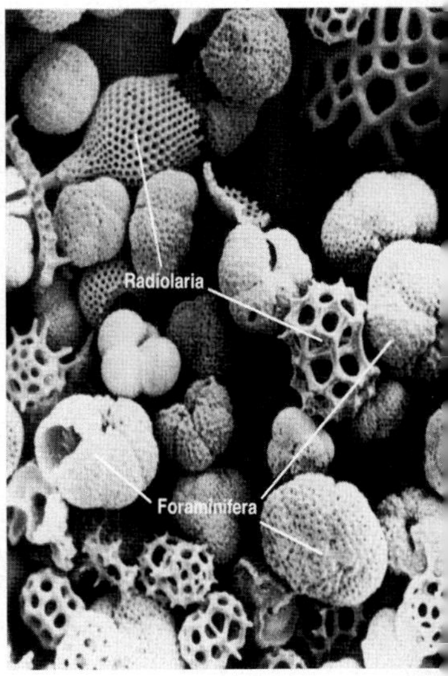

Sediments that blanket the ocean floor often contain large amounts of skeletal remains of microscopic organisms. (Deep Sea Drilling Project)

Abyssal Plains

Abyssal plains are deep, incredibly flat features; in fact, these regions are likely the most level places on Earth. The abyssal plain found off the coast of Argentina, for example, has less than 3 meters (10 feet) of relief over a distance exceeding 1300 kilometers (800 miles). The monotonous topography of abyssal plains is occasionally interrupted by the protruding summit of a partially buried volcanic peak.

Using *seismic profilers* (instruments that generate signals designed to penetrate far below the ocean floor), researchers have determined that abyssal plains owe their relatively featureless topography to thick accumulations of sediment that have buried an otherwise rugged ocean floor (see Figure 16.3). The nature of the sediment indicates that these plains consist primarily of fine sediments transported far out to sea by turbidity currents, deposits that have precipitated out of seawater, and shells and skeletons of microscopic marine organisms.

Table 16.1 Dimensions of Some Deep-Ocean Trenches

Trench	Depth (kilometers)	Average Width (kilometers)	Length (kilometers)
Aleutian	7.7	50	3700
Japan	8.4	100	800
Java	7.5	80	4500
Kurile-Kamchatka	10.5	120	2200
Mariana	11.0	70	2550
Central America	6.7	40	2800
Peru-Chile	8.1	100	5900
Philippine	10.5	60	1400
Puerto Rico	8.4	120	1550
South Sandwich	8.4	90	1450
Tonga	10.8	55	1400

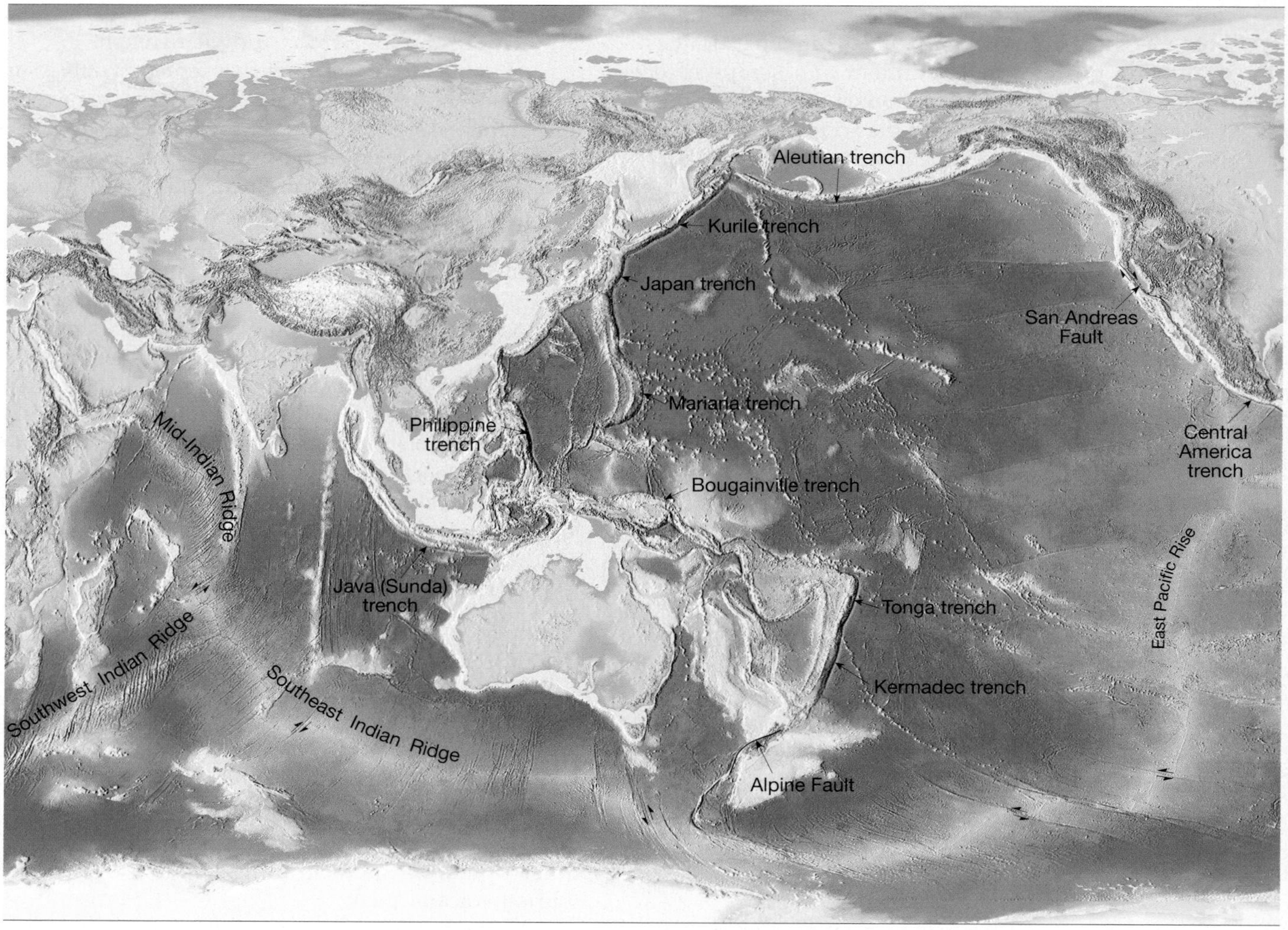

▲ **Figure 16.7** Distribution of the world's deep-ocean trenches.

Abyssal plains are found in all oceans. However, the Atlantic Ocean has the most extensive abyssal plains because it has few trenches to act as traps for sediment carried down the continental slope.

Seamounts, Guyots, and Oceanic Plateaus

Dotting the ocean floor are shield volcanoes called **seamounts,** which may rise hundreds of meters above the surrounding topography. It has been estimated that there are between 22,000 and 55,000 seamounts on the ocean floor—less than 2000 of these are active. Although these conical peaks are found on the floors of all the oceans, the greatest number have been identified in the Pacific. Furthermore, seamounts often form linear chains, or in some cases a more continuous volcanic ridge, not to be confused with mid-ocean ridges.

Some, like the Hawaiian Island–Emperor Seamount chain in the Pacific, which stretches from the Hawaiian Islands to the Aleutian trench, form over a volcanic hotspot in association with a mantle plume. Others are born near oceanic ridges, divergent plate boundaries where plates move apart. If the volcano grows large enough before being carried from the magma source by plate movement, the structure emerges as an island. Examples of volcanic islands in the Atlantic include the Azores, Ascension, Tristan da Cunha, and St. Helena.

During the time they exist as islands, some of these volcanoes are lowered to near sea level by weathering, mass wasting, running water, and wave action. Over a span of millions of years, the islands gradually sink and disappear below the water surface as the moving plate slowly carries them away from the elevated oceanic ridge or hot spot where they originated (see Box 16.1). These submerged, flat-topped seamounts are called **guyots** or **tablemounts.***

Mantle plumes have also generated several large **oceanic plateaus,** which resemble the flood basalt

*The term *guyot* is named after Princeton University's first geology professor. It is pronounced "GEE-oh" with a hard *g* as in "give."

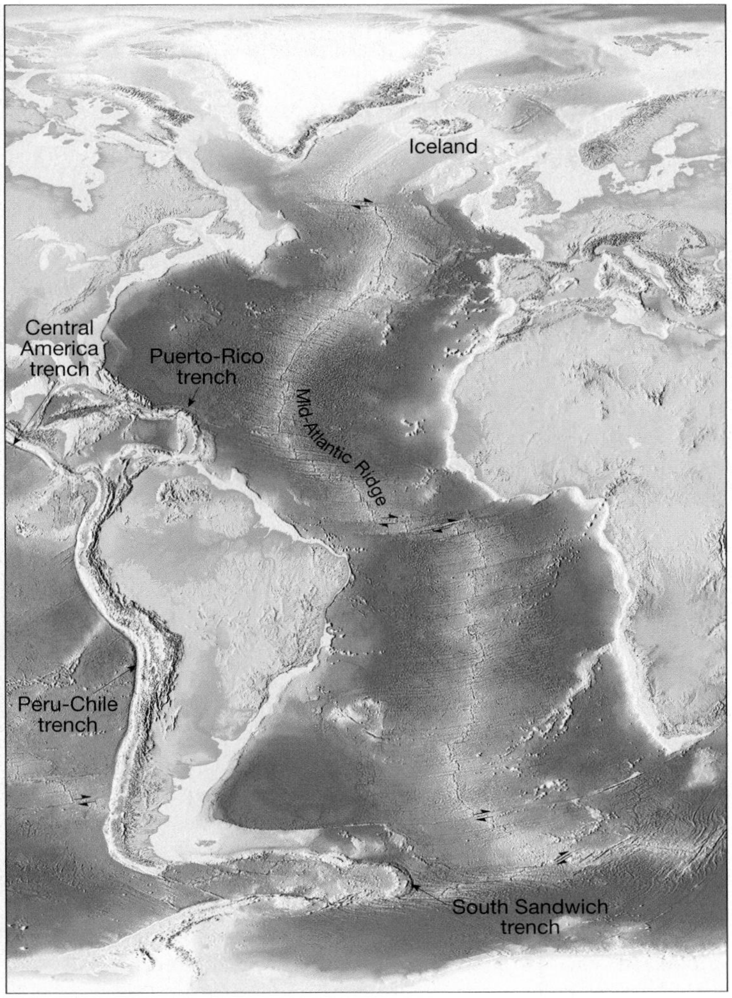

▲ Figure 16.7

provinces found on the continents. Examples of these extensive volcanic structures include the Ontong Java and Rockall plateaus, which formed from vast outpourings of fluid basaltic lavas onto the ocean floor (see Figure 17.15). Hence, oceanic plateaus are composed mostly of pillow basalt and other mafic rocks that in some cases exceed 30 kilometers in thickness.

Anatomy of the Oceanic Ridge

GEODe ESSENTIALS OF GEOLOGY

Origin and Evolution of the Ocean Floor

▼ Oceanic Ridges and Seafloor Spreading

Along well-developed divergent plate boundaries, the seafloor is elevated, forming a broad linear swell called the **oceanic ridge,** or **mid-ocean ridge.** Our knowledge of the oceanic ridge system comes from soundings taken of the ocean floor, core samples obtained from deep-sea drilling, visual inspection using deep-diving submersibles, and even firsthand inspection of slices of ocean floor that have been displaced onto dry land along convergent plate boundaries (Figure 16.8). An elevated position, extensive faulting and associated earthquakes, high heat flow, and numerous volcanic structures characterize the oceanic ridge.

The interconnected oceanic ridge system is the longest topographic feature on Earth's surface, exceeding 70,000 kilometers (43,000 miles) in length. Representing 20 percent of Earth's surface, the oceanic ridge winds through all major oceans in a manner similar to the seam on a baseball (Figure 16.9). The crest of this linear structure typically stands 2 to 3 kilometers above the adjacent deep-ocean basins and marks the plate margins where new oceanic crust is created.

Notice in Figure 16.9 that large sections of the oceanic ridge system have been named based on their locations within the various ocean basins. Ideally, ridges run along the middle of ocean basins, where they are called *mid-ocean* ridges. This holds true for the Mid-Atlantic Ridge, which is positioned in the middle of the Atlantic, roughly paralleling the margins of the continents on either side (Figure 16.9A). This is also true for the Mid-Indian Ridge, but note that the East Pacific Rise is displaced to the eastern side of the Pacific Ocean (Figure 16.9B,C).

The term *ridge* may be misleading, because these features are not narrow and steep as the term implies, but have widths of from 1000 to 4000 kilometers and the appearance of a broad elongated swell that exhibits various degrees of ruggedness. Furthermore, careful examination of Figure 16.7 shows that the ridge system is broken into segments that range from a few tens to hundreds of kilometers in length. Although each segment is offset from the adjacent segment, they are generally connected, one to the next, by a transform fault.

Oceanic ridges are as high as some mountains found on the continents, and thus they are often described as mountainous in nature. However, the similarity ends there. Whereas most continental mountains form when compressional forces fold and metamorphose thick sequences of sedimentary rocks along convergent plate boundaries, oceanic ridges form where tensional forces fracture and pull the ocean crust apart. The oceanic ridge consists of layers and piles of newly formed basaltic rocks that have been faulted into elongated blocks that are buoyantly uplifted.

Along the axis of some segments of the oceanic ridge system are deep down-faulted structures called **rift valleys** (Figure 16.10). These features may exceed

Did You Know?

In January 1960, U.S. Navy Lt. Don Walsh and explorer Jacques Piccard descended to the bottom of the Challenger Deep region of the Mariana Trench. More than five hours after leaving the surface, they reached the bottom at 10,912 meters (35,800 feet)—a record depth of human descent that has never been surpassed.

Iceland's largest fishing port was extensively damaged in 1973 by the eruption of Heimaey Volcano. (Sigurgeir Jonasson)

BOX 16.1

Explaining Coral Atolls—Darwin's Hypothesis

Coral *atolls* are ring-shaped structures that often extend to depths of several thousand meters below sea level (Figure 16.A). What causes atolls to form, and how do they attain such a great thickness?

Corals are colonial animals about the size of an ant that feed with stinging tentacles and are related to jellyfish. Most corals protect themselves by creating a hard external skeleton made of calcium carbonate. Where corals reproduce and grow over many centuries their skeletons fuse into large structures called *coral reefs*. Other corals—as well as sponges and algae—begin to attach to the reef, enlarging it further. Eventually fishes, sea slugs, octopus, and other organisms are attracted to these diverse and productive habitats.

Figure 16.A An aerial view of Tetiaroa Atoll in the Pacific. The light blue waters of the relatively shallow lagoon contrast with the dark blue color of the deep ocean surrounding the atoll. (Photo by Douglas Peebles Photography).

Corals require specific environmental conditions to grow. For example, reef-building corals grow best in waters with an average annual temperature of about 24°C (75°F). They cannot survive prolonged exposure to temperatures below 18°C (64°F) or above 30°C (86°F). In addition, reef-builders require an attachment site (usually other corals) and clear sunlit water. Consequently, the limiting depth of most active reef growth is only about 45 meters (150 feet).

The restricted environmental conditions required for coral growth create an interesting paradox: How can corals—which require warm, shallow, sunlit water no deeper than a few dozen meters to live—create thick structures such as coral atolls that extend into deep water?

The naturalist Charles Darwin was one of the first to formulate a hypothesis on the origin of atolls. From 1831 to 1836 he sailed aboard the British ship HMS *Beagle* during its famous circumnavigation of the globe. In various places that Darwin visited, he noticed a progression of stages in coral reef development from (1) a *fringing reef* along the margins of a volcano to (2) a *barrier reef* with a volcano in the middle to (3) an *atoll,* which consists of a continuous or broken ring of coral reef surrounded by a central lagoon (Figure 16.B). The essence of Darwin's hypothesis was that as a volcanic island slowly sinks, the corals continue to build the reef complex upward.

Darwin's hypothesis explained how coral reefs, which are restricted to shallow water, can build structures that now exist in much deeper water. During Darwin's time, however, there was no

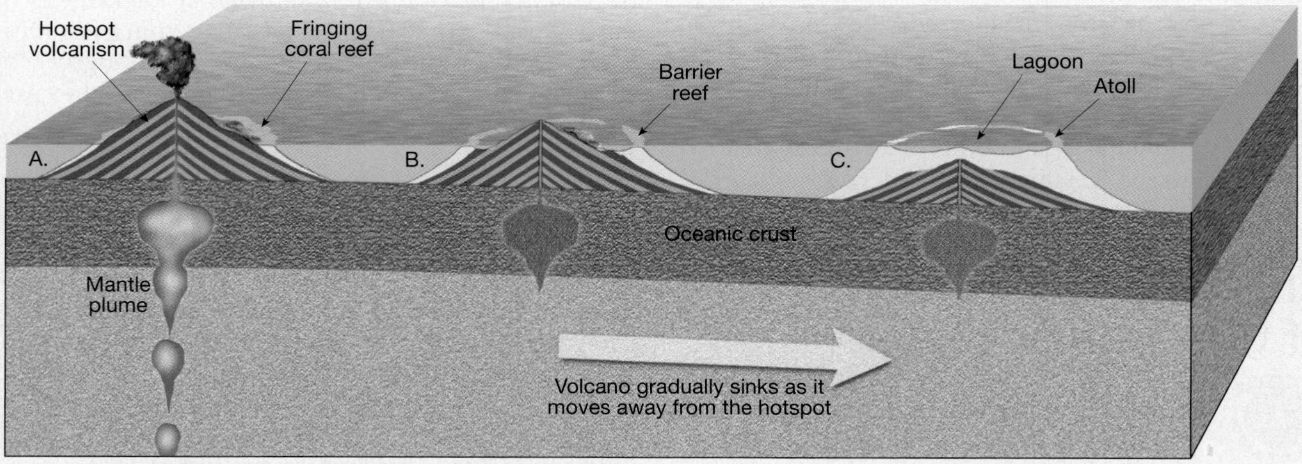

Figure 16.B Formation of a coral atoll due to the gradual sinking of oceanic crust and upward growth of the coral reef. A. A fringing coral reef forms around an active volcanic island. B. As the volcanic Island moves away from the region of hotspot activity it sinks, and the fringing reef gradually becomes a barrier reef. C. Eventually, the volcano is completely submerged and an atoll remains.

plausible mechanism to account for how an island might sink.

Today, plate tectonics helps explain how a volcanic island can become extinct and sink to great depths over long periods of time. Volcanic islands often form over a relatively stationary mantle plume, which causes the lithosphere to be buoyantly uplifted. Over a span of millions of years, these volcanic islands become inactive and gradually sink as the moving plate carries them away from the region of hotspot volcanism (Figure 16.B).

Drilling through atolls has revealed that volcanic rock does indeed underlie the oldest (and deepest) coral reef structures, confirming Darwin's hypothesis. Thus, atolls owe their existence to the gradual sinking of volcanic islands containing coral reefs that build upward through time.

50 kilometers in width and 2000 meters in depth. Because they contain faulted and tilted blocks of oceanic crust, as well as volcanic cones that have grown upon the newly formed seafloor, rift valleys usually exhibit rugged topography. The name *rift valley* has been applied to these features because they are so strikingly similar to continental rift valleys as exemplified by the East African Rift.

Topographically, the outermost flanks of most ridges are relatively subdued (except for isolated volcanic peaks) and rise very gradually (slope less than 1 degree) toward the ridge axis. Approaching the ridge crest, the topography becomes more rugged as volcanic structures, and faulted valleys that tend to parallel the ridge axis become more prominent. The most rugged topography is found on those ridges that exhibit large rift valleys.

▼ **Figure 16.8** The deep-diving submersible *Alvin* is 7.6 meters long, weighs 16 tons, has a cruising speed of 1 knot, and can reach depths as great as 4000 meters (13,000 feet). A pilot and two scientific observers are along during a normal 6- to 10-hour dive. *(Courtesy of Rod Catanach/Woods Hole Oceanographic Institution)*

Because of its accessibility to both American and European researchers, parts of the Mid-Atlantic Ridge have been studied in considerable detail (Figure 16.10). It is a broad, submerged structure standing 2500 to 3000 meters above the adjacent ocean basin floor. In a few places, such as Iceland, the ridge has actually grown above sea level (Figure 16.9). Throughout most of its length, however, this divergent plate boundary lies far below sea level. Another prominent feature of the Mid-Atlantic Ridge is its deep linear rift valley extending along the ridge axis. Using both surface ships and submersibles, as well as sophisticated side-scanning sonar equipment, "images" of this rift valley have been obtained for the benefit of current and future investigations. In places, this rift valley is more than 30 kilometers wide and bounded by walls that are about 1500 meters high. This makes it comparable to the deepest and widest part of Arizona's Grand Canyon.

This rifted area of Iceland sits astride the Mid-Atlantic Ridge. (George Gerster)

Origin of Oceanic Lithosphere

GEODe ESSENTIALS OF GEOLOGY

Origin and Evolution of the Ocean Floor

▼ Oceanic Ridges and Seafloor Spreading

Oceanic ridges represent constructive plate margins where new oceanic lithosphere originates. In fact, the greatest volume of magma (over 60 percent of Earth's total yearly output) is produced along the oceanic ridge system in association with seafloor spreading. As the plates diverge, fractures are created in the oceanic crust that are immediately filled with molten rock that wells up from the hot asthenosphere below. This molten material slowly cools to solid rock, producing new slivers of seafloor. This process occurs again and again, generating new lithosphere that moves from the ridge crest in a conveyor belt fashion.

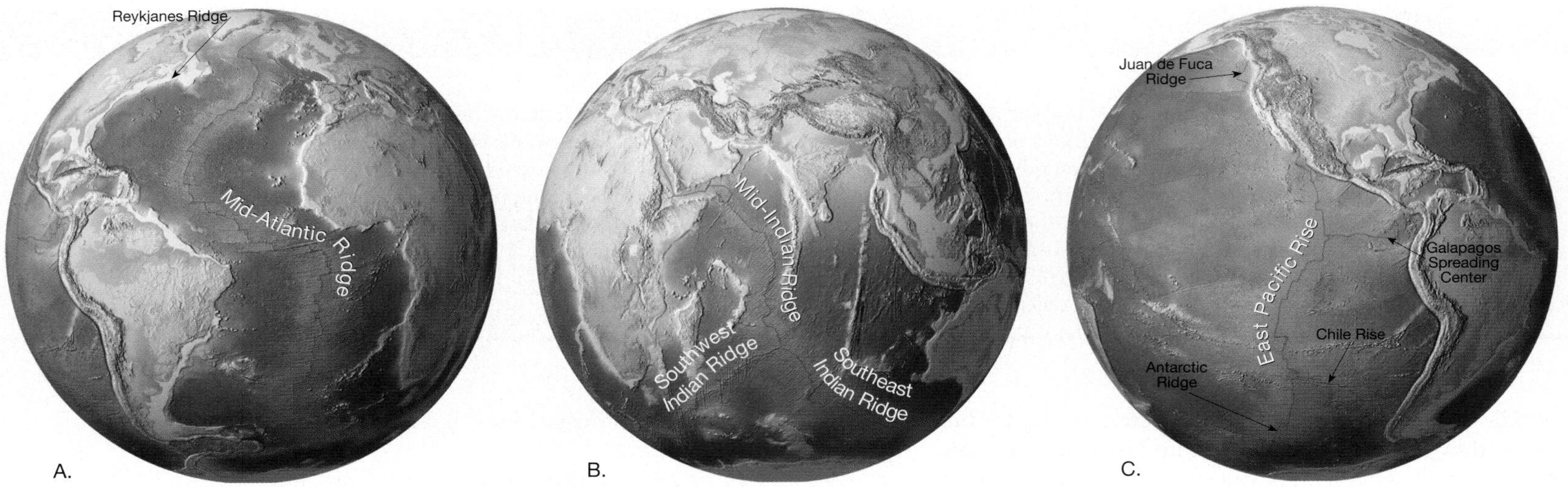

▲ **Figure 16.9** Distribution of the oceanic ridge system, which winds through all major ocean basins, like the seam on a baseball.

Seafloor Spreading

Harry Hess of Princeton University formulated the concept of seafloor spreading in the early 1960s. Later, geologists were able to verify Hess's view that seafloor spreading occurs along relatively narrow areas, located at the crests of oceanic ridges called **rift zones.** Here, below the ridge axis where the lithospheric plates separate, solid hot mantle rocks rise upward to replace the material that has shifted horizontally. Recall from Chapter 4 that as rock rises, it experiences a decrease in confining pressure and may undergo melting without the addition of heat. This process, called *decompression melting,* is how magma is generated along the ridge axis.

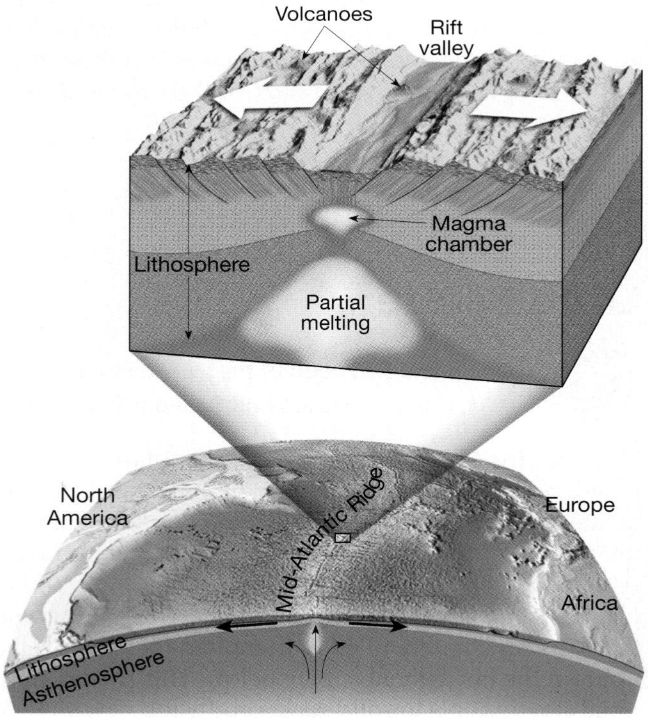

▲ **Figure 16.10** The axis of some segments of the oceanic ridge system contains deep downfaulted structures called *rift valleys.* Some may exceed 50 kilometers in width and 2000 meters in depth.

Partial melting of mantle rock produces basaltic magma having a composition that is surprisingly consistent along the entire length of the ridge system. This newly formed magma separates from the mantle rock from which it was derived, and rises toward the surface in the form of teardrop-shaped blobs. Although most of this magma collects in elongated reservoirs (magma chambers) located just beneath the ridge crest, about 10 percent eventually migrates upward along fissures to erupt as lava flows on the ocean floor (Figure 16.10). This activity continuously adds new basaltic rock to the plate margins, temporarily welding them together, only to be broken as spreading continues. Along some ridges, outpourings of bulbous lavas build submerged shield volcanoes (seamounts) as well as elongated lava ridges. At other locations, very fluid lavas create a more subdued topography.

During seafloor spreading, the magma that is injected into newly developed fractures forms dikes that cool from their outer borders inward toward their centers. Because the warm interiors of these newly formed dikes are weak, continued spreading produces new fractures that tend to split these young rocks roughly in half. As a result, new material is added equally to the two diverging plates. Consequently, new ocean floor grows symmetrically on each side of the centrally located ridge crest. Indeed, the ridge systems of the Atlantic and Indian oceans are located near the middle of these water bodies and is the reason they are called mid-ocean ridges. However, the East Pacific Rise is situated far from the center of the Pacific Ocean. Despite uniform spreading along the East Pacific Rise, much of the

the surface of a submarine lava flow is chilled quickly by seawater, it rarely travels more than a few kilometers before completely solidifying. The forward motion occurs as lava accumulates behind the congealed margin and then breaks through. This process occurs over and over, as molten basalt is extruded—like toothpaste out of a tightly squeezed tube. The result is tube-shaped protuberances resembling large bed pillows stacked one atop the other, hence the name **pillow basalts** (Figure 16.14). In some settings, pillow lavas may build into volcano-size mounds that resemble shield volcanoes, while in others they form elongated ridges. These structures will eventually be cut off from their supply of magma as they are carried away from the ridge crest by sea-floor spreading.

The lowest unit of the ocean crust develops from crystallization within the central magma chamber itself. The first minerals to crystallize are olivine, pyroxene, and occasionally chromite (chromium oxide), which fall through the magma to form a layered zone near the floor of the reservoir. The remaining magma tends to cool along the walls of the chamber and forms massive amounts of coarse-grained gabbro. This unit makes up the bulk of the oceanic crust, where it may account for as much as 5 of its 7-kilometer total thickness.

In this manner, the processes at work along the ridge system generate the entire sequence of rocks found in an ophiolite complex. Since the magma chambers are periodically replenished with fresh magma rising from the asthenosphere, oceanic crust is continuously being generated.

Interactions Between Seawater and Oceanic Crust

In addition to serving as a mechanism for the dissipation of Earth's internal heat, the interaction between seawater and the newly formed basaltic crust alters both the seawater and the crust. Because submarine lava flows are very permeable and the upper basaltic crust is highly fractured, seawater can penetrate to a depth of 2 kilometers. As seawater circulates through the hot crust, it is heated and alters the basaltic rock by a process called *hydrothermal* (hot water) *metamorphism.* One consequence of this activity is that the dark silicates found in basalt are often transformed into the mineral chlorite.

In addition to the basaltic crust being altered, so is the seawater. As the hot seawater circulates through the newly formed rock, it dissolves ions of silica, iron, copper, and other metals from the hot basalts. Once the water is heated to several hundred degrees, it buoyantly rises along fractures and eventually spews out at the surface. Studies conducted by submersibles along the Juan de Fuca Ridge have photographed these metallic-rich solutions as they gush from the seafloor to form particle-filled clouds called **black smokers.** As the hot liquid (about 350°C) mixes with the cold seawater, the dissolved minerals precipitate to form massive deposits of metallic sulfide minerals, some of which are economically important. Occasionally these deposits grow upward to form large chimneylike structures as tall as skyscrapers.

▼ **Figure 16.14** Pillow lava exposed along a sea cliff, Cape Wanbrow, New Zealand. Notice that each pillow shows an outer, rapidly cooled, dark glassy layer enclosing a dark gray basalt interior. *(Photo by G. R. Roberts)*

Did You Know?

In July of 1958, the largest wave ever recorded occurred in Lituya Bay, about 100 miles from Juneau, Alaska. A rockslide triggered by an earthquake dumped 90 million tons of rock into the bay, which created a huge *splash wave.* The wave swept over a ridge, uprooting and snapping tree trunks that had been 1740 feet above the bay.

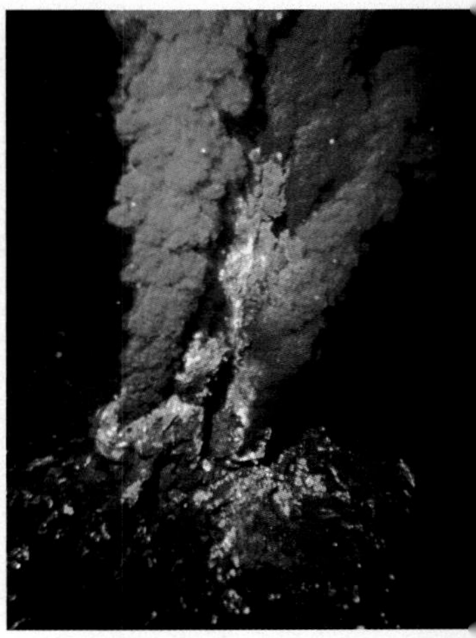

Black smoker chimney. (P. Arona)

Continental Rifting: The Birth of a New Ocean Basin

GEODe: Essentials of Geology — Origin and Evolution of the Ocean Floor ▼ The Formation of Ocean Basins

Why the supercontinent of Pangaea began to split apart nearly 200 million years ago is not known with certainty. Nevertheless, this event serves to illustrate that perhaps most ocean basins get their start when a continent begins to break apart. This clearly is the case for the Atlantic Ocean, which formed as the Americas drifted from Europe and Africa. It is also true for the Indian Ocean, which developed as Africa rifted from Antarctica and India.

Evolution of an Ocean Basin

The development of a new ocean basin begins with the formation of a **continental rift,** an elongated depression in which the entire thickness of the lithosphere has been deformed. Examples of continental rifts include the East African Rift, the Baikal Rift (south central Siberia), the Rhine Valley (northwestern Europe), the Rio Grand Rift, and the Basin and Range province in the western United States. It appears that continental rifts form in a variety of tectonic settings and may result in the breakup of a landmass.

Did You Know?

The circulation of seawater through the oceanic crust extracts heat from the volcanically produced material and is the main method by which the crust is cooled.

In those settings where rifting continues, the rift system will evolve into a young, narrow ocean basin, exemplified by the present-day Red Sea. Eventually, seafloor spreading results in the formation of a mature ocean basin bordered by rifted continental margins. The Atlantic Ocean is such a feature. What follows is a look at this model of ocean basin evolution using modern examples to represent the various stages of rifting.

East African Rift. An example of an active continental rift is the East African Rift, which extends through eastern Africa for approximately 3000 kilometers (2000 miles). Rather than being a single rift, the East African Rift consists of several somewhat interconnected rift valleys that split into an eastern and western section around Lake Victoria (see Figure 15.12, p. 348). Whether this rift will develop into a divergent boundary, where the Somali subplate separates from the continent of Africa, is still being debated. Nevertheless, the East African Rift is thought to characterize the initial stage in the breakup of a continent.

The most recent period of rifting began about 20 million years ago as upwelling in the mantle forcefully intruded the base of the lithosphere (Figure 16.15A).

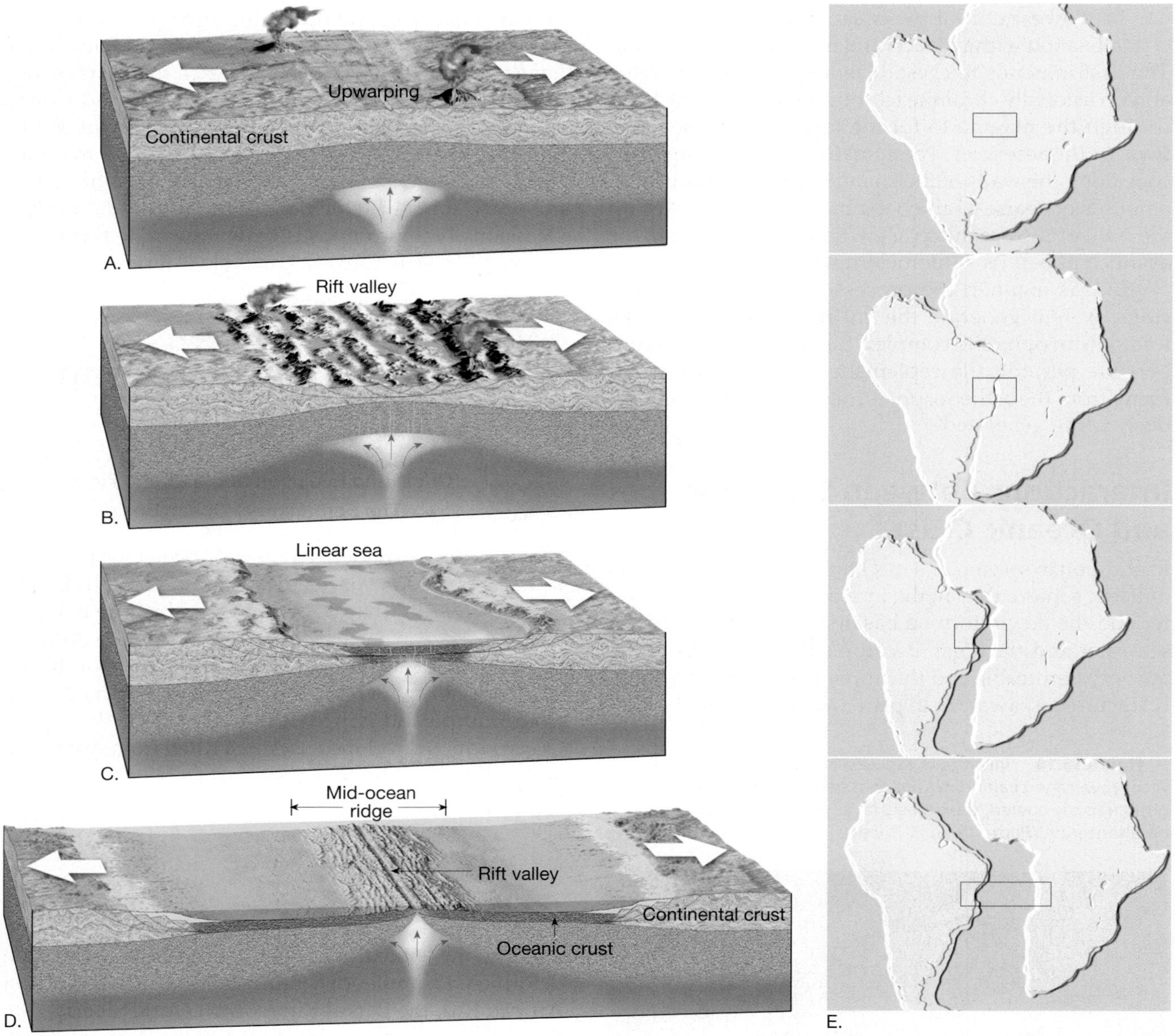

▲ **Figure 16.15** Formation of an ocean basin. **A.** Tensional forces and buoyant uplifting of the heated lithosphere cause the upper crust to be broken along normal faults, while the lower crust deforms by ductile stretching. **B.** As the crust is pulled apart, large slabs of rock sink, generating a rift zone. **C.** Further spreading generates a narrow sea. **D.** Eventually, an expansive ocean basin and ridge system are created. **E.** Illustration of the separation of South America and Africa to form the South Atlantic.

Buoyant uplifting of the heated lithosphere led to doming of the crust. As a consequence, the upper crust was broken along steep-angle normal faults, producing downfaulted blocks, or *grabens,* while the lower crust deformed by ductile stretching (Figure 16.15B).

In its early stage of formation, magma generated by decompression melting of the rising mantle plume intrudes the crust. Some of the magma migrates along fractures and erupts at the surface. This activity produces extensive basaltic flows within the rift as well as volcanic cones—some forming more than 100 kilometers from the rift axis. Examples include Mount Kilimanjaro, which is the highest point in Africa, rising almost 6000 meters (20,000 feet) above the Serengeti Plain, and Mount Kenya.

Red Sea. Research suggests that if extensional forces are maintained, a rift valley will lengthen and deepen, eventually extending out to the margin of the continent, thereby splitting it in two (Figure 16.15C). At this point, the continental rift becomes a narrow linear sea with an outlet to the ocean, similar to the Red Sea.

The Red Sea formed when the Arabian Peninsula rifted from Africa beginning about 30 million years ago. Steep fault scarps that rise as much as 3 kilometers above sea level flank the margins of this water body. Thus, the escarpments surrounding the Red Sea are similar to the cliffs that border the East African Rift. Although the Red Sea only reaches oceanic depths (up to 5 kilometers) in a few locations, symmetrical magnetic stripes indicate that typical seafloor spreading here has been taking place for the past 5 million years.

Atlantic Ocean. If spreading continues, the Red Sea will grow wider and develop an elevated oceanic ridge similar to the Mid-Atlantic Ridge (Figure 16.15D). As new oceanic crust is added to the diverging plates, the rifted continental margins move ever so slowly away from one another. As a result, the rifted continental margins that were once situated above the region of upwelling are displaced toward the interior of the growing plates. Consequently, as the continental lithosphere moves away from the source of heat, it cools, contracts, and subsides.

In time these continental margins will subside below sea level. Simultaneously, material eroded from the adjacent landmass will be deposited atop the faulted topography of the submerged continental margin. Eventually, this material will accumulate to form a thick, broad wedge of relatively undisturbed sediment and sedimentary rock. Recall that continental margins of this type are called *passive continental margins.* Examples of passive continental margins surround the Atlantic Ocean, including the east coasts of North and South America, as well as the coastal areas of Western Europe and Africa. Because passive margins are not associated with plate boundaries, they experience little volcanism and few earthquakes. Recall, however, that this was not the case when these lithospheric blocks made up the flanks of a continental rift.

Not all continental rift valleys develop into full-fledged divergent boundaries. Running through the central United States is a failed rift that extends from Lake Superior into central Kansas. This once active rift valley is filled with volcanic rock that was extruded onto the crust more than a billion years ago. Why one rift valley develops into an active plate boundary while others are abandoned is not yet known.

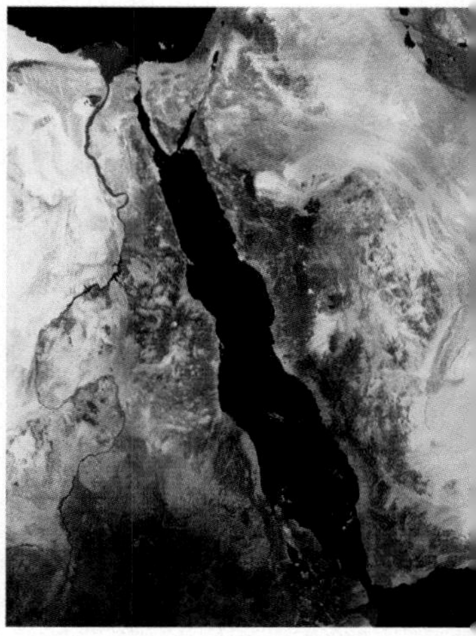

Satellite image of the Red Sea. (NASA)

Destruction of Oceanic Lithosphere

Although new lithosphere is continually being produced at divergent plate boundaries. Earth's surface area is not growing larger. In order to balance the amount of newly created lithosphere, there must be a process whereby plates are destroyed. Recall that this occurs along *convergent boundaries,* also called *subduction zones.*

Why Oceanic Lithosphere Subducts

The process of plate subduction is complex, and the ultimate fate of subducted plates is still being debated. What is known with some certainty is that a slab of oceanic lithosphere subducts because its overall density is greater than that of the underlying mantle. Recall that when ocean crust forms along a ridge, it is warm and buoyant, a fact that results in the ridge being elevated above the deep-ocean basins. However, as oceanic lithosphere moves away from the site of warm upwelling, it cools and thickens. After about 15 million years, an oceanic slab tends to be denser than the supporting asthenosphere. In parts of the western Pacific some oceanic lithosphere is nearly 180 million years old. This is the thickest and most dense in today's oceans. The subducting slabs in this region typically descend at angles approaching 90 degrees (Figure 16.16A). Sites where plates subduct at such steep angles are found in association with the Tonga, Mariana, and Kurile trenches.

When a divergent boundary is located near a subduction zone, the oceanic lithosphere is still young and therefore warm and buoyant. Hence, the angle of descent for these slabs is small (Figure 16.16B). It is even possible that oceanic lithosphere may be overridden by a continental landmass before it has cooled sufficiently to

Did You Know?

Communities of organisms reside around hydrothermal vents (black smokers) in dark, hot, sulfur-rich environments where photosynthesis cannot occur. The base of the food web is provided by bacteria-like organisms that use a process called chemosynthesis and heat energy from the vents to produce sugars and other foods that allow them and many other organisms to live in this extreme environment.

Did You Know?

Scientists at the Jet Propulsion Laboratory have predicted that a 1-kilometer-wide asteroid has a one-in-300 chance of plunging into the North Atlantic Ocean on March 16, 2880. If the event actually occurs, it would send a 300-foot wall of water crashing into the Eastern Seaboard of the United States.

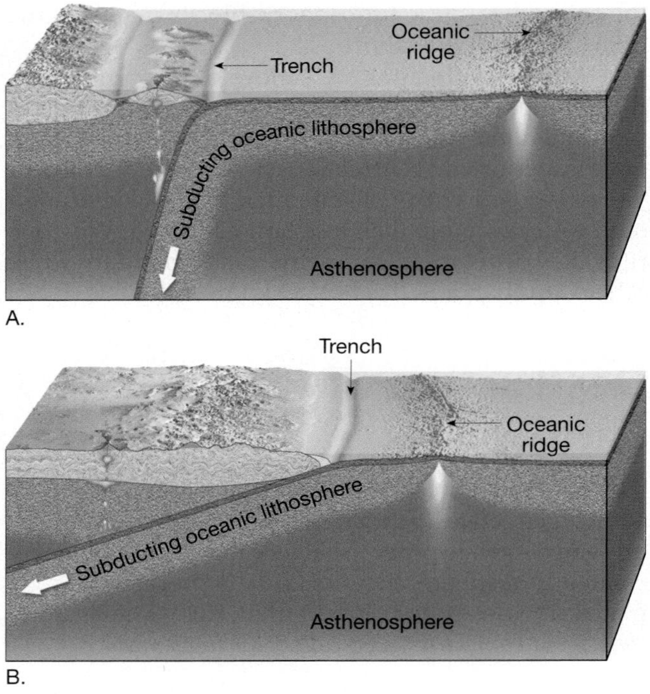

▲ **Figure 16.16** The angle at which oceanic lithosphere descends into the asthenosphere depends on its density. **A.** In parts of the Pacific, some oceanic lithosphere is older than 160 million years and typically descends into the mantle at angles approaching 90 degrees. **B.** Young oceanic lithosphere is warm and buoyant, hence it tends to subduct at a low angle.

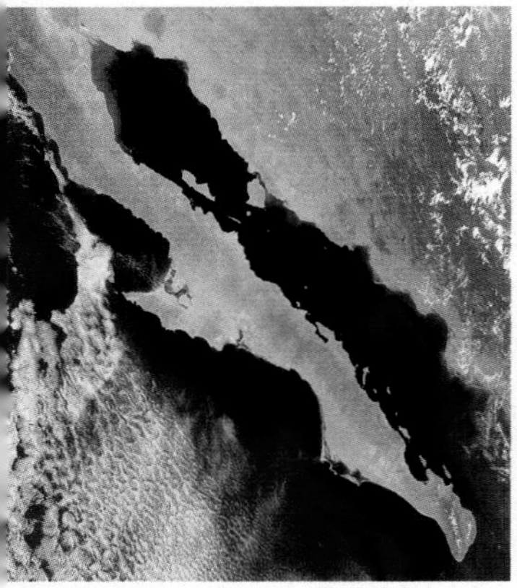

Satellite image showing the separation between the Baja Peninsula and North America. (NASA)

readily subduct. In this situation, the slab may be so buoyant that rather than plunging into the mantle, it moves horizontally beneath a block of continental lithosphere. This phenomenon is called **buoyant subduction.** Buoyant slabs are thought to eventually sink when they cool sufficiently, and their density increases.

It is important to note that it is the *lithospheric mantle,* located beneath the oceanic crust, that drives subduction. Even when the oceanic crust is quite old, its density is 3.0 g/cm^3, which is less than the underlying asthenosphere with a density of about 3.2 g/cm^3. Only because the cold lithospheric mantle is denser than the warmer asthenosphere, which supports it, does subduction occur.

Did You Know?

The Bering Sea is the most northerly marginal sea of the Pacific Ocean and connects to the Arctic Ocean through the Bering Straits. This water body is effectively cut off from the Pacific basins by the Aleutian Islands, which were created by volcanic activity associated with northward subduction of the Pacific basin.

Subducting Plates: The Demise of an Ocean Basin

Using magnetic stripes and fracture zones on the ocean floor, geologists began reconstructing the movement of plates over the past 200 million years. From this work they discovered that parts, or even entire ocean basins, have been destroyed along subduction zones. For example, during the breakup of Pangaea shown in Figure 15.A (p. 338) notice that the African plate rotates and moves northward. Eventually the northern margin of Africa collides with Eurasia. During this event, the floor of the intervening Tethys Ocean was almost entirely consumed into the mantle, leaving behind only a small remnant—the Mediterranean Sea.

Reconstructions of the breakup of Pangaea also helped investigators understand the demise of the Farallon plate—a large oceanic plate that once occupied much of the eastern Pacific basin. Prior to the breakup, the Farallon plate, plus one or two smaller plates, were situated opposite the Pacific plate on the eastern side of a divergent boundary located near the center of the Pacific basin. A modern remnant of this boundary, which generated both the Farallon and Pacific plates, is the East Pacific Rise.

Beginning about 180 million years ago, the Americas were propelled westward by seafloor spreading in the Atlantic. Hence, the convergent plate boundaries that formed along the west coasts of North and South America gradually migrated westward relative to the spreading center located in the Pacific. The Farallon plate, which was subducting beneath the Americas faster than it was being generated, got smaller and smaller (Figure 16.17B). As its surface area decreased, it broke into smaller pieces, some of which subducted entirely. The remaining fragments of the once mighty Farallon plate are the Juan de Fuca, Cocos, and Nazca plates.

As the Farallon plate shrank, the Pacific plate grew larger, encroaching on the American plates. About 30 million years ago, a section of the East Pacific Rise collided with the subduction zone that once lay off the coast of California (Figure 16.17B). As this spreading center subducted into the California trench, these structures were mutually destroyed and replaced by a newly generated transform fault system that accommodates the differential motion between the North American and Pacific plates. As more of the ridge was subducted, the transform fault system, which we now call the San Andreas Fault, propagated through western California (Figure 16.17). Farther north, a similar event generated the Queen Charlotte transform fault.

Consequently, much of the present boundary between the Pacific and North American plates lies along transform faults located within the continent. In the United States (outside of Alaska), the only remaining part of the extensive convergent boundary that once ran along the entire West Coast is the Cascadia subduction zone. Here the subduction of the Juan de Fuca plate has generated the volcanoes of the Cascade Range.

Today, the southern end of the San Andreas Fault connects to a young divergent boundary (an extension of the East Pacific Rise) that generated the Gulf of California. Because of this change in plate geometry,

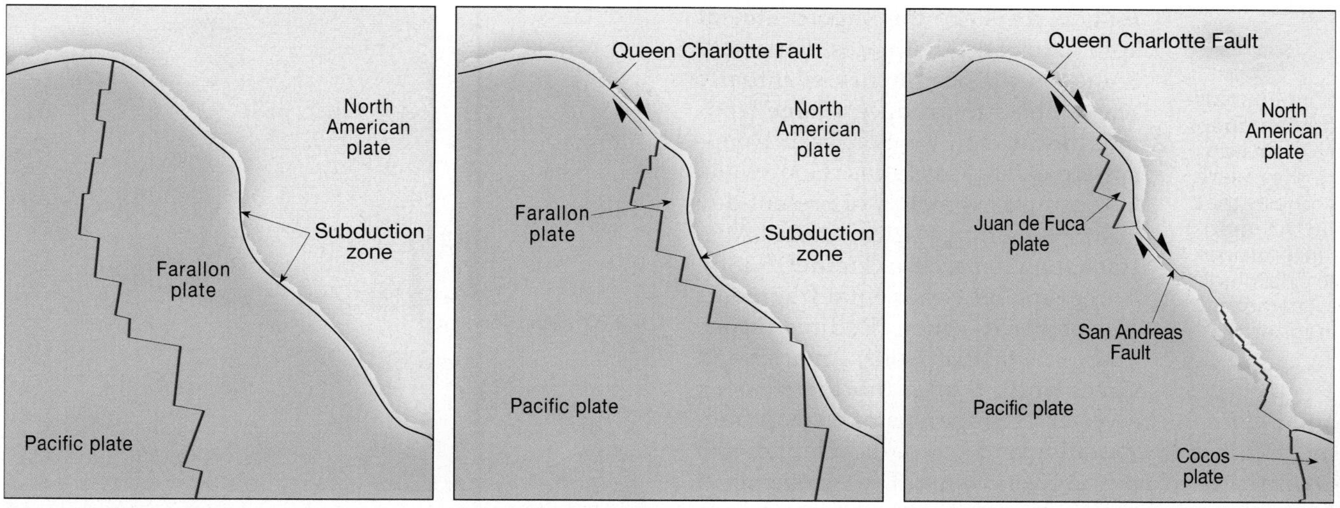

A. 56 million years ago B. 37 million years ago C. Today

▲ **Figure 16.17** Simplified illustration of the demise of the Farallon plate, which once ran along the western margin of the Americas. Because the Farallon plate was subducting faster than it was being generated, it got smaller and smaller. The remaining fragments of the once mighty Farallon plate are the Juan de Fuca, Cocos, and Nazca plates.

the Pacific plate has captured a sliver of North America (the Baja Peninsula) and is carrying it northwestward toward Alaska at a rate of about 6 centimeters per year.

Opening and Closing Ocean Basins: The Supercontinent Cycle

GEODe Essentials of Geology
Origin and Evolution of the Ocean Floor
▼ Pangaea: Formation and Breakup of a Supercontinent

Geologists are confident that plate tectonics has operated over the last 2 billion years, and perhaps even longer. The question that arises is "What came before Pangaea?" Pangaea was the most recent, but not the only supercontinent to exist in the geologic past. We can get some notion of what came before Pangaea by taking a closer look at the fate of this supercontinent.

Recall that Pangaea began to break up about 180 million years ago, and the pieces are still dispersing today. Crustal fragments from the breakup of Pangaea have already begun to reassemble to form a new supercontinent as evidenced by the collision of India with Asia. The idea that rifting and dispersal of one supercontinent is followed by a long period during which the fragments are gradually reassembled into a new supercontinent having a different configuration is called the **supercontinent cycle.*** We will look at the breakup of an earlier supercontinent and its reassembly into Pangaea as a way of examining the supercontinental cycle.

Before Pangaea

The plate motions that resulted in the breakup and dispersal of Pangaea are well documented. The dates when individual crustal fragments rifted from each other can be computed from the magnetic stripes left on the newly formed ocean floor. However, this technique cannot be used for reconstructing events prior to the breakup of Pangaea because much of the ocean crust from this time period has since been subducted. Nevertheless, geologists have been able to reconstruct the positions of continents in earlier times by using apparent polar-wandering paths, paleoclimatic data, and matching ancient geologic structures, such as mountain belts and rock units.

The earliest well-documented supercontinent, *Rodinia,* formed about 1 billion years ago. Although its reconstruction is still being researched, it is clear that Rodinia had a much different configuration than Pangaea (Figure 16.18A). During the period between 750 and 550

*The supercontinent cycle is sometimes called the *Wilson Cycle* after J. Tuzo Wilson, who first described the opening and closing of a proto-Atlantic basin.

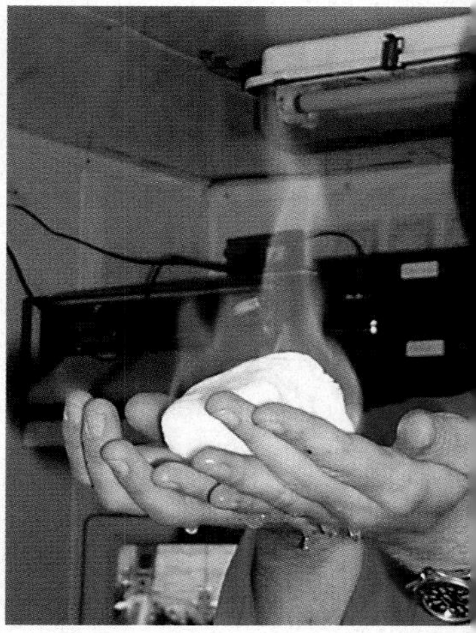

Gas hydrates are natural gas reservoirs in icelike crystalline solids found in submarine sediments. (GEOMAR Research Center)

Did You Know?

The Gulf of California, also known as the Sea of Cortez, formed over the past 6 million years by seafloor spreading. This 750-mile-long basin is located between the west coast of mainland Mexico and the Baja Peninsula.

Did You Know?

Sea turtles, salmon, and other animals travel great distances across the open ocean so that they can lay their eggs in the same location where they were hatched. Recent research suggests that these animals may use Earth's magnetic field for navigation. Interestingly, many animals that have "homing" ability also contain minuscule amounts of the magnetic mineral magnetite.

million years ago this supercontinent split apart and the pieces dispersed. Some of the fragments eventually reassembled to produce a large landmass located in the Southern Hemisphere called *Gondwana*. Gondwana was comprised mainly of present-day South America, Africa, India, Australia, and Antarctica (Figure 16.18B). Three smaller continental fragments also formed when Rodinia broke apart—*Laurentia* (North America and Greenland), *Siberia* (northern Asia), and *Baltica* (northwestern Europe). Later a small fragment, called *Avalonia* (England and parts of France and Spain), was rifted from Gondwana. The continents of Laurentia, Siberia, Baltica, and Avalonia began to collide about 430 million years ago, forming a landmass that straddled the equator, while the southern continent of Gondwana remained positioned over the South Pole (Figure 16.18C).

Pangaea began to take form over the next 100 million years as Gondwana migrated northward and collided with Laurentia and Baltica. As the developing supercontinent drifted northward, smaller fragments were added to Eurasia (Baltica and Siberia), and South America rammed into North America (Laurentia). By 230 million years ago, the supercontinent of Pangaea was nearly complete (Figure 16.18D). (Several crustal blocks, which today make up much of Southeast Asia, were never part of Pangaea.) Even before the last crustal fragments were added to Pangaea, North America and Africa began to rift apart. This event marks the beginning of the breakup and dispersal of this "newly" formed supercontinent.

Did You Know?

Today, coral reefs cover about 230,000 square miles of the world's marine areas. Many different stresses can destroy coral reefs, including human activity and rising sea levels due to global warming.

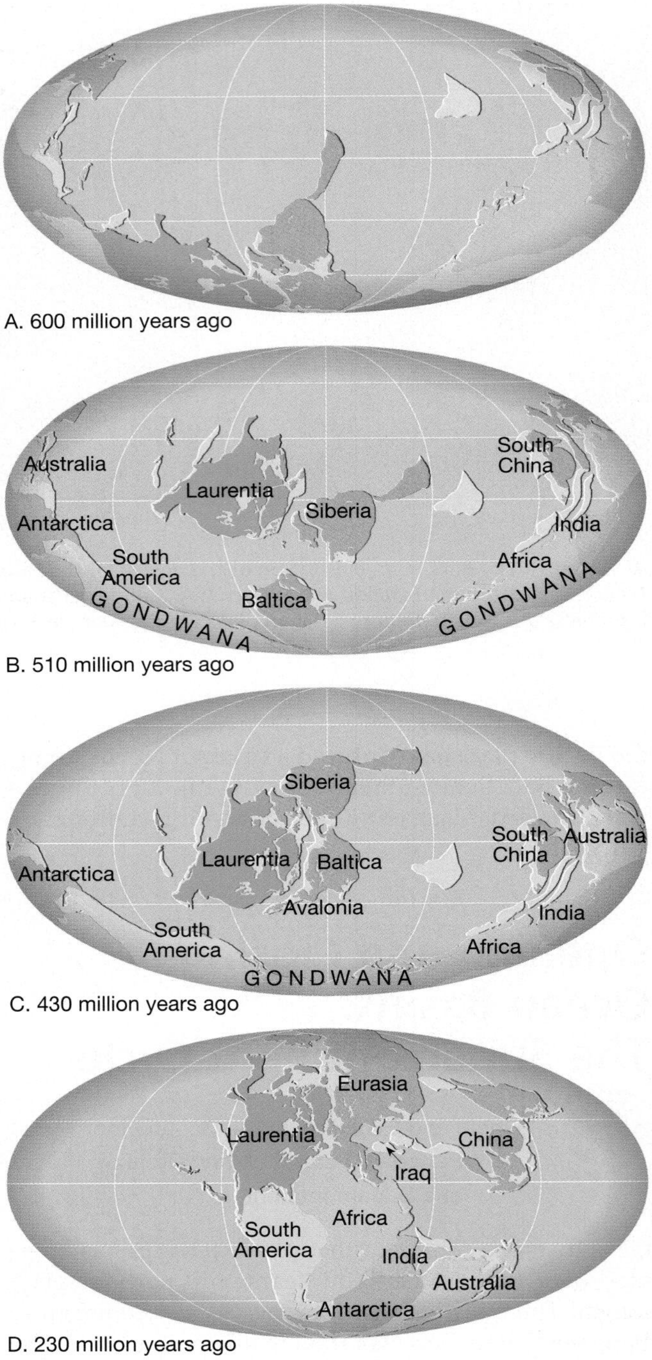

▲ **Figure 16.18** Sequence showing the breakup and dispersal of the supercontinent of Rodinia and the gradual reassembly of the fragments into the new supercontinent of Pangaea. Whenever the curved surface of Earth is drawn on a flat plane (map), some distortion results. The map used here, called a Mollweide projection, greatly distorts distances toward the edges but has very little distortion near the center. *(After C. Scotese, R. K. Bambach, C. Barton, R. VanderVoo, and A. Ziegler)*

The Chapter in Review

- *Ocean bathymetry is determined using echo sounders and multibeam sonars,* which bounce sonic signals off the ocean floor. Ship-based receivers record the reflected echoes and accurately measure the time interval of the signals. With this information, ocean depths are calculated and plotted to produce maps of ocean-floor topography. Recently, *satellite measurements* of the ocean surface have added data for mapping ocean-floor features.
- Oceanographers studying the topography of the ocean basins have delineated three major units: *continental margins, deep-ocean basins,* and *oceanic (mid-ocean) ridges.*
- The zones that collectively make up a *passive continental margin* include the *continental shelf* (a gently sloping, submerged surface extending from the shoreline toward the deep-ocean basin); *continental slope* (the true edge of the continent, which has a steep slope that leads from the continental shelf into deep water); and in regions where trenches do not exist, the relatively steep continental slope merges into a more gradual incline known as the *continental rise.* The continental rise consists of sediments that have moved downslope from the continental shelf to the deep-ocean floor.
- *Active continental margins* are located primarily around the Pacific Ocean in areas where the leading edge of a continent is overrunning oceanic lithosphere. Here sediment scraped from the descending oceanic plate is plastered against the continent to form a collection of sediments called an *accretionary wedge.* An active continental margin generally has a narrow continental shelf, which grades into a deep-ocean trench.
- The deep-ocean basin lies between the continental margin and the oceanic ridge system. Its features include *deep-ocean trenches* (long, narrow depressions that are the deepest parts of the ocean and are located where moving crustal plates descend into the mantle); *abyssal plains* (among the most level places on Earth, consisting of thick accumulations of sediments that were deposited atop the low, rough portions of the ocean floor by turbidity currents); *seamounts* (volcanic peaks on the ocean floor that originate near oceanic ridges or in association with volcanic hotspots); and *oceanic plateaus* (large flood basalt provinces similar to those found on the continents).
- *Oceanic (mid-ocean) ridges,* the sites of seafloor spreading, are found in all major oceans and represent more than 20 percent of Earth's surface. They are the most prominent features in the oceans, because they form an almost continuous swell that rises 2 to 3 kilometers above the adjacent ocean basin floor. Ridges are characterized by an *elevated position, extensive faulting,* and *volcanic structures* that have developed on newly formed oceanic crust. Most of the geologic activity associated with ridges occurs along a narrow region on the ridge crest, called the *rift zone,* where magma from the asthenosphere moves upward to create new slivers of oceanic crust. The topography of the various segments of the oceanic ridge is controlled by the rate of seafloor spreading.
- New oceanic crust is formed in a continuous manner by the process of seafloor spreading. The upper crust is composed of *pillow lavas* of basaltic composition. Below this layer are numerous interconnected dikes *(sheeted dike complex)* that are underlain by a thick layer of gabbro. This entire sequence is called an *ophiolite complex.*
- The development of a new ocean basin begins with the formation of a *continental rift* similar to the East African Rift. In those settings where rifting continues, a young, narrow ocean basin develops, exemplified by the Red Sea. Eventually, seafloor spreading creates an ocean basin bordered by rifted continental margins similar to the present-day Atlantic Ocean.
- Oceanic lithosphere subducts because its overall density is greater than the underlying asthenosphere. The subduction of oceanic lithosphere may result in the destruction of parts—or even entire—ocean basins. A classic example is the Farallon plate, most of which subducted beneath the American plates as they were displaced westward by seafloor spreading in the Atlantic.
- Rifting and dispersal of one supercontinent followed by a long period during which the fragments are gradually reassembled into a new supercontinent, having a different configuration, is called the *supercontinent cycle.*

Key Terms

abyssal plain (p. 373)
accretionary wedge (p. 373)
active continental margin (p. 372)
bathymetry (p. 368)
black smokers (p. 381)
buoyant subduction (p. 384)
continental margin (p. 370)
continental rift (p. 381)
continental rise (p. 372)
continental shelf (p. 370)
continental slope (p. 372)
deep-ocean basin (p. 373)
deep-ocean trench (p. 373)
deep-sea fan (p. 372)
echo sounder (p. 369)
guyot (p. 374)
mid-ocean ridge (p. 375)
oceanic plateau (p. 374)
oceanic ridge (p. 375)
ophiolite complex (p. 380)
passive continental margin (p. 370)
pillow basalts (p. 381)
rift valley (p. 375)
rift zone (p. 378)
seamount (p. 374)
seismic reflection profile (p. 370)
sheeted dike complex (p. 380)
sonar (p. 369)
supercontinent cycle (p. 385)
tablemount (p. 374)

Review Questions

1. Assuming that the average speed of sound waves in water is 1500 meters per second, determine the water depth if the signal sent out by an echo sounder requires 6 seconds to strike bottom and return to the recorder (see Figure 16.2)
2. Describe how satellites orbiting Earth can determine features on the seafloor without being able to directly observe them beneath several kilometers of seawater.
3. What are the three major topographic provinces of the ocean floor?
4. List the three major features that comprise a passive continental margin. Which of these features is considered a flooded extension of the continent? Which one has the steepest slope?
5. Describe the differences between active and passive continental margins. Be sure to include how various features relate to plate tectonics, and give a geographic example of each type of margin.
6. Why are abyssal plains more extensive on the floor of the Atlantic than on the floor of the Pacific?
7. How does a flat-topped *seamount*, or *guyot*, form?
8. Briefly describe the oceanic ridge system.
9. Although oceanic ridges can be as tall as some mountains found on the continents, how are these features different?
10. What is the source of magma for seafloor spreading?
11. What is the primary reason for the elevated position of the oceanic ridge system?
12. How does hydrothermal metamorphism alter the basaltic rocks that make up the seafloor? How is seawater changed during this process?
13. What is a *black smoker*?
14. Briefly describe the four layers of the ocean crust.
15. How does the *sheeted dike complex* form? What about the lower unit?
16. Name a place that exemplifies a *continental rift*.
17. Explain why oceanic lithosphere subducts even though the oceanic crust is less dense than the underlying asthenosphere.
18. What happened to the Farallon plate? Name the remnants of this plate.
19. Describe the *supercontinent cycle*.

Online Study Guide

The *Essentials of Geology* Web site uses the resources and flexibility of the Internet to aid in your study of the topics in this chapter. Written and developed by geology instructors, this site will help improve your understanding of geology. Visit **www.prenhall.com/lutgens** and click on the cover of *Essentials of Geology 9e* to find:

- Online review quizzes.
- Critical thinking exercises.
- Links to chapter-specific Web resources.
- Internet-wide key-term searches.

http://www.prenhall.com/lutgens

Hikers camping in Pakistan's Charakusa Valley with the bold peaks of the Karakoram Range in the background. *(Photo by Jimmy Chin/National Geographic/Getty)*

CHAPTER 17

Crustal Deformation and Mountain Building

Focus on Learning

To assist you in learning the important concepts in this chapter, you will find it helpful to focus on the following questions:

- What are the two basic types of rock deformation? What factors influence how rock deforms?
- What are the most common types of folds and faults? How does each type form?
- How are Aleutian- and Andean-type mountain building similar? How are they different?
- How does continental accretion relate to mountain building?
- How is the concept of isostasy related to mountain building?

Mountains provide some of the most spectacular scenery on our planet (Figure 17.1). This splendor has been captured by poets, painters, and songwriters alike. Geologists believe that at some time all continental regions were mountainous masses and have concluded that the continents grow by the addition of mountains to their flanks. Consequently, as geologists unravel the secrets of mountain formation, they also gain a deeper understanding of the evolution of Earth's continents. If continents do indeed grow by adding mountains to their flanks, how do geologists explain the existence of mountains (the Urals, for example) that are located in the interior of a landmass? To answer this and related questions, this chapter attempts to piece together the sequence of events believed to generate these lofty structures. We begin our look at mountain building by examining the process of rock deformation and the structures that result.

Rock Deformation

GEODe ESSENTIALS OF GEOLOGY

Crustal Deformation and Mountain Building: Part A
▼ Deformation

Every body of rock, no matter how strong, has a point at which it will fracture or flow. **Deformation** is a general term that refers to all changes in the original shape and/or size of a rock body. Most crustal deformation occurs along plate margins. Recall from Chapter 15 that the lithosphere consists of large segments (plates) that move relative to one another. Plate motions and the interactions along plate boundaries generate forces that cause rock to deform.

When rocks are subjected to forces (stresses) greater than their own strength, they begin to deform, usually by folding, flowing, or fracturing (Figure 17.2). It is easy to visualize how rocks break, because we normally think of them as being brittle. But how can rock masses be bent into intricate folds without being broken during the process? To answer this question, geologists performed laboratory experiments in which rocks were subjected to forces under conditions that simulated those existing at various depths within the crust.

Although each rock type deforms somewhat differently, the general characteristics of rock deformation were determined from these experiments. Geologists discovered that when stress is gradually applied, rocks first respond by deforming elastically. Changes that result from *elastic deformation* are recoverable; that is, like a rubber band, the rock will return to nearly its original size and shape when the force is removed. (As you saw in Chapter 14, the energy for most earthquakes comes from stored elastic energy that is released as rock snaps back to its original shape.)

Once the elastic limit (strength) of a rock is surpassed, it either flows (*ductile deformation*) or fractures (*brittle deformation*). The factors that influence the strength of a rock and thus how it will deform include temperature, confining pressure, rock type, and time.

Temperature and Confining Pressure

Rocks near the surface, where temperatures and confining pressures are low, tend to behave like a brittle solid and fracture once their strength is exceeded. This type of deformation is called **brittle failure** or **brittle deformation.** From our everyday experience, we know that glass objects, wooden pencils, china plates, and even our bones exhibit brittle failure once their

▼ **Figure 17.1** Mount Sneffels in the Colorado Rockies. *(Photo by Gavrel Jecan/Art Wolfe, Inc.)*

◀ **Figure 17.2** Folded sedimentary layers exposed on the face of Mount Kidd, Alberta, Canada. *(Photo by Peter French/DRK Photo)*

strength is surpassed. By contrast at depth, where temperatures and confining pressures are high, rocks exhibit *ductile* behavior. **Ductile deformation** is a type of solid-state flow that produces a change in the size and shape of an object without fracturing. Ordinary objects that display ductile behavior include modeling clay, beeswax, caramel candy, and most metals. For example, a copper penny placed on a railroad track will be flattened and deformed (without breaking) by the force applied by a passing train.

Ductile deformation of a rock—strongly aided by high temperature and high confining pressure—is somewhat similar to the deformation of a penny flattened by a train. Rocks that display evidence of ductile flow usually were deformed at great depth and may exhibit contorted folds that give the impression that the strength of the rock was akin to soft putty (Figure 17.2).

Rock Type

In addition to the physical environment, the mineral composition and texture of a rock greatly influence how it will deform. For example, crystalline rocks, such as granite, basalt, and quartzite, that are composed of minerals that have strong internal molecular bonds tend to fail by brittle fracture. By contrast, sedimentary rocks that are weakly cemented, or metamorphic rocks that contain zones of weakness, such as foliation, are more susceptible to ductile flow. Rocks that are weak and thus most likely to behave in a ductile manner when subjected to differential forces include rock salt, gypsum, and shale; limestone, schist, and marble are of intermediate strength. In fact, rock salt is so weak that it deforms under small amounts of differential stress and rises like stone pillars through beds of sediment that lie in and around the Gulf of Mexico.

Time

One key factor that researchers are unable to duplicate in the laboratory is how rocks respond to small amounts of force applied over long spans of *geologic time.* However, insights into the effects of time on deformation are provided in everyday settings. For example, marble benches have been known to sag under their own weight over a span of a hundred years or so, and wooden bookshelves may bend after being loaded with books for a relatively short period. In nature small stresses applied over long time spans surely play an important role in the deformation of rock. Forces that are unable to deform rock when initially applied may cause rock to flow if the force is maintained over an extended period of time.

Did You Know?

The tallest mountain in North America is the 20,321-foot-high Mount McKinley. The area surrounding the mountain was established as Mount McKinley National Park in 1917. In 1980 the park was enlarged and renamed Denali National Park and Preserve. Denali, the "High One" is the name used by native Athabascan people for this massive peak that crowns the 600-mile-long Alaska Range.

Folds

GEODe ESSENTIALS OF GEOLOGY

Crustal Deformation and Mountain Building: Part A
▼ Folds

During mountain building, flat-lying sedimentary and volcanic rocks are often bent into a series of wavelike undulations called **folds.** Folds in sedimentary strata are much like those that would form if you were to hold the ends of a sheet of paper and then push them together. In nature, folds come in a wide variety of sizes and configurations. Some folds are broad flexures in which rock units hundreds of meters thick have been slightly warped. Others are very tight microscopic structures found in metamorphic rocks. Size differences notwithstanding, most folds are the result of *compressional forces* that result in the shortening and thickening of the crust.

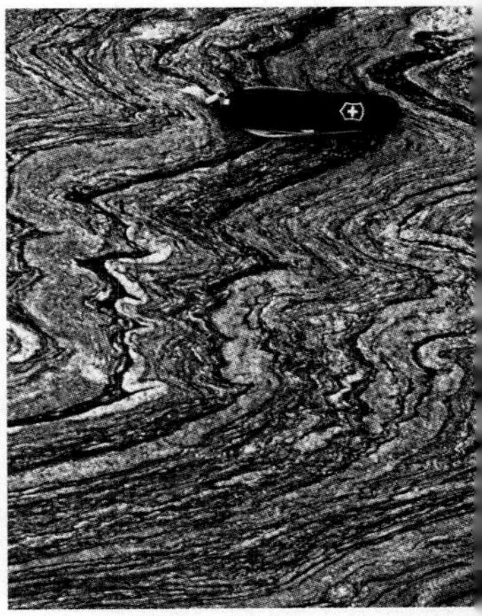

Rocks exhibiting the results of ductile behavior. (M. Miller)

Types of Folds

The two most common types of folds are anticlines and synclines (Figure 17.3). An **anticline** is most

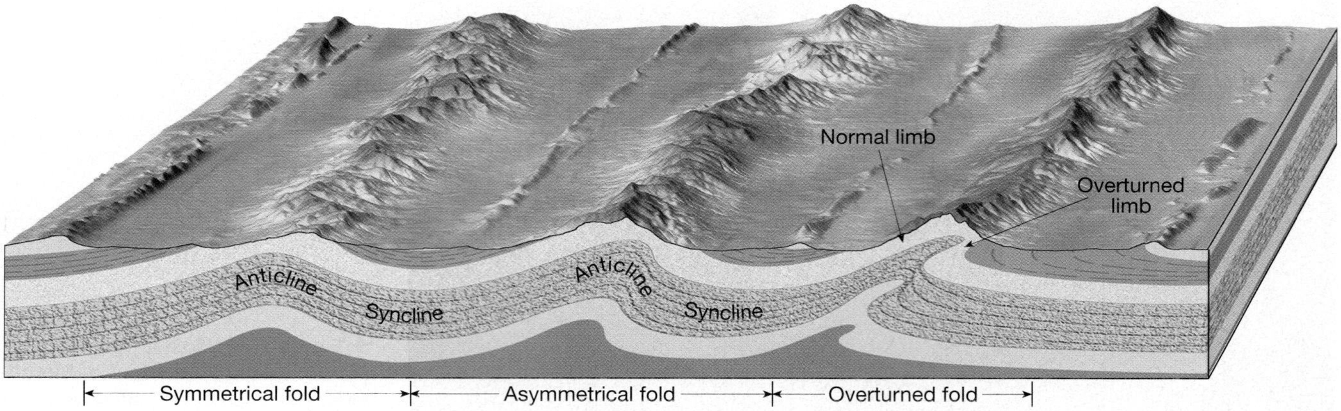

▲ **Figure 17.3** Block diagram of principal types of folded strata. The upfolded or arched structures are anticlines. The downfolds or troughs are synclines. Notice that the limb of an anticline is also the limb of the adjacent syncline.

commonly formed by the upfolding, or arching, of rock layers.* Often found in association with anticlines are downfolds, or troughs, called **synclines.** Notice in Figure 17.3 that the limb of an anticline is also a limb of the adjacent syncline.

Depending on their orientation, these basic folds are described as *symmetrical* when the limbs are mirror images of each other and *asymmetrical* when they are not. An asymmetrical fold is said to be *overturned* if one limb is tilted beyond the vertical (Figure 17.3). An overturned fold can also "lie on its side" so that a plane extending through the axis of the fold would be horizontal. These *recombent* folds are common in mountainous regions such as the Alps.

Folds do not continue forever; rather, their ends die out much like the wrinkles in cloth. Some folds *plunge* because the axis of the fold penetrates into the ground (Figure 17.4). As the figure shows, both anticlines and synclines can plunge. Figure 17.4C shows an example of

Did You Know?

Because winds tend to increase in strength with an increase in altitude, strong winds are common in high mountain settings. The highest wind speed recorded at a surface station is 231 miles per hour, measured at Mount Washington, New Hampshire.

*By strict definition, an anticline is a structure in which the oldest strata are found in the center. This most typically occurs when strata are upfolded. Further, a syncline is strictly defined as a structure in which the youngest strata are found in the center. This occurs most commonly when strata are downfolded.

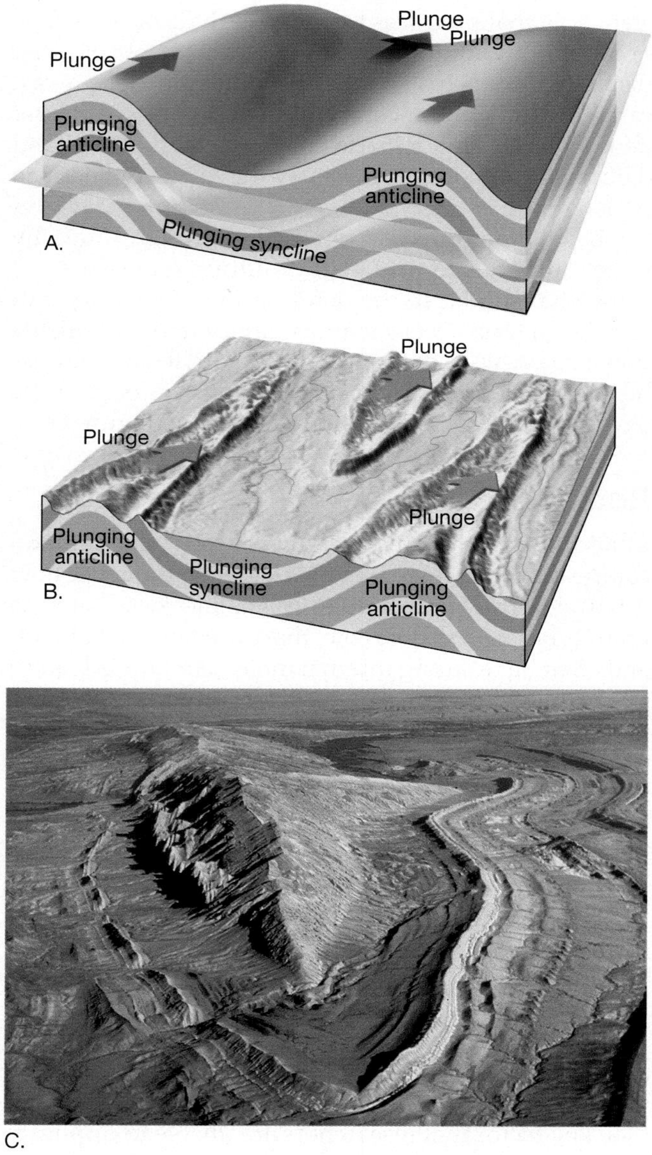

▶ **Figure 17.4** Plunging folds. A. Idealized view of plunging folds in which a horizontal surface has been added. B. View of plunging folds as they might appear after extensive erosion. Notice that in a plunging anticline the outcrop pattern "points" in the direction of the plunge, while the opposite is true of plunging synclines. C. Sheep Mountain, a doubly plunging anticline. *(Photo by John S. Shelton)*

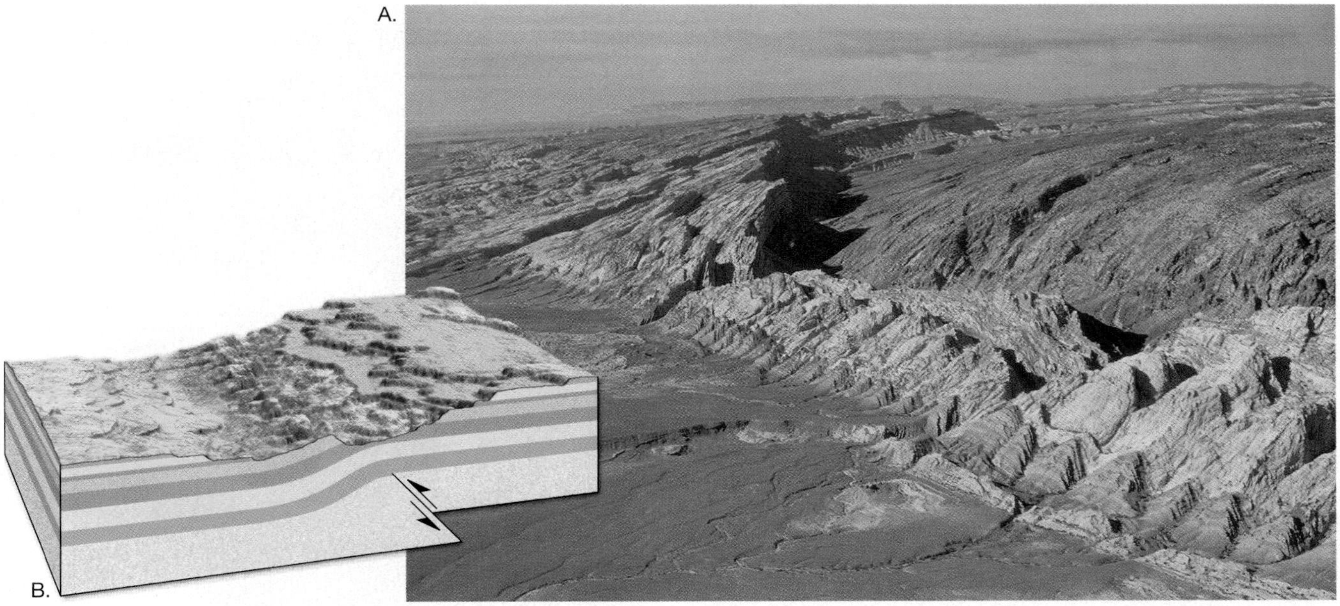

▲ **Figure 17.5** Monocline. A. Monocline located near Mexican Hat, Utah. *(Photo by Stephen Trimble)* B. Monocline consisting of bent sedimentary beds that were deformed by faulting in the bedrock below.

a plunging anticline and the pattern produced when erosion removes the upper layers of the structure and exposes its interior. Note that the outcrop pattern of an anticline points in the direction it is plunging, whereas the opposite is true for a syncline. A good example of the kind of topography that results when erosional forces attack folded sedimentary strata is found in the Valley and Ridge Province of the Appalachians.

Although we have separated our discussion of folds and faults, in the real world folds are generally intimately coupled with faults. Examples of this close association are broad, regional features called *monoclines.* Particularly prominent features of the Colorado Plateau, **monoclines** are large, steplike folds in otherwise horizontal sedimentary strata (Figure 17.5). These folds appear to be the result of the reactivating of steeply dipping fault zones located in basement rocks beneath the plateau. As large blocks of basement rock were displaced upward along ancient faults, the comparatively ductile sedimentary strata above responded by folding. On the Colorado Plateau, monoclines display a narrow zone of steeply inclined beds that flatten out to form the uppermost layers of large elevated areas, including the Zuni Uplift, Echo Cliffs Uplift, and San Rafael Swell. Displacement along these reactivated faults often exceeds 1 kilometer (0.6 mile).

Domes and Basins

Broad upwarps in basement rock may deform the overlying cover of sedimentary strata and generate large folds. When this upwarping produces a circular or elongated structure, the feature is called a **dome.** Downwarped structures having a similar shape are termed **basins.**

The Black Hills of western South Dakota are a large domed structure thought to be generated by upwarping. Here erosion has stripped away the highest portions of the upwarped sedimentary beds, exposing older igneous and metamorphic rocks in the center (Figure 17.6). Remnants of these once continuous sedimentary layers are visible, flanking the crystalline core of these mountains.

Syncline (left) and anticline (right) share a common limb.

Several large basins exist in the United States (Figure 17.7). The basins of Michigan and Illinois have very gently sloping beds similar to saucers. These basins are thought to be the result of large accumulations of sediment, whose weight caused the crust to subside.

Because large basins usually contain sedimentary beds sloping at very low angles, they are usually identified by the age of the rocks composing them. The youngest rocks are found near the center, and the oldest rocks are at the flanks. This is just the opposite order of a domed structure, such as the Black Hills, where the oldest rocks form the core.

Faults

GEODe: Crustal Deformation and Mountain Building: Part A
▼ Faults

Faults are fractures in the crust along which appreciable displacement has taken place. Occasionally, small faults can be recognized in road cuts where sedimentary beds

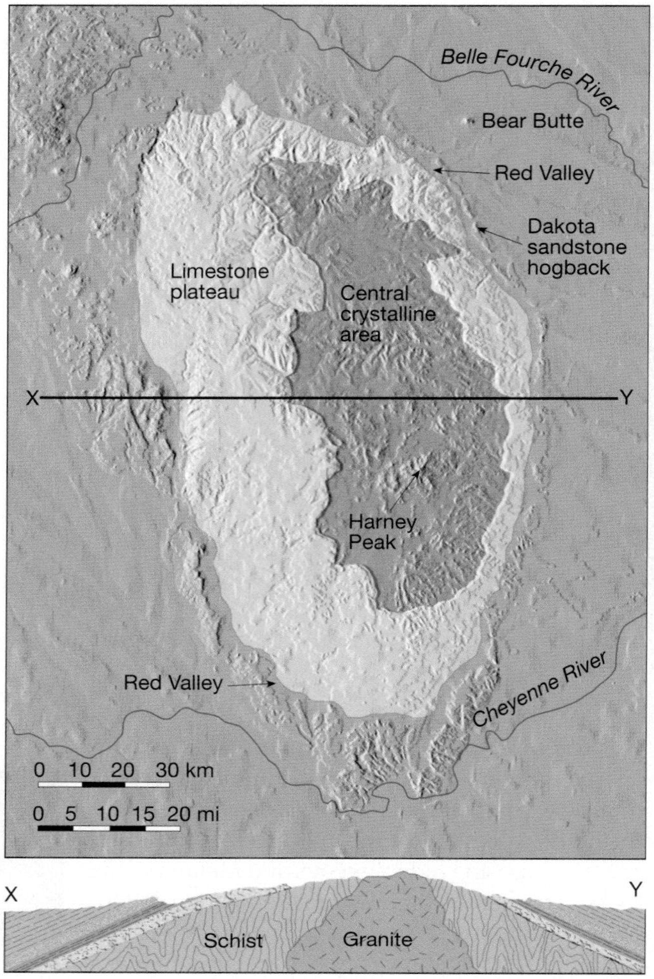

▲ **Figure 17.6** The Black Hills of South Dakota, a large domal structure with resistant igneous and metamorphic rocks exposed in the core.

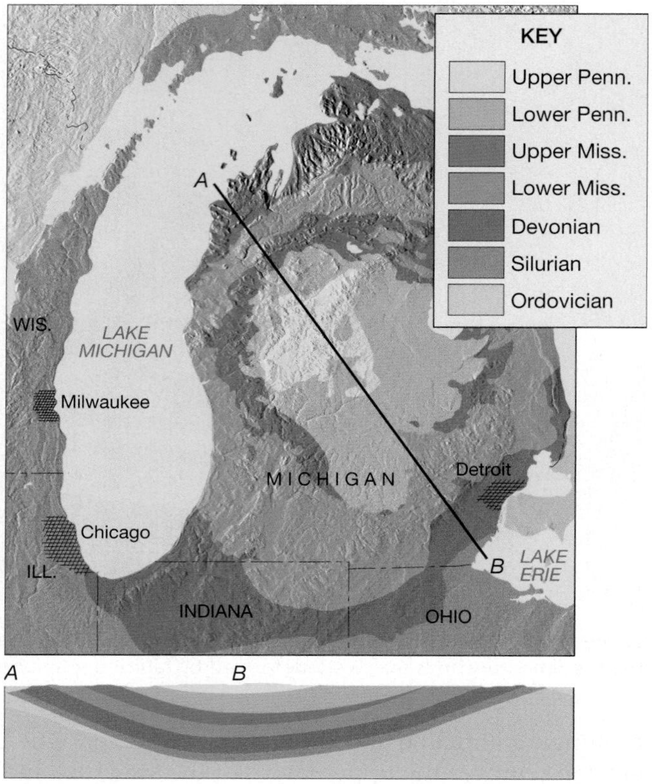

▲ **Figure 17.7** The bedrock geology of the Michigan Basin. Notice that the youngest rocks are centrally located, while the oldest beds flank this structure.

Recumbent folds in the Swiss Alps. (Mike Andrews)

have been offset a few meters, as shown in Figure 17.8. Faults of this scale usually occur as single discrete breaks. By contrast, large faults, like the San Andreas Fault in California, have displacements of hundreds of kilometers and consist of many interconnecting fault surfaces. These *fault zones* can be several kilometers wide and are often easier to identify from high-altitude photographs than at ground level.

Dip-Slip Faults

Faults in which the movement is primarily parallel to the inclination, or dip, of the fault surface are called **dip-slip faults.** Vertical displacements along dip-slip faults may produce long, low cliffs called **fault scarps.** Fault scarps are produced by displacements that generate earthquakes.

It has become common practice to call the rock surface that is immediately above the fault the *hanging wall* and to call the rock surface below, the *footwall.* This nomenclature arose from prospectors and miners who excavated shafts and tunnels along fault zones because these are frequently sites of ore deposits. In these tunnels, the miners would walk on the rocks below the mineralized fault zone (the footwall) and hang their lanterns on the rocks above (the hanging wall).

Two major types of dip-slip faults are *normal faults* and *reverse faults.*

▼ **Figure 17.8** Faulting caused the vertical displacement of these beds located near Kanab, Utah. Arrows show relative motion of rock units. *(Photo by Tom Bean/DRK Photo)*

Normal Faults. Dip-slip faults are classified as **normal faults** when the hanging wall block moves down relative to the footwall block (Figure 17.9A). Most normal faults have steep dips of about 60°, which tend to flatten out with depth. However, some dip-slip faults have much lower dips, with some approaching horizontal. Because of the downward motion of the hanging wall block, normal faults accommodate lengthening, or extension, of the crust.

Normal faulting is prevalent at divergent plate boundaries. Here, a central block called a **graben** is bounded by normal faults and drops as the plates separate (Figure 17.10). These grabens produce an elongated valley bounded by relatively uplifted structures called **horsts.** An excellent example of horst and graben topography is found in the Basin and Range Province, a region that encompasses Nevada and portions of surrounding states (Figure 17.10). Here the crust has been elongated and broken to create more than 200 relatively small mountain ranges. Averaging about 80 kilometers (about 50 miles) in length, the ranges rise 900 to 1500 meters (about 2950 to 4920 feet feet) above the adjacent down-faulted basins. Notice in Figure 17.10 that slopes of the normal faults in the Basin and Range Province decrease with depth and join together to form a nearly horizontal fault called a *detachment fault*. These detachment faults extend for several kilometers below the surface. Here they form a major boundary between the rocks below, which exhibit ductile deformation, and the rocks above, which demonstrate brittle deformation via faulting.

Fault motion provides geologists with a method of determining the nature of the forces at work within Earth. Normal faults indicate the existence of tensional forces that pull the crust apart. This "pulling apart" can be accomplished either by uplifting that causes the surfaces to stretch and break or by opposing horizontal forces.

Reverse and Thrust Faults. **Reverse faults** and **thrust faults** are dip-slip faults in which the hanging wall block moves up relative to the footwall block (Figure 17.9B and C). Reverse faults have dips greater than 45°, and thrust faults have dips less than 45°. Because the hanging wall block moves up and over the footwall block, reverse and thrust faults accommodate shortening of the crust.

Most high-angle reverse faults are small and accommodate local displacements in regions dominated by other types of faulting. Thrust faults, on the other hand, exist at all scales. In mountainous regions such as the Alps, Northern Rockies, Himalayas, and Appalachians, thrust faults have displaced strata as far as 50 kilometers (about 30 miles) over adjacent rock units. The result of this large-scale movement is that older strata end up overlying younger rocks.

Whereas normal faults occur in tensional environments, reverse and thrust faults result from strong

This satellite image shows the topography that results when erosional forces attack folded sedimentary strata. (LANDSAT image)

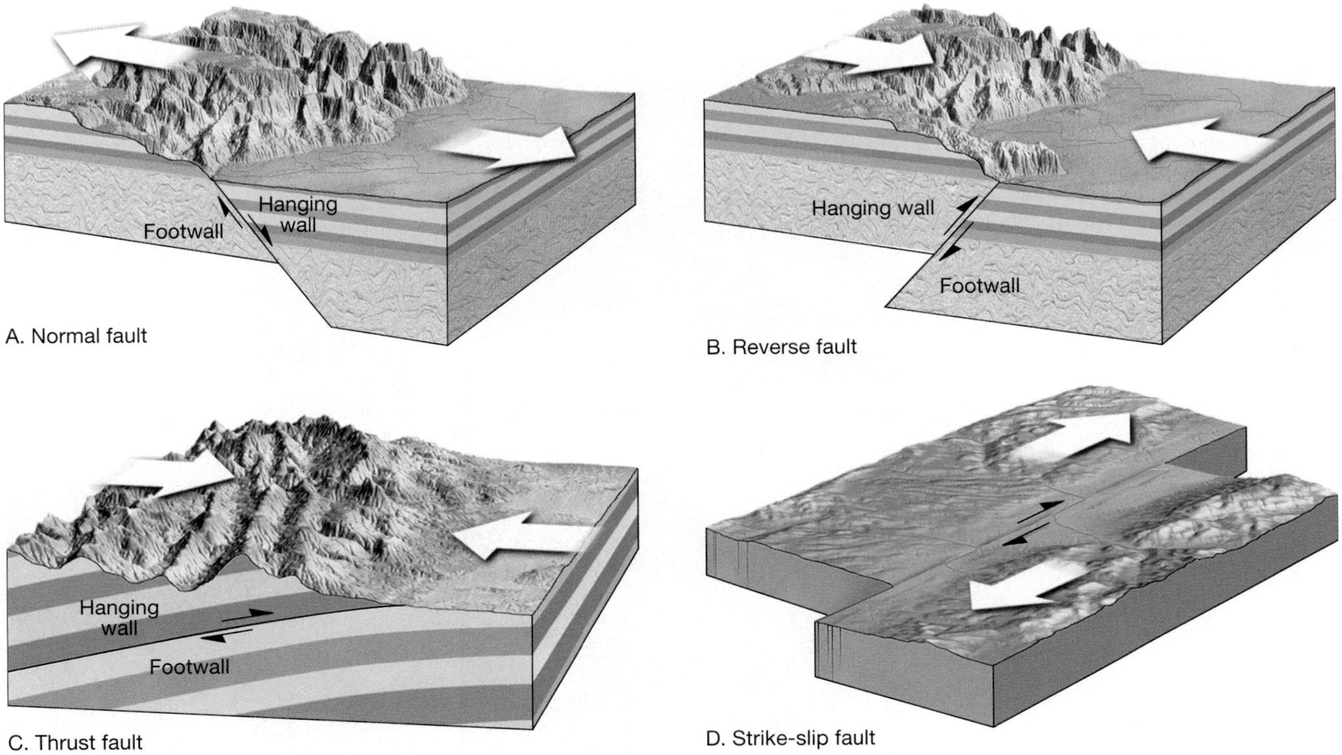

▲ **Figure 17.9** Block diagrams of four types of faults. **A.** Normal fault. **B.** Reverse fault. **C.** Thrust fault. **D.** Strike-slip fault.

Did You Know?

People have actually observed the creation of a fault scarp—and have lived to tell about it. In Idaho a large earthquake in 1983 created a 10-foot fault scarp that was witnessed by several people, many of whom were knocked off their feet. More often, though, fault scarps are noticed only after they form.

compressional stresses. In these settings, crustal blocks are displaced *toward* one another, with the hanging wall being displaced upward relative to the footwall. Thrust faulting is most pronounced in subduction zones and other convergent boundaries where plates are colliding. Compressional forces generally produce folds as well as faults and result in a thickening and shortening of the material involved.

Strike-Slip Faults

Faults in which the dominant displacement is horizontal and parallel to the trend, or strike, of the fault surface are called **strike-slip faults** (Figure 17.9D). Because of their large size and linear nature, many strike-slip faults produce a trace that is visible over a great distance. Rather than a single fracture along which movement takes place, large strike-slip faults consist of a zone of roughly parallel fractures. The zone may be up to several kilometers wide. The most recent movement, however, is often along a strand only a few meters wide, which may offset features such as stream channels (see Figure 17.A in Box 17.1). Furthermore, crushed and broken rocks produced during faulting are more easily eroded, often producing linear valleys or troughs that mark the locations of strike-slip faults.

Did You Know?

Many faults do not reach the surface, and hence do not produce a fault scarp during an earthquake. Faults of this type are termed *blind* or *hidden faults.* The 1994 Northridge earthquake in the Los Angeles area occurred along a blind trust fault.

The earliest scientific records of strike-slip faulting were made following surface ruptures that produced large earthquakes. One of the most noteworthy of these was the great San Francisco earthquake of 1906. During this strong earthquake, structures such as fences that were built across the San Andreas Fault were displaced as much as 4.7 meters (15 feet). Because the movement along the San Andreas causes the crustal block on the opposite side of the fault to move to the right as you face the fault, it is called *a right-lateral* strike-slip fault. The Great Glen fault in Scotland is a well-known example of a *left-lateral* strike-slip fault, which exhibits the opposite sense of displacement.

Many major strike-slip faults cut through the lithosphere and accommodate motion between two large crustal plates. This special kind of strike-slip fault is called a **transform fault.** Numerous transform faults cut the oceanic lithosphere and link offset segments of oceanic ridges. Others accommodate displacement between continental plates that move horizontally with respect to each other. One of the best-known transform faults is California's San Andreas Fault (see Box 17.1). This plate-bounding fault can be traced for about 950 kilometers (600 miles) from the Gulf of California to a point along the Pacific Coast north of San Francisco, where it heads out to sea. Ever since its formation, about

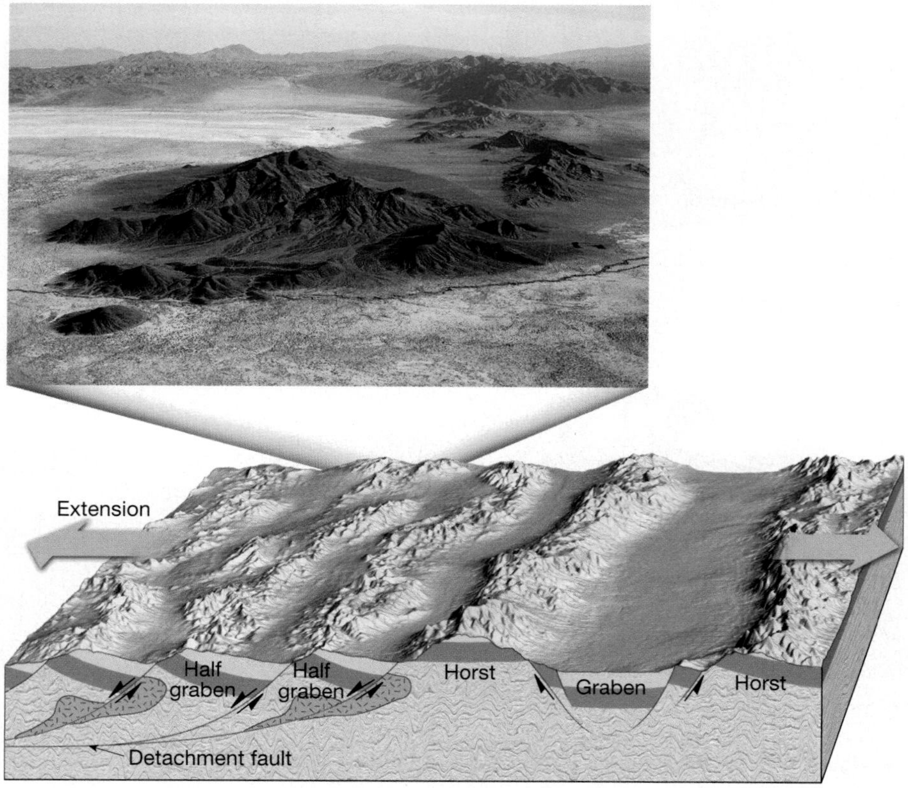

▶ **Figure 17.10** Normal faulting in the Basin and Range Province. Here, tensional stresses have elongated and fractured the crust into numerous blocks. Movement along these fractures has tilted the blocks producing parallel mountain ranges called fault-block mountains. *(Photo by Michael Collier)*

29 million years ago, displacement along the San Andreas Fault has exceeded 560 kilometers (340 miles). This movement has accommodated the northward displacement of southwestern California and the Baja Peninsula of Mexico in relation to the remainder of North America.

Joints

GEODe Crustal Deformation and Mountain Building: Part A
▼ Faults and Fractures

Among the most common rock structures are fractures called **joints.** Unlike faults, joints are fractures along which *no appreciable displacement* has occurred. Although some joints have a random orientation, most occur in roughly parallel groups (Figure 17.11).

We have already considered two types of joints. Earlier we learned that *columnar joints* form when igneous rocks cool and develop shrinkage fractures that produce elongated, pillarlike columns. Also recall that sheeting produces a pattern of gently curved joints that develop more or less parallel to the surface of large exposed igneous bodies such as batholiths. Here the jointing results from the gradual expansion that occurs when erosion removes the overlying load (see Figure 5.4, p. 119).

In contrast to the situations just described, most joints are produced when rocks in the outermost crust are deformed. Here forces associated with crustal movements cause the rock to fail by brittle fracture. For example, when folding occurs, rocks situated at the axes of the folds are elongated and pulled apart to produce tensional joints. Extensive joint patterns can also develop in response to relatively subtle and often barely perceptible regional upwarping and downwarping of the crust. In many cases, the cause for jointing at a particular locale is not readily apparent.

Many rocks are broken by two or even three sets of intersecting joints that slice the rock into numerous regularly shaped blocks. These joint sets often exert a strong influence on other geologic processes. For example, chemical weathering tends to be concentrated along joints, and in many areas groundwater movement and the resulting dissolution in soluble rocks is controlled by the joint pattern (Figure 17.11). Moreover, a system of joints can influence the direction that stream courses follow. The rectangular drainage pattern described in Chapter 9 is such a case.

▼ **Figure 17.11** Chemical weathering is enhanced along joints in sandstone, near Moab, Utah. *(Photo by Michael Collier)*

Fault scarp. (A. P. Trujillo)

Mountain Building

GEODe Crustal Deformation and Mountain Building: Part B
▼ Introduction

Like other people, geologists have been inspired more by Earth's mountains than by any other landforms (Figure 17.12, p. 402). Through extensive scientific exploration over the last 150 years, much has been learned about the internal processes that generate these often spectacular terrains. The name for the processes that collectively produce a mountain belt is **orogenesis.** The rocks comprising mountains provide striking visual evidence of the enormous compressional forces that have deformed large sections of Earth's crust and subsequently elevated them to their present positions. Although folding is often the most conspicuous sign of these forces, thrust faulting, metamorphism, and igneous activity are always present in varying degrees.

Mountain building has occurred during the recent geologic past in several locations around the world. These young mountainous belts include the American Cordillera, which runs along the western margin of the Americas from Cape Horn to Alaska and includes the Andes and Rocky Mountains; the Alpine-Himalaya chain, which extends from the Mediterranean through Iran to northern India and into Indochina; and the mountainous terrains of the western Pacific, which include volcanic island arcs such as Japan, the Philippines, and Sumatra. Most of these young mountain belts have come into existence within the last 100 million years. Some, including the Himalayas, began their growth as recently as 45 million years ago.

In addition to these relatively young mountain belts, several chains of older mountains exist on Earth as well.

Did You Know?

During Charles Darwin's famous voyage on the HMS *Beagle*, he experienced a strong earthquake that accompanied sudden uplift of the land around the Bay of Concepcion in Chile. From this and other observations, Darwin concluded that the uplift he witnessed "marked one step in the elevation of a mountain chain."

Did You Know?

New Zealander Edmund Hillary and Tenzing Norgay of Nepal were the first to reach the summit of Mount Everest on May 29, 1953. Not one to rest on his laurels, Hillary later went on to lead the first crossing of Antarctica.

BOX 17.1

The San Andreas Fault System

The San Andreas, the best-known and largest fault system in North America, first attracted wide attention after the great 1906 San Francisco earthquake and fire. Following this devastating event, geologic studies demonstrated that a displacement of as much as 5 meters (over 16 feet) along the fault had been responsible for the earthquake. It is now known that this dramatic event is just one of many thousands of earthquakes that have resulted from repeated movements along the San Andreas throughout its 29-million-year history.

Where is the San Andreas fault system located? As shown in Figure 17.A, it trends in a northwesterly direction for nearly 1300 kilometers (780 miles) through much of western California. At its southern end, the San Andreas connects with a spreading center located in the Gulf of California. In the north, the fault enters the Pacific Ocean at Point Arena, where it is thought to continue its northwesterly trend, eventually joining the Mendocino fracture zone. In the central section, the San Andreas is relatively simple and straight. However, at its two extremities, several branches spread from the main trace, so that in some areas the fault zone exceeds 100 kilometers (60 miles) in width.

Over much of its extent, a linear trough reveals the presence of the San Andreas Fault. When the system is viewed from the air, long narrow scars,

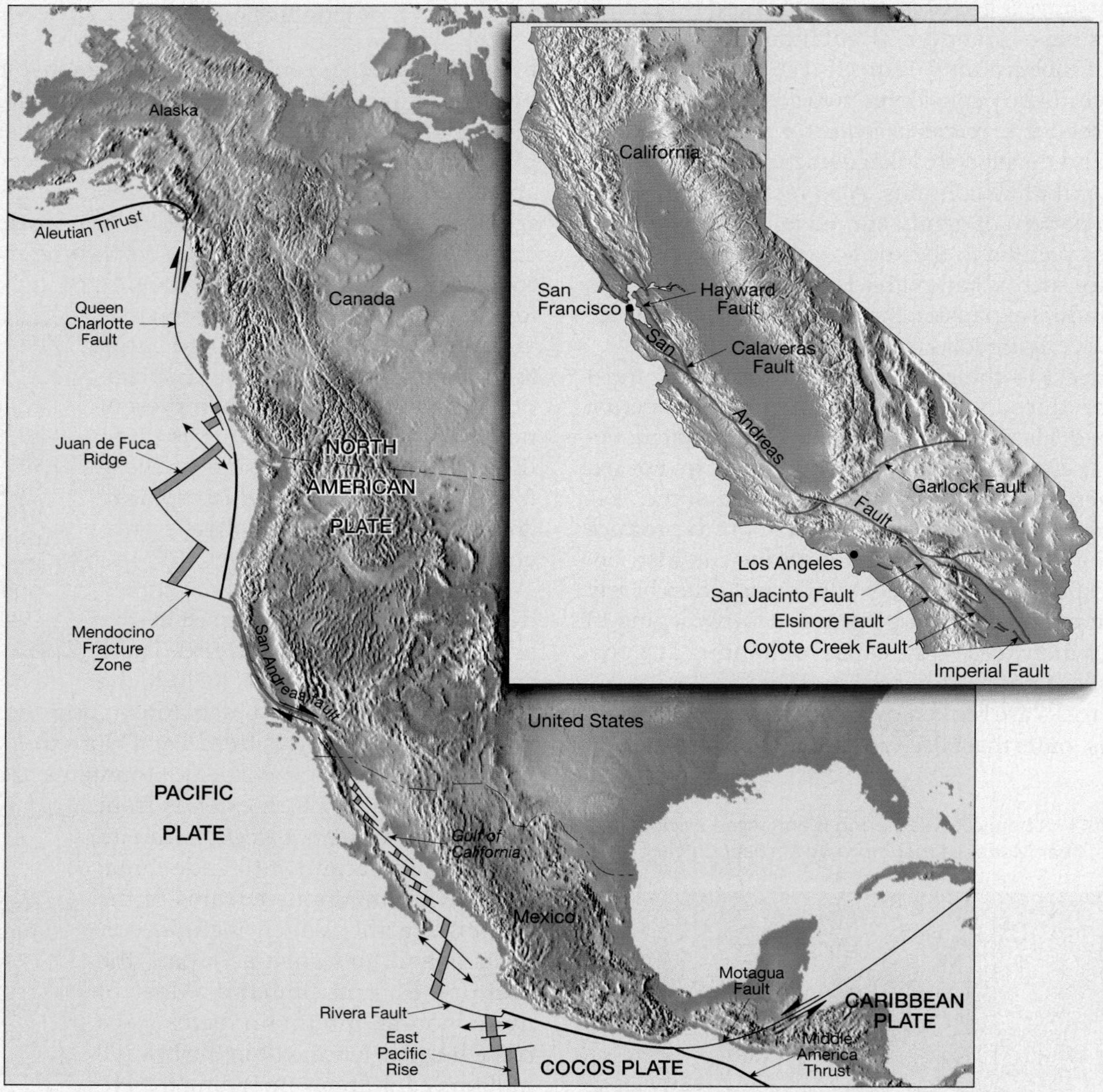

Figure 17.A Map showing the extent of the San Andreas fault system. Inset shows only a few of the many splinter faults that are part of this great fault system.

offset stream channels, and elongated ponds mark the trace in a striking manner. On the ground, however, surface expressions of the faults are much more difficult to detect. Some of the most distinctive landforms include long, straight escarpments, narrow ridges, and sag ponds formed by settling of blocks within the fault zone. Further, many stream channels characteristically bend sharply to the right where they cross the fault (Figure 17.B).

With the development of the theory of plate tectonics, geologists began to realize the significance of this great fault system. The San Andreas Fault is a transform boundary separating two crustal plates that move very slowly. The Pacific plate, located to the west, moves northwestward relative to the North American plate, causing earthquakes along the fault (Table 17.A).

The San Andreas is undoubtedly the most studied of any fault system in the world. Although many questions remain unanswered, geologists have learned that each fault segment exhibits somewhat different behavior. Some portions of the San Andreas exhibit a slow creep with little noticeable seismic activity. Other segments regularly slip, producing small earthquakes, whereas still other segments seem to store elastic energy for hundreds of years and rupture in great earthquakes. This knowledge is useful when assigning earthquake-hazard potential to a given segment of the fault zone.

Because of the great length and complexity of the San Andreas Fault, it is more appropriately referred to as a "fault system." This major fault system consists primarily of the San Andreas Fault and several major branches, including the Hayward and Calaveras faults of central California and the San Jacinto and Elsinore faults of southern California (Figure 17.A). These major segments, plus a vast number of smaller faults that include the Imperial Fault, San Fernando Fault, and the Santa Monica Fault, collectively accommodate the relative motion between the North American and Pacific plates.

Ever since the great San Francisco earthquake of 1906, when as much as 5 meters of displacement occurred, geologists have attempted to establish the cumulative displacement along this fault over its 29-million-year history. By matching rock units across the fault, geologists have determined that the total accumulated displacement from earthquakes and creep exceeds 560 kilometers (340 miles).

Figure 17.B Aerial view showing offset stream channel across the San Andreas Fault on the Carrizo Plain west of Taft, California. (Photo by Michael Collier/DRK Photo)

Table 17.A Major earthquakes on the San Andreas Fault system

Date	Location	Magnitude	Remarks
1812	Wrightwood, CA	7	Church at San Juan Capistrano collapsed, killing 40 worshippers.
1812	Santa Barbara channel	7	Churches and other buildings wrecked in and around Santa Barbara.
1838	San Francisco peninsula	7	At one time thought to have been comparable to the great earthquake of 1906.
1857	Fort Tejon, CA	8.25	One of the greatest U.S. earthquakes. Occurred near Los Angeles, then a city of 4000.
1868	Hayward, CA	7	Rupture of the Hayward fault caused extensive damage in San Francisco Bay area.
1906	San Francisco, CA	8.25	The great San Francisco earthquake. As much as 80 percent of the damage caused by fire.
1940	Imperial Valley	7.1	Displacement on the newly discovered Imperial fault.
1952	Kern County	7.7	Rupture of the White Wolf fault. Largest earthquake in California since 1906. Sixty million dollars in damages and 12 people killed.
1971	San Fernando Valley	6.5	One-half billion dollars in damages and 58 lives claimed.
1989	Santa Cruz Mountains	7.1	Loma Prieta earthquake. Six billion dollars in damages, 62 lives lost, and 3757 people injured.
1994	Northridge (Los Angeles area)	6.9	Over 15 billion dollars in damages, 51 lives lost, and over 5000 injured.

▲ **Figure 17.12** This peak is part of the Karakoram Range in Pakistan. The Karakoram are part of the Himalayan system. *(Photo by Art Wolfe)*

Although these older structures are deeply eroded and topographically less prominent, they clearly possess the same structural features found in younger mountains. Typical of this older group are the Appalachians in the eastern United States and the Urals in Russia.

Did You Know?

The lake that is the home of the legendary Loch Ness Monster is located along the Great Glen fault that divides Northern Scotland. Strike-slip movement along this fault zone shattered a wide zone of rock, that glacial erosion subsequently removed, to produce the elongated valley in which Loch Ness is situated. The broken rock was easily eroded, a fact that accounts for the great depth of Loch Ness—as much as 600 feet below sea level.

Over the years, several hypotheses have been put forward regarding the formation of Earth's major mountain belts. One early proposal suggested that mountains are simply wrinkles in Earth's crust, produced as the planet cooled from its original semimolten state. As Earth lost heat, it contracted and shrank. In response to this process, the crust was deformed similar to when the peel of an orange wrinkles as the fruit dries out. However, neither this nor any other early hypothesis was able to withstand careful scrutiny and had to be discarded.

Mountain Building at Subduction Zones

With the development of the theory of plate tectonics, a model for orogenesis with excellent explanatory power has emerged. According to this model, most mountain building occurs at convergent plate boundaries. Here, the subduction of oceanic lithosphere triggers partial melting of mantle rock, providing a source of magma that intrudes the crustal rocks that form the margin of the overlying plate. In addition, colliding plates provide the tectonic forces that fold, fault, and metamorphose the thick accumulations of sediments that have been deposited along the flanks of landmasses. Together, these processes thicken and shorten the continental crust, thereby elevating rocks that may have formed near the ocean floor, to lofty heights.

To unravel the events that produce mountains, researchers examine ancient mountain structures as well as sites where orogenesis is currently active. Of particular interest are active subduction zones, where lithospheric plates are converging. Here the subduction of oceanic lithosphere generates Earth's strongest earthquakes and most explosive volcanic eruptions, as well as playing a pivotal role in generating many of Earth's mountain belts.

The subduction of oceanic lithosphere gives rise to two different types of mountain belts. Where *oceanic lithosphere* subducts beneath an *oceanic plate,* an *island arc* and related tectonic features develop. Subduction beneath a *continental block,* on the other hand, results in the formation of a *continental volcanic arc* along the margin of the adjacent landmass. Plate boundaries that generate continental volcanic arcs are often referred to as *Andean-type plate margins.*

Island Arcs

Island arcs form where two oceanic plates converge and one is subducted beneath the other (Figure 17.13). This

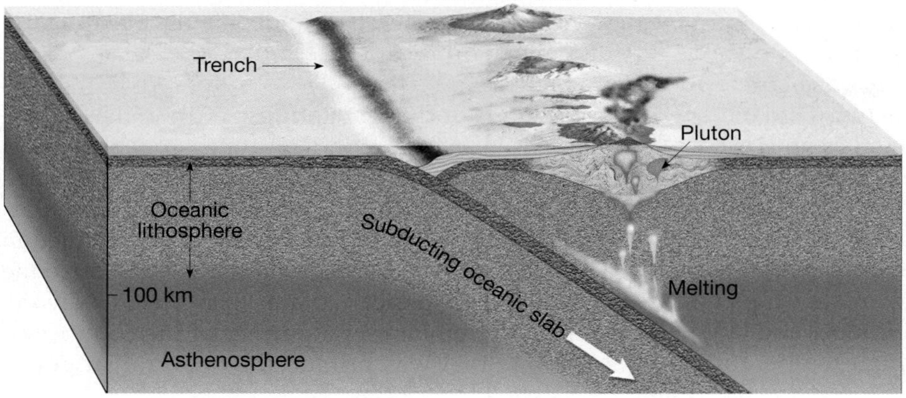

◄ **Figure 17.13** The development of a volcanic island arc by the convergence of two oceanic plates. Continuous subduction along these Aleutian-type convergent zones results in the development of thick units of continental-type crust.

activity results in partial melting of the mantle wedge located above the subducting plate and eventually leads to the growth of a volcanic island arc on the ocean floor. Because they are associated with subducting oceanic lithosphere, island arcs are typically found on the margins of an ocean basin, such as the Pacific—where the majority of volcanic island arcs are found. Examples of active island arcs include the Mariana, Kurile, Tonga, and Aleutian arcs.

Island arcs represent what are perhaps the simplest mountain belts. These structures result from the steady subduction of oceanic lithosphere, which may last for 100 million years or more. Somewhat sporadic volcanic activity, the emplacement of plutonic bodies at depth, and the accumulation of sediment that is scraped from the subducting plate gradually increase the volume of crustal material capping the upper plate. Some mature volcanic island arcs, such as Japan, appear to have been built upon a preexisting fragment of crustal material.

The continued development of a mature volcanic island arc can result in the formation of mountainous topography consisting of belts of igneous and metamorphic rocks. This activity, however, is viewed as just one phase in the development of a major mountain belt. As you will see later, some volcanic arcs are carried by a subducting plate to the margin of a large continental block, where they become involved in a major mountain-building episode.

Mountain Building Along Andean-Type Margins

Mountain building along continental margins involves the convergence of an oceanic plate and a plate whose leading edge contains continental crust. Exemplified by the Andes Mountains, an *Andean-type convergent zone* results in the formation of a continental volcanic arc and related tectonic features inland of the continental margin.

The first stage in the development of an idealized Andean-type mountain belt occurs prior to the formation of the subduction zone. During this period the continental margin is a **passive continental margin;** that is, it is not a plate boundary but a part of the same plate as the adjoining oceanic crust. The east coast of North America provides a present-day example of a passive continental margin. Here, as at other passive continental margins surrounding the Atlantic, deposition of sediment on the continental shelf is producing a thick wedge of shallow-water sandstones, limestones, and shales (Figure 17.14A). Beyond the continental shelf, turbidity currents are depositing sediments on the continental slope and rise.

Did You Know?

The highest city in the world, La Paz, Bolivia, is located 11,910 feet above sea level, more than twice as high as Denver, Colorado—the highest major city in the United States.

At some point the continental margin becomes active. A subduction zone forms and the deformation process begins (Figure 17.14B). A good place to examine an **active continental margin** is the west coast of South America. Here the Nazca plate is being subducted beneath the South American plate along the Peru–Chile trench. This subduction zone probably formed prior to the breakup of the supercontinent of Pangaea.

In an idealized Andean-type subduction, convergence of the continental block and the subducting oceanic plate leads to deformation and metamorphism of the continental margin. Once the oceanic plate descends to about 100 kilometers (60 miles), partial melting of mantle rock above the subducting slab generates magma that migrates upward (Figure 17.14B).

Thick continental crust greatly impedes the ascent of magma. Consequently, a high percentage of the magma that intrudes the crust never reaches the surface—instead, it crystallizes at depth to form plutons. Eventually, uplifting and erosion exhume these igneous bodies and associated metamorphic rocks. Once they are exposed at the surface, these massive structures are called *batholiths* (Figure 17.14C). Composed of numerous plutons, batholiths form the core of the Sierra Nevada in California and are prevalent in the Peruvian Andes.

Uplift along a fault has created this dramatic precipice in Greece. The polished and striated rock surface is called a slickenside. (James Jackson)

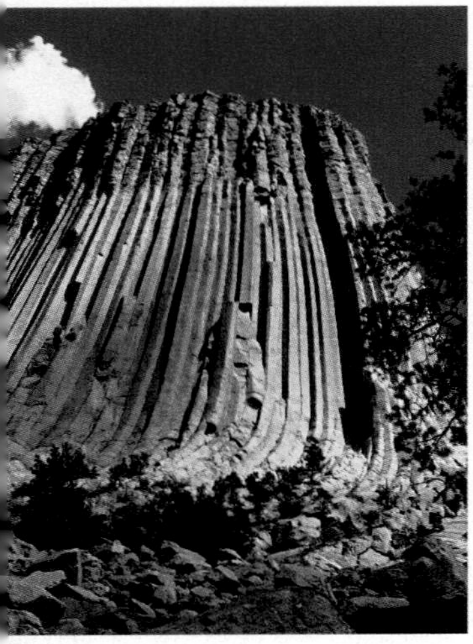

Devil's Tower, Wyoming, exhibits columnar jointing.

During the development of this continental volcanic arc, sediment derived from the land and scraped from the subducting plate is plastered against the landward side of the trench like piles of dirt in front of a bulldozer. This chaotic accumulation of sedimentary and metamorphic rocks with occasional scraps of ocean crust is called an **accretionary wedge** (Figure 17.14B). Prolonged subduction can build an accretionary wedge that is large enough to stand above sea level (Figure 17.14C).

Andean-type mountain belts are composed of two roughly parallel zones. The volcanic arc develops on the continental block. It consists of volcanoes and large intrusive bodies intermixed with high-temperature metamorphic rocks. The seaward segment is the accretionary wedge. It consists of folded, faulted sedimentary and metamorphic rocks (Figure 17.14C).

Sierra Nevada and Coast Ranges. One of the best examples of an inactive Andean-type orogenic belt is found in the western United States. It includes the Sierra Nevada and the Coast Ranges in California. These parallel mountain belts were produced by the subduction of a portion of the Pacific Basin under the western edge of the North American plate. The Sierra Nevada batholith is a remnant of a portion of the continental volcanic arc that was produced by several surges of

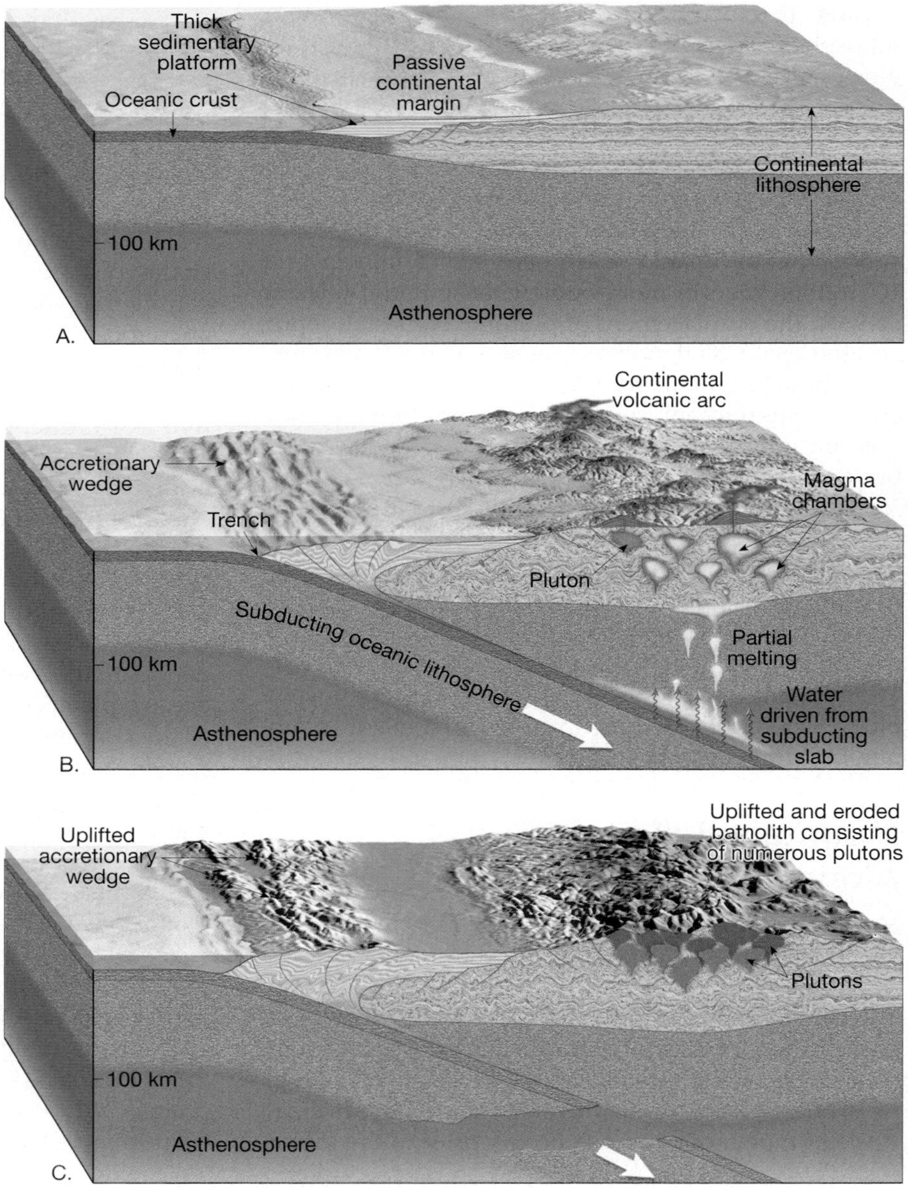

▶ **Figure 17.14** Mountain building along an Andean-type subduction zone. **A.** Passive continental margin with extensive wedge of sediments. **B.** Plate convergence generates a subduction zone, and partial melting produces a developing continental volcanic arc. Continued convergence and Igneous activity further deform and thicken the crust, elevating the mountain belt, while the accretionary wedge grows. **C.** Subduction ends and is followed by a period of uplift and erosion.

magma over tens of millions of years. Subsequent uplifting and erosion have removed most of the evidence of past volcanic activity and exposed a core of crystalline, igneous, and associated metamorphic rocks.

In the trench region, sediments scraped from the subducting plate, plus those provided by the eroding continental volcanic arc, were intensely folded and faulted into an accretionary wedge. This chaotic mixture of rocks presently constitutes the Franciscan Formation of California's Coast Ranges. Uplifting of the Coast Ranges took place only recently, as evidenced by the young unconsolidated sediments that still mantle portions of these highlands.

Collisional Mountain Ranges

GEODe ESSENTIALS OF GEOLOGY

Crustal Deformation and Mountain Building: Part B

▼ Continental Collisions / Crustal Fragments

As you have seen, when a slab of oceanic lithosphere subducts beneath a continental margin, an Andean-type mountain belt develops. If the subducting plate also contains a slab of continental lithosphere, continued subduction eventually carries the continental block to the trench. Oceanic lithosphere is relatively dense and readily subducts, but continental crust is composed of low-density material that is too buoyant to undergo subduction. Consequently, the arrival of the continental block at the trench results in a collision with the overriding continent. The result is crustal shortening and thickening to produce a mountain belt.

Mountain belts can develop as a result of the collision and merger of an island arc, or some other small crustal fragment with a continental block, as well as from the collision and joining of two or more continents.

Terranes and Mountain Building

The process of collision and accretion (joining together) of comparatively small crustal fragments to a continental margin has generated many of the mountainous regions rimming the Pacific. Geologists refer to these accreted crustal blocks as *terranes.* Simply, the term **terrane** refers to any crustal fragment that has a geologic history distinct from that of adjoining terranes. Terranes come in various shapes and sizes.

What is the nature of these crustal fragments, and from where do they originate? Research suggests that prior to their accretion to a continental block, some of the fragments may have been *microcontinents* similar to the present-day island of Madagascar, located east of Africa in the Indian Ocean. Many others were island arcs similar to Japan, the Philippines, and the Aleutian Islands. Still others are submerged crustal fragments, such as *oceanic plateaus,* which were created by massive outpourings of basaltic lavas associated with hot-spot activity (Figure 17.15).

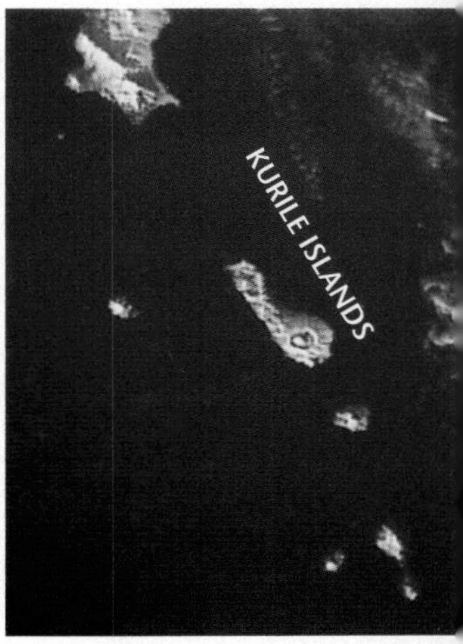

The volcanoes in this satellite image are part of an island arc known as the Kurile Islands. (NASA)

Accretion and Orogenesis. The widely accepted view is that as oceanic plates move, they carry embedded oceanic plateaus, volcanic island arcs, and microcontinents to an Andean-type subduction zone. When an oceanic plate contains a chain of small seamounts, these

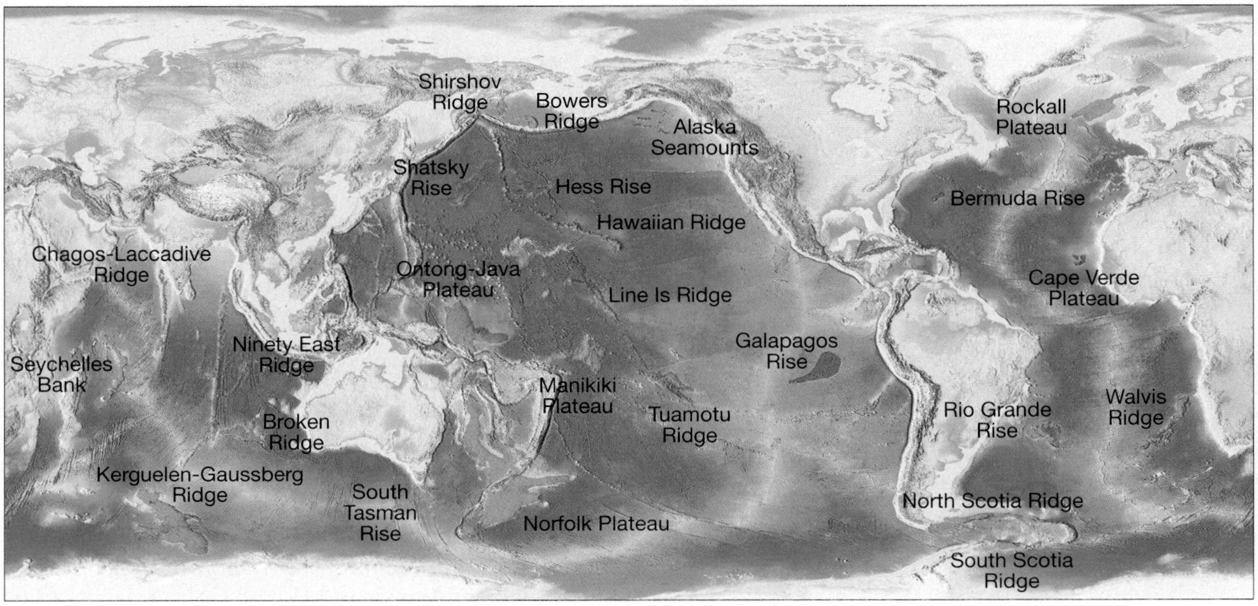

▲ **Figure 17.15** Distribution of present-day oceanic plateaus and other submerged crustal fragments. *(Data from Ben-Avraham and others)*

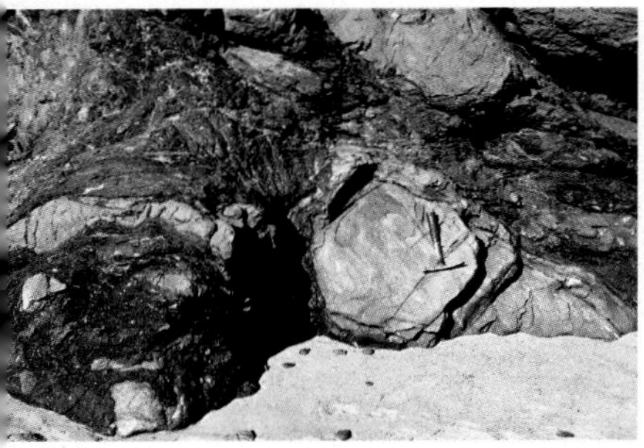

Accretionary wedge material, Franciscan formation, California. (M. Miller)

structures are generally subducted along with the descending oceanic slab. However, thick units of oceanic crust, such as the Ontong Java Plateau, or a mature island arc composed of abundant "light" igneous rocks may render the oceanic lithosphere too buoyant to subduct. In these situations, a collision between the crustal fragment and the continent occurs.

The sequence of events that occurs when a mature island arc reaches an Andean-type margin is shown in Figure 17.16. Because of its buoyancy, a mature island arc will not subduct beneath the continental plate. Instead, the upper portions of these thickened zones are peeled from the descending plate and thrust in relatively thin sheets upon the adjacent continental block. In some settings continued subduction may carry another crustal fragment to the continental margin. When this fragment collides with the continental margin, it displaces the accreted island arc further inland, adding to the zone of deformation and to the thickness and lateral extent of the continental margin.

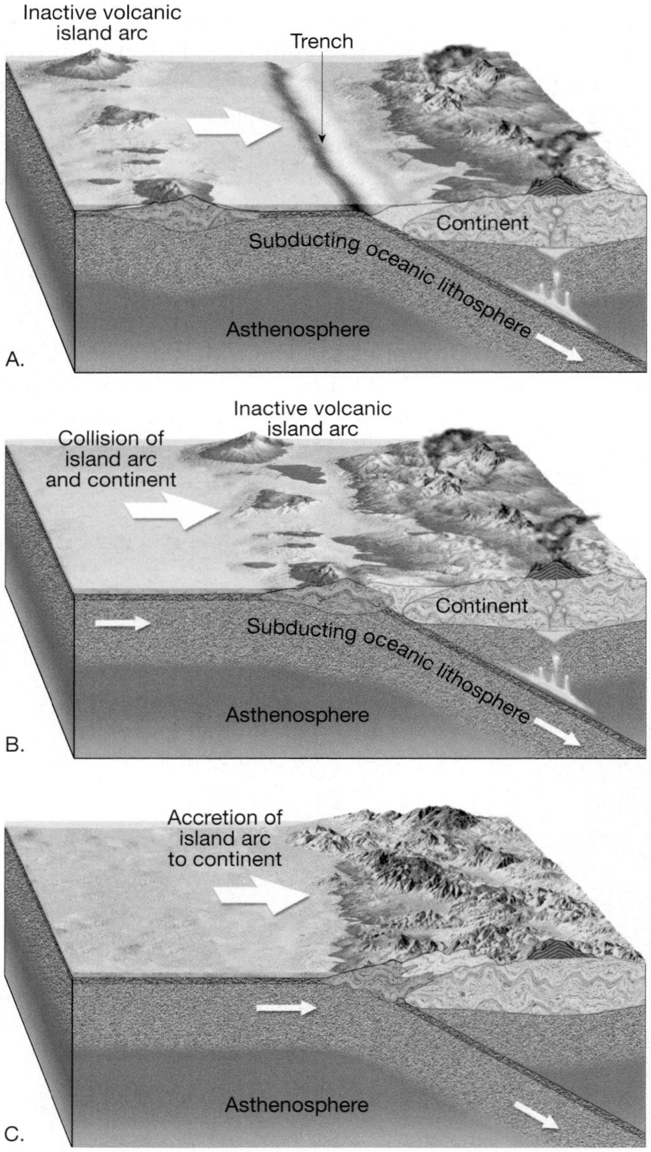

▲ **Figure 17.16** This sequence illustrates the collision of an inactive volcanic island arc with an Andean-type plate margin.

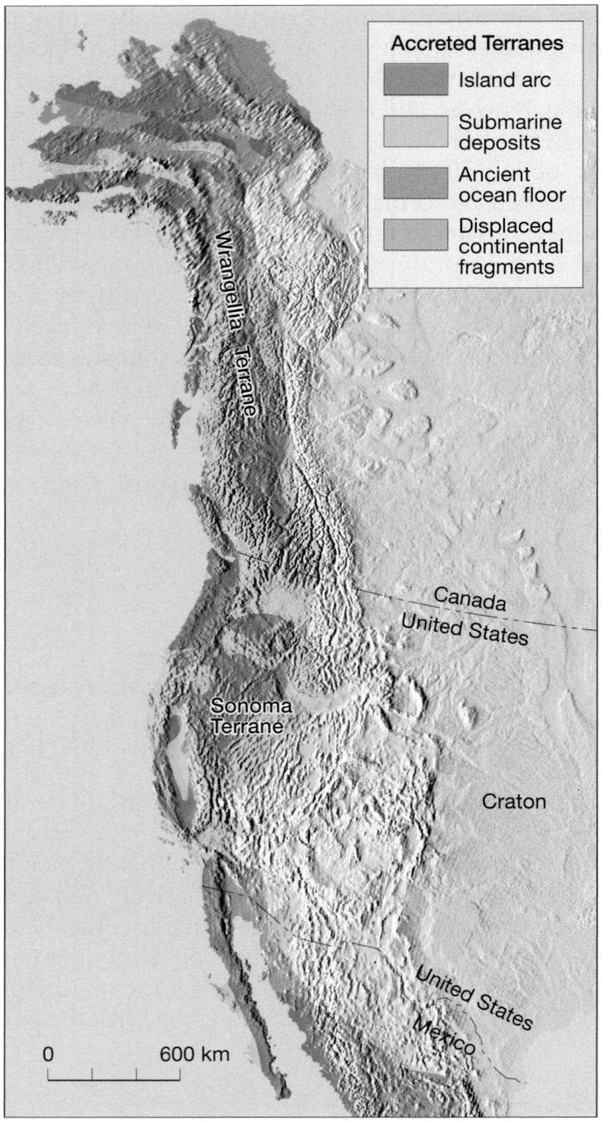

▲ **Figure 17.17** Map showing terranes thought to have been added to western North America during the past 200 million years. *(Redrawn after D. R. Hutchinson and others)*

The North American Cordillera. The idea that mountain building occurs in association with the accretion of crustal fragments to a continental mass arose principally from studies conducted in the North American Cordillera (Figure 17.17). Here it was determined that some mountainous areas, principally those in the orogenic belts of Alaska and British Columbia, contain fossil and paleomagnetic evidence indicating that these strata once lay nearer the equator.

It is now assumed that many of the other terranes found in the North American Cordillera were once scattered throughout the eastern Pacific, much as we find island arcs and oceanic plateaus distributed in the western Pacific today (see Figure 17.15). Since before the breakup of Pangaea, the eastern portion of the Pacific basin (Farallon plate) has been subducting under the western margin of North America. Apparently, this activity resulted in the piecemeal addition of crustal fragments to the entire Pacific margin of the continent—from Mexico's Baja Peninsula to northern Alaska (Figure 17.17). In a like manner, many modern microcontinents will eventually be accreted to active continental margins, producing new orogenic belts.

Continental Collisions

Continental collisions result in the development of mountains that are characterized by shortened and thickened crust. Thicknesses of 50 kilometers (30 miles) are common, and some regions have crustal thicknesses in excess of 70 kilometers (40 miles). In these settings, crustal thickening is achieved through folding and faulting.

We will take a closer look at two examples of collision mountains—the Himalayas and the Appalachians. The Himalayas are the youngest collision mountains on Earth and are still rising. The Appalachians are a much older mountain belt, in which active mountain building ceased about 250 million years ago.

The Himalayas. The mountain-building episode that created the Himalayas began roughly 45 million years ago when India began to collide with Asia. Prior to the breakup of Pangaea, India was part of a Southern Hemisphere landmass that also included Australia. Upon splitting from that continent, India moved rapidly, geologically speaking, a few thousand kilometers in a northward direction (see Figure 5.15, p 351).

The subduction zone that facilitated India's northward migration was located near the southern margin of Asia (Figure 17.18). Ongoing subduction along Asia's margin created an Andean-type plate margin that contained a well-developed volcanic arc and accretionary wedge. Eventually, the intervening ocean basin was consumed at the subduction zone and India collided with the Eurasian plate. The tectonic forces involved in the collision were immense and caused the more deformable materials located on the seaward edges of these landmasses to be highly folded and faulted (Figure 17.18). The shortening and thickening of the crust elevated great quantities of crustal material, thereby generating the spectacular Himalayan mountains (see Figure 17.12).

In addition to uplift, crustal thickening caused lower layers to become deeply buried and to experience

Scenic view of the Sierra Nevada. (Michael Collier)

Did You Know?

The term *terrane* is used to designate a distinct and recognizable series of rock formations that have been transported by plate tectonic processes. Don't confuse this with the term *terrain*, which describes the shape of the surface topography or "lay of the land." In other words, you might observe a terrane as part of the terrain in the Appalachian Mountains.

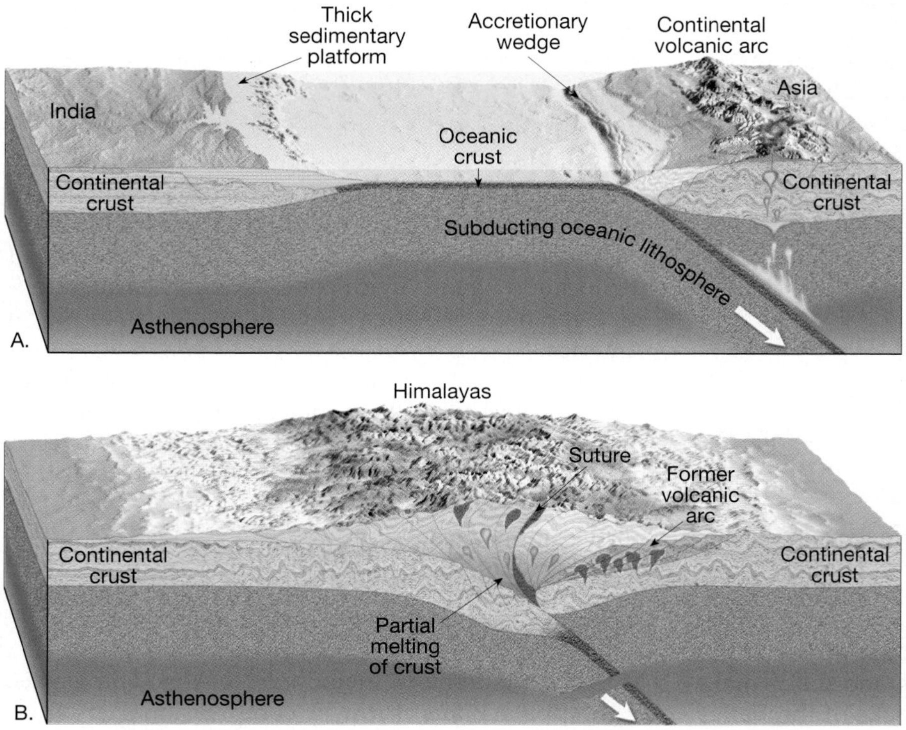

◀ **Figure 17.18** Illustration showing the collision of India with the Eurasian plate, producing the spectacular Himalayas.

The Blue Ridge Mountains are part of the Appalachians. (Carr Clifton)

elevated temperatures and pressures. Partial melting within the deepest and most deformed region of the developing mountain belt produced magma bodies that intruded and further deformed the overlying rocks. It is in such environments where the metamorphic and igneous core of a major mountain belt is generated.

The Appalachians. The Appalachian Mountains provide great scenic beauty near the eastern margin of North America from Alabama to Newfoundland. The orogeny that generated this extensive mountain system lasted a few hundred million years and was one of the stages in the assembling of Pangaea. Our simplified scenario begins roughly 600 million years ago when an ocean body, which predated the North Atlantic (referred to as the ancestral North Atlantic), began to close. Two subduction zones probably formed. One was located seaward of the coast of Africa and gave rise to a volcanic arc similar to those that presently rim the western Pacific. The other developed adjacent to a continental fragment that lay off the coast of North America, as shown in Figure 17.19A.

Did You Know?

Having an average altitude of over 16,000 feet, the Tibetan Plateau is the highest plateau on Earth. Its average elevation is greater than the highest peaks in the lower 48 states.

Between 450 and 500 million years ago, the marginal sea located between this crustal fragment and North America began to close. The ensuing collision deformed the continental shelf and sutured the crustal fragment to the North American plate. The metamorphosed remnants of the continental fragment are recognized today as the crystalline rocks of the Blue Ridge and western Piedmont regions of the Appalachians (Figure 17.19B). In addition to the pervasive metamorphism, igneous activity placed numerous plutonic bodies along the continental margin, particularly in New England.

Did You Know?

During the assembly of Pangaea, the landmasses of Europe and Siberia collided to produce the Ural Mountains. Long before the discovery of plate tectonics, this extensively eroded mountain chain was regarded as the boundary between Europe and Asia.

A second episode of mountain building occurred about 400 million years ago. The continued closing of the ancestral North Atlantic resulted in the collision of the developing volcanic arc with North America (Figure 17.19C). Evidence for this event is visible in the Carolina State Belt of the eastern Piedmont, which contains metamorphosed sedimentary and volcanic rocks characteristic of an island arc.

The final orogeny occurred somewhere between 250 and 300 million years ago, when Africa collided with North America. This event displaced and further deformed the shelf sediments and sedimentary rocks that had once flanked the eastern margin of North America (Figure 17.19D). Today these folded and thrust-faulted sandstones, limestones, and shales make up the largely unmetamorphosed rocks of the Valley and Ridge Province.

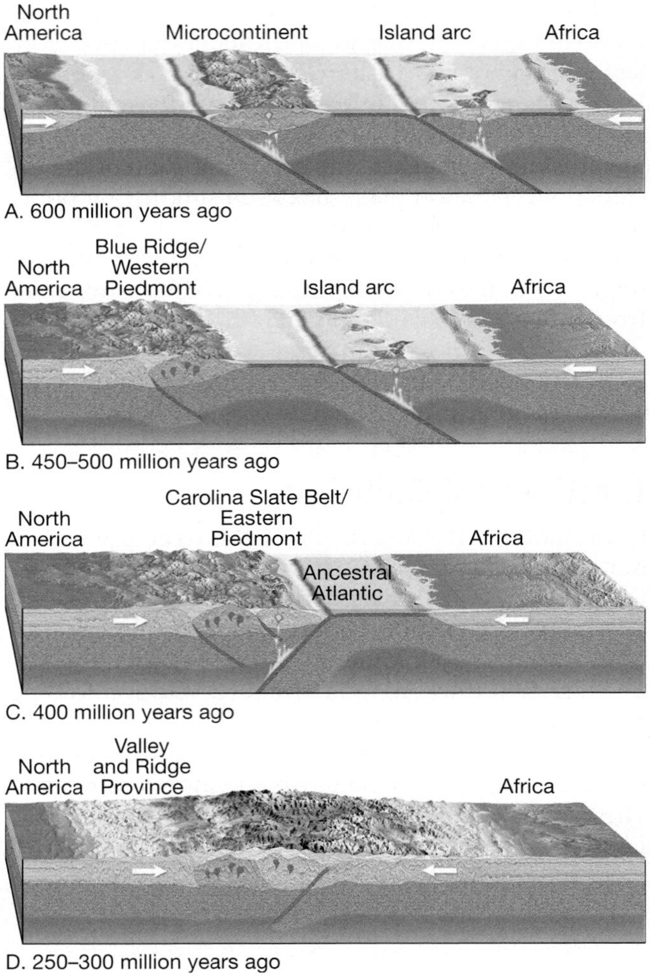

▲ **Figure 17.19** These simplified diagrams depict the development of the southern Appalachians as the ancient North Atlantic was closed during the formation of Pangaea. Three separate stages of mountain-building activity spanned more than 300 million years. *(After Zve Ben-Avraham, Jack Oliver, Larry Brown, and Frederick Cook)*

Geologically speaking, shortly after the formation of the Appalachian Mountains, the newly formed supercontinent of Pangaea began to break into smaller fragments. Because the zone of rifting occurred east of the location where Africa collided with North America, a remnant of Africa remains "welded" to North America.

Other mountain ranges that exhibit evidence of continental collisions include the Alps and the Urals. The Alps are thought to have formed as a result of a collision between Africa and Europe. The Urals, on the other hand, formed during the assembly of Pangaea when northern Europe and northern Asia collided.

Fault-Block Mountains

Most mountain belts, including the Alps, Himalayas, and Appalachians, form in compressional environments, as evidenced by the predominance of large thrust faults and folded strata. However, other tectonic processes, such as continental rifting, can also produce uplift and the formation of topographic mountains. The mountains that form in these settings, termed **fault-block mountains,** are bounded by high-angle normal faults that gradually flatten with depth. Most fault-block mountains form in response to broad uplifting, which causes elongation and faulting. Such a situation is exemplified by the fault blocks that rise high above the rift valleys of East Africa.

Mountains in the United States in which faulting and gradual uplift have contributed to their lofty stature include the Sierra Nevada of California and the Grand Tetons of Wyoming. Both are faulted along their eastern flanks, which were uplifted as the blocks tilted downward to the west. Looking west from Owens Valley, California, and Jackson Hole, Wyoming, the eastern fronts of these ranges (the Sierra Nevada and the Tetons, respectively) rise over 2 kilometers (1.2 miles) making them two of the most imposing mountain fronts in the United States.

One of Earth's largest regions of fault-block mountains is the Basin and Range Province. This region extends in a roughly north to south direction for nearly 3000 kilometers (2000 miles) and encompasses all of Nevada and portions of the surrounding states, as well as parts of southern Canada and western Mexico. Here, the brittle upper crust has literally been broken into hundreds of fault blocks. Tilting of these faulted structures (half-grabens) gave rise to nearly parallel mountain ranges, averaging about 80 kilometers (50 miles) in length, which rise above adjacent sediment-laden basins (see Figure 17.10).

Extension in the Basin and Range Province began about 20 million years ago and appears to have "stretched" the crust as much as twice its original width. Figure 17.20 shows a rough outline of the boundaries of the western states before and after this period of extension. High heat flow in the region, three times average, and several episodes of volcanism provide strong evidence that mantle upwelling caused doming of the crust, which in turn contributed to extension in the region.

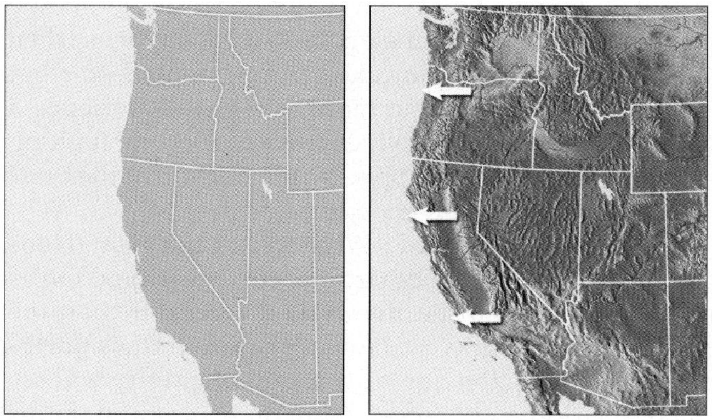

▲ **Figure 17.20** Extension in the Basin and Range Province has "stretched" the crust in some locations by as much as twice its original width. Shown here is a rough outline of the western states before (left) and after (right) extension.

The Grand Tetons of Wyoming are an example of fault-block mountains.

Vertical Movements of the Crust

In addition to the large crustal displacements driven mainly by plate tectonics, gradual up-and-down motions of the continental crust are observed at many locations around the globe. Although much of this vertical movement occurs along plate margins and is associated with active mountain building, some of it is not.

Evidence for crustal uplift occurs along the west coast of the United States. When the elevation of a coastal area remains unchanged for an extended period, a wavecut platform develops (see Figure 13.10, p. 294). In parts of California, ancient wave-cut platforms can now be found as terraces hundreds of meters above sea level. Such evidence of crustal uplift is easy to find; unfortunately, the reason for uplift is not always as easy to determine.

Did You Know?

The greatest accumulations of snow generally occur in mountainous regions. In the United States, the record for a single season was in Mount Baker ski area north of Seattle, Washington, where 1140 inches (95 feet!) of snow fell during the winter of 1998–99.

Isostasy

Early workers discovered that Earth's less-dense crust floats on top of the denser and deformable rocks of the mantle. The concept of a floating crust in gravitational balance is called **isostasy.** One way to grasp the concept of isostasy is to envision a series of wooden blocks of different heights floating in water, as shown in Figure 17.21. Note that the thicker wooden blocks float higher than the thinner blocks.

Similarly, many mountain belts stand high above the surrounding terrain because of crustal thickening. These compressional mountains have buoyant crustal "roots" that extend deep into the supporting material below, just like the thicker wooden blocks shown in Figure 17.21.

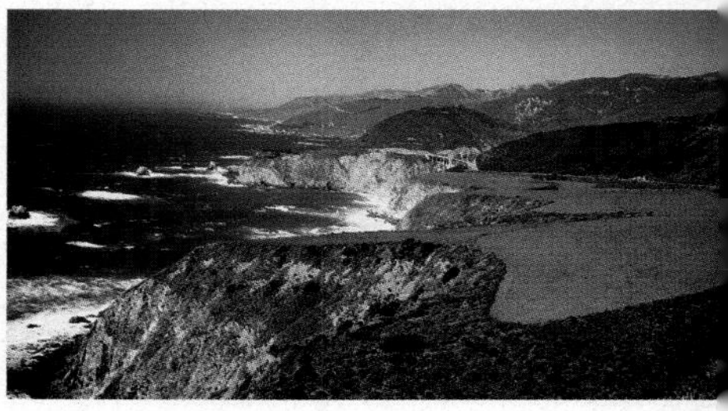

This terrace (flat grassy area) originated as a wave-cut platform that was later uplifted above sea level.

Visualize what would happen if another small block of wood were placed atop one of the blocks in Figure 17.21. The combined block would sink until a new isostatic (gravitational) balance was reached. However, the top of the combined block would actually be higher than before, and the bottom would be lower. This process of establishing a new level of gravitational equilibrium is called **isostatic adjustment.**

Applying the concept of isostatic adjustment, we should expect that when weight is added to the crust, it will respond by subsiding, and when weight is removed, the crust will rebound. (Visualize what hap-

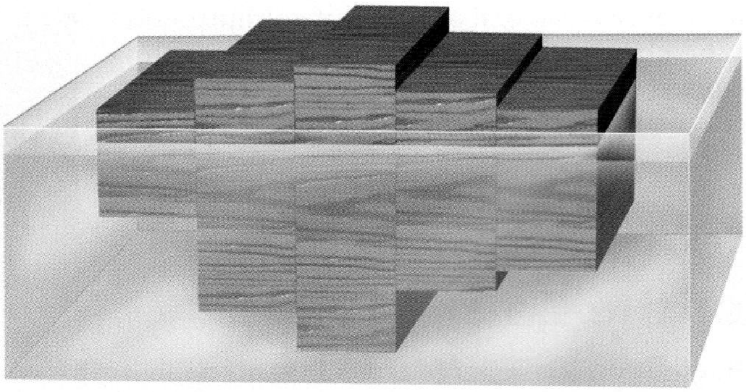

▲ **Figure 17.21** This drawing illustrates how wooden blocks of different thicknesses float in water. In a similar manner, thick sections of crustal material float higher than thinner crustal slabs.

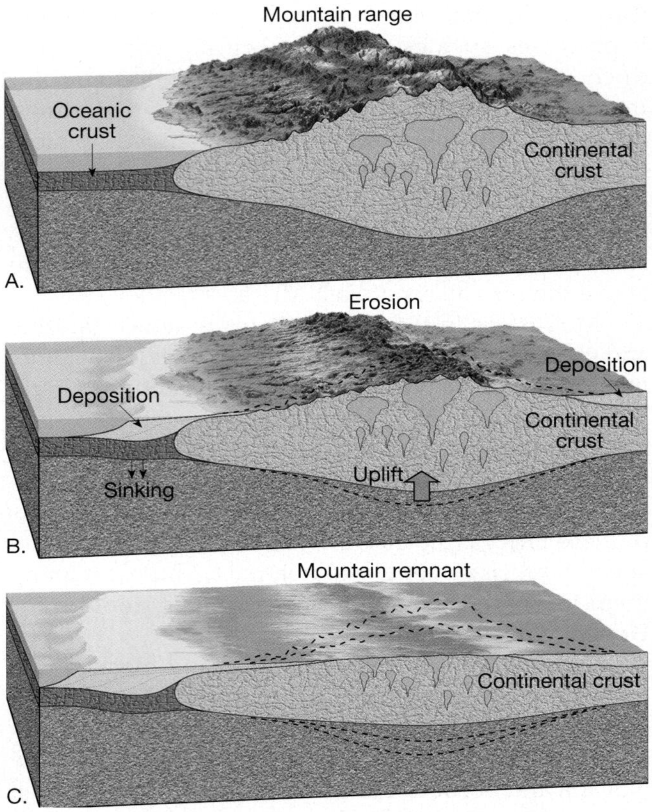

▲ **Figure 17.22** This sequence illustrates how the combined effect of erosion and isostatic adjustment results in a thinning of the crust in mountainous regions. **A.** When mountains are young, the continental crust is thickest. **B.** As erosion lowers the mountains, the crust rises in response to the reduced load. **C.** Erosion and uplift continue until the mountains reach "normal" crustal thickness.

Did You Know?

Since the first successful ascent to the summit of Mount Everest in 1953, thousands have set out to match this feat. Unfortunately, more than 100 people have perished in their effort. In addition, debris has become a major problem because waste, such as empty oxygen bottles, does not disintegrate at high altitudes and is impossible to bury.

pens to a ship as cargo is being loaded and unloaded.) Evidence for crustal subsidence followed by crustal rebound is provided by Ice Age glaciers. When continental ice sheets occupied portions of North America during the Pleistocene epoch, the added weight of 3-kilometer-thick (nearly 2-mile-thick) masses of ice caused downwarping of Earth's crust by hundreds of meters. In the 8000 years since the last ice sheet melted, uplifting of as much as 330 meters (1000 feet) has occurred in Canada's Hudson Bay region, where the thickest ice had accumulated.

One of the consequences of isostatic adjustment is that as erosion lowers the summits of mountains, the crust will rise in response to the reduced load (Figure 17.22). However, each episode of isostatic uplift is somewhat less than the elevation loss due to erosion. The processes of uplifting and erosion will continue until the mountain block reaches "normal" crustal thickness. When this occurs, the mountains will be eroded to near sea level, and the once deeply buried interior of the mountain will be exposed at the surface. In addition, as mountains are worn down, the eroded sediment is deposited on adjacent landscapes, causing these areas to subside (Figure 17.22).

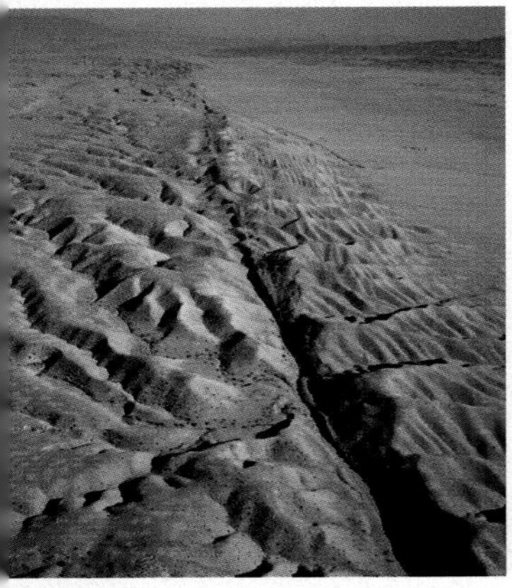

Aerial view of San Andreas Fault. (D. Parker/Photo Researchers)

How High Is Too High?

Where compressional forces are great, such as those driving India into Asia, mountains such as the Himalayas result. But is there a limit on how high a mountain can rise? As mountaintops are elevated, gravity-driven processes such as erosion and mass wasting accelerate, carving the deformed strata into rugged landscapes. Just as important, however, is the fact that gravity also acts on the rocks within these mountainous masses. The higher the mountain, the greater the downward force on the rocks near the base. (Visualize a group of cheerleaders at a sporting event building a human pyramid.) At some point the rocks deep within the developing mountain, which are comparatively warm and weak, will begin to flow laterally, as shown in Figure 17.23. This is analogous to what happens when a ladle of very thick pancake batter is poured on a hot griddle. As a result, the mountain will experience a **gravitational collapse,** which involves normal faulting and subsidence in the upper, brittle portion of the crust and ductile spreading at depth.

You then might ask: What keeps the Himalayas standing? Simply, the horizontal compressional forces that are driving India into Asia are greater than the vertical force of gravity. However, once India's northward trek ends, the downward pull of gravity will become the dominant force acting on this mountainous region.

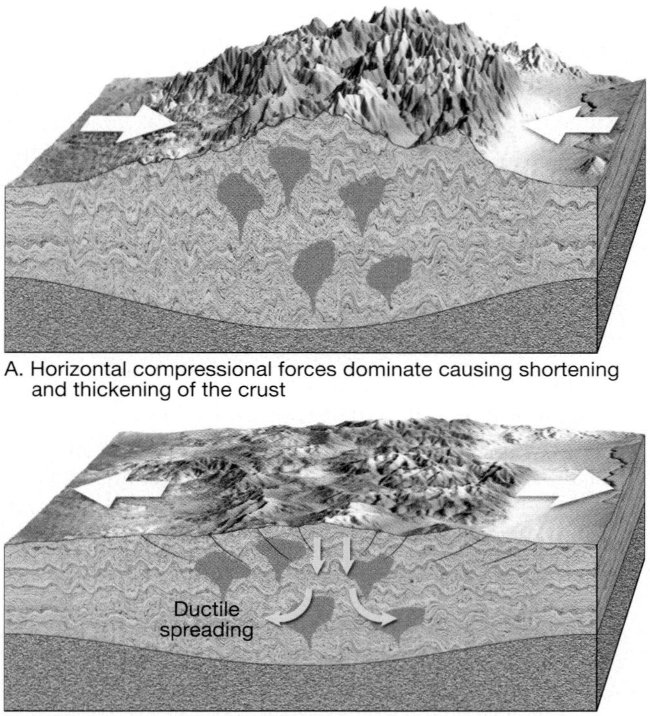

A. Horizontal compressional forces dominate causing shortening and thickening of the crust

B. Gravitational forces dominate resulting in stretching and thinning of the crust

◀ **Figure 17.23** Block diagram of a mountain belt that is collapsing under its own "weight." Gravitational collapse involves normal faulting in the upper, brittle portion of the crust and ductile spreading at depth.

Chapter Summary

- *Deformation* refers to changes in the shape and/or volume of a rock body. Rocks deform differently depending on the environment (temperature and confining pressure), the composition of the rock, and the length of time stress is maintained. Rocks first respond by deforming *elastically* and will return to their original shape when the stress is removed. Once their elastic limit (strength) is surpassed, rocks either deform by ductile flow or they fracture. *Ductile deformation* is a solid-state flow that results in a change in size and shape of rocks without fracturing. Ductile deformation occurs in a high-temperature/high-pressure environment. In a near-surface environment, most rocks deform by *brittle failure.*

- Among the most basic geologic structures associated with rock deformation are *folds* (flat-lying sedimentary and volcanic rocks bent into a series of wavelike undulations). The two most common types of folds are *anticlines,* formed by the upfolding, or arching, of rock layers, and *synclines,* which are downfolds. Most folds are the result of horizontal *compressional stresses. Domes* (upwarped structures) and *basins* (downwarped structures) are circular or somewhat elongated folds formed by vertical displacements of strata.

- Faults are fractures in the crust along which appreciable displacement has occurred. Faults in which the movement is primarily vertical are called *dip-slip faults.* Dip-slip faults include both *normal* and *reverse faults.* Low-angle reverse faults are called *thrust faults.* Normal faults indicate *tensional stresses* that pull the crust apart. Along divergent boundaries tensional stresses can cause a central block called a *graben,* bounded by normal faults, to drop as the plates separate.

- Reverse and thrust faulting indicate that *compressional forces* are at work. Large *thrust faults* are found along subduction zones and other convergent boundaries where plates are colliding.

- *Strike-slip faults* exhibit mainly horizontal displacement parallel to the fault surface. Large strike-slip faults, called *transform faults,* accommodate displacement between plate boundaries. Most transform faults cut the oceanic lithosphere and link divergent boundaries. The San Andreas Fault cuts the continental lithosphere and accommodates the northward displacement of southwestern California.

- *Joints* are fractures along which no appreciable displacement has occurred. Joints generally occur in groups with roughly parallel orientations and are the result of brittle failure of rock units located in the outermost crust.

- The name for the processes that collectively produce a mountain system is *orogenesis.* Most mountains consist of roughly parallel ridges of folded and faulted sedimentary and volcanic rocks, portions of which have been strongly metamorphosed and intruded by younger igneous bodies.

- Subduction of oceanic lithosphere under a continental block gives rise to an *Andean-type plate margin* that is characterized by a continental volcanic arc and associated igneous plutons. In addition, sediment derived from the land, as well as material scraped from the subducting plate, becomes plastered against the landward side of the trench, forming an *accretionary wedge*. An excellent example of an inactive Andean-type mountain belt is found in the western United States and includes the Sierra Nevada and the Coast Range in California.

- Mountain belts can develop as a result of the collision and merger of an island arc, oceanic plateau, or some other small crustal fragment to a continental block. Many of the mountain belts of the North American Cordillera, were generated in this manner.

- Continued subduction of oceanic lithosphere beneath an Andean-type continental margin will eventually close an ocean basin. The result will be a *continental collision* and the development of compressional mountains that are characterized by shortened and thickened crust as exhibited by the Himalayas. The development of a major mountain belt is often complex, involving two or more distinct episodes of mountain building. Continental collisions have generated many mountain belts, including the Alps, Urals, and Appalachians.

- Although most mountains form along convergent plate boundaries, other tectonic processes, such as continental rifting can produce uplift and the formation of topographic mountains. The mountains that form in these settings, termed *fault-block mountains,* are bounded by high-angle normal faults that gradually flatten with depth. The Basin and Range Province in the western United States consists of hundreds of faulted blocks that give rise to nearly parallel mountain ranges that stand above sediment-laden basins.

- Earth's less dense crust floats on top of the denser and deformable rocks of the mantle, much like wooden blocks floating in water. The concept of a floating crust in gravitational balance is called *isostasy.* Most mountainous topography is located where the crust has been shortened and thickened. Therefore, mountains have deep crustal roots that isostatically support them. As erosion lowers the peaks, *isostatic adjustment* gradually raises the mountains in response. The processes of uplifting and erosion will continue until the mountain block reaches "normal" crustal thickness. Gravity also causes elevated mountainous structures to collapse under their own "weight."

Key Terms

accretionary wedge (p. 404)
active continental margin (p. 403)
anticline (p. 393)
basin (p. 395)
brittle failure (brittle deformation) (p. 392)
deformation (p. 392)
dip-slip fault (p. 396)
dome (p. 395)
ductile deformation (p. 393)
fault (p. 395)
fault-block mountains (p. 406)
fault scarp (p. 396)
fold (p. 393)
graben (p. 397)
gravitational collapse (p. 410)
horst (p. 397)
isostasy (p. 409)
isostatic adjustment (p. 409)
joint (p. 401)
monocline (p. 395)
normal fault (p. 397)
orogenesis (p. 401)
passive continental margin (p. 403)
reverse fault (p. 397)
strike-slip fault (p. 398)
syncline (p. 394)
terrane (p. 405)
thrust fault (p. 397)
transform fault (p. 398)

Review Questions

1. What is rock *deformation*?
2. How is *brittle deformation* different from *ductile deformation*?
3. List three factors that determine how rocks will behave when exposed to stresses that exceed their strength. Briefly explain the role of each.
4. Distinguish between *anticlines* and *synclines, domes* and *basins, anticlines* and *domes.*
5. How is a *monocline* different from an *anticline*?
6. The Black Hills of South Dakota are a good example of what type of structural feature?
7. Contrast the movements that occur along normal and reverse faults. What type of force is indicated by each fault?
8. Is the fault shown in Figure 17.8 a normal or reverse fault?
9. Describe a *horst* and a *graben.* Explain how a graben valley forms, and name one.
10. What type of faults are associated with fault-block mountains?
11. How are reverse faults different from thrust faults? In what way are they the same?
12. The San Andreas Fault is an excellent example of a _____ fault.
13. How are joints different from faults?
14. In the plate tectonics model, which type of plate boundary is most directly associated with mountain building?

15. Briefly describe the development of a volcanic island arc.

16. The formation of mountainous topography at a volcanic island arc, such as Japan, is considered just one phase in the development of a major mountain belt. Explain.

17. What is an *accretionary wedge*? Briefly describe its formation.

18. What is a *passive margin*? Give an example. Then give an example of an *active continental margin*.

19. In what way are the Sierra Nevada and the Andes similar?

20. How can the Appalachian Mountains be considered a collision-type mountain range when the nearest continent is thousands of kilometers away?

21. How does the plate tectonics theory help explain the existence of fossil marine life in rocks atop mountains formed by continental collisions?

22. Define the term *terrane*. How is it different from the term *terrain*?

23. In addition to microcontinents, what other structures are thought to be carried by the oceanic lithosphere and eventually accreted to a continent?

24. Briefly describe the major differences between the evolution of the Appalachian Mountains and the North American Cordillera.

25. Compare the processes that generate fault-block mountains to those associated with most other major mountain belts.

26. Give one example of evidence that supports the concept of crustal uplift.

27. What happens to a floating object when weight is added? Subtracted? How does this principal apply to changes in the elevation of mountains? What term is applied to the adjustment that causes crustal uplift of this type?

28. How does the formation and melting of Pleistocene ice sheets support the idea that the lithosphere tries to remain in isostatic balance?

Online Study Guide

The *Essentials of Geology* Web site uses the resources and flexibility of the Internet to aid in your study of the topics in this chapter. Written and developed by geology instructors, this site will help improve your understanding of geology. Visit **www.prenhall.com/lutgens** and click on the cover of *Essentials of Geology 9e* to find:

- Online review quizzes.
- Critical thinking exercises.
- Links to chapter-specific Web resources.
- Internet-wide key-term searches.

http://www.prenhall.com/lutgens

The strata exposed in the Grand Canyon contain clues to hundreds of millions of years of Earth history. This view is from the North Rim. *(Photo by Carr Clifton)*

CHAPTER 18

Geologic Time

Focus on Learning

To assist you in learning the important concepts in this chapter, you will find it helpful to focus on the following questions:

- What are the two types of dates used by geologists to interpret Earth history?
- What are the laws, principles, and techniques used to establish relative dates?
- What are fossils? What are the conditions that favor the preservation of organisms as fossils?
- How are fossils used to correlate rocks of similar ages that are in different places?
- What is radioactivity, and how are radioactive isotopes used in radiometric dating?
- What is the geologic time scale, and what are its principal subdivisions?
- Why is it difficult to assign reliable numerical dates to samples of sedimentary rock?

Paleontologist working with dinosaur remains.
(Rich Frishman)

In the late eighteenth century, James Hutton recognized the immensity of Earth history and the importance of time as a component in all geological processes. In the nineteenth century, others effectively demonstrated that Earth had experienced many episodes of mountain building and erosion, which must have required great spans of geologic time. Although these pioneering scientists understood that Earth was very old, they had no way of knowing its true age. Was it tens of millions, hundreds of millions, or even billions of years old? Rather, a geologic time scale was developed that showed the sequence of events based on relative dating principles. What are these principles? What part do fossils play? With the discovery of radioactivity and radiometric dating techniques, geologists now can assign fairly accurate dates to many of the events in Earth history. What is radioactivity? Why is it a good "clock" for dating the geologic past?

Geology Needs a Time Scale

In 1869, John Wesley Powell, who was later to head the U.S. Geological Survey, led a pioneering expedition down the Colorado River and through the Grand Canyon (Figure 18.1). Writing about the rock layers that were exposed by the downcutting of the river, Powell said that "the canyons of this region would be a Book of Revelations in the rock-leaved Bible of geology." He was undoubtedly impressed with the millions of years of Earth history exposed along the walls of the Grand Canyon (see chapter-opening photo).

Powell realized that the evidence for an ancient Earth is concealed in its rocks. Like the pages in a long and complicated history book, rocks record the geological events and changing life forms of the past. The book, however, is not complete. Many pages, especially in the early chapters, are missing. Others are tattered, torn, or smudged. Yet enough of the book remains to allow much of the story to be deciphered.

Interpreting Earth history is a prime goal of the science of geology. Like a modern-day sleuth, the geologist must interpret clues found preserved in the rocks. By studying rocks, especially sedimentary rocks, and the features they contain, geologists can unravel the complexities of the past.

Geological events by themselves, however, have little meaning until they are put into a time perspective. Studying history, whether it be the Civil War or the Age of Dinosaurs, requires a calendar. Among geology's major contributions to human knowledge is the *geologic time scale* and the discovery that Earth history is exceedingly long.

Relative Dating—Key Principles

GEODe ESSENTIALS OF GEOLOGY
Geologic Time
▼ Relative Dating—Key Principles

The geologists who developed the geologic time scale revolutionized the way people think about time and perceive our planet. They learned that Earth is much older

A.

B.

▲ **Figure 18.1** **A.** Start of the expedition from Green River station. A drawing from Powell's 1875 book. **B.** Major John Wesley Powell, pioneering geologist and the second director of the U.S. Geological Survey. *(Courtesy of the U.S. Geological Survey, Denver)*

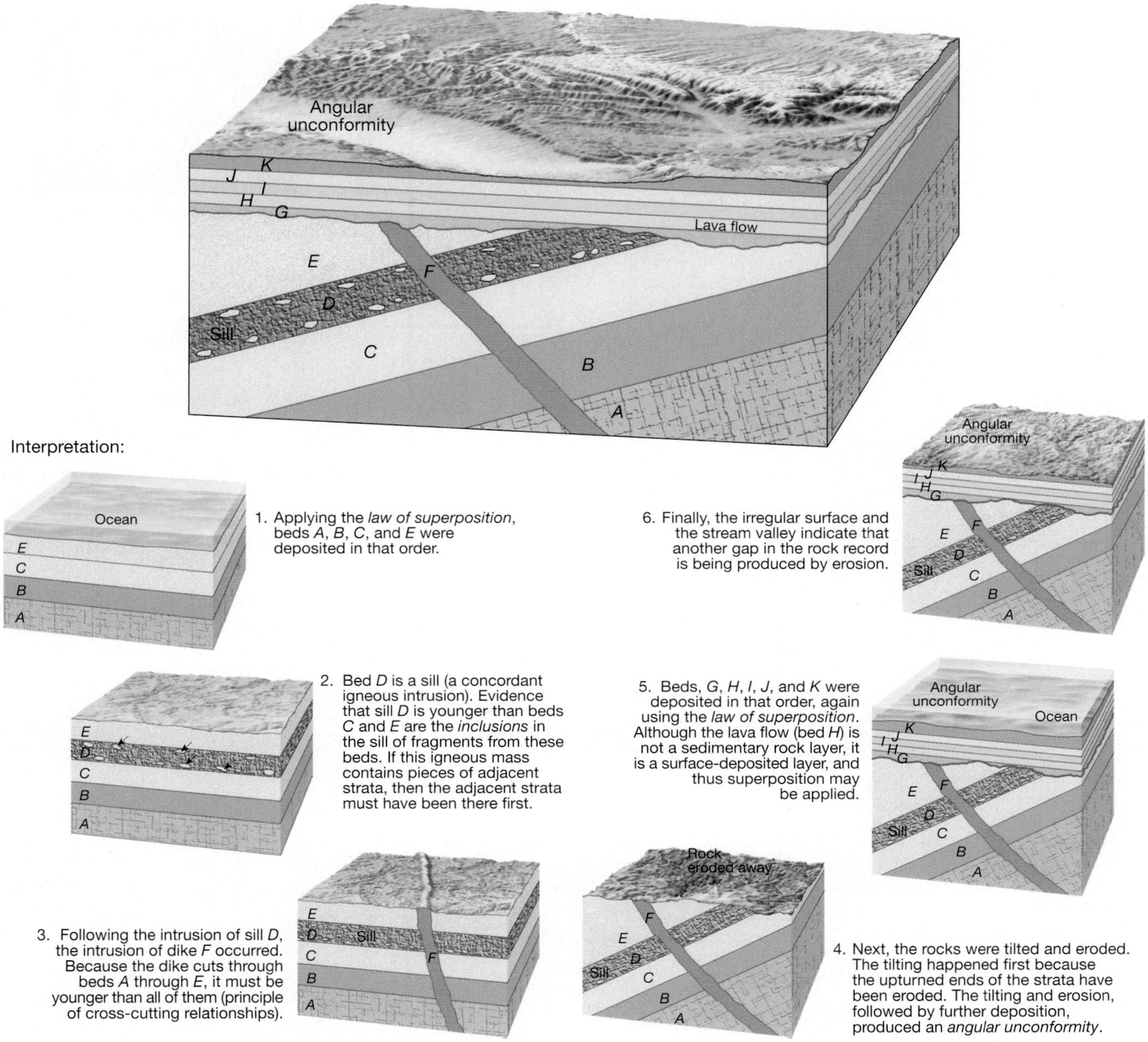

Angular unconformity
K
J
I
H
G
Lava flow
E
F
D
Sill
C
B
A
Interpretation:
Ocean
E
C
B
A
1. Applying the *law of superposition*, beds *A*, *B*, *C*, and *E* were deposited in that order.
E
D
C
B
A
2. Bed *D* is a sill (a concordant igneous intrusion). Evidence that sill *D* is younger than beds *C* and *E* are the *inclusions* in the sill of fragments from these beds. If this igneous mass contains pieces of adjacent strata, then the adjacent strata must have been there first.
3. Following the intrusion of sill *D*, the intrusion of dike *F* occurred. Because the dike cuts through beds *A* through *E*, it must be younger than all of them (principle of cross-cutting relationships).
E
D
Sill
C
B
F
A
Rock eroded away
F
E
D
Sill
C
B
A
4. Next, the rocks were tilted and eroded. The tilting happened first because the upturned ends of the strata have been eroded. The tilting and erosion, followed by further deposition, produced an *angular unconformity*.
5. Beds, *G*, *H*, *I*, *J*, and *K* were deposited in that order, again using the *law of superposition*. Although the lava flow (bed *H*) is not a sedimentary rock layer, it is a surface-deposited layer, and thus superposition may be applied.
Angular unconformity
Ocean
K
J
I
H
G
F
E
D
Sill
C
B
A
6. Finally, the irregular surface and the stream valley indicate that another gap in the rock record is being produced by erosion.
Angular unconformity
K
J
I
H
G
E
F
D
Sill
C
B
A

▲ **Figure 18.8** Geologic cross-section of a hypothetical region.

Swamps, an Age of Reptiles, and an Age of Mammals. These "ages" pertain to groups that were especially plentiful and characteristic during particular time periods. Within each of the "ages," there are many subdivisions based, for example, on certain species of trilobites and certain types of fish, reptiles, and so on. This same succession of dominant organisms, never out of order, is found on every continent.

Once fossils were recognized as time indicators, they became the most useful means of correlating rocks of similar age in different regions. Geologists pay particular attention to certain fossils called **index fossils.** These fossils are widespread geographically and are limited to a short span of geologic time, so their presence provides an important method of matching rocks of the same age. Rock formations, however, do not always contain a specific index fossil. In such situations, groups of fossils are used to establish the age of the bed. Figure 18.11 illustrates how a group of fossils can be used to date rocks more precisely than could be accomplished by the use of only one of the fossils.

In addition to being important and often essential tools for correlation, fossils are important environmental indicators. Although much can be deduced about past environments by studying the nature and characteristics of sedimentary rocks, a close

Did You Know?

People frequently confuse paleontology and archaeology. Paleontologists study fossils and are concerned with *all* life forms in the geologic past. By contrast, archaeologists focus on the material remains of past human life. These remains include both the objects used by people long ago, called *artifacts*, and the buildings and other structures associated with where people lived, called *sites*.

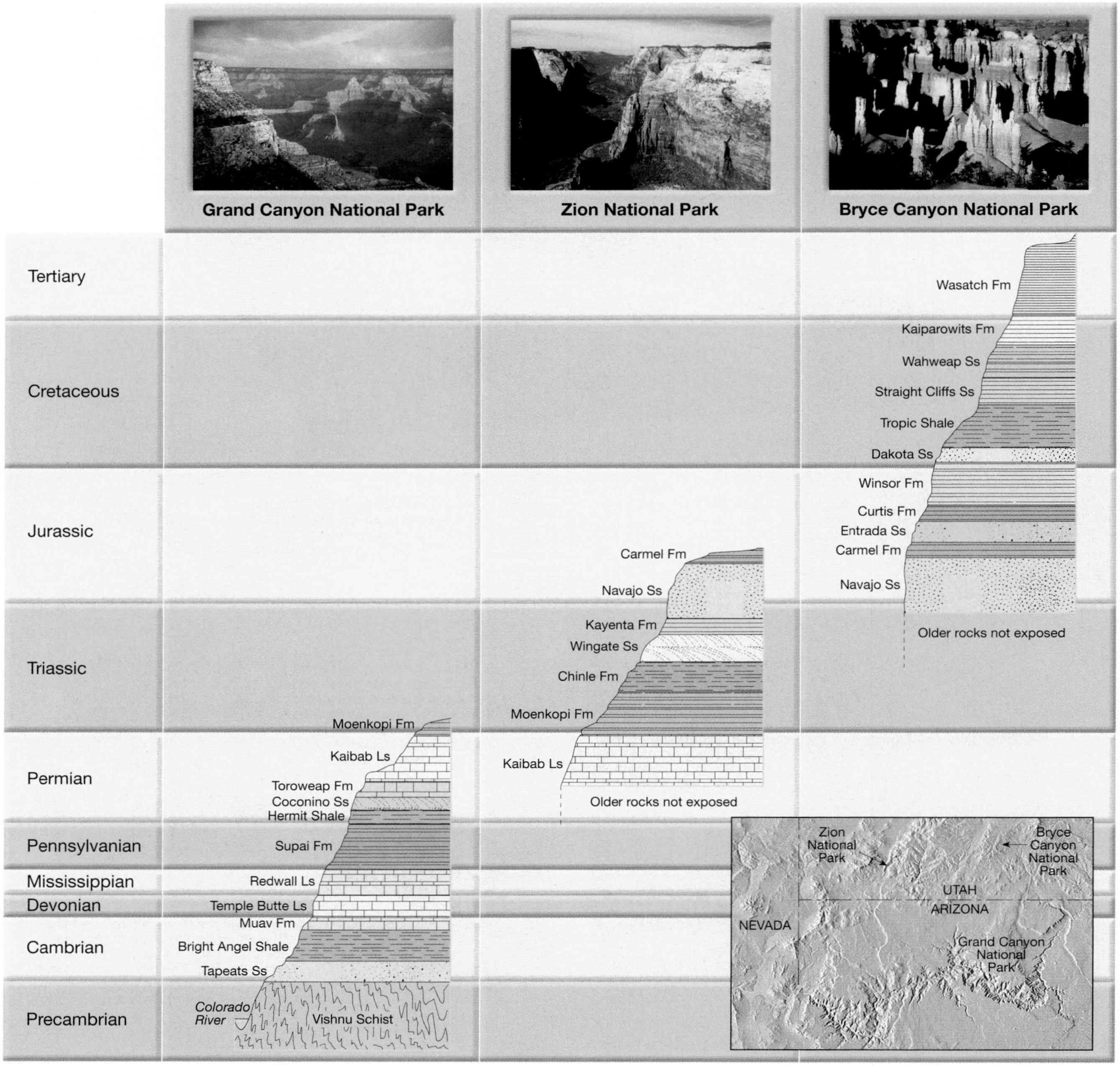

▲ **Figure 18.9** Correlation of strata at three locations on the Colorado Plateau reveals the total extent of sedimentary rocks in the region. *(After U.S. Geological Survey; Photos by E. J. Tarbuck)*

examination of any fossils present can usually provide a great deal more information.

For example, when the remains of certain clam shells are found in limestone, the geologist can assume that the region was once covered by a shallow sea, because that is where clams live today. Also, by using what we know of living organisms, we can conclude that fossil animals with thick shells capable of withstanding pounding and surging waves must have inhabited shorelines. In contrast, animals with thin, delicate shells probably indicate deep, calm offshore waters. Hence, by looking closely at the types of fossils, the approximate position of an ancient shoreline may be identified.

Further, fossils can indicate the former temperature of the water. Certain present-day corals require warm and shallow tropical seas like those around Florida and the Bahamas. When similar corals are found in ancient limestones, they indicate that a Florida-like marine environment must have existed when they were alive. These examples illustrate how fossils can help unravel the complex story of Earth history.

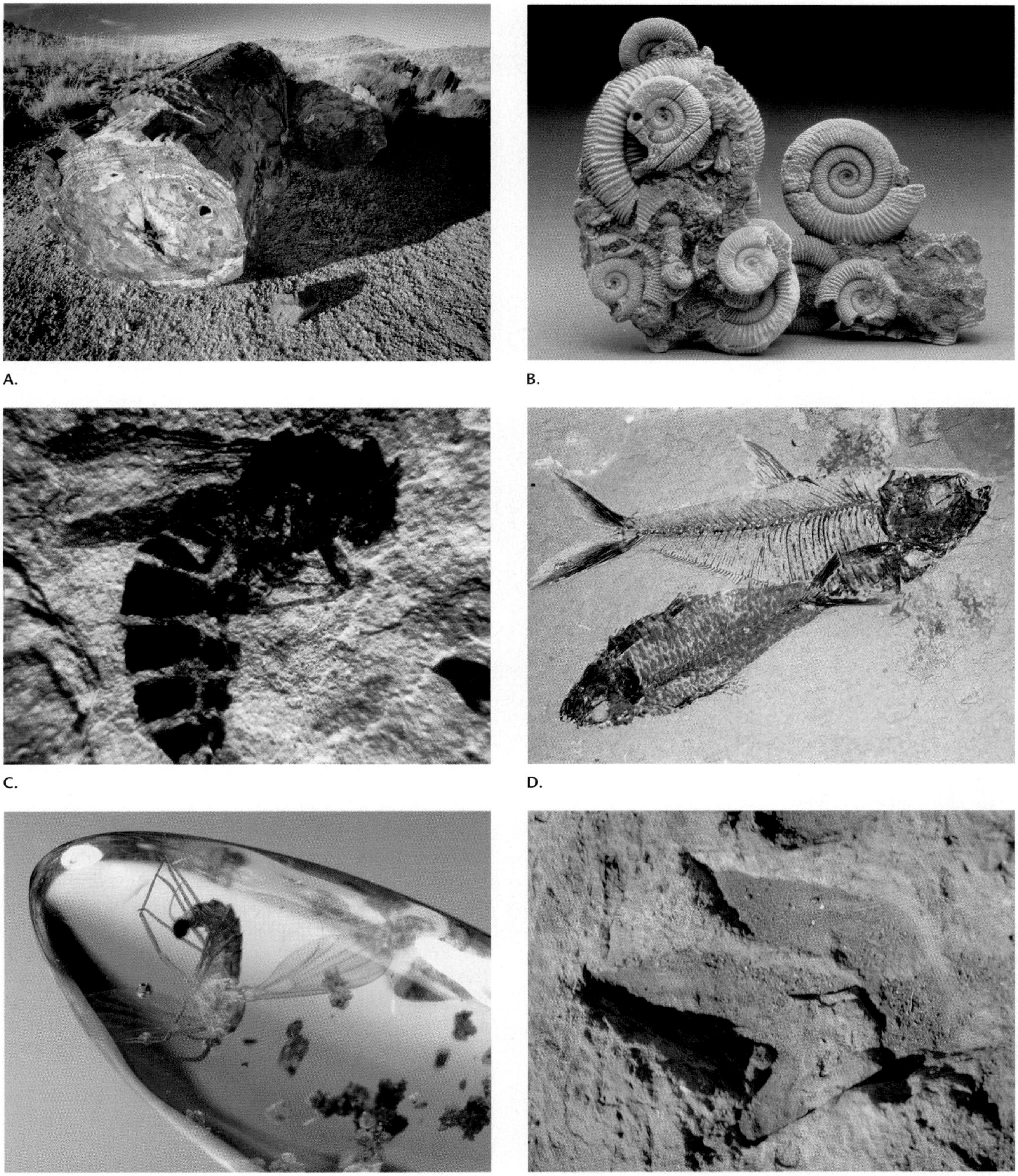

A. B. C. D. E. F.

▲ **Figure 18.10** There are many types of fossilization. Six examples are shown here. **A.** Petrified wood in Petrified Forest National Park, Arizona. **B.** Natural casts of shelled invertebrates. **C.** A fossil bee preserved as a thin carbon film. **D.** Impressions are common fossils and often show considerable detail. **E.** Insect in amber. **F.** Dinosaur footprint in fine-grained limestone near Tuba City, Arizona. *(Photo A by David Muench; Photos B, D, and F by E. J. Tarbuck; Photo C courtesy of the National Park Service; Photo E by Breck P. Kent)*

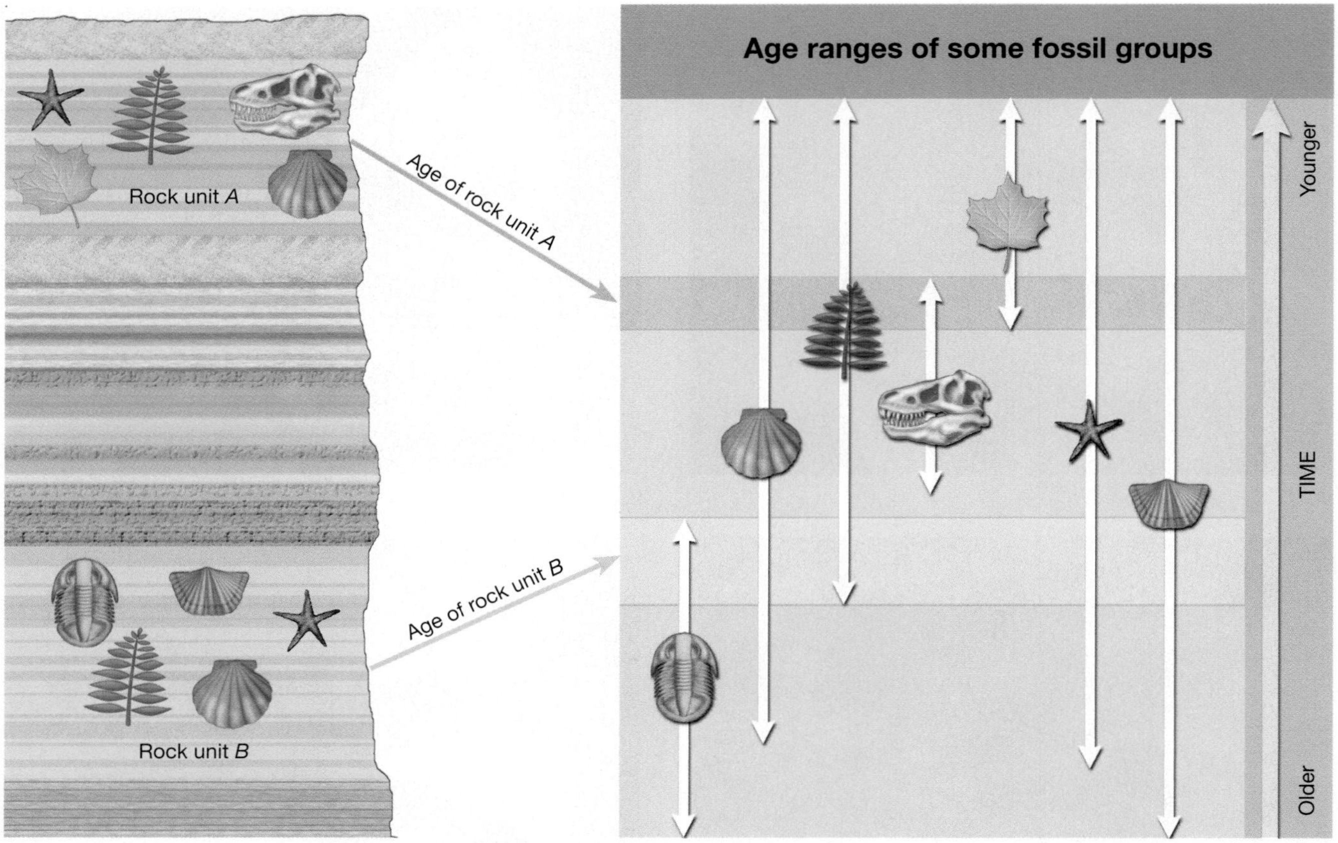

▲ **Figure 18.11** Overlapping ranges of fossils help date rocks more exactly than using a single fossil.

Dating with Radioactivity

GEODe Geologic Time ▼ Dating with Radioactivity

In addition to establishing relative dates by using the principles described in the preceding sections, it is also possible to obtain reliable numerical dates for events in the geologic past. For example, we know that Earth is about 4.5 billion years old and that the dinosaurs became extinct about 65 million years ago. Dates that are expressed in millions and billions of years truly stretch our imagination because our personal calendars involve time measured in hours, weeks, and years. Nevertheless, the vast expanse of geologic time is a reality, and it is radiometric dating that allows us to measure it. In this section you will learn about radioactivity and its application in radiometric dating.

This trilobite fossil illustrates mold and cast.

Reviewing Basic Atomic Structure

Recall from Chapter 2 that each atom has a *nucleus* containing protons and neutrons and that the nucleus is orbited by electrons. *Electrons* have a negative electrical charge, and *protons* have a positive charge. A *neutron* is actually a proton and an electron combined, so it has no charge (it is neutral).

The *atomic number* (each element's identifying number) is the number of protons in the nucleus. Every element has a different number of protons and thus a different atomic number (hydrogen = 1, carbon = 6, oxygen = 8, uranium = 92, etc.). Atoms of the same element always have the same number of protons, so the atomic number stays constant.

Practically all of an atom's mass (99.9 percent) is in the nucleus, indicating that electrons have virtually no mass at all. So, by adding the protons and neutrons in an atom's nucleus, we derive the atom's *mass number.* The number of neutrons can vary, and these variants, or *isotopes,* have different mass numbers.

To summarize with an example, uranium's nucleus has always 92 protons, so its atomic number is always 92. But its neutron population varies, so uranium has three isotopes: uranium-234 (protons + neutrons = 234), uranium-235, and uranium-238. All three isotopes are mixed in nature. They look the same and behave the same in chemical reactions.

Radioactivity

The forces that bind protons and neutrons together in the nucleus usually are strong. However, in some isotopes, the nuclei are unstable because the forces binding protons and neutrons together are not strong enough. As a result, the nuclei spontaneously break apart (decay), a process called **radioactivity.**

What happens when unstable nuclei break apart? Three common types of radioactive decay are illustrated in Figure 18.12 and are summarized as follows:

1. Alpha particles (α particles) may be emitted from the nucleus. An alpha particle is composed of 2 protons and 2 neutrons. Consequently, the emission of an alpha particle means that the mass number of the isotope is reduced by 4 and the atomic number is decreased by 2.
2. When a beta particle (β particle), or electron, is given off from a nucleus, the mass number remains unchanged because electrons have practically no mass. However, because the electron has come from a neutron (remember, a neutron is a combination of a proton and an electron), the nucleus contains one more proton than before. Therefore, the atomic number increases by 1.
3. Sometimes an electron is captured by the nucleus. The electron combines with a proton and forms an additional neutron. As in the last example, the mass number remains unchanged. However, as the nucleus now contains one less proton, the atomic number decreases by 1.

An unstable (radioactive) isotope of an element is called the *parent.* The isotopes resulting from the decay of the parent are the *daughter products.* Figure 18.13 provides an example of radioactive decay. Here it can be seen that when the radioactive parent, uranium-238 (atomic number 92, mass number 238) decays, it follows a number of steps, emitting 8 alpha particles and 6 beta particles before finally becoming the stable daughter product lead-206 (atomic number 82, mass number 206). One of the unstable daughter products produced during this decay series is radon. Box 18.1 examines the hazards associated with this radioactive gas.

Did You Know?

Even when organisms die and their tissues decay, the organic compounds (hydrocarbons) of which they were made may survive in sediments. These are called *chemical fossils.* Commonly, these hydrocarbons form oil and gas, but some residues can persist in the rock record and be analyzed to determine from which kind of organisms they are derived.

Certainly among the most important results of the discovery of radioactivity is that it provided a reliable means of calculating the ages of rocks and minerals that contain particular radioactive isotopes. The procedure

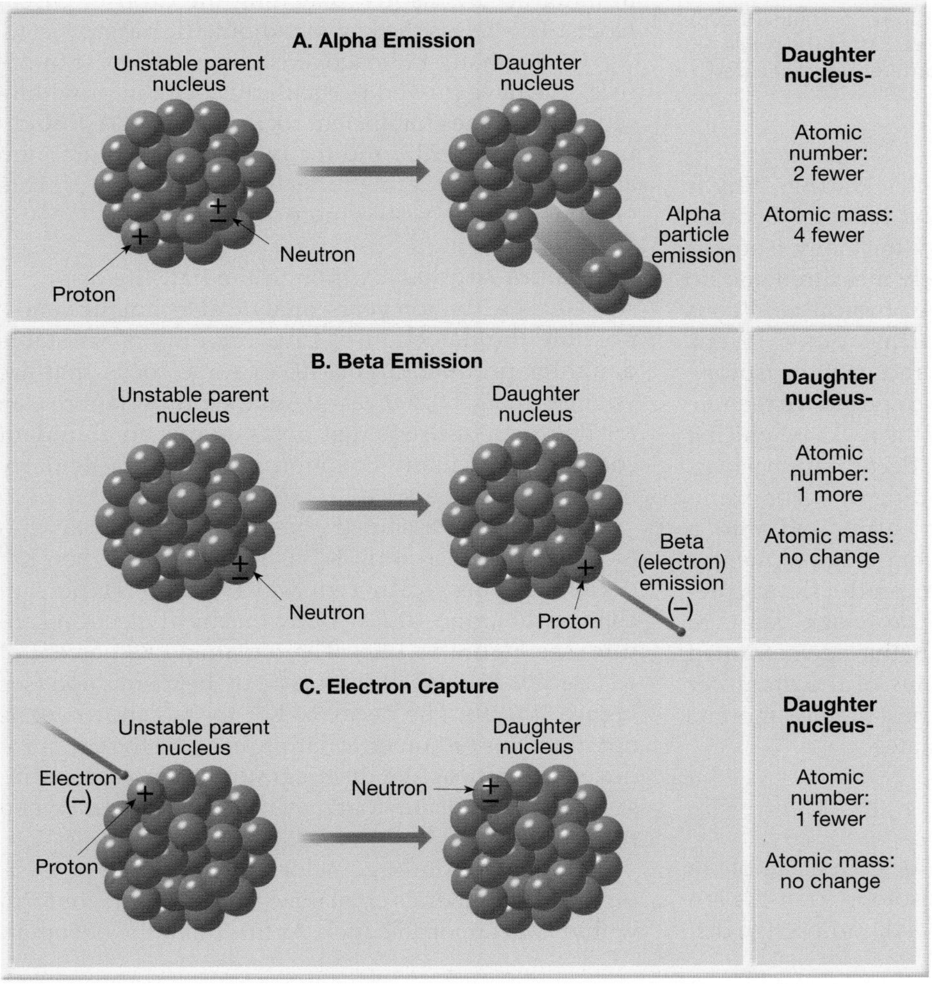

◄ **Figure 18.12** Common types of radioactive decay. Notice that in each case the number of protons (atomic number) in the nucleus changes, thus producing a different element.

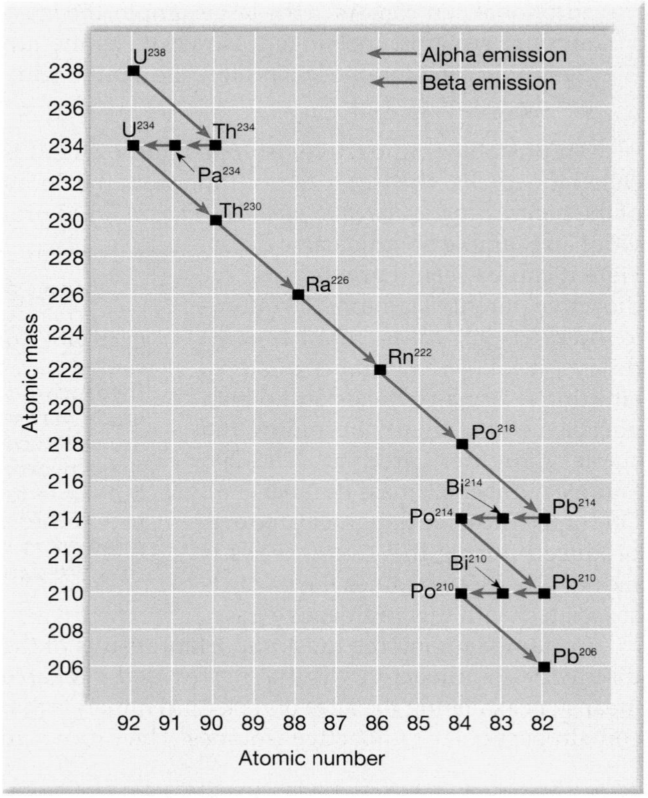

▲ **Figure 18.13** The most common isotope of uranium (U-238) is an example of a radioactive decay series. Before the stable end product (Pb-206) is reached, many different isotopes are produced as intermediate steps.

Did You Know?

Numerical dates of the fossil record show that life began in the ocean approximately 3.8 billion years ago.

Coprolite is fossil dung.

is called **radiometric dating.** Why is radiometric dating reliable? Because the rates of decay for many isotopes have been precisely measured and do not vary under the physical conditions that exist in Earth's outer layers. Therefore, each radioactive isotope used for dating has been decaying at a fixed rate ever since the formation of the rocks in which it occurs, and the products of decay have been accumulating at a corresponding rate. For example, when uranium is incorporated into a mineral that crystallizes from magma, there is no lead (the stable daughter product) from previous decay. The radiometric "clock" starts at this point. As the uranium in this newly formed mineral disintegrates, atoms of the daughter product are trapped and measurable amounts of lead eventually accumulate.

Half-Life

The time required for half of the nuclei in a sample to decay is called the **half-life** of the isotope. Half-life is a common way of expressing the rate of radioactive disintegration. Figure 18.14 illustrates what occurs when a radioactive parent decays directly into its stable daughter product. When the quantities of parent and daughters are equal (ratio 1:1), we know that one half-life has transpired. When one quarter of the original parent atoms remain and three quarters have decayed to the daughter product, the parent/daughter ratio is 1:3, and we know that two half-lives have passed. After three half-lives, the ratio of parent atoms to daughter atoms is 1:7 (one parent for every seven daughter atoms).

If the half-life of a radioactive isotope is known and the parent/daughter ratio can be measured, the age of the sample can be calculated. For example, assume that the half-life of a hypothetical unstable isotope is 1 million years and the parent/daughter ratio in a sample is 1:15. Such a ratio indicates that four half-lives have passed and that the sample must be 4 million years old.

Radiometric Dating

Notice that the *percentage* of radioactive atoms that decay during one half-life is always the same: 50 percent. However, the *actual number* of atoms that decay with the passing of each half-life continually decreases. Thus, as the percentage of radioactive parent atoms declines, the proportion of stable daughter atoms rises, with the increase in daughter atoms just matching the drop in parent atoms. This fact is the key to radiometric dating.

Of the many radioactive isotopes that exist in nature, five have proved particularly useful in providing radiometric ages for ancient rocks (Table 18.1). Rubidium-87, thorium-232, and the two isotopes of uranium are used only for dating rocks that are millions of years old, but potassium-40 is more versatile.

Potassium-Argon. Although the half-life of potassium-40 is 1.3 billion years, analytical techniques make possible the detection of tiny amounts of its stable daughter product, argon-40, in some rocks that are younger than 100,000 years. Another important reason for its frequent use is that potassium is an abundant constituent of many common minerals, particularly micas and feldspars.

Although potassium (K) has three natural isotopes—K^{39}, K^{40}, and K^{41}—only K^{40} is radioactive. When K^{40} decays, it does so in two ways. About 11 percent changes to argon-40 (Ar^{40}) by means of electron capture (see Figure 18.12C). The remaining 89 percent of K^{40} decays to calcium-40 (Ca^{40}) by beta emission (see Figure 18.12B). The decay of K^{40} to Ca^{40}, however, is not useful for radiometric dating, because the Ca^{40} produced by radioactive disintegration cannot be distinguished from calcium that may have been present when the rock formed.

The potassium-argon clock begins when potassium-bearing minerals crystallize from a magma or form within a metamorphic rock. At this point the new min-

BOX 18.1

Radon

Richard L. Hoffmann, Ph.D.*

Radioactivity is defined as the spontaneous emission of atomic particles and/or electromagnetic waves from unstable atomic nuclei. For example, in a sample of uranium-238, unstable nuclei decay and produce a variety of radioactive progeny or daughter products, as well as energetic forms of radiation (Table 18.A). One of its radioactive decay products is radon—a colorless, odorless, invisible gas.

Radon gained public attention in 1984 when a worker in a Pennsylvania nuclear power plant set off radiation alarms—not when he left work, but as he first arrived there. His clothing and hair were contaminated with radon decay products. Investigation revealed that his basement at home had a radon level 2800 times the average level in indoor air. The home was located along a geological formation known as the Reading Prong—a mass of uranium-bearing rock that runs from near Reading, Pennsylvania, to near Trenton, New Jersey.

Originating in the radio decay of traces of uranium and thorium found in almost all soils, radon isotopes (Rn-222 and Rn-220) are continually renewed in an ongoing, natural process. Geologists estimate that the top 6 feet of soil from an average acre of land contains about 50 pounds of uranium (about 2 to 3 parts per million); some types of rocks contain more. Radon is continually generated by the gradual decay of this uranium. Because uranium has a half-life of about 4.5 billion years, radon will be with us forever.

Radon itself decays, having a half-life of only about four days. Its decay products (except lead-206) are all radioactive solids that adhere to dust particles, many of which we inhale. During prolonged exposure to a radon-contaminated environment, some decay will occur while the gas is in the lungs, thereby placing the radioactive radon progeny in direct contact with delicate lung tissue. Steadily accumulating evidence indicates radon to be a significant cause of lung cancer second only to smoking.

A house with a radon level of 4.0 picocuries per liter of air has about eight to nine atoms of radon decaying every minute in every liter of air. The Environmental Protection Agency (EPA) suggests that indoor radon levels be kept below this level. The EPA risk estimates are conservative; they are based on an assumption that one would spend 75 percent of a 70-year time span (about 52 years) in the contaminated space, which most people would not.

Once radon is produced in the soil, it diffuses throughout the tiny spaces between soil particles. Some radon ultimately reaches the soil surface, where it dissipates into the air. Radon enters buildings and homes through holes and cracks in basement floors and walls. Radon's density is greater than air, so it tends to remain in basements during its short decay cycle.

The source of radon is as enduring as its generation mechanism within Earth; radon will never go away. However, cost-effective mitigation strategies are available to reduce radon to acceptable levels, generally without great expense.

Table 18.A Decay products of uranium-238.

Some Decay Products of Uranium 238	Decay Particle Produced	Half-Life
Uranium-238	alpha	4.5 billion years
Radium-226	alpha	1600 years
Radon-222	**alpha**	**3.82 days**
Polonium-218	alpha	3.1 minutes
Lead-214	beta	26.8 minutes
Bismuth-214	beta	19.7 minutes
Polonium-214	alpha	1.6×10^{-4} second
Lead-210	beta	20.4 years
Bismuth-210	beta	5.0 days
Polonium-210	alpha	138 days
Lead-206	none	stable

*Dr. Hoffmann, late Professor of Chemistry, Illinois Central College.

erals will contain K^{40} but will be free of Ar^{40}, because this element is an inert gas that does not chemically combine with other elements.

As time passes, the K^{40} steadily decays by electron capture. The Ar^{40} produced by this process remains trapped within the mineral's crystal lattice. Because no Ar^{40} was present when the mineral formed, all of the daughter atoms trapped in the mineral must have come from the decay of K^{40}. To determine a sample's age, the K^{40}/Ar^{40} ratio is measured precisely and the known half-life for K^{40} applied.

Sources of Error. It is important to realize that an accurate radiometric date can be obtained only if the mineral remained a closed system during the entire period since its formation. A correct date is not possible unless there was neither the addition nor loss of parent or daughter isotopes. This is not always the case. In fact, an important limitation of the potassium-argon method arises from the fact that argon is a gas and it may leak from minerals, throwing off measurements. Indeed, losses can be significant if the rock is subjected to relatively high temperatures.

Of course, a reduction in the amount of Ar^{40} leads to an underestimation of the rock's age. Sometimes temperatures are high enough for a sufficiently long period that all argon

Did You Know?

Although movies and cartoons have depicted humans and dinosaurs living side by side, this was never the case. Dinosaurs flourished during the Mesozoic era and became extinct about 65 million years ago. Humans and their close ancestors did not appear on the scene until the late Cenozoic, more than 60 million years *after* the demise of dinosaurs.

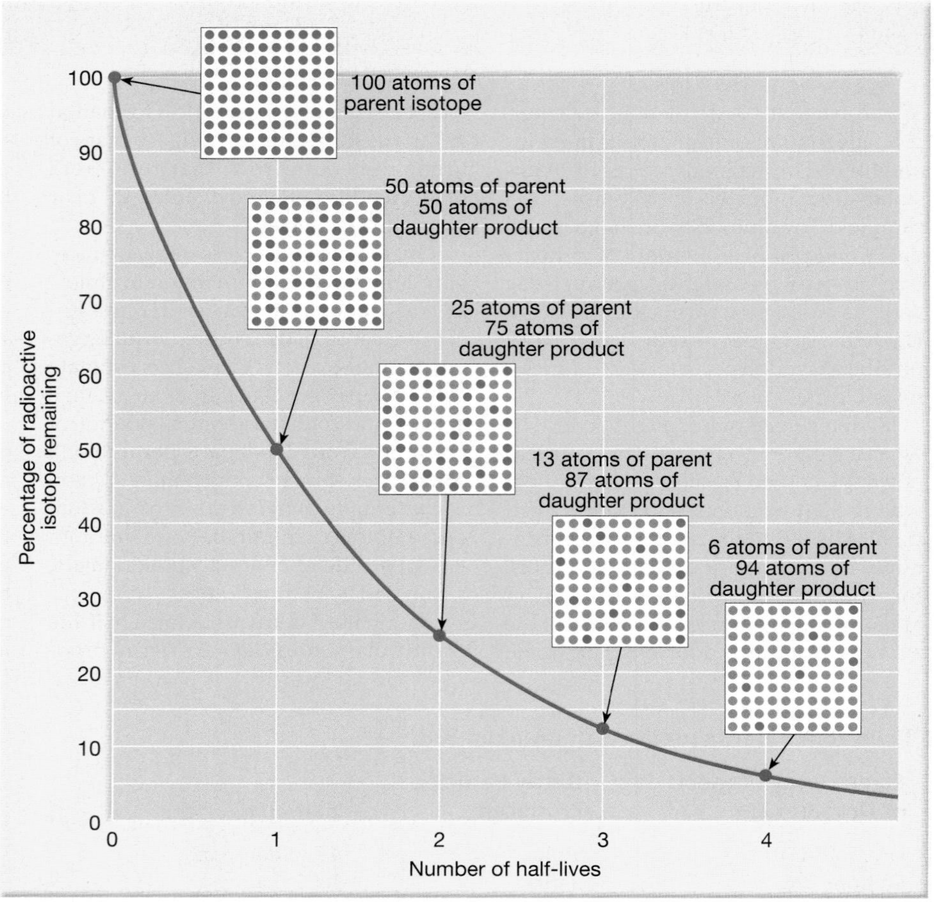

▶ **Figure 18.14** The radioactive-decay curve shows change that is exponential. Half of the radioactive parent remains after one half-life. After a second half-life, one-quarter of the parent remains, and so forth.

Did You Know?

One common precaution against sources of error in radiometric dating is the use of cross checks. Often this simply involves subjecting a sample to two different methods. If the two dates agree, the likelihood is high that the date is reliable. If an appreciable difference is found, other cross checks must be employed to determine which, if either, is correct.

escapes. When this happens, the potassium-argon clock is reset, and dating the sample will give only the time of thermal resetting, not the true age of the rock. For other radiometric clocks, a loss of daughter atoms can occur if the rock has been subjected to weathering or leaching. To avoid such a problem, one simple safeguard is to use only fresh, unweathered material and not samples that may have been chemically altered.

Table 18.1 Radioactive isotopes frequently used in radiometric dating.

Radioactive Parent	Stable Daughter Product	Currently Accepted Half-life Values
Uranium-238	Lead-206	4.5 billion years
Uranium-235	Lead-207	713 million years
Thorium-232	Lead-208	14.1 billion years
Rubidium-87	Strontium-87	47.0 billion years
Potassium-40	Argon-40	1.3 billion years

Dating with Carbon-14

To date very recent events, carbon-14 is used. Carbon-14 is the radioactive isotope of carbon. The process is often called **radiocarbon dating.** Because the half-life of carbon-14 is only 5730 years, it can be used for dating events from the historic past as well as those from very recent geologic history (see Box 18.2). In some cases carbon-14 can be used to date events as far back as 70,000 years.

Carbon-14 is continuously produced in the upper atmosphere as a consequence of cosmic-ray bombardment. Cosmic rays, which are high-energy particles, shatter the nuclei of gas atoms, releasing neutrons. Some of the neutrons are absorbed by nitrogen atoms (atomic number 7), causing their nuclei to emit a proton. As a result, the atomic number decreases by 1 (to 6), and a different element, carbon-14, is created (Figure 18.15A). This isotope of carbon quickly becomes incorporated into carbon dioxide, which circulates in the atmosphere and is absorbed by living matter. As a result, all organisms contain a small amount of carbon-14, including yourself.

While an organism is alive, the decaying radiocarbon is continually replaced, and the proportions of car-

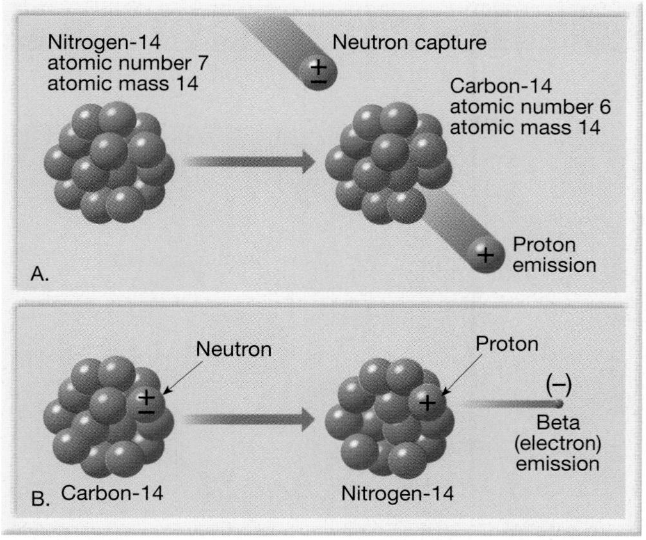

▲ **Figure 18.15** **A.** Production and **B.** decay of carbon-14. These sketches represent the nuclei of the respective atoms.

bon-14 and carbon-12 remain constant. Carbon-12 is the stable and most common isotope of carbon. However, when any plant or animal dies, the amount of carbon-14 gradually decreases as it decays to nitrogen-14 by beta emission (Figure 18.15B). By comparing the proportions of carbon-14 and carbon-12 in a sample, radiocarbon dates can be determined. It is important to emphasize that carbon-14 is only useful in dating organic materials, such as wood, charcoal, bones, flesh, and even cloth made of cotton fibers.

Although carbon-14 is useful in dating only the last small fraction of geologic time, it has become a very valuable tool for anthropologists, archeologists, and historians, as well as for geologists who study very recent Earth history. In fact, the development of radiocarbon dating was considered so important that the chemist who discovered this application, Willard F. Libby, received a Nobel Prize.

Importance of Radiometric Dating

Bear in mind that although the basic principle of radiometric dating is simple, the actual procedure is quite complex. The analysis that determines the quantities of parent and daughter must be painstakingly precise. In addition, some radioactive materials do not decay directly into the stable daughter product. As you saw in Figure 18.13, uranium-238 produces 13 intermediate unstable daughter products before the 14th and final daughter product, the stable isotope lead-206, is produced.

Radiometric dating methods have produced literally thousands of dates for events in Earth history. Rocks from several localities have been dated at more than 3 billion years, and geologists realize that still older rocks exist. For example, a granite from South Africa has been dated at 3.2 billion years—and it contains inclusions of quartzite. (Remember that inclusions are older than the rock containing them.) Quartzite itself is a metamorphic rock that originally was the sedimentary rock sandstone. Sandstone, in turn, is the product of the lithification of sediments produced by the weathering of existing rocks. Thus, we have a positive indication that much older rocks existed.

Did You Know?

Dating with carbon-14 is useful to archaeologists and historians as well as geologists. For example, University of Arizona researchers recently used carbon-14 to determine the age of the Dead Sea Scrolls, considered among the great archaeological discoveries of the twentieth century. Parchment from the scrolls dates between 150 B.C. and 5 B.C. Portions of the scrolls contain dates that match those determined by the carbon-14 measurements.

Radiometric dating has vindicated the ideas of James Hutton, Charles Darwin, and others who inferred that geologic time must be immense. Indeed, modern dating methods have proved that there has been enough time for the processes we observe to have accomplished tremendous tasks.

The Geologic Time Scale

GEODe ESSENTIALS OF GEOLOGY
Geologic Time
▼ Geologic Time Scale

Geologists have divided the whole of geologic history into units of varying magnitude. Together they comprise the **geologic time scale** of Earth history (Figure 18.16). The major units of the time scale were delineated during the nineteenth century, principally by workers in western Europe and Great Britain. Because radiometric dating was unavailable at that time, the entire time scale was created using methods of relative dating. It was only in the twentieth century that radiometric dating permitted numerical dates to be added.

Radiometric dates for some of the lunar samples collected by astronauts exceeded 4 billion years. (NASA)

Structure of the Time Scale

The geologic time scale subdivides the 4.5-billion-year history of Earth into many different units and provides a meaningful time frame within which the events of the geologic past are arranged. As shown in Figure 18.16, **eons** represent the greatest expanses of time. The eon that began about 540 million years ago is the **Phanerozoic,** meaning *visible life.* It is an appropriate description because the rocks and deposits of the Phanerozoic eon contain abundant fossils that document major evolutionary trends.

Another glance at the time scale reveals that the Phanerozoic eon is divided into **eras.** The three eras

BOX 18.2

Using Tree Rings to Date and Study the Recent Past

If you look at the top of a tree stump or at the end of a log, you will see that it is composed of a series of concentric rings. Each of these *tree rings* becomes larger in diameter outward from the center (Figure 18.A). Every year in temperate regions, trees add a layer of new wood under the bark. Characteristics of each tree ring, such as size and density, reflect the environmental conditions (especially climate) that prevailed during the year when the ring formed. Favorable growth conditions produce a wide ring; unfavorable ones produce a narrow ring. Trees growing at the same time in the same region show similar tree-ring patterns.

Because a single growth ring is usually added each year, the age of the tree when it was cut can be determined by counting the rings. If the year of cutting is known, the age of the tree and the year in which each ring formed can be determined by counting back from the outside ring.* This procedure can be used to determine the dates of recent geologic events. For example, the minimum number of years since a new land surface was created by a landslide or a flood. The dating and study of annual rings in trees is called *dendrochronology.*

Figure 18.A Each year a growing tree produces a layer of new cells beneath the bark. If the tree is felled and the trunk examined (or if a core is taken, to avoid cutting the tree), each year's growth can be seen as a ring. Because the amount of growth (thickness of a ring) depends upon precipitation and temperature, tree rings are useful records of past climates. (Photo by Stephen J. Krasemann/DRK Photo)

To make the most effective use of tree rings, extended patterns known as ring chronologies are established. They are produced by comparing the patterns of rings among trees in an area. If the same pattern can be identified in two samples, one of which has been dated, the second sample can be dated from the first by matching the ring pattern common to both. This technique, called *cross dating,* is illustrated in Figure 18.B. Tree-ring

*Scientists are not limited to working with trees that have been cut down. Small, nondestructive core samples can be taken from living trees.

Precambrian rocks are often buried beneath thousands of feet of younger strata. The ancient Precambrian rocks at the bottom of the Grand Canyon are an example.

within the Phanerozoic are the **Paleozoic** ("ancient life"), the **Mesozoic** ("middle life"), and the **Cenozoic** ("recent life"). As the names imply, the eras are bounded by profound worldwide changes in life forms. Each era is subdivided into **periods.** The Paleozoic has seven, the Mesozoic three, and the Cenozoic two. Each of these 12 periods is characterized by a somewhat less profound change in life forms as compared with the eras. The eras and periods of the Phanerozoic, with brief explanations of each, are shown in Table 18.2.

Finally, periods are divided into still smaller units called **epochs.** As you can see in Figure 18.16, seven epochs have been named for the periods of the Cenozoic. The epochs of other periods, however, are not usually referred to by specific names. Instead, the terms *early, middle,* and *late* are generally applied to the epochs of these earlier periods.

Precambrian Time

Notice that the detail of the geologic time scale does not begin until about 540 million years ago, the date for the beginning of the Cambrian period. The nearly 4 billion years prior to the Cambrian is divided into three eons, the *Hadean,* the *Archean,* and the *Proterozoic.* It is also common for this vast expanse of time to simply be referred to as the **Precambrian.** Although it represents about 88 percent of Earth history, the Precambrian is not divided into nearly as many smaller time units as the Phanerozoic eon.

The quantity of information that geologists have deciphered about Earth's past is somewhat analogous to the detail of human history. The further back we go, the less we know. Certainly more data and information exist about the past 10 years than for the first decade of the twentieth

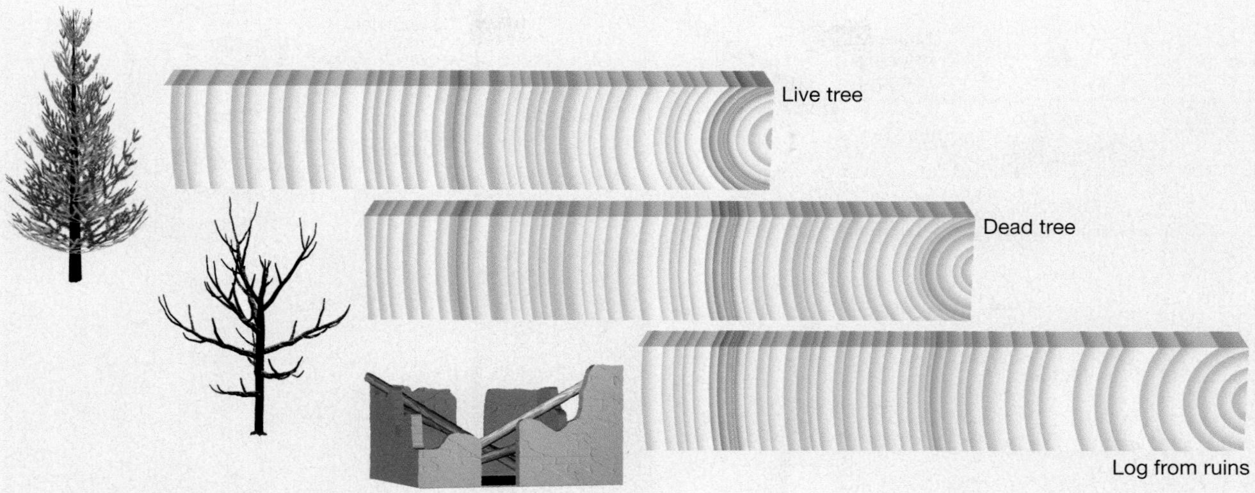

Figure 18.B Cross dating is a basic principle in dendrochronology. Here it was used to date an archaeological site by correlating tree-ring patterns for wood from trees of three different ages. First, a tree-ring chronology for the area is established using cores extracted from living trees. This chronology is extended further back in time by matching overlapping patterns from older, dead trees. Finally, cores taken from beams inside the ruin are dated using the chronology established from the other two sites.

chronologies extending back for thousands of years have been established for some regions. To date a timber sample of unknown age, its ring pattern is matched against the reference chronology.

Tree-ring chronologies are unique archives of environmental history and have important applications in such disciplines as climate, geology, ecology, and archaeology. For example, tree rings are used to reconstruct climate variations within a region for spans of thousands of years prior to human historical records. Knowledge of such long-term variations is of great value in making judgments regarding the recent record of climate change.

In summary, dendrochronology provides useful numerical dates for events in the historic and recent prehistoric past. Moreover, because tree rings are a storehouse of data, they are a valuable tool in the reconstruction of past environments.

century; the events of the nineteenth century have been documented much better than the events of the first century A.D.; and so on. So it is with Earth history. The more recent past has the freshest, least disturbed, and most observable record. The further back in time the geologist goes, the more fragmented the record and clues become.

Difficulties in Dating the Geologic Time Scale

Although reasonably accurate numerical dates have been worked out for the periods of the geologic time scale (Figure 18.16), the task is not without difficulty. The primary problem in assigning numerical dates to units of time is the fact that not all rocks can be dated by radiometric methods. Recall that for a radiometric date to be useful, all minerals in the rock must have formed at about the same time. For this reason, radioactive isotopes can be used to determine when minerals in an igneous rock crystallized and when pressure and heat created new minerals in a metamorphic rock.

However, samples of sedimentary rock can only rarely be dated directly by radiometric means. A sedimentary rock might include particles that contain radioactive isotopes, but the rock's age cannot be accurately determined, because the grains making up the rock are not the same age as the rock in which they occur. Rather, the sediments have been weathered from rocks of diverse ages.

Radiometric dates obtained from metamorphic rocks might also be difficult to interpret, because the age of a particular mineral in a metamorphic rock does not necessarily represent the time when the rock initially

A useful numerical date for this conglomerate is not possible, because the gravel particles that compose it were derived from rocks of diverse ages.

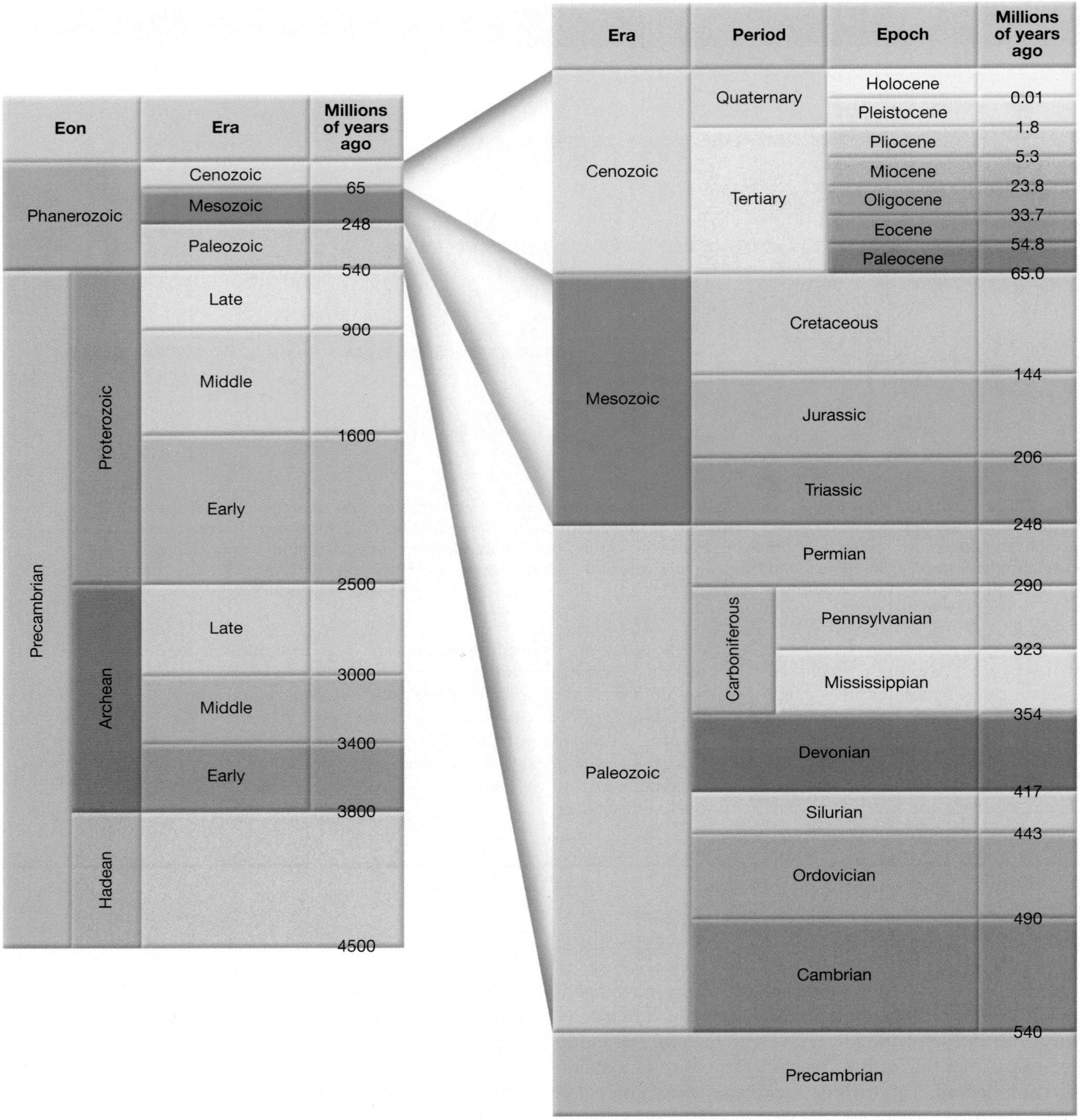

▲ **Figure 18.16** The geologic time scale. The numerical dates were added long after the time scale had been established using relative dating techniques. *(Data from Geological Society of America)*

Table 18.2 Major divisions of geologic time.

Cenozoic Era (Age of Recent Life)	Quaternary period	The several geologic eras were originally named Primary, Secondary, Tertiary, and Quaternary. The first two names are no longer used; Tertiary and Quaternary have been retained but used as period designations.
	Tertiary period	
Mesozoic Era (Age of Middle Life)	Cretaceous period	Derived from the Latin word for chalk (creta) and first applied to extensive deposits that form white cliffs along the English Channel.
	Jurassic period	Named for the Jura Mountains, located between France and Switzerland, where rocks of this age were first studied.
	Triassic period	Taken from word "trias" in recognition of the threefold character of these rocks in Europe.
Paleozoic Era (Age of Ancient Life)	Permian period	Named after the province of Perm, Russia, where these rocks were first studied.
	Pennsylvanian period*	Named for the state of Pennsylvania where these rocks have produced much coal.
	Mississippian period*	Named for the Mississippi River Valley where these rocks are well exposed.
	Devonian period	Named after Devonshire County, England, where these rocks were first studied.
	Silurian period	Named after Celtic tribes, the Silures and the Ordovices, that lived in Wales during the Roman Conquest.
	Ordovician period	
	Cambrian period	Taken from Roman name for Wales (Cambria), where rocks containing the earliest evidence of complex forms of life were first studied.
Precambrian		The time between the birth of the planet and the appearance of complex forms of life. About 88 percent of Earth's 4.5 billion-year history falls into this span.

**Outside of North America, the Mississippian and Pennsylvanian periods are combined into the Carboniferous period.*
Source: U.S. Geological Survey.

formed. Instead, the date could indicate any one of a number of subsequent metamorphic phases.

If samples of sedimentary rocks rarely yield reliable radiometric ages, how can numerical dates be assigned to sedimentary layers? Usually the geologist must relate them to datable igneous masses, as in Figure 18.17. In this example, radiometric dating has determined the ages of the volcanic ash bed within the Morrison Formation and the dike cutting the Mancos Shale and Mesaverde Formation. The sedimentary beds below the ash are obviously older than the ash, and all the layers above the ash are younger (principle of superposition). The dike is younger than the Mancos Shale and the Mesaverde Formation but older than the Wasatch Formation because the dike does not intrude the Tertiary rocks (cross-cutting relationships).

From this kind of evidence, geologists estimate that a part of the Morrison Formation was deposited about 160 million years ago, as indicated by the ash bed. Further, they conclude that the Tertiary period began after the intrusion of the dike, 66 million years ago. This is one example of literally thousands that illustrate how datable materials are used to *bracket* the various episodes in Earth history within specific time periods. It shows the necessity of combining laboratory dating methods with field observations of rocks.

Creta is the Latin word for chalk. The Cretaceous period of the Mesozoic era was named for the chalk deposits along the English Channel.

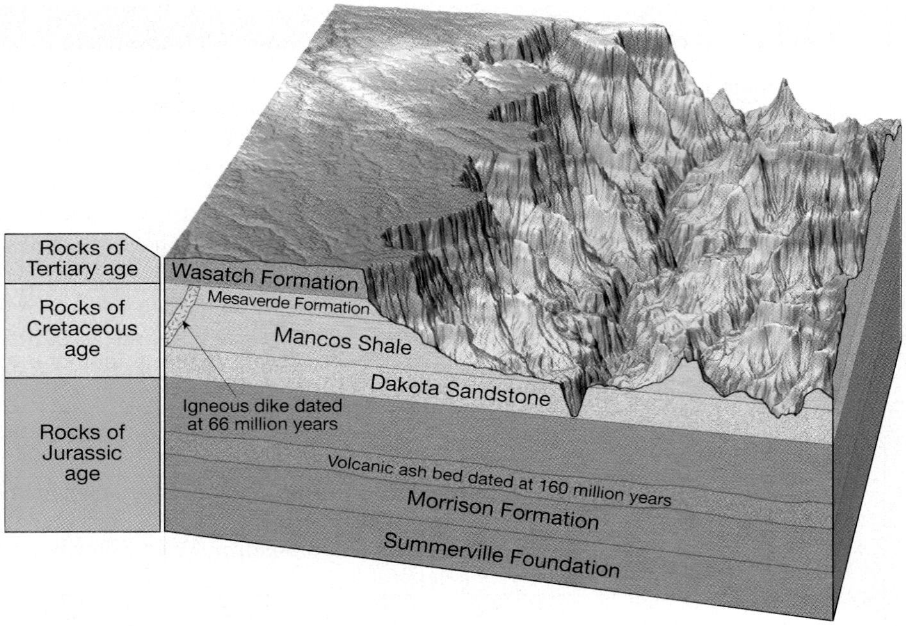

▲ **Figure 18.17** Numerical dates for sedimentary layers are usually determined by examining their relationship to igneous rocks. *(After U.S. Geological Survey)*

The Chapter in Review

- The two types of dates used by geologists to interpret Earth history are (1) *relative dates,* which put events in their *proper sequence of formation,* and (2) *numerical dates,* which pinpoint the *time in years* when an event occurred.
- Relative dates can be established using the *law of superposition* (in an underformed sequence of sedimentary rocks or surface-deposited igneous rocks, each bed is older than the one above, and younger than the one below), *principle of original horizontality* (most layers are deposited in a horizontal position), *principle of cross-cutting relationships* (when a fault or intrusion cuts through another rock, the fault or intrusion is younger than the rocks cut through), and *inclusions* (the rock mass containing the inclusion is younger than the rock that provided the inclusion).
- *Unconformities* are gaps in the rock record. Each represents a long period during which deposition ceased, erosion removed previously formed rocks, and then deposition resumed. The three basic types of unconformities are *angular unconformities* (tilted or folded sedimentary rocks that are overlain by younger, more flat-lying strata), *disconformities* (the strata on either side of the unconformity are essentially parallel), and *nonconformities* (where a break separates older metamorphic or intrusive igneous rocks from younger sedimentary strata).
- *Correlation,* matching rocks of similar age in different areas, is used to develop a geologic time scale that applies to the entire Earth.
- *Fossils* are the remains or traces of prehistoric life. The special conditions that favor preservation are *rapid burial* and the possession of *hard parts* such as shells, bones, or teeth.
- Fossils are used to *correlate* sedimentary rocks from different regions by using the rocks' distinctive fossil content and applying the *principle of fossil succession.* It is based on the work of *William Smith* in the late 1700s and states that fossil organisms succeed one another in a definite and determinable order, and therefore any time period can be recognized by its fossil content. The use of *index fossils,* those that are widespread geographically and are limited to a short span of geologic time, provides an important method for matching rocks of the same age.
- Each atom has a nucleus containing *protons* (positively charged particles) and *neutrons* (neutral particles). Orbiting the nucleus are negatively charged *electrons.* The *atomic number* of an atom is the number of protons in the nucleus. The *mass number* is the number of protons plus the number of neutrons in an atom's nucleus. *Isotopes* are variants of the same atom, but with a different number of neutrons, and hence a different mass number.
- *Radioactivity* is the spontaneous breaking apart (decay) of certain unstable atomic nuclei. Three common types of radioactive decay are (1) emission of *alpha particles* from the nucleus, (2) emission of *beta particles* (or electrons) from the nucleus, and (3) *capture of electrons* by the nucleus.

- An unstable *radioactive isotope,* called the *parent,* will decay and form stable *daughter products.* The length of time for half of the nuclei of a radioactive isotope to decay is called the *half-life* of the isotope. If the half-life of the isotope is known, and the parent/daughter ratio can be measured, the age of a sample can be calculated. An accurate radiometric date can only be obtained if the mineral containing the radioactive isotope remained a closed system during the entire period since its formation.
- The *geologic time scale* divides Earth's history into units of varying magnitude. It is commonly presented in chart form, with the oldest time and event at the bottom and the youngest at the top. The principal subdivisions of the geologic time scale, called *eons,* include the *Hadean, Archean, Proterozoic* (together, these three eons are commonly referred to as the *Precambrian*), and, beginning about 540 million years ago, the *Phanerozoic.* The Phanerozoic (meaning "visible life") eon is divided into the following *eras: Paleozoic* ("ancient life"), *Mesozoic* ("middle life"), and *Cenozoic* ("recent life").
- A significant problem in assigning numerical dates to units of time is that *not all rocks can be radiometrically dated.* A sedimentary rock may contain particles of many ages that have been weathered from different rocks that formed at various times. One way geologists assign numerical dates to sedimentary rocks is to relate them to datable igneous masses, such as dikes and volcanic ash beds.

Key Terms

angular unconformity (p. 419)
Cenozoic era (p. 432)
conformable (p. 418)
correlation (p. 419)
cross-cutting relationships, principle of (p. 418)
disconformity (p. 419)
eon (p. 431)
epoch (p. 431)
era (p. 431)
fossil (p. 420)
fossil succession, principle of (p. 422)
geologic time scale (p. 431)
half-life (p. 428)
inclusion (p. 418)
index fossil (p. 423)
Mesozoic era (p. 432)
nonconformity (p. 419)
numerical date (p. 417)
original horizontality, principle of (p. 417)
paleontology (p. 420)
Paleozoic era (p. 432)
period (p. 432)
Phanerozoic eon (p. 431)
Precambrian (p. 432)
radioactivity (p. 427)
radiocarbon dating (p. 430)
radiometric dating (p. 427)
relative dating (p. 417)
superposition, law of (p. 417)
unconformity (p. 418)

Questions for Review

1. Distinguish between *numerical* and *relative dating.*
2. What is the *law of superposition*? How are *cross-cutting relationships* used in relative dating?
3. When you observe an outcrop of steeply inclined sedimentary layers, what principle allows you to assume that the beds were tilted *after* they were deposited?
4. Refer to Figure 18.4 and answer the following questions:
   a. Is fault *A* older or younger than the sandstone layer?
   b. Is dike *A* older or younger than the sandstone layer?
   c. Was the conglomerate deposited before or after fault *A*?
   d. Was the conglomerate deposited before or after fault *B*?
   e. Which fault is older, *A* or *B*?
   f. Is dike *A* older or younger than the batholith?
5. A mass of granite is in contact with a layer of sandstone. Using a principle described in this chapter, explain how you might determine whether the sandstone was deposited on top of the granite, or whether it was intruded from below after the sandstone was deposited.
6. Distinguish among angular *unconformity, disconformity,* and *nonconformity.*
7. What is meant by the term *correlation*?
8. Describe several types of fossils. What organisms have the best chance of being preserved as fossils?
9. Describe William Smith's important contribution to the science of geology.
10. Why are fossils such useful tools in correlation?
11. In addition to being important aids in dating and correlating rocks, how else are fossils helpful in geologic investigations?

12. If a radioactive isotope of thorium (atomic number 90, mass number 232) emits 6 alpha particles and 4 beta particles during the course of radioactive decay, what is the atomic number and mass number of the stable daughter product?

13. Why is radiometric dating a reliable method of dating the geologic past?

14. A hypothetical radioactive isotope has a half-life of 10,000 years. If the ratio of radioactive parent to stable daughter product is 1:3, how old is the rock containing the radioactive material?

15. Briefly describe why tree rings might be helpful in studying the recent geologic past (see Box 18.2).

16. In order to provide a reliable radiometric date, a mineral must remain a closed system from the time of its formation until the present. Why is this true?

17. What precautions are taken to ensure reliable radiometric dates?

18. To make calculations easier, let us round the age of Earth to 5 billion years.
    a. What fraction of geologic time is represented by recorded history? (Assume 5000 years for the length of recorded history.)
    b. The first abundant fossil evidence does not appear until the beginning of the Cambrian period (540 million years ago). What percentage of geologic time is represented by abundant fossil evidence?

19. What subdivisions make up the geologic time scale?

20. Briefly describe the difficulties in assigning numerical dates to layers of sedimentary rock.

21. Figure 18.18 is a block diagram of a hypothetical area in the American Southwest. Place the lettered features in the proper sequence, from oldest to youngest. Identify an angular unconformity and a nonconformity.

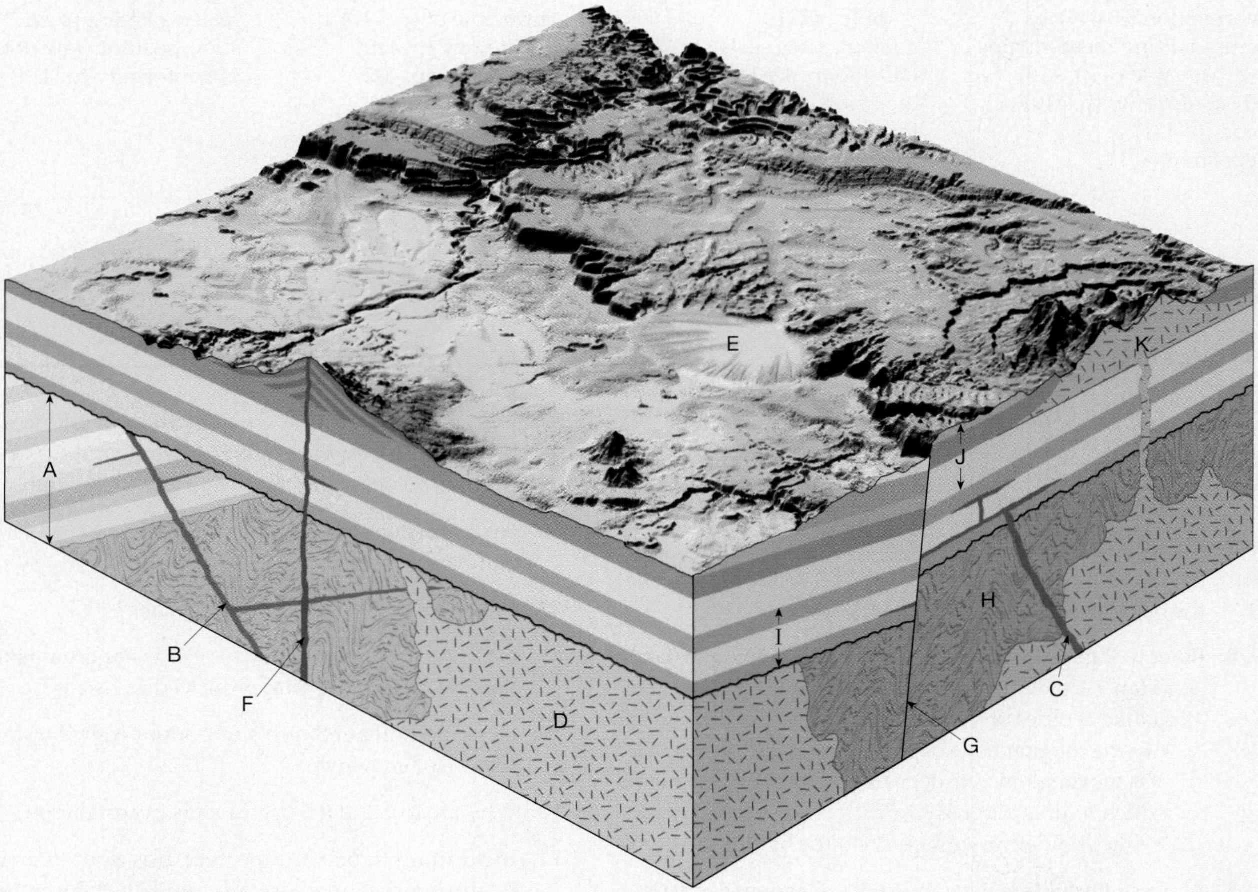

▲ **Figure 18.18** Use this block diagram in conjunction with Review Question 21.

Online Study Guide

The *Essentials of Geology* Web site uses the resources and flexibility of the Internet to aid in your study of the topics in this chapter. Written and developed by geology instructors, this site will help improve your understanding of geology. Visit **www.prenhall.com/lutgens** and click on the cover of *Essentials of Geology 9e* to find:

- Online review quizzes.
- Critical thinking exercises.
- Links to chapter-specific Web resources.
- Internet-wide key-term searches.

http://www.prenhall.com/lutgens

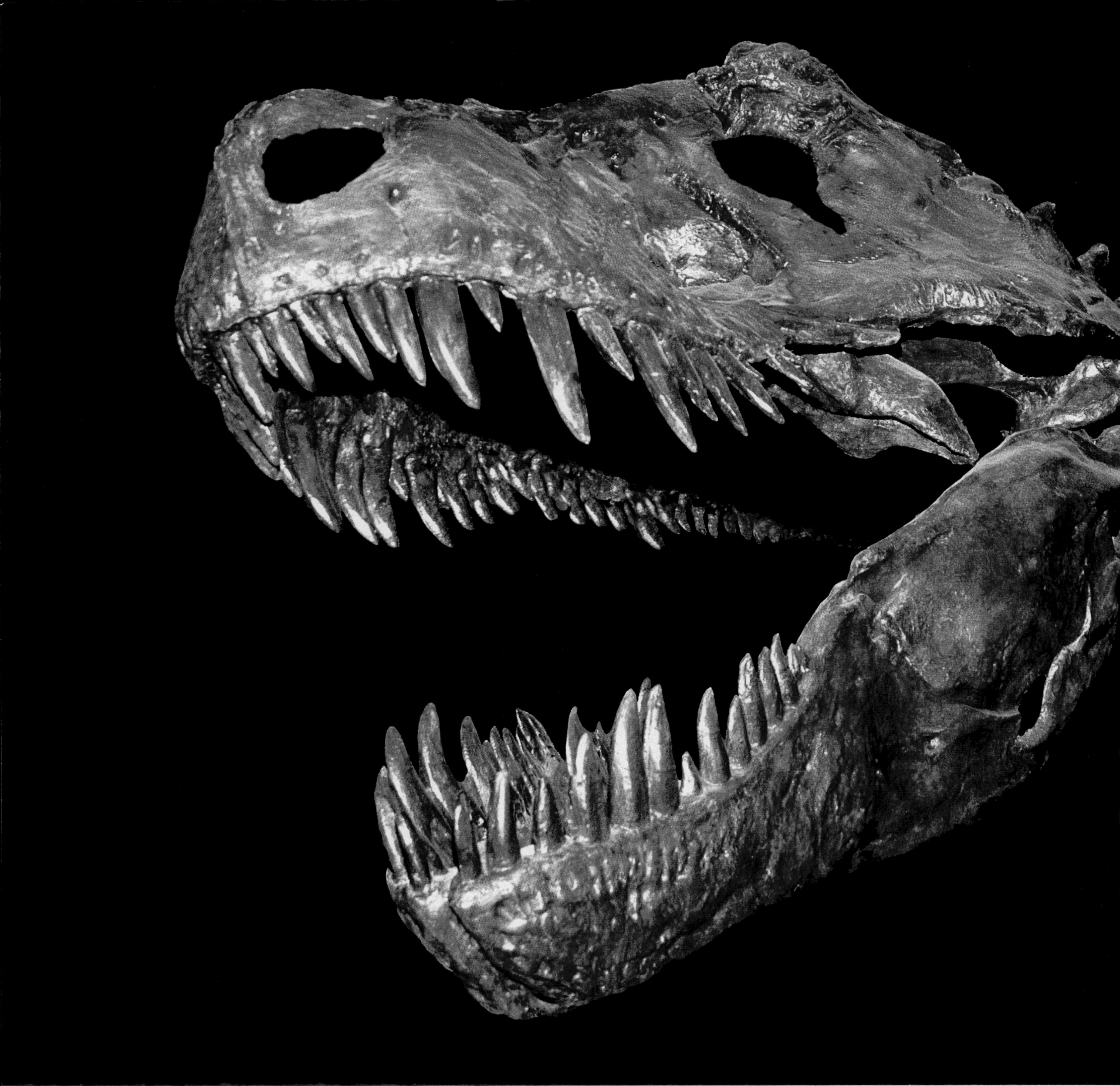

Tyrannosaurus Rex **on display at the Smithsonian Museum.**
(Photo by Russ Merne/Alamy Images)

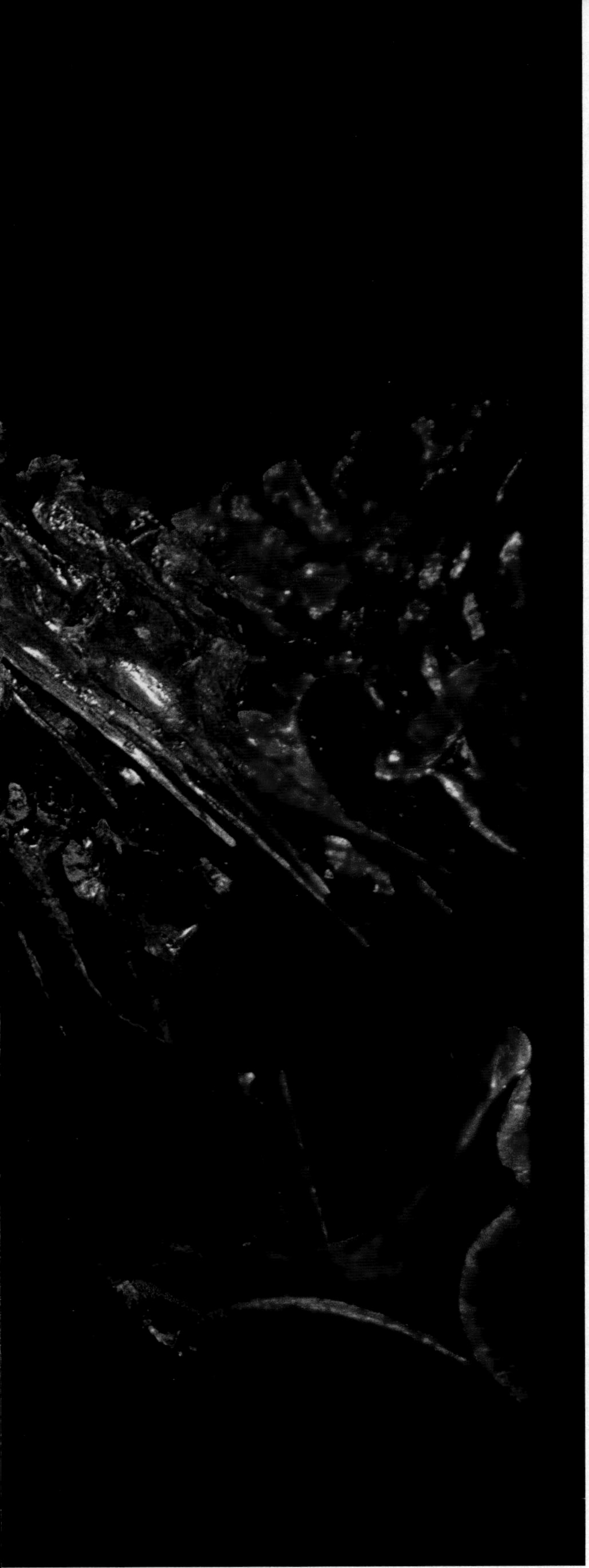

CHAPTER 19

Earth History: A Brief Summary

Focus on Learning

To assist you in learning the important concepts in this chapter, you will find it helpful to focus on the following questions:

- How did Earth's atmosphere evolve and change through time?
- What were the principal geological and biological events of Earth history?
- How have the continents changed through time?

Glaciers are one of the forces that have modified our planet during its long history.

Earth has a long and complex history. The splitting and colliding of continents has resulted in the formation of new ocean basins and the creation of Earth's great mountain ranges. Furthermore, the nature of life on our planet has experienced dramatic changes through time.

Many of the changes on planet Earth occur at a "snail's pace," generally too slow for people to perceive. Thus, human awareness of evolutionary change is fairly recent. Evolution is not confined to life forms, for all Earth's "spheres" have evolved together: the atmosphere, hydrosphere, geosphere, and biosphere. Examples are evolutionary changes in the air we breathe, evolution of the world's oceans, the rise of mountains, ponderous movements of crustal plates, the comings and goings of vast ice sheets, and the evolution of a vast array of life forms (Figure 19.1). As each facet of Earth has evolved, it has powerfully influenced the others.

As you learned in Chapter 18, geologists have many tools at their disposal for interpreting the clues about Earth's past. Using these tools, and clues that are contained in the rock record, geologists have been able to unravel many of the complex events of the geological past. The goal of this chapter is to provide a brief overview of the history of our planet and its life forms (Figure 19.2). We will begin nearly 4.5 billion years ago with the formation of Earth and its atmosphere. Then we will describe how our physical world assumed its present form and how Earth's inhabitants changed through time.

Early Evolution of Earth

Recall from Chapter 1 the nebular hypothesis, which proposes that Earth, along with the rest of our solar system, condensed from a cloud of dust and gases. As material accumulated to form Earth (and for a short period afterward), the high-velocity impact of nebular debris and the decay of radioactive elements caused the temperature of our planet to steadily increase. During this time of intense heating, Earth became hot enough that iron and nickel began to melt. Melting produced liquid blobs of heavy metal that sank toward the center of the planet. This process occurred rapidly on the scale of geologic time and produced Earth's dense iron-rich core.

The early period of heating resulted in another process of chemical differentiation, whereby melting formed buoyant masses of molten rock that rose toward the surface where they solidified to produce a primitive crust. These rocky materials were enriched in oxygen and "oxygen-seeking" elements, particularly silicon and aluminum, along with lesser amounts of calcium, sodium, potassium, iron, and magnesium. In addition, some heavy metals such as gold, lead, and uranium, which have low melting points or were highly soluble in the ascending molten masses, were scavenged from Earth's interior and concentrated in the developing crust. This early period of chemical segregation established the three basic divisions of Earth's interior—the iron-rich *core,* the thin *primitive crust,* and Earth's largest layer, called the *mantle,* which is located between the core and crust.

▼ **Figure 19.1** Paleontologists study ancient life. Here scientists are excavating the remains of *Albertasaurus,* a large carnivore similar to *Tyrannosaurus rex,* that lived during the late Cretaceous Period. The excavation site is near Red Deer River, Alberta, Canada. *(Photo by Richard T. Nowitz/Science Source/Photo Researchers, Inc.)*

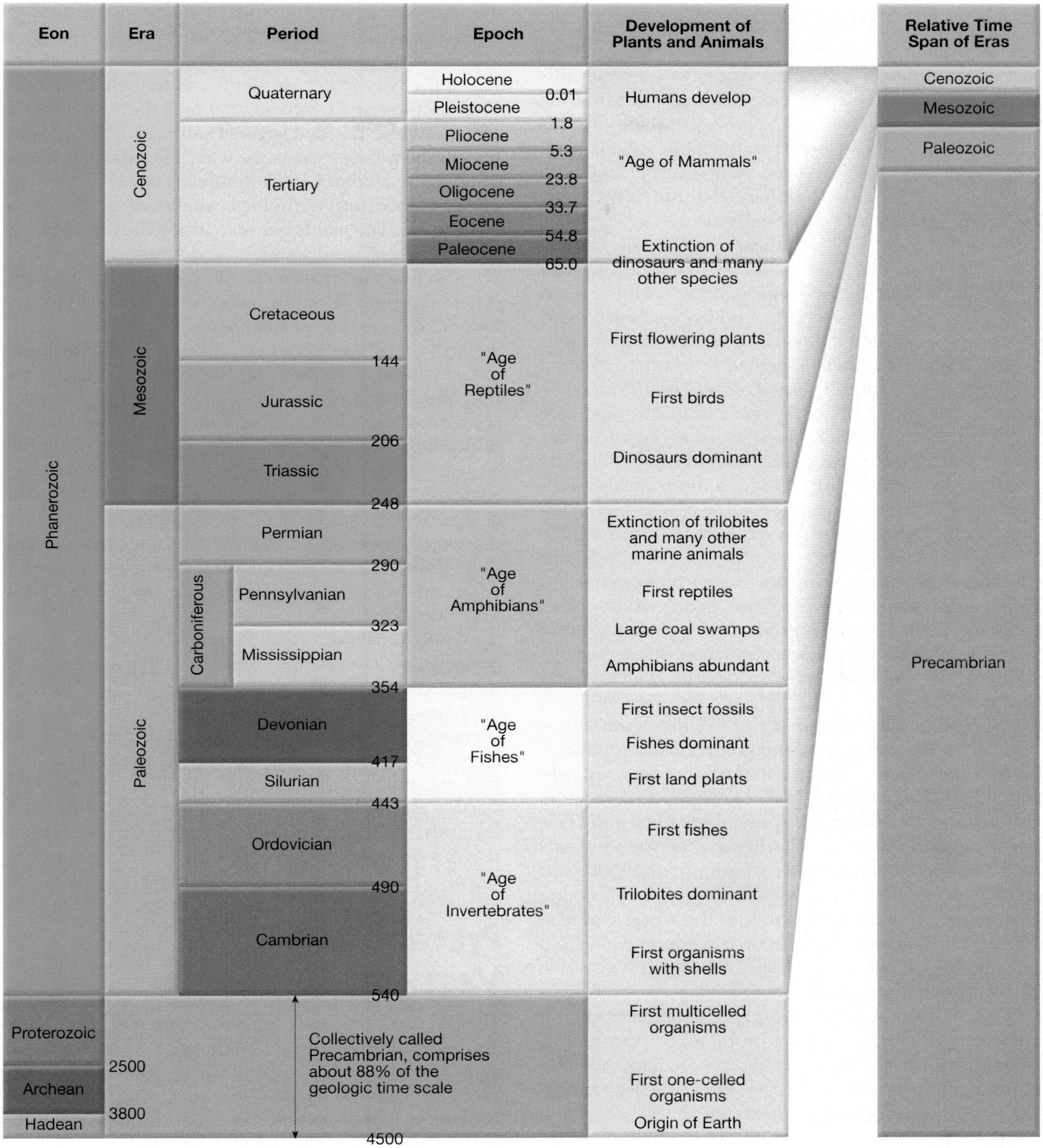

Eon
Era
Period
Epoch
Development of Plants and Animals
Relative Time Span of Eras
Phanerozoic
Cenozoic
Quaternary
Holocene
0.01
Pleistocene
1.8
Humans develop
Tertiary
Pliocene
5.3
Miocene
23.8
"Age of Mammals"
Oligocene
33.7
Eocene
54.8
Paleocene
65.0
Extinction of dinosaurs and many other species
Mesozoic
Cretaceous
First flowering plants
144
"Age of Reptiles"
Jurassic
First birds
206
Triassic
Dinosaurs dominant
248
Paleozoic
Permian
Extinction of trilobites and many other marine animals
290
Carboniferous
Pennsylvanian
"Age of Amphibians"
First reptiles
323
Large coal swamps
Mississippian
Amphibians abundant
354
Devonian
"Age of Fishes"
First insect fossils
Fishes dominant
417
Silurian
First land plants
443
Ordovician
First fishes
490
"Age of Invertebrates"
Trilobites dominant
Cambrian
First organisms with shells
540
Proterozoic
First multicelled organisms
Collectively called Precambrian, comprises about 88% of the geologic time scale
2500
Archean
First one-celled organisms
3800
Hadean
Origin of Earth
4500
Cenozoic
Mesozoic
Paleozoic
Precambrian

▲ **Figure 19.2** The geologic time scale. Numbers on the time scale represent time in millions of years before the present. These dates were added long after the time scale had been established using relative dating techniques. The Precambrian accounts for about 88 percent of geologic time. *(Data from Geological Society of America)*

Earth's Primitive Atmosphere

An important consequence of this early period of chemical differentiation is that large quantities of gaseous materials were allowed to escape from Earth's interior, as happens today during volcanic eruptions. By this process a primitive atmosphere gradually evolved. It is on this planet, with this atmosphere, that life as we know it came into existence.

Today, the air you breathe is a stable mixture of 78 percent nitrogen, 21 percent oxygen, about 1 percent argon (an inert gas), and trace gases like carbon dioxide and water vapor. But our planet's original atmosphere, several billion years ago, was far different.

Earth's very earliest atmosphere probably was swept away into space by the *solar wind,* a vast stream of particles emitted by the Sun. As Earth slowly cooled, a more enduring atmosphere formed. The molten surface solidified into a crust, and gases that had been dissolved in the molten rock were gradually released, a process called **outgassing.** Outgassing continues today from hundreds of active volcanoes worldwide. Thus, geologists hypothesize that Earth's original atmosphere was made up of gases similar to those released in volcanic emissions today: water vapor, carbon dioxide, nitrogen, and several trace gases.

As the planet continued to cool, the water vapor condensed to form clouds, and great rains commenced. At first, the water evaporated in the hot air before reaching the ground or quickly boiled away upon contacting the surface, just like water sprayed on a hot grill. This accelerated the cooling of Earth's crust. When the surface had cooled below water's boiling point (100°C or 212°F), torrential rains slowly filled low areas, forming the oceans. This reduced not only the water vapor in the air but the amount of carbon dioxide as well, for it became dissolved in the water. What remained was a nitrogen-rich atmosphere.

If Earth's primitive atmosphere resulted from volcanic outgassing, we have a problem, because volcanoes do not emit free oxygen. Where did the very significant percentage of oxygen in our present atmosphere (nearly 21 percent) come from?

Did You Know?

The names of several periods on the geologic time scale refer to places that have prominent strata of that age. For example, the Cambrian period is taken from the Roman name for Wales (Cambria). The Permian is named for the province of Perm in Russia, while the Jurassic period gets its name from the Jura Mountains located between France and Switzerland.

Plants have strongly influenced the composition of our atmosphere.

Earth's Atmosphere Evolves

The major source of oxygen is green plants. Put another way, *life itself* has strongly influenced the composition of our present atmosphere. Plants did not just adapt to their environment; they actually influenced it, dramatically altering the composition of the entire planet's atmosphere by using carbon dioxide and releasing oxygen. This is a good example of how Earth operates as a giant system in which living things interact with their environment.

How did plants come to alter the atmosphere? The key is the way in which plants create their own food. They employ *photosynthesis,* in which they use light energy to synthesize food sugars from carbon dioxide and water. The process releases a waste gas, oxygen. Those of us in the animal kingdom rely on oxygen to metabolize our food, and we in turn exhale carbon dioxide as a waste gas. The plants use this carbon dioxide for more photosynthesis, and so on, in a continuing system.

The first life forms on Earth, probably bacteria, did not need oxygen. Their life processes were geared to the earlier, oxygenless atmosphere. Even today, many *anaerobic* bacteria thrive in environments that lack free oxygen. Later, primitive plants evolved that used photosynthesis and released oxygen. Slowly, the oxygen content of Earth's atmosphere increased. The Precambrian rock record suggests that much of the first free oxygen did not remain free, because it combined with (oxidized) other substances dissolved in water, especially iron. Iron has tremendous affinity for oxygen, and the two elements combine to form iron oxides at any opportunity. As a result, most of Earth's iron-ore deposits are from the middle Precambrian (1.2 to 2.5 billion years ago).

Then, once the available iron satisfied its need for oxygen, substantial quantities of oxygen accumulated in the atmosphere. By the beginning of the Paleozoic era, about 4 billion years into Earth's existence (after seven-eighths of Earth's history had elapsed), the fossil record reveals abundant ocean-dwelling organisms that require oxygen to live. Hence, the composition of Earth's atmosphere has evolved together with its life forms, from an oxygenless envelope to today's oxygen-rich environment.

Precambrian Time: Vast and Enigmatic

The Precambrian encompasses immense geological time, from Earth's distant beginnings 4.5 billion years ago until the start of the Cambrian period, about 4 billion years later. Thus, the Precambrian spans about 88 percent of Earth's history. To get a visual sense of the proportion, look at the right side of Figure 19.2, which shows the relative time span of eras. Our knowledge of this ancient time is sketchy, for much of the early rock record has been obscured by the very Earth processes you have been studying, especially plate tectonics, erosion, and deposition.

Untangling the long, complex Precambrian rock record is a formidable task, and it is far from complete. Most Precambrian rocks are devoid of fossils, which hinders correlation of rocks. Rocks of this great age are metamorphosed and deformed, extensively eroded, and obscured by overlying strata. Indeed, Precambrian history is written in scattered, speculative episodes, like a long book with many missing chapters.

Precambrian History

Looking at Earth from the space, astronauts see plenty of ocean (71 percent) and much less land area (29 percent). Over large expanses of the continents, orbiting space scientists gaze upon many Paleozoic, Mesozoic, and Cenozoic rock surfaces, but fewer Precambrian surfaces. This demonstrates the law of superposition: Precambrian rocks in these regions are buried from view beneath varying thicknesses of more recent rocks. Here, Precambrian rocks "peek" through the surface in places where younger strata are extensively eroded, as in the Grand Canyon and in some mountain ranges. However, on each continent, large core areas of Precambrian rocks dominate the surface mostly as deformed crystalline metamorphic and igneous rocks. These areas are called **shields** because they roughly resemble a warrior's shield in shape.

Figure 19.3 shows the distribution of Precambrian shields worldwide. In North America (including Greenland), the Canadian Shield encompasses 7.2 million square kilometers (2.8 million square miles), an area 10 times larger than the state of Texas.

Landmasses of the Precambrian. Based on radiometric dates, the oldest fragments of continental crust so far uncovered formed between 3.8 and 3.5 billion years ago. Once formed, these small continental blocks grew both in thickness and in lateral extent by adding material along their margins. Here, subduction of oceanic lithosphere generated volcanic arcs and igneous plutons that increased the thickness of crust. In addition, as these continents moved about the globe, continental collisions "squeezed" sediments that had been eroded from these landmasses and "plastered" them on the continental margins.

Beginning about 3 billion years ago the first large blocks of continental crust became stable and moved about Earth without being appreciably deformed. The collision and accretion of these continental fragments formed even larger continental units. Today these crustal platforms form the stable interiors of our modern continents.

About 1 billion years ago most of the large continental blocks assembled into a supercontinent that we have come to call *Rodinia.* Although its size and shape are still being researched, Rodinia clearly had a much different configuration than Pangaea (Figure 19.4). One noticeable difference is that North America was located near the center of this ancient supercontinent.

Between 800 and 600 million years ago Rodinia gradually split apart and the pieces dispersed. By the close of the Precambrian, many of the fragments had reassembled to produce a large landmass located in the Southern Hemisphere called *Gondwana.* Gondwana was comprised mainly of present-day South America, Africa, India, Australia, and Antarctica (Figure 19.5). Other continental fragments also formed

Precambrian rocks exposed at the bottom of the Grand Canyon.

Did You Know?

The earliest evidence for life on Earth comes from hydrocarbon residues of bacteria (prokaryotes) in metamorphosed sedimentary rocks in Greenland and date from about 3.8 billion years ago.

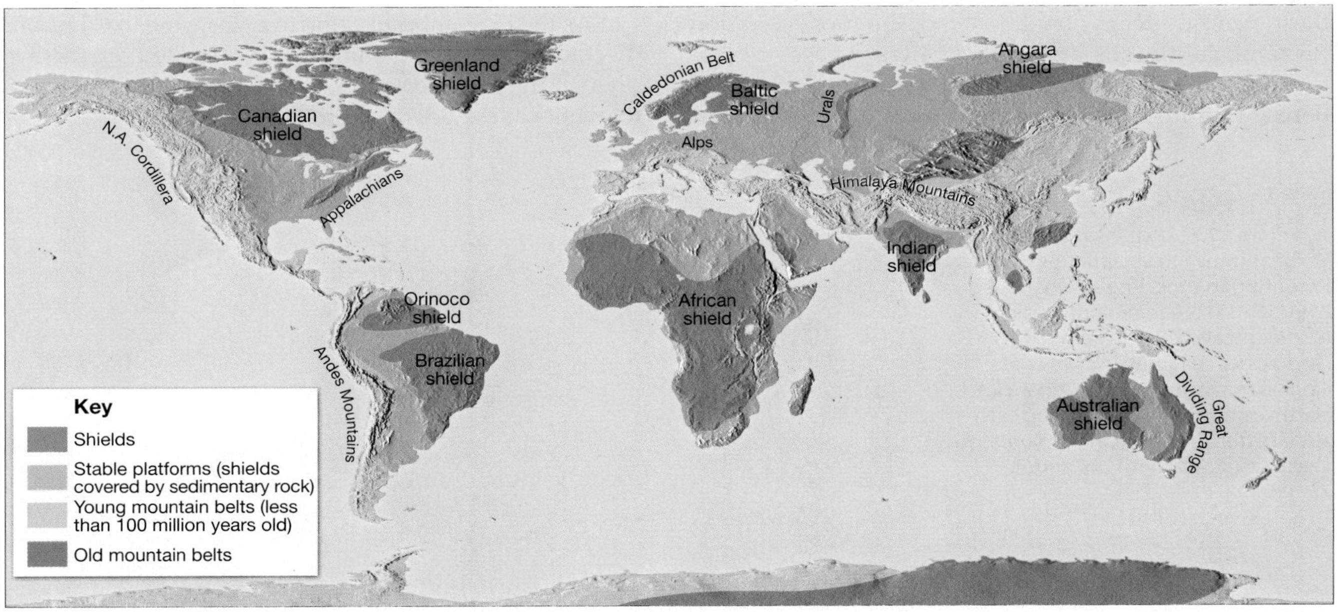

▲ **Figure 19.3** The stable interiors of continents contain expansive, flat regions composed of deformed Precambrian crystalline rocks called shields.

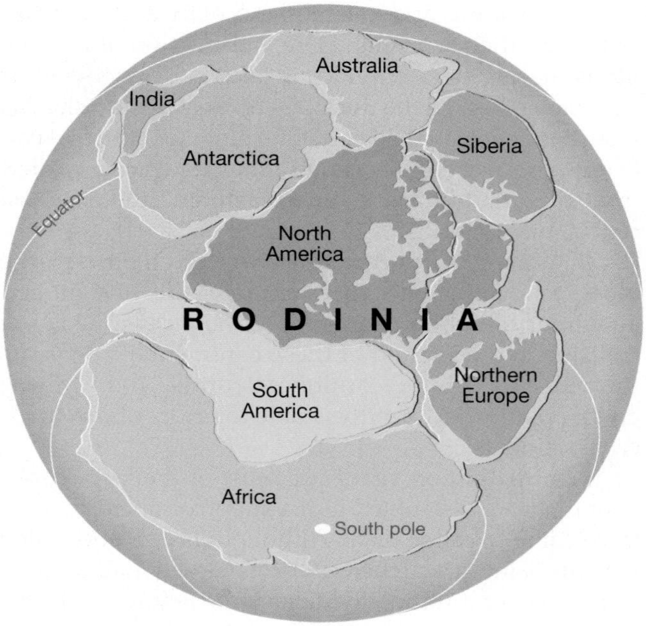

▲ **Figure 19.4** Simplified drawing of one of several possible configurations of the supercontinent Rodinia. For clarity, the continents are drawn with somewhat modern shapes, which was not the situation 1 billion years ago. *(After P. Hoffman, J. Rogers, M. Santosh, and others)*

when Rodinia broke apart—North America, Siberia, and northern Europe. We will consider the fate of these Precambrian landmasses later in the chapter.

Precambrian Fossils

Precambrian fossils are disappointing if you are expecting to see fascinating plants and large animals, for these had not yet evolved. Instead, the most common Precambrian fossils are **stromatolites.** These are distinctively layered mounds or columns of calcium carbonate (Figure 19.6). Stromatolites are not the remains of actual organisms, but are material deposited by algae. Evidence of the origin of these ancient fossils is the close similarity they have to deposits made by modern algae.

Many of these most ancient fossils are preserved in *chert,* a hard, dense chemical sedimentary rock. Chert must be very thinly sliced and studied under powerful microscopes to observe bacteria and algae fossils within it. Precambrian microfossils, encased in chert, have been found at several locations worldwide. Two notable areas are in southern Africa, where the rocks date to more than 3.1 billion years, and in the Gunflint Chert (named for its use in flintlock rifles) of Lake Superior, which dates to 1.7 billion years. In both places, bacteria and blue-green algae have been discovered. The fossils are of the most primitive organisms, *prokaryotes.* Their cells lack organized nuclei, and they reproduce asexually.

More advanced organisms, *eukaryotes,* have cells that contain nuclei. Eukaryotes are among billion-year-old fossils discovered at Bitter Springs in Australia, such as green algae. Unlike prokaryotes, eukaryotes reproduce sexually, which means that genetic material is exchanged between organisms. This reproductive mode permits greatly increased genetic variation. Thus, development of eukaryotes may have dramatically increased the rate of evolutionary change on our planet.

Plant fossils date from the middle Precambrian, but animal fossils came a bit later, in the late Precambrian. Many of these fossils are *trace fossils,* meaning that they are not of the animals themselves, but evidence of their activities, such as trails and wormholes. Areas in Australia and Newfoundland have yielded hundreds of fossil impressions of soft-bodied creatures. Most, if not all, of the Precambrian fauna lacked shells, which would develop as protective armor during the Paleozoic.

As the Precambrian came to a close, the fossil record discloses diverse and complete multicelled organisms. This set the stage for more complex plants and animals to evolve at the dawn of the Paleozoic era.

▶ **Figure 19.5** Reconstruction of Earth as it may have appeared in late Precambrian time. The southern continents were joined into a single landmass called Gondwana. Other landmasses that were not part of Gondwana include North America, northwestern Europe, and northern Asia. *(After C. Scotese, R. K. Bambach, C. Barton, R. Vander Voo, and A. Ziegler)*

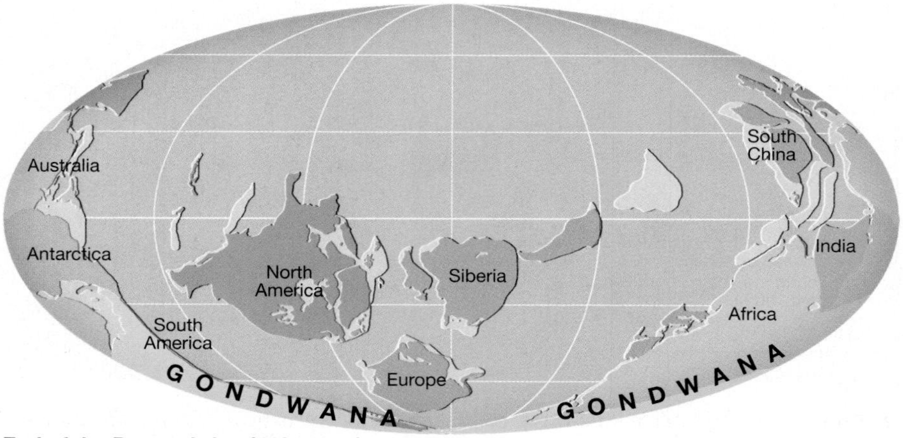

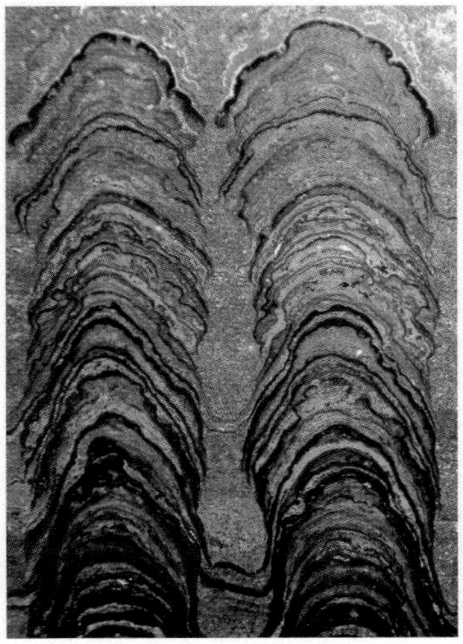

A.

B.

▲ **Figure 19.6** Stromatolites are among the most common Precambrian fossils. **A.** Precambrian fossil stromatolites composed of calcium carbonate deposited by algae in the Helena Formation, Glacier National Park. *(Photo by Ken M. Johns/Photo Researchers, Inc.)* **B.** Modern stromatolites growing in shallow saline seas, western Australia. *(Photo by Bill Bachman/Photo Researchers, Inc.)*

Paleozoic Era: Life Explodes

The span since the close of the Precambrian encompasses 540 million years and is divided into three eras: Paleozoic, Mesozoic, and Cenozoic. The Paleozoic era, which lasted about 292 million years, is by far the longest of the three. Seven periods make up the Paleozoic era (see Figure 19.2).

The beginning of the Paleozoic is marked by the appearance of the first life forms with hard parts such as shells, scales, bones, or teeth (Figure 19.7). Hard parts greatly enhanced the chance of an organism being preserved in the fossil record. Therefore, our knowledge of life's diversification improves greatly from the Paleozoic onward. This diversity is demonstrated in Figure 19.8.

Abundant fossils have allowed geologists to construct a far more detailed time scale for the most recent one-eighth of geologic time than for the preceding seven-eights (the Precambrian). Moreover, because every organism is associated with a particular environment, the greatly improved fossil record provided invaluable information for deciphering ancient environments.

Paleozoic History

As the Paleozoic opened, North America was a land with no living things, plant or animal. There were no Appalachian or Rocky Mountains; the continent was largely barren lowlands. Several times during the early Paleozoic, shallow seas moved inland and then receded from the interior of the continent. Deposits of clean sandstones, used today to make glass, mark the shorelines of these shallow seas in the mid-continent.

The Paleozoic era was marked with a series of collisions that gradually joined together North America, Europe, Siberia, and other smaller crustal fragments (Figure 19.9). These events eventually generated a large northern continent called *Laurasia*. At this time, the vast southern continent of Gondwana encompassed five continents—South America, Africa, Australia, Antarctica, India, and perhaps portions of China. Evidence of an extensive continental glaciation places this landmass near the South Pole! By the end of the Paleozoic, Gondwana had migrated northward to collide with Laurasia, culminating in the formation of the supercontinent of Pangaea.

Formation of Pangaea. The accretion of Pangaea spans over 200 million years and resulted in the creation of several mountain belts (Figure 19.9). This time period saw the collision of northern Europe (mainly Norway and Scotland) with Greenland to produce the Caledonian Mountains. At roughly the same time, at least two microcontinents collided with and deformed the sediments that had accumulated along the eastern margin of North America. This event was an early phase in the formation of the Appalachian Mountains.

By the late Paleozoic, the joining of northern Asia (Siberia) and Europe created the Ural Mountains. Northern China is also thought to have accreted to Asia by the end of the Paleozoic, whereas southern China may not have joined Asia until sometime after Pangaea had begun to rift apart in the early Mesozoic era. (Recall that India did not accrete to Asia until about 45 million years ago.)

This leaflike animal is one of more than 100,000 unique Paleozoic fossils found in the Burgess Shale. (National Museum of Natural History)

Did You Know?

Fossil evidence suggests that the first organisms with relatively complex cells that contain a nucleus (called eukaryotes) appeared about 2.7 billion years ago.

A. B.

▲ **Figure 19.7** Fossils of common Paleozoic life forms. **A.** Natural cast of a trilobite. Trilobites dominated the Paleozoic ocean, scavenging food from the bottom. **B.** Extinct coiled cephalopods. Like their modern descendants, these were highly developed marine organisms. *(Photos courtesy of E. J. Tarbuck)*

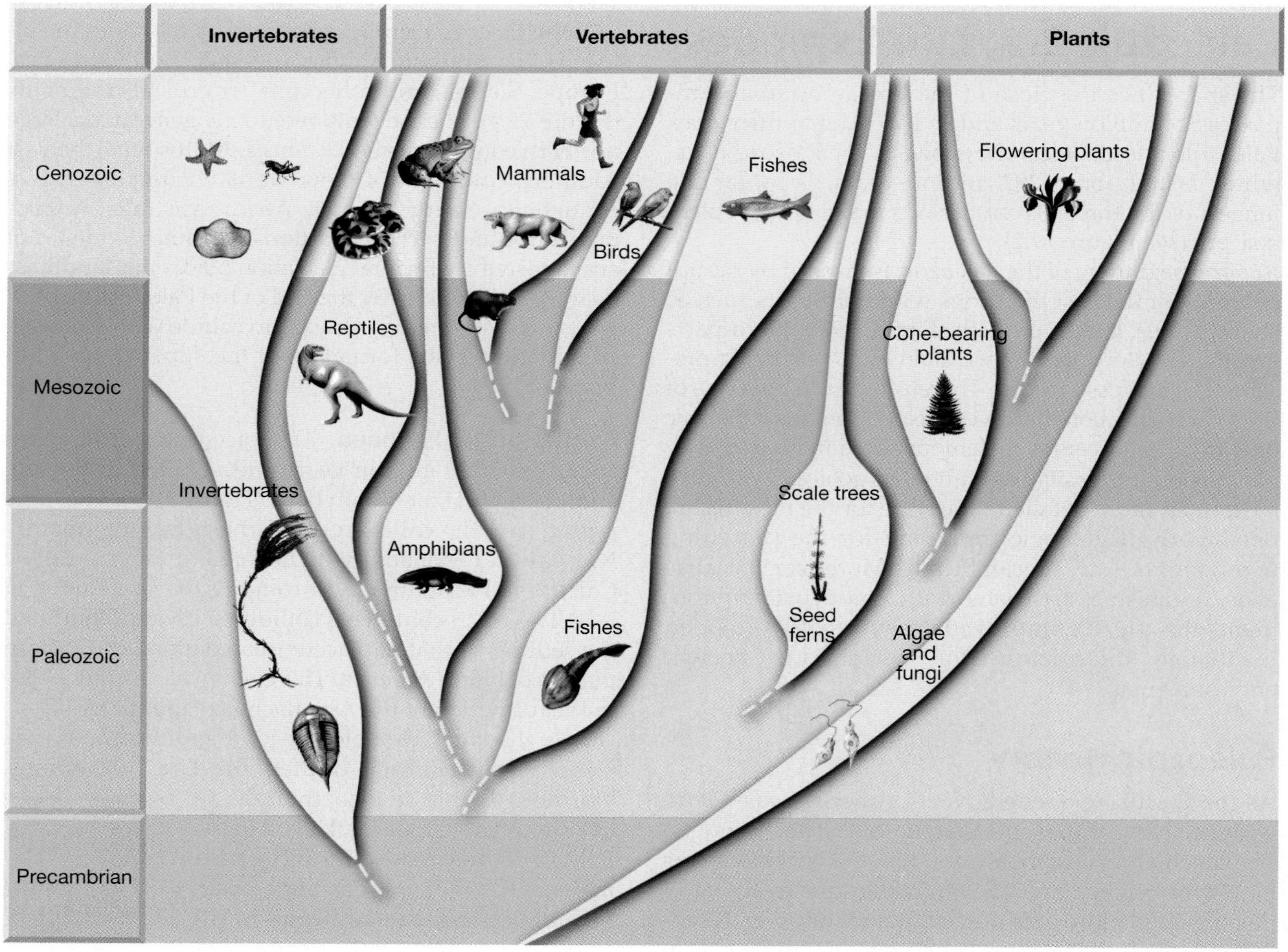

▲ **Figure 19.8** This chart indicates the times of appearance and relative abundance of major groups of organisms. The wider the band, the more abundant the group.

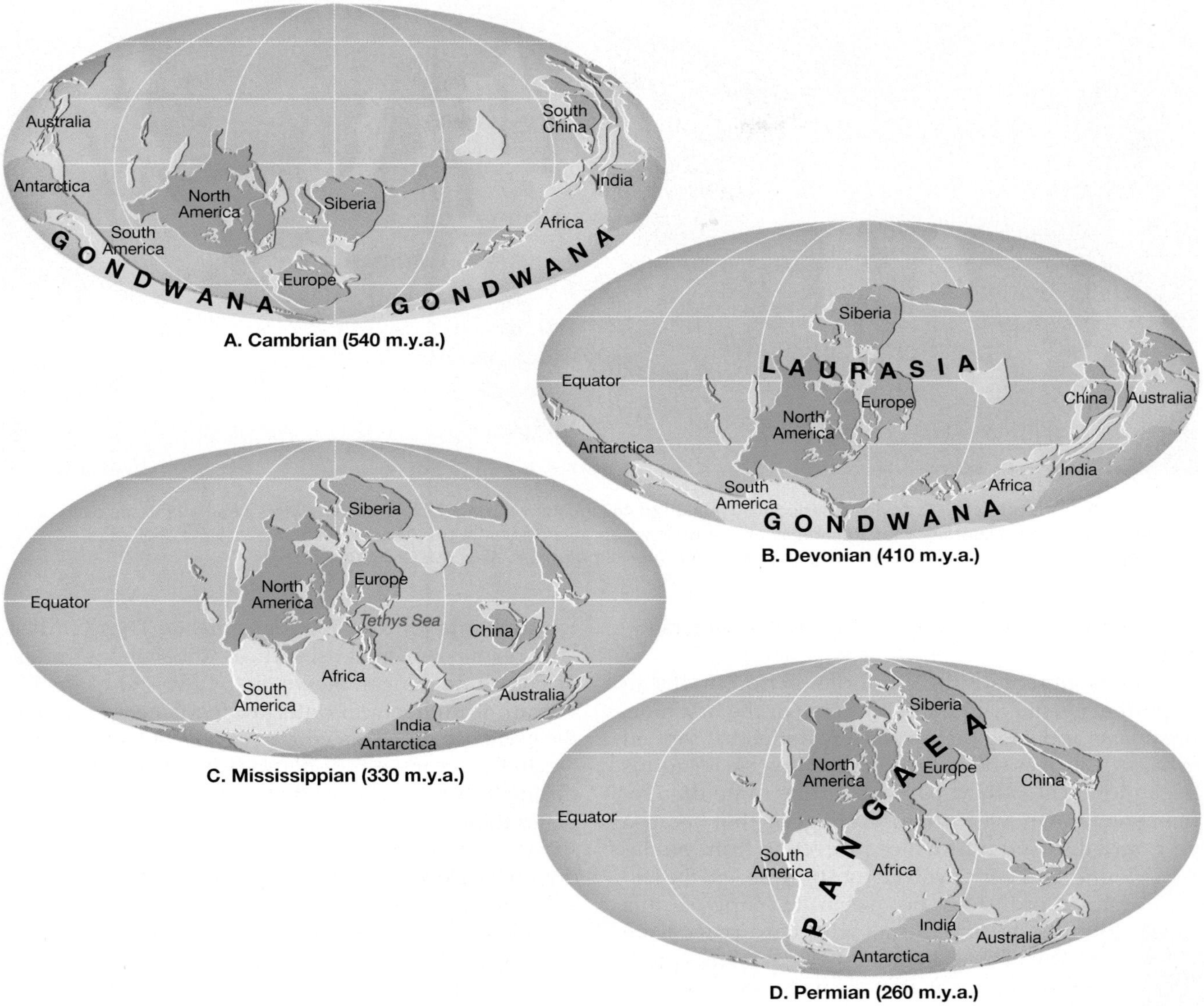

▲ **Figure 19.9** During the late Paleozoic, plate movements were joining together the major landmasses to produce the supercontinent of Pangaea. *(After C. Scotese, R. K. Bambach, C. Barton, R. VanderVoo, and A. Ziegler)*

Pangaea reached its maximum size about 260 million years ago as Africa (and Gondwana) collided with North America (Figure 19.9D). This event marked the final episode of growth in the long history of the Appalachian Mountains.

Early Paleozoic Life

Life in early Paleozoic time was restricted to the seas. Vertebrates had not yet evolved, so life consisted of several invertebrate groups (shown in Figure 19.10). The Cambrian period was the golden age of *trilobites.* More than 600 genera of these mud-burrowing scavengers flourished worldwide. By Ordovician times, *brachiopods* outnumbered the trilobites. Brachiopods are among the most widespread Paleozoic fossils and, except for one modern group, are now extinct. Although the adults lived attached to the seafloor, the young larvae were free-swimming. This mobility accounts for the group's wide geographic distribution.

The Ordovician also marked the appearance of abundant *cephalopods*—mobile, highly developed mollusks that became the major predators of their time. The descendants of cephalopods include the modern squid, octopus, and nautilus. Cephalopods were the first truly large organisms on Earth (Figure 19.10). Whereas the largest trilobites seldom exceeded 30 centimeters (12 inches) in length and the biggest brachiopods were no more than about 20 centimeters (8 inches) across, one species of

Did You Know?

Although the possession of hard parts greatly enhances the likelihood of organisms being preserved as a fossil, there have been rare occasions when large numbers of soft-bodied organisms were preserved. One of the best examples is the Burgess Shale, located in the Canadian Rockies, where more than 100,000 unique fossils have been uncovered.

▲ **Figure 19.10** During the Ordovician period (490 million to 443 million years ago), the shallow waters of an inland sea over central North America contained an abundance of marine invertebrates. Shown in this reconstruction are straight-shelled cephalopods, trilobites, brachiopods, snails, and corals. *(© The Field Museum, Neg. #GEO80820c, Chicago)*

cephalopod reached a length of nearly 10 meters (30 feet).

The beginning of the Cambrian period marks an important event in animal evolution. For the first time, organisms appeared that secreted material that formed *hard parts,* such as shells. Why several diverse life forms began to develop hard parts about the same time remains unknown. One proposal suggests that because an external skeleton provides protection from predators, hard parts evolved for survival. Yet, the fossil record does not seem to support this hypothesis. Organisms with hard parts were plentiful in the Cambrian period, whereas predators such as cephalopods were not abundant until the Ordovician period, some 70 million years later.

The modern Appalachians are the remnants of a massive mountain belt that formed in the Paleozoic era.

Whatever the answer, hard parts clearly served many useful purposes and aided adaptations to new ways of life. Sponges, for example, developed a network of fine interwoven silica spicules that allowed them to grow larger and more erect, capable of extending above the seafloor in search for food. Mollusks (clams and snails) secreted external shells of calcium carbonate that protected them and allowed body organs to function in a more controlled environment. The successful trilobites developed an exoskeleton of a protein called *chitin,* which permitted them to burrow through soft sediment in search of food.

Late Paleozoic Life

During most of the late Paleozoic, organisms diversified dramatically. Some 400 million years ago, plants that had adapted to survive at the water's edge began to move inland, becoming *land plants.* These earliest land plants were leafless, vertical spikes about the size of your index finger. However, by the end of the Devonian, 40 million years later, the fossil record indicates the existence of forests with trees tens of meters high.

In the oceans, armor-plated fishes that had evolved during the Ordovician continued to adapt. Their armor plates thinned to lightweight scales that increased their speed and mobility (Figure 19.11). Other fishes evolved during the Devonian, including primitive sharks that had a skeleton made of cartilage and bony fishes, the groups to which many modern fishes belong. Because of this, the Devonian period is often called the "Age of Fishes."

By late Devonian time, two groups of bony fishes—the lung fish and the lobe-finned fish—became adapted to land environments. Not unlike their modern relatives, these fishes had primitive lungs that supplemented their breathing through gills. It is believed that the lobe-finned fish occupied tidal flats or small ponds and that in times of drought, they may have used their bony fins to "walk" from dried-up pools in search of other ponds. Through time, the lobe-finned fish began to rely more on their lungs and less on their gills. By late Devonian time, they had evolved into true air-breathing amphibians with fishlike heads and tails.

Modern amphibians, like frogs, toads, and salamanders, are small and occupy limited biological niches. But conditions during the late Paleozoic were ideal for these newcomers to the land. Plants and insects, which were their main diet, already were very abundant and large. Having only minimal competition from other land dwellers, the amphibians rapidly diversified. Some groups took on roles and forms that were more similar to modern reptiles, such as crocodiles, than modern amphibians.

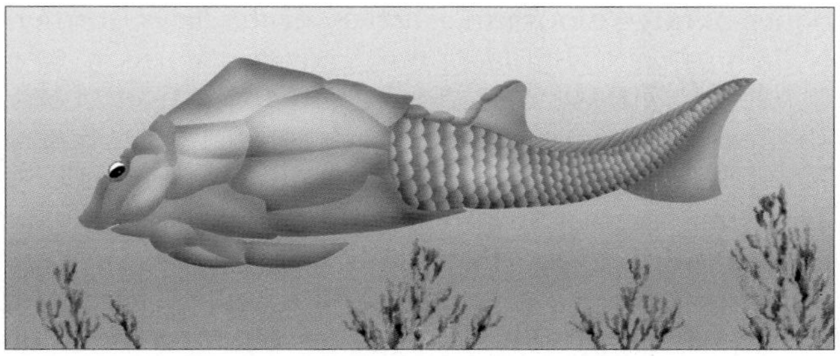

▲ **Figure 19.11** These placoderms or "plate-skinned" fish were abundant during the Devonian (417 million to 354 million years ago). *(Drawing after A. S. Romer)*

By the Pennsylvanian period, large tropical swamps extended across North America, Europe, and Siberia (Figure 19.12). Trees approached 30 meters (100 feet) tall, with trunks more than 1 meter across. The coal deposits that fueled the Industrial Revolution and which provide a substantial portion of our electric power today, originated in these vast swamps. Further, it was in the lush coal swamp environment of the late Paleozoic in which amphibians evolved quickly into a variety of species.

The Great Paleozoic Extinction. The Paleozoic ended at a time when Earth's major landmasses were joined to form the supercontinent of Pangaea (see Figure 19.9D). This redistribution of land and water and changes in the elevations of landmasses brought pronounced changes in world climates. Broad areas of the northern continents became elevated above sea level, and the climate grew drier. These changes apparently triggered extinctions of many species on land and sea.

By the close of the Paleozoic, 75 percent of the amphibian families had disappeared, and plants declined in number and variety. Although many amphibian groups became extinct, their descendants, the reptiles, would become the most successful and advanced animals on Earth. Marine life was not spared. At least 80 percent, and perhaps as much as 95 percent, of marine life disappeared. Many marine invertebrates that had been dominant during the Paleozoic, including all the remaining trilobites as well as some types of corals and brachiopods, failed to adapt to the widespread environmental changes.

The late Paleozoic extinction was the greatest of at least five mass extinctions to occur over the past 500 million years. Each extinction wreaked havoc with the existing biosphere, wiping out large numbers of species. In each case, however, the survivors formed new biological communities that were more diverse than their predecessors. Thus, mass extinctions actually invigorated life on Earth, as the few hardy survivors eventually filled more niches than the ones left by the victims.

The cause of the great Paleozoic extinction is uncertain. The climate changes brought about by the formation of Pangaea and the associated drop in sea level undoubtedly stressed many species. In addition, at least 2 million cubic kilometers of lava flowed across Siberia to produce what is called the Siberian Traps. Perhaps debris from these eruptions blocked incoming

Did You Know?

You are probably most familiar with animals that have a vertebra, such as humans, dogs, fish, and frogs. However, 98 percent of the major animal groups are invertebrates. Examples of the latter group include insects, corals, clams, and starfish.

Did You Know?

The Devonian period, which began about 417 million years ago, is known as the "Age of the Fishes." During this time the lobe-finned fishes developed muscular fins packed with bone that enabled them to crawl. This group, whose modern relatives include lungfish and coelacanths, gave rise to the first amphibians.

▼ **Figure 19.12** Restoration of a Pennsylvanian-age coal swamp (323 million to 290 million years ago). Shown are scale trees (left), seed ferns (lower left), and scouring rushes (right). Also note the large dragonfly. *(© The Field Museum, Neg. #GEO85637c, Chicago. Photographer John Weinstein.)*

Did You Know?

The Mesozoic era is often characterized as the "Age of the Reptiles" because many major reptile groups were the dominant forms of animal life. The dinosaurs occupied the land while predators, such as the fast-swimming *ichthyosaurs*, took to the seas.

sunlight, or perhaps enough sulfuric acid was emitted to make the seas virtually uninhabitable. A recently discovered impact crater off the northwest coast of Australia has been cited as a potential contributing cause of the late Paleozoic extinction. Whatever caused this extinction, it is clear that without it a very different population of organisms would inhabit our planet today.

Mesozoic Era: Age of the Dinosaurs

Spanning about 183 million years, the Mesozoic era is divided into three periods: the Triassic, Jurassic, and Cretaceous. The Mesozoic era witnessed the beginning of the breakup of the supercontinent Pangaea. Also in this era, organisms that had survived the great Permian extinction began to diversify in spectacular ways. On land, dinosaurs became dominant and remained unchallenged for over 100 million years.

This Tyrannosaurus stands on display in Chicago's Field Museum. (Don and Pat Valenti / DRK)

Early geologists recognized a profound difference between the fossils in Permian strata and those in younger Triassic rocks, as if someone had drawn a bold line to separate the two time periods. Clearly one half of the fossil groups that occurred in late Paleozoic rocks were missing in Mesozoic rocks. On this basis, it was decided to separate the Paleozoic and Mesozoic at the Permian–Triassic boundary.

Mesozoic History

The Mesozoic era began with much of the world's land above sea level. In fact, in North America no period exhibits a more meager marine sedimentary record than the Triassic period. Of the exposed Triassic strata, most are red sandstones and mudstones that lack fossils and contain features indicating a terrestrial environment.

As the Jurassic period opened, the sea invaded western North America. Adjacent to this shallow sea, extensive continental sediments were deposited on what is now the Colorado Plateau. The most prominent is the Navajo Sandstone, a windblown, white quartz sandstone that in places approaches a thickness of 300 meters (1000 feet). These massive dunes indicate that a major desert occupied much of the American Southwest during early Jurassic times.

A well-known Jurassic deposit is the Morrison Formation, within which is preserved the world's richest storehouse of dinosaur fossils. Included are fossilized bones of huge dinosaurs, such as *Apatosaurus* (formerly *Brontosaurus*), *Brachiosaurus*, and *Stegosaurus*.

As the Jurassic period gave way to the Cretaceous, shallow seas once again invaded much of western North America, the Atlantic, and Gulf coastal regions. This created great swamps like those of the Paleozoic era, forming Cretaceous coal deposits that are very important economically in the western United States and Canada. For example, on the Crow Indian reservation in Montana, there are nearly 20 billion tons of high-quality coal of Cretaceous age.

A major event of the Mesozoic era was the breakup of Pangaea (Figure 19.13). A rift developed between what is now North America and western Africa, marking the birth of the Atlantic Ocean. As Pangaea gradually broke apart, the westward-moving North American plate began to override the Pacific plate. This tectonic event marks the beginning of a continuous wave of deformation that moved inland along the entire western margin of North America. By Jurassic times, subduction of the Pacific plate had begun to produce the chaotic mixture of rocks that exists today in the Coast Ranges of California. Further inland, igneous activity was widespread, and for nearly 60 million years, huge masses of magma rose to within a few miles of the surface before crystallizing. The remnants of this intrusive activity include the granitic rocks of the Sierra Nevada as well as the Idaho batholith, and British Columbia's Coast Range batholith.

Tectonic activity that began in the Jurassic continued throughout the Cretaceous, ultimately forming the vast Northern Rockies. Compressional forces moved huge rock units in a shinglelike fashion toward the east. Throughout much of the western margin of North America, older rocks were thrust eastward over younger strata, for a distance exceeding 150 kilometers (90 miles).

As the Mesozoic came to an end, the southern ranges of the Rocky Mountains formed. This mountain-building event, called the Laramide Orogeny, occurred when tectonic forces uplifted large blocks of Precambrian rocks in Wyoming, Colorado, and New Mexico.

Mesozoic Life

As the Mesozoic era dawned, its life forms were the survivors of the great Paleozoic extinction. These organisms diversified in many ways to fill the biological voids created at the close of the Paleozoic (Figure 19.14). On land, conditions favored those that could adapt to drier climates. Among plants, the gymnosperms were one such group. Unlike the first plants to invade the land, the seed-bearing gymnosperms did not depend on free-standing water for fertilization. Consequently, these plants were not restricted to a life near the water's edge.

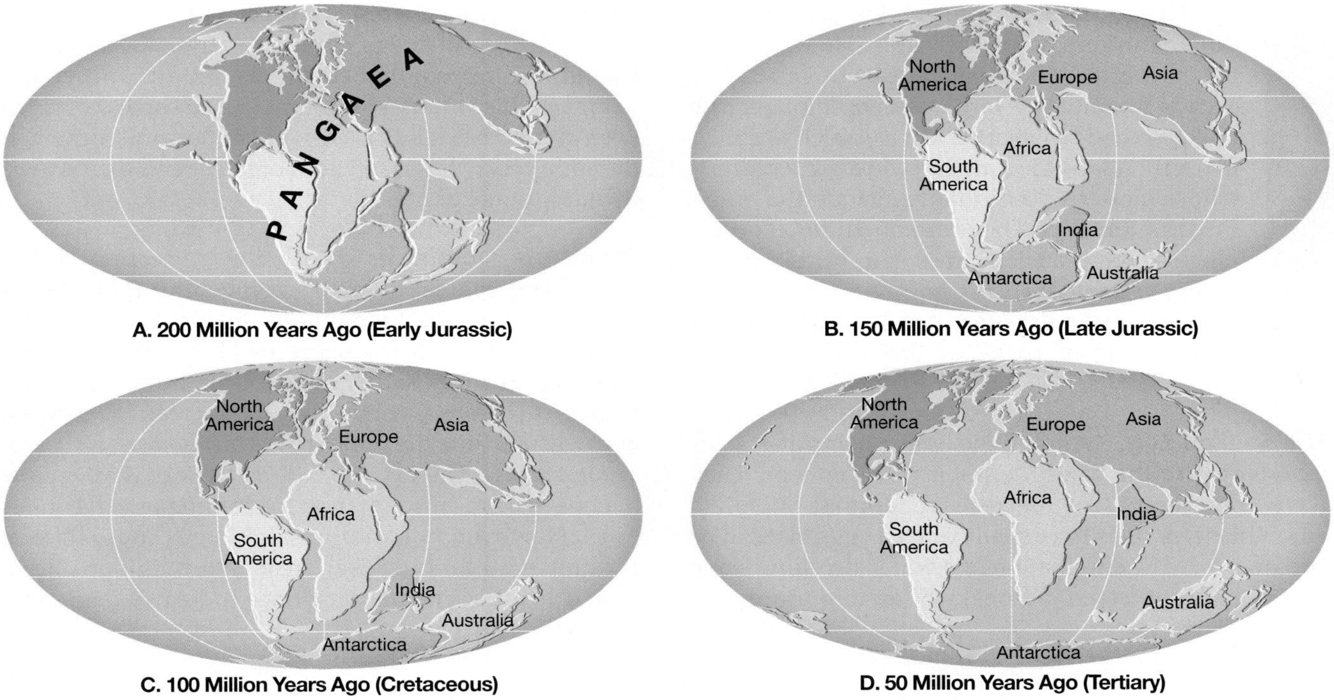

▲ **Figure 19.13** The breakup of the supercontinent of Pangaea began and continued throughout the Mesozoic era. *(After D. Walsh and C. Scotese)*

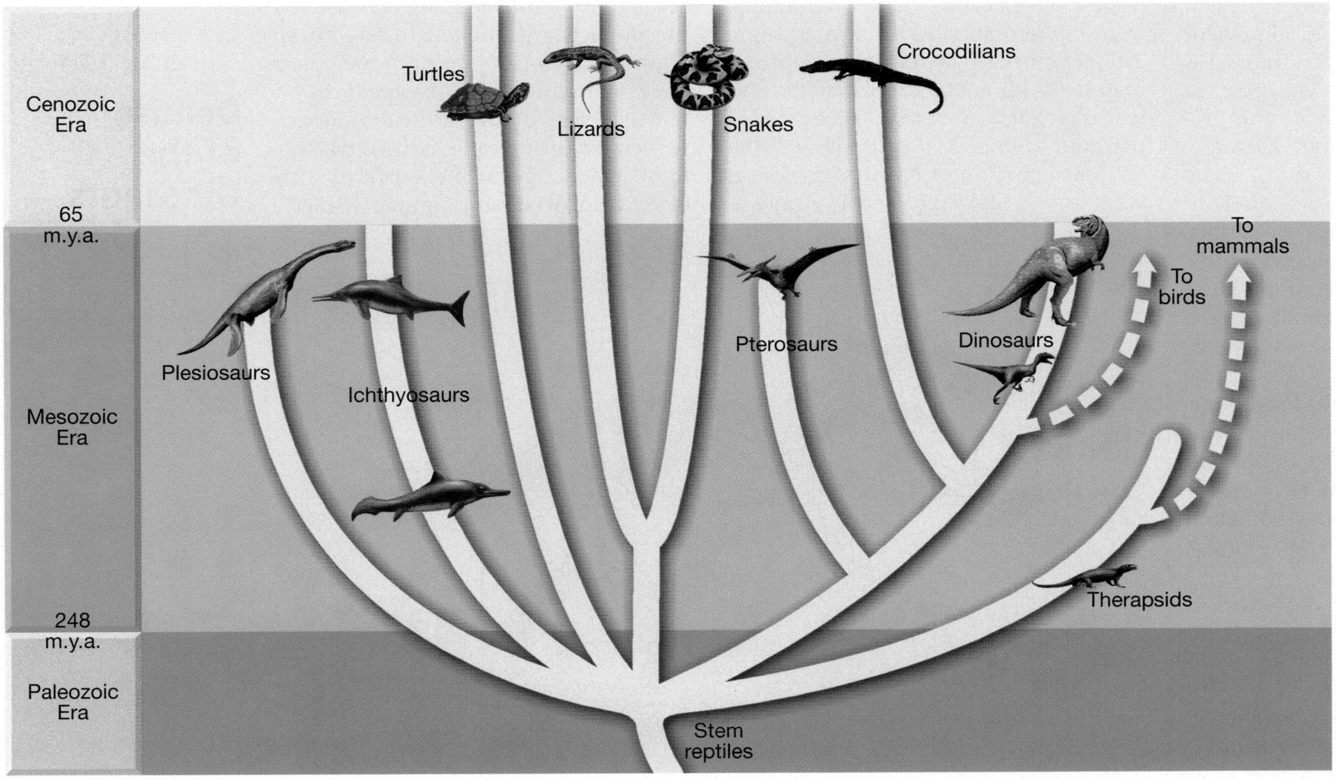

▲ **Figure 19.14** Adaptive radiation of major reptile groups during the Mesozoic era—"the Golden Age of Reptiles." Dinosaurs, pterosaurs (flying reptiles), and large marine reptiles all became extinct by the end of the Mesozoic.

Petrified logs of Triassic age in Arizona's Petrified Forest National Park. (David Muench)

The gymnosperms quickly became the dominant trees of the Mesozoic. They included the cycads, the conifers, and the ginkgoes. The cycads resembled a large pineapple plant. The ginkgoes had fan-shaped leaves, much like their modern relatives. Largest were the conifers, whose modern descendants include the pines, firs, and junipers. The best-known fossil occurrence of these ancient trees is in northern Arizona's Petrified Forest National Park. Here, huge petrified logs lie exposed at the surface, having been weathered from rocks of the Triassic Chinle Formation.

The Shelled Egg. Among the animals, reptiles readily adapted to the drier Mesozoic environment. They were the first true terrestrial animals. Unlike amphibians, reptiles have shell-covered eggs that can be laid on land. The elimination of a water-dwelling stage (like the tadpole stage in frogs) was an important evolutionary step. Of interest is the fact that the watery fluid within the reptilian egg closely resembles seawater in chemical composition. Because the reptile embryo develops in this watery environment, the shelled egg has been characterized as a "private aquarium" in which the embryos of these land vertebrates spend their water-dwelling stage of life.

Dinosaurs Dominate. With the perfection of the shelled egg, reptiles quickly became the dominant land animals. They continued this dominance for more than 160 million years. Most awesome of the Mesozoic reptiles were the dinosaurs. Some of the huge dinosaurs were carnivorous (*Tyrannosaurus*), whereas others were herbivorous (like the ponderous *Apatosaurus,* formerly *Brontosaurus*). The extremely long neck of *Apatosaurus* may have been an adaptation for feeding on tall conifer trees. However, not all dinosaurs were large. In fact, certain small forms closely resembled modern fleet-footed lizards. Further, evidence indicates that some

BOX 19.1

Demise of the Dinosaurs

The boundaries between divisions on the geologic time scale represent times of significant geological and/or biological charge. Of special interest is the boundary between the Mesozoic era ("middle life") and Cenozoic era ("recent life"), about 65 million years ago. Around this time, about three-quarters of all plant and animal species died out in a *mass extinction.* This boundary marks the end of the era in which dinosaurs and other reptiles dominated the landscape and the beginning of the era when mammals become very important (Figure 19.A). Because the last period of the Mesozoic is the Cretaceous (abbreviated K to avoid confusion with other "C" periods), and the first period of the Cenozoic is the Tertiary (abbreviated T), the time of this mass extinction is called the *Cretaceous-Tertiary* or *KT boundary.*

The extinction of the dinosaurs is generally attributed to this group's inability to adapt to some radical change in environmental conditions. What event could have triggered the rapid extinction of the dinosaurs—one of the most successful groups of land animals ever to have lived?

The most strongly supported hypothesis proposes that about 65 million years ago our planet was struck by a large carbonaceous meteorite, a relic from the formation of the solar system. The errant mass of rock was approximately 10 kilometers in diameter and was traveling at about 90,000 kilometers per hour at impact. It collided with the southern portion of North America in what is now Mexico's Yucatán Peninsula but at the time was a shallow tropical sea (Figure 19.B). The energy released by the impact is estimated to have been equivalent to 100 million megatons (*mega* = million) of high explosives.

For a year or two after the impact, suspended dust greatly reduced the sunlight reaching Earth's surface. This caused global cooling ("impact winter")

Figure 19.A Dinosaurs dominated the Mesozoic landscape until their extinction at the close of the Cretaceous period. This skeleton of *Tyrannosaurus* stands on display in New York's Museum of Natural History. (Photo by Gail Mooney/CORBIS Photo)

▲ **Figure 19.15** Fossils of the great flying Pteranodon have been recovered from Cretaceous chalk deposits located in Kansas. Although Pteranodon had a wingspan of 7 meters (22 feet), flying reptiles with twice this wingspan have been discovered in west Texas.

dinosaurs, unlike their present-day reptile relatives, were warm-blooded.

The reptiles made one of the most spectacular adaptive radiations in all of Earth history (Figure 19.14). One group, the pterosaurs, took to the air. These "dragons of the sky" possessed huge membranous wings that allowed them rudimentary flight (Figure 19.15). Another group of reptiles, exemplified by the fossil *Archaeopteryx,* led to more successful flyers: the birds. Whereas some reptiles took to the skies, others returned to the sea, including the fish-eating plesiosaurs and ichthyosaurs. These reptiles became proficient swimmers, but they retained their reptilian teeth and breathed by means of lungs.

Fossil skull of a huge crocodile of Mesozoic age. (Project Exploration)

At the close of the Mesozoic, many reptile groups became extinct. Only a few types survived to recent times, including the turtles, snakes, crocodiles, and lizards. The huge land-dwelling dinosaurs, the marine plesiosaurs, and the flying pterosaurs all are known only through the fossil record. What caused this great extinction (see Box 19.1)?

and inhibited photosynthesis, greatly disrupting food production. Long after the dust settled, carbon dioxide, water vapor, and sulfur oxides that had been added to the atmosphere by the blast remained. If significant quantities of sulfate aerosols formed, their high reflectivity would have helped to perpetuate the cooler surface temperatures for a few more years. Eventually sulfate aerosols leave the atmosphere as acid precipitation. By contrast, carbon dioxide has a much longer residence time in the atmosphere. Carbon dioxide is a *greenhouse gas,* a gas that traps a portion of the radiation emitted by Earth's surface. With the aerosols gone, the enhanced greenhouse effect caused by the carbon dioxide would have led to a long-term rise in average global temperatures. The likely result was that some of the plant and animal life that had survived the initial environmental assault finally fell victim to stresses associated with global cooling, followed by acid precipitation and global warming.

The extinction of the dinosaurs opened up habitats for the small mammals that survived. These new habitats, along with evolutionary forces, led to the development of the large mammals that occupy our modern world.

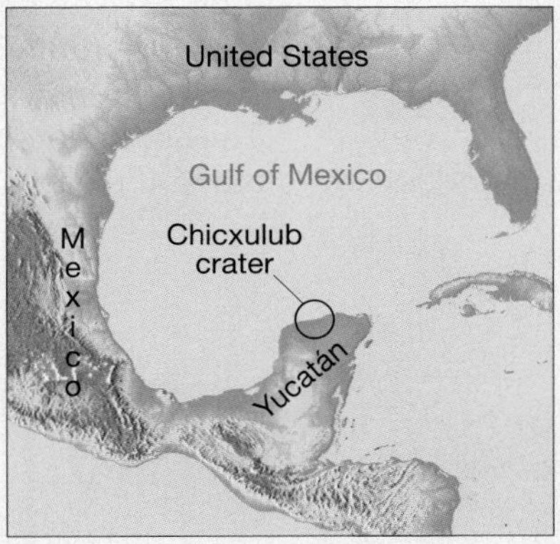

Figure 19.B Chicxulub crater is a giant impact crater that formed about 65 million years ago and has since been filled with sediments. About 180 kilometers in diameter, Chicxulub crater is regarded by some researchers to be the impact site that resulted in the demise of the dinosaurs.

What evidence points to such a catastrophic collision 65 million years ago? First, a thin layer of sediment nearly 1 centimeter thick has been discovered at the KT boundary, worldwide. This sediment contains a high level of the element *iridium,* rare in Earth's crust but found in high proportions in stony meteorites. Could this layer be the scattered remains of the meteorite that was responsible for the environmental changes that led to the demise of many reptile groups?

Despite growing support, some scientists disagree with the impact hypothesis. They suggest instead that huge volcanic eruptions may have led to a breakdown in the food chain. To support this hypothesis, they site enormous outpourings of lavas in the Deccan Plateau of northern India about 65 million years ago.

Whatever caused the KT extinction, we now have a greater appreciation of the role of catastrophic events in shaping the history of our planet and the life that occupies it. Could a catastrophic event having similar results occur today? This possibility may explain why an event that occurred 65 million years ago has captured the interest of so many.

Did You Know?

Dinosaurs were not the only large animals to roam Earth in the distant past. One example is a crocodile known as *Sarcosuchus imperator.* This huge river dweller weighed 8 metric tons (more than 17,600 pounds!) and was about 12 meters (40 feet) long—as long as *Tyrannosaurus rex* and much heavier. Its jaws were roughly as long as an adult human.

Cenozoic Era: Age of Mammals

The Cenozoic era, or *"era of recent life,"* encompasses the last 65 million years of Earth history. It is the *"post-dinosaur era,"* the time of mammals, including humans. It is during this span that the physical landscapes and life forms of our modern world came into being. The Cenozoic era represents a much smaller fraction of geologic time than either the Paleozoic or the Mesozoic. Although shorter, it nevertheless possesses a rich history, because the completeness of the geologic record improves as time approaches the present. The rock formations of this time span are more widespread and less disturbed than those of any preceding time.

Marine reptiles like this ichthyosaur *were the most spectacular of sea animals.* (Chip Clark)

The Cenozoic era is divided into two periods of very unequal duration—the Tertiary period and the Quaternary period. The Tertiary period includes five epochs and embraces about 63 million years, practically all of the Cenozoic era. The Quaternary period consists of two epochs that represent only the last 2 million years of geologic time.

Cenozoic North America

Most of North America was above sea level throughout the Cenozoic era. However, the eastern and western margins of the continent experienced markedly contrasting events, because of their different relationships with plate boundaries. The Atlantic and Gulf coastal regions, far removed from an active plate boundary, were tectonically stable. Western North America, in contrast, was the leading edge of the North American plate. As a result, plate interactions during the Cenozoic gave rise to many events of mountain building, volcanism, and earthquakes in the West.

Eastern North America. The stable continental margin of eastern North America was the site of abundant marine sedimentation. The most extensive deposition surrounded the Gulf of Mexico, from the Yucatán Peninsula to Florida. Here, the great buildup of sediment caused the crust to downwarp and produced numerous faults. In many instances, the faults created traps in which oil and natural gas accumulated. Today, these and other petroleum traps are the most economically important resource in Cenozoic strata of the Gulf Coast, as evidenced by the Gulf's well-known offshore drilling platforms.

By early Cenozoic time, most of the original Appalachians had been eroded to a low plain. Then, by the mid-Cenozoic, isostatic adjustments raised the region once again, rejuvenating its rivers. Streams eroded with renewed vigor, gradually sculpturing the surface into its present-day topography (Figure 19.16). The sediments from all of this erosion were deposited along the eastern margin of the continent, where they

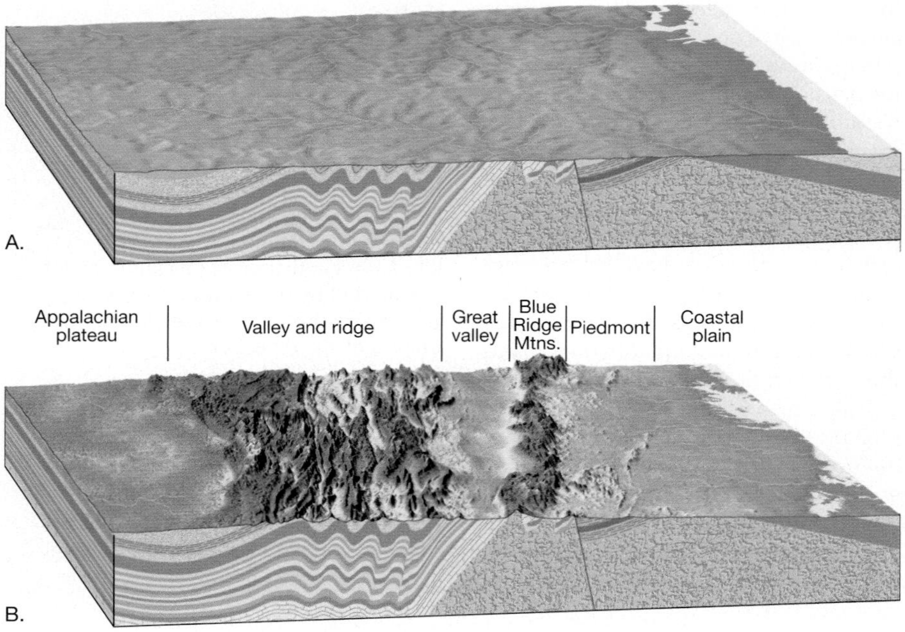

▶ **Figure 19.16** The formation of the modern Appalachian Mountains. **A.** The original Appalachians eroded to a low plain. **B.** The recent upwarping and erosion of the Appalachians began nearly 30 million years ago to produce the present topography.

attained a thickness of many kilometers. Today, portions of the strata deposited during the Cenozoic are exposed as the gently sloping Atlantic and Gulf coastal plains.

Western North America. In the West, the Laramide Orogeny that built the southern Rocky Mountains was coming to an end. As erosion lowered the mountains, the basins between uplifted ranges filled with sediments. Eastward, a great wedge of sediment from the eroding Rockies was building, creating the Great Plains.

Beginning in the Miocene epoch, a broad region from northern Nevada into Mexico experienced crustal extension that created more than 150 fault-block mountain ranges. Today, they rise abruptly above the adjacent basins, creating the Basin and Range Province (see Chapter 17).

As the Basin and Range Province was forming, the entire western interior of the continent was gradually uplifted. This uplift re-elevated the Rockies and rejuvenated many of the West's major rivers. As the rivers became incised, many spectacular gorges were formed, including the Grand Canyon of the Colorado River, the Grand Canyon of the Snake River, and the Black Canyon of the Gunnison River.

Volcanic activity was also common in the West during much of the Cenozoic. Beginning in the Miocene epoch, great volumes of fluid basaltic lava flowed from fissures in portions of present-day Washington, Oregon, and Idaho. These eruptions built the extensive (1.3 million square kilometers) Columbia Plateau. Immediately west of the Columbia Plateau, volcanic activity was quite different. Here, more viscous magmas with higher silica contents erupted explosively, creating a chain of stratovolcanoes from northern California to the Canadian border, some of which remain active—such as Mount St. Helens.

A final episode of folding occurred in the West in late Tertiary time, creating the Coast Ranges that stretch along the Pacific Coast. Meanwhile, the Sierra Nevada were faulted and uplifted along their eastern flank, creating the imposing mountain front we know today.

As the Tertiary period drew to a close, the effects of mountain building, volcanic activity, isostatic adjustments, and extensive erosion and sedimentation had created a physical landscape very similar to the configuration of today. All that remained of Cenozoic time was the final 2-million-year episode called the Quaternary period. During this most recent (and current) phase of Earth history, in which humans evolved, the action of glacial ice and other erosional agents added the finishing touches.

Cenozoic Life

Mammals replaced reptiles as the dominant land animals in the Cenozoic. Angiosperms (flowering plants with covered seeds) replaced gymnosperms as the dominant land plants. Marine invertebrates took on a modern look. Microscopic animals called *foraminifera* became especially important. Today, foraminifera are among the most intensely studied of all fossils because their widespread occurrence makes them invaluable in correlating Tertiary sediments. Tertiary strata are very important to the modern world, for they yield more oil than rocks of any other age.

Did You Know?

Bats and humans are both mammals, but they have developed very different physical structures and lifestyles. One clue that they are related is their pentadactyl (five-fingered) limbs. Humans have evolved hands for grasping, whereas bats have proportionally much longer "fingers" that support their wings for flying.

The Cenozoic is often called the "Age of Mammals," because these animals came to dominate land life. It could also be called the "Age of Flowering Plants," for the angiosperms enjoy a similar status in the plant world. As a result of advances in seed fertilization and dispersal, angiosperms experienced rapid development and expansion as the Mesozoic drew to a close.

Did You Know?

The first modern humans (*Homo sapiens*) began their migration out of Africa roughly 100,000 years ago. After migrating across Asia, they entered the Americas across the Bering land bridge about 15,000 years ago.

Development of the flowering plants, in turn, strongly influenced the evolution of both birds and mammals that feed on seeds and fruits. For example, during the middle Tertiary, grasses, which are angiosperms, developed rapidly and spread over the plains. This fostered the emergence of herbivorous (plant-eating) mammals that were mainly grazers. In turn, the development and spread of grazing animals established the setting for the evolution of the large carnivorous mammals that preyed upon them.

Mammals Replace Reptiles. Primitive mammals emerged in the late Triassic, about the same time as the dinosaurs. Yet throughout the period of dinosaur dominance, mammals remained in the background as small and inconspicuous animals. It was only at the close of the Mesozoic era, when most large reptiles became extinct, that mammals came into their own as the dominant land animals. The transition is a major example in the fossil record of the replacement of one dominant group by another.

Mount Rainier, one of several large Cenozoic volcanoes in the Pacific Northwest.

Mammals are distinct from reptiles in that they give birth to live young (which they suckle on milk) and they are warm-blooded. This latter adaptation allowed mammals to lead more active and diversified lives than reptiles because they could survive in cold regions and search for food during any

Did You Know?

The La Brea tar pits, located in downtown Los Angeles, are famous because they contain a rich and very well preserved assemblage of mammals that roamed southern California from 40,000 to 80,000 years ago. Tar pits form when crude oil seeps to the surface and the light portion evaporates, leaving behind sticky pools of heavy tar.

season or time of day. (Most reptiles are dormant during cold weather.) Other mammalian adaptations included the development of insulating body hair and a more efficient heart and lungs.

It is worth noting that perfect boundaries cannot be drawn between mammalian and reptilian traits. One minor group of mammals, the *monotremes,* lay eggs. The two species in this group, the duck-billed platypus and the spiny anteater, are found only in Australia. Moreover, although modern reptiles are cold-blooded, some paleontologists believe that dinosaurs may have been warm-blooded.

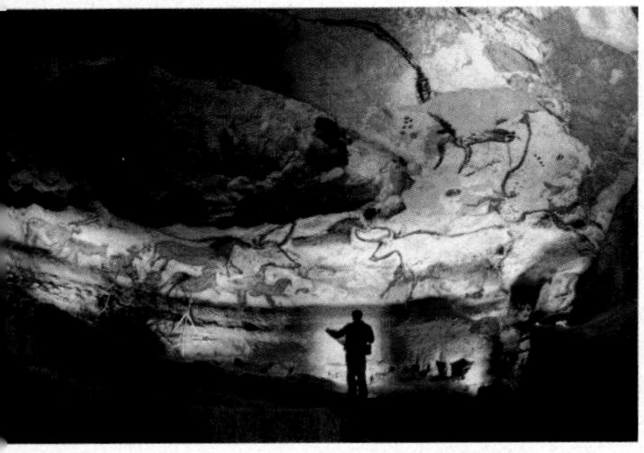

Cave painting of animals that early humans encountered about 17,000 years ago. (National Geographic)

With the demise of the large Mesozoic reptiles, Cenozoic mammals diversified rapidly. The many forms that exist today evolved from small primitive mammals that were characterized by short legs, flat five-toed feet, and small brains. Their development and specialization took four principal directions: (1) increase in size, (2) increase in brain capacity, (3) specialization of teeth to better accommodate their diet, and (4) specialization of limbs to better equip the animal for a particular lifestyle or environment.

Marsupial and Placental Mammals. Two groups of mammals, the *marsupials* and the *placentals,* evolved and diversified during the Cenozoic. The groups differ principally in their modes of reproduction. Young marsupials are born live but at a very early stage of development. At birth, the tiny and immature young enter the mother's pouch to suckle and complete their development. Examples are kangaroos, opossums, and koalas.

Placental mammals, conversely, develop within the mother's body for a much longer period, so birth occurs after the young are comparatively mature. Most modern mammals are placental, including humans.

Today, marsupials are found primarily in Australia, where they underwent a separate evolutionary expansion during the Cenozoic, largely isolated from placental mammals.

In South America, primitive marsupials and placentals coexisted in isolation for about 40 million years after the breakup of Pangaea. Evolution and specialization of both groups continued undisturbed until about 3 million years ago when the Panamanian landbridge emerged, connecting the two American continents. This event permitted the exchange of fauna between the two continents. Monkeys, armadillos, sloths, and opossums arrived in North America, while various types of horses, bears, rhinos, camels, and wolves migrated southward. Many animals that had

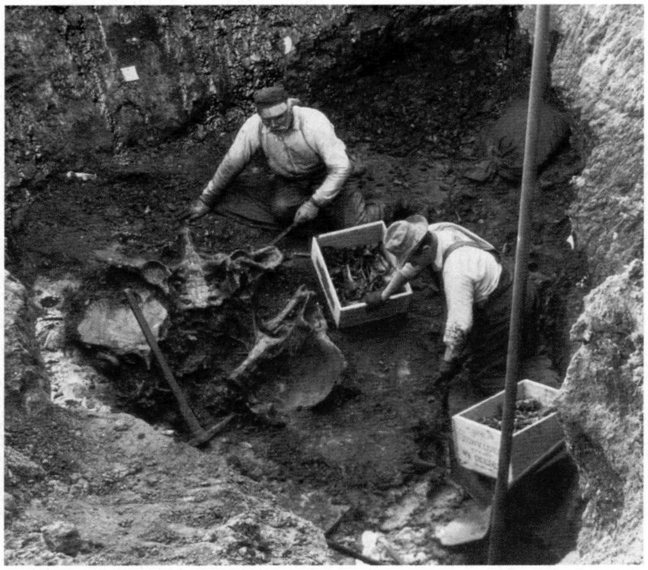

A.

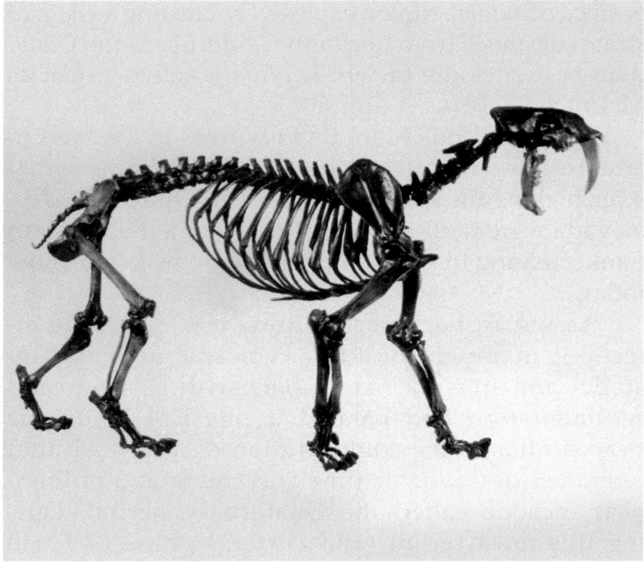

B.

▲ **Figure 19.17** La Brea tar pits in Los Angeles, California. **A.** Fossils being excavated from the thick tar in 1914. **B.** Skeleton of an extinct saber-toothed cat. These bones came from the La Brea tar pits. *(Photo courtesy of the George C. Page Museum)*

been unique to South America disappeared completely after this event, including hoofed mammals, rhino-sized rodents, and a number of carnivorous marsupials. Because this period of extinction coincided with the formation of the Panamanian land bridge, it was believed that the advanced carnivores from North America were responsible. Recent research, however, suggests that other factors, including climatic changes, may have played a role.

Large Mammals and Extinction. As you have seen, mammals diversified rapidly during the Cenozoic era. One tendency was for some groups to become very large. For example, by the Oligocene epoch a hornless rhinoceros that stood nearly 5 meters (16 feet) high had evolved. It is the largest land mammal known to have existed. As time approached the present, many other types evolved to a large size as well; more, in fact, than now exist. Many of these large forms were common as recently as 11,000 years ago. However, a wave of late Pleistocene extinctions rapidly eliminated these animals from the landscape.

In North America, the mastodon and mammoth, both huge relatives of the elephant, became extinct. In addition, saber-toothed cats, giant beavers, large ground sloths, horses, camels, giant bison, and others died out (Figure 19.17). In Europe, late Pleistocene extinctions included woolly rhinos, large cave bears, and the Irish elk. The reason for this recent wave of large animal extinctions puzzles scientists. These animals had survived several major glacial advances and interglacial periods, so it is difficult to ascribe these extinctions to climatic change. Some scientists believe that early humans hastened the decline of these mammals by selectively hunting large forms.

Angiosperms (flowering plants) became dominant in the Cenozoic era.

The Chapter in Review

- The decay of radioactive elements and heat released by colliding particles aided the melting of Earth's interior, allowing the denser elements, principally iron and nickel, to sink to its center. As a result of this *differentiation,* Earth's interior consists of shells or spheres of materials, each having distinct properties.
- Earth's primitive atmosphere probably consisted of water vapor, carbon dioxide, nitrogen, and several trace gases that were released in volcanic emissions, a process called *outgassing.* The first life forms on Earth, probably *anaerobic bacteria,* did not need oxygen. As life evolved, plants, through the process of *photosynthesis,* used carbon dioxide and water and released oxygen into the atmosphere. Once the available iron on Earth was oxidized (combined with oxygen), substantial quantities of oxygen accumulated in the atmosphere. About 4 billion years into Earth's existence, the fossil record reveals abundant ocean-dwelling organisms that require oxygen to live.
- The *Precambrian* spans about 88 percent of Earth's history, beginning with the formation of Earth about 4.5 billion years ago and ending approximately 540 million years ago with the diversification of life, which marks the start of the Paleozoic era. It is the least understood span of Earth's history. On each continent, there is a core area of Precambrian rocks called a *shield.* The most common middle Precambrian fossils are *stromatolites.* Microfossils of bacteria and blue-green algae, both primitive *prokaryotes* whose cells lack organized nuclei, have been found in a hard, dense, chemical sedimentary rock called *chert. Eukaryotes,* with cells containing organized nuclei, are among billion-year-old fossils discovered in Australia. Plant fossils date from the middle Precambrian, but animal fossils came a bit later, in the late Precambrian.
- The beginning of the Paleozoic is marked by the *appearance of the first life forms with hard parts* such as shells. Therefore, abundant Paleozoic fossils occur, and a far more detailed record of Paleozoic events can be constructed. As the Paleozoic opened, the vast continent of Gondwana was situated in the Southern Hemisphere, while ancestral North America, northern Asia, and northwestern Europe were located nearer the equator. Gradually the northern continents accreted to form Laurasia, which then collided with Gondwana to generate the supercontinent of Pangaea. Life in the

early Paleozoic was restricted to the seas and consisted of several invertebrate groups. During the Paleozoic, organisms diversified dramatically. Insects and plants moved onto the land, and amphibians evolved. By the Pennsylvanian period, large tropical swamps, which became the major coal deposits of today, extended across North America, Europe, and Siberia. At the close of the Paleozoic, altered climatic conditions contributed to one of the most dramatic biological declines in all of Earth's history.

- The *Mesozoic era,* literally the *era of middle life,* is often called the "Age of Reptiles." Early in the Mesozoic; much of the land was above sea level. However, by the middle Mesozoic, seas invaded western North America. As Pangaea began to break up, the westward-moving North American plate began to override the Pacific plate, causing crustal deformation along the entire western margin of the continent. Organisms that had survived extinction at the end of the Paleozoic began to diversify in spectacular ways. *Gymnosperms* (cycads, conifers, and ginkgoes) became the dominant trees of the Mesozoic because they could adapt to the drier climates. Reptiles became the dominant land animals. The most awesome of the Mesozoic reptiles were the *dinosaurs.* At the close of the Mesozoic, many large reptiles, including the dinosaurs, became extinct.
- The *Cenozoic era,* or *era of recent life,* begins approximately 65 million years ago and continues today. It is the *time of mammals,* including humans. The widespread, less disturbed rock formations of the Cenozoic provide a rich geologic record. Most of North America was above sea level throughout the Cenozoic. Owing to their different relations with tectonic plate boundaries, the eastern and western margins of the continent experienced contrasting events. The stable eastern margin was the site of abundant sedimentation as isostatic adjustment raised the Appalachians, causing streams to erode with renewed vigor, depositing their sediment along the continental margin. In the West, building of the Rocky Mountains was coming to an end, the Basin and Range Province was forming, and volcanic activity was extensive. The Cenozoic is often called the *"Age of Mammals"* because these animals replaced the reptiles as the dominant land life. Two groups of mammals, the marsupials and the placentals, evolved and expanded during this era. One tendency was for some mammal groups to become very large. However, a wave of late *Pleistocene* extinctions rapidly eliminated these animals from the landscape. Some scientists believe that early humans hastened their decline by selectively hunting the larger animals. The Cenozoic could also be called the *"Age of Flowering Plants."* As a source of food, flowering plants (angiosperms) strongly influenced the evolution of both birds and herbivorous (plant-eating) mammals throughout the Cenozoic era.

Key Terms

outgassing (p. 444) shields (p. 445) stromatolites (p. 446)

Questions for Review

1. Briefly describe the events that are believed to have led to the formation of our atmosphere.

2. What is the major source of free oxygen in Earth's atmosphere?

3. Explain why Precambrian history is more difficult to decipher than more recent geological history.

4. Match the following words and phrases to the most appropriate time span. Select among the following: Precambrian, Paleozoic, Mesozoic, Cenozoic.

a. Pangaea came into existence.
b. Bacteria and blue-green algae preserved in chert.
c. The era that encompasses the least amount of time.
d. Shields.
e. "Age of dinosaurs."
f. Formation of the original northern Appalachian Mountains.
g. Mastodons and mammoths.
h. "Age of Trilobites."
i. Triassic, Jurassic, and Cretaceous.
j. Coal swamps extended across North America, Europe, and Siberia.
k. Gulf Coast oil deposits formed.
l. Formation of most of the world's major iron-ore deposits.
m. Massive sand dunes covered large portions of the Colorado Plateau region.
n. "Age of Fishes."

o. Cambrian, Ordovician, and Silurian.
p. Pangaea began to break apart.
q. "Age of Mammals."
r. Animals with hard parts first appeared in abundance.
s. Gymnosperms were the dominant trees.
t. Columbia Plateau formed.
u. Stromatolites are among its more common fossils.
v. Fault-block mountains form in the Basin and Range region.

5. Briefly discuss two proposals that attempt to explain why several groups developed hard parts at the beginning of the Cambrian period. Do these proposals appear to provide a satisfactory explanation? Why or why not?

6. List some differences between amphibians and reptiles. List differences between reptiles and mammals.

7. Describe one hypothesis that attempts to explain the extinction of the dinosaurs.

8. Contrast the eastern and western margins of North America during the Cenozoic era in terms of their relationships to plate boundaries.

Online Study Guide

The *Essentials of Geology* Web site uses the resources and flexibility of the Internet to aid in your study of the topics in this chapter. Written and developed by geology instructors, this site will help improve your understanding of geology. Visit **www.prenhall.com/lutgens** and click on the cover of *Essentials of Geology 9e* to find:

- Online review quizzes.
- Critical thinking exercises.
- Links to chapter-specific Web resources.
- Internet-wide key-term searches.

http://www.prenhall.com/lutgens

APPENDIX

A Metric and English Units Compared

Units

1 kilometer (km)	=	1000 meters (m)
1 meter (m)	=	100 centimeters (cm)
1 centimeter (cm)	=	0.39 inch (in.)
1 mile (mi)	=	5280 feet (ft)
1 foot (ft)	=	12 inches (in.)
1 inch (in.)	=	2.54 centimeters (cm)
1 square mile (mi^2)	=	640 acres (a)
1 kilogram (kg)	=	1000 grams (g)
1 pound (lb)	=	16 ounces (oz)
1 fathom	=	6 feet (ft)

Conversions

When you want to convert:	Multiply by:	To find:
Length		
inches	2.54	centimeters
centimeters	0.39	inches
feet	0.30	meters
meters	3.28	feet
yards	0.91	meters
meters	1.09	yards
miles	1.61	kilometers
kilometers	0.62	miles
Area		
square inches	6.45	square centimeters
square centimeters	0.15	square inches
square feet	0.09	square meters
square meters	10.76	square feet
square miles	2.59	square kilometers
square kilometers	0.39	square miles
Volume		
cubic inches	16.38	cubic centimeters
cubic centimeters	0.06	cubic inches
cubic feet	0.028	cubic meters
cubic meters	35.3	cubic feet
cubic miles	4.17	cubic kilometers
cubic kilometers	0.24	cubic miles
liters	1.06	quarts
liters	0.26	gallons
gallons	3.78	liters

Masses and Weights

ounces	28.33	grams
grams	0.035	ounces
pounds	0.45	kilograms
kilograms	2.205	pounds

Temperature

When you want to convert degrees Fahrenheit (°F) to degrees Celsius (°C), subtract 32 degrees and divide by 1.8.

When you want to convert degrees Celsius (°C) to degrees Fahrenheit (°F), multiply by 1.8 and add 32 degrees.

When you want to convert degrees Celsius (°C) to kelvins (K), delete the degree symbol and add 273. When you want to convert kelvins (K) to degrees Celsius (°C), add the degree symbol and subtract 273.

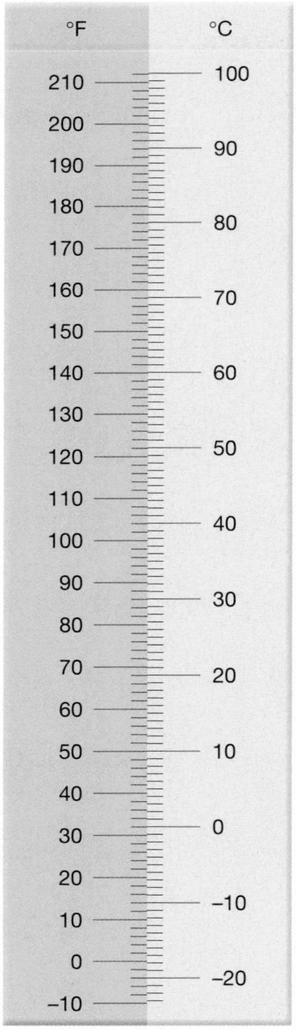

Figure A.1 A comparison of Fahrenheit and Celsius temperature scales.

APPENDIX

B Topographic Maps

A map is a representation on a flat surface of all or a part of Earth's surface drawn to a specific scale. Maps are often the most effective means for showing the locations of both natural and human structures, their sizes, and their relationships to one another. Like photographs, maps readily display information that would be impractical to express in words.

While most maps show only the two horizontal dimensions, geologists, as well as other map users, often require that the third dimension—elevation—be shown on maps. Maps that show the shape of the land are called **topographic maps.** Although various techniques may be used to depict elevations, the most accurate method involves the use of contour lines.

Contour Lines

A **contour line** is a line on a map representing a corresponding imaginary line on the ground that has the same elevation above sea level along its entire length. While many map symbols are pictographs, resembling the objects they represent, a contour line is an abstraction that has no counterpart in nature. It is, however, an accurate and effective device for representing the third dimension on paper.

Some useful facts and rules concerning contour lines are listed as follows. This information should be studied in conjunction with Figure B.1.

1. Contour lines bend upstream or upvalley. The contours form V's that point upstream, and in the upstream direction the successive contours represent higher elevations. For example, if you were standing on a stream bank and wished to get to the point at the same elevation directly opposite you on the other bank without stepping up or down, you would need to walk upstream along the contour at that elevation to where it crosses the stream bed, cross the stream, and then walk back downstream along the same contour.
2. Contours near the upper parts of hills form closures. The top of a hill is higher than the highest closed contour.
3. Hollows (depressions) without outlets are shown by closed, hatched contours. Hatched contours are contours with short lines on the inside pointing downslope.
4. Contours are widely spaced on gentle slopes.
5. Contours are closely spaced on steep slopes.
6. Evenly spaced contours indicate a uniform slope.
7. Contours usually do not cross or intersect each other, except in the rare case of an overhanging cliff.
8. All contours eventually close, either on a map or beyond its margins.
9. A single high contour never occurs between two lower ones, and vice versa. In other words, a change in slope direction is always determined by the repetition of the same elevation either as two different contours of the same value or as the same contour crossed twice.
10. Spot elevations between contours are given at many places, such as road intersections, hill summits, and lake surfaces. Spot elevations differ from control elevation stations, such as benchmarks, in not being permanently established by permanent markers.

Relief

Relief refers to the difference in elevation between any two points. Maximum relief refers to the difference in elevation between the highest and lowest points in the area being considered. Relief determines the **contour interval,** which is the difference in elevation between succeeding contour lines that is used on topographic maps. Where relief is low, a small contour interval, such as 10 or 20 feet, may be used. In flat areas, such as wide river valleys or broad, flat uplands, a contour interval of 5 feet is often used. In rugged mountainous terrain, where relief is many hundreds of feet, contour intervals as large as 50 or 100 feet are used.

Scale

Map **scale** expresses the relationship between distance or area on the map to the true distance or area on Earth's surface. This is generally expressed as a ratio or fraction, such as 1:24,000 or 1/24,000. The numerator, usually 1, represents map distance, and the denominator, a large number, represents ground distance. Thus, 1:24,000 means that a distance of 1 unit on the map represents a distance of 24,000 such units on the surface of Earth. It does not matter what the units are.

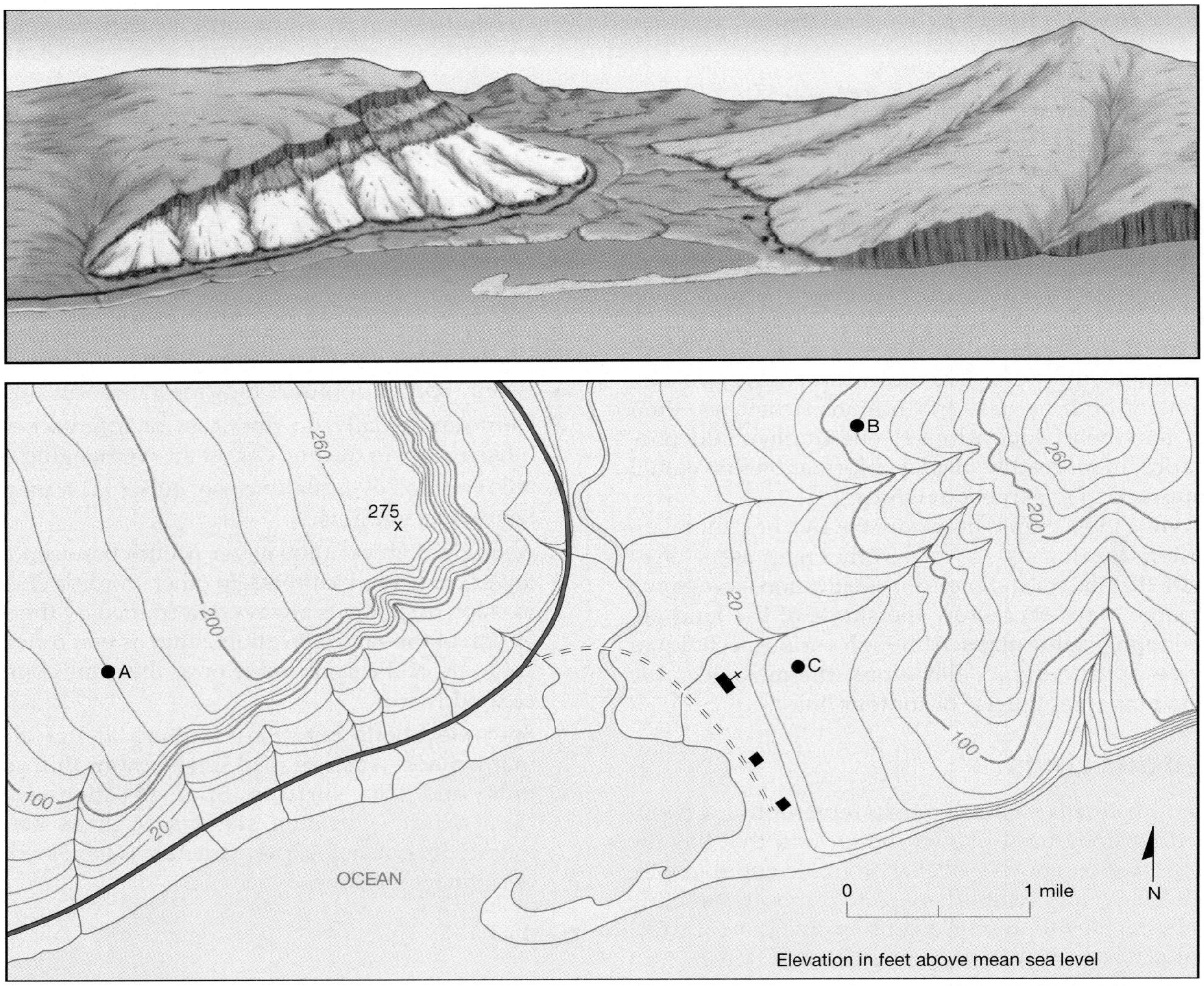

Figure B.1 Perspective view of an area and a contour map of the same area. These illustrations show how features are depicted on a topographic map. The upper illustration is a perspective view of a river valley and the adjoining hills. The river flows into a bay, which is partly enclosed by a hooked sandbar. On either side of the valley are terraces through which streams have cut gullies. The hill on the right has a smoothly eroded form and gradual slopes, whereas the one on the left rises abruptly in a sharp precipice, from which it slopes gently, and forms an inclined plateau traversed by a few shallow gullies. A road provides access to a church and the two houses situated across the river from a highway that follows the seacoast and curves up the river valley. The lower illustration shows the same features represented by symbols on a topographic map. The contour interval (vertical distance between adjacent contours) is 20 feet. *(After U.S. Geological Survey)*

Often, the graphic or bar scale is more useful than the fractional scale, because it is easier to use for measuring distances between points. The graphic scale (Figure B.2) consists of a bar divided into equal segments, which represent equal distances on the map. One segment on the left side of the bar is usually divided into smaller units to permit more accurate estimates of fractional units.

Topographic maps, which are also referred to as *quadrangles,* are generally classified according to publication scale. Each series is intended to fulfill a specific type of map need. To select a map with the proper scale for a particular use, remember that large-scale maps show more detail and small-scale maps show less detail. The sizes and scales of topographic maps published by the U.S. Geological Survey are shown in Table B.1.

Color and Symbol

Each color and symbol used on U.S. Geological Survey topographic maps has significance. Common topographic map symbols are shown in Figure B.3. The meaning of each color is as follows:

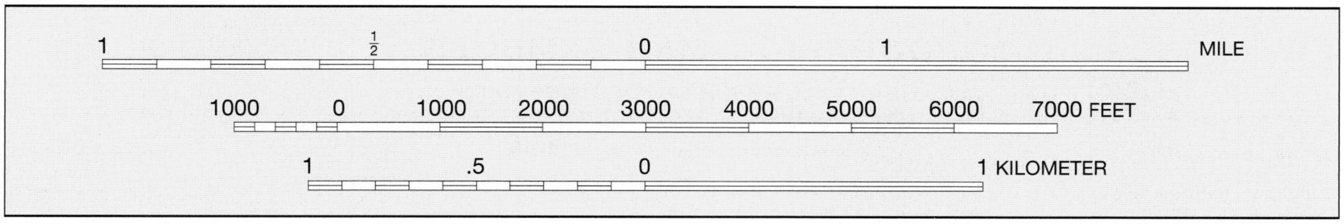

Figure B.2 Graphic scale.

Blue—water features
Black—works of humans, such as homes, schools, churches, roads, and so forth
Brown—contour lines
Green—woodlands, orchards, and so forth
Red—urban areas, important roads, public land subdivision lines

Table B.1 National topographic maps

Series	Scale	1 inch Represents	Standard Quadrangle Size (latitude-longitude)	Quadrangle Area (square miles)	Paper Size E-W N-S Width Length (inches)
$7\frac{1}{2}$-minute	1:24,000	2000 feet	$7\frac{1}{2}' \times 7\frac{1}{2}'$	49–70	22×27
Puerto Rico $7\frac{1}{2}$-minute	1:20,000	about 1667 feet	$7\frac{1}{2}' \times 7\frac{1}{2}'$	71	$29\frac{1}{2} \times 32\frac{1}{2}$
15-minute	1:62,500	nearly 1 mile	$15' \times 15'$	197–282	17×21
Alaska 1:63,360	1:63,360	1 mile	$15' \times 20' - 36'$	207–281	18×21
U.S. 1:250,000	1:250,000	nearly 4 miles	$1° \times 2°$	4580–8669	34×22
U.S. 1:1,000,000	1:1,000,000	nearly 16 miles	$4° \times 6°$	73,734–102,759	27×27

Source: U.S. Geological Survey

TOPOGRAPHIC MAP SYMBOLS
VARIATIONS WILL BE FOUND ON OLDER MAPS
Primary highway, hard surface
Secondary highway, hard surface
Light-duty road, hard or improved surface
Unimproved road
Road under construction, alinement known
Proposed road
Dual highway, dividing strip 25 feet or less
Dual highway, dividing strip exceeding 25 feet
Trail
Railroad: single track and multiple track
Railroads in juxtaposition
Narrow gage: single track and multiple track
Railroad in street and carline
Bridge: road and railroad
Drawbridge: road and railroad
Footbridge
Tunnel: road and railroad
Overpass and underpass
Small masonry or concrete dam
Dam with lock
Dam with road
Canal with lock
Buildings (dwelling, place of employment, etc.)
School, church, and cemetery
Cem
Buildings (barn, warehouse, etc.)
Power transmission line with located metal tower
Telephone line, pipeline, etc. (labeled as to type)
Wells other than water (labeled as to type)
Oil
Gas
Tanks: oil, water, etc. (labeled only if water)
Water
Located or landmark object; windmill
Open pit, mine, or quarry; prospect
Shaft and tunnel entrance
Horizontal and vertical control station:
Tablet, spirit level elevation
BM Δ 5653
Other recoverable mark, spirit level elevation
Δ 5455
Horizontal control station: tablet, vertical angle elevation
VABM Δ 9519
Any recoverable mark, vertical angle or checked elevation
Δ 3775
Vertical control station: tablet, spirit level elevation
BM × 957
Other recoverable mark, spirit level elevation
× 954
Spot elevation
× 7369
× 7369
Water elevation
670
670
Boundaries: National
State
County, parish, municipio
Civil township, precinct, town, barrio
Incorporated city, village, town, hamlet
Reservation, National or State
Small park, cemetery, airport, etc.
Land grant
Township or range line, United States land survey
Township or range line, approximate location
Section line, United States land survey
Section line, approximate location
Township line, not United States land survey
Section line, not United States land survey
Found corner: section and closing
Boundary monument: land grant and other
Fence or field line
Index contour
Intermediate contour
Supplementary contour
Depression contours
Fill
Cut
Levee
Levee with road
Mine dump
Wash
Tailings
Tailings pond
Shifting sand or dunes
Intricate surface
Sand area
Gravel beach
Perennial streams
Intermittent streams
Elevated aqueduct
Aqueduct tunnel
Water well and spring
Glacier
Small rapids
Small falls
Large rapids
Large falls
Intermittent lake
Dry lake bed
Foreshore flat
Rock or coral reef
Sounding, depth curve
10
Piling or dolphin
Exposed wreck
Sunken wreck
Rock, bare or awash; dangerous to navigation
Marsh (swamp)
Submerged marsh
Wooded marsh
Mangrove
Woods or brushwood
Orchard
Vineyard
Scrub
Land subject to controlled inundation
Urban area

Figure B.3 U.S. Geological Survey topographic map symbols. (Variations will be found on older maps.)

APPENDIX
C
Landforms of the Conterminous United States

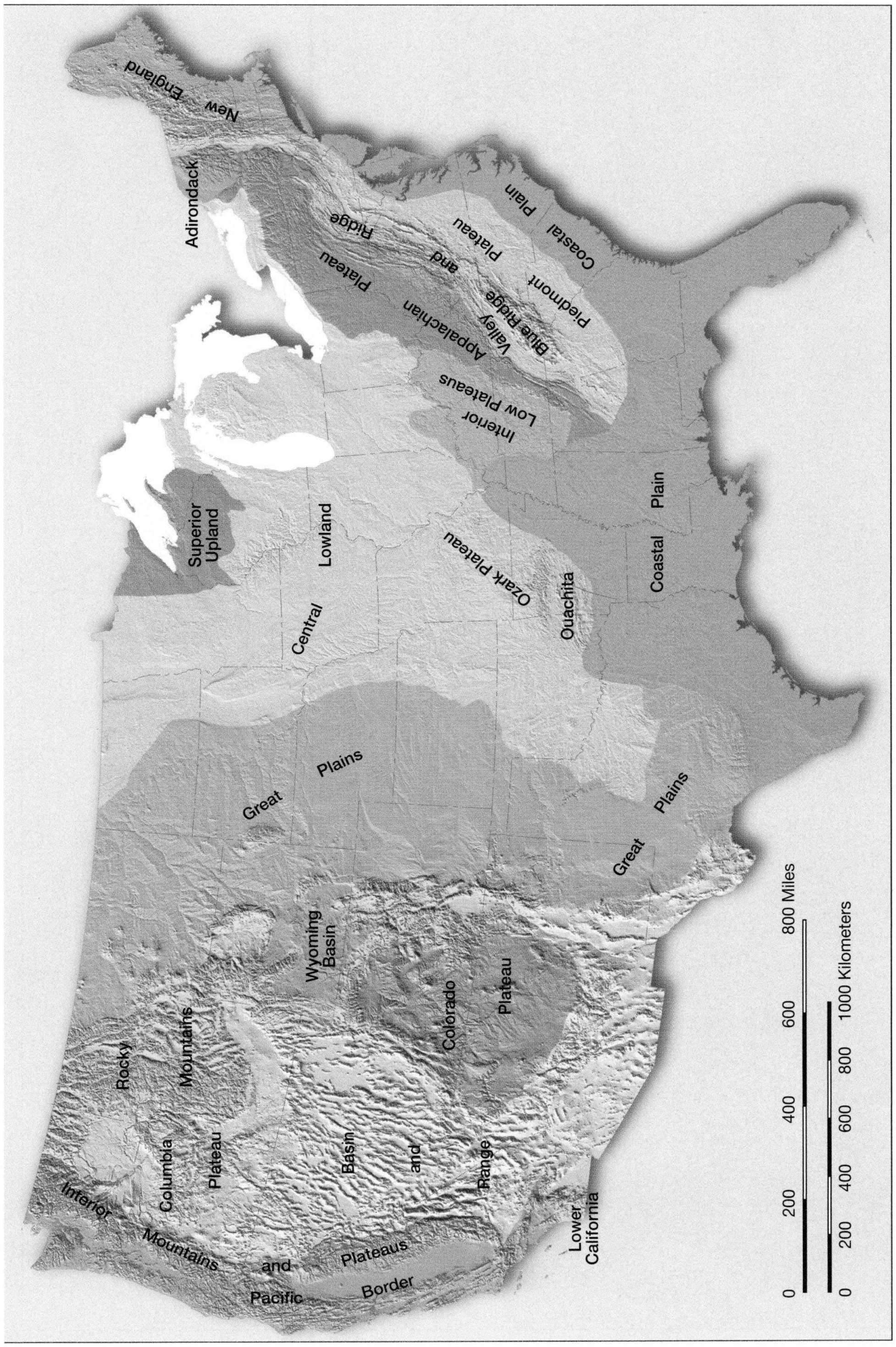

Figure C.1 Outline map showing major physiographic provinces of the United States.

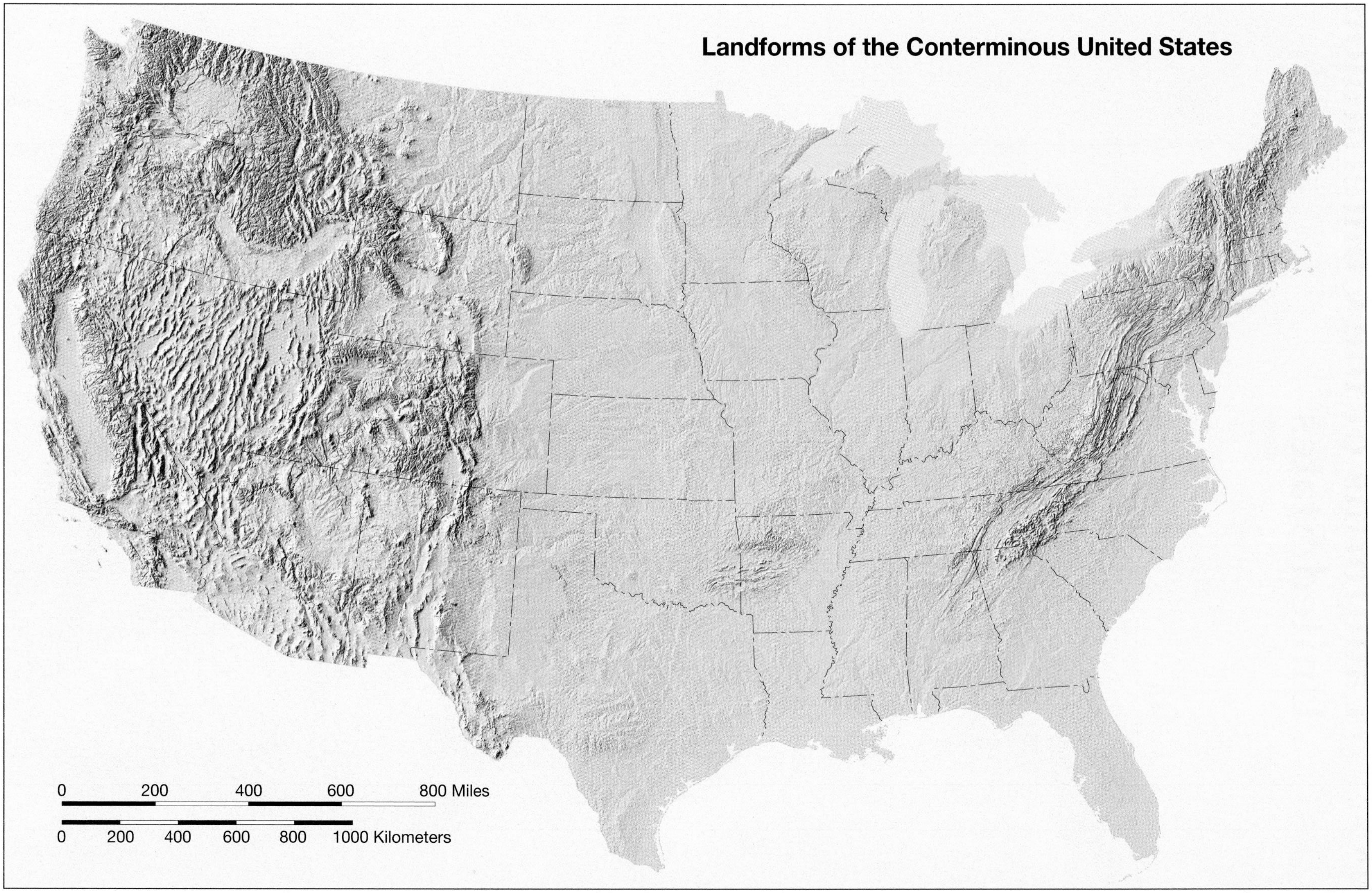

Figure C.2 Digital shaded relief landform map of the United States. *(Data provided by the U.S. Geological Survey)*

Glossary

Aa A type of lava flow that has a jagged, blocky surface.
Ablation A general term for the loss of ice and snow from a glacier.
Abrasion The grinding and scraping of a rock surface by the friction and impact of rock particles carried by water, wind, or ice.
Abyssal plain Very level area of the deep-ocean floor, usually lying at the foot of the continental rise.
Accretionary wedge A large wedge-shaped mass of sediment that accumulates in subduction zones. Here sediment is scraped from the subducting oceanic plate and accreted to the overriding crustal block.
Active continental margin Usually narrow and consisting of highly deformed sediments. They occur where oceanic lithosphere is being subducted beneath the margin of a continent.
Active layer The zone above the permafrost that thaws in summer and refreezes in winter.
Aftershock A smaller earthquake that follows the main earthquake.
Alluvial fan A fan-shaped deposit of sediment formed when a stream's slope is abruptly reduced.
Alluvium Unconsolidated sediment deposited by a stream.
Alpine glacier See *Valley glacier.*
Andesitic Igneous rocks having a mineral makeup between that of granite and basalt, after the common volcanic rock andesite. See also *Intermediate.*
Angle of repose The steepest angle at which loose material remains stationary without sliding downslope.
Angular unconformity An unconformity in which the older strata dip at an angle different from that of the younger beds.
Anthracite A hard, metamorphic form of coal that burns clean and hot.
Anticline A fold in sedimentary strata that resembles an arch.
Aphanitic texture A texture of igneous rocks in which the crystals are too small for individual minerals to be distinguished with the unaided eye.
Aquifer Rock or sediment through which groundwater moves easily.
Aquitard An impermeable bed that hinders or prevents groundwater movement.
Archean eon The second eon of Precambrian time. The eon following the Hadean eon and preceding the Proterozoic eon. It extends between about 3.8 and 2.5 billion years ago.
Arête A narrow, knifelike ridge separating two adjacent glaciated valleys.
Arkose A feldspar-rich sandstone.
Artesian well A well in which the water rises above the level where it was initially encountered.
Asthenosphere A subdivision of the mantle situated below the lithosphere. This zone of weak material exists below a depth of about 100 kilometers and in some regions extends as deep as 700 kilometers. The rock within this zone is easily deformed.
Atmosphere The gaseous portion of a planet; the planet's envelope of air. One of the traditional subdivisions of Earth's physical environment.
Atoll A continuous or broken ring of coral reef surrounding a central lagoon.
Atom The smallest particle that exists as an element.
Atomic mass unit A mass unit equal to exactly one-twelfth the mass of a carbon-12 atom.
Atomic number The number of protons in the nucleus of an atom.
Atomic weight The average of the atomic masses of isotopes of a given element.
Aureole A zone or halo of contact metamorphism found in the host rock surrounding an igneous intrusion.
Backshore The inner portion of the shore, lying landward of the high-tide shoreline. It is usually dry, being affected by waves only during storms.
Back swamp A poorly drained area on a floodplain, resulting when natural levees are present.
Bar Common term for sand and gravel deposits in a stream channel.
Barchan dune A solitary sand dune shaped like a crescent with its tips pointed downwind.
Barchanoid dunes Dunes forming scalloped rows of sand oriented at right angles to the wind. This form is intermediate between isolated barchans and extensive waves of transverse dunes.
Barrier island A low, elongated ridge of sand that parallels the coast.
Basal slip A mechanism of glacial movement in which the ice mass slides over the surface below.
Basalt An aphanitic igneous rock of mafic composition.
Basaltic Term used to describe igneous rocks that contain abundant dark (ferromagnesian) minerals and about 50 percent silica.
Base level The level below which a stream cannot erode.
Basin A circular downfolded structure.
Batholith A large mass of igneous rock that formed when magma was emplaced at depth, crystallized, and was subsequently exposed by erosion.
Baymouth bar A sandbar that completely crosses a bay, sealing it off from the main body of water.
Beach An accumulation of sediment found along the landward margin of the ocean or a lake.
Beach face The wet, sloping surface that extends from the berm to the shoreline.
Beach drift The transport of sediment in a zigzag pattern along a beach. It is caused by the uprush of water from obliquely breaking waves.
Beach nourishment Large quantities of sand are added to the beach system to offset losses caused by wave erosion. By building beaches seaward, beach quality and storm protection are both improved.
Bed See *Strata.*
Bedding plane A nearly flat surface separating two beds of sedimentary rock. Each bedding plane marks the end of one deposit and the beginning of another having different characteristics.
Bed load Sediment rolled along the bottom of a stream by moving water, or particles rubbed along the ground surface by wind.
Belt of soil moisture A zone in which water is held as a film on the surface of soil particles and may be used by plants or withdrawn by evaporation. The uppermost subdivision of the zone of aeration.
Berm The dry, gently sloping zone on the backshore of a beach at the foot of the coastal cliffs or dunes.
Biochemical Describing a type of chemical sediment that forms when material dissolved in water is precipitated by water-dwelling organisms. Shells are common examples.
Biogenous sediment Seafloor sediments consisting of material of marine-organic origin.
Bituminous coal The most common form of coal, often called soft, black coal.
Block lava Lava having a surface of angular blocks associated with material having andesitic and rhyolitic compositions.
Blowout (deflation hollow) A depression excavated by wind in easily eroded materials.
Body wave A seismic wave that travels through Earth's interior.
Bottomset bed A layer of fine sediment deposited beyond the advancing edge of a delta and then buried by continuous delta growth.

Bowen's reaction series A concept proposed by N. L. Bowen that illustrates the relationship between magma and the minerals crystallizing from it during the formation of igneous rock.
Braided stream A stream consisting of numerous intertwining channels.
Breakwater A structure protecting a nearshore area from breaking waves.
Breccia A sedimentary rock composed of angular fragments that were lithified.
Brittle deformation Deformation that involves the fracturing of rock. Associated with rocks near the surface.
Burial metamorphism Low-grade metamorphism that occurs in the lowest layers of very thick accumulations of sedimentary strata.
Caldera A large depression typically caused by collapse of the summit area of a volcano following a violent eruption.
Caliche A hard layer, rich in calcium carbonate, that forms beneath the *B* horizon in soils of arid regions.
Calving Wastage of a glacier that occurs when large pieces of ice break off into water.
Capacity The total amount of sediment a stream is able to transport.
Capillary fringe A relatively narrow zone at the base of the zone of aeration. Here water rises from the water table in tiny threadlike openings between grains of soil or sediment.
Cap rock A necessary part of an oil trap. The cap rock is impermeable and hence keeps upwardly mobile oil and gas from escaping at the surface.
Catastrophism The concept that Earth was shaped by catastrophic events of a short-term nature.
Cavern A naturally formed underground chamber or series of chambers most commonly produced by solution activity in limestone.
Cementation One way in which sedimentary rocks are lithified. As material precipitates from water that percolates through the sediment, open spaces are filled and particles are joined into a solid mass.
Cenozoic era A time span on the geologic time scale beginning about 65 million years ago following the Mesozoic era.
Chemical sedimentary rock Sedimentary rock consisting of material that was precipitated from water by either inorganic or organic means.
Chemical weathering The processes by which the internal structure of a mineral is altered by the removal and/or addition of elements.
Cinder cone A rather small volcano built primarily of pyroclastics ejected from a single vent.
Cirque An amphitheater-shaped basin at the head of a glaciated valley produced by frost wedging and plucking.
Clastic A sedimentary rock texture consisting of broken fragments of preexisting rock.
Cleavage The tendency of a mineral to break along planes of weak bonding.
Coast A strip of land that extends inland from the coastline as far as ocean-related features can be found.
Coastline The coast's seaward edge; the landward limit of the effect of the highest storm waves on the shore.
Col A pass between mountain valleys where the headwalls of two cirques intersect.
Color A phenomenon of light by which otherwise identical objects may be differentiated.
Column A feature found in caves that is formed when a stalactite and stalagmite join.
Columnar joints A pattern of cracks that forms during cooling of molten rock to generate columns.
Compaction A type of lithification in which the weight of overlying material compresses more deeply buried sediment. It is most important in fine-grained sedimentary rocks such as shale.
Competence A measure of the largest particle a stream can transport; a factor dependent on velocity.
Composite cone A volcano composed of both lava flows and proclastic material.
Compound A substance formed by the chemical combination of two or more elements in definite proportions and usually having properties different from those of its constituent elements.
Concordant A term used to describe intrusive igneous masses that form parallel to the bedding of the surrounding rock.
Conduit A pipelike opening through which magma moves toward Earth's surface. It terminates at a surface opening called a vent.
Cone of depression A cone-shaped depression immediately surrounding a well.
Conformable layers Rock layers that were deposited without interruption.
Conglomerate A sedimentary rock consisting of rounded, gravel-size particles.
Contact metamorphism Changes in rock caused by the heat of a nearby magma body.
Continental drift A hypothesis, credited largely to Alfred Wegener, that suggested all present continents once existed as a single supercontinent. Further, beginning about 200 million years ago, the supercontinent began breaking into smaller continents, which then drifted to their present positions.
Continental margin See *Active continental margin* and *Passive continental margin.*
Continental rise The gently sloping surface at the base of the continental slope.
Continental shelf The gently sloping submerged portion of the continental margin, extending from the shoreline to the continental slope.
Continental slope The steep gradient that leads to the deep-ocean floor and marks the seaward edge of the continental shelf.
Continental volcanic arc Mountains formed in part by igneous activity associated with the subduction of oceanic lithosphere beneath a continent. Examples include the Andes and the Cascades.
Convergent plate boundary A boundary in which two plates move together, resulting in oceanic lithosphere being thrust beneath an overriding plate, eventually to be reabsorbed into the mantle. It can also involve the collision of two continental plates to create a mountain system.
Core Located beneath the mantle, it is Earth's innermost layer. The core is divided into an outer core and an inner core.
Correlation Establishing the equivalence of rocks of similar age in different areas.
Covalent bond A chemical bond produced by the sharing of electrons.
Crater The depression at the summit of a volcano, or that which is produced by a meteorite impact.
Creep The slow downhill movement of soil and regolith.
Crevasse A deep crack in the brittle surface of a glacier.
Cross-bedding Structure in which relatively thin layers are inclined at an angle to the main bedding. Formed by currents of wind or water.
Cross-cutting A principle of relative dating. A rock or fault is younger than any rock (or fault) through which it cuts.
Crust The very thin, outermost layer of Earth.
Crystal An orderly arrangement of atoms.
Crystal form The external appearance of a mineral as determined by its internal arrangement of atoms.
Crystalline texture See *Nonclastic texture.*
Crystallization The formation and growth of a crystalline solid from a liquid or gas.
Curie point The temperature above which a material loses its magnetization.
Cut bank The area of active erosion on the outside of a meander.
Cutoff A short channel segment created when a river erodes through the narrow neck of land between meanders.
Dark silicate Silicate minerals containing ions of iron and/or magnesium in their structure. They are dark in color and have a higher specific gravity than nonferromagnesian silicates.
Daughter product An isotope resulting from radioactive decay.
Debris flow A relatively rapid type of mass wasting that involves a flow of soil and regolith containing a large amount of water. Also called *mudflows.*
Decompression melting Melting that occurs as rock ascends due to a drop in confining pressure.
Deep-focus earthquake An earthquake focus at a depth of more than 300 kilometers.
Deep-ocean basin The portion of seafloor that lies between the continental margin and the oceanic ridge system. This region comprises almost 30 percent of Earth's surface.

Deep-ocean trench See *Trench.*
Deflation The lifting and removal of loose material by wind.
Deformation General term for the processes of folding, faulting, shearing, compression, or extension of rocks as the result of various natural forces.
Delta An accumulation of sediment formed where a stream enters a lake or ocean.
Dendritic pattern A stream system that resembles the pattern of a branching tree.
Density The weight per unit volume of a particular material.
Desert One of the two types of dry climate; the drier of the dry climates.
Desert pavement A layer of coarse pebbles and gravel created when wind removes the finer material.
Detachment fault A low-angle fault that represents a major boundary between unfaulted rocks below that exhibit ductile deformation and rocks above that exhibit brittle deformation via faulting.
Detrital sedimentary rocks Rocks that form from the accumulation of materials that originate and are transported as solid particles derived from both mechanical and chemical weathering.
Diagenesis A collective term for all the chemical, physical, and biological changes that take place after sediments are deposited and during and after lithification.
Dike A tabular-shaped intrusive igneous feature that cuts through the surrounding rock.
Dip The angle at which a rock layer is inclined from the horizontal. The direction of dip is at a right angle to the strike.
Dip-slip fault A fault in which the movement is parallel to the dip of the fault.
Discharge The quantity of water in a stream that passes a given point in a given period of time.
Disconformity A type of unconformity in which the beds above and below are parallel.
Discontinuity A sudden change with depth in one or more of the physical properties of the material making up Earth's interior. The boundary between two dissimilar materials in Earth's interior as determined by the behavior of seismic waves.
Discordant A term used to describe plutons that cut across existing rock structures, such as bedding planes.
Disseminated deposit Any economic mineral deposit in which the desired mineral occurs as scattered particles in the rock but in sufficient quantity to make the deposit an ore.
Dissolved load The portion of a stream's load carried in solution.
Distributary A section of a stream that leaves the main flow.
Diurnal tide A tide characterized by a single high and low water height each tidal day.
Divergent plate boundary A boundary in which two plates move apart, resulting in upwelling of material from the mantle to create new seafloor.
Divide An imaginary line that separates the drainage of two streams; often found along a ridge.
Dome A roughly circular, upfolded structure.
Drainage basin The land area that contributes water to a stream.
Drawdown The difference in height between the bottom of a cone of depression and the original height of the water table.
Drift See *Glacial drift.*
Drumlin A streamlined asymmetrical hill composed of glacial till. The steep side of the hill faces the direction from which the ice advanced.
Dry climate A climate in which the yearly precipitation is less than the potential loss of water by evaporation.
Ductile deformation A type of solid-state flow that produces a change in the size and shape of a rock body without fracturing. Occurs at depths where temperatures and confining pressures are high.
Dune A hill or ridge of wind-deposited sand.
Earthflow The downslope movement of water-saturated, clay-rich sediment. Most characteristic of humid regions.
Earthquake Vibration of Earth produced by the rapid release of energy.
Echo sounder An instrument used to determine the depth of water by measuring the time interval between emission of a sound signal and the return of its echo from the bottom.
Elastic deformation Nonpermanent deformation in which rock returns to its original shape when the stress is released.
Elastic rebound The sudden release of stored strain in rocks that results in movement along a fault.
Electron A negatively charged subatomic particle that has a negligible mass and is found outside the atom's nucleus.
Element A substance that cannot be decomposed into simpler substances by ordinary chemical or physical means.
Eluviation The washing out of fine soil components from the *A* horizon by downward-percolating water.
Emergent coast A coast where land formerly below the sea level has been exposed either by crustal uplift or a drop in sea level or both.
End moraine A ridge of till marking a former position of the front of the glacier.
Energy levels Spherically shaped negatively charged zones that surround the nucleus of an atom.
Eon The largest time unit on the geologic time scale, next in order of magnitude above *era.*
Epicenter The location on Earth's surface that lies directly above the focus of an earthquake.
Epoch A unit of the geological time scale that is a subdivision of a *period.*
Era A major division on the geologic time scale; eras are divided into shorter units called *periods.*
Erosion The incorporation and transportation of material by a mobile agent, such as water, wind, or ice.
Eruption column Buoyant plumes of hot ash-laden gases that can extend thousands of meters into the atmosphere.
Esker Sinuous ridge composed largely of sand gravel deposited by a stream flowing in a tunnel beneath a glacier near its terminus.
Estuary A partially enclosed coastal water body that is connected to the ocean. Salinity here is measurably reduced by the freshwater flow of rivers.
Evaporite A sedimentary rock formed of material deposited from solution by evaporation of water.
Evapotranspiration The combined effect of evaporation and transpiration.
Exfoliation dome Large, dome-shaped structure, usually composed of granite, formed by sheeting.
Exotic stream A permanent stream that traverses a desert and has its source in well-watered areas outside the desert.
External process Process such as weathering, mass wasting, or erosion that is powered by the Sun and transforms solid rock into sediment.
Extrusive Igneous activity that occurs at Earth's surface.
Fall A type of movement common to mass-wasting processes that refers to the free falling of detached individual pieces of any size.
Fault A break in a rock mass along which movement has occurred.
Fault-block mountain A mountain formed by the displacement of rock along a fault.
Fault creep Slow, gradual displacement along a fault that occurs relatively smoothly and with little noticeable seismic activity.
Fault scarp A cliff created by movement along a fault. It represents the exposed surface of the fault prior to modification by weathering and erosion.
Felsic A term derived from *fel*dspar and *si*lica (quartz). It is a term used to describe granitic igneous rocks.
Ferromagnesian silicate See *Dark silicate.*
Fetch The distance that the wind has traveled across the open water.
Fiord A steep-sided inlet of the sea formed when a glacial trough was partially submerged.
Fissure A crack in rock along which there is a distinct separation.
Fissure eruption An eruption in which lava is extruded from narrow fractures or cracks in the crust.
Flood The overflow of a stream channel that occurs when discharge exceeds the channel's

capacity. The most common and destructive geologic hazard.
Flood basalts Flows of basaltic lava that issue from numerous cracks or fissures and commonly cover extensive areas to thicknesses of hundreds of meters.
Floodplain The flat, low-lying portion of a stream valley subject to periodic inundation.
Flow A type of movement common to mass-wasting processes in which water-saturated material moves downslope as a viscous fluid.
Fluorescence The absorption of ultraviolet light, which is reemitted as visible light.
Focus (earthquake) The zone within Earth where rock displacement produces an earthquake.
Fold A bent layer or series of layers that were originally horizontal and subsequently deformed.
Foliated A texture of metamorphic rocks that gives the rock a layered appearance.
Foliation A term for a linear arrangement of textural features often exhibited by metamorphic rocks.
Foreset bed An inclined bed deposited along the front of a delta.
Foreshocks Small earthquakes that often precede a major earthquake.
Foreshore That portion of the shore lying between the normal high- and low-water marks; the intertidal zone.
Fossil The remains or traces of organisms preserved from the geologic past.
Fossil fuel General term for any hydrocarbon that may be used as a fuel, including coal, oil, natural gas, bitumen from tar sands, and shale oil.
Fossil succession Fossil organisms succeed one another in a definite and determinable order, and any time period can be recognized by its fossil content.
Fossil magnetism See *Paleomagnetism.*
Fractional crystallization The process that separates magma into components having varied compositions and melting points.
Fracture (mineral) One of the basic physical properties of minerals. It relates to the breakage of minerals when there are no planes of weakness in the crystalline structure. Examples include conchoidal, irregular, and splintery.
Fracture (rock) Any break or rupture in rock along which no appreciable movement has taken place.
Fracture zone Linear zone of irregular topography on the deep-ocean floor that follows transform faults and their inactive extensions.
Fragmental texture See *Pyroclastic texture.*
Frost wedging The mechanical breakup of rock caused by the expansion of freezing water in cracks and crevices.
Fumarole A vent in a volcanic area from which fumes or gases escape.
Gaining stream Streams that gain water from the inflow of groundwater through the streambed.
Geology The science that examines Earth, its form and composition, and the changes it has undergone and is undergoing.
Geosphere The solid Earth; one of Earth's four basic spheres.
Geothermal energy Natural steam used for power generation.
Geothermal gradient The gradual increase in temperature with depth in the crust. The average is 30°C per kilometer in the upper crust.
Geyser A fountain of hot water ejected periodically from the ground.
Glacial budget The balance, or lack of balance, between accumulation at the upper end of a glacier, and loss at the other end.
Glacial drift An all-embracing term for sediments of glacial origin, no matter how, where, or in what shape they were deposited.
Glacial erratic An ice-transported boulder that was not derived from the bedrock near its present site.
Glacial striations Scratches and grooves on bedrock caused by glacial abrasion.
Glacial trough A mountain valley that has been widened, deepened, and straightened by a glacier.
Glacier A thick mass of ice originating on land from the compaction and recrystallization of snow. The ice shows evidence of past or present flow.
Glass (volcanic) Natural glass produced when molten lava cools too rapidly to permit crystallization. Volcanic glass is a solid composed of unordered atoms.
Glassy texture A term used to describe the texture of certain igneous rocks, such as obsidian, that contain no crystals.
Gneissic texture The texture displayed by the metamorphic rock *gneiss* in which dark and light silicate minerals have separated, giving the rock a banded appearance.
Gondwanaland The southern portion of Pangaea consisting of South America, Africa, Australia, India, and Antarctica.
Graben A valley formed by the downward displacement of a fault-bounded block.
Graded bed A sediment layer characterized by a decrease in sediment size from bottom to top.
Graded stream A stream that has the correct channel characteristics to maintain the exact velocity required to transport the material supplied to it.
Gradient The slope of a stream; generally measured in feet per mile.
Granitic Igneous rocks composed mainly of light-colored silicates (quartz and feldspar) are said to have this composition.
Gravitational collapse The gradual subsidence of mountains caused by lateral spreading of weak material located deep within these structures.
Greenhouse effect Carbon dioxide and water vapor in a planet's atmosphere absorb and reradiate infrared wavelengths, effectively trapping solar energy and raising the temperature.
Groin A short wall built at a right angle to the seashore to trap moving sand.
Groundmass The matrix of smaller crystals within an igneous rock that has porphyritic texture.
Ground moraine An undulating layer of till deposited as the ice front retreats.
Groundwater Water in the zone of saturation.
Guyot A submerged flat-topped seamount.
Hadean eon The first eon on the geologic time scale. The eon ending 3.8 billion years ago that preceded the Archean eon.
Half-life The time required for one half of the atoms of a radioactive substance to decay.
Hanging valley A tributary valley that enters a glacial trough at a considerable height above the floor of the trough.
Hardness A mineral's resistance to scratching and abrasion.
Head The vertical distance between the recharge and discharge points of a water table. Also, the source area or beginning of a valley.
Headward erosion The extension upslope of the head of a valley due to erosion.
Historical geology A major division of geology that deals with the origin of Earth and its development through time. Usually involves the study of fossils and their sequence in rock beds.
Hogback A narrow, sharp-crested ridge formed by the upturned edge of a steeply dipping bed of resistant rock.
Horn A pyramid-like peak formed by glacial action in three or more cirques surrounding a mountain summit.
Horst An elongated, uplifted block of crust bounded by faults.
Hot spot A proposed concentration of heat in the mantle capable of introducing magma that in turn extrudes onto Earth's surface. The intraplate volcanism that produced the Hawaiian Islands is one example.
Hot spring A spring in which the water is 6° to 9°C (10°C to 15°F) warmer than the mean annual air temperature of its locality.
Humus Organic matter in soil produced by the decomposition of plants and animals.
Hydrogenous sediment Seafloor sediments consisting of minerals that crystallize from seawater. The principal example is manganese nodules.
Hydrologic cycle The unending circulation of Earth's water supply. The cycle is powered by energy from the Sun and is characterized by continuous exchanges of water among the oceans, the atmosphere, and the continents.
Hydrolysis A chemical-weathering process in which minerals are altered by chemically reacting with water and acids.
Hydrosphere The water portion of our planet; one of the traditional subdivisions of Earth's physical environment.

Hydrothermal metamorphism Chemical alterations that occur as hot, ion-rich water circulates through fractures in rock.

Hydrothermal solution The hot, watery solution that escapes from a mass of magma during the latter stages of crystallization. Such solutions may alter the surrounding country rock and are frequently the source of significant ore deposits.

Hypothesis A tentative explanation that is then tested to determine if it is valid.

Ice cap A mass of glacial ice covering a high upland or plateau and spreading out radially.

Ice-contact deposit An accumulation of stratified drift deposited in contact with a supporting mass of ice.

Ice sheet A very large, thick mass of glacial ice flowing outward in all directions from one or more accumulation centers.

Igneous rock A rock formed by the crystallization of molten magma.

Immature soil A soil lacking horizons.

Impact metamorphism Metamorphism that occurs when meteorites strike Earth's surface.

Incised meander Meandering channel that flows in a steep, narrow valley. These meanders form either when an area is uplifted or when base level drops.

Inclusion A piece of one rock unit contained within another. Inclusions are used in relative dating. The rock mass adjacent to the one containing the inclusion must have been there first in order to provide the fragment.

Index fossil A fossil that is associated with a particular span of geologic time.

Index mineral A mineral that is a good indicator of the metamorphic environment in which it formed. Used to distinguish different zones of regional metamorphism.

Inertia Objects at rest tend to remain at rest, and objects in motion tend to stay in motion unless either is acted upon by an outside force.

Infiltration The movement of surface water into rock or soil through crack and pore spaces.

Infiltration capacity The maximum rate at which soil can absorb water.

Inner core The solid, innermost layer of Earth, about 1216 kilometers (754 miles) in radius.

Inselberg An isolated mountain remnant characteristic of the late stage of erosion in a mountainous region.

Intensity (earthquake) A measure of the degree of earthquake shaking at a given locale based on the amount of damage.

Interface A common boundary where different parts of a system interact.

Interior drainage A discontinuous pattern of intermittent streams that do not flow to the ocean.

Intermediate Compositional category for igneous rocks found near the middle of Bowen's reaction series, mainly amphibole and the intermediate plagioclase feldspars.

Intermediate focus An earthquake focus at a depth of between 60 and 300 kilometers.

Internal process A process such as mountain building or volcanism that derives its energy from Earth's interior and elevates Earth's surface.

Intraplate volcanism Igneous activity that occurs within a tectonic plate away from plate boundaries.

Intrusive rock Igneous rock that formed below Earth's surface.

Ion An atom or molecule that possesses an electrical charge.

Ionic bond A chemical bond between two oppositely charged ions formed by the transfer of valence electrons from one atom to another.

Island arc See *Volcanic island arc.*

Isostasy The concept that Earth's crust is floating in gravitational balance upon the material of the mantle.

Isotopes Varieties of the same element that have different mass numbers; their nuclei contain the same number of protons but different numbers of neutrons.

Joint A fracture in rock along which there has been no movement.

Kame A steep-sided hill composed of sand and gravel originating when sediment collected in openings in stagnant glacial ice.

Kame terrace A narrow, terracelike mass of stratified drift deposited between a glacier and an adjacent valley wall.

Karst A topography consisting of numerous depressions called *sinkholes.*

Kettle holes Depressions created when blocks of ice become lodged in glacial deposits and subsequently melt.

Laccolith A massive, concordant igneous body intruded between preexisting strata.

Lahar Mudflows on the slopes of volcanoes that result when unstable layers of ash and debris become saturated and flow downslope, usually following stream channels.

Laminar flow The movement of water particles in straight-line paths that are parallel to the channel. The water particles move downstream without mixing.

Lateral moraine A ridge of till along the sides of a valley glacier composed primarily of debris that fell to the glacier from the valley walls.

Laterite A red, highly leached soil type found in the tropics and rich in oxides of iron and aluminum.

Laurasia The northern portion of Pangaea consisting of North America and Eurasia.

Lava Magma that reaches Earth's surface.

Lava dome A bulbous mass associated with an old-age volcano, produced when thick lava is slowly squeezed from the vent. Lava domes may act as plugs to deflect subsequent gaseous eruptions.

Lava tube Tunnel in hardened lava that acts as a horizontal conduit for lava flowing from a volcanic vent. Lava tubes allow fluid lavas to advance great distances.

Law of superposition In any undeformed sequence of sedimentary rocks or surface-deposited igneous materials, each layer is older than the one above it and younger than the one below.

Leaching The depletion of soluble materials from the upper soil by downward-percolating water.

Light silicate Silicate minerals that lack iron and/or magnesium. They are generally lighter in color and have lower specific gravities than dark silicates.

Lithification The process, generally by cementation and/or compaction, of converting sediments to solid rock.

Lithosphere The rigid outer layer of Earth, including the crust and upper mantle.

Loess Deposits of windblown silt, lacking visible layers, generally buff colored, and capable of maintaining a nearly vertical cliff.

Longitudinal dunes Long ridges of sand oriented parallel to the prevailing wind; these dunes form where sand supplies are limited.

Longitudinal profile A cross section of a stream channel along its descending course from the head to the mouth.

Longshore current A nearshore current that flows parallel to the shore.

Losing stream Streams that lose water to the groundwater system by outflow through the streambed.

Lower mantle See *Mesosphere.*

Luster The appearance or quality of light reflected from the surface of a mineral.

Mafic Because basaltic rocks contain a high percentage of ferromagnesian minerals, they are also called mafic (from *ma*gnesium and *fe*rrum, the Latin name for iron).

Magma A body of molten rock found at depth, including any dissolved gases and crystals.

Magnetic reversal A change in Earth's magnetic field from normal to reverse or vice versa.

Magnetic time scale The detailed history of Earth's magnetic reversals developed by establishing the magnetic polarity of lava flows of known age.

Magnetometer A sensitive instrument used to measure the intensity of Earth's magnetic field at various points.

Magnitude (earthquake) The total amount of energy released during an earthquake.

Manganese nodules A type of hydrogenous sediment scattered on the ocean floor, consisting mainly of manganese and iron, and usually containing small amounts of copper, nickel, and cobalt.

Mantle The 2885-kilometer (1789-mile) thick layer of Earth located below the crust.

Mantle plume A mass of hotter-than-normal mantle material that ascends toward the surface, where it may lead to igneous activity. These plumes of solid yet mobile material may originate as deep as the core-mantle boundary.

Mass number The sum of the number of neutrons and protons in the nucleus of an atom.
Mass wasting The downslope movement of rock, regolith, and soil under the direct influence of gravity.
Meander A looplike bend in the course of a stream.
Meander scar A floodplain feature created when an oxbow lake becomes filled with sediment.
Mechanical weathering The physical disintegration of rock, resulting in smaller fragments.
Medial moraine A ridge of till formed when lateral moraines from two coalescing valley glaciers join.
Melt The liquid portion of magma excluding the solid crystals.
Mesosphere The part of the mantle that extends from the core-mantle boundary to a depth of 660 kilometers. Also known as the lower mantle.
Mesozoic era A time span on the geologic time scale between the Paleozoic and Cenozoic eras—from about 248 million to 65 million years ago.
Metallic bond A chemical bond present in all metals that may be characterized as an extreme type of electron sharing in which the electrons move freely from atom to atom.
Metamorphic rock Rock formed by the alteration of preexisting rock deep within Earth (but still in the solid state) by heat, pressure, and/or chemically active fluids.
Metamorphism The changes in mineral composition and texture of a rock subjected to high temperature and pressure within Earth.
Metasomatism A significant change in the chemical composition of a rock, usually by the addition or removal of ions in solution.
Mid-ocean ridge See *Oceanic ridge.*
Migmatite A rock exhibiting both igneous and metamorphic rock characteristics. Such rocks may form when light-colored silicate minerals melt and then crystallize, while the dark silicate minerals remain solid.
Mineral A naturally occurring, inorganic crystalline material with a unique chemical structure.
Mineral resource All discovered and undiscovered deposits of a useful mineral that can be extracted now or at some time in the future.
Modified Mercalli intensity scale A 12-point scale developed to evaluate earthquake intensity based on the amount of damage to various structures.
Mohorovičić discontinuity (Moho) The boundary separating the crust and the mantle, discernible by an increase in seismic velocity.
Mohs scale A series of 10 minerals used as a standard in determining hardness.
Moment magnitude A more precise measure of earthquake magnitude than the Richter scale that is derived from the amount of displacement that occurs along a fault zone.
Monocline A one-limbed flexure in strata. The strata are usually flat lying or very gently dipping on both sides of the monocline.
Mouth The point downstream where a river empties into another stream or water body.
Mud crack A feature in some sedimentary rocks that forms when wet mud dries out, shrinks, and cracks.
Mudflow See *Debris flow.*
Natural levees The elevated landforms composed of alluvium that parallel some streams and act to confine their waters, except during floodstage.
Neap tide The lowest tidal range, occurring near the times of the first and third quarters of the Moon.
Nearshore zone The zone of a beach that extends from the low-tide shoreline seaward to where waves break at low tide.
Nebular hypothesis A model for the origin of the solar system that assumes a rotating nebula of dust and gases that contracted to produce the Sun and planets.
Neutron A subatomic particle found in the nucleus of an atom. The neutron is electrically neutral, with a mass approximately equal to that of a proton.
Nonclastic A term for the texture of sedimentary rocks in which the minerals form a pattern of interlocking crystals.
Nonconformity An unconformity in which older metamorphic or intrusive igneous rocks are overlain by younger sedimentary strata.
Nonferromagnesian silicate See *Light silicate.*
Nonfoliated texture Metamorphic rocks that do not exhibit foliation.
Nonmetallic mineral resource Mineral resource that is not a fuel or processed for the metals it contains.
Nonrenewable resource Resource that forms or accumulates over such long time spans that it must be considered as fixed in total quantity.
Normal fault A fault in which the rock above the fault plane has moved down relative to the rock below.
Normal polarity A magnetic field the same as that which presently exists.
Nucleus The small, heavy core of an atom that contains all of its positive charge and most of its mass.
Nuée ardente Incandescent volcanic debris that is buoyed up by hot gases and moves downslope in an avalanche fashion.
Numerical date Date that specifies the actual number of years that have passed since an event occurred.
Oceanic plateau An extensive region on the ocean floor composed of thick accumulations of pillow basalts and other mafic rocks that in some cases exceed 30 kilometers in thickness.
Oceanic ridge A continuous mountainous ridge on the floor of all the major ocean basins and varying in width from 500 to 5000 kilometers (300 to 3000 miles). The rifts at the crests of these ridges represent divergent plate boundaries.
Octet rule Atoms combine in order that each may have the electron arrangement of a noble gas; that is, the outer energy level contains eight electrons.
Offshore zone The relatively flat submerged zone that extends from the breaker line to the edge of the continental shelf.
Oil trap A geologic structure that allows for significant amounts of oil and gas to accumulate.
Ore Usually a useful metallic mineral that can be mined at a profit. The term is also applied to certain nonmetallic minerals such as fluorite and sulfur.
Original horizontality Layers of sediment are generally deposited in a horizontal or nearly horizontal position.
Orogenesis The processes that collectively result in the formation of mountains.
Outer core A layer beneath the mantle about 2270 kilometers (1410 miles) thick that has the properties of a liquid.
Outgassing The release of gases dissolved in molten rock.
Outlet glacier A tongue of ice normally flowing rapidly outward from an ice cap or ice sheet, usually through mountainous terrain to the sea.
Outwash plain A relatively flat, gently sloping plain consisting of materials deposited by meltwater streams in front of the margin of an ice sheet.
Oxbow lake A curved lake produced when a stream cuts off a meander.
Oxidation The removal of one or more electrons from an atom or ion. So named because elements commonly combine with oxygen.
Pahoehoe A lava flow with a smooth-to-ropy surface.
Paleomagnetism The natural remnant magnetism in rock bodies. The permanent magnetization acquired by rock that can be used to determine the location of the magnetic poles and the latitude of the rock at the time it became magnetized.
Paleontology The systematic study of fossils and the history of life on Earth.
Paleozoic era A time span on the geologic time scale between the Precambrian and Mesozoic eras—from about 540 million to 248 million years ago.
Pangaea The proposed supercontinent that 200 million years ago began to break apart and form the present landmasses.
Parabolic dune A sand dune similar in shape to a barchan dune except that its tips point into the wind. These dunes often form along coasts that have strong onshore winds, abundant sand, and vegetation that partly covers the sand.
Paradigm Theory that is held with a very high degree of confidence and is comprehensive in scope.

Parasitic cone A volcanic cone that forms on the flank of a larger volcano.
Parent material The material upon which a soil develops.
Parent rock The rock from which a metamorphic rock formed.
Partial melting The process by which most igneous rocks melt. Because individual minerals have different melting points, most igneous rocks melt over a temperature range of a few hundred degrees. If the liquid is squeezed out after some melting has occurred, a melt with a higher silica content results.
Passive continental margin A margin that consists of a continental shelf, continental slope, and continental rise. They are *not* associated with plate boundaries and therefore experience little volcanism and few earthquakes.
Pater noster lakes A chain of small lakes in a glacial trough that occupy basins created by glacial erosion.
Pegmatite A very coarse-grained igneous rock (typically granite) commonly found as a dike associated with a large mass of plutonic rock that has smaller crystals. Crystallization in a water-rich environment is believed to be responsible for the very large crystals.
Pegmatitic texture A texture of igneous rocks in which the interlocking crystals are all larger than one centimeter in diameter.
Perched water table A localized zone of saturation above the main water table created by an impermeable layer (aquitard).
Peridotite An igneous rock of ultramafic composition thought to be abundant in the upper mantle.
Period A basic unit of the geologic calendar that is a subdivision of an *era.* Periods may be divided into smaller units called *epochs.*
Periodic table The tabular arrangement of the elements according to atomic number.
Permafrost Any permanently frozen subsoil. Usually found in the subarctic and arctic regions.
Permeability A measure of a material's ability to transmit water.
Phaneritic texture An igneous rock texture in which the crystals are roughly equal in size and large enough so that individual minerals can be identified with the unaided eye.
Phanerozoic eon That part of geologic time represented by rocks containing abundant fossil evidence. The eon extending from the end of the Proterozoic eon (about 540 million years ago) to the present.
Phenocryst Conspicuously large crystals in a porphyry that are imbedded in a matrix of finer-grained crystals (the groundmass).
Physical geology A major division of geology that examines the materials of Earth and seeks to understand the processes and forces acting upon Earth's surface from below.
Piedmont glacier A glacier that forms when one or more valley glaciers emerge from the confining walls of mountain valleys and spread out to create a broad sheet in the lowlands at the base of the mountains.
Pillow lava Basaltic lava that solidifies in an underwater environment and develops a structure that resembles a pile of pillows.
Pipe A vertical conduit through which magmatic materials have passed.
Placer Deposit formed when heavy minerals are mechanically concentrated by currents, most commonly streams and waves. Placers are sources of gold, tin, platinum, diamonds, and other valuable minerals.
Plastic deformation Permanent deformation that results in a change in size and shape through folding or flowing.
Plastic flow A type of glacial movement that occurs within the glacier, below a depth of approximately 50 meters, in which the ice is not fractured.
Plate One of numerous rigid sections of the lithosphere that moves as a unit over the material of the asthenosphere.
Plate tectonics The theory that proposes Earth's outer shell consists of individual plates, which interact in various ways and thereby produce earthquakes, volcanoes, mountains, and the crust itself.
Playa The flat central area of an undrained desert basin.
Playa lake A temporary lake in a playa.
Pleistocene epoch An epoch of the Quaternary period beginning about 1.6 million years ago and ending about 10,000 years ago. Best known as a time of extensive continental ice sheets.
Plucking The process by which pieces of bedrock are lifted out of place by a glacier.
Pluton A structure that results from the emplacement and crystallization of magma beneath Earth's surface.
Pluvial lake A lake formed during a period of increased rainfall. For example, this occurred in many nonglaciated areas during periods of ice advance elsewhere.
Point bar A crescent-shaped accumulation of sand and gravel deposited on the inside of a meander.
Polar wandering hypothesis As the result of paleomagnetic studies in the 1950s, researchers proposed that either the magnetic poles migrated greatly through time or the continents had gradually shifted their positions.
Polymorphs Two or more minerals having the same chemical composition but different crystalline structures. Exemplified by the diamond and graphite forms of carbon.
Porosity The volume of open spaces in rock or soil.
Porphyritic texture An igneous rock texture characterized by two distinctively different crystal sizes. The larger crystals are called *phenocrysts,* and the matrix of smaller crystals is termed the *groundmass.*
Porphyroblastic texture A texture of metamorphic rocks in which particularly large grains (porphyroblasts) are surrounded by a fine-grained matrix of other minerals.
Porphyry An igneous rock with a porphyritic texture.
Pothole A depression formed in a stream channel by the abrasive action of the water's sediment load.
Precambrian All geologic time prior to the Paleozoic era.
Principal shells See *Energy levels.*
Principle of fossil succession Fossil organisms succeed one another in a definite and determinable order, and any time period can be recognized by its fossil content.
Principle of original horizontality Layers of sediment are generally deposited in a horizontal or nearly horizontal position.
Proterozoic eon The eon following the Archean and preceding the Phanerozoic eon. It extends between 2.5 billion and 540 million years ago.
Proton A positively charged subatomic particle found in the nucleus of an atom.
P wave The fastest earthquake wave; travels by compression and expansion of the medium.
Pyroclastic flow A highly heated mixture, largely of ash and pumice fragments, traveling down the flanks of a volcano or along the surface of the ground.
Pyroclastic material The volcanic rock ejected during an eruption. Pyroclastics include ash, bombs, and blocks.
Pyroclastic texture An igneous rock texture resulting from the consolidation of individual rock fragments that are ejected during a violent eruption.
Radial drainage A system of streams running in all directions away from a central elevated structure, such as a volcano.
Radioactivity The spontaneous decay of certain unstable atomic nuclei.
Radiocarbon (carbon-14) The radioactive isotope of carbon produced continuously in the atmosphere and used in dating events as far back as 75,000 years.
Radiometric dating The procedure of calculating the absolute ages of rocks and minerals containing certain radioactive isotopes.
Rainshadow desert A dry area on the lee side of a mountain range. Many middle-latitude deserts are of this type.
Rapids A part of a stream channel in which the water suddenly begins flowing more swiftly and turbulently because of an abrupt steepening of the gradient.
Recessional moraine An end moraine formed as the ice front stagnated during glacial retreat.
Rectangular pattern A drainage pattern that develops on jointed or fractured bedrock and is characterized by numerous right-angle bends.

Refraction A change in direction of waves as they enter shallow water. The portion of the wave in shallow water is slowed, which causes the wave to bend and align with the underwater contours.
Regional metamorphism Metamorphism associated with large-scale mountain building.
Regolith The layer of rock and mineral fragments that nearly everywhere covers Earth's land surface.
Relative dating Rocks are placed in their proper sequence or order. Only the chronological order of events is determined.
Renewable resource A resource that is virtually inexhaustible or that can be replenished over relatively short time spans.
Reserve Already identified deposits from which minerals can be extracted profitably.
Reservoir rock The porous, permeable portion of an oil trap that yields oil and gas.
Residual soil Soil developed directly from the weathering of the bedrock below.
Reverse fault A fault in which the material above the fault plane moves up in relation to the material below.
Reverse polarity A magnetic field opposite to that which presently exists.
Richter scale A scale of earthquake magnitude based on the motion of a seismograph.
Ridge push A mechanism that may contribute to plate motion. It involves the oceanic lithosphere sliding down the oceanic ridge under the pull of gravity.
Rift A region of Earth's crust along which divergence (separation) is taking place.
Ripple marks Small waves of sand that develop on the surface of a sediment layer by the action of moving water or air.
Roche moutonnée An asymmetrical knob of bedrock formed when glacial abrasion smoothes the gentle slope facing the advancing ice sheet and plucking steepens the opposite side as the ice overrides the knob.
Rock A consolidated mixture of minerals.
Rock avalanche The very rapid downslope movement of rock and debris. These rapid movements may be aided by a layer of air trapped beneath the debris, and they have been known to reach speeds in excess of 200 kilometers per hour.
Rock cleavage The tendency of rock to split along parallel, closely spaced surfaces. These surfaces are often highly inclined to the bedding planes in the rock.
Rock cycle A model that illustrates the origin of the three basic rock types and the interrelatedness of Earth's materials and processes.
Rock flour Ground-up rock produced by the grinding effect of a glacier.
Rockslide The rapid slide of a mass of rock downslope along planes of weakness.
Runoff Water that flows over the land rather than infiltrating into the ground.
Saltation Transportation of sediment through a series of leaps or bounces.
Salt flat A white crust on the ground produced when water evaporates and leaves its dissolved materials behind.
Schistosity A type of foliation characteristic of coarser-grained metamorphic rocks. Such rocks have a parallel arrangement of platy minerals such as the micas.
Scoria Hardened lava that has retained the vesicles produced by escaping gases.
Scoria cone See *Cinder cone.*
Sea arch An arch formed by wave erosion when caves on opposite sides of a headland unite.
Seafloor spreading The hypothesis first proposed in the 1960s by Harry Hess that suggested that new oceanic crust is produced at the crests of mid-ocean ridges, which are the sites of divergence.
Seamount An isolated volcanic peak that rises at least 1000 meters (3300 feet) above the deep-ocean floor.
Sea stack An isolated mass of rock standing just offshore, produced by wave erosion of a headland.
Seawall A barrier constructed to prevent waves from reaching the area behind the wall. Its purpose is to defend property from the force of breaking waves.
Secondary enrichment The concentration of minor amounts of metals that are scattered through unweathered rocks into economically valuable concentrations by weathering processes.
Secondary (S) wave A seismic wave that involves oscillation perpendicular to the direction of propagation.
Sediment Unconsolidated particles created by the weathering and erosion of rock, by chemical precipitations from solution in water, or from the secretions of organisms, and transported by water, wind, or glaciers.
Sedimentary rock Rock formed from the weathered products of preexisting rocks that have been transported, deposited, and lithified.
Seiche The rhythmic sloshing of water in lakes, reservoirs, and other smaller enclosed basins. Some seiches are initiated by earthquake activity.
Seismic sea wave A rapidly moving ocean wave generated by earthquake activity and capable of inflicting heavy damage in coastal regions.
Seismogram The record made by a seismograph.
Seismograph An instrument that records earthquake waves.
Seismology The study of earthquakes and seismic waves.
Settling velocity The speed at which a particle falls through a still fluid. The size, shape, and specific gravity of particles influence settling velocity.
Shadow zone The zone between 105 and 140 degrees distance from an earthquake epicenter that direct waves do not penetrate because of refraction by Earth's core.
Shallow-focus earthquake An earthquake focus at a depth of less than 60 kilometers.
Shear Stress that causes two adjacent parts of a body to slide past one another.
Sheet flow Runoff moving in unconfined thin sheets.
Sheeting A mechanical weathering process characterized by the splitting off of slablike sheets of rock.
Shelf break The point at which a rapid steepening of the gradient occurs, marking the outer edge of the continental shelf and the beginning of the continental slope.
Shield A large, relatively flat expanse of ancient metamorphic rock within the stable continental interior.
Shield volcano A broad, gently sloping volcano built from fluid basaltic lavas.
Shore Seaward of the coast, this zone extends from the highest level of wave action during storms to the lowest tide level.
Shoreline The line that marks the contact between land and sea. It migrates up and down as the tide rises and falls.
Silicate Any one of numerous minerals that have the silicon-oxygen tetrahedron as their basic structure.
Silicon-oxygen tetrahedron A structure composed of four oxygen atoms surrounding a silicon atom that constitutes the basic building block of silicate minerals.
Sill A tabular igneous body that was intruded parallel to the layering of preexisting rock.
Sinkhole A depression produced in a region where soluble rock has been removed by groundwater.
Slab-pull A mechanism that contributes to plate motion in which cool, dense oceanic crust sinks into the mantle and "pulls" the trailing lithosphere along.
Slaty cleavage The type of foliation characteristic of slates in which there is a parallel arrangement of fine-grained metamorphic minerals.
Slide A movement common to mass-wasting processes in which the material moving downslope remains fairly coherent and moves along a well-defined surface.
Slip face The steep, leeward surface of a sand dune that maintains a slope of about 34 degrees.
Slump The downward slipping of a mass of rock or unconsolidated material moving as a unit along a curved surface.
Snowfield An area where snow persists year-round.
Snowline Lower limit of perennial snow.
Soil A combination of mineral and organic matter, water, and air; that portion of the regolith that supports plant growth.

Soil horizon A layer of soil that has identifiable characteristics produced by chemical weathering and other soil-forming processes.

Soil profile A vertical section through a soil showing its succession of horizons and the underlying parent material.

Soil Taxonomy A soil classification system consisting of six hierarchical categories based on observable soil characteristics. The system recognizes 12 soil orders.

Solifluction Slow, downslope flow of water-saturated materials common to permafrost areas.

Solum The *O*, *A*, and *B* horizons in a soil profile. Living roots and other plant and animal life are largely confined to this zone.

Solution The change of matter from the solid or gaseous state into the liquid state by its combination with a liquid.

Sorting The degree of similarity in particle size in sediment or sedimentary rock.

Specific gravity The ratio of a substance's weight to the weight of an equal volume of water.

Speleothem A collective term for the dripstone features found in caverns.

Spheroidal weathering Any weathering process that tends to produce a spherical shape from an initially blocky shape.

Spit An elongated ridge of sand that projects from the land into the mouth of an adjacent bay.

Spring A flow of groundwater that emerges naturally at the ground surface.

Spring tide The highest tidal range; occurs near the times of the new and full moons.

Stable platform That part of the craton that is mantled by relatively undeformed sedimentary rocks and underlain by a basement complex of igneous and metamorphic rocks.

Stalactite The icicle-like structure that hangs from the ceiling of a cavern.

Stalagmite The columnlike form that grows upward from the floor of a cavern.

Star dune Isolated hill of sand that exhibits a complex form and develops where wind conditions are variable.

Steppe One of the two types of dry climate. A marginal and more humid variant of the desert that separates the desert from bordering humid climates.

Stock A pluton similar to but smaller than a batholith.

Strata Parallel layers of sedimentary rock.

Stratified drift Sediments deposited by glacial meltwater.

Stratovolcano See *Composite cone.*

Streak The color of a mineral in powdered form.

Stream A general term to denote the flow of water within any natural channel. Thus, a small creek and a large river are both streams.

Stream valley The channel, valley floor, and the sloping valley walls of a stream.

Stress The force per unit area acting on any surface within a solid.

Striations The multitude of fine parallel lines found on some cleavage faces of plagioclase feldspars but not present on orthoclase feldspar.

Striations (glacial) Scratches or grooves in a bedrock surface caused by the grinding action of a glacier and its load of sediment.

Strike The compass direction of the line of intersection created by a dipping bed or fault and a horizontal surface. Strike is always perpendicular to the direction of dip.

Strike-slip fault A fault along which the movement is horizontal.

Stromatolite Structures that are deposited by algae and that consist of layered mounds or columns of calcium carbonate.

Subduction The process of thrusting oceanic lithosphere into the mantle along a convergent zone.

Subduction zone A long, narrow zone where one lithospheric plate descends beneath another.

Submarine canyon A seaward extension of a valley that was cut on the continental shelf during a time when sea level was lower, or a canyon carved into the outer continental shelf, slope, and rise by turbidity currents.

Submergent coast A coast whose form is largely the result of the partial drowning of a former land surface either due to a rise of sea level or subsidence of the crust, or both.

Subsoil A term applied to the *B* horizon of a soil profile.

Superposition, law of In any undeformed sequence of sedimentary rocks, each bed is older than the one above and younger than the one below.

Surf A collective term for breakers; also the wave activity in the area between the shoreline and the outer limit of breakers.

Surface soil The upper portion of a soil profile consisting of the *O* and *A* horizons.

Surface waves Seismic waves that travel along the outer layer of Earth.

Surge A period of rapid glacial advance; surges are typically sporadic and short-lived.

Suspended load The fine sediment carried within the body of flowing water or air.

S wave An earthquake wave, slower than a P wave, that travels only in solids.

Swells Wind-generated waves that have moved into an area of weaker winds or calm.

Syncline A linear downfold in sedimentary strata; the opposite of anticline.

System A group of interacting or interdependent parts that form a complex whole.

Tablemount See *Guyot.*

Tabular Describing a feature such as an igneous pluton having two dimensions that are much longer than the third.

Talus An accumulation of rock debris at the base of a cliff.

Tarn A small lake in a cirque.

Tectonics The study of the large-scale processes that collectively deform Earth's crust.

Temporary (local) base level The level of a lake, resistant rock layer, or any other base level that stands above sea level.

Terminal moraine The end moraine marking the farthest advance of a glacier.

Terrace A flat, benchlike structure produced by a stream that was left elevated as the stream cut downward.

Terrane A crustal block bounded by faults whose geologic history is distinct from the histories of adjoining crustal blocks.

Terrigenous sediment Seafloor sediment derived from terrestrial weathering and erosion.

Texture The size, shape, and distribution of the particles that collectively constitute a rock.

Theory A well-tested and widely accepted view that explains certain observable facts.

Thermal metamorphism See *Contact metamorphism.*

Thrust fault A low-angle reverse fault.

Tide Periodic change in the elevation of the ocean's surface.

Till Unsorted sediment deposited directly by a glacier.

Tillite A rock formed when glacial till is lithified.

Tombolo A ridge of sand that connects an island to the mainland or to another island.

Topset bed An essentially horizontal sedimentary layer deposited on top of a delta during floodstage.

Transform fault A major strike-slip fault that cuts through the lithosphere and accommodates motion between two plates.

Transform fault boundary A boundary in which two plates slide past one another without creating or destroying lithosphere.

Transpiration The release of water vapor to the atmosphere by plants.

Transported soil Soils that form on unconsolidated deposits.

Transverse dunes A series of long ridges oriented at right angles to the prevailing wind; these dunes form where vegetation is sparse and sand is very plentiful.

Travertine A form of limestone ($CaCO_3$) that is deposited by hot springs or as a cave deposit.

Trellis drainage A system of streams in which nearly parallel tributaries occupy valleys cut in folded strata.

Trench An elongate depression in the seafloor produced by bending of oceanic crust during subduction.

Tsunami The Japanese word for a seismic sea wave.

Turbidite Turbidity current deposit characterized by graded bedding.

Turbidity current A downslope movement of dense, sediment-laden water created when sand and mud on the continental shelf and slope are dislodged and thrown into suspension.

Turbulent flow The movement of water in an erratic fashion often characterized by swirling, whirlpool-like eddies. Most streamflow is of this type.

Ultimate base level Sea level; the lowest level to which stream erosion could lower the land.

Ultramafic Compositional category for igneous rocks made up almost entirely of ferromagnesian minerals (mostly olivine and pyroxene).

Unconformity A surface that represents a break in the rock record; caused by erosion or nondeposition.

Uniformitarianism The concept that the processes that have shaped Earth in the geologic past are essentially the same as those operating today.

Valence electron The electrons involved in the bonding process; the electrons occupying the highest principal energy level of an atom.

Valley glacier A glacier confined to a mountain valley, which in most instances had previously been a stream valley.

Valley train A relatively narrow body of stratified drift deposited on a valley floor by meltwater streams that issue from the terminus of a valley glacier.

Vein deposit A mineral filling a fracture or fault in a host rock. Such deposits have a sheetlike, or tabular, form.

Ventifact A cobble or pebble polished and shaped by the sandblasting effect of wind.

Vesicles Spherical or elongated openings on the outer portion of a lava flow that were created by escaping gases.

Vesicular texture A term applied to aphanitic igneous rocks that contain many small cavities, called *vesicles.*

Viscosity A measure of a fluid's resistance to flow.

Volatiles Gaseous components of magma dissolved in the melt. Volatiles will readily vaporize (form a gas) at surface pressures.

Volcanic Pertaining to the activities, structures, or rock types of a volcano.

Volcanic island arc A chain of volcanic islands generally located a few hundred kilometers from a trench where there is active subduction of one oceanic plate beneath another.

Volcanic bomb A streamlined pyroclastic fragment ejected from a volcano while the fragment is still molten.

Volcanic neck An isolated, steep-sided, erosional remnant consisting of lava that once occupied the vent of a volcano.

Volcano A mountain formed from lava and/or pyroclastics.

Wash A desert stream course that is typically dry except for brief periods immediately following rainfall.

Water gap A pass through a ridge or mountain in which a stream flows.

Water table The upper level of the saturated zone of groundwater.

Wave-cut cliff A seaward-facing cliff along a steep shoreline formed by wave erosion at its base and by mass wasting.

Wave-cut platform A bench or shelf along a shore at sea level, cut by wave erosion.

Wave height The vertical distance between the trough and crest of a wave.

Wave length The horizontal distance separating successive crests or troughs.

Wave period The time interval between the passage of successive crests at a stationary point.

Weathering The disintegration and decomposition of rock at or near the surface of the Earth.

Welded tuff A pyroclastic deposit composed of particles fused together by the combination of heat still contained in the deposit after it has come to rest and by the weight of overlying material.

Well An opening bored into the zone of saturation.

Wind gap An abandoned water gap. These gorges typically result from stream piracy.

Xenolith An inclusion of unmelted country rock in an igneous pluton.

Xerophyte A plant highly tolerant of drought.

Yazoo tributary A tributary that flows parallel to the main stream because a natural levee is present.

Zone of accumulation The part of a glacier characterized by snow accumulation and ice formation. The outer limit of this zone is the *snowline.*

Zone of aeration Area above the water table where openings in soil, sediment, and rock are not saturated but are filled mainly with air.

Zone of fracture The upper portion of a glacier consisting of brittle ice.

Zone of saturation Zone where all open spaces in sediment and rock are completely filled with water.

Index

Aa flow, 87
Ablation, 248, 249
Abrasion, 250, 276–77, 290–91, 293
Abyssal plain, 24, 373–74
Accessory minerals, 173
Accretion, continental, 405–7
Accretionary wedge, 372, 373, 404, 407
Acid precipitation, 123, 157
Active continental margins, 370, 372–73, 403
Aerosols, 109
Aftershocks, 312
Agassiz, Louis, 261
Agate, 145
Aggregate, 34, 36
Air pollution, 157
Alaska pipeline, 197, 198
Aleutian-type subduction, 403
Alluvial channels, 208–10
Alluvial fan, 193, 195, 214, 273, 274
Alluvium, 208
Alpha particle, 427, 428
Alpine Fault of New Zealand, 28
Alpine glacier. *See* Valley glacier
Altisols, 129
Amber, 421, 425
Amphiboles, 47, 49, 349
Amplification of seismic waves, 321
Anaerobic bacteria, 444
Andean-type subduction, 402, 403–5
Andesite, 64, 65, 67, 68
Andesitic composition, 64
Andesitic igneous rocks, 63, 65, 67–68
Andisols, 129
Angiosperms, 457, 458
Angle of repose, 188, 189
Angular unconformity, 419, 420, 421, 422
Antarctic Ice Sheet, 245
Antarctic ice shelves, 245–46, 248
Anthracite, 148, 149, 172
Anticlines, 393–95
Aphanitic texture, 60, 61, 65
Appalachian Mountains, 408, 456
Aquifer, 228, 232
Aquitard, 228, 230
Aragonite (mother-of-pearl), 50, 139
Aral Sea, 272
Arête, 252–53
Arid climates. *See* Desert
Aridosols, 129
Aristotle, 3, 4
Arkose, 143, 151
Artesian springs, 230
Artesian wells, 230–31, 232
Artificial cutoffs, 219
Artificial levee, 218–19
Asbestos, 51
Ash, volcanic, 88–89
Assimilation, 73–74
Asthenosphere, 19, 20, 328, 329, 342, 348
Atlantic coast, 299–300
Atlantic Ocean, 383
Atmosphere, 9, 10, 444
Atoll, 376–77
Atom, 37–38, 39
Atomic mass unit, 39
Atomic number, 36, 37, 426
Atomic structure, 37, 426
Augite, 49
Aureole, 175
Avalanches, 191, 192. *See also* Landslide; Mass wasting
Avalonia, 386

Backshore, 286, 287
Backswamp, 214
Backwash, 290, 291
Bajada, 273, 274
Baltica, 386
Barchan dune, 278, 280
Barchanoid dune, 279, 280
Barrier island, 295–96, 299–300
Barrier reef, 376
Bars, 208, 213, 295
Basal slip, 247
Basalt, 14, 19, 48, 65, 68, 69
  flood, 99
  pillow, 380, 381
Basaltic composition, 63–64
Basaltic igneous rocks, 63, 65, 68, 69
Basalt plateaus, 98–99
Base level, 210, 211, 212–13
Basin and Range, 273–74, 409, 457
Basins, 395, 397
  deep-ocean, 24, 373–75
  drainage, 203–4, 218
  ocean, 22–24, 381–86
Batholith, 102–3, 404
Bathymetry, 368, 369–70
Bauxite, 132–33
Baymouth bar, 295, 296, 297, 298
Beach, 286–88
  sand movement on, 291–93
Beach drift, 292–93, 294
Beach face, 286–88
Beach nourishment, 299, 300
Bedding plane, 151
Bed load, 207, 275
Bedrock channels, 208
Beds, 151, 155
Belt of soil moisture, 225
Berms, 286, 287
Beta particle, 427, 428
Big Bang, 16
Biochemical sediment, 144
Biological activity, weathering and, 119
Biosphere, 9, 10, 125–27
Biotite, 47, 49
Bird-foot delta, 213, 216
Bituminous coal, 148, 149
Black Hills, 395, 396
Black smokers, 176, 381
Block lavas, 87
Blocks, volcanic, 89
Blowout, 275–76
Body waves, 314
Bombs, volcanic, 88, 89
Bottomset beds, 214
Bowen, N.L., 71, 72
Bowen's reaction series, 71–73
Brachiopods, 449, 450
Braided stream, 209–10
Breakwater, 298
Breccia, 140, 144, 172, 177–78
Brittle failure (deformation), 392–93
Brittle rocks, 166
Bronowski, Jacob, 7
Bronze Age, 34
Building materials, 154
Buoyant subduction, 384
Burial metamorphism, 177
Burrows (fossils), 422

Calcareous tufa, 234
Calcite, 49–50
Calcium carbonate, 144, 145
Caldera, 89, 90, 97–98, 99
Calving, 248, 249
Cambrian period, 449, 450
Capacity, 207–8
Cape Hatteras Lighthouse, moving, 289
Cap rock, 158
Carbon-14, 430–31
Carbonates, 43, 49, 52
Carbon cycle, 146
Carbon dioxide, atmospheric, 146
  global warming and, 11–12
Carbonic acid, 120, 146, 236
Carbonization, 421, 425
Casts (fossils), 421, 425
Catastrophism, 3–4
Caverns, 144, 224, 225, 236–37, 240
Cementation, 139
Cenozoic era, 432, 434, 435, 448, 453, 454, 456–59
Cephalopods, 449–50
Chalk, 136, 145
Challenger, H.M.S., 368–69
Chamberlin, T.C., 342
Channel, stream, 205, 208–10
Channel deposits, 208–10
Channelization, 219
Chemical bonds, 37–39
Chemically active fluids, metamorphism and, 167–68
Chemical pollution, 132
Chemical sedimentary rock, 140, 144–49
Chemical weathering, 116, 118, 119–22, 123
Chemosynthesis, 383
Chert, 145–47, 446
Chesterfield, Lake, 239
Chikyu, 354
Chitin, 450
Chrysotile, 51
Cinder cone, 91–92
Cinders, 89
Circular orbital motion of waves, 288–90
Circum-Pacific belt, 316
Cirque, 252–53, 254
Clastic texture, 149–50
Clay, 48–49, 120, 140
Cleavage, 41–42, 169–70
Climate
  ancient, 340–42
  dry, 268–73

Climate (*cont.*)
glacial ice and data on past, 263
soil formation, 125
weathering rate, 122–23
Climate change, 108–9, 451
Closed systems, 11
Coal, 34, 139, 148–49, 155, 156–58
Coastal wetlands, decline in, 215, 301
Coastal zone, 286–88. *See also* Shorelines
Coastline, 286, 287
Coast Range (California), 404–5
Collapse pits, 97
Color, 40
Columbus, 10
Columnar joints, 101, 103, 401
Compaction, 139, 148
Competence, 207–8
Composite cone, 92–95
Compound, 34
Conchoidal fracture, 42, 43
Concordant plutons, 100
Conduit (pipe), volcanic, 89, 99–100
Cone of depression, 229
Confining pressure, 166, 392–93
Conformable layers, 418
Conglomerate, 140, 143, 144, 151
Contact metamorphism, 165, 166, 175–76
Continental accretion, 405–7
Continental collisions, 407–8
Continental-continental convergence, 349, 350–51
Continental crust, formation of, 18. *See also* Crust
Continental divide, 204
Continental drift, 24, 336–42
Continental environment, 150, 152–53
Continental margins, 22–23, 370–73, 383, 403–5, 406
Continental rifts, 107, 109, 346–47, 381–85
Continental rise, 371, 372
Continental shelf, 23, 370–72
Continental slope, 23–24, 371, 372
Continental volcanic arc, 105, 106, 349, 350, 372, 373, 402, 404, 407
Continents, features of, 21–23
Continuous reaction series, 71, 72
Convergent plate boundaries, 25–27, 28, 343, 345, 347–51
destruction of oceanic lithosphere at, 383–85
igneous activity at, 105, 106
mountain building, 402–5
Coprolites, 422
Coquina, 145
Coral reef, 144–45, 147, 376–77, 386
Core, 18, 19, 20, 328, 329, 330, 331, 442
Correlation, 419–20, 422–26
Covalent bond, 38, 39
Crater, 89
Crater Lake-type calderas, 97, 98, 99
Creep, 191, 196–97
Crest of wave, 288, 289, 290
Cretaceous period, 452, 453, 454
Cretaceous-Tertiary (KT) boundary, 454
Crevasses, 247, 249
Critical settling velocity, 208
Cross-beds, cross-bedding, 151, 155, 278, 279
Cross-cutting relationships, 418, 419, 435
Cross dating, 432, 433
Crust, 18, 19, 20, 21–24, 327, 328, 330–31
deformation, 392–401
most common elements in, 44
oceanic, structure of, 380–81. *See also* Ocean floor
primitive, 18, 442
vertical movements of, 409–11
Crustal fragments, accretion of, 405–7
Crustal uplift, 409–10
Crystal form, 34, 40, 42, 150
Crystallization, 14–16, 59–60
Crystal settling, 73, 74
Curie point, 355
Current(s)
glaciation and ocean, 263
longshore, 292–93, 294
tidal, 304–5
turbidity, 152, 155, 372
Current ripple marks, 151
Cut bank, 208, 209
Cutoff, 209, 219

Dam-failure floods, 218
Dams, flood-control, 219
Darcy, Henry, 229
Darcy's law, 229
Dark (ferromagnesian) silicates, 46, 49, 62–63
Darwin, Charles, 376–77, 401
Debris flow, 190, 193–96
lahars, 96–97, 193–96
Decompression melting, 70, 71, 108, 378, 383
Deep-layer model of mantle convection, 361, 362
Deep-ocean basin, 24, 373–75
Deep-ocean trench, 24, 348–49, 350, 373, 374–75
Deep Sea Drilling Project, 352–53
Deep-sea fan, 372
Deflation, 275–76, 277
Deformation, 392–401
Delta, 213, 214, 215, 216
tidal, 295, 305
Dendritic pattern, 214–15, 217
Dendrochronology, 432–33
Density, 42–43
Deposition
environment of, 150
glacial deposits, 254–57
stream, 208–10, 213–14
wind deposits, 277–81
Depositional landforms, 213–14
Desert, 268–74
Basin and Range, 273–74
common misconceptions, 270–73
distribution and causes, 268–70
geologic processes, 270–73
rainshadow, 270
Desert pavement, 276, 277
Destructive plate margins. *See* Convergent plate boundaries
Detachment fault, 397
Detrital sedimentary rock, 139, 140–44, 149
Devonian period, 449, 450, 451
Diagenesis, 139
Diamonds, 76, 77, 100, 178
Differential stress, 166–67
Differential weathering, 114, 123
Dike, 100–101, 102
Dinosaurs, 452–55, 457, 458. *See also* Mesozoic era
Diorite, 65, 67–68
Dip-slip faults, 396–98
Discharge, 205–6
Disconformities, 419, 421
Discontinuous reaction series, 71, 72
Discordant plutons, 100
Disseminated deposits, 76–77
Dissolved load, 207
Distributaries, 213, 214
Divergent plate boundary, 25, 26, 28, 105–9, 343–47
Divide, 204
Dolomite, 49
Dolostone, 49–50, 145
Domes, 395
Double refraction, 43
Drainage basin, 203–4, 218
Drake, Edwin, 150, 157
Drawdown, 229
Drift
beach, 292–93, 294
continental, 24, 336–42
glacial, 254–55
Dripstone, 144, 237
Drumlins/drumlin fields, 257, 258
Dry climate, 268–73. *See also* Desert
Ductile deformation, 392, 393
Ductile rocks, 167
Dune, sand, 277–79, 280
Dust Bowl, 131, 275, 276, 278

Earth
distribution of land and water, 368
dynamic, 24–28
early evolution of, 16–19
interior, 12, 18–24, 327–31, 442
origin of, 16–18
spheres, 8–10
Earth-centered model of universe, 8
Earthflow, 190, 196
Earth history, 441–61
atmosphere, 444
Cenozoic era, 432, 434, 435, 448, 453, 454, 456–59
early evolution, 442–44
Mesozoic era, 432, 434, 435, 448, 452–55, 457
Paleozoic era, 432, 434, 435, 447–52, 453
Precambrian, 432–33, 434, 435, 444–47, 448
Earthquakes, 2, 309–33
cause, 312
defined, 311
destruction, 317, 320–24
Earth's interior and, 327–31
in eastern U.S., 318
epicenter, 316, 320
faults, 311–12
intensity, 317
magnitude, 315, 317–20
mass wasting, 189
notable, 326
plate tectonics, 312, 352

prediction, 324–27
waves, 314–15, 321
Earth system, 11, 12–13, 123–24
Earth system science, 11
East African Rift, 109, 346, 348, 382–83
Ebb currents, 304
Ebb deltas, 305
Eccentricity, 261
Echo sounder, 369
Eggs, shell-covered, 454
Elastic deformation, 392
Elastic rebound, 312, 313
El Chichón, 109
Electron capture, 427, 428
Electrons, 37–39, 426
Elements, 35–37, 38, 44, 68
Eluviation, 127
Emergent coasts, 302
End moraine, 255–56, 257, 258
Energy, Earth system, 12
Energy levels, 37
Energy resources, 155–58, 234–36
Entisols, 129
Environment of deposition, 150
Eon, 431, 434
Ephemeral stream, 271, 274
Epicenter, 316, 320
Epoch, 432, 434
Equatorial low, 269
Era, 431–32, 434
Erastosthenes, 10
Erosion, 116, 129–32
glacial, 249–53
shoreline, 294, 295, 296–302
stream, 207, 210
wave, 290–91
wind, 131, 275–77
Eruption columns, 84
Esker, 257, 258
Estuaries, 302, 303
Etna, Mount, 108
Eukaryotes, 446
Evaporites, 147–48, 150
Evapotranspiration, 202
Exfoliation dome, 118, 119
External processes, 116
Extinction
dinosaurs, 454–55
large mammals of Pleistocene, 459
mass, 454–55
Paleozoic, 451–52
Extrusive rocks, 58

Fall, 190–91
Farallon plate, 384–85
Fathoms, 24
Fault-block mountains, 398, 408–9
Fault breccia, 172, 177–78
Fault creep, 312, 327
Fault gouge, 178
Faults, 311–12, 395–401
San Andreas, 27–28, 312–14, 324, 326–27, 352, 384, 398–401
Fault scarps, 396, 398, 401
Fault zones, 396
metamorphism along, 177–78
Feedback mechanisms, 11–12, 263
Feldspar, 46–48, 63, 120–21
Felsic igneous rocks, 63, 64–67
Ferromagnesian silicates, 46, 49, 62–63
Fetch, 288
Fiord, 252, 253
Fire, earthquakes and, 324
Fissility, 141
Fissure eruptions, 98–99, 100, 102
Fissures, 98
Flash flood, 217, 218
Flint, 145
Flood basalts, 99
Flood control, 218–20
Flood currents, 304
Flood deltas, 305
Floodplain, 212–13, 216, 219
Floods, 212, 215–20
Flow, 191
glacial ice, 245, 246
sheet, 207
streamflow, 204–6
Focus, 311, 312
Folds, 393–95
Foliated metamorphic rocks, 172–74
Foliated textures, 169–71
Foliation, 168–69
Footwall, 396
Foraminifera, 457
Foreset beds, 214
Foreshocks, 312
Foreshore, 286, 287
Fossil(s), 5–6, 151, 156, 420–26
conditions favoring preservation, 422
continental drift evidence, 337–39
correlation, 422–26
dinosaur, 452
index, 423
Paleozoic, 447, 448
Precambrian, 446–47
trace, 446
types, 420–22, 425
Fossil fuels, 155–58
Fossil succession, principle of, 5–6, 422–23
Fracture, 42
Fracture zone, 352
Fragmental texture, 62
Fringing reef, 376
Frost wedging, 117, 118
Fumaroles, 90

Gabbro, 19, 65, 68, 69, 380, 381
Gaining streams, 226, 227
Gases dissolved in magma, 84–86
Gastrolith, 422
Gelisols, 129
Gems, 39, 40
Geologic time, 4–6, 415–39
dating with radioactivity, 426–31
as factor in deformation, 393
relative dating, 5, 6, 416–19
Geologic time scale, 5–6, 416, 431–36, 443
Geology, 2–4
historical, 2
historical notes, 3–4
physical, 2
science of, 2–3
Geosphere, 10
Geothermal energy, 234–36
Geothermal gradient, 69–70, 165–66
Geyser, 108, 234, 235
Geyserite, 234
Glacial budget, 248–49
Glacial drift, 254–55
Glacial erratic, 254–55, 257
Glacial ice, climate data stored in, 263
Glacial striations, 250
Glacial trough, 251, 252, 253
Glacier, 243–65, 410
budget, 248–49
causes of glaciation, 260–63
definition, 244
deposits, 254–57
erosion, 249–53
Ice Age, 257–60, 410
indirect effects, 259–60
movement, 8, 247–49
types, 245–46
water stored as ice, 245
Glassy texture, 60, 61–62
Global Positioning System (GPS), 358–59
Global warming, 11–12, 301
Glomar Challenger, 353, 354
Glossary, 469–81
Gneiss, 171, 172, 174
Gneissic texture, 170–71, 174
Gold, 38
Gondwana/Gondwanaland, 386, 445, 449
Graben, 383, 397, 398
Graded beds, 151, 155
Gradient, 205
Grand Canyon, 5
Granite, 64–67, 120–21
Granitic igneous rocks, 63, 64–67
Granodiorite, 19
Graphite, 173
Gravitational collapse, 410–11
Gravity, specific, 42–43
Graywacke, 143
Greenhouse gases, 12, 262–63
Groin, 298, 299
Groundmass, 61
Ground moraine, 256, 258
Ground subsidence, 232, 235, 323
Groundwater, 223–41
artesian, 230–31, 232
contamination, 232–33, 234
definition, 225
distribution, 224–25, 226–27
environmental problems, 231–33, 234
geologic work, 236–40
importance, 224
interaction with streams, 226–27
movement, 227–28
springs, 228–29, 230, 234
water table, 225, 226–27
wells, 229–31, 232
Gulf coast, 299–300
Gullies, 130, 207
Gutenberg, Beno, 329–30
Guyot, 374
Gymnosperms, 452–54
Gypsum, 50, 147

Hadean eon, 432
Half-life, 428
Halides, 49, 52
Halite, 50, 147
Hanging valley, 251, 252
Hanging wall, 396
Hardness, 41
Hardpan, 127
Hard stabilization, 298
Hawaiian-type calderas, 97–98
Hays, J.D., 262
Head/headwaters (stream), 206, 207, 208

Heat
  generating magma from solid rock, 69–70
  metamorphic agent, 165–66
Hess, Harry, 378
High-resolution multibeam sonar, 369–70
Hillary, Edmund, 401
Himalaya Mountains, 407–8
Historical geology, 2
Histosols, 129
Horizon (soil), 127–28, 129
Horn, 251, 252–53
Hornblende, 47, 49
Hornfels, 172, 175–76
Horst, 397, 398
Hot spots, 106, 107, 110–11, 354–55
Hot spring, 108, 234
Humus, 124, 127
Hutton, James, 4, 419
Hydrated minerals, 168
Hydraulic conductivity, 229
Hydrologic cycle, 12, 13, 202–3
Hydrosphere, 9–10, 202, 224
Hydrothermal metamorphism, 76, 165, 176, 381
Hydrothermal solutions, 76–77, 176
Hydrothermal vents, 383
Hypothesis, 6–7

Ice Age, 257–60, 410
Iceberg, 248, 249
Ice cap, 246
Ice cores, climate data in, 263
Ice-jam floods, 218
Ice sheet, 245
Ice shelves, 245–46
Igneous activity, plate tectonics and, 104–11
Igneous rock(s), 15, 16, 19, 57–79
  composition, 62–64, 71–73, 82
  resources, 75–77
  textures, 60–62
  types, 64–69
Impactiles, 178
Impact metamorphism, 178, 180
Impression (fossil), 421, 425
Inceptisols, 129
Incised meander, 213
Inclusions, 418, 420, 423
Index fossils, 423
Index minerals, 178–80
Indonesian earthquake, tsunami damage from 2004, 322–23
Industrial minerals, 154
Infiltration, 202
Inner core, 20, 21, 329, 330
Inselberg, 273, 274, 277
Integrated Ocean Drilling Program (IODP), 354
Intensity scales, 317
Interface, 12, 286
Interior drainage, 272, 273
Intermediate composition, 64
Intermediate igneous rocks, 63, 65, 67–68
Internal processes, 116. *See also* Mountain building; Volcanoes
Intraplate volcanism, 106, 107, 109–11
Intrusive activity, 100–103
Intrusive rocks, 58
Ionic bond, 38, 39
Ions, 38
Iridium, 455
Island arc. *See* Volcanic island arc
Island archipelagos, 105
Isostasy, 409–10
Isostatic adjustment, 409–10
Isotopes, 39–40, 426–27, 428, 430

Jasper, 145
JOIDES Resolution, 353, 354
Joints, 101, 103, 401
Jurassic period, 452, 453

Kame, 257, 258
Karst topography, 237–40
Kettle, 256–57, 258
Kilauea, 88, 90, 91, 92
Kimberlite, 77

Laccoliths, 101–2
Lahar, 96–97, 193–96
Lakes, 209, 210, 239, 252, 260, 274
Laminae, 141
Laminar flow, 204
Landform development, mass wasting and, 187–88
Landslide, 186–87, 189–90, 195, 323, 324. *See also* Mass wasting
Land subsidence, 232, 235, 323
Lapilli, 89
Laramide Orogeny, 452, 457
Lateral moraine, 255, 258
Laterite, 126
Laurasia, 447, 449
Laurentia, 386
Lava, 58
Lava flows, 86–87
Lava plateaus, 98–99
Law, scientific, 9
Layered model of mantle convection, 361
Leaching, 126, 127, 133
Lehmann, Inge, 330
Levees, 213–14, 215, 218–19
Lightfoot, John, 4
Light (nonferromagnesian) silicates, 46–49, 63
Light-year, 16
Lignite, 148, 149
Limestone, 49–50, 144–45, 146, 147, 236
Liquefaction, 189, 321–22, 323
Lithification, 16, 139
Lithosphere, 19, 20, 24, 328, 342, 348. *See also* Plate tectonics
  oceanic, 377–80, 383–85
Local base level, 210
Loess, 277, 279–81
Longitudinal dune, 279, 280
Longitudinal profile, 207
Longshore current, 292–93, 294
Losing streams, 226–27
Louisiana coastal wetlands, 215
Lower mantle, 19, 20, 21, 328–29
Low-latitude deserts, 269
Luster, 40

Mafic (basaltic) igneous rocks, 63, 65, 68, 69
Magma, 14–16, 58–60, 69–75. *See also* Seafloor spreading; Volcanoes
  compositions, 86
  gas content of, 84–86
  viscosity, 82–84, 86
Magma mixing, 74
Magmatic differentiation, 73, 74, 84
Magmatic segregation, 76
Magnetic reversals, 356–58
Magnetic time scale, 357
Magnetometers, 357
Magnitude (earthquake), 315, 317–20
Mammals, 456–59
Mantle, 18, 19, 20, 328–29, 330, 331, 349, 351, 384, 442
  lower, 19, 20, 21, 328–29
Mantle convection, 359, 361–62
Mantle plume, 106, 107, 110, 354, 361–62
Marble, 171, 172, 174
Marine environment, 150, 152–53
Marine terrace, 294
Marsupials, 458–59
Mass extinction, 454–55
Massive plutons, 100
Mass number, 39, 426
Mass wasting, 116, 185–99
  classification, 190–91
  controls and triggers, 188–90
  defined, 187
  landform development, 187–88
  rate of movement, 191
  slow movements, 196–98
  types, 192–98
Matrix, 143
Matthews, D.H., 357
Meanders, 208, 210, 212–13
Mechanical weathering, 116–19
Medial moraine, 255
Melt, 58–59
Melting
  decompression, 70, 71, 108, 378, 383
  partial, 58, 74–75, 105, 108, 349, 351, 378, 408
Mercalli, Guiseppe, 317
Mesosaurus, 339, 340
Mesosphere, 19, 20, 21, 328–29
Mesozoic era, 432, 434, 435, 448, 452–55, 457
Metaconglomerate, 172
Metallic bonding, 39
Metallic luster, 40
Metallic minerals, 75
Metamorphic grade, 165, 178–80
Metamorphic rock, 15, 16, 163–83
  classification, 172
  environments, 175–78
  foliated, 172–74
  nonfoliated, 174–75
  textures, 168–71, 178
Metamorphism, 164–65
  agents, 165–68
  types of, 76, 165, 175–77, 178, 180, 381
  zones of, 175, 178–80
Metasomatism, 168
Meteorites, 18, 178, 331
Mica schists, 173
Microcline feldspar, 48
Microcontinents, 405
Mid-Atlantic Ridge, 375, 377
Middle-latitude deserts, 269–70
Mid-ocean ridge, 24, 342, 343–46, 352, 368, 375–81
Migmatite, 172, 180
Milankovitch, Milutin, 261–62
Minerals, 14, 33–55
  characteristics, 34
  composition, 34–40
  defined, 34
  family of, 45